The Human Microbiome in Health & Disease

The Human MICROBIOME in Health & Disease

An Introduction

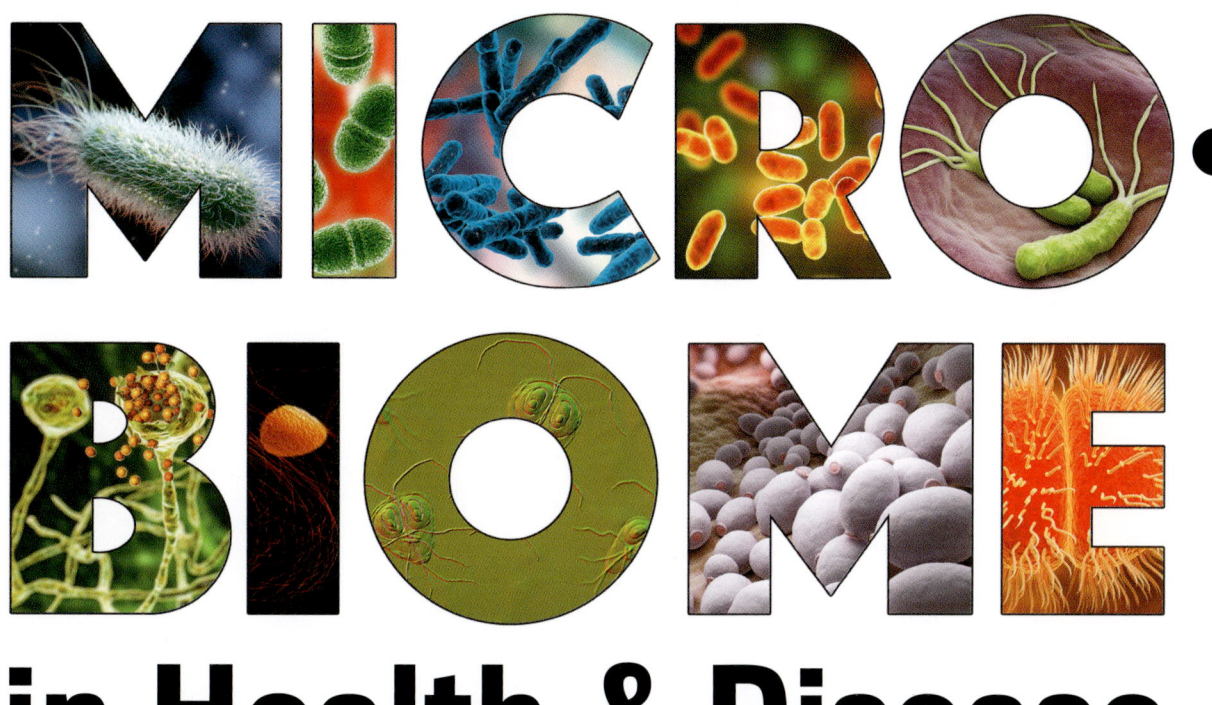

Margaret Riley
University of Massachusetts Amherst

Princeton University Press
Princeton and Oxford

Copyright © 2026 by Princeton University Press

Princeton University Press is committed to the protection of copyright and the intellectual property our authors entrust to us. Copyright promotes the progress and integrity of knowledge created by humans. By engaging with an authorized copy of this work, you are supporting creators and the global exchange of ideas. As this work is protected by copyright, any reproduction or distribution of it in any form for any purpose requires permission; permission requests should be sent to permissions@press.princeton.edu. Ingestion of any IP for any AI purposes is strictly prohibited.

Published by Princeton University Press
41 William Street, Princeton, New Jersey 08540
99 Banbury Road, Oxford OX2 6JX

press.princeton.edu

GPSR Authorized Representative: Easy Access System Europe - Mustamäe tee 50, 10621 Tallinn, Estonia, gpsr.requests@easproject.com

All Rights Reserved

ISBN (pbk.) 9780691243214
ISBN (e-book) 9780691243252

Library of Congress Control Number: 2025930778
British Library Cataloging-in-Publication Data is available

Editorial: Sydney Carroll and Johannah Walkowicz
Production Editorial: Mark Bellis
Text and Cover Design: Wanda España
Production: Jacquie Poirier
Publicity: William Pagdatoon

Cover images from left to right: iStock.com/peterschreiber.media, Pasieka / Science Source, Kateryna Kon/Shutterstock, iStock.com/Dr_Microbe, iStock.com/iLexx, iStock.com/Dr_Microbe, Power and Syred/ Science Source, iStock.com/Dr_Microbe, iStock.com/iLexx, iStock.com/Gilnature.

This book has been composed in Sabon 10/12

Printed in China

10 9 8 7 6 5 4 3 2 1

To the unseen, yet indispensable, microbial companions within and around us, whose intricate worlds inspire endless curiosity and shape our health. And to my family, friends, and former students, for their unwavering support and steadfast belief in this journey of discovery.

Brief Contents

CHAPTER 1	The Ancient Origin of Microbes	1
CHAPTER 2	A Brief History of Microbiome Research	27
CHAPTER 3	The Human Holobiont	51
CHAPTER 4	Generating Microbiome Data	85
CHAPTER 5	Analyzing Microbiome Data	115
CHAPTER 6	Applying Microbiome Analysis	145
CHAPTER 7	Mother's First Gift	175
CHAPTER 8	The Microbiome and the Brain	209
CHAPTER 9	The Microbiome and Immunity	235
CHAPTER 10	Microbiome Dysbiosis	253
CHAPTER 11	The Microbiome and Obesity	273
CHAPTER 12	Allergic Diseases and the Microbiome	295
CHAPTER 13	Our Evolving Microbiome	321
CHAPTER 14	The Microbiome of the Built Environment	345
CHAPTER 15	Taking Charge of Your Microbiome	373

Contents

	Preface xix	
CHAPTER 1	**The Ancient Origin of Microbes**	1
	1.1 **IN THE BEGINNING**	2
	Origin of Life	3
	The Very First Cell	5
	BOX 1.1 RESEARCH IN ACTION Building Life from Scratch—The Quest for a Protocell	6
	Competition Drives Diversification	7
	Anaerobic versus Aerobic Respiration	7
	Photosynthesis Evolves	8
	Endosymbiosis and the Origin of Eukaryotes	10
	1.2 **THE GREAT TREE OF LIFE**	12
	A Molecular Tree of Life	13
	The Three Domains of Life	14
	The Tiniest Microbes	16
	1.3 **MAKING THE INVISIBLE VISIBLE**	17
	The First Sightings of Bacteria	17
	Culturing the Invisible	18
	Extremophiles, Life on the Edge	18
	1.4 **THE MICROBES WITHIN US**	19
	A Universe of Microbes within Us	19
	How Much of You Is Human?	20
	1.5 **OUR MICROBIOMES, OUR HEALTH**	21
	Microbiomes and Human Nutrition	22
	Microbial Metabolites, Key to Human Health	22
	Reflections on Your Microbiome	22
CHAPTER 2	**A Brief History of Microbiome Research**	27
	2.1 **OUR FIRST VIEW OF MICROBES**	27
	Spontaneous Generation, or Not?	29
	BOX 2.1 Redi's Exemplary Experimental Design	30

		Modern Experimental Design	30
		The Germ Theory of Disease	31
		Koch's Postulates	32
		The Discovery of Antibiotics	34
		BOX 2.2 Discovery of Penicillin	34
	2.2	**THE GOLDEN AGE OF MICROBIOLOGY**	35
		Discovering Colonization Resistance	36
		Not Just Germs	37
		Anaerobic Culturing Methods	37
	2.3	**GENOMICS AND BIOINFORMATICS**	38
		Environmental Metagenomics	38
		Fecal Microbiota Transplantation	39
		Germ-Free Mice	40
	2.4	**BRINGING IT ALL TOGETHER**	41
		Fluorescing Microbes	42
		The Father of Microbiome Research	43
		Human Microbiome Project	45
		No One Else Has Your Exact Microbiome	45
		The Normal Human Microbiome	45

CHAPTER 3 The Human Holobiont — 51

3.1	**THE HUMAN HOLOBIONT**	51
3.2	**THE MANY HUMAN MICROBIOMES**	52
	A Healthy versus Normal Microbiome	52
	Microbiome in Names and Numbers	52
	Microbial Taxonomy	53
	The Core Microbiome	54
	Primary Functions of the Core	55
	Hallmarks of a Healthy Microbiome	56
3.3	**THE MICROBES IN OUR GI TRACT**	58
	Crowdsourcing for Carbs	59
	Cooperation and Conflict in the Large Intestine	61
	The Friendly Gut Phageome	63
	BOX 3.1 Communication between the Large Intestine Microbiome and Immune System	64
	Dysbiosis of the Large Intestine Microbiome	64
	BOX 3.2 Bacteriophages Protect Epithelial Cells in the Large Intestine	65
	Inflammatory Bowel Disease	66
	BOX 3.3 RESEARCH IN ACTION Fecal Amino Acids and Dysbiosis—Unlocking Crohn's Disease Therapies	68
3.4	**THE MICROBES IN OUR MOUTH**	69
	Archaeal Syntrophy	69
	Fungal Diversity of Unknown Function	69
	Creation of Dental Plaque	70
	The Diverse Roles of the Oral Microbiome	70

	Our Evolving Oral Microbiota	70
	Oral Microbiome Dysbiosis	71
	BOX 3.4 The Gum Microbiome and Gum Disease	72
3.5	**THE MICROBES ON OUR SKIN**	73
	A Nutritional Desert	73
	Primary Functions of the Skin Microbiome	74
	Skin Microbiome Dysbiosis	75
	Acne and Your Skin Microbes	75
3.6	**SAMPLING YOUR OWN MICROBIOME**	76
	Identifying a Testable Hypothesis	76
	BOX 3.5 Bacterial Growth Media Protocol	77
	Preparing Nutrient Media	77
	Experimental Controls Are Key	77
	Sampling Your Skin	78
	BOX 3.6 Culturing Your Skin Microbiome	79

CHAPTER 4 Generating Microbiome Data — 85

4.1	**AN OPPORTUNITY AND MANY CHALLENGES**	86
	Training to Become a Microbiome Scientist	86
	Designing a Microbiome Study	86
	Using Model Organisms When Practical	87
	A Testable Hypothesis Is Key	87
	Experimental Variables	89
	Exploring the Primary Literature	89
	Performing a Literature Search	90
4.2	**DESIGNING OUR STUDY**	90
	Choosing Our Subjects	91
	Statistical Power Is Key	91
	Testing for Statistical Significance	91
	The Power of Our Sample Size	94
	Taking Control of Our Experiment	95
	Cross-Sectional versus Longitudinal Study Design	96
	Experimental Data versus Metadata	96
4.3	**ENTERING THE EXPERIMENTAL PHASE**	97
	Obtaining Our Samples	97
	DNA Chemistry	98
	Extracting Metagenomic DNA	99
	Amplifying Metagenomic DNA	101
	Billions of Amplification Products	102
	High-Throughput DNA Sequencing	104
4.4	**THE "OMICS"**	106
	Metatranscriptomics	106
	Metaproteomics	107
	Metabolomics	109
	Spatial Omics	109

CHAPTER 5 Analyzing Microbiome Data — 115

- 5.1 **THE DATA ANALYSIS PIPELINE** — 115
  - The Raw Data — 117
  - Garbage In, Garbage Out — 118
- 5.2 **RECONSTRUCTION OF GENES AND GENOMES** — 118
  - Locating Open Reading Frames — 119
  - Predicting Sequence Functions — 120
- 5.3 **MICROBIOME DATA ANALYSIS** — 121
  - Online Data Analysis Package Programs — 121
  - Operational Taxonomic Units (OTUs) — 121
  - Assigning Taxonomic Identities — 121
  - Species Accumulation Curves — 124
- 5.4 **SPECIES DIVERSITY MEASURES** — 126
  - Alpha Diversity — 126
  - Alpha Diversity Indices — 129
  - Faith's Phylogenetic Diversity Metric — 132
  - Beta Diversity — 135
  - Bray-Curtis Dissimilarity Metric — 135
  - Unique Fraction Metric — 137
- 5.5 **VISUALIZING DIVERSITY ESTIMATES** — 139
  - Principal Component Analysis — 139

CHAPTER 6 Applying Microbiome Analysis — 145

- 6.1 **THE SCIENTIFIC PRIMARY LITERATURE** — 146
- 6.2 **DISSECTING A RESEARCH ARTICLE** — 148
  - Abstract — 150
  - Introduction — 152
  - Results — 153
  - Methods — 154
  - Data Analysis — 154
  - Discussion — 157
- 6.3 **INTRODUCTION TO QIITA** — 158
  - The Basics — 158
  - Diving Deeper into Qiita — 163
  - Qiita-Based Taxonomic Distribution Analysis — 164
  - Qiita-Based Alpha Diversity Analysis — 166
  - Qiita-Based PCA Analysis — 169
- 6.4 **BEYOND QIITA** — 170

CHAPTER 7 Mother's First Gift — 175

- 7.1 **THE MICROBIOME OF THE FEMALE REPRODUCTIVE TRACT** — 176
  - Vaginal Community State Types — 177
  - Primary Functions of the Vaginal Microbiome Types — 178
  - Dysbiosis of the Vaginal Microbiome Types — 179

	7.2	**THE MOTHER'S MICROBIOME DURING PREGNANCY**	179
		What We Can Learn from the Mouse about a Mother's Microbiome during Pregnancy	180
		Immune Interactions between the Developing Fetus and the Maternal Microbiome	180
		The Maternal Impact on Development of the Fetal Immune System	182
	7.3	**THE BIRTHING PROCESS AND THE NEWBORN MICROBIOME**	186
		Vaginal Delivery	186
		Cesarean Section	186
	7.4	**THE INFANT'S CORE MICROBIOME**	187
		Structuring the Infant's Core Microbiome	187
		Wave after Wave of Microbial Colonization	189
	7.5	**BEYOND THE GUT MICROBIOME**	190
		The Newborn's Skin Microbiome	190
	7.6	**THE WONDER OF MOTHER'S MILK**	191
		The Composition of Breast Milk	191
	7.7	**TRANSITIONING TO SOLID FOODS**	194
	7.8	**ENVIRONMENTAL IMPACTS ON THE INFANT'S MICROBIOME**	194
	7.9	**HEALTH IMPACTS OF A NEWBORN'S DYSBIOTIC MICROBIOME**	196
		Antibiotics	196
		BOX 7.1 RESEARCH IN ACTION Antibiotics and the Newborn Gut—Long-Term Impacts on Microbiome Development	197
		Malnutrition	197
		Allergic Diseases	198
		Obesity	199
		Diabetes	200
	7.10	**MICROBIOME-BASED THERAPIES**	201
		Probiotics	201
		Vaginal Microbiome Transplant	202
		Fecal Microbiota Transplant	202
		Oral Probiotics	203
		Oral Prebiotics	203

CHAPTER 8 The Microbiome and the Brain — 209

	8.1	**THE NERVOUS SYSTEM**	210
	8.2	**THE MATERNAL MICROBIOME AND NEURAL DEVELOPMENT**	212
		BOX 8.1 RESEARCH IN ACTION Maternal Microbiota—Shaping the Fetal Brain's Biochemistry	215
		Formation of the Blood-Brain Barrier	216
	8.3	**THE MICROBIOTA-GUT-BRAIN AXIS**	217
		MGBA Communication via the Endocrine Pathway	218
		MGBA Communication via the Neural Pathway	220
		MGBA Communication via the Immune Pathway	221
		MGBA Communication via Autophagy	222
	8.4	**GUT MICROBIOTA AND NEUROPSYCHIATRIC DISORDERS**	222
		Depression	223
		Autism Spectrum Disorder	226
		Parkinson's Disease	227

8.5	**GUT MICROBIOME-BASED THERAPIES**	228
	Probiotic Therapies	229
	Prebiotic Therapies	229
8.6	**THE EVOLVED DEPENDENCE OF OUR MICROBIOTA**	229

CHAPTER 9 The Microbiome and Immunity — 235

9.1	**COMPONENTS AND ACTIVATION OF THE IMMUNE SYSTEM**	236
	Primary Components of the Immune System	236
	Activating the Immune System	237
	BOX 9.1 Components of the Innate and Adaptive Immune Systems	238
9.2	**THE GUT MICROBIOME AND IMMUNE RESPONSE**	239
9.3	**THE MUCOSAL FIREWALL INFLUENCES IMMUNE FUNCTION**	239
9.4	**IMMUNE SYSTEM AND EXTRA-INTESTINAL MICROBIOME CROSS TALK**	239
	Skin	240
	Lung	240
	Liver	241
9.5	**INFLUENCE OF THE MATERNAL MICROBIOTA DURING FETAL DEVELOPMENT**	242
	Immune Cells in Fetal Development	242
	Distinguishing Self from Non-Self	243
	Maternal Inflammation and Infection and the Fetus	243
9.6	**INFLUENCE OF THE MATERNAL AND FETAL MICROBIOTA DURING THE INFANT YEARS**	244
	Breast Milk	244
	Immunoglobulin A	245
9.7	**IMMUNE DISEASES AND DYSBIOSIS OF THE GUT MICROBIOME**	245
	Rheumatoid Arthritis	246
	Cardiometabolic Disease	247
	Cancer	248

CHAPTER 10 Microbiome Dysbiosis — 253

10.1	**DEFINING EUBIOSIS AND DYSBIOSIS**	254
	The Eubiotic Microbiome	254
	The Dysbiotic Microbiome	254
	More Is Not Necessarily Better	255
10.2	**AN ECOLOGICAL PERSPECTIVE ON THE MICROBIOME-HOST RELATIONSHIPS**	255
	Ecological Interactions between the Gut and Its Microbiome	255
	The Microbial Leash Metaphor	256
	Host Control of Microbiomes in Small versus Large Intestines	257
	Host Control of a Microbiome-Based Disease	258
	Microbial Control of the Gut Microbiota	259
	Dysbiosis, the Cause or Consequence of Disease?	260
10.3	**A FRAMEWORK FOR ASSESSING MICROBIOTA-DISEASE ASSOCIATIONS**	260
	Polymicrobial Synergy and Dysbiosis	261
	Polymicrobial Synergy in Gum Disease	262
	BOX 10.1 A Microbiome-Based Model of Periodontal Disease	263

		BOX 10.2 RESEARCH IN ACTION Breaking the Single-Species Myth in Periodontal Disease	264
		Shared Dysbiosis	265
		Is the Term *Dysbiosis* Still Useful?	266
		Koch's Postulates Applied to Microbiome-Health Associations	267
		BOX 10.3 A Comparison of the Original and Ecological Koch's Postulates	268

CHAPTER 11 The Microbiome and Obesity — 273

- **11.1 THE OBESITY EPIDEMIC** — 274
  - Losing Weight Is Hard to Do — 275
- **11.2 THE MICROBIOME OF OBESITY** — 276
  - Obese versus Lean Gut Microbiomes — 277
  - Microbial Guilds at Work in Obesity — 279
  - Not Just Your Gut Microbes — 281
- **11.3 METABOLIC MARKERS OF WEIGHT LOSS** — 281
  - Rendering Bile Acids Impotent — 283
  - Triggers of Fat Storage — 283
  - Mediating Low-Grade Inflammation — 284
- **11.4 THE GUT MICROBIOME AND WEIGHT LOSS** — 285
  - BOX 11.1 RESEARCH IN ACTION Microbiome Predictors—Biomarkers for Weight Loss Success — 286
- **11.5 DIET-BASED APPROACHES** — 287
  - Phytochemicals — 287
  - Fermented Foods — 288
  - Probiotics — 288
  - Prebiotics — 289
  - Fecal Microbiota Transplantation — 290

CHAPTER 12 Allergic Diseases and the Microbiome — 295

- **12.1 THE ALLERGIC RESPONSE** — 295
  - The Allergic Cascade — 296
- **12.2 A CRITICAL WINDOW OF IMMUNE TRAINING TO PREVENT ALLERGIC DISEASE** — 297
- **12.3 EPIGENETIC CHANGES AND ALLERGIC DISEASE** — 298
- **12.4 IMPACT OF THE MATERNAL MICROBIOME ON ALLERGIC REACTIONS** — 299
- **12.5 THE IMPACT OF THE ENVIRONMENT IN ALLERGIC DISEASE** — 300
  - The Response of Gut Microbes to Environmental Triggers of Allergic Disease — 301
  - The Farm Effect — 302
  - The Atopic March — 303
- **12.6 THE OLD FRIENDS AND BIODIVERSITY HYPOTHESES** — 303
- **12.7 THE ROLE OF ANTIBIOTICS IN ALLERGIC DISEASE** — 304
- **12.8 THE IMPACT OF THE MICROBIOME ON ALLERGIC DISEASE** — 305
  - Asthma — 306
  - Atopic Dermatitis — 309
  - Food Allergy — 311
- **12.9 MICROBIOME-BASED THERAPEUTICS FOR ALLERGIC DISEASES** — 313
  - Oral Immunotherapy — 313
  - Synbiotics — 314

		Fecal Microbiota Transplantation	314
		Faecalibacterium, *Lachnospira*, *Veillonella*, and *Rothia*	314
		Inulin	315
	12.10	A CIRCLE OF CAUSALITY	315

CHAPTER 13 Our Evolving Microbiome — 321

13.1	WHERE DID OUR MICROBIOME COME FROM?	322
	Our Closest Living Relatives	324
	Looking to Our Evolutionary Siblings	326
	Coprolites	326
	Dental Plaque	327
	Oral Microbiome	328
	BOX 13.1 Expansion of the Human Brain May Have Been Triggered by a Starch-Rich Diet	329
	Microbes from the Middle Ages	330
	Bacteriophages	330
13.2	THE INDUSTRIALIZATION OF OUR MICROBIOME	331
	The Hygiene Hypothesis	333
	The Disappearing-Microbiota Hypothesis	334
	Antibiotics	335
	Diet	335
	The Consequences of the Missing Microbiota	336
	Recolonizing a Vacated Niche	337
13.3	THE MICROBIOME AND THE MISSING-HERITABILITY PROBLEM	339
13.4	REACQUIRING OUR ANCIENT MICROBIOTA	339
	BOX 13.2 RESEARCH IN ACTION Lactose Digestion and the Microbiome—An Evolutionary Link	340

CHAPTER 14 The Microbiome of the Built Environment — 345

14.1	WHAT IS THE BUILT ENVIRONMENT?	346
	The Earliest Human-Built Environments	346
	The Earliest Homes Appear	346
14.2	WHAT IS THE MICROBIOME OF THE BUILT ENVIRONMENT?	347
	Health Impacts of the MoBE	348
	Controlling the MoBE	348
	Physical Factors Influence the MoBE	349
	Early Studies of the MoBE	349
	Human Microbial Clouds	350
14.3	MICROBIOLOGY OF THE BUILT ENVIRONMENT	350
	Constituents of the MoBE	351
	The Skin and Oral Microbiomes Contribute Most to the MoBE	351
	The Virome	352
	Plant Microbiomes	352

14.4 BE FACTORS THAT INFLUENCE THE MoBE — 353
- Humidity and Mold — 355
- Ventilation and Microbial Spread — 355
- Light Influences Which Microbes Survive — 355
- Indoor Plumbing — 355
- Cleaning Practices Impact the MoBE — 356

14.5 THE IMPACT OF THE MoBE ON HEALTH — 356
- Sick Building Syndrome — 356
- The Farm Effect — 357
- BOX 14.1 RESEARCH IN ACTION The Amish Advantage—How Dust Exposure Reduces Asthma Risk — 358

14.6 TRACKING MICROBES IN THE BUILT ENVIRONMENT — 358
- Tracking Hospital Pathogens — 359
- The Neonatal Intensive Care Unit — 360
- The Hospital MoBE — 361
- The Hospital Resistome — 362

14.7 MICROBIAL METABOLOMICS AND THE BE — 362
- Metabolites in the BE — 363
- Volatile Organic Compounds and the BE — 364

14.8 THE FUTURE MoBE — 365
- COVID-19 and the MoBE — 365
- Build Back Better — 365
- The Rewilding Hypothesis — 366
- Nature-Based Solutions — 366

CHAPTER 15 Taking Charge of Your Microbiome — 373

15.1 IS YOUR GUT MICROBIOME HEALTHY? — 374

15.2 SEEKING PROFESSIONAL ADVICE — 375
- Consulting a Medical Professional — 375
- Gut Microbiome Index — 376
- BOX 15.1 RESEARCH IN ACTION Gut Microbiome Health Index—A Predictor of Disease Probability — 377
- Consulting a Naturopath — 377

15.3 THE DIY APPROACH — 378
- Assessing Stool Quality — 378
- Assessing Gut Transit Time — 379
- Assessing Gut Microbiome Composition — 380

15.4 DEFINING A HEALTHY MICROBIOME — 384
- Fiber-Fermenting Bacteria — 385

15.5 THE HEALTHY PLATE — 387

15.6 MICROBIOME-BASED THERAPEUTICS — 388
- Probiotics — 389
- Prebiotics — 392
- Synbiotics — 393

15.7 MICROBIOME RECOVERY FOLLOWING ANTIBIOTIC USE   394
15.8 LET FOOD BE THY MEDICINE   396

Glossary   401
References   417
Index   435

Preface

The human microbiome—a dynamic and intricate ecosystem within and on our bodies—has captivated both the scientific community and the public. It is a realm where cutting-edge research intersects with everyday health, revealing profound connections between microbial life and our well-being. This textbook aims to bridge these discoveries with a structured understanding to empower students, educators, and researchers to navigate this transformative field.

The journey to creating this book began with a simple question: *How do microbes shape us?* I have been continually amazed by the microbiome's influence on human digestion, immunity, behavior, and even our evolutionary history. Yet, the more we uncover, the more we realize how much remains to be explored.

This book is designed to serve as a comprehensive resource, integrating foundational concepts with the latest research. Each chapter builds on core principles while delving into advanced topics, from microbial metagenomics to the interplay between diet, environment, and microbial communities. My hope is that this book will inspire readers to see microbes not merely as microscopic entities, but as vital partners in health and disease. Whether you are a student beginning your exploration of the microbiome, a seasoned researcher delving into its complexities, or an educator shaping the next generation of scientists, this text is for you.

As we journey into this microbial frontier, I invite you to explore, question, and envision the possibilities that await. Our expanding knowledge of the human microbiome is undoubtedly transforming the future of health and medicine.

ACKNOWLEDGMENTS

Let me begin by acknowledging the microbial communities that inhabit and shape our world, constantly reminding us of the interconnectedness of all life. It is my hope that this book contributes, even in a small way, to the growing appreciation and understanding of their vital roles.

To my students, past and present: your curiosity, questions, and enthusiasm have inspired me to think more deeply and communicate more clearly. You are the reason this textbook exists, and I hope it serves as a valuable resource in your academic and professional journeys. I am particularly grateful to Ally Brookhart and Sean Sullivan, for their invaluable editorial and writing assistance in the development of this textbook. Their expertise, dedication, and thoughtful contributions have greatly enhanced the clarity and depth of this work.

To the editorial and publishing teams at Princeton University Press: thank you for your professionalism, patience, and expertise in transforming my manuscript into this polished textbook. Your meticulous attention to detail and steadfast support were invaluable throughout this process.

Reviewers who provided feedback on specific chapters include Dobrusia Bialonska (University of North Georgia, Dahlonega), Heather A. Bruns (University of Alabama at Birmingham Heersink School of Medicine), Wen-Hsing Cheng (Mississippi State University), Nora Demers (Florida Gulf Coast University), Patrick M. Erwin

(University of North Carolina Wilmington), Sara Flood (Utah Valley University), Jack Gilbert (University of California San Diego), Bryan Hsu (Virginia Tech), Braden McFarland (University of Alabama at Birmingham Heersink School of Medicine), Oleg Paliy (Wright State University), and Mitch Walkowicz (University of Massachusetts Amherst).

Finally, to my husband: your unwavering faith in me and constant encouragement have been the foundation of this project. Thank you for always being my greatest cheerleader and for reminding me, through your love and support, that I could accomplish this dream.

The Ancient Origin of Microbes

1

CHAPTER CONTENTS

1.1 In the Beginning
1.2 The Great Tree of Life
1.3 Making the Invisible Visible
1.4 The Microbes within Us
1.5 Our Microbiomes, Our Health

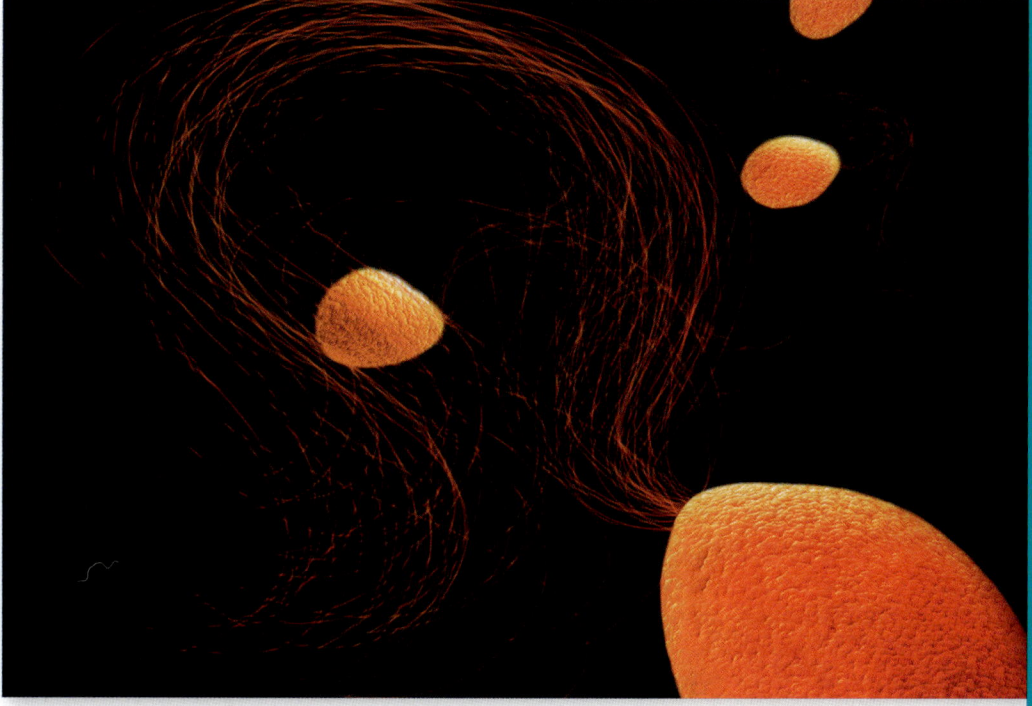

Hi there! My name is *Pyrococcus furiosus*. No fears, I am not a furious microbial monster. I am simply an extremophilic, hyperthermophilic archaeon that thrives in extremely hot environments. Maybe now you would prefer me to simply be furious! It isn't that hard to figure me out. I love hot! I mean I really, really love hot. My optimal growth temperature is a mere 100°C (or 212°F). I am also "allergic to oxygen," meaning I need to live in anaerobic environments, such as near hydrothermal vents. In fact, I was first found in waters near Italy, hanging out in a vent. Why should you care about little 'ole me? Well, I am a chemoorganotroph, meaning I break down sulfur to obtain energy. In the process I produce hydrogenases and amylases that are extremely heat-stable and efficient, which makes them valuable for some of your human industrial applications. So, a little kudos to me, please! (Photo from Power and Syred / Science Source)

Before we begin our exploration of the human microbiome, we must first develop an understanding of microorganisms, also called microbes—those minute creatures, far too small to be seen by the naked eye, that are both the creators and constituents of a breathtaking spectrum of microbiomes found on Earth. As you will learn, microorganisms emerged on our planet shortly after its origin and have spent over 4 billion years adapting to every conceivable environment our planet has to offer, including us! This first chapter provides an overview of the origins and diversification of microbes on Earth, with a special emphasis on what makes microbes so unique among life on our planet.

"If you don't like bacteria, you're on the wrong planet."
—Stewart Brand (Brand, 2014)

1.1 IN THE BEGINNING

If you could peer back in time to the birth of our planet, some 4.5 billion years ago (bya), what might you find? Certainly nothing even remotely resembling the Earth of today. Our young planet had no oceans, although there were plenty of volcanoes spewing out magma, water vapor, and gasses. It had no free oxygen in its atmosphere and no protective **ozone layer**, which is the thin layer of the Earth's atmosphere that absorbs most of the sun's harmful ultraviolet light. It would have been an exceedingly hot place—imagine a surface temperature upwards of 2,000° Celsius (3,632° Fahrenheit). An artist's rendition of early Earth shows a planet that does not appear even remotely hospitable to life (**Figure 1.1**).

Or was it? In fact, some of the earliest signs of life appear in 3.7 bya rock, formed when our planet was just beginning to cool from its volcanic origin (Dodd et al., 2017). Some of this ancient rock has survived the ages and paints a fascinating picture of early life. The dark gray peaks in the cross section of sedimentary rock shown in **Figure 1.2A** have tentatively been identified as fossilized microbial mats, also known as **stromatolites**, which are mounds of layers of lime-secreting bacteria and trapped sediment. Stromatolites were the only biological structures on Earth until about 540 million years ago (mya), and they can still be found in certain lagoons in Australasia (**Figure 1.2B**). In other words, regardless of how inhospitable early Earth might look to us, by 3.7 bya Earth was already teeming with life!

The word **microbe** literally means "small life," from the Greek words *mikros* and *bios*. Microbes are small life forms that are usually too small to be seen without magnification. As we shall learn, they represent the greatest diversity of life on our planet. Although most of us are aware microbes exist, we may be unaware that they appeared very early in Earth's history and have remained the dominant life forms ever since. Exploring present-day **hydrothermal vents** in the seafloor provides valuable clues about how these earliest life forms flourished in the extreme environments of our young planet. Heated, mineral-rich water flows out of these seafloor vents, and it supports untold numbers of **chemolithotrophs**, which are bacteria that harvest energy from the minerals and chemicals that spew from the vents and release compounds that other microorganisms then use for food. Fossils of hydrothermal vents have been discovered in rock as old as 3.8 bya (Cavalazzi et al., 2021).

Figure 1.1 Early Earth This artist's rendition provides a glimpse of what early Earth may have looked like. Our planet coalesced just over 4.5 billion years ago from cosmic debris. Transient oceans and lakes existed from the start, although they had been repeatedly vaporized by the massive meteorites that showered our planet back then. The environment of the planet had settled down by about 3.8 million years ago, when the earliest rocks appear in the fossil record in what is now southeast Greenland, and the planet might have looked as this artist portrays it. (Photo © Don Dixon)

(A)  (B)

Each stromatolite is built up from many thin layers of different bacterial species living together.

Figure 1.2 Ancient Microbial Fossils (A) The earliest fossil evidence of microbial communities. The layering in this rock is very likely due to biological activity. Cyanobacteria form mats of cells that secrete sticky substances that trap sediments in the surrounding water. Over time, these sediments form a mat and then new layers of Cyanobacteria attach. Layers of volcanic ash compacted against these structures, preserving them in the Greenland fossil record for the past ~3.7 billion years. Small fossils like these, buried under billions of years of collected rock, allow us to learn more about life in the distant past. (B) A cluster of living stromatolites from Shark Bay, Australia. There are very few such structures remaining on Earth. (A photo from Muséum de Toulouse, CC BY-SA 4.0, via Wikimedia Commons; B photo from Paul Harrison, CC BY-SA 3.0, via Wikimedia Commons)

One microbial species commonly found in vents, *Methanopyrus kandleri*, uses hydrogen gas as a food source and releases methane as a waste product. This process is known as **methanogenesis**, and it is one of the most ancient forms of energy production. The name of this microbe describes its fondness for extreme environments; *methanopyrus* literally means "methane fire," which is highly appropriate as it can grow in temperatures up to 122°C (252°F), the highest temperature known to be compatible with life. Consider that water boils at 100°C; with this in mind, we can begin to imagine how life emerged on what we had previously considered to be an inhospitable early Earth.

Origin of Life

If we can't rewind the tape of time and return to early Earth, can we ever learn about life's origins? In 1953, a young scientist, Stanley Miller, and his mentor, Harold Urey, showed us the way by answering the question: Could the complex organic molecules necessary for life be created under the conditions of our planet billions of years ago? Miller and Urey designed a glass chamber in which they could create conditions that were believed to mimic those on early Earth (**Figure 1.3**). Starting with simple ingredients, such as heat, which would have been provided by the Earth's molten core; an electrical charge to mimic lightning; water (H_2O); and an early atmosphere made of methane (CH_4), hydrogen (H_2), and ammonia (NH_3) gasses, Miller and Urey showed that complex **organic molecules** could be created from what was a predominately inorganic planet. Organic molecules are primarily made of carbon atoms bonded with hydrogen and other elements and are of biological origin. All living things on Earth are composed of organic molecules. In contrast, inorganic compounds are substances that do not contain both carbon and hydrogen. Hydrogen atoms are contained in many inorganic compounds, such as water (H_2O) and the hydrochloric acid (HCl) produced by your stomach. In contrast, only a handful of inorganic compounds contain carbon atoms. Carbon dioxide (CO_2) is one of the few examples. Miller and Urey showed that with heat, electricity, and simple inorganic ingredients, complex organic molecules, such as amino acids, could be produced. Amino acids are

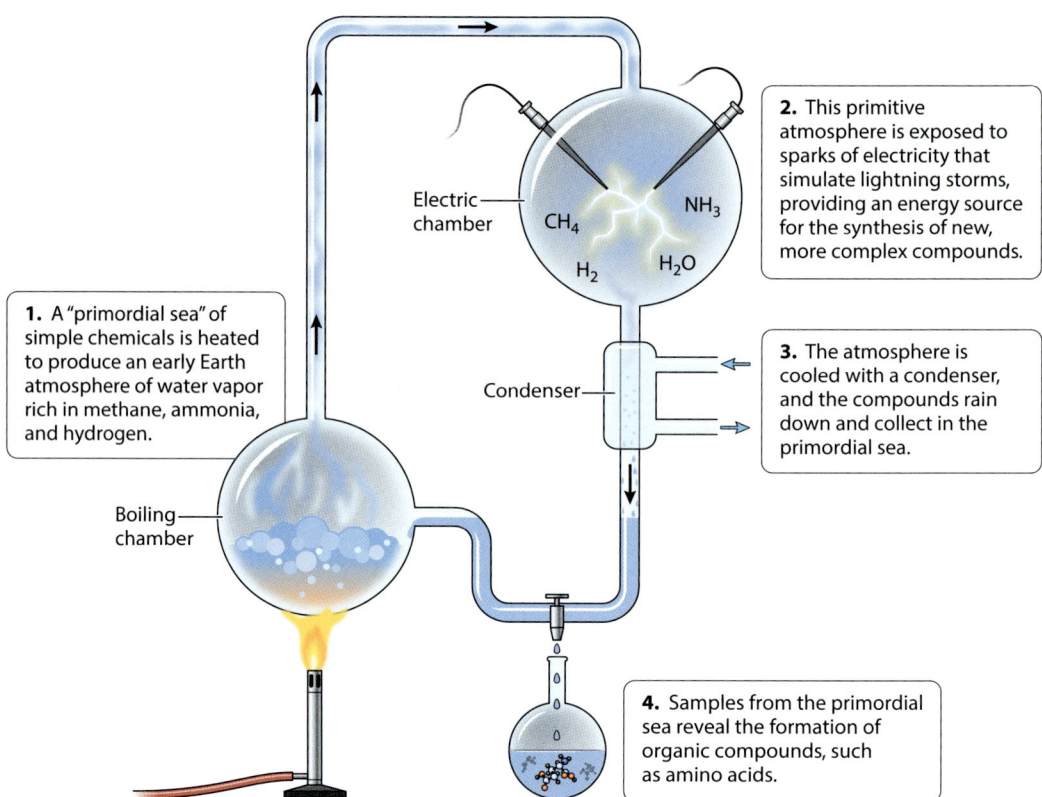

Figure 1.3 The Miller-Urey Origin of Life Experiment This experimental apparatus was designed to simulate the origin of organic compounds on early Earth.

the building blocks of **proteins**, the workhorses of cells that carry out many biological functions.

Miller and Urey's findings were extraordinary for several reasons. First, their data suggested that life could have arisen from the simple ingredients present in the "primordial soup" found on early Earth. We now know that many of the essential building blocks of life, such as amino acids and **nucleotides** (the key ingredients of **deoxyribonucleic acid**, or **DNA**), would have rapidly accumulated from simple inorganic constituents. Furthermore, this was the very first experiment in what was to emerge as a rich and exciting field of **abiogenesis**, or the study of the creation of life from nonlife. Their publication helped transform studies of the origin of life into a respectable field of research.

By the 1990s many scientists agreed that at least two functions were required for cellular life to emerge from a nonliving precursor: a means to physically separate the cell's internal functions from the environment (a membrane), and the ability to generate offspring, which involves copying the genetic information and producing daughter cells (replication). **Figure 1.4** provides an overview of how a cell's genetic information, DNA, is transcribed into RNA, which is then used to produce proteins. This theory is known as the **Central Dogma of Molecular Biology**, and there were vigorous debates about which element (DNA, RNA, or protein) came first. One argument emphasized the role of genetics and inheritance (**replication argument**), that is, DNA or RNA was first on the scene. A competing view proposed that creating the cells' structure came first (**membrane argument**), that is, proteins creating the structure of a cell came first. It was at this moment that a particularly innovative thinker entered the field. Jack Szostak, who won a Nobel Prize for his work on **chromosome structure**, which refers to the way DNA is packaged in a cell, decided to retool his lab and to focus on an entirely new research question—the origin of life. Szostak was

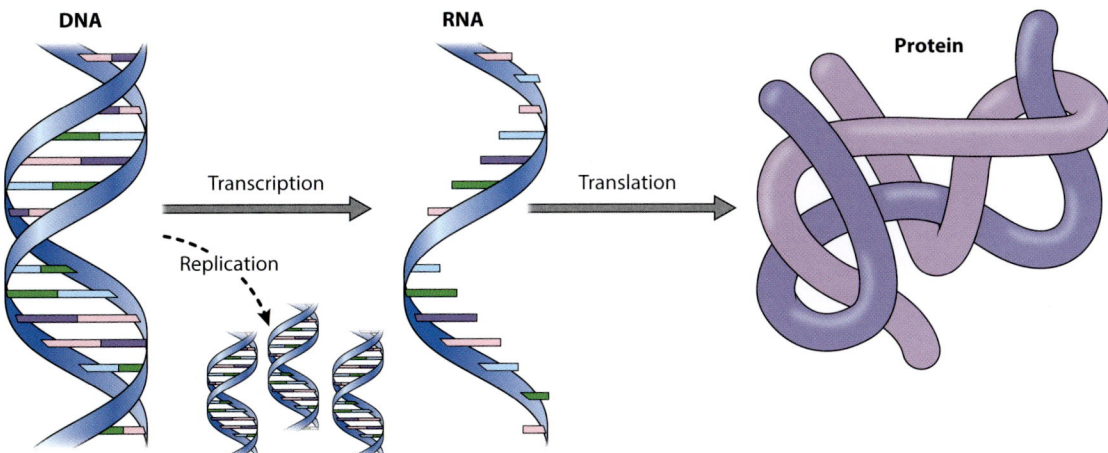

Figure 1.4 Central Dogma The Central Dogma of Molecular Biology describes the fundamental biological process by which proteins are built. In most cells, the genetic information is encoded in the DNA, which can be transcribed into a messenger molecule known as RNA. This RNA is then translated into proteins, using complex cellular machinery to "read" the RNA sequence and build the corresponding protein structure. The Central Dogma is essential to our understanding of early Earth because it illustrates the connection between genetic information and cellular processes.

fascinated with the origins debate, and he set off to create a primitive cell, or **protocell**, that would permit him to experimentally explore how the first cell might have evolved (**Box 1.1**).

The Very First Cell

Szostak knew that fatty acids could transition from small spheres (or **micelles**) into multilayered membranes as the pH of the local environment goes from a basic to a more neutral state, so he decided to simply add more membrane-forming molecules (**fatty acids**) to the mix and see what happened (see Box 1.1 and Figure 1.5B). The team added fatty acids, some of which inserted themselves into the cell's membrane. This spontaneous growth process transformed the small spheres into long filamentous vesicles, which could be induced to divide when agitated and then to re-form cells when the agitation stopped. This elegantly simple protocell appears to possess one of the key characteristics of life—a cellular structure that could make copies of itself.

Szostak had proven that the earliest cells could have created a protected environment in which metabolism could take place. Next, he tested whether the RNA fragments located in these protocells were able to replicate, or make copies of themselves, which would permit the identical genetic information to be passed on when the protocell divides. Given that RNA is capable of both replicating itself and performing enzymatic activities, it is often considered the likely ancestor to our own DNA-based mode of inheritance. However, RNA requires high concentrations of magnesium, which can destroy the delicate membranes of the cell. Szostak found conditions that protected the membrane but still provided sufficient magnesium to permit RNA replication. These experiments provide us with a membrane-bound genetic system that is capable of self-replication and growth—two of the hallmarks of life. All done in test tubes in a laboratory!

Since this revolutionary experiment, Szostak and many others continue to dive ever deeper into questions about the origin of life (Mann, 2021). One current focus is on planetary habitability, or the potential for planets to develop and sustain life. A second research area examines the environmental conditions required to produce **biomolecules** (such as carbohydrates, lipids, nucleic acids, and proteins) in concentrations that permit metabolism. A third focus is on determining the ways in which the precursors to DNA and RNA might have assembled and replicated. These studies are just beginning to answer some fundamental questions about the origin of life (Mann, 2021). Szostak himself notes, "Many challenges remain before we will be close to a full understanding of the origin of life, so the future of research in this field is brighter than ever!" (Szostak, 2017).

BOX 1.1. RESEARCH IN ACTION
Building Life from Scratch—The Quest for a Protocell

Researchers in the Szostak lab made their first protocell out of self-replicating genetic material, in this case a fragment of RNA. **Figure 1.5A** shows the organization of this protocell and reveals the inner compartment created by this structure and the fragments of RNA floating within. This internal environment would have permitted the cell to carry out key functions, such as metabolism, which allows the cell to transform food into energy. However, the first cells would have had none of the machinery needed for their own growth and division. The researchers hypothesized that the coupled growth and division of protocells could be achieved using conditions likely to have been present on Earth over 3 billion years ago.

❖ **Experiment.** A protocell is placed in dilute acid, such that the interior of the protocell is slightly more acidic than the solution. Osmosis will cause water to enter the protocell, resulting in large (~4 mm in diameter) vesicles. Fatty acids are then added to the mixture and modest shear forces are provided (**Figure 1.5B**).

❖ **Results.** The growth of small protocells is achieved by placing them in a solution where liquid permeates the membrane, resulting in the transformation of initially spherical vesicles into long threadlike vesicles that can divide into multiple daughter vesicles.

❖ **Conclusion.** This experiment shows that protocells can be created, enlarged, and replicated in the laboratory, suggesting that similar processes might have occurred under the prebiotic conditions of early Earth.

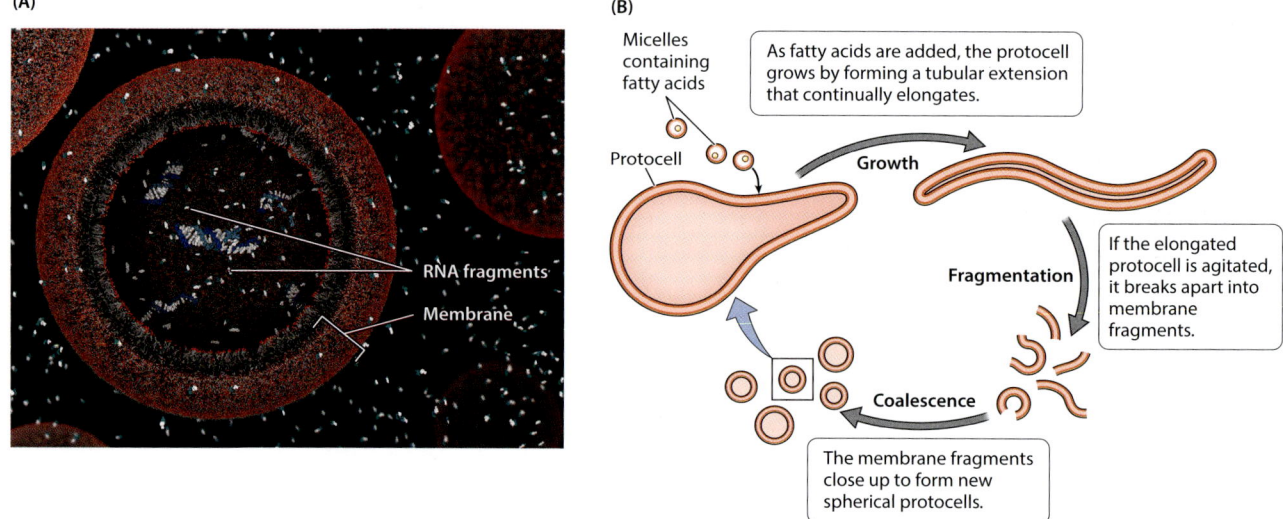

Figure 1.5 The Protocell (A) A computer-generated image of the type of protocell created by the Szostak lab. The protocell is spherical but is shown in cross section here so the inside can be seen. The lipid membrane (red outer circle) provides an internal environment for the protocell to store and replicate its genetic material and undergo metabolic processes to generate energy. Noticeably lacking are more-complex cellular structures that you may already be familiar with, such as a nucleus or mitochondrion. The protocell is surrounded by a "primordial soup" consisting of inorganic and organic molecules. Most of these were small, but some were more complex, such as RNA fragments. (B) The proposed cyclical process of protocell membrane growth and division. The cell incorporates micelles that cause its size to increase until it reaches a point where agitation results in its splitting open. The resulting fragments of the original protocell then reconfigure into new cells. This series of events is a precursor to the modern cell cycle. (A photo courtesy Janet Iwasa, Szostak Laboratory, Harvard Medical School and Massachusetts General Hospital; B after Zhu and Szostak, 2009.)

Competition Drives Diversification

Szostak's research shows us how an ancestral life form could have emerged on early Earth. However, these primitive processes were inefficient. Each time a new protocell was formed, a new RNA fragment would have been captured in the cell, which would have encoded completely different functions, or none at all. We envision the cycle shown in Figure 1.5B repeating itself billions, if not trillions, of times. Some cells captured RNA that encoded novel functions, and those cells might have survived longer and had a greater likelihood of "reproducing," which at this point means that the protocell divided into two daughter cells that share the same RNA fragment. Imagine a primitive ocean filled with trillions upon trillions of protocells. Those that had features that resulted in the production of more copies would consume more ingredients, which would then not be available for others.

The process just described is known as **natural selection**, and it is one of the most powerful forces affecting life on Earth, through which populations of living organisms adapt to the ever-changing environment. Organisms more adapted to their environment are more likely to survive and pass on the genes that contributed to their success. This process, natural selection, causes species to change and diverge over time. Individuals in a population are naturally variable, meaning that they all differ in some ways. This variation means that some individuals have traits better suited to the present environment than others. Individuals with **adaptive traits**—traits that give them some advantage—are more likely to survive and reproduce. These individuals then pass the adaptive traits on to their offspring. Over time, these advantageous traits become more common in the population. Through this process of natural selection, favorable traits are transmitted through generations, and organisms adapt to their environment.

Overwhelming evidence shows us that all **extant** species (meaning that they are alive today) are related, having descended from a common ancestral protocell. We call this extinct organism the **Last Universal Common Ancestor**, or **LUCA**. LUCA was very likely a single-celled **autotroph**, which means it was able to make its own energy and relied on available inorganic compounds as a food source. It is envisioned that LUCA engaged in **chemolithoautotrophy**, meaning it obtained energy by oxidizing inorganic compounds (like hydrogen or sulfur) and fixing carbon dioxide to produce organic molecules. Its genetic material was almost certainly DNA, and it employed RNA molecules, such as tRNA and mRNA, in translating the information encoded in its DNA into proteins.

Heterotrophs would have emerged next, which are organisms that lack the ability to make their own organic compounds. Instead, heterotrophs obtain their energy by breaking down complex organic molecules, such as carbohydrates, fats, and proteins, which they acquire from other organisms—either by eating plants, animals, or decomposing organic matter. As heterotrophs reproduced and became more numerous, they would have rather quickly consumed the organic compounds being produced by autotrophs, resulting in selection for organisms capable of using alternative foods.

Anaerobic versus Aerobic Respiration

These earliest heterotrophs evolved on a planet with an atmosphere composed of methane, ammonia, and hydrogen cyanide, which derived primarily from the gasses emitted from volcanoes. Free oxygen was present at only trace levels. Therefore, the earliest Earth ecosystems existed in an **anoxic** world, devoid of oxygen, and the microbial communities present were supported by anaerobic respiration. Cellular respiration is the process by which cells break down sugar and turn it into energy, which is then used to perform work at the cellular level. The most primitive form happens in the absence of oxygen and is called anaerobic **respiration** (**Figure 1.6A**). Early anaerobic microbes used chemicals to derive energy for respiration by mediating the oxidation and reduction of inorganic compounds in their environments. For example, methanogens obtain their energy from hydrogen (H_2) and carbon dioxide (CO_2) and release methane (CH_4) as a waste product, hence their name. Similarly, sulfate-reducing

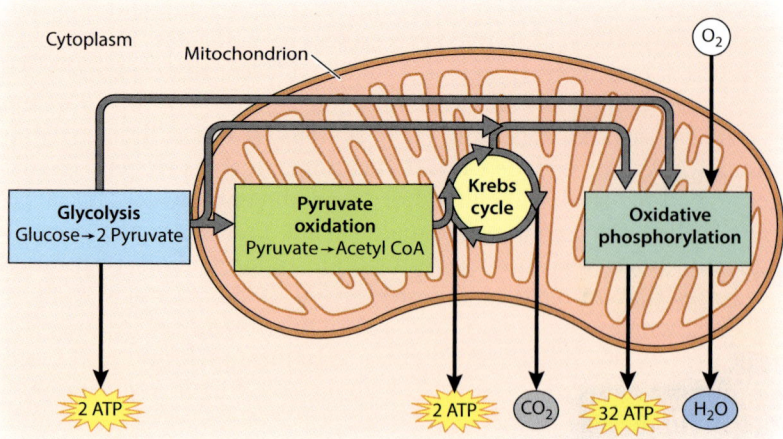

Figure 1.6 Cellular Respiration Cellular respiration is the process by which cells release energy by breaking down sugar molecules, such as glucose. (A) Anaerobic respiration, the most primitive form of respiration on Earth, is how cells convert the stored energy of glucose into adenosine triphosphate (ATP) in the absence of free oxygen. It provides energy to the cells very rapidly. (B) Aerobic respiration is the process through which cells break down the glucose molecule to convert its stored biochemical energy into ATP in the presence of oxygen.

microbes feed on sulfate. These **chemoautotrophs**, which use chemicals for energy, would have had an enormous supply of inorganic chemicals to feed on and are still commonly found in environments rich in inorganic compounds, such as near deep-sea hydrothermal vents.

The anaerobic **biosphere** of early Earth, that is, the regions of the planet occupied by living organisms, was less energetically active than our present-day aerobic biosphere—in other words, energy flow from chemicals into and between microbes was slow, roughly 5% of the rate of energy conversion found in our current biosphere. Life forms would have been engaged in intensive competition for the limited energy sources, which would have driven the process of natural selection, resulting in novel approaches to finding and harvesting energy.

Some microbial lineages evolved the ability to use oxygen as an energy source, which we refer to as aerobic respiration (**Figure 1.6B**). This novel form of respiration converts glucose or other organic molecules into energy in the form of **ATP**, or **adenosine triphosphate**, which is essential for various cellular functions. ATP uses the energy stored in its phosphate bonds to power chemical reactions. It is often referred to as the "currency" of the cell. Although anaerobic respiration also produces ATP, aerobic respiration is much more efficient, and it produces ATP much more quickly. This is because oxygen is an excellent electron acceptor for the chemical reactions involved in generating ATP.

Photosynthesis Evolves

One of the truly great metabolic innovations involved the ability to harness the sun's energy, a process called **photosynthesis**. The earliest photosynthetic organisms evolved specialized pigments capable of extracting energy directly from sunlight. These pigments captured the sun's energy and used it to transform carbon dioxide and water into carbohydrates (food) and oxygen (waste product) (**Figure 1.7A**). The first organisms capable of photosynthesis were the ancestors of the modern-day **Cyanobacteria**, a phylum of bacteria also known as blue-green algae. Thanks to photosynthesis, these organisms no longer needed to rely on a limited pool of organic

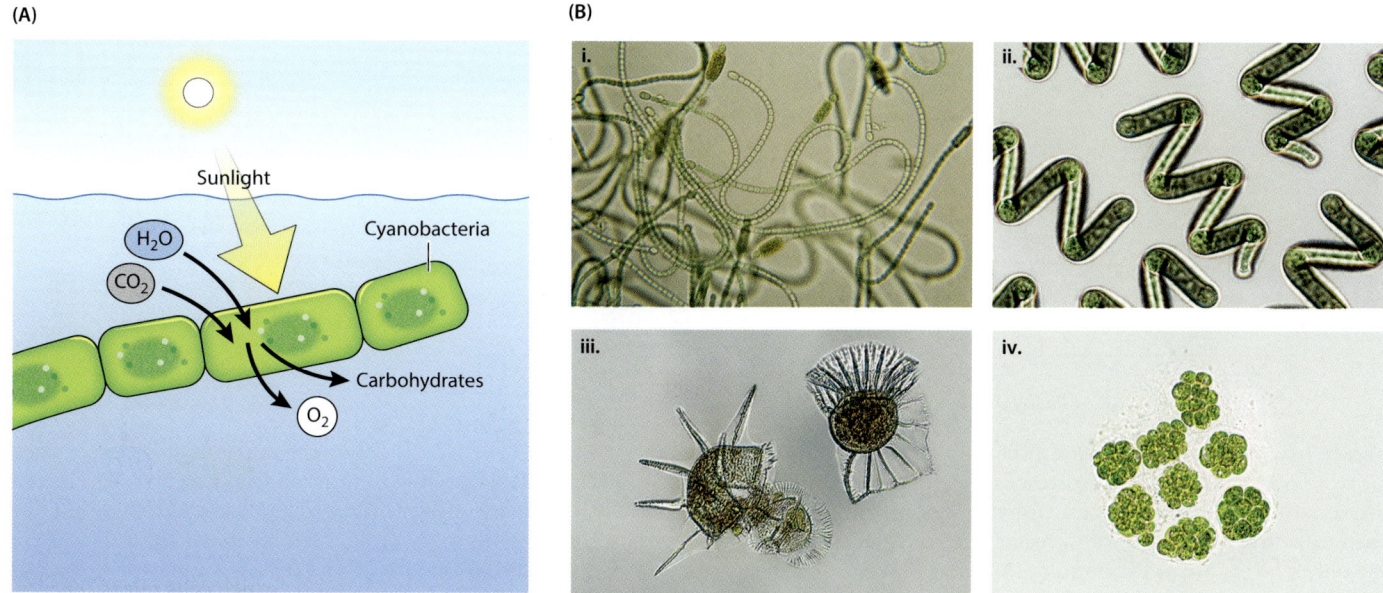

Figure 1.7 Cyanobacterial Photosynthesis and Diversity (A) Cyanobacteria use the energy of sunlight to drive photosynthesis, a process where the energy of light is used to synthesize organic compounds from carbon dioxide and water, resulting in oxygen as a waste product. (B) Some of the diverse types of cyanobacteria. Left column: Blue green algae (top), Dinophysis algae (bottom). Right column: Spirulin (top), Pandorina (bottom). (B photos from [i] istock.com/Nnehring; [ii, iii, iv] iStock.com/Elif Bayraktar)

molecules or engage in the far slower process of extracting energy from chemicals and could instead get their energy directly from the sun, which offered them a profound selective advantage. Descendants of these very first photosynthetic cells can be found in almost any water source you examine today. **Figure 1.7B** provides a snapshot of some of the stunning and diverse members of this ancient lineage.

Cyanobacteria played a key role in transforming early Earth's biosphere. Every time a cell broke down a molecule of carbon dioxide, it would release a molecule of oxygen as waste. Imagine trillions upon trillions of cells, each puffing out oxygen over the millennia. At first this free oxygen was captured by minerals, which we see as massive iron oxide (or rust) deposits in the geological record about 2.5 bya (**Figure 1.8**). Once these minerals were saturated with oxygen, the excess began to accumulate in the atmosphere. This period in Earth's history is referred to as the **Great Oxidation Event** (**GOE**), in which the atmosphere was transformed into one rich in oxygen, like Earth's atmosphere today, which is 78% nitrogen (N_2), 21% oxygen (O_2), 0.93% argon (Ar), 0.04% carbon dioxide (CO_2), and trace levels of other chemicals. Figure 1.8 shows the dramatic impact of the GOE on levels of atmospheric oxygen on Earth.

The rising levels of oxygen resulted in one of the first mass **extinction** events on our planet. A mass extinction event is identified when species go extinct faster than new species evolve, defined as about 75% of the world's species being lost in less than 3 million years. Oxygen is toxic to anaerobic bacteria, which do not possess mechanisms to protect their enzymes from oxidants, and thus, most did not survive this period of atmospheric transformation. A lucky few found ways to avoid the oxygen. For example, it is likely that the ancestors of modern-day methanotrophs, microorganisms that produce methane (CH_4) as a by-product of their metabolism, would have continued to flourish in the so-called dead zones in the ocean (areas where the levels of oxygen remain low) and deep in the ocean floor. Our fossil record of that time is limited, and given the microscopic size of the organisms, we are forced to infer features of these ancient life forms.

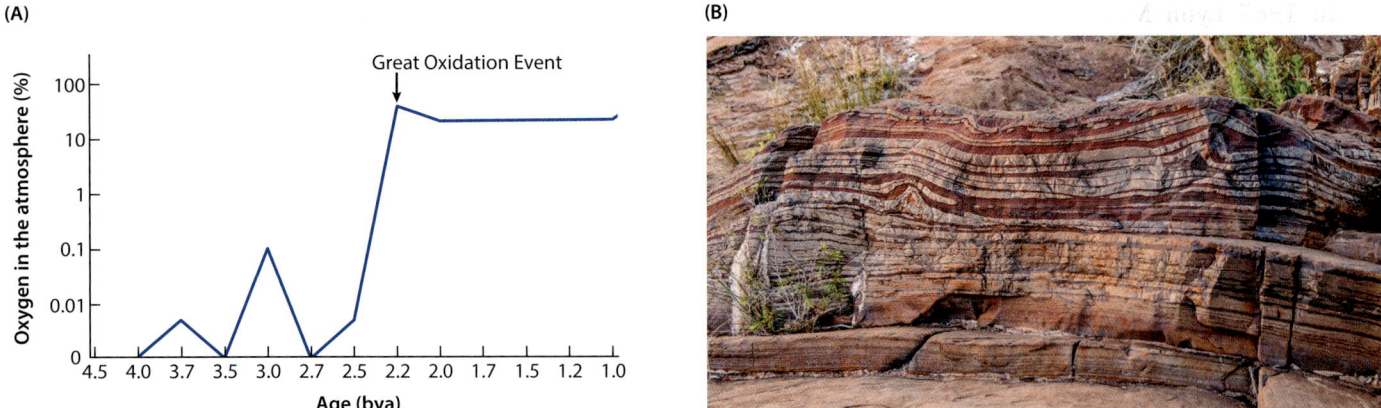

Figure 1.8 The Great Oxidation Event (A) Timeline of atmospheric oxygen levels on early Earth. Note the dramatic increase, labeled Great Oxidation Event, that corresponds with the saturation of minerals with oxygen, resulting in iron oxide sediments. (B) The red inserts of sedimentary rock are rust deposits providing evidence of the Great Oxidation Event. Rust is the common name for the chemicals that result when iron reacts with oxygen and water. Sedimentary rock is built by layering different rocks and soils, where the oldest layers are at the bottom. (A after R.A. White III 2020; oxygen data were provided by Dr. Sean Crowe [University of British Columbia] with permission, D.E. Canfield 2005, C. Dupraz and P. T. Visscher 2005, T. W. Lyons et al. 2014, and S.A. Crowe et al., 2013; B photo from Graeme Churchard from Bristol, UK, CC BY 2.0, via Wikimedia Commons)

Endosymbiosis and the Origin of Eukaryotes

With the rise in atmospheric oxygen and the advent of aerobic respiration, large, complex multicellular organisms first appear in the fossil record. Multicellularity has several obvious advantages over single-celled life forms. One of the earliest selective pressures for it may have been related to the fact that a group of cells presents a great challenge for a predator. As cells group together, their survival rate increases. Further, multicellular organisms can have longer lifespans—the organism survives even when individual cells die. Finally, multicellularity also permits increasing complexity by allowing differentiation of cell types, or tissue specificity (Pentz et al., 2020). These changes paved the way for evolution of circulatory and respiratory systems and intestines that break down food sources and extract nutrients from them.

For the first half of the history of life on Earth, single-celled **prokaryotes**, whose genetic information is found floating in the cell's cytoplasm, were the sole inhabitants. However, sometime around 2 bya, a new type of cellular life form arose: the **eukaryotes**, whose DNA is enclosed in a protective membrane called the **nucleus**. **Figure 1.9** provides a comparison of a simple prokaryote and a more complex eukaryotic cell.

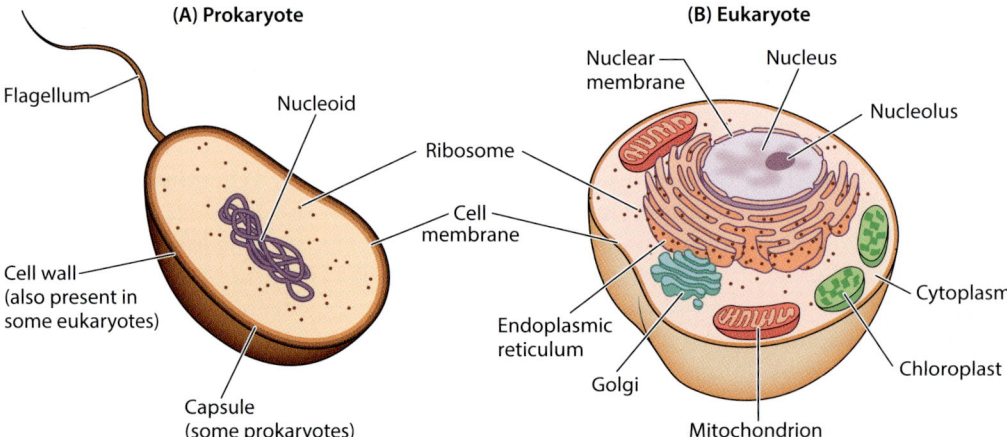

Figure 1.9 Prokaryote versus Eukaryote Cellular Complexity This diagram illustrates the similarities and differences between prokaryotic and eukaryotic cells. Both contain genetic material, a cell membrane, and ribosomes. Eukaryotic cells also contain membrane-bound organelles, such as the nucleus, mitochondria, and Golgi body, whereas prokaryotic cells do not.

In 1967 Lynn Margulis, a microbiologist and evolutionary biologist at the University of Massachusetts, proposed that the eukaryotic cell was the result of a chance fusion between two prokaryotes. An ancestral prokaryotic host cell engulfed, but didn't digest, a second prokaryotic cell, one capable of aerobic metabolism (**Figure 1.10A** shows this with a eukaryotic cell). The engulfed cell, or **endosymbiont**, provided its host with the ability to use oxygen to release energy stored in nutrients. In turn, the host cell protected the endosymbiont from predators. Over time, a **symbiotic** relationship, which refers to a close, long-term interaction between two different species, where at least one of the species benefits from the relationship, developed between the two organisms to the point that neither could survive on its own. This endosymbiotic event is immortalized in eukaryotic cells by the presence of the **mitochondrion**, which is the descendent of that ancient, engulfed aerobic symbiont and now serves as the energy factory in nearly all eukaryotic cells today.

Margulis's idea was largely ridiculed, and some 15 journals rejected her research findings before they were published (Sagan, 1967). She spent much of her career defending the hypothesis until enough experimental evidence was garnered to support its recognition as a valid theory. In fact, it is now clear that a series of symbiotic events (**serial endosymbiosis**) occurred. One endosymbiosis resulted in eukaryotic cells possessing a mitochondrion, which became the cell's energy factory (**Figure 1.10B**). Plant cells went even further, with chloroplasts resulting from a fusion of a heterotrophic bacterium with a photosynthetic cyanobacterium (**Figure 1.10C**). Chloroplasts are the membrane-bound organelles in plants and algae where photosynthesis takes place. Margulis was an extraordinary scientist, one who remained steadfast in her then-revolutionary belief that eukaryotic origins could be found. When questioned about the controversy surrounding her proposal of endosymbiosis, she replied, "I don't consider my ideas controversial. I consider them right" (Teresi, 2011).

With the advent of the eukaryotic cell, the diversification of life took on a whole new dimension. A tidal wave of biological diversification occurred about 540 mya. This period, known as the **Cambrian Explosion**, was literally that, an explosion of macroscopic life forms that appear all at once in the fossil record during the geological period known as the Cambrian. What was previously a planet dominated by microscopic prokaryotes is now rich with complex macroscopic, multicellular life

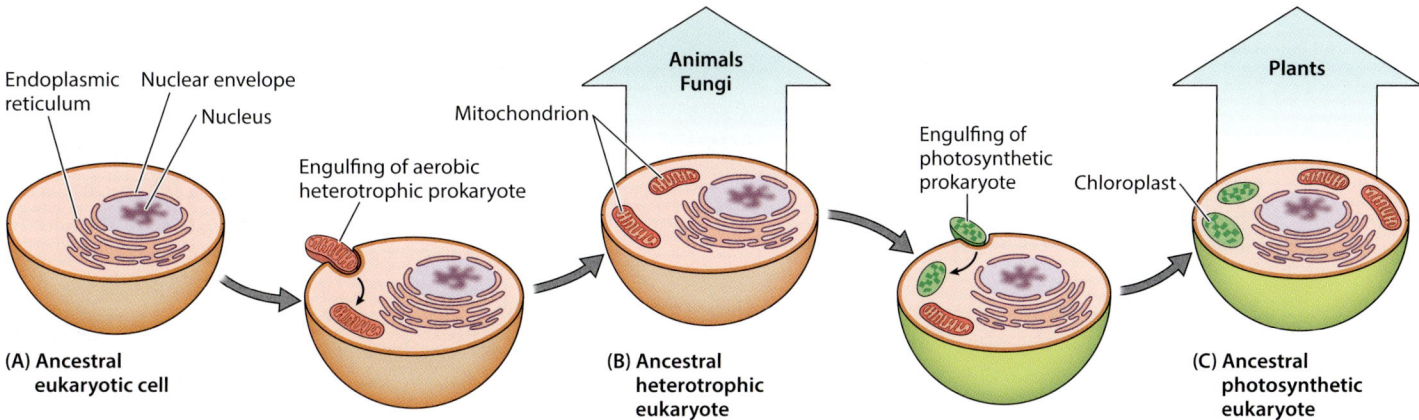

Figure 1.10 The First Endosymbiotic Events Imagine an ancestral eukaryotic cell (A), similar to a present-day amoeba. It engaged in phagocytosis, gaining energy from ingested organic matter, such as prokaryotic cells. The endosymbiotic theory posits that in several instances, the ingested cells survived and developed a symbiotic relationship with the host. Mitochondria (in B) and chloroplasts (in C) were the result of this process and were capable of aerobic respiration or photosynthesis, respectively.

1.2 THE GREAT TREE OF LIFE

Now that our planet is teeming with microscopic and macroscopic life, we need a system to name all this diversity. In 1735, Carolus Linnaeus proposed a hierarchical scheme of classification that started with the most inclusive groupings, **kingdoms**, and descended into smaller and smaller subgroups, ultimately ending with a **species** name. Linnaeus would assign each species a unique two-word Latin name, or **binomial**, such as *Homo sapiens*, the binomial for humans. It consists of the species designation (*sapiens* or "wise man") preceded by the **genus** (*Homo*). Genera were grouped into **families**, families into superfamilies, and so on until the level of kingdom was reached. **Figure 1.11** shows a portion of the hierarchical levels of the Linnaean classification system and provides an example of how the human species is classified. Beyond the level of order, humans are members of the class Mammalia, the phylum Chordata, and the kingdom Animalia.

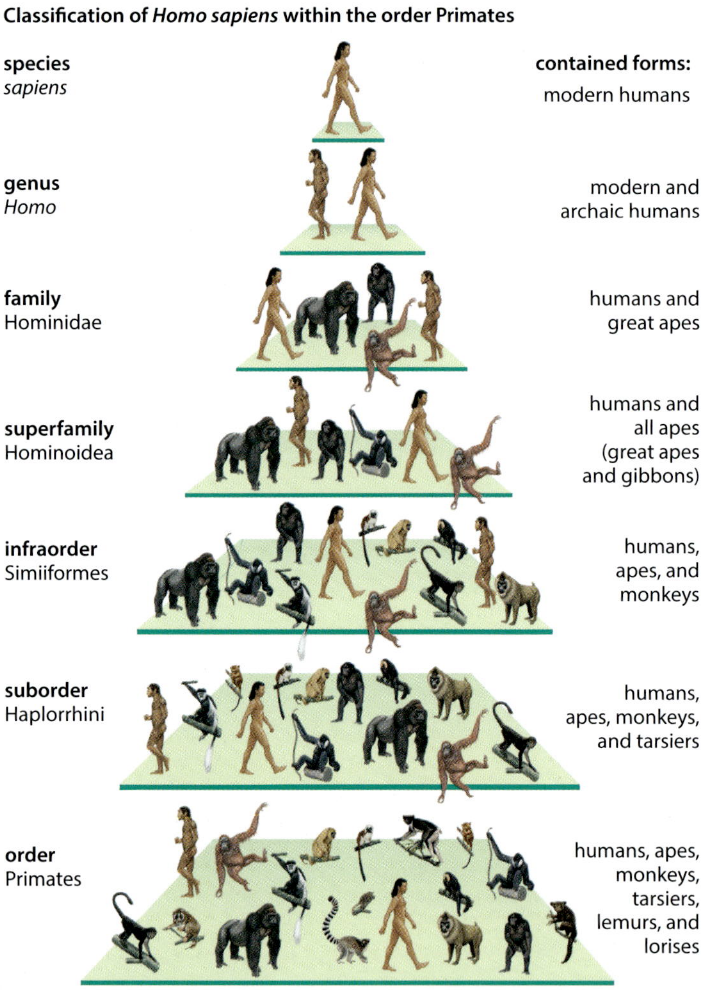

Although Linnaeus sought to classify organisms based upon similarities, his methods often resulted in clusters that reflected evolutionary relationships, which we can represent in a phylogenetic tree. A **phylogenetic tree** (also **phylogeny** or **evolutionary tree**) is a branching diagram showing the evolutionary relationships among organisms based upon similarities and differences in their physical or genetic characteristics. In 1859, when Charles Darwin published his thesis on the origin of species, he introduced the concept of a great **tree of life** (**ToL**) connecting all living and extinct life forms to a common ancestor (Darwin, 1859). He envisioned an ever-growing tree whose root is our common ancestor (LUCA), with the branches representing distinct lineages terminating in foliage, which represent the species. **Figure 1.12A** shows Darwin's illustration of his tree of life. He went so far as to describe the fallen limbs and leaves as those extinct lineages that we know only from the fossil record: "Buds give rise by growth to fresh buds, and these, if vigorous, branch out and overtop on all sides many a feebler branch, so by generation I believe it has been with the great Tree of Life, which fills with its dead and broken branches the crust of the earth, and covers the surface with its ever branching and beautiful ramifications" (Darwin, 1859).

The ToL envisioned by Darwin was transformed over the next hundred years as more and more organisms were discovered, described, and added. **Figure 1.12B** provides a version of the ToL popular in the mid-20th century, called the five-kingdom ToL, with animals, plants, fungi, protists, and monera identified as the five categories, or kingdoms, of life. Animals and plants are obvious; however, you may be less familiar with the other kingdoms. **Fungi** refers to spore-

Classification of *Homo sapiens* within the order Primates

		contained forms:
species *sapiens*		modern humans
genus *Homo*		modern and archaic humans
family Hominidae		humans and great apes
superfamily Hominoidea		humans and all apes (great apes and gibbons)
infraorder Simiiformes		humans, apes, and monkeys
suborder Haplorrhini		humans, apes, monkeys, and tarsiers
order Primates		humans, apes, monkeys, tarsiers, lemurs, and lorises

Figure 1.11 Hierarchical Classification This image shows a portion of the Linnaean classification of humans, or *Homo sapiens*. The broadest level of Linnaeus's classification system is kingdom. The kingdom Animalia includes all animals, including humans. The groups become more specific as classification continues. Humans are in the genus *Homo*, which contains modern humans as well as now-extinct humans, such as Neanderthals. A species' name consists of its genus name followed by its species name, which is specific to it, so humans are given the name *Homo sapiens*. (Photo from Universal Images Group North America LLC/Alamy Stock Photo)

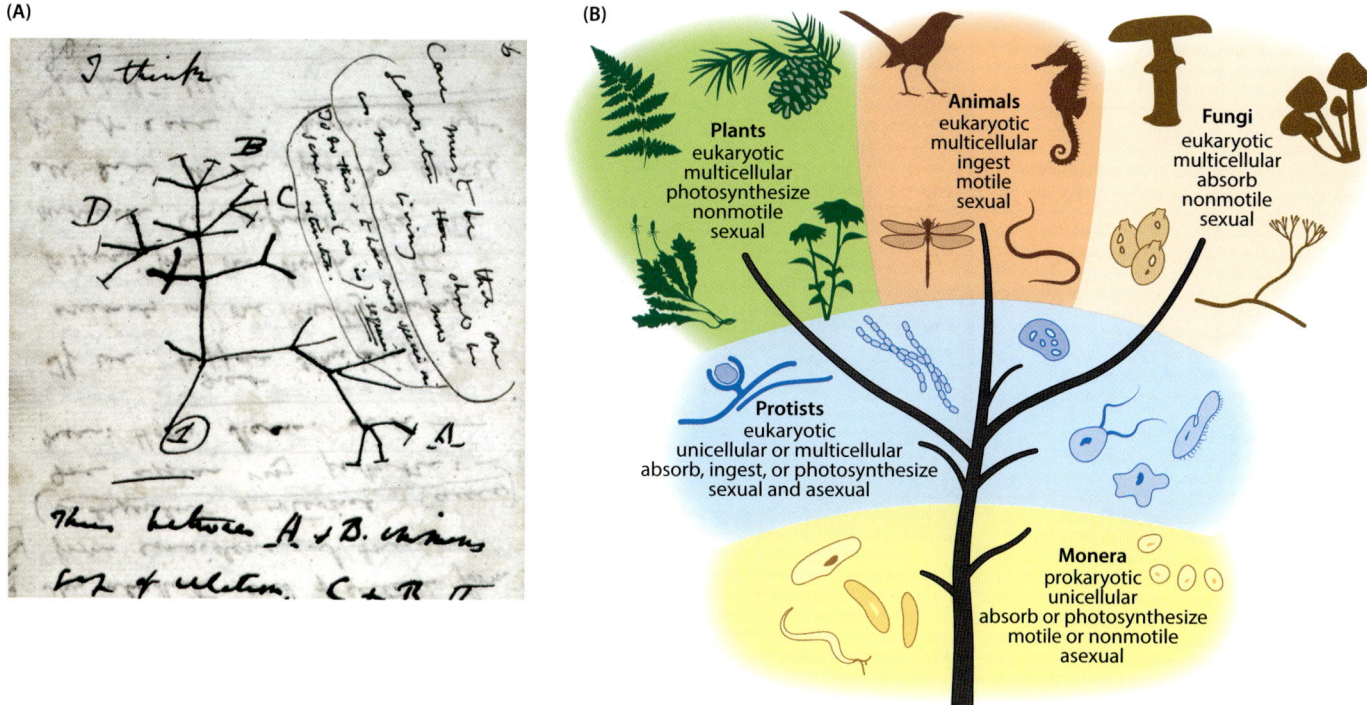

Figure 1.12 Darwin's Tree of Life (A) Darwin's illustration of the tree of life, which was first drawn in one of his notebooks in 1837. The base of the tree, which is labeled by the number 1, represents the cenancestor (or LUCA), and the ends of the branches represent species. A version of this illustration was included in his landmark book on evolution, On the Origin of Species, which was published 12 years after Darwin's original tree drawing. (B) The five-kingdom ToL was widely used until molecular technology became advanced enough to permit us a window into the incredible diversity of Protista and Monera, which we now recognize as the prokaryotes. Monera, which includes bacteria and other prokaryotes in this ToL, is at the base. Monera was seen as a more primitive group, from which more-advanced multicellular life evolved. The uppermost branches of the tree represent plants, animals, and fungi. Scientists could easily observe these large, multicellular life forms, so their diversity was better understood and took up most of the branches of the tree. (A illustration reproduced by kind permission of the Syndics of Cambridge University Library.)

producing organisms that feed on organic matter, including molds, yeast, mushrooms, and toadstools. **Protists** are single-celled eukaryotic organisms, such as protozoa or simple algae. **Monera** is the kingdom into which prokaryotes, such as bacteria, are placed. This view of life's diversity focuses your attention first and foremost on the macroscopic organisms and suggests that protists and monera are somewhat more primitive and less diverse. In truth, scientists at that time couldn't make sense of the evolutionary relationships among monera, simply because they didn't have **phenotypes**, or observable characteristics, to compare.

A Molecular Tree of Life

In 1977, Carl Woese tackled this formerly intractable problem, inferring the evolutionary relationships among the monera, or prokaryotes (Woese & Fox, 1977). Lacking visible physical traits with which to classify microorganisms, Woese turned to molecules. He chose the ribosome, which is a complex of RNA and associated proteins that functions to synthesize proteins, and which is one of the most ancient and well-conserved biochemical structures shared by all life. This means that any two species' ribosomes are similar, even if the species are not otherwise closely related. The ribosome is essentially a mini factory that translates genetic information into proteins

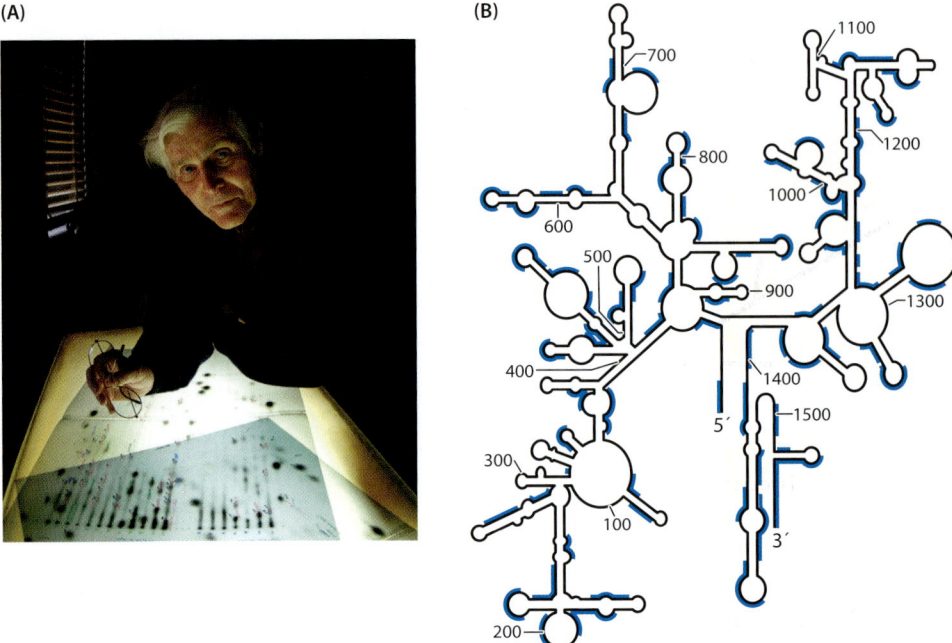

Figure 1.13 Carl Woese and a Molecular-Based Tree of Life (A) Photograph of Carl Woese peering at a radiograph that shows the ribosomal fragments of a microorganism's 16S rRNA separated based upon electrical charge. (B) Two-dimensional structure of the 16S rRNA molecule with regions indicated in blue that are cleaved during the RNA digestion procedure employed by Woese. (A photo courtesy of Jason Lindsey, University of Illinois Urbana-Champaign)

(**Figure 1.13**). All life forms use this same fundamental process for making proteins, so they all share at least some portions of the ribosomal RNA–protein complex.

Woese proposed that by comparing portions of the ribosomal complex among all life forms, it would be possible to group organisms in the same manner that they had been grouped previously using physical traits, or phenotypes. Within the ribosomal complex are subunits made of RNA and protein. The RNA molecules within those subunits are named based upon their weight in Svedberg units, such as 16S and 18S. All organisms, even those from across the three domains of life, have ribosomal subunits. Woese used information obtained by cleaving the RNA sequences of these ribosomal subunits and comparing the resulting fragments to estimate how closely related two organisms are. Pairs of taxa that are more similar in their ribosomal fragments are inferred to be more closely related. The number of differences between the ribosomal RNA fragments then serves as a measure of the amount of evolutionary time that separates a pair of taxa. These evolutionary distances can be used to create a phylogeny.

The Three Domains of Life

Woese first focused on a subunit of the ribosome (the 16S subunit) that is present in all bacteria. He produced fragments of the 16S ribosomal RNA (rRNA) for a diverse sample of what he thought of as bacteria and immediately noticed something striking. There was one cluster of fragments that was quite different from all the others. The organisms represented by that cluster were methanogens, prokaryotic cells that produce methane as a waste product. Woese quickly realized the significance of this finding: methanogens were not bacteria, but something completely different. He then employed an additional subunit of the ribosome (the 18S subunit), which is related to the 16S subunit but is found in eukaryotes, so that he could include eukaryotes in his clustering process. Although methanogens looked superficially like bacteria, their ribosomes reveal a very different ancestry. To his surprise the ribosomal fragments of methanogens were more like those found in eukaryotes than in bacteria. Woese named this new lineage **Archaea**, which is a Latin term meaning "primitive."

Based upon these results, Woese created a new ToL, which required a higher level of organization than the five kingdoms. He identified and named three groups within a higher level of biological relationships: Eukarya (animals, plants, fungi, and

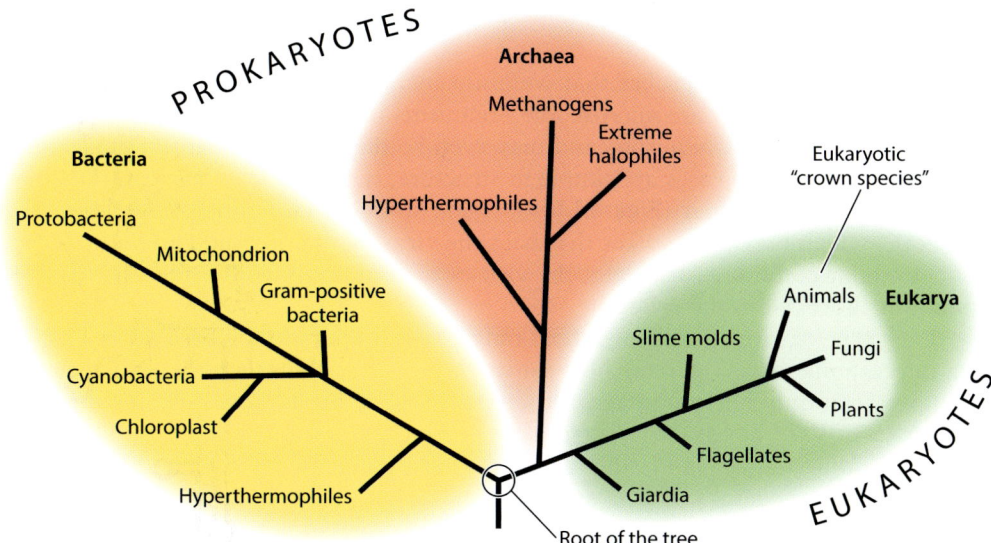

Figure 1.14 The Ribosomal RNA–Based Tree of Life This phylogenetic tree was developed using data from rRNA sequences. While eukaryotes made up most of the five-kingdom-view-based tree, they are only a small portion of the modern tree of life. Monera was found to include two distinct domains: Bacteria and Archaea. Although archaeans are microorganisms like bacteria, they are actually more closely related to eukaryotes, like us, than they are to bacteria! (After M. T. Madigan and M. Martinko, 2006.)

protists), Bacteria, and Archaea (Figure 1.14). The discovery of Archaea stimulated both enormous interest and intense skepticism at first. However, as more lineages of Archaea were identified, it became clear that it did, indeed, represent a novel and ancient branch on the ToL. Table 1.1 summarizes some of the similarities and differences observed between members of the three domains. The prokaryotes, which encompass members of the domains Archaea and Bacteria, share certain characteristics, such as size and a lack of intracellular organelles, while the eukaryotes appear to be chimeras, sharing key characteristics with both archaeans and bacteria. If we think back to the endosymbiotic theory, these patterns of similarities and differences begin to make sense. Eukaryotes, which were created through a series of endosymbiotic events, may very well have been derived from an ancestral archaean host that harbored a bacterial endosymbiont.

Woese's breakthrough was momentous for several reasons. First, by focusing on the ribosome, he had identified a way to compare all cellular life. Second, Woese revealed our ignorance of one of the three main branches of life, the Archaea. Further, he showed us that microbes occupy a dominant place in Earth's biodiversity. If we compare the five-kingdom and three-domain views of biodiversity, we see a fundamental shift from a view of life in which the eukaryotic **crown species** (plants, animals, and fungi) dominate, to one in which these eukaryotes are in the minority (see Figures 1.12B and 1.14). Woese himself described how unsettling this new view of life's diversity truly was: "Imagine walking out in the countryside and not being able to tell a snake from a cow from a mouse from a blade of grass, that's been the level of our ignorance" (Blakeslee, 1996).

Table 1.1 Comparison of Domains

	EUKARYA	BACTERIA	ARCHAEA
Cell type	Eukaryotic	Prokaryotic	Prokaryotic
Chromosomes	Linear	Circular	Circular
Membrane-bound organelles	Yes	No	No
Nuclear envelope	Yes	No	No
RNA polymerase	Many	One	Many
Cell wall composition	Not always present Plants—cellulose Fungi—chitin	Peptidoglycan	Lacks peptidoglycan
Cell membrane composition	Ester linked lipid with proteins (straight chain)	Ester linked lipids with D-glycerol (straight chain)	Ester linked lipids with L-glycerol (branched chain)

The Tiniest Microbes

There is one group of microbes that were not included in Woese's molecular tree of life, the viruses. Viruses are microscopic organisms that require a living cell, or host, to multiply. They are ubiquitous and may even be the most abundant biological entities on our planet. Viruses are simple in structure, with a genetic material (DNA or RNA) and a protein coat (**Figure 1.15A**). Some sport an additional outer layer, the envelope, which may have spikes that help the virus latch onto and enter a host cell. If the cellular conditions are right, the viruses then multiply within their host, often killing the host cell in the process.

Each type of virus has its own **host range**, which refers to the breadth of hosts it can infect. Some have a narrow host range; for example, *Variola virus*, which causes smallpox, can only infect humans. Other viruses have broad host ranges; for example, SARS-CoV-2, the causative agent of COVID-19, may infect hundreds of different hosts, including humans and other primates, bats, pangolins, ferrets, and camels.

Viruses are generally not given species names, so they don't fit neatly into the Linnaean classification system. In fact, many scientists don't consider them to be alive! They lack some of the basic features we think of when we attempt to define life, such as being cellular, maintaining homeostasis (or a stable internal state), growing, and making or acquiring energy. They do, however, replicate—using the host's replication machinery—and they adapt to their environment. Whether they are alive or not, viruses are one of the most abundant and diverse forms of microorganisms on Earth. They are categorized according to various characteristics they possess, includ-

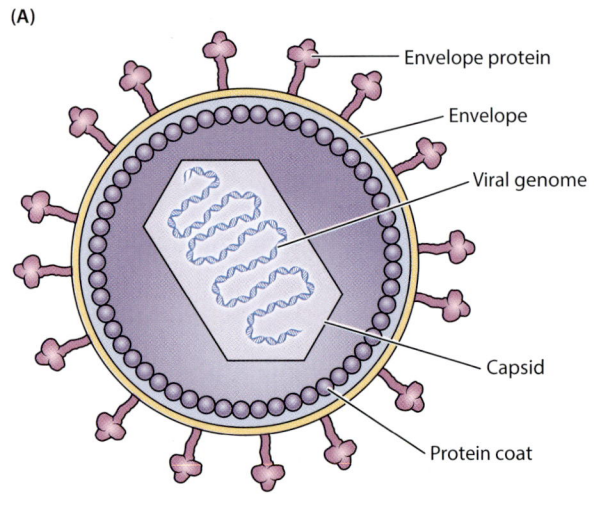
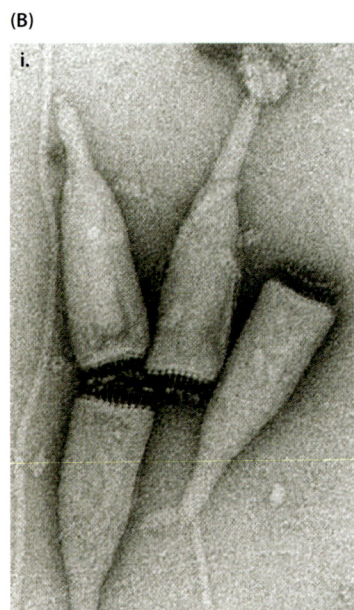
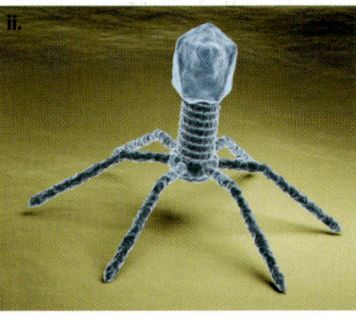
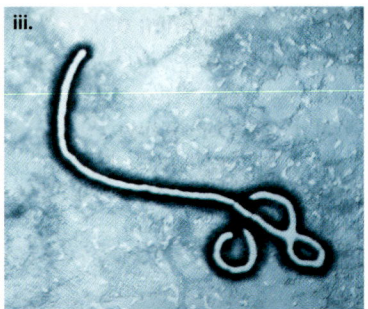

Figure 1.15 Viral Structure and Diversity (A) Most viruses are enclosed by an envelope embedded with proteins, which help the virus enter a host cell. A virus may have a DNA or RNA genome, which may be protected by a capsid. (B) A variety of different viral structures: [i] *Acidianus* bottle-shaped virus (colorized electron micrograph image), [ii] Bacteriophage on a bacterial cell (computer generated image), [iii] Ebola virus (microscopic view), and [iv] SARS-CoV-2 (computer generated image). (B photos from [i] ICTV International Committee on Taxonomy of Viruses, David Prangishvili, Mart Krupovic, Andrew M. Kropinski, Stuart G. Siddell, CC BY-SA 4.0, via Wikimedia Commons; [ii] extender_01/Shutterstock; [iii] iStock.com/Nixx photography; [iv] iStock.com/Naeblys)

ing their shape and size, the type of genetic material they possess (DNA or RNA), and whether they have an envelope layer. **Figure 1.15B** illustrates the major types of viruses.

It is challenging to identify the origin of viruses, as they don't leave fossils. In addition, some viruses can insert their genetic material into their hosts' genomes, which makes it difficult to untangle viral from host evolutionary histories. Since viruses do not share homologous genes or proteins with members of the three domains (Bacteria, Archaea, and Eukarya), we are not able to place them onto one or more branches of the ToL, leaving their relationships with other life forms in question.

1.3 MAKING THE INVISIBLE VISIBLE

With Woese's transformation of the ToL, microbes took center stage in our understanding of the diversity of life for the first time. In fact, according to Woese, microbes are the core of life on Earth: "If you wiped all multicellular life-forms off the face of the earth, microbial life might shift a tiny bit, if microbial life were to disappear, that would be it—instant death for the planet" (Blakeslee, 1996).

Before the 16S rRNA ToL revolution, we hadn't appreciated the immense diversity of microbes on our planet. In large part this was due to their seemingly simple morphology, which resulted in our tendency to group these simple life forms together. In the five-kingdom view of life, we see the microbial lineages clustered in two pools at the base of the tree (see Figure 1.12B). These pools represent the protists and monera (Bacteria and Archaea) with virtually no branches to represent what we now know is an incredible diversity of microscopic life.

We have known that microbes exist for over 400 years, ever since Robert Hooke invented the first microscope and explored the detailed structure of all sorts of biological entities, such as sponges, seaweed, and wood. Of particular interest here are his observations of mold. He describes its appearance on numerous decaying substances and notes that these creatures "will not be unworthy of our more serious speculation and examination" (Hooke, 1665). In short, Hooke was describing a microorganism's appearance for the first time.

The First Sightings of Bacteria

Inspired by Hooke, Antonie van Leeuwenhoek developed an even more powerful microscope and explored numerous samples from his own body, such as stool. In 1677, he reported to the British Royal Society that he had discovered over 1,000 "**animalcules**," or little animals, that differed from one location in the body to another (**Figure 1.16**) (van Leeuwenhoek, 1677). When he examined scrapings from his teeth, van Leuwenhoek noted, "I then most always saw, with great wonder, that in the said matter there were many very little living animalcules, very prettily a-moving. The biggest sort . . . had a very strong and swift motion and shot through the water (or spittle) like a pike does through the water. The second sort . . . oft-times spun round like a top . . . and these were far more in number" (van Leeuwenhoek, 1677). These were the very first observations of living bacteria ever recorded, and they inspired the development of an entirely new field of study, **microbiology**, or the branch of science that deals with microorganisms. Van Leuwenhoek is considered the father of microbiology, and from the

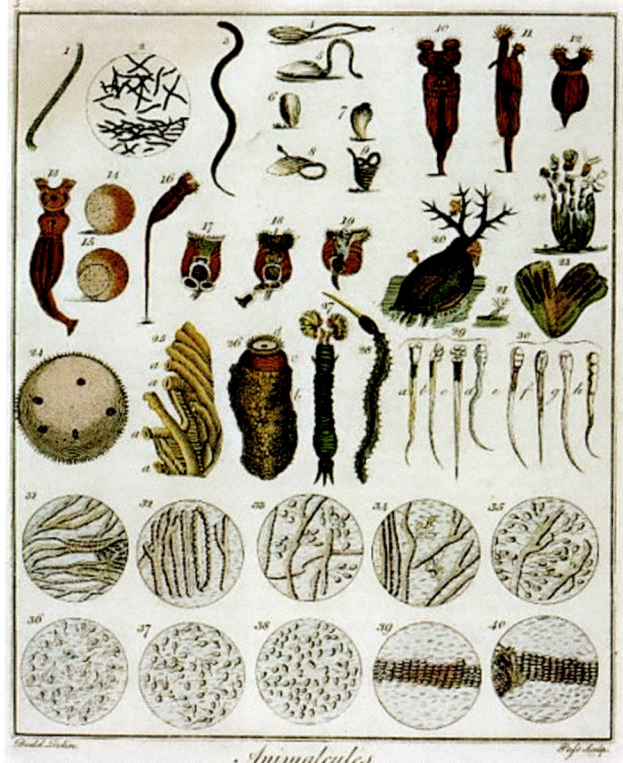

Figure 1.16 Animalcules Antonie van Leeuwenhoek was the first person to record observations of the microbiome. He obtained microbiome samples from various body parts and viewed them under a microscope. The "animalcules," or microorganisms, he saw are illustrated in this figure. (Photo from The Picture Art Collection/Alamy Stock Photo)

late 1600s to present day, scientists have been exploring the rich diversity of microbes on Earth.

Culturing the Invisible

Ever since the invention of the microscope, microbiologists have developed a rich toolbox with which to further explore microscopic life forms. The most common approach is to **culture** the cells, which allows them to grow and divide until there are enough for us to see. The basic procedure is straightforward. Say you want to see some of the microorganisms present in a nearby pond. You start with a sample of pond water and spread a drop of it on a rich growth medium. Each cell lands on a unique spot on the growth medium. If its requirements for growth are present, it grows and divides in this spot, and its daughter cells then replicate and eventually form a visible "colony" of hundreds of thousands of identical cells (**Figure 1.17**). In our pond water sample, we might find 50 or more different types of microbes growing on the food source we provide.

By altering the nutrients offered in growth medium to meet different species' growth requirements, scientists have identified several thousand prokaryotic and protist species. However, that seemingly impressive number pales in comparison with the number that actually inhabit the pond water. If we were to apply Woese's molecular methods of comparing all the 16S rDNA present in our pond water sample, we might find several thousand microbial species. This discrepancy between what we can grow in artificial media and what microscopic life is present in a sample is known as the **great plate count anomaly**, and it hindered progress in microbiology for decades. We simply didn't know what (or how much) we didn't know! For example, it is common knowledge that urine is sterile, unless you have a urinary tract infection. And yet, if you take a sample of supposedly sterile urine from a bladder and sequence the 16S rDNA present, you will find a wealth of different microbes have made urine, or the bladder, their home. For every novel environment we sample, we identify an ever-greater breadth and depth of microbial diversity.

Extremophiles, Life on the Edge

With the advent of molecular tools for identifying microbes, microbiologists engaged in an expansive hunt for novel microorganisms. We now know that microbes exist in

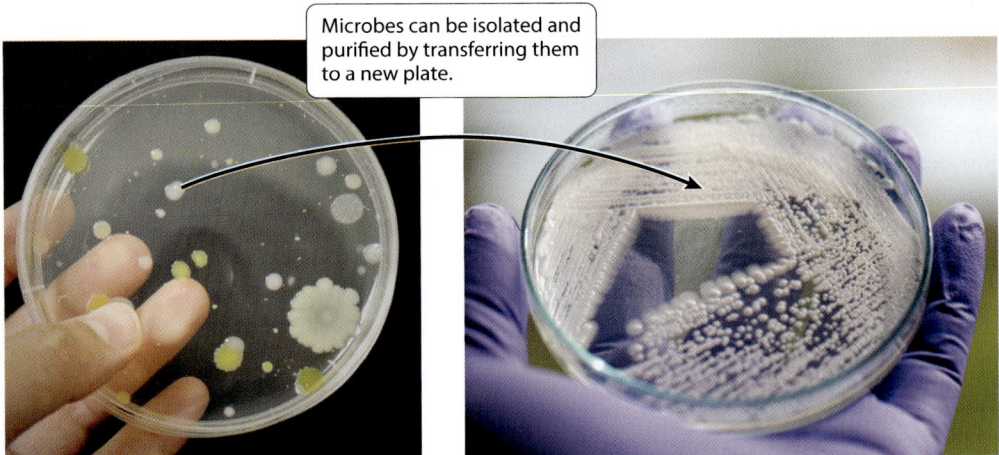

Figure 1.17 Bacterial Culture To isolate a genetically identical group of bacteria, a sample can be spread across a nutrient-filled petri dish to isolate individual cells, which can grow and divide to form visible colonies (A). Every member of a colony is a descendant of the first individual cell that landed on that spot on the petri dish. To obtain a pure culture of each individual cell from the original sample, cells from a colony are transferred to a fresh petri dish and grown in isolation (B). (Photos from [left] iStock.com/aorphoto; [right] iStock.com/Sinhyu)

some of Earth's most extreme environments. Some thrive in ice or salt, in the most acidic or basic conditions, living in organic solvents, consuming heavy metals and even toxic waste. Such **extremophiles** have been found in every imaginable, and even the most unimaginable, conditions on Earth. In every extreme environment investigated, a variety of organisms have been shown to not only tolerate the conditions there, but often require them to survive. Table 1.2 shows just a sliver of the extreme environments where extremophiles have been identified so far.

The term *extremophile* means "lover of the extreme," and the Archaea domain is where most extremophiles are found. In fact, when archaeans were unveiled to the world, they were thought of as extremophile weirdos. We now know that archaeans can readily adapt to extreme conditions, which may be due, in part, to the composition of their cell membrane. All cells have a plasma membrane made of a phospholipid bilayer, which evolved from the lipid-based protocell membrane we discussed earlier. The archaeans employ ether bonds in that bilayer, while bacteria and eukaryotes use ester bonds. This distinction is important because ether bonds are more resistant to chemical activity, which permits archaeal cells to survive in more extreme environments.

Some archaeans are among the most extremely thermophilic (heat tolerant), acidophilic (acid tolerant), alkaliphilic (base tolerant), and halophilic (salt tolerant) microorganisms known. Figure 1.18 shows the location where extremophiles were first discovered, in the hot springs of Yellowstone National Park. The genus *Picrophilus*, a member of Archaea, includes the most acidophilic organisms known, which can grow at a pH of 0.06, which is more acidic than hydrochloric acid. Despite their heat-loving reputation, archaeans are also found in very cold places, like Arctic seawater. Aside from our fascination with how extremophiles adapt to their extreme environments, this relatively unknown domain of life is particularly important to humans, due to its position on the ToL. Eukaryotes share a more recent common ancestor with Archaea than they do with Bacteria. Archaeans are our sister lineage, and there is so much more we must learn from them about them, and thus our own place in the biosphere.

Table 1.2 Types of Extreme Environments

Hot springs
Deep sea hydrothermal vents
Salt lakes
Polar regions
Volcanic areas
Acidic mine drainage
Deserts
Environments with high radiation levels

1.4 THE MICROBES WITHIN US

We now understand that microbes have a long and rich evolutionary history on Earth, one that is essentially as old as the planet itself. They continuously adapt to novel environments, invent new methods of energy capture, and in the process, have transformed our planet. Given this central role of microbes in the biosphere, it may be less surprising to learn that microbes have also adapted to living in and on us. We refer to these invisible residents as members of our **microbiome** (from the Greek terms *micro* meaning "small" and *bios* meaning "life"). The formal definition of a microbiome refers to a characteristic microbial community occupying a defined habitat that has certain properties. We can find microbiomes essentially everywhere we look—in our gut, in the soil surrounding the roots of a plant, in clouds, and even in the plume from a hydrothermal vent.

A Universe of Microbes within Us

The term *microbiome* refers to both the microorganisms present and the functions they provide, while the term **microbiota** refers simply to which species are present. For example, our

Figure 1.18 Extremophiles at Yellowstone National Park Extremophiles were first discovered in Yellowstone National Park's hot springs, where the water regularly reaches 189°F. The thermophiles that live in the hot springs give the pool its ring of colors. To survive at such high temperatures, these bacteria have evolved very stable membranes and proteins. One of these proteins, Taq polymerase, is now used in an important technique for creating copies of DNA, known as the polymerase chain reaction, or PCR. Taq is able to maintain its structure and function even at the high temperatures required for PCR. We can thank extremophiles for our ability to perform PCR for COVID-19 testing, gene sequencing, forensic testing, and more! (Photo from Framalicious/Shutterstock)

gut microbiome is home to approximately 300 to 500 species of microbes, collectively called the gut microbiota. These members together with the functions they provide, such as digesting some of the food we ingest, are called our gut microbiome. Each microbiome is integrated into its host or ecosystem and is crucial for the proper functioning and health of the organism(s) in that niche.

Our goal in this textbook is to explore what microbes are present in humans, what functions they encode in their genomes, and how those functions impact us, their human hosts, in both healthy and diseased states. This knowledge may force us to redefine what it means to be human. Rather than consider ourselves as distinct biological entities, separate from all other life forms, we must now acknowledge that humans, indeed all multicellular organisms, are composed of numerous complex ecosystems each consisting of a mixture of their own and microbial cells. This new entity, the human with all its microbiomes, is referred to as the **holobiont**, a term derived from the Greek *hólos* or "whole" and *biont* for "unit of life." The term was coined by Lynn Margulis in the 1990s as she was exploring the endosymbiotic origin of eukaryotes. Her intent was to provide a term that would acknowledge the key role of symbiotic relationships in the evolution and diversification of multicellular eukaryotic organisms, such as when an ancestral prokaryotic cell gave rise to mitochondria or chloroplasts. However, the term is equally appropriate to refer to a human body with its invisible microbial symbionts that, as you will learn, provide the key to our health while at the same time serving as the harbingers of certain diseases.

Each of us consists of about 30 trillion human cells, which carry our genetic blueprint and the machinery required to translate that information into what becomes the visible "us." These cells form collections of tissues and organs, which play critical roles in keeping our bodies functioning. For example, skin serves as our frontline defense against invading pathogens, while the heart provides the force required to ensure all of our cells receive the oxygen-rich blood they require. For several thousand years physicians and scientists have explored our cells, tissues, and organs in their quest to understand what makes us uniquely human, what keeps us healthy, and what can go wrong in our bodies to cause disease and death.

We have long known that bacteria and viruses could invade our bodies and cause illness; however, they were considered temporary intruders that our bodies, or the medications we took, would fight to eliminate. In just the past 20 years we have gained an entirely new perspective on the important role microorganisms play in keeping our bodies healthy, leading some to argue that the microbiome should be considered the 11th critical organ, equal in importance to our brain! Let's explore this new organ and learn a bit about its role in keeping us healthy.

How Much of You Is Human?

It's estimated that we have about 35 trillion microbes in and on our bodies—about 5 trillion more than the number of human cells! This count excludes viruses, whose numbers may dwarf the human and microbial cell counts combined. Those numbers translate into a weight of just over 1 kg (2.5 pounds), with a volume of about 1.5 liters (6 cups) of cells. That's nearly half a gallon of microbes per human!

Our body hosts numerous, distinct microbiomes (**Figure 1.19**). We have an oral microbiome in our mouth, one that covers our skin, another in our urinary tract, one in our gut, and even one deep in our lungs. There are far more fine-tuned distinctions we could make. For example, the microbes that inhabit the surface of our tongue are distinct from those that live under our gums, which are different from those that live attached to our teeth, and so on.

These distinct microbial communities also vary greatly in their cell densities. Blood is a virtual microbial desert, while the large intestine contains one of the densest microbial communities on Earth (Bojanova & Bordenstein, 2016)! While the precise number of microbes may differ, each microbiome is highly diverse, with over

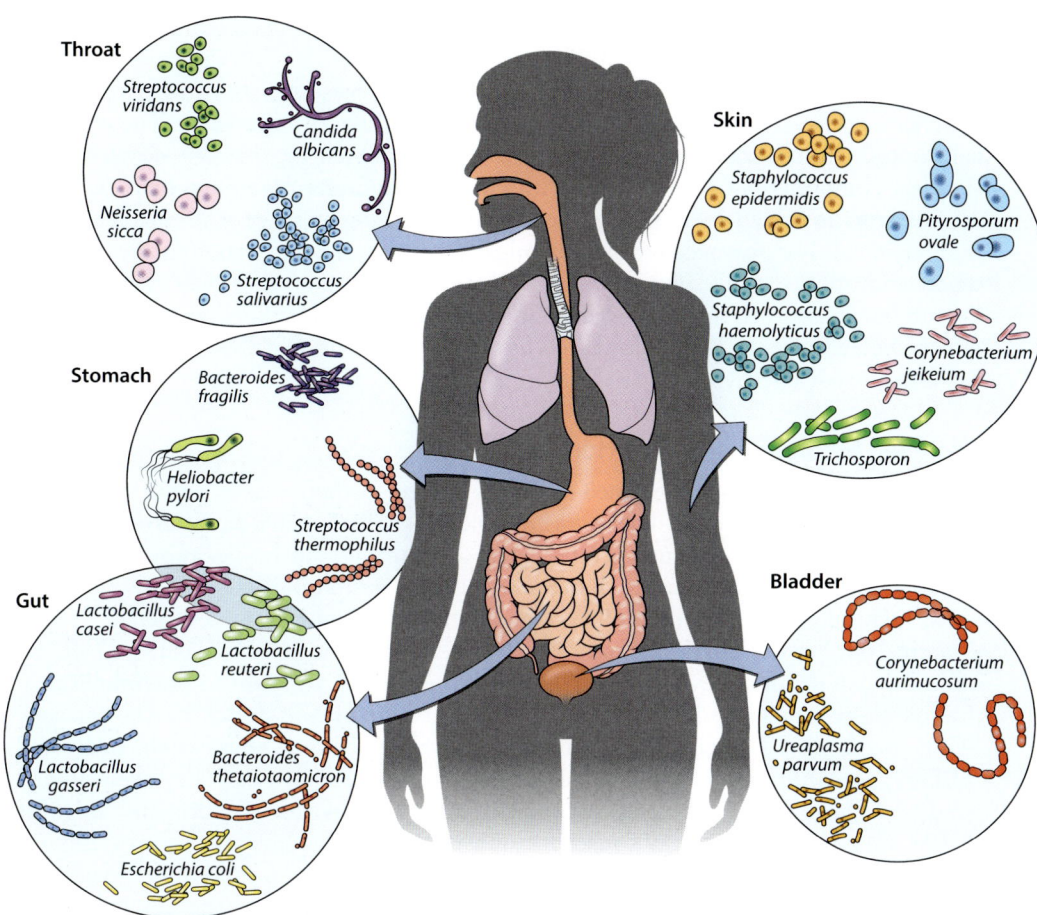

Figure 1.19 So Many Human Microbiomes The human microbiome includes many different microbial communities, each with its own unique composition of species and role in maintaining our health. (After V. D. Appanna, 2018.)

300 distinct bacterial species identified in the human gut microbiome alone (Almeida et al., 2021).

Even more compelling than their sheer numbers is the fact that the genetic information our microbiomes encode far exceeds our own. The human genome encodes 20,000 genes, while our microbiomes provide an additional 45 million, each encoding functions with the potential to impact us, their host. For example, if not for genes carried by certain species of bacteria, we would not be able to digest most of the fiber we consume.

1.5 OUR MICROBIOMES, OUR HEALTH

The rapidly growing field of microbiome science is revealing the complex roles these fellow travelers serve in human health. There is now overwhelming evidence that most functions of our body, such as growth, development, and metabolism, depend on our microbiome. Our immune system is trained first by our mother's microbiome during pregnancy and then by our own microbiome, particularly during the first few years of life. Dysfunctions in the gut microbiome are associated with several autoimmune diseases such as arthritis, fibromyalgia, and multiple sclerosis. Our gut microbiome also plays a role in several intestinal conditions, such as inflammatory bowel disease (IBD) and irritable bowel syndrome (IBS), while obesity is often associated with an imbalance in the members of our gut microbiome.

Microbiomes and Human Nutrition

Another example of the key role our microbiome serves is in nutrition. Sugars and starches are two classes of carbohydrates synthesized by all organisms. The plants we eat contain thousands of different carbohydrates, which are broken down to their simplest components to provide us with energy. The human genome has fewer than 20 **enzymes** involved in digesting carbohydrates. Enzymes are proteins that act as biological catalysts by accelerating chemical reactions. Those carbohydrates we can't digest end up in the large intestine, where our microbiome takes over. The microbes in our gut encode thousands of carbohydrate-digesting enzymes in their genomes, which they employ to break down, or ferment, carbohydrates that are not digestible by humans, for energy.

Microbial Metabolites, Key to Human Health

One outcome of the microbiome's digestive efforts is their waste, some of which is essential for human health. These waste molecules, also known as **by-products**, serve key roles in our nutrition and metabolism. For example, our bodies require vitamins, which are organic compounds that are essential for maintaining various body systems, including the immune and nervous systems. You might have learned that the vitamins our bodies need can only be obtained from the food we eat. In fact, our microbes can produce several key vitamins for us. Many vitamins are **metabolites**, or intermediaries, produced during the fermentation of fibrous foods by the microbes living in our gut. Bacteria in the microbiome also produce **short-chain fatty acids (SCFAs)**, which are fatty acids with fewer than six carbon atoms. They are primarily produced through the fermentation of dietary fibers by gut bacteria in the colon. SCFAs are an essential energy source for our intestinal cells. It is an elegant symbiosis: our gut provides an energy-rich environment that supports an incredible diversity of microbial life, while that life, in turn, provides us with some of the key ingredients required to ensure our health.

Reflections on Your Microbiome

Let's think about our microbiomes from a slightly different perspective. As you walk from one lecture hall to another, passing people who may look very different from you—in height, weight, skin, or eye color—consider this fact: your genome differs by about 0.1% from any other human genome. Regardless of how different you look, you are nearly identical in terms of your DNA content. Now, look again at those passing by, and imagine that you can see the members of their microbiomes as easily as you see their facial features. Each person's microbiome differs by as much as 90% in terms of the species present, not to mention the genetic repertoires those species possess.

All these facts are causing us to reconsider how we think of ourselves as uniquely "us." Traditional explanations for what makes an individual unique focus on our brain or the contents of our genome. However, as you will learn, our microbial residents communicate directly with our brain, and they provide far more gene functions than does our own genome. We are realizing that humans are not discrete entities of human cells and genes; rather, each of us is a consortium of thousands of organisms that result in a functioning, hopefully healthy, human. Indeed, it takes a microbial village to be a human!

Take a moment to reflect on what this new understanding of our microbial partnerships means to you. Does it scare you (or gross you out) to imagine the astronomical numbers of microbes in and on your body? Do you get excited about the genetic potential we carry inside us? Or do your thoughts turn to the role these microbes have played in our evolutionary history? Perhaps you wonder if you can take advantage of them to improve your health. Simply said, we are not alone, and it can feel empowering to understand that you have a fair bit of help in keeping your body healthy.

CHECK YOUR UNDERSTANDING

1. Approximately when do we think life emerged on this planet?
   a. 4 billion years ago
   b. 1 million years ago
   c. 0.5 million years ago
   d. 1,000 years ago

2. The Miller-Urey experiment was designed to test whether
   a. early Earth's conditions could be mimicked.
   b. organic molecules could be created under early Earth conditions.
   c. inorganic molecules could create life.
   d. life could be created in a glass chamber.

3. Which represents the Central Dogma of Molecular Biology?
   a. RNA → DNA → protein
   b. Membrane → DNA → protein
   c. DNA → RNA → protein
   d. DNA → membrane → protein

4. The advanced protocell created by Szostak's lab was essentially a
   a. membrane-bound cell containing DNA.
   b. fragment of RNA that could replicate itself.
   c. cellular structure that could make copies of itself.
   d. cellular structure that was unable to replicate itself.

5. Natural selection occurs when
   a. an individual organism gains new, advantageous traits during its lifespan.
   b. individuals with advantageous traits are better able to survive and reproduce, and those traits become more common in the population over time.
   c. random events result in organisms better able to survive and reproduce.
   d. a population of organisms survives to reproduce.

6. Hydrothermal vents provide a rich nutrient source that some of the earliest life forms likely took advantage of.
   a. True
   b. False

7. What are deep-sea hydrothermal vents?
   a. Magma transmitted from the Earth's core
   b. The ocean's equivalent of geysers
   c. Very hot plumes of air at the bottom of the ocean
   d. Underwater volcanoes

8. The two competing arguments about the origin of life are the replication argument and the cell division argument.
   a. True
   b. False

9. The protocell membrane was created with
   a. DNA.
   b. RNA.
   c. fatty acids.
   d. proteins.

10. What's the difference between autotrophs and heterotrophs?
    a. An autotroph makes its own food.
    b. A heterotroph makes its own food.
    c. A heterotroph uses the sun's energy to fuel itself.
    d. An autotroph uses the sun's energy to fuel itself.

11. How did the Great Oxidation Event affect life?
    a. Anaerobic life largely went extinct.
    b. Aerobic life largely went extinct.
    c. It created the rust deposits found in some sedimentary rocks.
    d. It enabled anaerobic life to flourish.

12. Identify 2 characteristics of eukaryotes not found in prokaryotes.
    a. Cell membranes, flagella
    b. Nuclei, mitochondria
    c. Nuclei, flagella
    d. Golgi bodies, cell membranes

13. Lynn Margulis proposed that eukaryotic cells came from a chance fusion of 2 protists.
    a. True
    b. False

14. How did cyanobacteria transform Earth's atmosphere?
    a. By producing methane
    b. By consuming all the existing oxygen
    c. By producing oxygen
    d. By consuming all the existing carbon dioxide

15. LUCA was the very first organism.
    a. True
    b. False

16. What technology allowed the microbiome to be viewed for the first time?
    a. Telescope
    b. Microscope
    c. Electron microscope
    d. 16S ribosomal sequence

17. What genus are humans members of?
    a. Eukarya
    b. Sapiens
    c. Mammalia
    d. *Homo*

18. Which kingdom were prokaryotes a part of in the 5-kingdom view of life?
    a. Monera
    b. Fungi
    c. Protists
    d. Bacteria

19. What did Carl Woese use to infer relationships between prokaryotes?
    a. Whole genome sequencing
    b. Phenotypic observations
    c. Metabolic pathways
    d. 16S rRNA

20. What are the 3 domains of life?
    a. Eukarya, Prokarya, and Monera
    b. Eukarya, Bacteria, and Archaea
    c. Fungi, Protista, and Bacteria
    d. Eukarya, Bacteria, and Protista

21. What is the cause of the great plate anomaly?
    a. Some bacteria have RNA genomes.
    b. Many bacteria cannot be cultured using available techniques.
    c. It is difficult to find bacteria in the environment.
    d. It is impossible to isolate a single species from a sample.

22. Extremophiles are microbes that survive in intense conditions, such as very high or low temperatures.
    a. True
    b. False

23. Which human microbiome is less dense than the others?
    a. Gut microbiome
    b. Oral microbiome
    c. Blood microbiome
    d. Skin microbiome

24. A human and their microbiome have about the same number of enzymes involved in digesting carbohydrates.
    a. True
    b. False

25. Vitamins, short-chain fatty acids, and other metabolites are produced when certain microbes digest which compounds in food?
    a. Simple sugars
    b. Fatty acids
    c. Lipids
    d. Fibers

Answers: 1A, 2B, 3C, 4C, 5B, 6A, 7B, 8B, 9C, 10A, 11A, 12B, 13B, 14C, 15B, 16B, 17D, 18A, 19D, 20B, 21B, 22A, 23C, 24B, 25D

DIVING DEEPER

1. Why were deep-sea hydrothermal vents advantageous locations for early life?
2. How did Miller and Urey show that the organic molecules necessary for life could form from inorganic material?
3. What were the two competing views about the origin of life, and what did Jack Szostak's protocell reveal?
4. What's the difference between autotrophs and heterotrophs?
5. How did the Great Oxidation Event affect life?
6. Can you explain three differences and three similarities between prokaryotes and eukaryotes?
7. According to Lynn Margulis's endosymbiotic theory, how did eukaryotic cells acquire mitochondria and chloroplasts?
8. Identify three differences between the five-kingdoms and three-domains views of life's diversity.
9. Why is the ribosome a good tool to use for inferring the tree of life?
10. Why was Woese's use of 16S rDNA sequencing revolutionary?
11. What technology allowed the microbiome to be viewed for the first time?
12. Why are bacterial culture techniques limited, and what technology solves this problem?
13. Can you give examples of the environments that extremophiles are able to live in?
14. What is a virus's host range?
15. Why can't viruses be placed on the tree of life, and how are they different from Bacteria, Archaea, and Eukarya?
16. What's the difference between the microbiome and microbiota?

17. Lynn Margulis introduced the term holobiont to explain what?
18. Can you list five microbiomes found in/on humans?
19. How does the human microbiome vary by body part?
20. Why is the microbiome necessary for carbohydrate digestion?
21. What are the two main metabolites bacteria produce as waste, and why are they important for human health?

DISCUSSING AND REFLECTING

1. Lynn Margulis's serial endosymbiosis theory was a harbinger of the discovery of the microbiome. Explain what is meant by that statement.
2. Woese's impact on our understanding of biodiversity has been enormous. Describe the key features of biodiversity that we were ignorant about before Woese's research revealed the three-domain tree of life.
3. What can extremophiles tell us about the origin of life on Earth and the possibility of life existing on other planets?
4. Reflection. Carl Woese said, "If you wiped all multicellular life-forms off the face of the earth, microbial life might shift a tiny bit, if microbial life were to disappear, that would be it—instant death for the planet" (Blakeslee, 1996). How do you feel now that you know the importance of microbes, and how does this affect your view of life on this planet?

RECOMMENDED READINGS

Popular Science Reviews

O'Donnell, E. (2019, June 7). How Life Began: Jack Szostak's Pursuit of the Biggest Questions on Earth. *Harvard Magazine*, 40–79.

Quammen, D. (2018, August 13). The Scientist Who Scrambled Darwin's Tree of Life. *New York Times*, 34.

Popular Science Book

Sagan, D. (2012). *Lynn Margulis: The Life and Legacy of a Scientific Rebel*. Chelsea Green.

Scientific Reviews

Gray, M. W. (2017). Lynn Margulis and the Endosymbiont Hypothesis: 50 Years Later. *Molecular Biology of the Cell*, 28(10), 1285–1287. https://doi.org/10.1091/mbc.e16-07-0509

Nasir, A., Romero-Severson, E., & Claverie, J.-M. (2020). Investigating the Concept and Origin of Viruses. *Trends in Microbiology*, 28(12), 959–967. https://doi.org/10.1016/j.tim.2020.08.003

A Brief History of Microbiome Research

CHAPTER CONTENTS

2.1 Our First View of Microbes

2.2 The Golden Age of Microbiology

2.3 Genomics and Bioinformatics

2.4 Bringing It All Together

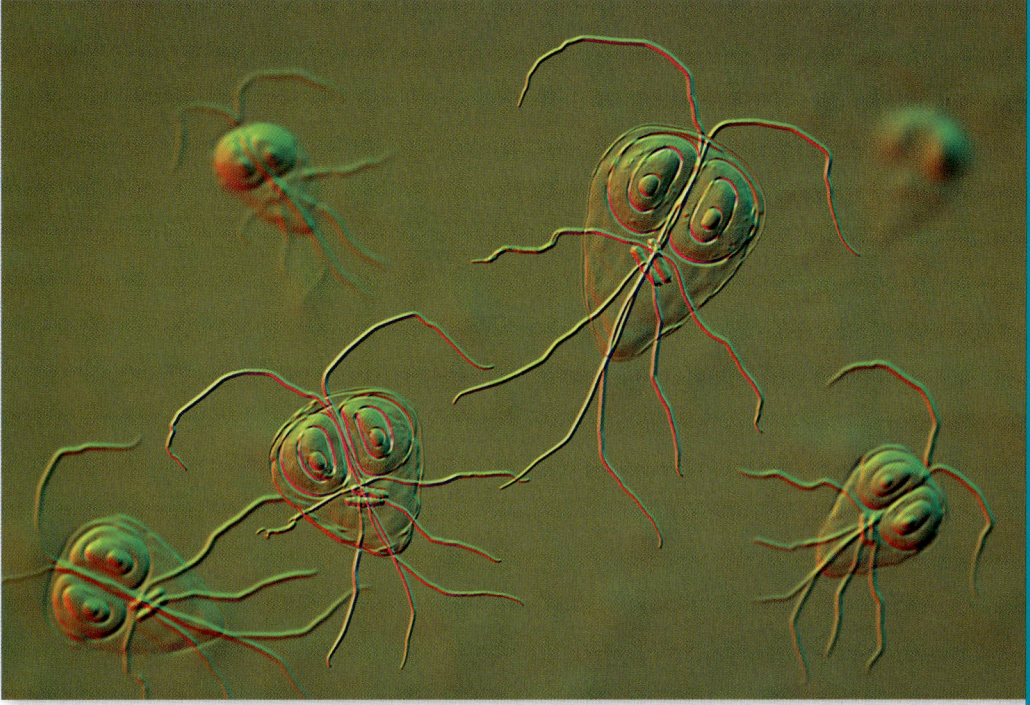

"Hey there, human! *Giardia intestinalis* here—just dropping by to stir things up a bit. I'm all about adventure, especially in your small intestine, where I hitch a ride after you sip from that refreshing mountain stream (or maybe just an unlucky glass of water). Once I get settled, I love disrupting your digestion—think greasy stools, cramps, and some serious bloating. Sure, your gut microbes don't love me, and your immune system tries to show me the door, but hey, a microbe's gotta live, right? Don't worry; I usually don't stick around long . . . unless you're up for a repeat visit!" (Photo from iStock.com/Dr_Microbe)

Microorganisms have been our most intimate partners since our species first evolved. In this chapter we will learn about the physicians and scientists who, since as early as the 17th century, have been developing the tools required to make this invisible partnership come to light. Our focus will include understanding the technological advances required to see these fascinating microbes, learning about the biochemical processes they engage in, and introducing the enormous DNA sequence datasets being generated to aid in this exploration. We will identify some of the most influential microbiome scientists and explore how they are revealing the role of the microbiome in human health and disease.

"Every person is a collective, a vast and complex gathering of interdependent life."
—Guy Harrison (Harrison, 2021)

2.1 OUR FIRST VIEW OF MICROBES

The field of microbiome science is young. However, at least one curious individual was exploring the diversity of microbes in his own body as early as the 17th century. As mentioned in chapter 1, the first observations of the microbiome were recorded by Antonie van Leeuwenhoek, the owner of a fabric shop in Holland. In his spare time, he indulged in his hobby: grinding glass lenses. Van Leeuwenhoek made a pair of eyeglasses that magnified 270 times, 10 times more powerful than any other lens of the day. He was a deeply curious individual, and as he made his fine lenses, he began to use these tools to explore the microscopic world around him. He magnified just about anything he could find, including samples from his own body. Imagine his surprise when he scraped plaque from his teeth and found more than 1,000 tiny organisms he called **animalcules**. What Leeuwenhoek saw were bacteria, along with single-celled eukaryotes known as protists. In fact, he was the very first person to ever view a bacterium!

Figure 2.1 shows the very first drawing of a bacterium, which van Leeuwenhoek identified from a solution of pepper in water that had turned cloudy. He also examined samples from different parts of his body, such as tooth plaque and stool, and saw that they contained different microbes (see Figure 1.16). His writings show just how delighted he was to find these organisms, and he spent considerable time observing their behaviors and interactions. In a letter written to the Royal Society of London he noted, "I found floating therein divers earthy particles, and some green streaks, spirally wound serpent-wise, and orderly arranged, after the manner of the copper or tin worms, which distillers use to cool their liquors as they distil over. The whole circumference of each of these streaks was about the thickness of a hair of one's head. . . . All consisted of very small green globules joined together: and there were very many small green globules as well."

Although van Leeuwenhoek's observations generated much discussion in the 17th century among the members of the Royal Society, which was the epicenter of biological research at the time, we must wait another two centuries for the next significant advance in microbiome science. In part, this long delay was due to the skepticism many scientists expressed about van Leeuwenhoek's findings. This period of disdain for all things microbial is clearly seen in Carolus Linnaeus's influential contemporaneous work on biological classification. We learned in chapter 1 about the hierarchical categories he created for organizing all organisms into groups and then subgroups upon subgroups until the level of species was reached. Linnaeus was aware of Leeuwenhoek's discoveries and yet dumped all microbes into a single phylum, Vermes (worms), and genus, *Chaos* (formless). Current estimates are that, where Linnaeus saw formless worms, we currently recognize over 1,300 bacterial phyla and 5 archaeal phyla, at the very least!

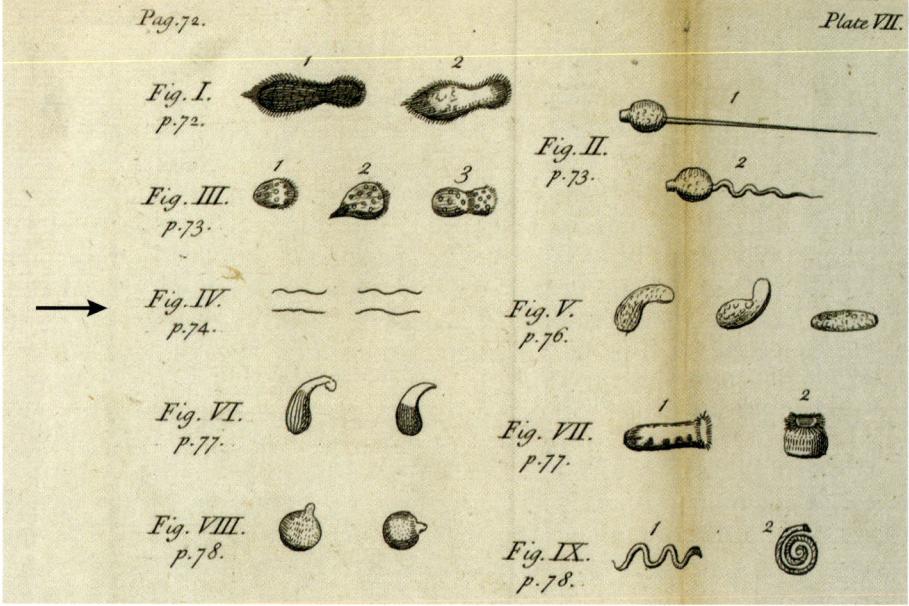

Figure 2.1 Oral Microbes Observed by Antonie van Leeuwenhoek In this figure you see a sketch made from observing a sample from a bottle of pepper water that turned cloudy. This drawing is believed to contain the first sketch of bacteria (Fig IV). The field of microbiology was born! (Photo from H. Baker, 1743)

Figure 2.2 **The Butterfly Amoeba** Joseph Leidy was an excellent artist, as well as a research scientist. He produced magnificent teaching charts that presented the diversity of microbes he found in animals. Shown here is a portion of a plate from Joseph Leidy's 1879 monograph showing some of his drawings of *Hyalosphenia papilio*, the butterfly gut amoeba. (Photo from Pictorial Press Ltd/Alamy Stock Photo)

It wasn't until Joseph Leidy, a 19th-century American scientist, published *A Flora and Fauna within Living Animals* that microbiology truly blossomed as a research endeavor (Leidy, 1853). Leidy made an intensive study of the organisms living in the guts of insects. His illustrations are extraordinary in their clarity, and he went well beyond simple descriptions, exploring how these organisms developed and where in the host they were found (**Figure 2.2**). Gazing at dissected insect guts through his microscope, Leidy saw hundreds of small specks pouring out of the intestine: "Myriads of the living occupants escape, reminding one of the turning out of a multitude of persons from the door of a crowded meetinghouse" (Leidy, 1881).

Leidy used this knowledge to identify the microbial origins of certain diseases, such as trichinosis, which results from eating undercooked meat infected with the larvae of the worm *Trichinella*. He even proposed that meat should be fully cooked to prevent infection. Provocatively, he also proposed that the organisms residing in the insect's guts were not the result of spontaneous generation, which smacked squarely against the dominant view of the time (Ward, 1923).

Spontaneous Generation, or Not?

As early as 400 BC, Greek philosophers believed that some life forms arose spontaneously from nonliving matter, particularly decaying matter. The **miasma theory**, also known as **spontaneous generation**, held that certain diseases originated from "bad air" or particles rising from decomposing matter, and if you ventured too close, you might be infected. The plague, responsible for as many as 200 million deaths in Europe during the Middle Ages, was thought to have spread in this way. You may have seen images of plague doctors wearing their distinctive uniforms with masks bearing long protuberances into which they would place herbs and perfumes to filter out the miasma. Although this theory was eventually proven wrong, many of the observations were correct, like the fact that decaying matter has a distinctive odor, which we now know is from microorganisms breaking down organic matter.

In the mid-1600s one scientist challenged the concept of spontaneous generation. Francesco Redi, an Italian physician and naturalist, was studying maggots, which are the larval form of the fly. In Redi's time, most scientists believed that maggots were spontaneously generated in rotting foods. Redi had a different idea. He hypothesized that the maggots arose from the eggs laid by flies, and he devised a simple, ingenious

BOX 2.1. REDI'S EXEMPLARY EXPERIMENTAL DESIGN

An **experiment** is an investigation in which a hypothesis is scientifically tested. In Francesco Redi's experiment he wanted to know whether flies were the source of the maggots that appeared on rotting meat (**Figure 2.3**).

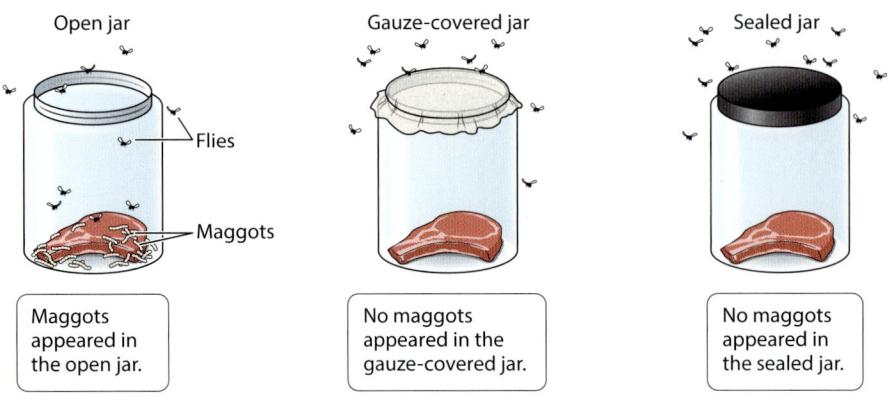

Figure 2.3 The Design of Redi's Maggot Experiment

experiment to test his idea (Redi, 1909). He placed meat inside flasks that were open, sealed, or covered with gauze. Maggots grew in the open flasks, where flies could enter and lay eggs, but not in the sealed or gauze-covered flasks, which flies could not enter.

Modern Experimental Design

Redi's experiment was one of the first examples of modern-day experimental design, in which one develops a testable **hypothesis** (or idea), identifies one specific factor (or **experimental treatment**) they wish to test, and then ensures that their sample size is adequate to permit the hypothesis to be evaluated (Sant, 2019). **Box 2.1** introduces the concept of experimental design and provides a description of the most important features of a well-designed experiment, using Redi's experiment as an example. Despite his well-executed experiments, the theory of spontaneous generation survived for several more centuries. However, there was a twist; by the 18th century the idea of spontaneous generation was only applied to microscopic forms of life, rather than animals, plants, and fungi large enough to see with the naked eye.

It was not until the mid-19th century that the miasma theory was finally debunked. The French Academy of Sciences held a competition to either prove or disprove spontaneous generation. In a deceptively simple experiment, Louis Pasteur provided indisputable proof against the miasma theory (**Figure 2.4**) (Pasteur, 1882). He took two swan-neck flasks containing rich broth that he sterilized by boiling, which ensures no living microorganisms are present. The S-shaped neck prevents air from carrying microbes into the flask, where they could grow in the broth. He left one flask intact and broke the neck of the other. The broth in the first flask remained clear, whereas the broth in the broken-neck flask became cloudy with microbial growth. This proved that the microbial growth was caused by microbes in the air entering the flask with the broken neck, which did not occur in the intact flask. He hypothesized that in the unbroken flask, the S curve prevented microbes from reaching the broth; instead they settled in the flask's neck. When Pasteur then tipped the flask with the unbroken neck and the broth reached the bend in the neck, it became cloudy with microbial growth as well. Pasteur's experiment clearly proved that

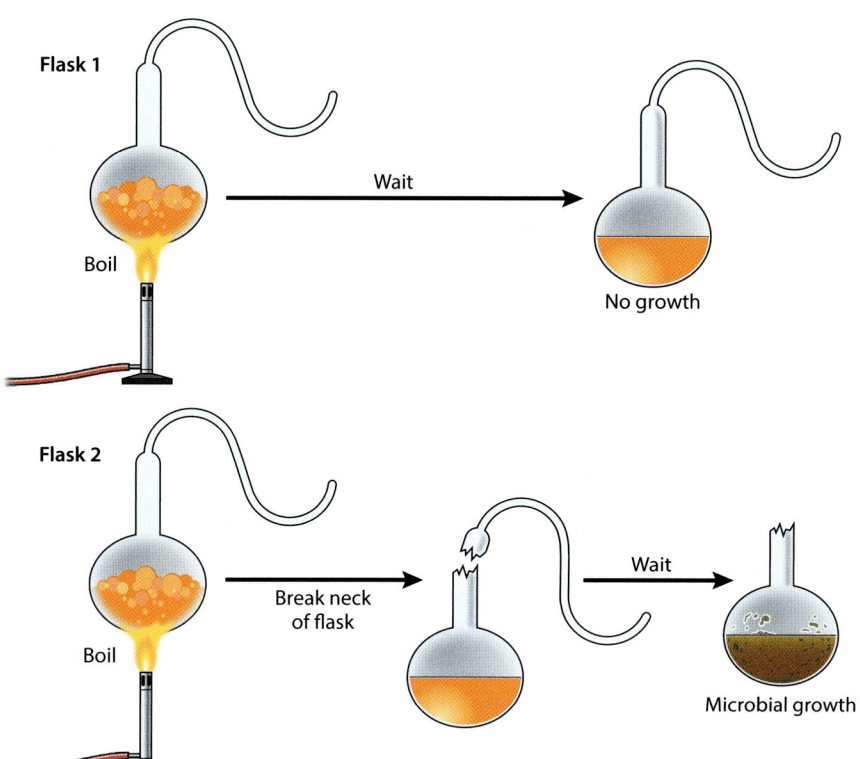

Figure 2.4 Pasteur's Spontaneous Generation Experiment Pasteur filled two swan-neck flasks with broth and boiled them. He broke the swan neck of one of the flasks, which allowed microbes to enter the flask and replicate, whereas there was no growth in the other flask. This proved that the circulating air had provided the microbes that grew in the flask with the broken neck.

spontaneous generation was not necessary to explain his observations. Further, he demonstrated that microorganisms are found everywhere, even in the air we breathe. Or, in his own words, "*Omne vivum ex vivo*," which translates to "life only comes from life" (Pasteur, 1864). Pasteur was awarded the Jecker medal from the French Academy of Sciences in recognition for his experiments refuting spontaneous generation.

The Germ Theory of Disease

The **germ theory of disease** states that certain microorganisms (germs) can invade humans and other living hosts and that their growth and reproduction within their hosts can cause disease. Some of the first evidence for the germ theory was provided by an unlikely source, the silkworm! Silk fiber is produced from the cocoons the silkworm larvae encase themselves in before metamorphosing into adult moths (**Figure 2.5**) France had been a center of silk production since the 18th century, and the demand for this luxury item exploded in the 19th century. Large silkworm farms were constructed, and with the huge populations, the worms soon became overrun with disease, the cause of which was unknown. Some of the earliest therapies involved burning incense because farmers believed that bad odors made silkworms sick.

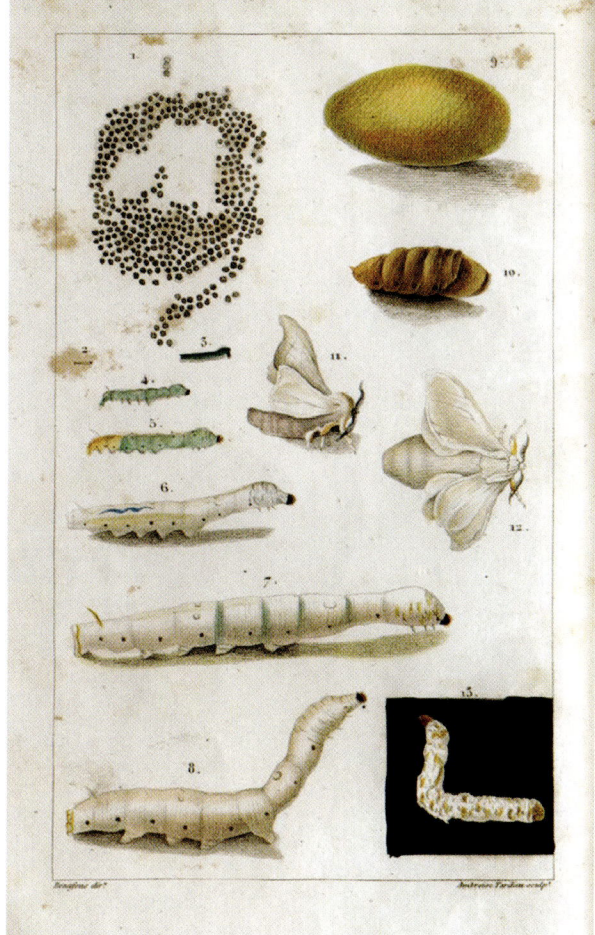

Figure 2.5 Developmental Stages of the Silkworm Stages of the development of the silkworm, one of the world's most famous insects, which produces silk during its caterpillar stage of life. (From S. Julien and M. Bonafous, 1840)

Agostino Bassi, an insect scientist, spent 25 years studying silkworm illnesses and determined that one of the diseases was the result of a fungal infection. This was the first microorganism to be recognized as an agent of disease. Bassi showed that the disease could be transmitted by moving fungal spores from a dead silkworm into a healthy one. The title of the resulting manuscript was "Disease that affects silkworms and on the means of freeing therefrom even the most devastated breeding establishments. Herein, the fundamental discovery that microbes (namely, 'an extraneous germ') can cause disease is interwoven with practical tips for rearing silkworms." Bassi further proposed that disease in humans and animals was also caused by microorganisms. His work reached Louis Pasteur, who turned to the silkworm and explored several other diseases before championing his germ theory of infectious disease. Imagine, a simple silkworm is credited with changing our perception of human health.

Over the next hundred years, the young field of **microbiology**, or the study of microorganisms, was focused on the identification of microbes responsible for disease. In the early 1900s, the three leading causes of death were pneumonia, tuberculosis, and diarrhea, all due to infectious agents. Surgery was still a highly risky proposition, with the potential of dying from a surgical infection often outweighing the benefits of the surgery.

Koch's Postulates

As early as 1890, the German physician and microbiologist Robert Koch advanced the provocative idea that microbes caused certain animal diseases. Koch was study-

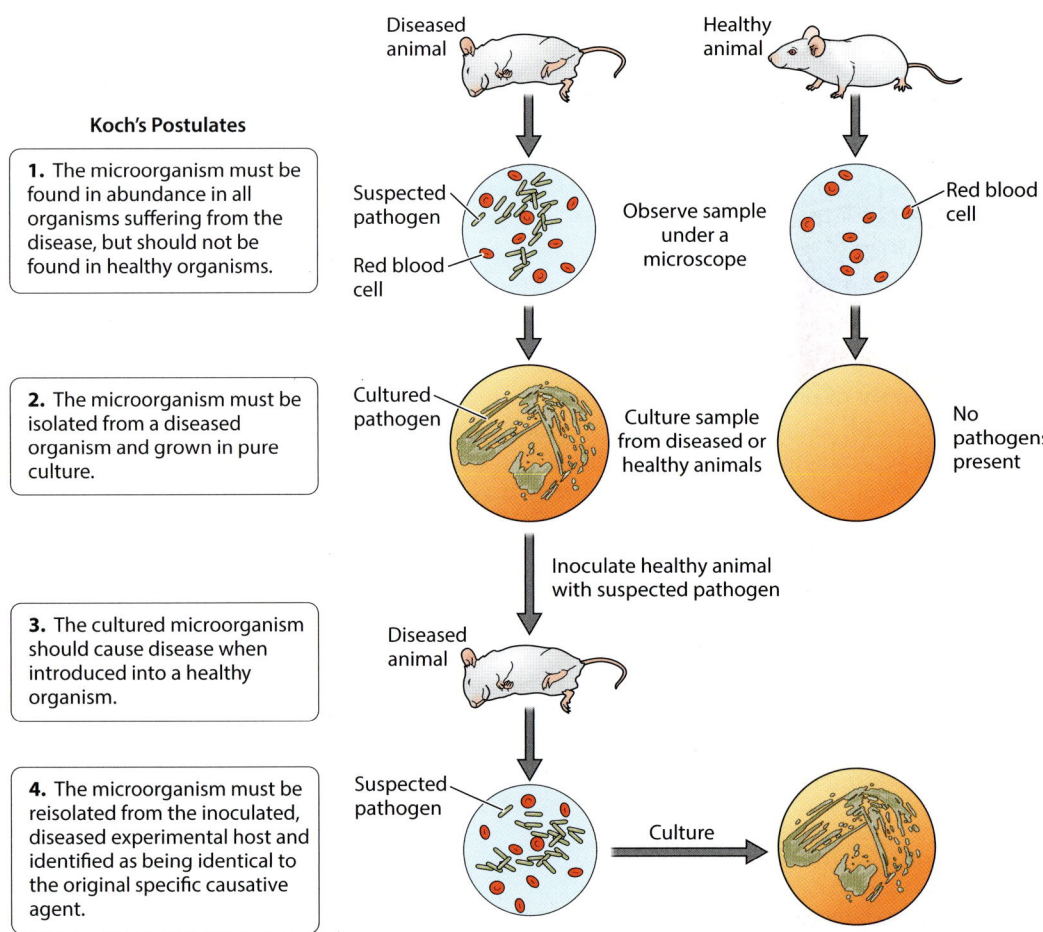

Figure 2.6 **Illustration of Koch's Postulates** (Adapted from Mike Jones, https://en.wikipedia.org/wiki/File:Koch%27s_Postulates.svg. CC BY-SA 3.0.)

ing the disease known as anthrax. Hoofed animals, such as cows and sheep, tend to get this disease, and by the time they display symptoms, such as staggering and trembling, death comes quickly. Koch examined the blood of cows that had died of anthrax and saw bacteria. He suspected these bacteria had caused the disease, and to prove his hypothesis, he infected mice with the blood from the dead cows, which, as he predicted, caused the mice to develop anthrax. This led to the creation of **Koch's postulates** (Figure 2.6), which are four criteria to determine whether a particular microbe (or germ) causes a particular disease (Koch, 1876). As originally stated, the four criteria are (1) the microorganism must be found in diseased but not healthy individuals; (2) the microorganism must be cultured from diseased individuals; (3) inoculation of healthy individuals with the cultured microorganism must result in the disease; and, finally, (4) the microorganism must be reisolated from inoculated, diseased individuals and matched to the original microorganism. Koch's postulates have been critically important in establishing the criteria whereby the scientific community agrees that a microorganism causes a disease.

It is now obvious that certain microbes can cause human disease. Although infectious diseases are not the focus of this textbook, it is worth identifying some of the more common ones found in the US (chlamydia, influenza, and *Staphylococcus aureus* skin infection; Figure 2.7A) and some of the deadliest in the world (lower respiratory tract infections, diarrheal diseases, and HIV/AIDS; Figure 2.7B). We are used to thinking of bacteria solely as germs, which negatively impact our health. However, it is important to remember that microbes are not purposely trying to harm humans, but, rather, just trying to survive. This distinction was beautifully described in the 1970s by Lewis Thomas, an American physician, poet, and research scientist. According to Thomas, "disease usually results from inconclusive negotiations for symbiosis, an overstepping of the line by one side or the other, a biological misinterpretation of the borders" (Thomas, 1978). As we discussed in chapter 1, the term

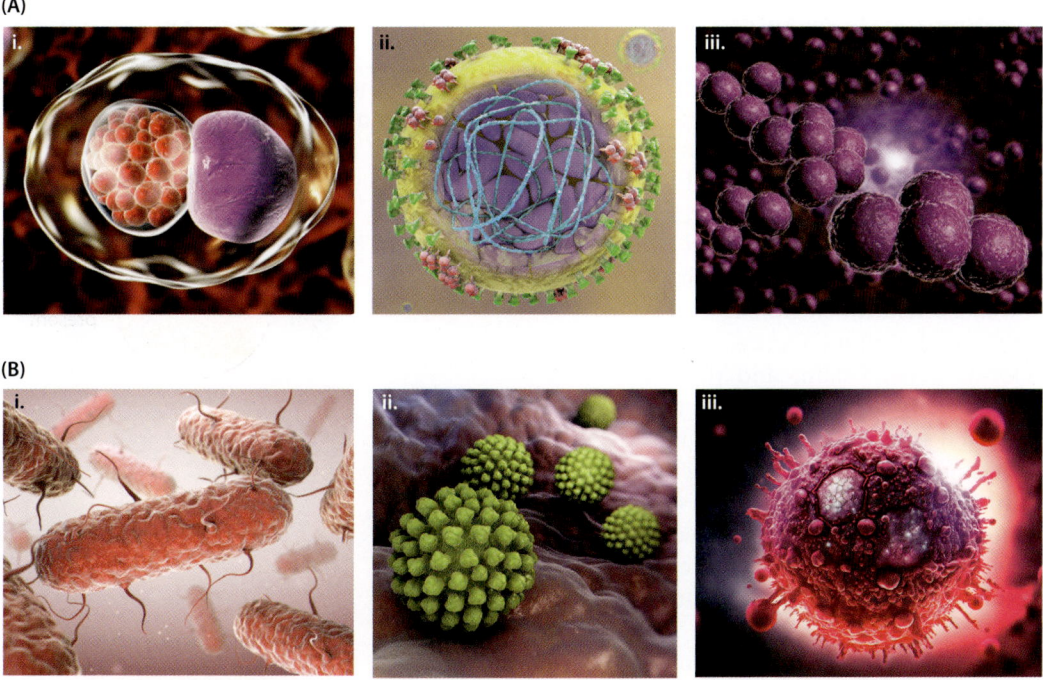

Figure 2.7 The Most Common and Most Deadly Infectious Diseases (A) Most common infectious diseases in the US: chlamydia (i), influenza (ii), and *S. aureus* skin infections (iii). (B) The deadliest diseases, by case numbers, in the world: lower respiratory tract infections (i), diarrheal diseases (ii), and HIV/AIDS (iii). (A photos from Kateryna Kon/Shutterstock [i]; Manu5, CC BY-SA 4.0, via Wikimedia Commons [ii]; Ezume Images/Shutterstock [iii]; B photos from iStock/iLexx, [i and ii]; iStock/CIPhotos [iii])

symbiosis refers to different organisms living in close physical association, typically to the advantage of both. We now know that we are the host of untold numbers of bacterial symbionts that play a critical role in keeping us healthy, and only rarely create disease.

The Discovery of Antibiotics

One critical response to our rapidly growing knowledge about the causes of infectious disease was an equally intense hunt for therapeutic interventions. Without any understanding of the microbiome's presence in our own bodies at that time, people believed microbes, particularly bacteria, were exclusively harmful to human health. Thus, scientists looked for any treatments that could kill bacteria. In 1928, Alexander Fleming discovered one such treatment: penicillin, the first true antibiotic used in human health. **Antibiotics** are chemical substances produced by microorganisms that kill or inhibit the growth of bacteria. **Box 2.2** describes Alexander Fleming's accidental discovery of penicillin, which forever changed the course of human health (Gaynes, 2017). In a world essentially at the mercy of **germs**, the term used for viruses, bacteria, and other microorganisms that cause disease, finding a superweapon that could kill germs was a game-changer. After the discovery of penicillin, scientists rushed to identify more antibiotics. The period of discovery from 1950 to 1960 is known as the golden age of antibiotics, as 50% of the compounds we still use to treat bacterial infections were discovered then.

BOX 2.2. DISCOVERY OF PENICILLIN

Sir Alexander Fleming was a Scottish physician and scientist working in a London hospital. His job was to study wound infections. In September 1928, Fleming returned to his laboratory after being away for a month and noticed that a culture of *Staphylococcus aureus* he had left out on his workbench had become contaminated with a mold (**Figure 2.8**), later identified as *Penicillium notatum*. He also discovered that the colonies of staphylococci surrounding this mold had been destroyed. Fleming called the substance responsible "mold juice" at first and then eventually named it penicillin after the mold that produced it. He determined that it was a chemical substance and tried to purify it from the mold without success.

It was Howard Florey, Ernst Chain, and their colleagues at Oxford University who first purified penicillin. They experimented with an array of different culture vessels, such as bedpans and butter churns, and eventually turned their laboratory into a penicillin production factory. In 1940, Florey showed that the penicillin they were producing could protect mice against disease resulting from a lethal infection of *Streptococcus*.

On February 12, 1941, a 43-year-old policeman, Albert Alexander, became the first patient to be treated with penicillin. He had a life-threatening infection that his physicians could not treat. They injected penicillin, and he improved rapidly, within days. However, there was very little penicillin available, and before the physicians could give him more, Alexander died. The onset of WWII drove a more rapid push to produce penicillin for use on the battlefield.

Sir Fleming commented on his discovery later in his life, saying, "When I woke up just after dawn on September 28, 1928, I certainly didn't plan to revolutionize all medicine by discovering the world's first antibiotic, or bacteria killer. But I suppose that was exactly what I did." (Tan and Tatsumura, 2015).

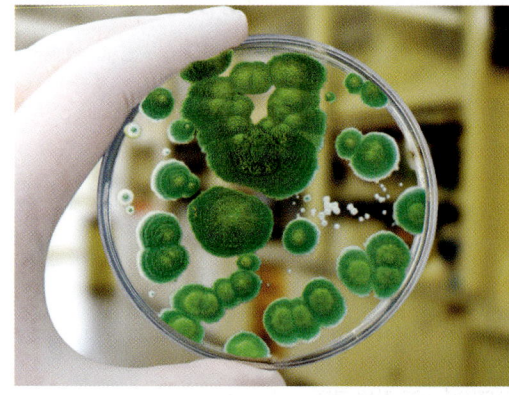

Figure 2.8 Penicillin Mold This petri dish shows the penicillin mold growing in green. (Photo from Kateryna Kon/Shutterstock)

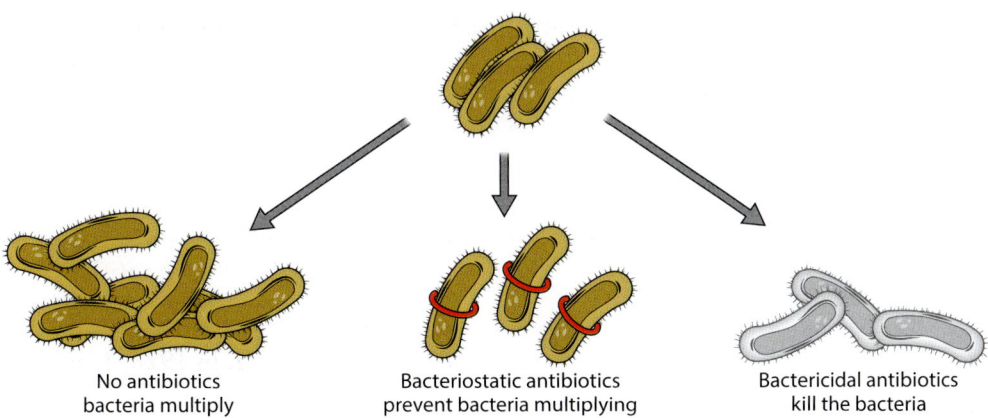

Figure 2.9 Bacteriostatic versus Bactericidal Modes of Action Bacteriostatic antibiotics inhibit the bacterial cellular processes, while bactericidal antibiotics kill the bacteria.

Any substance that inhibits the growth of or kills a bacterial cell can be called an antibiotic. These drugs are employed in medicine to target bacterial infections on or in the body, such as tetanus, bacterial pneumonia, strep throat, and cholera. There are two main ways in which antibiotics target bacteria (**Figure 2.9**). Antibiotics that kill the bacteria are known as bactericidal, while those that simply inhibit their growth are known as bacteriostatic. Although antibiotics are powerful treatments for eliminating bacterial infections, they are not effective against viruses, which is why you aren't prescribed antibiotics for a cold or the flu. In comparison, antiseptics such as hydrogen peroxide and rubbing alcohol are used to sterilize surfaces of living tissue, while disinfectants such as bleach and chlorine are used to sterilize nonliving surfaces.

As we learn about the importance of our microbiome in human health, we will realize that the golden age of antibiotics, although still golden, had a faint shadow lurking behind it right from the start. We now recognize, if a bit late, that these superweapons not only kill many of our most challenging infectious agents, they also simultaneously decimate some of the most helpful members of our microbiome, leading to a plethora of new diseases that we will touch on in later chapters. The ecology of our microbiome is something that we are just now beginning to understand. One key ecological principle is that any niche abhors a vacuum. If we eliminate microorganisms from a microbial community, such as the one we find in our gut, something will invade this now-empty niche. Unfortunately, it is often a pathogen that wins that race! The faint shadow lurking in the background during the golden age of antibiotics is now recognized as a full-fledged human health imperative. We must find ways to reduce or eliminate our reliance on antibiotics to treat infectious diseases. Their broad spectrum of activity, which means they kill both bacterial friend and foe alike, makes them an untenable human health solution in the long run. We will discuss in chapter 8 how misuse and overuse of antibiotics has resulted in the emergence of highly drug-resistant bacterial pathogens, requiring that we quickly return to the blackboard and rethink the future of anti-infective treatments.

2.2 THE GOLDEN AGE OF MICROBIOLOGY

This intense focus on microorganisms and the diseases they caused in the mid- to late-1800s is understandable. It was the start of the golden age of microbiology in which we learned for the first time that microbes are responsible for some of the deadliest diseases, including plague, tuberculosis, and cholera. Scientists enjoyed enormous success in their early investigations of infectious agents for one very simple reason: most human pathogens grow in the presence of oxygen, making it easy to culture them. Additionally, there was a growing body of scientists exploring the potential

Figure 2.10 Theodor Escherich and His Namesake Theodor Escherich was a German-Austrian pediatrician and a professor. In 1886, he published a monograph on the relationship of intestinal bacteria to the physiology of digestion in infants. It was also the publication where Escherich described "bacterium coli commune," which was later renamed in his honor as *Escherichia coli*. (Photos from Unknown Retouched by Lichtspiel, Public domain, via Wikimedia Commons [left]; ymd2881/Shutterstock [right])

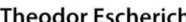

Theodor Escherich

health benefits of some of our invisible companions. While this focus on infection continues today, there was a growing body of scientists exploring the potential health benefits of some of our invisible companions.

In 1885, Theodor Escherich, a German-Austrian pediatrician and professor, was focused on describing the microbes present in the guts of infants (**Figure 2.10**). One of his most famous discoveries was the bacterial species *Escherichia coli*, which was later named in his honor (Escherich, 1989). *E. coli* went on to become a model bacterium in microbiology, meaning that it has served as the focus of intense research, such that we know more about *E. coli* than we do any other cellular organism. Its ability to grow rapidly (doubling in numbers every 20 minutes) in an inexpensive medium has ensured its endurance as a research model. By the way, *E. coli* is present in most human guts and plays a role in digesting food and producing vitamin K.

Escherich's work was followed by Alfred Nissle, who was an army physician during World War I. Soldiers were dying not only from bullet wounds and poison gas inhalation, but also from infectious disease. One particularly common disease during the war was a severe form of diarrhea known as dysentery. Nissle obtained a stool sample from a convalescing soldier who had been fighting in a region heavily contaminated with a pathogenic bacterium, *Shigella*, known to cause dysentery. This soldier was not visibly suffering from his infection, and Nissle hypothesized that he might have a strain of *E. coli* resident in his gut that prevented *Shigella* from increasing in numbers and causing symptoms.

Discovering Colonization Resistance

Wait a minute, let's think about that last statement. Here was a doctor, working during the early 20th century, well before we knew about the good bacteria that live in our guts, proposing that one patient hadn't succumbed to dysentery because he had a second type of bacteria in his gut that protected him from *Shigella* pathogens. This is an example of a scientist clearly thinking outside the box! Nissle carefully observed his patients, and contrary to the accepted knowledge at the time, he proposed something truly innovative: that there was an ecology in the gut, and bacteria were competing to survive in it. In this one target patient, Nissle proposed, perhaps there was a nonpathogenic bacterium present that simply outcompeted the pathogenic *Shigella* and thus ensured the patient remained healthy.

Nissle's studies showed that there was such a strain, later named in his honor—*E. coli* Nissle 1917—capable of competing against certain pathogenic gut bacteria. Nissle's research revealed the concept of **colonization resistance**, in which one microbe (in this case *E. coli* Nissle 1917) prevents the establishment of a pathogen (such as *Shigella*) in the same niche (in this case the large intestine) (**Figure 2.11**) (Nissle, 1925). In other words, if the nonpathogenic strain *E. coli* Nissle 1917 is present in a gut, it can inhibit the invasion of *Shigella*. This same strain of *E. coli* remains in use today as an oral treatment for diarrhea and can reduce the timeline of an infection by half (Henker et al., 2007).

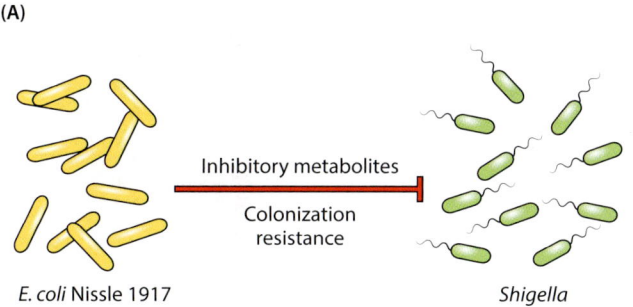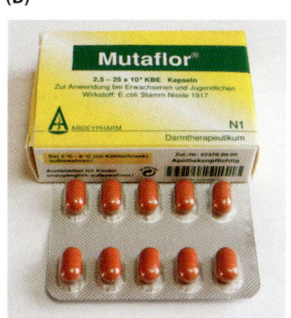

Figure 2.11 Colonization Resistance (A) This figure shows how a resident microbe, such as the *E. coli*, can give a host resistance to an invading pathogen just by being present in the microbiome. *E. coli* Nissle 1917 produces inhibitory metabolites that prevent *Shigella* bacteria from becoming established in the gut microbiome, preventing infection. (B) Preparations of the Nissle strain are still used to treat intestinal infections. (B photo from Cofiant Images/Alamy Stock Photo)

Not Just Germs

By the early 1900s scientists were starting to realize that the microbes in our gut were not just the cause of disease, but in fact played critical roles in our health. As more methods to grow members of our microbiota were developed, scientists began to explore the diversity of microbes throughout our bodies. Those in our mouth had largely been ignored for the two centuries following Leeuwenhoek's observations of bacteria in his tooth plaque. However, it turns out that oral microbes are easy to collect, and many are **aerobic**, growing in the presence of oxygen, which makes them far easier to work with than the mostly **anaerobic** bacteria, which are sensitive to oxygen, that reside in our gut. Between the 1920s and 1950s, an American dentist, Joseph Appleton, cataloged the bacteria sampled from the mouths of his patients and noticed that the microbes present changed depending upon age, diet, the season, and the presence of disease. These studies helped transform dentistry from its once-marginalized place in medicine, into a rigorous field of scientific inquiry.

Anaerobic Culturing Methods

In contrast to the aerobic bacteria focused on by Appleton, numerous bacteria thrive in **anoxic** (oxygen-depleted) environments such as soil sediments or the guts of animals. These anaerobes are extremely sensitive to oxygen. Historically, it has been difficult to study anaerobes in the lab because oxygen in air stresses or kills them. The development of methods to grow anaerobes was an important stepping stone in microbiology, one that paved the way for the discovery of a multitude of new bacteria. Many of these discoveries could not have been possible without a technical breakthrough that was achieved in the 1940s. Robert Hungate was a pioneering **microbial ecologist**, a scientist studying the interactions between microbes and their environment. He was investigating **cellulose-degrading bacteria** that can break down cellulose, an insoluble carbohydrate found in plant cell walls. His focus was on bacteria found in the cow's rumen, one of its several stomachs, that break down grass and provide the cow with energy. Hungate was frustrated that he couldn't observe the cellulose-degrading bacteria that he knew were present, which pushed him to invent an anaerobic growth method called the **roll-tube** (Figure 2.12).

The roll-tube procedure is quite complicated and requires a series of steps to ensure that no oxygen remains in the growth medium, which would kill anaerobic microbes. Hungate started by creating a growth medium that contained cellulose. To kill any other microbes that might be present in the medium, he used an **autoclave** to apply steam under pressure. The high heat and high-pressure conditions sterilized the growth medium, so no microbes

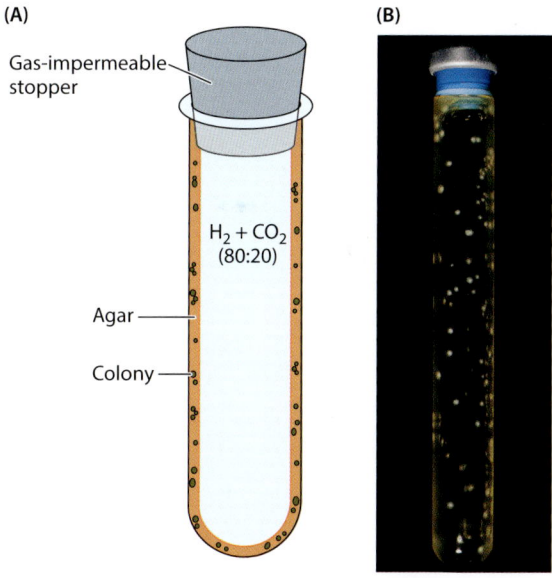

Figure 2.12 The Hungate Roll-Tube Because our atmosphere is rich in oxygen, special preparations must be made to study anaerobic bacteria in the lab. The roll-tube technique allows for a thin layer of nutrient agar to be exposed to an oxygen-free environment, allowing the bacteria of interest to thrive (A). The tube in (B) was inoculated with fungal spores from cultures of fresh dung samples collected from goats at the Santa Barbara Zoo. Numerous anaerobic bacterial colonies can be seen. (B photo from C. Haitjema, et al., 2014. Anaerobic Gut Fungi: Advances in Isolation, Culture, and Cellulolytic Enzyme Discovery for Biofuel Production. *Biotechnology and Bioengineering*. 111. 10.1002/bit.25264. © 2014 Wiley Periodicals, Inc.)

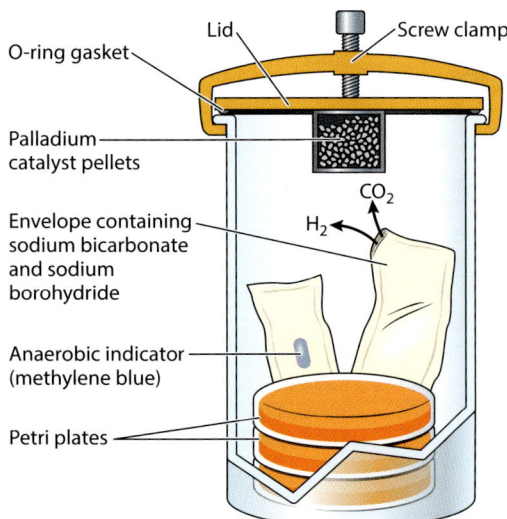

Figure 2.13 The GasPak A GasPak™ jar is used to create an anaerobic environment, which means an atmosphere free of oxygen, in which organisms that thrive under such conditions may be cultivated in the laboratory. The chemicals sodium bicarbonate and sodium borohydride are placed in the jar, where they react with water to produce hydrogen and carbon dioxide gas. (Adapted from Tortora, Funke, and Case, 2003.)

survived. Hungate then poured the now-sterile medium into sterile rubber-stoppered tubes. Next, he introduced a stream of nitrogen and carbon dioxide gasses into the tube to eliminate any oxygen. Molten agar was then added to the tube, which was rolled horizontally as the agar solidified into a thin layer that covered the inner surface of the tube and the growth medium. Using a needle and syringe, the bacteria being cultured were then injected through the rubber stopper. The method, now known as the **Hungate technique**, is still in use to this day.

Hungate's efforts paid off, and in 1950 he published both the roll-tube technique and a description of the cellulose-degrading organisms he was able to successfully culture from the cow's rumen (Hungate, 1950). Several simpler approaches to anaerobic growth have since been developed, such as the **GasPak**™ (**Figure 2.13**). This commercial product is a packet containing a dry powder that, when mixed with water and kept in an airtight vessel, produces an atmosphere free of oxygen gas. This simple system made anaerobes, such as those found in our gut microbiome, much easier to grow in laboratories, opening the door to new areas of microbiome research.

2.3 GENOMICS AND BIOINFORMATICS

Microbiologists spent the next several decades exploring the anaerobic diversity that was finally culturable. We learned an extraordinary amount about specific microbes, usually bacteria, that were involved in the metabolism of complex carbohydrates in our gut. However, everything changed with the advent of the field of **genomics**, which is the branch of molecular biology concerned with the structure, function, evolution, and mapping of genomes. Since the discovery by Watson and Crick of the structure of DNA in the 1950s, there has been an intense focus on improving our ability to extract DNA from an organism and then determine the precise sequence of nucleotide bases (adenine, guanine, thymine, and cytosine) present. Together with the refinement of sequencing technologies was the development of ever more efficient and sophisticated computational algorithms to make sense of the enormous sequence databases being generated.

Environmental Metagenomics

Once these advances were in place, it became possible to apply these same technologies to the challenge of determining the complete DNA content of an environmental sample, with its complex mixture of organisms and their genomes. Jo Handelsman, a microbial ecologist at the University of Wisconsin, coined the term **metagenomics** to refer to the analysis of all the DNA isolated from an environmental sample and is considered the mother of this field of research (Handelsman et al., 1988). Metagenomics is distinct from genomics in that the latter focuses on an individual or species as the target of study, while the former includes all life present in a sample. Metagenomic analysis aims at addressing one or more of the following questions: Who is there, or what is the diversity and abundance of community members? What are they doing, or what is the metabolic potential of the community and its members? Why are they there, or what are the ecological relationships between members of the community?

The metagenomic process involves extracting DNA from a sample, such as soil or feces, and determining all the DNA sequences present. **Figure 2.14** provides a simplified example of such a study. The result is a compilation of the DNA sequences present in that sample without the need for complex culturing techniques. Early metagenomic studies revealed that most of the microbial diversity in an environmen-

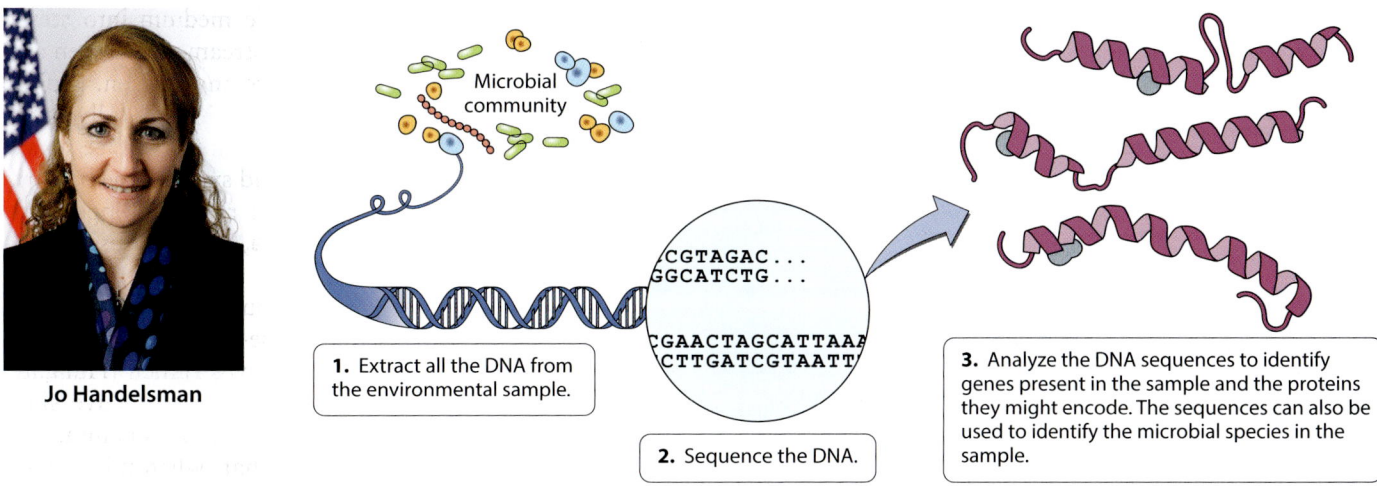

Jo Handelsman

Figure 2.14 Metagenomic Study Jo Handelsman is considered by many to be the mother of metagenomics. Although these studies seem complicated, they are fairly easy to carry out in the lab. You simply obtain a sample from the environment, extract the microbial DNA, sequence it, and analyze the resulting data to identify who is present, at what frequency, and what they are doing! (Photo from Chuck Kennedy/White House official photo, Public domain, via Wikimedia Commons, Illustration after Gilbert and Dupont, 2011.)

tal sample was unculturable, even when using our novel anaerobic culturing tricks. These studies also told us that, even after spending decades to learn how to culture more diverse microbes, we still had no idea at all about the actual abundance and diversity of microbes on Earth. This knowledge inspired a rebirth in the development of even more culturing techniques. Now that we know such extraordinary microbial diversity is present, can we tease these formerly invisible microbes into growing in our labs? This is a critical step if we wish to isolate the bacteria to study them more directly. We have certainly made great headway in this regard, and it is now possible to cultivate some 50% of the organisms in some environmental samples.

Fecal Microbiota Transplantation

A further milestone in microbiome science occurred in 1958 when physician Ben Eiseman and colleagues reported a novel treatment for the disease pseudomembranous enterocolitis (Eiseman et al., 1958). Enterocolitis is caused by a bacterium, *Clostridioides difficile*, which can cause debilitating diarrheal symptoms. The colon becomes inflamed, and there is the formation of "pseudomembranes" consisting of cellular debris. Eiseman had several patients with this disease, and he noticed that their stool contained the bacterium *Staphylococcus aureus*, which was considered a possible cause of the disease. He also noted that these patients had been treated extensively with antibiotics, and he proposed that the disease may be the result of the disruption of the gut microflora by the drugs. He came up with the idea that if stool were transferred from healthy individuals into the colon of a sick patient, the microbes present in the transferred stool could eliminate or at least substantially decrease the density of *S. aureus*. So, Eiseman performed what may have been the first **fecal microbiota transplant** (**FMT**) in Western medicine—and it worked! The stool in the patients following the FMT procedure no longer contained *S. aureus*. Although it was not known at the time, FMT also resulted in the elimination of the true cause of disease, *C. difficile*. Eiseman suggested that one or more of the microbes present in the healthy guts of the donors displaced the colitis-causing pathogen. Although FMT appears, at first glance, to be a very strange approach to treating infectious disease, it really isn't that different from how Nissle used the probiotic strain *E. coli* Nissle 1917 to displace *Shigella* in patients with dysentery.

Since this first inspired application, FMT has been subjected to numerous clinical trials for use in treating recurrent *C. difficile* infections and is now considered a

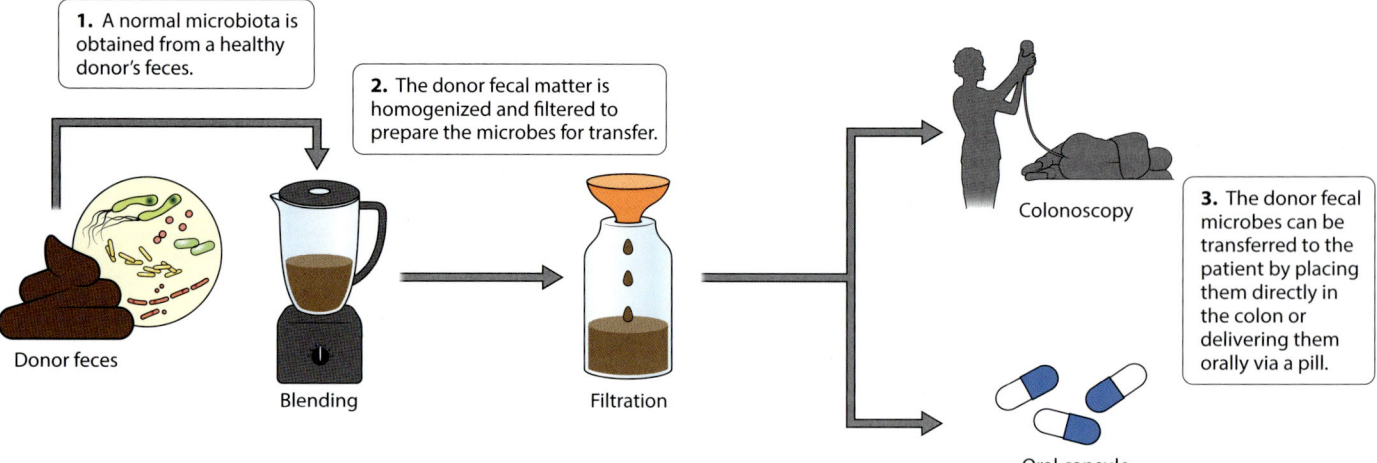

Figure 2.15 Fecal Microbiota Transfer The schematic diagram of the FMT procedure, which involves fecal sample procurement and preparation and transfer into the target patient. (Adapted from Wang et al., 2019.)

treatment of choice (**Figure 2.15**). The microbes from the donor, once introduced into the patient's colon, appear to normalize the gut microbial community. Several mechanisms may be involved, such as the production of antimicrobial peptides by the healthy microbes, which kill bacterial pathogens; the inhibition of germination of **spores** (a dormant form of *C. difficile*); and competition between bacteria for nutrients. FMT is now under investigation for other conditions, such as colitis, which is a chronic digestive disease that results in inflammation of the inner lining of the colon. It has even shown promise in addressing symptoms of obesity. FMT from lean donors was introduced into obese males, and there was an associated improvement in glucose metabolism and alterations in intestinal microbiota composition, which we'll discuss further in chapter 3 (Napolitano & Covasa, 2020).

These studies inspired a new perspective on our microbiomes and their potential use in therapeutic applications. Unlike the previous era, when scientists looked for treatments to kill bacteria, scientists now saw bacteria as the potential treatment for disease. When Eisemen carried out the very first FMT, he had no idea what microbes were present in the donor's feces. However, the power of the approach spurred scientists to find tools that would allow them to learn more about these microbial "good guys."

Germ-Free Mice

In the 1960s, another crucial tool was incorporated into the microbiome research tool kit that allowed us to manipulate a microbiome and measure the impact of those changes on the host. Germ-free animals were already well established in studies aimed at learning how intestinal microbes contributed to a host's nutritional needs. **Gnotobiotic** (germ-free) animals are born and raised under sterile conditions—they have no microbiomes. At birth, they are removed from the mother by Cesarean section and live in tightly controlled isolation chambers with germ-free foster mothers. These animals, usually mice, rats, or guinea pigs, have a lifespan similar to that of conventional animals, although they grow more slowly and have some physiological differences.

In 1965, Russell Schaedler, a professor at Rockefeller University, published an article that described his pivotal work on establishing a defined and stable microbiome in germ-free mice, using cultures from the guts of conventional mice, essentially employing FMT (Schaedler et al., 1965). Schaedler and colleagues inoculated food with bacterial cultures isolated from the guts of conventional mice. After only one week on this special diet, the germ-free mice had bacterial diversity in the gastrointestinal tract comparable to that of the donor mice. Although the speed of this colonization was surprising at the time, we now understand that a similarly rapid process happens during the birthing process. A mouse, or human, at birth is transformed from a sterile vessel to a microbial fermentation chamber in very short order!

Schaedler and his students eventually created proprietary mixtures of bacterial species that were sold to companies that supply gnotobiotic mice. Soon, many research labs in the US engaged in gnotobiotic mouse studies of the microbiome, employing these same strain mixtures to create defined and stable gut microbiota. By adding a set strain mixture to germ-free mice, scientists could make many mice have the same microbiome, which was a useful control for experiments. Studies aimed at investigating the role of the microbiome on the host could now be directly compared.

2.4 BRINGING IT ALL TOGETHER

In 1999 David Relman, an infectious disease physician and professor at Stanford University and a pioneer in modern microbiome science, repeated, in essence, the experiment first carried out several hundred years earlier by Leeuwenhoek. Relman scraped plaque from his teeth, but rather than observe the sample under a microscope as Leeuwenhoek had done, he sent part of the sample to a clinical microbiology lab, where they employed a variety of growth media to identify all of the bacteria that would grow. He used the remaining sample to extract and sequence the 16S **ribosomal DNA (rDNA)** (the gene that encodes ribosomal RNA), employing what we now consider a standard metagenomic sequencing screen. Recall from chapter 1 that Carl Woese employed the 16S ribosomal protein in his quest to incorporate microorganisms into a tree of life in the 1970s. He fragmented the protein and then compared fragmentation patterns between organisms. By the 1990s DNA sequencing and related molecular biology technologies had improved, and we could now rapidly sequence the 16S ribosomal encoding gene (rDNA) itself (**Figure 2.16**). These sequences

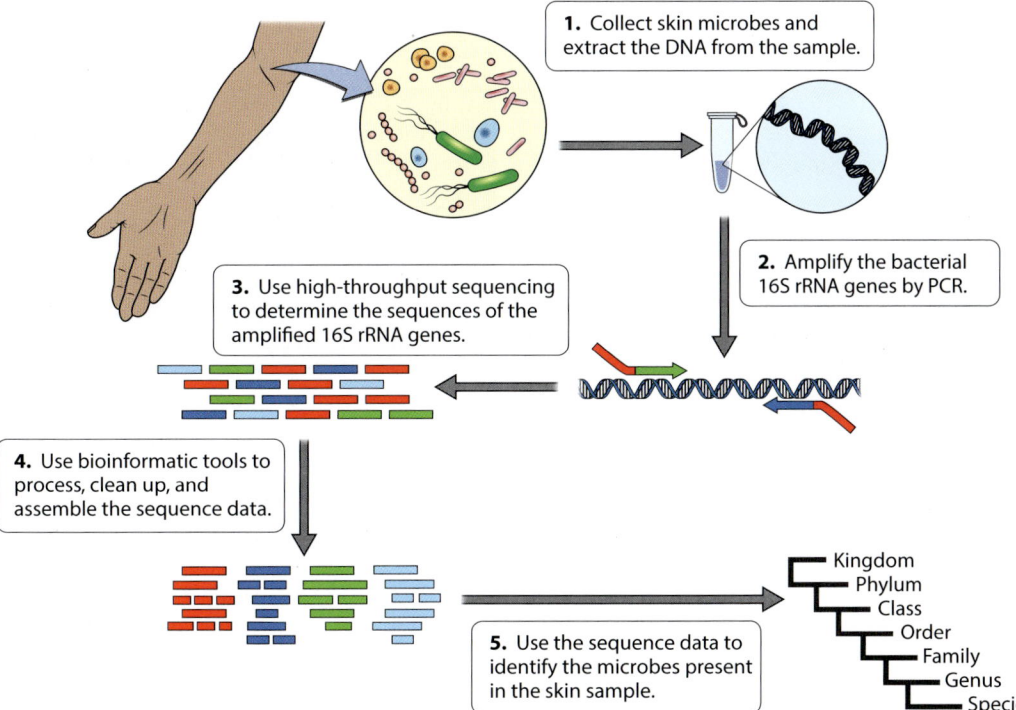

Figure 2.16 16S rDNA Sequencing for Microorganism Identification Schematic illustration of basic workflow for skin 16S rRNA gene-based sequencing. (After Jo et al., 2016 and Kong, 2011.)

provided vastly more information about the organisms present than did the RNA fragment patterns employed by Woese.

Relman's initial goal was to compare the bacterial census results obtained with these two different methods: conventional culturing versus molecular probing (Proctor et al., 2018). Although metagenomic screens were by then being applied to a diverse sample of natural environments, no one had yet used them to target the human microbiome and directly compare the results with those from culturing methods. What Relman found was that, even though the mouth was the most well-studied of human microbiomes at that time, 10 times more microbial species were identified with the metagenomic methods, confirming what had been suspected—that our prior reliance on culturing had vastly underestimated microbial diversity (Proctor et al., 2018). Further, many of the species identified by rDNA sequencing had never been seen before. This was the first application of metagenomics to the human body—revealing a diverse, and mostly undiscovered, microbiota. According to Relman, "looking at some scrapings from human teeth (they were actually my own teeth) . . . the sequencing approach revealed far more, and was an order of magnitude more sensitive, and informative" (Relman, 2019).

Fluorescing Microbes

Jessica Welch, a microbial ecologist from the Marine Biological Laboratory in Woods Hole, has since taken Relman's oral microbiome cataloging effort one step further. She also took a dental plaque sample, but this time she introduced 12 unique molecular tags into the sample, each of which had a different fluorescent color bound to it. These tags were designed to recognize and bind to different bacterial species (Welch et al., 2016). Using a microscope with fluorescent lights, Welch was able to see the spatial organization of the bacteria within the dental plaque matrix for the very first time. Welch and her colleagues named the plaque image produced "hedgehog," and it certainly has some resemblance to that adorable spiky mammal (**Figure 2.17**).

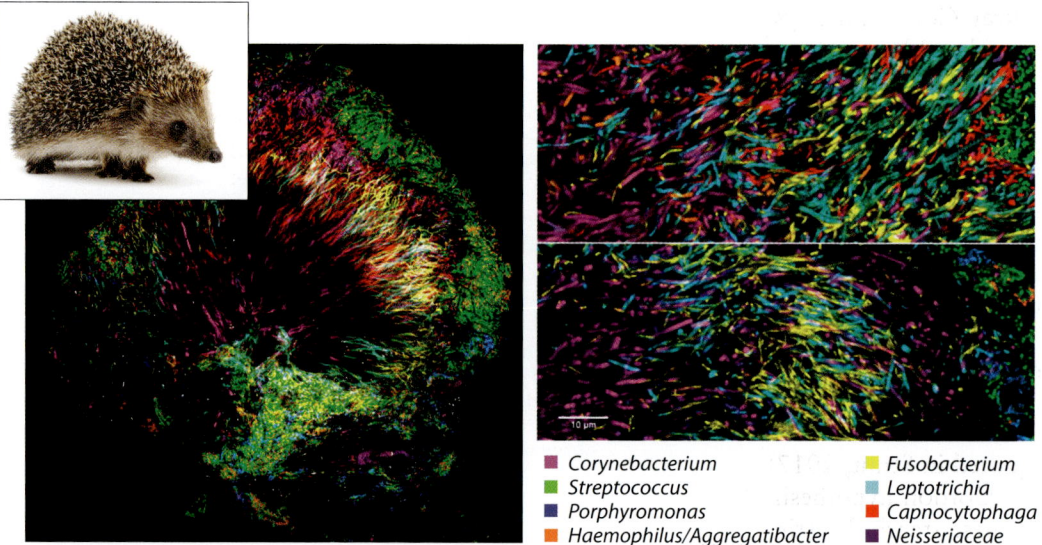

Figure 2.17 Fluorescing Bacteria in Dental Plaque This fluorescent hedgehog structure on the left is a reconstruction of the bacteria found in dental plaque. The images on the right are higher-magnification views. Note that the scale bar = 10 μm; for reference, an *E. coli* cell is typically 1–2 μm long. Each of the different colors is associated with a different bacterial genus, as indicated in the key. Understanding where bacteria are located in an environment can provide insight into their evolutionary niches, and what effects they may have on their human hosts. The adorable mammal at upper left is the plaque structure's namesake, the hedgehog. (Photo inset from iStock.com/Voren1; fluorescent images from J. L. Mark Welch et al., 2016. Biogeography of a human oral microbiome at the micron scale. *Proc Natl Acad Sci USA*. 2016 Feb 9;113(6): E791–800.)

Rather than being randomly distributed, the different types of bacteria in the mouth are organized in specific ways. Filamentous bacteria, which grow end to end and resemble long strands of spaghetti, interlock with each other to form a mesh that serves as the foundation of this microenvironment. The filaments run the length of the plaque, while other types of bacteria are found restricted to certain regions within this mesh. The hedgehog structure reveals the complexity of the oral microbiome and shows us how organization can emerge from micron-scale interactions of the microbial community members.

Since the mid-1990s, microbiomes from every imaginable location on the human body have been sampled using metagenomic methods and the microbiota have been identified. This period has provided one amazing discovery after another—far beyond the sheer numbers of microbes in our bodies is the fact that most of these organisms had never been encountered before! Imagine a naturalist such as Darwin exploring the coastline of South America in the 1800s. The diversity of unique plants and animals would have been nearly overwhelming to him. These past 20 years have marked a similar period of discovery and excitement for microbiologists as we have revealed the extraordinary, previously unknown diversity of microbes that are part of what makes us human.

There is, however, one major drawback to sequence- and fluorescence-based approaches in cataloging microbial diversity. These molecular methods are superb at telling you which microbes are present, but far less useful in identifying what they are doing. If you imagine yourself, again, as a naturalist, it's like being given a list of the DNA sequences of species present in a rainforest but not knowing what they look like or what they do. How would you figure out what a sloth or monkey does from DNA alone? As microbiomes from every nook and cranny of the human body were being discovered, one scientist turned his attention to figuring out precisely what they were doing.

The Father of Microbiome Research

Jeffrey Gordon, a physician and professor at Washington University in St. Louis, is considered by many to be the father of modern microbiome research. He harnessed the power of germ-free mice to experimentally explore the functional roles of microorganisms in the human gut. Early on, he and a former graduate student, Lynn Bry, focused on the development of the intestine in conventional versus germ-free mice. Rather than focusing on the diversity of the intestinal contents, Gordon and Bry were interested in their functional roles. They discovered that certain microbes in the guts of mice produce enzymes that permit them to digest starchy foods. If those microbes are absent, such as in the guts of germ-free mice, the starchy food sources remain undigested. Gordon and Bry also found that microbes in mice were somehow inducing the host's intestinal cells to produce sugars for the microbes to eat! The sterile mice had no such sugar production, unless they were given fecal microbiota transplants from the conventional mice. According to Gordon, "Lynn found evidence that gut microbes are directing the host to serve them a meal of complex carbohydrates" (quoted in Strait, 2017).

Gordon hypothesized that individuals who tend toward obesity may have more of these starch-digesting microbes, which are able to extract more energy from their food sources than the gut microbes in lean individuals can. In other words, if a lean and an obese individual were to consume the same diet, the obese individual would extract more calories from their food than the lean individual, simply due to the types of microbes present in their guts. To test this hypothesis, Gordon and his colleagues employed a mouse with a genetic mutation that resulted in obesity, and they compared its gut microbiota with that of a mouse without this mutation. The mice were born from the same mother, lived in identical environments, and ate the same diet, so the mutation that led to weight gain was the only difference. The obese mice developed a different composition of gut bacteria, suggesting that there was an interaction between the obesity mutation and the gut microbiota (Turnbaugh et al., 2006).

These findings resulted in quite a media frenzy; perhaps microbes caused obesity! The composition of the gut microbiome affected the weight of mice, but would this work in humans as well? To find out, Gordon next turned his attention to genetically identical human siblings, one of whom was thin and the other obese. He used 16S rRNA gene sequencing to identify the microbes present in the twins' guts and showed that although the twins were genetically identical, their microbiomes were not. The levels of microbial diversity were lower in the gut of the obese twin, and there were differences in the numbers of microbiome-encoded genes related to metabolism (Ridaura et al., 2013).

Although these differences showed that the microbiome is associated with weight, it didn't prove that differences in the microbiome caused weight gain. It was possible that some other factor, like diet, was causing the differences in the twins' weight and that these weight differences then caused changes in the gut microbiome's composition. To determine if the microbiome could be causing changes in weight, Gordon's team transplanted stool samples from the twins into germ-free genetically identical mice and revealed that obesity is transmissible through stool (**Figure 2.18**). In other words, mice that had received the obese twin's microbiota showed an increase in body fat (Ridaura et al., 2013).

Equally interesting, when the obese and lean mice were housed together, the microbes associated with leanness were able to establish themselves in the guts of the obese mice, which then lost weight. Gordon's research revealed clearly for the first time that our gut microbiome plays a significant role in the development of obesity, and that role is transmissible. "Being able to see ourselves as a splendid collection of interacting human and microbial parts teaches us that we do not travel through life alone, unaccompanied," Gordon says. "There is a microbial dimension to our development that offers an expanded view of the 'self'" (quoted in Strait, 2017).

Gordon and his colleagues continue to produce fascinating stories about our microbiomes. One recent project involves the creation of snack foods that contain fibers that deliberately change the gut microbiome in ways linked to health. Gordon started this project feeding prototype snacks to mice, and his team has since engaged several rounds of human volunteers. The overweight volunteers ate snacks with specific combinations of fibers that select for microbes involved in metabolizing fiber. Gordon's team has shown that subsequent shifts in the participants' microbiomes were linked to changes in certain blood proteins in ways that could improve health over the long term (Delannoy-Bruno et al., 2021). Just imagine, healthy snacks that promote a healthy microbiome and that result in weight loss and improved health!

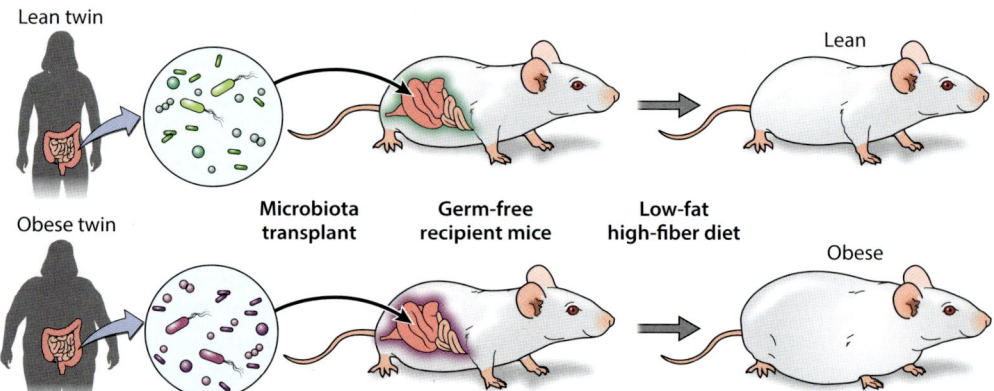

Figure 2.18 **Obesity in Germ-Free Mice** Germ-free mice were inoculated with microbiota from obese or lean human twins. The mouse microbiomes took on the characteristics of the donors. (After Ridaura et al., 2013.)

Human Microbiome Project

As we enter the 21st century, microbiome research is truly taking off. Prior to 2003, there were essentially no publications that specifically referred to the microbiome. In 2010 there were 67, and by 2020 over 1,000 publications reported on the human microbiome. In part, this rapid increase was the result of significant investment made by the US **National Institutes of Health** (**NIH**). The NIH, a part of the US Department of Health and Human Services, is the nation's medical research agency, responsible for funding the research that makes important discoveries to improve health and save lives. In 2007, the NIH created the **Human Microbiome Project** (**HMP**), with an initial goal of providing a census of the content of the healthy human microbiome.

The first phase of the project (HMP1) took a systematic approach to identifying the microbiota associated with health. Scientists participating in this effort examined the microbiomes of 242 healthy individuals at 15 (male) to 18 (female) body sites. The microbiota at these sites were quite diverse—and it was quickly realized that there wasn't one single human microbiome, but rather numerous microbial communities adapted to different regions of our bodies. The process of niche adaptation is so deliberate that, for example, your skin microbiome is more like someone else's skin microbiome than it is like other microbiomes on your own body. We also learned that after the first few years of life, one's microbiomes become relatively stable in composition. Finally, regardless of which microbiome sites are sampled, the microbes present carry out many of the same metabolic functions. In other words, certain functions are performed in every microbiome on your body, but they may not be carried out by the same species.

The HMP1 study, and subsequent efforts, have generated an enormous dataset of microbiome data from healthy adults. There are over 10 terabytes (TB) of information, equivalent to 10 times one trillion (10^{12}) bytes of DNA sequence data available for public use, making it the largest set of microbiome data from any source. It includes the microbial community composition from five body regions, chosen primarily for their roles in human health (nasal cavity, oral cavity, skin, gastrointestinal tract, and urogenital tract) (NIH, 2024) https://www.hmpdacc.org/hmp/.

No One Else Has Your Exact Microbiome

Over the course of defining what they considered to be the healthy human microbiome, researchers made numerous critical scientific discoveries (Proctor et al., 2019). One of the most fundamental of these is that everyone's microbiome is different from all other human microbiomes. In healthy adults, microbiomes are unique to each individual and relatively stable over time. This one finding instantly helped us understand why we respond differently to clinical treatments—our microbes aren't the same and thus may interact with drugs differently. However, these findings also simultaneously created confusion—if healthy humans each have a unique microbiota, then how do we define a healthy microbiome? As discussed below, the solution is that we refer to a **normal**, rather than healthy, microbiome. At this point we must address an issue that relates to science equity and equality. The HMP did not target the diversity of the human species to decipher the normal human microbiome. It might more reasonably be called the WHMP, or White HMP, for its almost exclusive engagement of white participants. Fortunately, subsequent studies are addressing this critical weakness of the study design.

The Normal Human Microbiome

While the first stage of this massive, government-funded project helped us understand the normal microbiome of white individuals, the second phase (HMP2) aimed to study how the microbiome changes depending on certain health conditions. This phase is called the integrated HMP (iHMP), and it sought to create datasets of host and microbiome properties over time (L. Proctor & iHMP Research Network Consortium,

2014). The conditions studied include (1) pregnancy and preterm birth, (2) onset of inflammatory bowel disease (IBD), and (3) onset of type 2 diabetes. The studies followed the changes in the hosts and microbiomes associated with these conditions over time, including changes in microbial community composition, metabolic profiles, gene expression, and protein profiles, as well as host-specific properties such as genetic, epigenomic, antibody, and cytokine profiles. All tools and data resulting from these efforts are freely available and serve as a central resource for the research community. We will return to the results produced by HMP-funded studies in subsequent chapters.

Although the human microbiome is still considered a very young field, the explosion of interest in it has been extraordinary. The HMP provided one of the first infusions of funding to permit the microbiome research committee to work together and develop sound methods of data generation that could be compared between studies. Many labs have continued this work well beyond the original, rather limited goals. Particularly prolific contributors include Eran Elinav and colleagues at the Weizmann Institute, whose lab is leading the charge to discover the role of the microbiome in a variety of human disorders, including metabolism, cancer, IBD, autoimmunity, and neurodegeneration, and recently published a report on the impact of smoking on several human microbiome sites (**Figure 2.19**). Jeffrey Gordon and his colleagues at Washington University in St. Louis continue explorations of the functional changes in the microbiome associated with disease, particularly obesity and malnutrition. Gordon also ushered in a new era of microbiome-based therapeutics and preventive medicine. Rob Knight's lab at University of California, San Diego, has produced many of the software tools and laboratory techniques that have enabled high-throughput microbiome science. Jack Gilbert, also at UC San Diego, has helped create an entirely new field of microbiome science focused on the built environment,

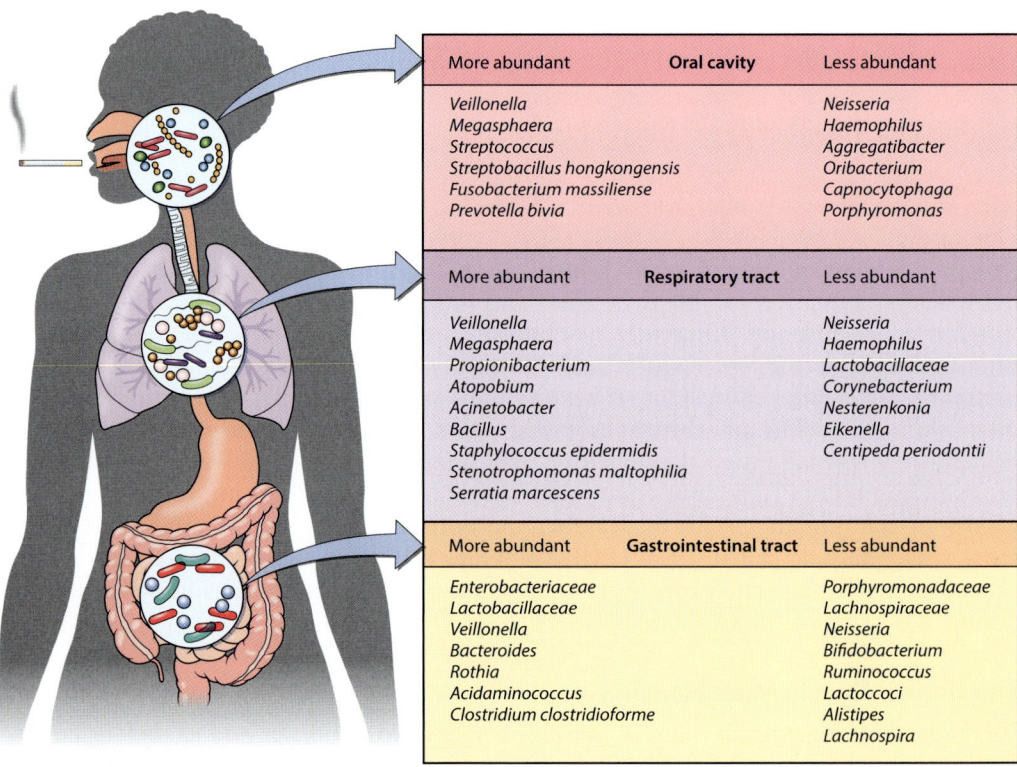

Figure 2.19 Microbiome Disturbance Induced by Smoking Smoking causes changes in the population sizes of different microbes in the oral cavity and respiratory and gastrointestinal tracts. Some microbes become more abundant, while others become less abundant. (After Shapiro et al., 2022.)

which comprises all human-created structures and environments. These investigators are leading the way in providing the information required to begin to better understand how our microbiomes influence our health and respond to disease.

By now you have learned about the evolutionary history of microbes, how and when they were discovered, and the amazing technologies scientists have invented for their study. We also understand what a microbiome is and some of its most general features, such as being unique to each person, varying by location / body site, being relatively stable over time, and playing an outsized role in human health. We have reached a point where it is time to literally dive into the human microbiome and find out who is there and what they are doing. That is precisely the focus of chapter 3.

CHECK YOUR UNDERSTANDING

1. What is Joseph Leidy's *A Flora and Fauna within Living Animals*?
   a. A travel guide to the Amazon rainforest
   b. A book of illustrations and descriptions of insects' microbiomes
   c. A field guide about plants in the western US
   d. A book about the role of microbes in nitrogen fixation of plant roots

2. According to the miasma theory, or spontaneous generation, where did disease come from?
   a. Bad air
   b. Bacteria
   c. Maggots
   d. Insect eggs

3. A hypothesis must be
   a. proven to be correct.
   b. explainable.
   c. significant to the world.
   d. testable.

4. Which of the following is true about the open flask in Redi's experiment?
   a. It is part of the treatment group.
   b. It serves as a negative control.
   c. Maggots did not grow in it.
   d. It was used to prove spontaneous generation.

5. Why did the liquid inside the intact swan-neck flask not become cloudy in Pasteur's experiment?
   a. Bad air was not able to get into the flask.
   b. The liquid was contaminated.
   c. Air was not able to carry microbes into the flask.
   d. The liquid inside did not support microbial growth.

6. What do we call the theory that microorganisms cause disease?
   a. Germ theory
   b. Bacterial waste theory
   c. Miasma theory
   d. Spontaneous generation

7. Which is not one of Koch's postulates?
   a. If you give the pathogen to a healthy organism, it will become sick.
   b. The pathogen can be isolated from a diseased organism.
   c. The pathogen is only found in diseased organisms.
   d. The pathogen can be transmitted through the air or fecal matter.

8. What are Koch's postulates used for?
   a. To identify potential causes of a disease
   b. To prove a particular agent causes a disease
   c. To understand the mechanism of disease on a cellular level
   d. To explain how a disease is transmitted

9. What do antibiotics kill?
   a. Bacteria
   b. Viruses
   c. Any microorganism
   d. Some eukaryotes

10. 1950–1960 was called the golden age of antibiotics because many antibiotics were discovered then.
    a. True
    b. False

11. The strain *E. coli* Nissle 1917 prevents Shigella from overtaking the gut and causing dysentery. What is this an example of?
    a. Germ theory of disease
    b. Fecal microbiota transplant
    c. Miasma theory
    d. Colonization resistance

12. Bacteria in the oral microbiome are easy to culture because most are
    a. anaerobic.
    b. aerobic.
    c. autotrophic.
    d. heterotrophic.

13. What is the difference between genomics and metagenomics?
    a. Genomics is the study of an organism's DNA, and metagenomics is the study of an organism's DNA and the epigenetic modifications it encodes.
    b. Genomics is the study of transcription, and metagenomics is the study of transcriptional regulators.
    c. Genomics is the study of an organism's DNA, and metagenomics is the study of all the DNA present in a community.
    d. Genomics is the study of differences between individuals' DNA, and metagenomics is the study of differences between species' DNA.

14. How many of the microbial species that we can sequence are currently culturable?
    a. None
    b. About half
    c. The great majority
    d. All

15. What is the process of transferring stool from a healthy individual into a sick patient to treat their disease called?
    a. Fecal microbiota transplant
    b. Pseudomembranous therapy
    c. Colonization treatment
    d. Gut growth transfer

16. FMT has been tested in clinical trials for the treatment of which disease?
    a. Recurrent *C. difficile* infection
    b. Dysentery caused by *Shigella*
    c. Colitis
    d. None

17. How long did it take for the gut microbiota to be established in Russell Schaedler's germ-free mice?
    a. One day
    b. One week
    c. One month
    d. One year

18. What does 16S rDNA encode?
    a. Butyrate
    b. Shiga toxin
    c. RNA polymerase
    d. Ribosomal RNA

19. Microbial species are intermixed and evenly distributed throughout the oral microbiome.
    a. True
    b. False

20. Jeffrey Gordon and Lynn Bry found that microbiota could induce host cells to produce sugars for the microbiota to consume.
    a. True
    b. False

21. What was the first goal of the Human Microbiome Project?
    a. To understand how microbiome composition changes over time
    b. To identify all species found in the human gut microbiome
    c. To characterize healthy adults' microbiome compositions at several body sites
    d. To compare microbes present in healthy and diseased adults

22. Data generated as part of the Human Microbiome Project must be publicly accessible.
    a. True
    b. False

23. What conditions were studied in the second phase of the Human Microbiome Project?
    a. Ulcerative colitis and Crohn's disease
    b. *C. difficile* infection
    c. Cancer, autoimmune disease, and allergies
    d. Pregnancy and preterm birth, IBD, and type 2 diabetes

Answers: 1B, 2A, 3D, 4B, 5C, 6A, 7D, 8B, 9A, 10A, 11D, 12B, 13C, 14B, 15A, 16A, 17B, 18D, 19B, 20A, 21C, 22A, 23D

DIVING DEEPER

1. What two scientists first recorded observations of the microbiome?
2. What is the miasma theory, also known as spontaneous generation?
3. How did Redi's and Pasteur's experiments disprove the miasma theory?
4. What are Koch's four postulates?
5. Think back to chapter 1—what is the difference between aerobic and anaerobic microbes?
6. What type of bacteria could first be cultured using the roll-tube procedure? What was the significance of this development?
7. What is the difference between genomics and metagenomics?

8. Can you envision what sorts of problems might arise with FMT?

9. Identify four changes that occur in germ-free mouse physiology.

10. What are the benefits of generating a community profile using a metagenomic instead of a culture-based approach?

11. Describe some of the shortfalls of a metagenomic approach to microbiome analyses. What are some ways that we might avoid them?

12. What fundamental insight into the nature of obesity has Jeffrey Gordon's research provided?

13. What are the two phases of the Human Microbiome Project and their objectives?

14. Why is the realization that everybody has a unique microbiome important to physicians?

DISCUSSING AND REFLECTING

1. Early in the chapter we discuss the role of the microbiome in our digestive system. How might our diets and digestive processes be different if we had no microbes in our gut?

2. In the section Fecal Microbiota Transplantation, we introduced the idea of colonization resistance, where the presence of a specific species can confer resistance to a pathogenic bacterium. Why do you think bacteria would have mechanisms for antagonizing other bacteria? What evolutionary benefits might this offer?

3. Gordon's group demonstrated that obesity is linked to the microbes present in an individual's microbiome. What other health characteristics do you think could depend directly on microbiome composition? Why?

4. Reflection. Jeffrey Gordon said, "Lynn found evidence that gut microbes are directing the host to serve them a meal of complex carbohydrates." How do you feel about the concept that microbes are, literally, directing you to eat certain foods?

RECOMMENDED READINGS

Popular Science Review

Zhang, L. (2019). *Beginner's Guide to Bioinformatics Tools for Analyzing Microbiome Data*. Genomics Core at NYU CGSB. https://gencore.bio.nyu.edu/beginners-guide-to-bioinformatic-tools-for-analyzing-microbiome-data/

Scientific Reviews

Allaband, C., McDonald, D., Vázquez-Baeza, Y., Minich, J. J., Tripathi, A., Brenner, D. A., Loomba, R., Smarr, L., Sandborn, W. J., Schnabl, B., Dorrestein, P., Zarrinpar, A., & Knight, R. (2019). Microbiome 101: Studying, Analyzing, and Interpreting Gut Microbiome Data for Clinicians. *Clinical Gastroenterology and Hepatology*, 17(2), 218–230. https://doi.org/10.1016/j.cgh.2018.09.017

Peeters, J., Thas, O., Shkedy, Z., Kodalci, L., Musisi, C., Owokotomo, O. E., Dyczko, A., Hamad, I., Vangronsveld, J., Kleinewietfeld, M., Thijs, S., & Aerts, J. (2021). Exploring the Microbiome Analysis and Visualization Landscape. *Frontiers in Bioinformatics*, 1, 774631. https://doi.org/10.3389/fbinf.2021.774631

The Human Holobiont

3

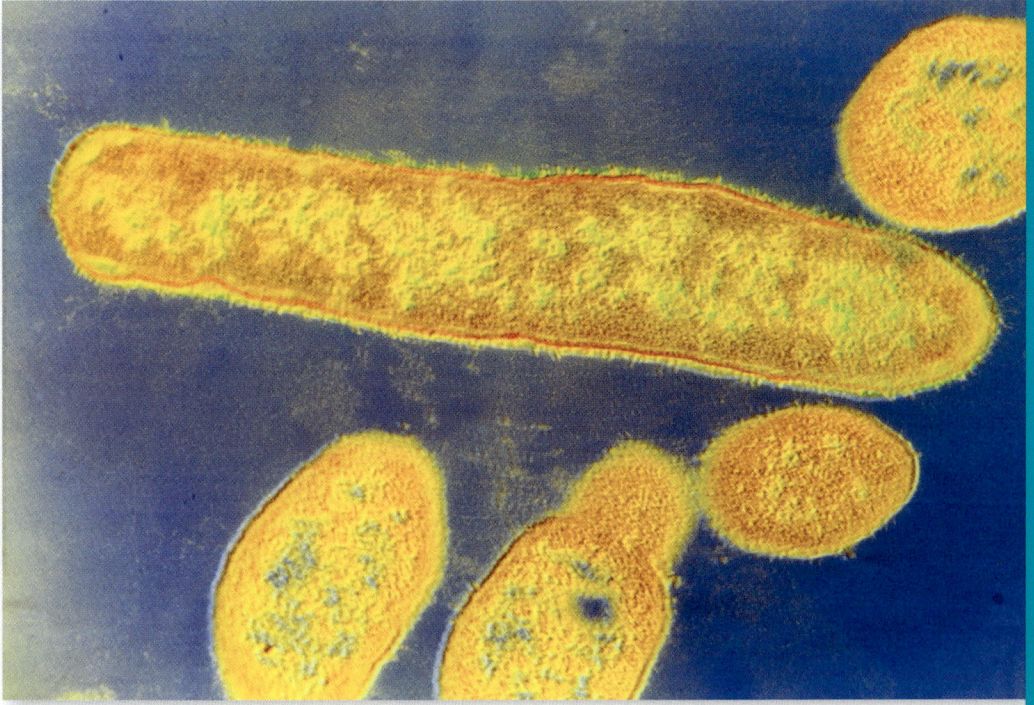

CHAPTER CONTENTS

- 3.1 The Human Holobiont
- 3.2 The Many Human Microbiomes
- 3.3 The Microbes in Our GI Tract
- 3.4 The Microbes in Our Mouth
- 3.5 The Microbes on Our Skin
- 3.6 Sampling Your Own Microbiome

Permit me to introduce myself, *Bacteroides fragilis*. I was given this name in 1898, and I honestly don't know why it was determined I was fragile. My claim to fame is that I am the most abundant member of my genus (Bacteroides) in the human gut, by a long shot! I was delighted to learn that scientists have finally noted that my role in human health is very much based upon location, as one scientist noted—"*B. fragilis*, as with real estate, it's location, location, location" (Vandamme et al., 1995). When I reside in the gut, I am a real team player, and I employ a complex system to sense what you are eating and tailor my metabolism accordingly. However, when I stray, I can cause horrible abscesses in the abdomen, brain, liver, and pretty much wherever I end up. So, don't mess around with me. (Photo from CNRI/Science Source)

This chapter marks the beginning of our journey into the very heart of the human microbiome. We will consider what we mean by the terms *holobiont* and *microbiome* and then learn about the structure and function of several of our major microbiomes, those found in the gut, in the mouth, and on the skin. Our focus will then turn from what constitutes a "normal" or healthy microbiome to some of the many ways things can go wrong with these delicate ecological relationships, and which may result in disease. But first, let's explore the extraordinary microbial symbionts that we host.

"The microbiome is the sum of our experiences throughout our lives: the genes we inherited, the drugs we took, the food we ate, the hands we shook. It is unlikely to yield one-size-fits-all solutions to modern maladies."
—Ed Yong (Yong, 2014)

3.1 THE HUMAN HOLOBIONT

Now that we know we are the host to an extraordinary diversity of "others"—such as bacteria, archaea, viruses, and microscopic eukaryotes—our view of ourselves has forever changed. Some call this assemblage of life forms the human superorganism. However, the term *holobiont* was created to capture this new perspective more precisely. Lynn Margulis, the same scientist who proposed what was then considered a radical theory for eukaryote origins (**endosymbiosis**) (see Figure 1.10), popularized the term *holobiont* in the 1990s. According to Margulis, a **holobiont** is a community of a host and the species living in or on it, which form a discrete ecological unit. The key is that these life forms work synergistically as a collective, which means that they interact and cooperate to produce a combined effect greater than the sum of their separate effects. That, in a nutshell, describes the relationship of a human and its microbiome!

Many of us have spent a lifetime focused on ourselves: how we feel, what we eat, what mood we are in, questioning whether we need to gain or lose weight, or if we are as healthy as we can be. Few have considered that the answers to most, if not all, of those questions are highly dependent on the members of our microbiome. It seems high time to put these life partners into the limelight and focus on who they are, what they are doing, and how their actions influence us, their host. The remainder of this chapter will introduce you to the key microbial constituents of our microbiome.

3.2 THE MANY HUMAN MICROBIOMES

It is perhaps the right moment to pause and reflect on the fact that when we refer to our microbiome, we actually mean microbio*mes*, plural. As we will learn, there are well-characterized, discrete, microbiomes in our gut, mouth, skin, and pretty much any site in or on our body. Their members differ, their primary functions differ, and the roles they play in human health may differ as well. We can further identify unique microbiomes within each of these larger entities. As you may recall from chapter 1, our mouth hosts an oral microbiome, but the **microbiota**, or members of the microbiome, differ depending on where in the mouth you look. We have a "below the gums" oral microbiome, a "top of the tongue" oral microbiome, and so on. There are simply too many microbiomes upon microbiomes for us to attempt to delineate them all. Thus, we will focus first on the generic human microbiome, which is the sum of all of the site-specific microbiomes in and on our bodies, and then turn to several of the major site-specific microbiomes, such as those found in the gut and mouth and on the skin.

A Healthy versus Normal Microbiome

Our primary goal in this chapter is to highlight the microbiome from the perspective of a person without known disease. Given the enormous diversity in microbiota found from one person to the next, it remains challenging to characterize a "healthy" microbiome. Additionally, many people can have underlying diseases or conditions that aren't identified by researchers, making it challenging to even characterize what we mean by "healthy." Instead, microbiomes in individuals without obvious disease are now described as "normal," which is a term that suggests there is some sort of average composition of each site-specific microbiome. When you hear the term **normal microbiome**, it is referring to the consensus microbiome as deduced from a large sample of individuals without known disease.

Before we learn the scientific names of some of our microbiota, let's agree that the most important outcome from reading this chapter is for you to develop a basic understanding of the key players and their respective roles in our major microbiomes. As you have already learned, over a thousand microbial species have been identified in the human gut microbiome alone. Learning about them all would be overwhelming and not very useful. Instead, we will identify the major phyla of microbes involved and only identify genera and species when appropriate.

Microbiome in Names and Numbers

There is some bookkeeping we need to address before we dive too deeply into our microbiomes. **Figure 3.1** provides a color-coded key to the most common microbial phyla and genera identified in a large sample of human gut microbiomes, along with

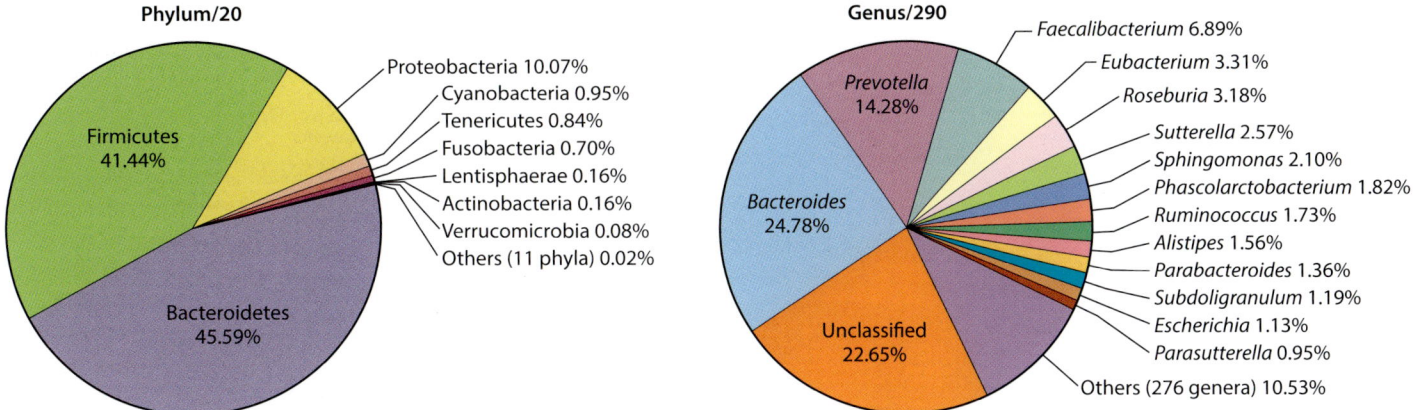

Figure 3.1 The Major Phyla and Genera of the Human Gut Microbiome The most common bacterial phyla and genera found in the human gut microbiome are shown. (After Yang et al., 2020.)

their relative frequencies (Yang et al., 2020). All are members of the Bacteria domain of life, as they make up most of the cellular microbial diversity in our microbiome. The three most abundant phyla are **Bacteroidetes**, **Firmicutes**, and **Proteobacteria**, and we will refer to them frequently in the following chapters. The genera within these phyla that are present in the gut are more varied from one individual to the next. However, we will introduce several of them as well as we explore the various human microbiomes.

Let's turn our focus from the taxa present in the gut microbiome (i.e., the microbiota) to their genetic contributions. The human holobiont has a composite genome of roughly 9 million unique genes, with the majority (8 million) contributed by bacteria. The paltry human-specific contribution of 22,000 genes seems like an afterthought—and yet those relatively few genes encode the organism we recognize as human! The take-home message is that bacteria rule our microbiome in terms of species count as well as in the number of unique genes they contribute. The more than 8 million bacterial genes represent some 93% of the total human holobiont genome.

Microbial Taxonomy

We will be employing Latin species names to identify some of our microbial partners. We were introduced to the Linnaean classification system in chapter 1, but let's have a quick refresher, with a snapshot of the classification of one very famous member of our microbiome, *Escherichia coli* (**Figure 3.2**). We see that *E. coli*—by the way, once we have identified a **genus**, we are then free to abbreviate that genus name in future use—is contained within the domain Bacteria, the phylum Proteobacteria, and then a series of ever more restrictive subgroups (class, order, and family) until you reach the ranks of genus (*Escherichia*) and **species** (*E. coli*). The genus name is always capitalized, while the species designation is not, and both are italicized.

There will be times when we require an even more specific name than is provided by a species-level designation. We use the term **strain** to identify a genetic variant or subtype within a species. For example, *E. coli* is a normal member of the human gut microbiome and plays a role in synthesizing key vitamins we require. However, there are strains of *E. coli* that can cause disease. For example, *E. coli* O157:H7 is a strain of *E. coli* responsible for numerous disease outbreaks, often originating from contaminated beef. That strain possesses genetic information that results in release of a toxin (Shiga toxin), which disrupts protein synthesis in the human epithelial cells lining the intestinal mucosa, leading to cell death, sloughing of the mucosa, and eventual bloody diarrhea.

Figure 3.2 Linnaean Classification of *Escherichia coli* The Linnaean classification system is based upon a series of nested groupings, starting with the most inclusive, domains, and ending with the most specific, species.

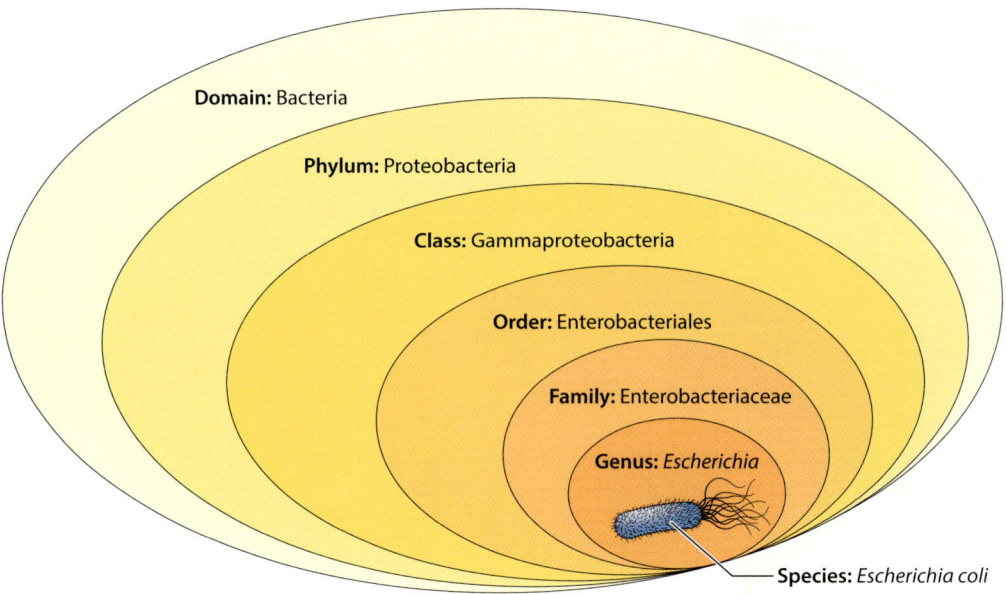

The Core Microbiome

We wrapped up chapter 2 with a short description of the HMP, or the Human Microbiome Project. A primary goal of the first phase of the HMP (HMP1) was to identify the microbes present in a normal microbiome—in other words, the core taxa that are shared by all, or most, seemingly healthy humans. Understanding the normal microbiome is essential for determining its role in disease, because it allows scientists to compare microbiomes between healthy and unhealthy hosts and identify differences that are associated with, and may even contribute to or cause, the diseased state. From this concerted effort we learned that the primary members of this core include species from three bacterial phyla: Bacteroidetes, Firmicutes, and Proteobacteria (see Figure 3.1). Each of us will sport different members of these core phyla, and the individual genera, species, and strains will vary over time due to numerous factors, such as a host's genome, health status, immune system, lifestyle, diet, and the like.

Since everyone's microbiome can have such different species compositions, how do we define a normal one? Some argue that the species names don't matter, it is the functions that they provide that are relevant, since we care more about what the organisms do than who they are. Although the species present vary greatly between individuals, similar functions are generally present. In other words, the functions of most microbiomes are more highly conserved than the taxa (Dogra et al., 2020). These findings suggest that microbiome functions are more **resilient** than taxa, which means that they are more likely to remain the same or return to their original state after an environmental perturbation or change. In part, this is because there are often numerous species that can perform the same, or similar, functions. Perhaps a more relevant description of a core microbiome, then, is a list of the gene families, **metabolic modules** (conserved sequences of chemical transformations), and **regulatory pathways** (which turn genes on and off) provided by members of a microbiome, rather than a simple list of the species present.

You can think of a person's microbiome like a baseball team—the players might change, but the team will still always have a pitcher, catcher, etc. (**Figure 3.3**). Further, even though different teams have different players, the positions they play stay the same, much like how functions are conserved across different individuals' microbiomes. If we trade a pitcher, we know what that player will do for the team, which is to pitch. It's more useful to define a baseball team as a list of positions than as a list

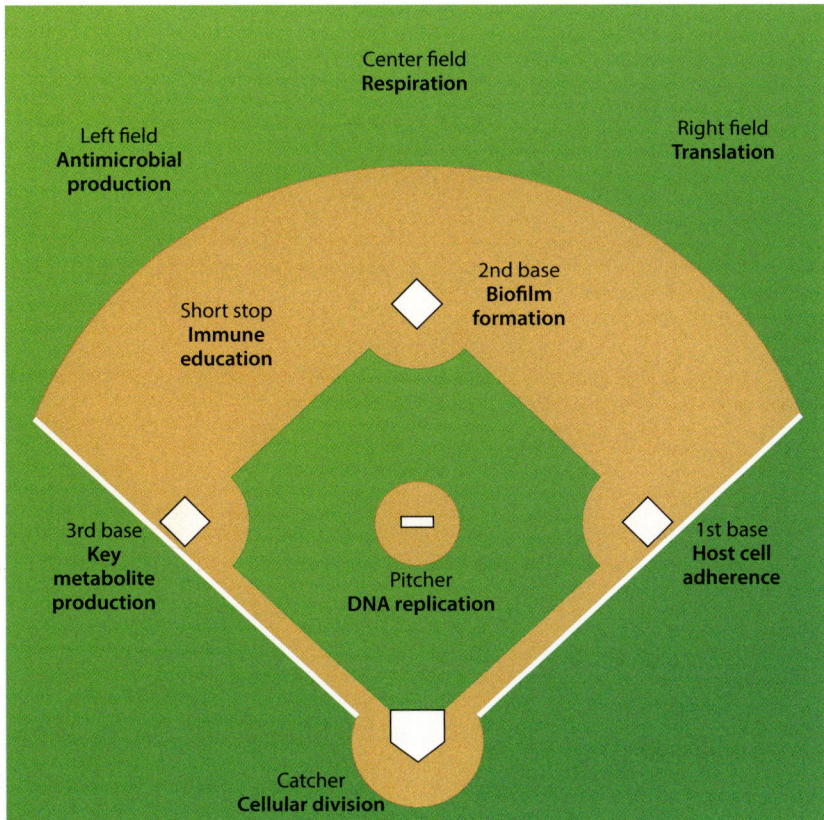

Figure 3.3 Microbiomes as a Baseball Team Metaphor Several of the major functions provided by the human core microbiome are identified as if they were members of a baseball team.

of players' names, since the specific players will vary. Similarly, we may define the human microbiome as the functions it performs, such as the breakdown of food, vitamin production, cholesterol synthesis, and immune system modulation, instead of a list of who performs those functions.

Primary Functions of the Core

If we continue in this vein, our core microbiome will include functions from the following three categories: (1) **housekeeping tasks** necessary for microbes to exist regardless of their host or the environment, such as the ability to replicate DNA, divide, and harness energy; (2) **processes specific to human-associated adaptations**, such as the ability to adhere to host cell surfaces or the production of metabolites, such as short-chain fatty acids; and (3) **specialized functions** that differ by body site, such as fiber metabolism in the gut, or biofilm formation on teeth (Neu et al., 2021). A normal microbiome requires an assemblage of microbial species able to carry out these foundational sets of functions.

Table 3.1 provides a slightly different way of thinking about the functions of the core microbiome. This perspective includes a consideration of how the microbiota changes over time (**temporal core**). It also identifies an **ecological core** as those microbes that interact with each other and the host to create a community. The **functional core** is defined as the microbial members that provide essential biological functions to the host, and finally, a **host-adapted core** refers to those species that have evolved with the host and whose presence results in an increase in host fitness. Certainly, from this perspective, a single species can be part of multiple different functional cores.

If we define the core microbiome based on a list of functions, we can greatly simplify our view of the community. Organizing by function allows us to group

Table 3.1 Functions of the Core Microbiome

TERM	DEFINITION	CRITERIA
Common core	The component of the microbiome that is found across a considerable proportion of hosts within a defined host population or species	High prevalence/occupancy frequency across host population/species. Can be identified using occupancy-abundance curves
Temporal core	A temporally stable or predictable component of the microbiota	Taxa that demonstrate stable or predictable dynamics over time, either within a single host or across host population/species
Ecological core	The component of the microbiome that is disproportionally important for shaping the organization and diversity of the ecological community	Removal or introduction results in large cascading effects on ecological structure and diversity. May increase community stability
Functional core	The component of the microbiome that performs essential biological functions to the host, usually in respect to their biochemical, physiological, or ecological services to the host	A set of genes or taxa that are linked to a measurable facet or host function. Natural variation in host function does not affect host fitness
Host-adapted core	A set of microbes that has coevolved with the host species of subpopulation and whose presence increases host fitness in at least some ecological contexts	Taxa that are linked to a measurable facet of host function. Natural variation in host function affects host fitness in at least some ecological contexts. Are not functionally redundant (other taxa cannot perform same function)

Source: From Risely, 2000. CC BY 4.0 International Deed

multiple species together that fill a similar role. Within a diverse microbiota, even if certain species are lost, the core functions will still be provided by those taxa that remain. **High microbiome diversity**, meaning that there are numerous species present, is usually positively correlated with the health of the host and with the stability of the microbiota over time, as the important functions continue to be carried out even when the microbiome is perturbed. In contrast, **low microbiome diversity**, the presence of relatively few taxa, may result in the loss of key functions and result in disease states that can range from allergies and diabetes to cancer.

Hallmarks of a Healthy Microbiome

Stability, **resistance**, and **resilience** are key characteristics of any ecosystem. Although constantly exposed to external factors, such as the foods we consume or the medications we take, our microbiome can both resist perturbations, such as the invasion of a pathogen, and regain equilibrium afterwards. We call this capacity of self-modulation resilience. Unless perturbed, a mature microbial community exists in a stable state, with somewhat predictable interactions and functions, and can remain intact for years.

When we employ the term *stable*, it is critical to note that it does not mean "unchanging." A perturbation can push the microbiome from a stable state towards an unstable state. The microbiome is then able to recover—by reattaining its original state or by reaching a new stable state. The rate of recovery or transition into a new state will depend on what are called **tipping points**, which are factors such as particular stressors, changes in diet, introduction of a medication, etc. This dichotomy between stability and dynamism is challenging to capture in microbiome studies, as it requires very fine-scale and long-term investigations.

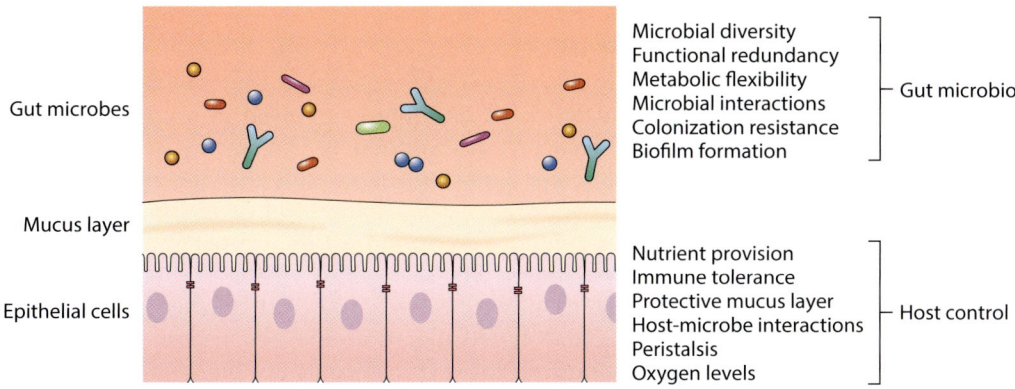

Figure 3.4 Hallmarks of a Healthy Microbiome-Host Relationship A healthy gut microbiota is usually diverse, which ensures that it possesses features—such as functional redundancy and metabolic flexibility—to ensure it is resilient to change. The host contributes to maintaining a healthy gut microbiota through functions such as nutrient provisioning and immune tolerance.

The resilience of a microbiome seems to be governed by both the microbiome and host (Dogra et al., 2020). The top half of **Figure 3.4** illustrates characteristics of the gut microbiome that ensure resilience, including microbial diversity, functional redundancy, microbial interactions, and colonization resistance. The bottom half of Figure 3.4 shows how the host exerts its own set of selective pressures, such as the foods it consumes, the sensitivity of its immune response, and the type of medications it takes, all of which help determine which microbes can persist. Experimental studies have shown that when a microbiome is more diverse, it responds less severely to impacts such as a change in diet or the use of antibiotics (Liu et al., 2022).

A normal gut microbiome usually recovers diversity within a few weeks to months after many types of perturbations. However, this is not always the case. For instance, a decrease in microbiome diversity may clear the way for microbes already present in the community to become detrimental, though they weren't before the microbiome was perturbed. One example is the fungal species *Candida albicans*, a common member of our gut microbiome. A recent study revealed that certain gut bacteria can feed on sugars found in the fungal cell walls. When perturbed by antibiotic use, these same bacteria release starchy subunits that promote the movement of *C. albicans* from the intestine. This can result in life-threatening bloodstream infections (Pérez, 2021). Other microbes that can be both friend and foe include a genus of Proteobacteria (*Escherichia*) and of Firmicutes (*Clostridioides*). Members of each of these genera are found in healthy individuals but may cause disease when the microbiome is disturbed or when these microbes gain access to parts of the body where they do not normally reside. For example, we previously learned that certain strains of *E. coli*, such as O157:H7, can cause bloody diarrhea when they replace resident *E. coli* strains—often after the host consumes undercooked, contaminated beef. *Clostridioides difficile* can cause diarrhea and colitis (inflammation of the colon) when their abundance increases, such as when the diversity of the gut microbiome plummets following antibiotic use.

As we visit three of our most influential site-specific microbiomes, our goal is not to catalog all the species present, but rather to highlight some of the core members and the critical functions they perform in maintaining human health. We will become fluent with some of the central players in the complex dynamic that occurs as our microbiome responds to and influences our body. Further, we will learn that there is a repeating pattern of core microbiome functions regardless of body site: those involved in defense of the host from invading microbes, those required in microbial growth and proliferation, and those that influence the roles of human cells and organs, such as functions provided by our immune and nervous systems.

3.3 THE MICROBES IN OUR GI TRACT

Our **gastrointestinal** (**GI**) **tract** includes all the organs involved in food and liquids being swallowed, digested, and absorbed and ultimately leaving the body as feces. The GI tract includes the mouth, where digestion begins with chewing and the secretion of saliva, which contains enzymes that break down food. It continues with the esophagus—a muscular tube that connects the mouth to the stomach, pushing food down through a process called peristalsis. The stomach is where food is mixed with gastric acids and enzymes that break down proteins and other nutrients. The small intestine is the primary site of human-based digestion and absorption of nutrients, while the large intestine is where our microbes digest foods not digested by the host, and where water and electrolytes are absorbed from undigested food. The rectum is the final section of the large intestine that stores undigested matter (stool) until it is ready to be expelled.

The large intestine hosts the densest microbial community in our body, with over 1 trillion microbial cells per centimeter. In fact, the large intestine is the home to the densest microbial ecosystems ever observed on Earth (Senghor et al., 2018)! What is the secret to this extraordinary concentration of microbes? The answer is quite simple—surface area. The inner surfaces of the large intestine are highly convoluted, which makes it challenging to measure; however, one estimate suggests there is 32 m^2 of inner surface area in the large intestine alone (or 344 ft^2) (Helander and Fandriks, 2014) (**Figure 3.5**). Microbial communities coat the lining and fill the interior space of the large intestine with microbial cell densities 10 to 100 times higher than at any other body site.

Over 1,000 microbial species have been identified in the human gut microbiome, although the average person hosts a more modest 200 to 500 species. These microbial residents are primarily members of the prokaryotic domains Bacteria and Archaea, which consume, produce, and exchange hundreds of metabolites. The gut microbiota encodes over 22 million genes (Almeida et al., 2019). This rich genetic resource provides functions that we depend upon for digestion and overall health. Together, these microbes and the genes they encode create a complex ecosystem that helps maintain GI tract **homeostasis**, or a stable physiological state.

Our gut microbes play a key role in helping us to digest indigestible foods; produce essential vitamins; detoxify drugs, heavy metals, and other pollutants; chemically modify bile and cholesterol; protect against the invasion of pathogens; and perhaps most critically, train and modulate our immune system. These beneficial microbes allow our guts to continue functioning normally despite being exposed to continuous change, in terms of the varied foods we eat, medications we take, or our levels of stress. When homeostasis is perturbed, a state we call **dysbiosis**, such as

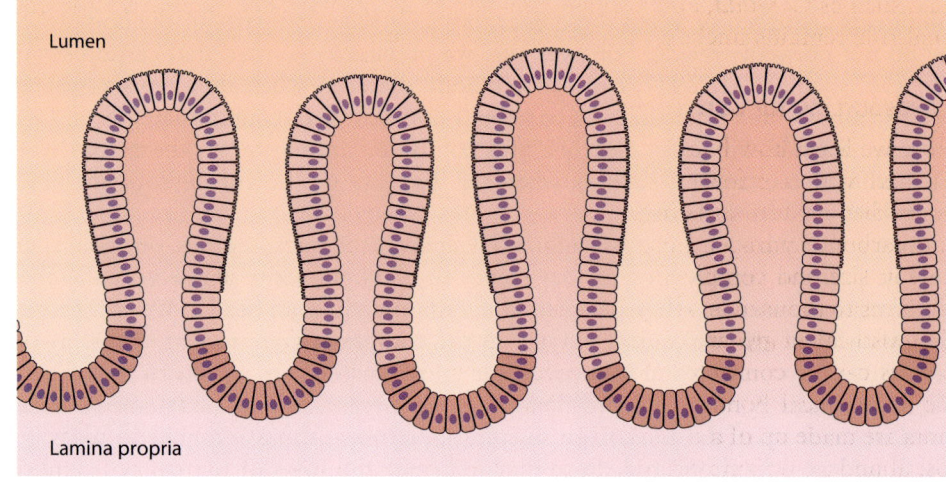

Figure 3.5 The Intestines' Incredible Surface Area The surface of the large intestine is convoluted, providing substantial surface area on which dense populations of microbes take up residence. The lumen is the interior space within the intestines. The lamina propria is the loose layer of connective tissue that lies just under the epithelial cells lining the intestines.

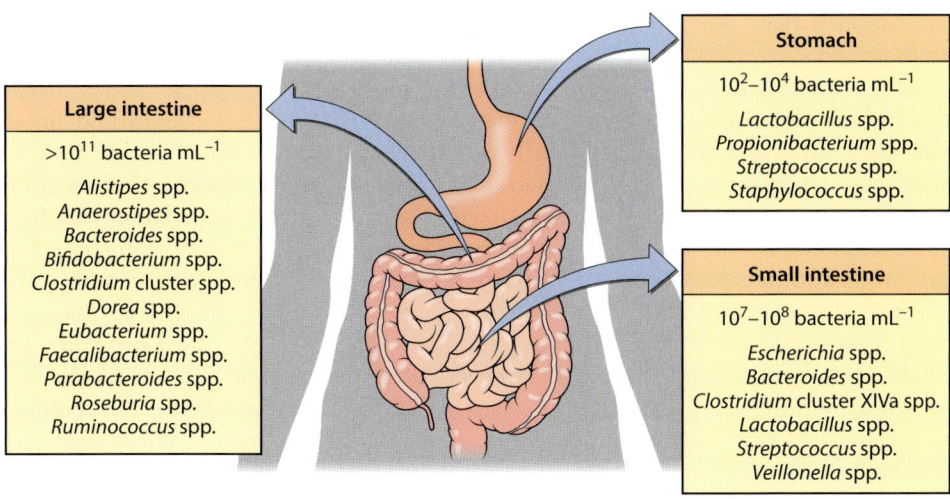

Figure 3.6 Common Species in the GI Tract The abundance of microbes varies greatly along the GI tract, as do the types of microbes that thrive in the distinct environments provided. The stomach, which is highly acidic and inhospitable to many microbes, has the lowest density of bacteria. The small intestine has a moderate level of diversity; however, peristalsis and the production of bile acids limit who can survive along its length. The large intestine has the highest abundance and greatest diversity of bacterial species. (After Tuohy and Scott, 2015; Tap et al., 2009; Zoetendal et al., 2012; Delgado et al., 2013; and Walker et al., 2014.)

when we take antibiotics or experience high levels of stress, some of these same microbes can impact our health in more negative ways, such as producing toxins, reducing the production of beneficial metabolites, and triggering inflammation.

The density and composition of the microbiome differ along the GI tract (**Figure 3.6**). The mouth will be considered separately, and the esophagus is only involved in a brief, transient interaction with foods. The stomach is home to a unique microbiome composed of very few microbial species, which is unsurprising given that it has a very low pH, ranging from 1.5 to 3. The small intestine hosts progressively more-dense populations of microbes the closer you are to the large intestine. The first stretch of the small intestine is rich in bile and has a rapid flow of contents, and consequently it hosts few microbes until it nears the large intestine, where transit slows, and the environment is well suited for hundreds of different species to exist. The large intestine wins the prize for hosting the densest human microbiome ever documented, boasting over 1 trillion microbial cells per centimeter of its length.

Let's return for a moment to Figure 3.1, which provides a pie chart showing the frequency of the dominant bacterial phyla in the large intestine, with the more common genera identified as well. Two of the phyla, Firmicutes and Bacteroidetes, represent about 90% of the large intestine microbiota and are bacterial. Comparatively little is known about the resident archaeans, except that most are methane producers, and they serve a key role in the efficient digestion of complex sugars. Even rarer are members of the microscopic eukaryotes, whose presence is limited to fungi, generally yeast, such as *Candida*, and a single-celled, parasitic protist, *Blastocystis*, whose functional role remains unclear (Pérez, 2021).

Crowdsourcing for Carbs

Before we learn how bacteria in the large intestine keep our digestive system working, let's start with a refresher on carbohydrates, which represent the bulk of a normal human diet. **Dietary carbohydrates** consist of sugars, starches, and fibers that are built from carbon, hydrogen, and oxygen atoms. Members of this heterogeneous family range in size and complexity. The simplest sugars are **monosaccharides**, such as glucose. Tens to thousands of monosaccharides can be joined together to form **polysaccharides** (also called **glycans**), resulting in **starches** and **fibers**. The sugar units in polysaccharides can be connected in complex ways through **glycosidic linkages**, which are a type of chemical bond that requires digestive enzymes to break apart. Animals and plants are made up of a diverse array of polysaccharides, such as **cellulose**, which is the most abundant organic molecule on the planet and a common component of plant cell walls. Spoiler alert—humans do not encode the enzymes required to digest cellulose, and most other complex carbohydrates, without the intervention of our microbiome!

When we consume starch, its breakdown is initiated in the oral cavity through chewing and the enzymatic action of salivary amylases. Digestion continues in the small intestine, which has epithelial cells covered with microvilli that anchor digestive enzymes and help increase the surface area for absorption of nutrients. The remaining starch is resistant to human digestive enzymes and is known as **resistant starch**. A mutual dependence has evolved between humans and the symbiotic bacteria in their large intestines that can access these energy-rich sources. The complete metabolism of resistant starch by the large intestine microbiome is a team effort that involves a group of primary starch degraders and secondary starch scavengers (**Figure 3.7**). Primary degraders (fermenters), such as *Ruminococcus bromii* and *Bifidobacterium adolescentis*, initiate the breakdown of resistant starch using anaerobic fermentation and release substrates, such as glucose and oligosaccharides, that can be used by other microbial species (Figure 3.7A). The physical structure of the carbohydrate granule changes during this bacterial fermentation process, which makes it more accessible to other bacteria, such as *Eubacterium rectale* or *Bacteroides thetaiotaomicron* (Figure 3.7B). A variety of simple sugars are released during fermentation and support other species, including *Lactobacillus reuteri* and *Escherichia coli,* that cannot degrade starch but can scavenge malto-oligosaccharides (Figure 3.7C). The precise metabolic interactions within this food web are likely dictated by the type of starch and the glycan degradation strategy employed by individual bacteria.

One of the main products, or **metabolites**, of anaerobic digestion of resistance starch is **short-chain fatty acids** (**SCFAs**) (Figure 3.7D). These key acids, which we are unable to make on our own, include acetate, propionate, and butyrate, and they are known to improve gut health in several ways, which we will discuss further below. Microbes gain energy from the degradation of SCFAs, primarily in the form of **adenosine triphosphate** (**ATP**), which is the primary source of energy for cellular use and storage. ATP is found in all known life forms and is often called the "molecular unit of currency" of intracellular energy transfer. It is used to power many cellular processes, such as chemical synthesis and building proteins.

The phylum Bacteroidetes accounts for a major fraction of the microbiota of the large intestine, and their genomes possess over 300 genes involved in starch binding and utilization. In fact, these microbes encode an almost unimaginable 60,000-plus carbohydrate-degrading enzymes. So, to feed the cells lining the large intestine, we require a minimum of four different microbes (or functions), each contributing a particular step to accomplish the degradation and fermentation of starch, which results in energy for the host in the form of SCFAs and, ultimately, ATP. Starch metabolism is truly a crowdsourcing activity.

Although the Firmicutes are also essential members of the large intestine microbiota, their presence can often result in intestinal problems. Think of these bacteria as hyperactive children after consuming sugar-filled junk food. They cause your body to crave even more sugar—so you consume more junk food—and the Firmicutes increase in density. It is a problematic cycle that can result in the production of fat and inflammation in the large intestine, which slows down your body's metabolism. Early in our evolutionary history, Firmicutes would have been critical for human survival, as fat stores are essential during periods when food is scarce. However, our diets have changed since the dawn of agriculture in 10,000 BC. The Firmicutes have been presented with ever-more resources, and the result is an increase in fat production. In fact, we might be tempted to blame the current obesity epidemic in the industrialized countries on Firmicutes. However, it is not as simple as that!

If the microbiome of your large intestine has more Firmicutes than Bacteroidetes, it may be hard to lose weight. As we noted in chapter 2, if microbes from the large intestine of an obese human are transplanted into the large intestine of a mouse, the mouse gains weight (Turnbaugh, 2017). Obese people tend to have more Firmicutes in their large intestines, which results in genes being turned on that increase the risk for other illnesses, such as diabetes and heart disease. It is a vicious cycle that results

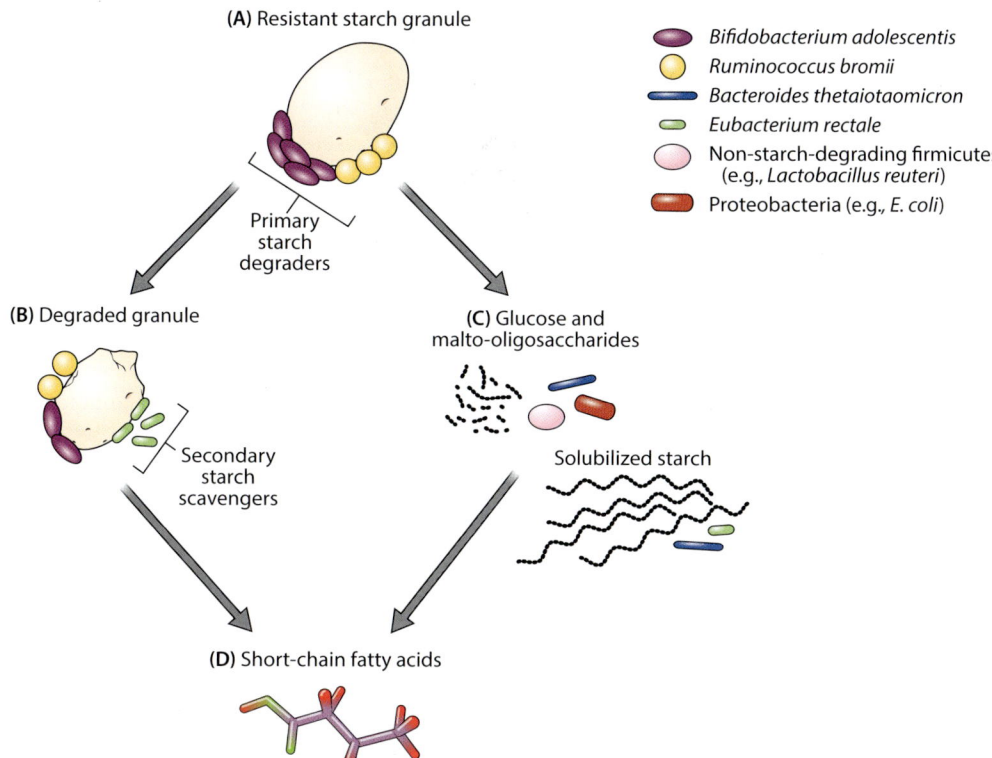

Figure 3.7 Interactions between Starch Granules, the Microbiome, and Host Epithelial Cells (A) Starch granules provided by the host require breakdown by microbially produced enzymes such as those produced by the primary starch degraders. (B) The partially degraded starch granules are then accessible for further degradation by secondary starch scavengers. (C) Degradation releases simple sugars that other microbes can consume. (D) A key metabolite released during these processes is short-chain fatty acids, which some bacteria consume but which are also taken up by the epithelial cells in the large intestine and contribute to host health. (After Cerqueira et al., 2020.)

in serious health implications that we will explore more fully in future chapters. For now, it is enough to know that members of these two critical bacterial phyla, Bacteroidetes and Firmicutes, play an oversized role in maintaining a healthy gut.

Cooperation and Conflict in the Large Intestine

The microbiome of the large intestine provides two primary benefits to the human host: metabolic and protective. With respect to **metabolism**, as mentioned above, carbohydrate **fermentation** by members of the Bacteroidetes, Actinobacteria, and Firmicutes results in several key metabolites, including SCFAs, but it also results in gasses, such as hydrogen, methane, and carbon dioxide, and vitamins. The SCFAs are used by the host as an energy source, and the gasses are removed in the breath or as flatulence. Thus, our microbiome helps us extract more energy from our food than we would be able to on our own.

The three main SCFAs produced by microbes in the large intestine are acetate, propionate, and butyrate, in **molar** ratios of 60:20:20 (*molar* simply refers to how many molecules of each acid are in a specified volume). SCFAs serve as energy sources for intestinal cells, messenger molecules that allow the microbiome to communicate with the host, and much, much more (**Figure 3.8**). **Butyrate**, a key SCFA, also activates the conversion of noncarbohydrate substances into glucose—all of which contributes to keeping levels of energy stable in the intestines. **Propionate** is converted to glucose and provides additional energy for the intestinal cells, some of which is sent to the liver, where it plays a role in the breakdown of noncarbohydrate substances

Figure 3.8 The Key Effects of SCFAs in the Large Intestine Acetate is able to cross the blood-brain barrier and has a direct appetite-suppressing role in the central nervous system. Propionate increases leptin (the main hormone involved in hunger and satiety regulation) levels, leading to a decrease in energy intake. Butyrate acts mainly locally and is a main energy substrate for colonocytes. Butyrate has also been shown, along with propionate, to regulate the division and proliferation of the large intestinal mucosa, with a possible role in preventing colon cancer, as well as other types of cancers. (© 2020 Silva, Bernardi and Frozza. 2020. CC-BY 4.0.)

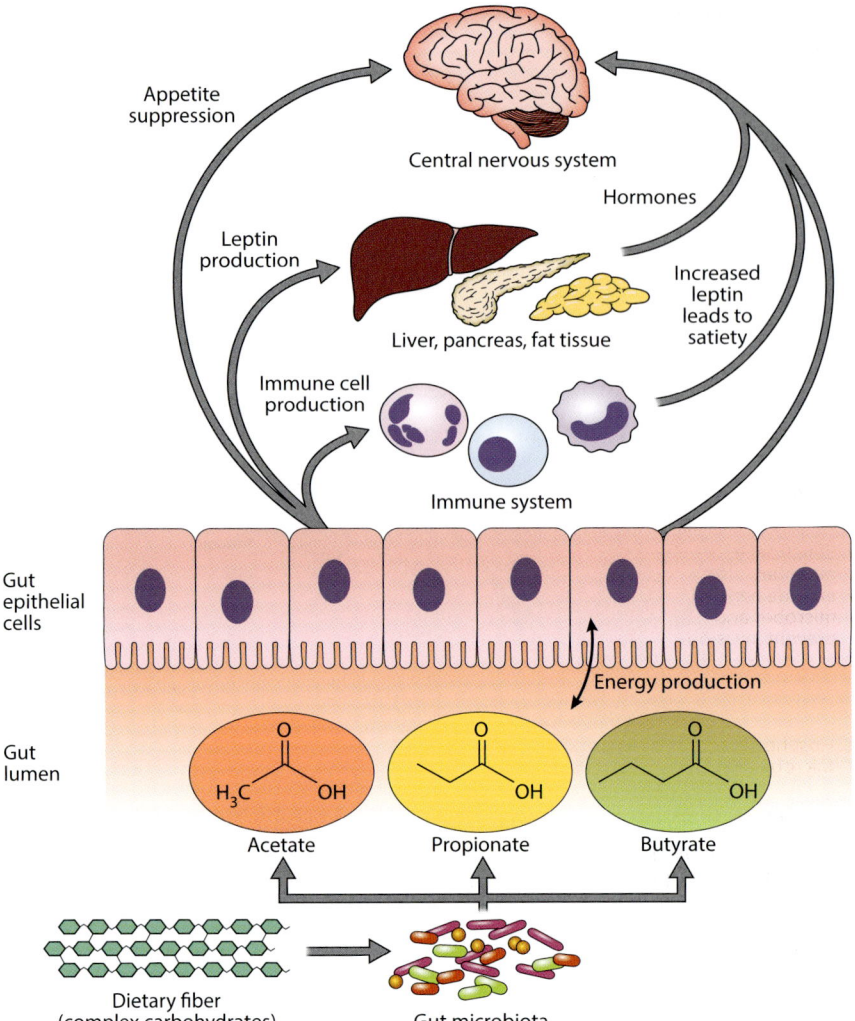

into glucose. Propionate is also a **satiety signal**, telling you that you are not hungry, which it accomplishes by interactions with fatty acid receptors on the walls of intestinal cells and adipocytes. Finally, there is **acetate**, the most abundant SCFA, which serves multiple purposes, such as feeding other bacteria in the large intestine, contributing to cholesterol metabolism and fatty acid production, and serving in appetite regulation (van der Hee & Wells, 2021).

The microbiome of the large intestine also serves to prevent invasion by pathogens, and there is a never-ending war between microbial residents and invaders. Our microbiomes have evolved complex strategies that involve both competition and cooperation to ensure homeostasis, and thus prevent invasion by potential pathogens. Microbial residents of the large intestine contribute to **colonization resistance** by preventing the attachment of pathogens to epithelial cells in the large intestine. *Lactobacillus* (Firmicutes) outcompetes potential pathogens, in part due to its production of numerous antimicrobial compounds, such as lactic acid. *Bifidobacterium* (Actinobacteria) also plays a key role in colonization resistance, but in this case simply by eating fast! These bacteria consume nutrients faster than invading pathogens, and thus they outcompete them. **Figure 3.9** illustrates this duality, otherwise known as war and peace in the microbiome. Beneficial microbes can be our benefactors, but when starved, they can become agents of infection. According to Athena Aktipis, at Arizona State University, whose research is the focus of Figure 3.9, "our gut microbes are not just passive recipients of the food that we eat—they evolve and change in response to what we feed our bodies. And there are certain foods that lead to resource

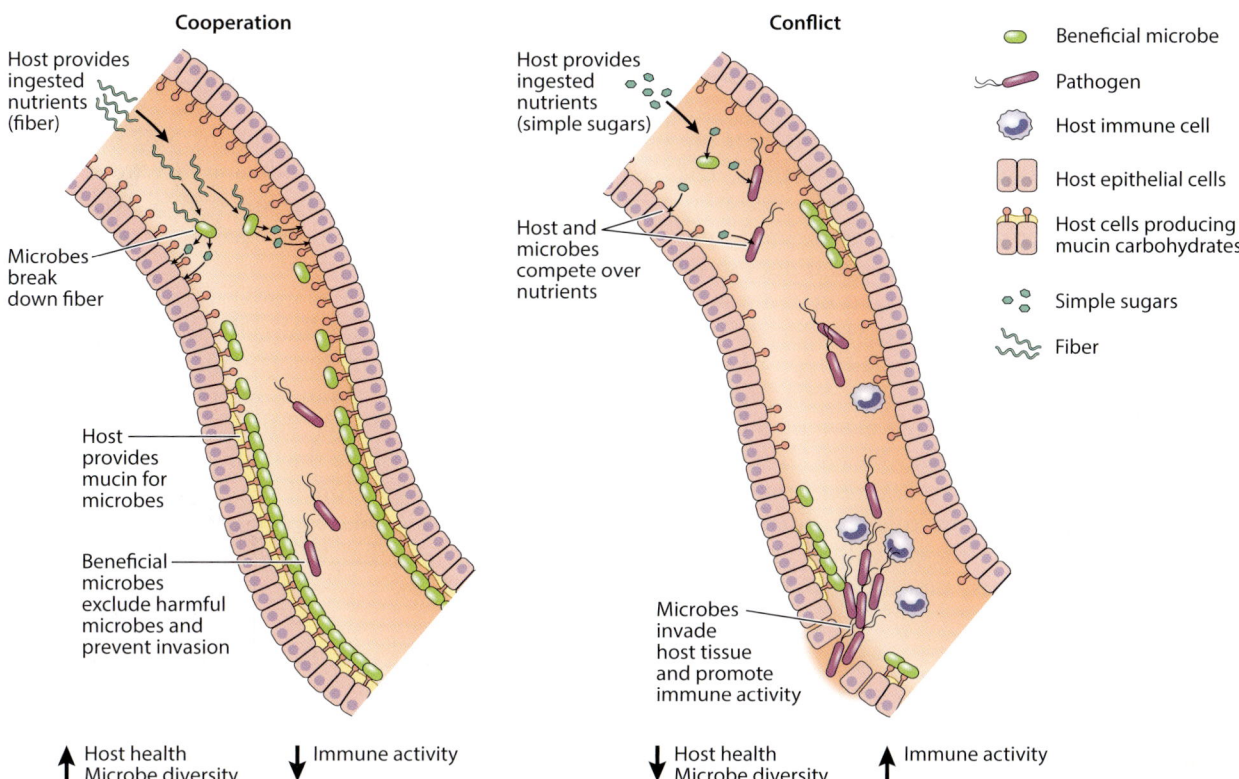

Figure 3.9 Cooperation and Conflict in the Gut Microbiome Beneficial microbes (green) may outcompete pathogens (purple) for resources, preventing infection in the host. This cooperation strategy increases host health and microbiome diversity. However, if beneficial microbes are not given the proper resources, they may not be able to outcompete the pathogens, potentially leading to infection and a host immune response.

sharing between us and our microbes, while other foods can lead to conflict and resource competition between our bodies and our microbes."

One further beneficial function of the microbiome of the large intestine is its role in ensuring our immune system develops and functions properly. Although scientists are not certain precisely how microbes train our immune system, it appears to be an amazing form of chemical communication that permits the adaptation of our immune system to our ever-changing resident microbiota (**Box 3.1**). The immune system plays a role in recognizing and tolerating beneficial microbes, while the microbes serve to both train the immune system and respond to invading pathogens. When these two organs (immune system and large intestine microbiome) work in harmony, the body responds rapidly to pathogens, while tolerating our beneficial microbes. If your microbiota is depleted due to poor diet, the use of antibiotics, or chemotherapy, your immune system will be immediately impacted.

The Friendly Gut Phageome

There is one highly significant component of our large intestine microbiome that we have not yet mentioned—the beneficial viruses present in untold but extremely high numbers in our intestinal **lumen**. The types of viruses you find in the gut are not the ones that we think of as causing flu or COVID-19. Instead, much of the gut viral diversity is composed of **bacteriophages**, or just **phages**, which literally means "bacteria eaters."

We mentioned above that a resilient microbiome can recover from perturbations. Recent studies suggest that bacteriophages may contribute to this function. Healthy individuals possess a very large set of phages that are unique to them, and a much smaller subset of shared phages (<5%) that are present in humans all around the

BOX 3.1. COMMUNICATION BETWEEN THE LARGE INTESTINE MICROBIOME AND IMMUNE SYSTEM

The microbiome in your large intestine and your immune system are in constant communication. The immune system acts to prevent microbial penetration of the mucus layer protecting the epithelium, and thus penetration to the rest of the body. It does so by directing mucus production and secretion of antibodies, such as immunoglobulin A (IgA), that stick to that mucus. IgA is the most abundant type of antibody in the body, and it serves to protect the mucosal tissues from microbial invasion and maintain immune homeostasis with the microbiota. Beneficial bacteria in the large intestine, such as *Bifidobacterium* and *Lactobacillus*, play a significant role in regulating our immune system, inhibiting the growth of pathogens by competing for resources, supporting the development of immune cells, fighting against inflammation, protecting the gut barrier, producing metabolic products, and contributing to the fine-tuning of immune responses. SCFAs produced by bacterial fermentation in the gut serve as a link between the microbiota and the immune system. SCFAs are an important energy source for the gut's epithelial cells and are essential for regulating the cellular functioning and turnover of the intestinal barrier. Many microbial metabolites cross the epithelial barrier into the bloodstream to other body tissues, where they influence the development, maturation, and function of immune cells, including neutrophils, macrophages, and T lymphocytes (immune cells that kill pathogens) in different organs (**Figure 3.10**).

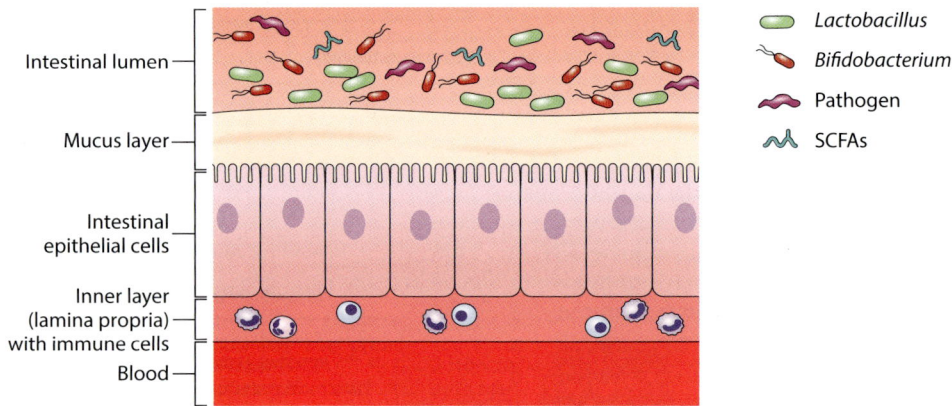

Figure 3.10 Chemical Communication between Microbes and the Gut Microbes in the gut lumen communicate with immune cells, which the immune cells direct to pass into the mucus layer. These same microbes communicate with other members of the microbiota and thus limit the invasion of potential pathogens.

world. The presence of these shared phages appears to be associated with host health and was named the **healthy gut phageome** (Townsend et al., 2021). One group of researchers envisions a process through which phages might control the presence and abundance of particular bacteria in the large intestine (Barr et al., 2013) (**Box 3.2**).

Dysbiosis of the Large Intestine Microbiome

Dysbiosis of the large intestine microbiome is an "imbalance" in the resident microbial community. This imbalance could be due to the gain or loss of community members or changes in their relative abundances. Dysbiosis can result from a variety of environmental events, such as a sudden dietary change; new medications, such as antibiotics, that affect your intestinal flora; and high levels of stress or anxiety, which can weaken your immune system.

We now recognize that the host itself can also directly impact gut microbiome homeostasis (Byndloss et al., 2018). **Figure 3.12** shows one of the mechanisms, which involves antimicrobe-producing Paneth cells located in the gut epithelial layer. The

BOX 3.2. BACTERIOPHAGES PROTECT EPITHELIAL CELLS IN THE LARGE INTESTINE

Bacteriophages engage in a critical protective process in our guts, which is to prevent the attachment of bacteria to the cells lining our large intestines. A model for how this occurs is provided in **Figure 3.11** and involves the following five stages. (1) Mucus is produced and secreted by the underlying epithelium. (2) Phages bind to the mucus glycoproteins. (3) Phage adherence creates an antimicrobial layer that reduces bacterial attachment to and colonization of the mucus, which in turn lessens epithelial cell death. (4) Mucus-adherent phages are more likely to encounter bacterial hosts and thus can serve as a first line of defense against invading pathogens. (5) Continual sloughing of the outer mucus (including the phages) provides a dynamic mucosal environment.

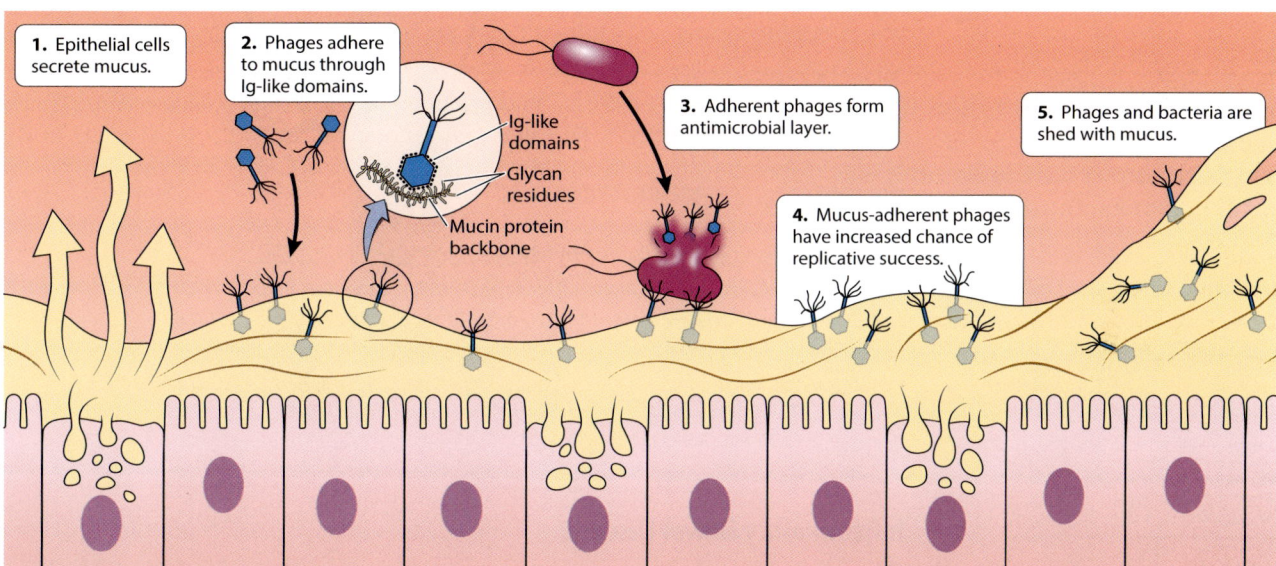

Figure 3.11 Model for Bacteriophage Protection of Gut Epithelial Cells Bacteriophages bind to mucus lining gut epithelial cells, which serves to increase phage-bacterial interactions and results in a robust line of defense against invading pathogens. (After Barr et al., 2013.)

production of the antimicrobials limits which microbes can approach the mucin layer. Under stress, Paneth cells may cease peptide production, which, in turn, permits various microbes to access the epithelial cells and induce inflammation. Colonocytes, epithelial cells lining the colon, consume SCFAs produced by the **obligate anaerobic** bacteria that make up most of the large intestine microbiota. Obligate anaerobes can only live in environments that lack oxygen. As colonocytes consume SCFAs, they also consume oxygen, which results in an oxygen gradient in the lumen. Low levels of oxygen select for more anaerobes that then produce more SCFAs—and the healthful anaerobic fermentation cycle continues. Inflammation and epithelial repair can shift the colonocytes towards a different form of SCFA metabolism, this time one that does not consume oxygen. With oxygen levels not being depleted by the colonocytes, oxygen accumulates in the lumen. The presence of increasing levels of oxygen causes a shift in the microbiome from obligate anaerobes to **facultative anaerobes**, which can metabolize with or without oxygen present. This shift results in dysbiosis of the gut microbiome and is associated with several human diseases, such as **colitis**, which is a chronic digestive disease characterized by inflammation of the inner lining of the colon. The presence of the facultative anaerobes triggers an immune response that results in inflammation.

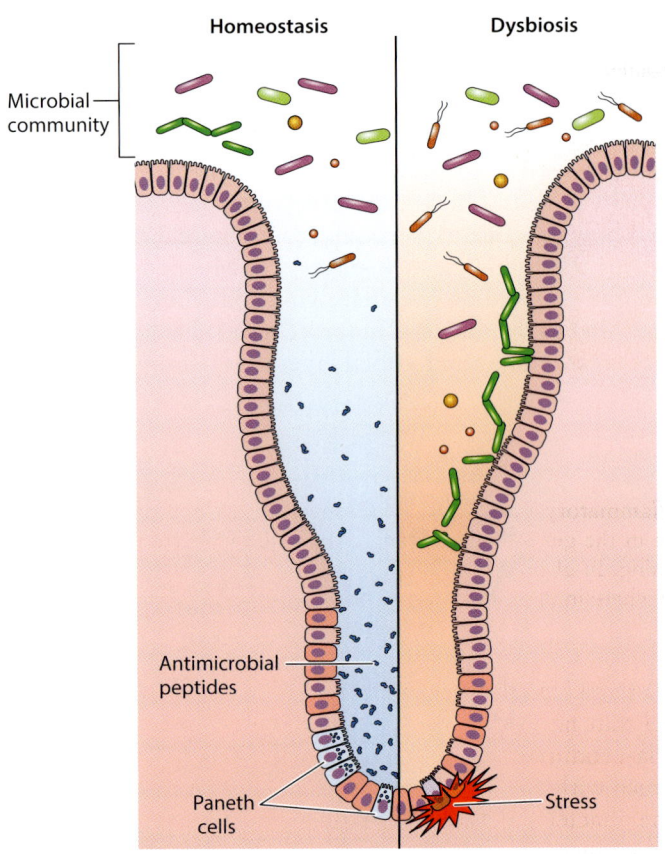

Figure 3.12 Paneth Cells Contribute to Microbial Homeostasis Antimicrobial peptides secreted by Paneth cells in the small-intestinal crypts check bacterial growth in proximity to the epithelial surface, thereby maintaining homeostasis in the small intestine. (After Byndloss et al., 2018.)

Inflammatory Bowel Disease

We will tackle several of the diseases resulting from large intestine microbiome dysbiosis in greater detail in future chapters, but for now let's briefly examine several of the more common ones. **Inflammatory bowel disease** (**IBD**) is a chronic (lifelong) disease affecting over 3 million individuals in the US. IBD results in painful inflammation in the intestines that is caused by a combination of factors, including genes, diet, and gut microbiome composition. It can take years to properly diagnose IBD and even longer to treat it, which may involve anti-inflammatory medications to reduce inflammation and surgery to remove diseased parts of the bowel.

The two most common forms of IBD are Crohn's disease and ulcerative colitis, which differ in where in the gut they create inflammation (**Figure 3.13**). With **Crohn's disease**, inflammation occurs anywhere in the GI tract, while an individual with colitis has inflammation only in the colon. In both cases, the host becomes intolerant of members of their intestinal microbiota, the immune system is triggered, and the result is an alteration of the intestinal epithelial barrier. Several bacterial species normally associated with a healthy intestinal microbiome then begin to penetrate the epithelial surface and damage the host cells.

It has been proposed that the large intestine microbiome plays a direct role in IBD. In one study, researchers explored whether specific bacterial metabolites could be altering the microbiome, thereby leading to Crohn's disease (Ni et al., 2017). **Box 3.3** describes the details of the first phase of the study. The goal was to analyze fecal samples from people with Crohn's disease and compare them with samples from healthy individuals to identify **metabolites**, such as SCFAs and amino acids, that differ between these two participant groups. Most of the metabolites that were found at higher levels in the samples from people with Crohn's disease were amino acids. Knowing that many types of bacteria will produce amino acids by breaking down a nitrogen-rich compound called urea, the scientists focused on an enzyme,

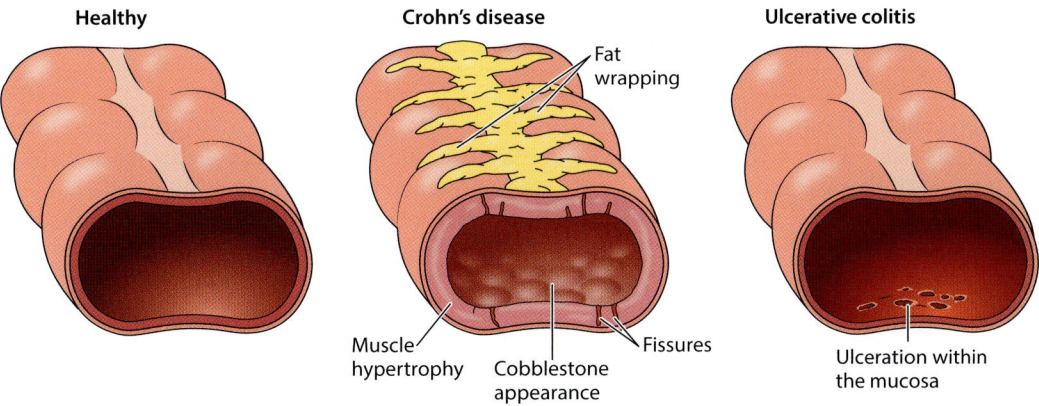

Figure 3.13 Inflammatory Bowel Disease Crohn's disease and ulcerative colitis are the result of inflammation in the gut epithelial lining, which allows bacteria to invade and damage the gut epithelial lining. In ulcerative colitis, this occurs in the colon, whereas in Crohn's disease, it can occur anywhere in the GI tract.

urease, that is critical for this process (Ni et al., 2017). The researchers established a mouse model that harbors bacteria that produce higher levels of urease to determine if bacteria-produced urease might play a role in the development of Crohn's. They treated mice with antibiotics to deplete their microbiomes, then inoculated the mice with either bacteria lacking urease or bacteria engineered to produce urease and allowed the mice to naturally re-establish their microbiomes over the next month. They found that the two groups of mice developed significantly different microbiomes: the mice initially inoculated with the urease-producing bacteria were more likely to develop microbiomes that contained relatively greater numbers of bacteria associated with poor health. These results suggest that urease may play a role in exacerbating inflammation in the gut by disrupting the microbiome. This research also points to urease as a possible target for therapy, either by directly inhibiting its production or by manipulating the microbiome's bacterial components to decrease the amount of urease in the intestines. According to Gary Wu, a physician at the Perelman School of Medicine at the University of Pennsylvania, "the outcomes of this study . . . will be an important first step in building a technology platform to engineer a beneficial composition of the gut microbiota for the treatment of inflammatory bowel diseases" (Wu, 2017).

Another common disease resulting from gut dysbiosis is colorectal cancer, which begins with the formation of small noncancerous cell clusters, or polyps. The likelihood of developing these polyps is often associated with a diet rich in red meat and fat. The intestinal microbiota has been identified as a key factor in colorectal cancer, and several mechanisms have been proposed, including the formation of biofilms that secrete genotoxins, which are chemical agents that can cause DNA damage and lead to cancer (Silva et al., 2021). Bacteria such as *Bacteroides* (Bacteroidetes) have been found in association with an increased risk of colorectal cancer. In contrast, *Lactobacillus* and *Eubacterium* (Firmicutes) are associated with decreased risks (Lucas et al., 2017). Thus, although we tend to think of Bacteroidetes as the "good guys" in the gut and Firmicutes as the hyperactive, sugar-craving troublemakers, their roles are switched in this situation. The moral of this story is that members of a microbiota play numerous roles, which are highly dependent upon the other members they find themselves with—what is considered a good bacterium in one setting may be something quite nasty in another!

BOX 3.3. RESEARCH IN ACTION
Fecal Amino Acids and Dysbiosis—Unlocking Crohn's Disease Therapies

- **Hypothesis.** The presence of amino acids in fecal microbiota from patients with Crohn's disease positively correlates with disease activity.

- **Experiment.** Fecal samples were obtained from patients with Crohn's disease and from healthy volunteers whose gut microbiota had previously been established. The metabolites in their feces were examined using liquid chromatography–mass spectrometry, which is a method that employs mass-to-charge ratios to identify substances.

- **Results.** There were 341 known metabolites detected in the feces. Levels of amino acids were

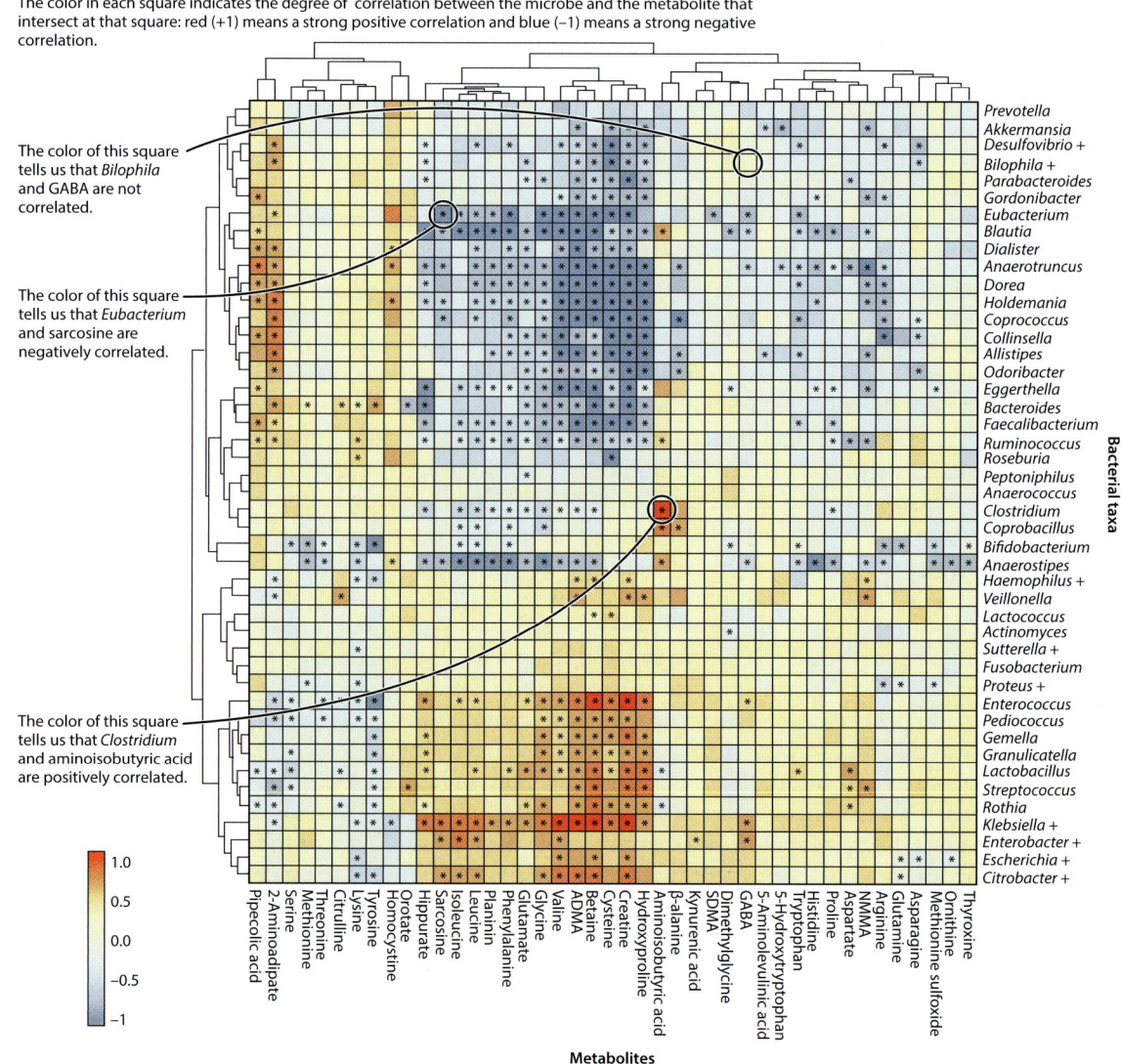

Figure 3.14 Bacterial Abundance Associated with Crohn's Disease This matrix shows the level of correlation between amino acid metabolites identified in fecal samples and the bacterial taxa present in the microbiomes of healthy volunteers and individuals with Crohn's disease. Boxes colored orange indicate a positive correlation between the bacterial taxa—indicated on the *y*-axis—and the metabolites—indicated on the *x*-axis. There is a positive association between the taxa related to amino acid metabolites and the severity of symptoms in patients with Crohn's disease. (From Ni et al., A role for bacterial urease in gut dysbiosis and Crohn's disease. *Sci Transl Med* 2017 Nov 15; 9 (416): eaah6888. Reprinted with permission from AAAS.)

significantly associated with the patients with Crohn's disease and with disease severity. Dietary intake of protein was explored as a possible explanation for the association between fecal amino acids and Crohn's disease, but there was no significant difference between the healthy and ill participants. Fecal amino acids and their derivatives were also positively associated with taxa in the Proteobacteria phylum, such as *Escherichia*, *Klebsiella*, *Haemophilus*, and *Proteus*, all of which have been associated with intestinal microbiota dysbiosis in patients with IBD (**Figure 3.14**).

❖ **Conclusions.** Fecal amino acids are associated with Crohn's disease and positively correlated with increasing disease activity. These data suggest that one treatment option involves the manipulation of the dysbiotic microbiome to reduce or eliminate key members of the Proteobacteria, which are found at elevated densities in the intestines of Crohn's patients.

3.4 THE MICROBES IN OUR MOUTH

Let's now turn to our second-richest microbiome after the large intestine, the oral microbiome. The mouth offers a wealth of microenvironments, with the hard surfaces of the teeth and the soft oral mucosa, all hydrated in saliva that helps to transport nutrients between cells. Unlike the large intestine, which hosts a predominately anaerobic microbiome, the mouth has sufficient oxygen to support a thriving aerobic bacterial community as well. The four most common bacterial phyla in the mouth are Firmicutes, Bacteroidetes, Proteobacteria, and Actinobacteria. Note that these same phyla were dominant in the large intestine. Well over 700 species of bacteria have been identified in the mouth to date, although the average oral microbiome hosts a mere 250 or so.

Archaeal Syntrophy

Archaeal diversity is far more limited in the mouth, with one genus, *Methanobrevibacter*, identified as the dominant archaeal representative. Members of this genus serve a **syntrophic** function, which refers to the phenomenon of one species living off the metabolites of another. For example, *Bacteroides thetaiotaomicron* assists the host by fermenting dietary polysaccharides, whereas *Methanobrevibacter smithii* consumes the resulting end-stage fermentation products and triggers further fermentation by *B. thetaiotaomicron*. Under certain conditions, *B. thetaiotaomicron* and *M. smithii* even form interspecies granules, just as has been observed for syntrophic partnerships between microbes in soil (Catlett et al., 2022).

Fungal Diversity of Unknown Function

There are several microbial eukaryotes present in the oral cavity, including protists, such as *Entamoeba* and *Trichomonas*, and over 85 genera of fungi, including *Candida*, *Aspergillus*, and *Fusarium*. We know virtually nothing about the role of these microbial eukaryotes in the normal oral microbiome. It has been suggested that they are commensals, which are microbes that live on or in a host without causing harm, though both *Entamoeba* and *Trichomonas* have been implicated as parasites involved in periodontitis, or gum disease (Yaseen et al., 2021). Given that gum disease is so common, however, it is possible that the presence of these protists is simply correlated with oral disease, rather than a cause of it.

Eukaryotic cells produce cell-surface receptors that bacteria attach to. Indeed, numerous bacterial species have evolved appendages, such as pili, that specifically facilitate attachment to these receptors. These interactions are exquisitely specific, like the interlocking of matching puzzle pieces. Due to this receptor attachment specificity, the bacteria present in a microbiome can vary greatly by location. For example, in the oral microbiome, one side of the mouth will host different species from the other, and teeth have different bacteria above and below the gums and on their interior versus exterior surfaces.

Creation of Dental Plaque

Some of the earliest colonizers of the oral cavity are members of the genus *Streptococcus*, which remain a dominant member for life (Nath et al., 2021). As teeth emerge, they acquire a protective coat of glycoprotein, which is rapidly colonized by microbes; this results in the formation of a **biofilm** community we call **dental plaque**. We learned in chapter 2 that the microbiome that develops on teeth is composed of spatially structured members whose functions are dependent on a complex web of symbiotic interactions (see Figure 2.17). This matrix of bacteria is known as a biofilm and its members create acidic and anaerobic microenvironments in which *Corynebacterium* (Actinobacteria) thrive. This foundational member of plaque binds to tooth enamel and provides a framework for additional members of the dental community. It is attached so firmly to teeth that toothbrushes and mouthwashes have little impact. *Streptococcus* lives deep at the base of a *Corynebacterium* forest and produces carbon dioxide, which *Corynebacterium* needs to grow. *Streptococcus* also consumes sugars and produces lactate and hydrogen peroxide, which inhibits the growth of many other microbes. *Aggregatibacter* (Pseudomonadota) has enzymes to detoxify peroxide, and it can use lactate as a food source, so it happily inhabits the plaque canopy. This region is semianaerobic due to *Streptococcus* metabolism, which provides a home to anaerobic species like *Fusobacterium*. "These results show the very best of what can happen when you look at something well known—plaque—in a new way," said Michael Fischbach from Stanford University. "The degree of organization in the community is beyond my wildest dreams" (Yong, 2016).

The Diverse Roles of the Oral Microbiome

The oral microbiome performs numerous functions that help keep our mouth (and body) healthy. Certain members are involved in transport, such as ferrying minerals from saliva to the surface of teeth to help in remineralization, or the movement of oxygen to the gums and soft tissue. Others are involved in nutrient cycling or helping digest certain foods, and as happens in the large intestine, they hydrolyze indigestible polysaccharides into SCFAs that we can then metabolize for energy. Still others protect us from pathogen invasion or eliminate waste products. Even more surprising, certain members of the oral microbiome play a role in our perception of taste!

As key components of nutrient cycling, some members of the oral microbiome can reduce nitrate to nitrite, which is then transformed into nitric oxide in the bloodstream. Levels of nitric oxide are indicators of cardiovascular health, as the chemical plays a role in regulating blood pressure. Inhibition of nitric oxide signaling can result in increased blood pressure, obesity, and heart disease. Activities as simple as the use of antibacterial mouthwash can result in increased blood pressure through the inhibitory effect the mouthwash has on the oral microbiome (Bescos et al., 2020). So perhaps think twice when considering that bit of oral hygiene!

As in the gut microbiome, members of the oral microbiota play a key role preventing pathogen invasion. They are highly competitive in colonizing niches and thus prevent invaders from establishing. Some oral bacteria, such as *Streptococcus* and *Actinomyces* (Actinobacteria) can inhibit the growth of *Porphyromonas* (Bacteroidetes), which is involved in gum disease. Others, such as *Lactococcus* (Firmicute), produce their own antimicrobials, while still others, such as *Streptococcus* and *Lactobacillus*, produce alkaline metabolic by-products that buffer the acid produced by bacteria involved in cavity production.

Our Evolving Oral Microbiota

Historical shifts in the human diet have resulted in significant changes in the members of the oral microbiome. The emergence of agriculture resulted in a more acidic diet rich in meat, dairy products, and grains, which led to significant changes to the oral microbiota, including the increased presence of acid-producing and acid-tolerant organisms. One fascinating study investigated changes in the oral microbiome spanning 100,000 years (Yates et al., 2021). The authors analyzed dental biofilms of humans and Neanderthals (scraped from fossilized teeth) and compared them with dental microbiomes of chimpanzees, gorillas, and howler monkeys. Their data, which involve a comparison of the taxa present in the dental plaque of each primate taxon studied, show major differences between the oral microbiomes of humans and chimps, but relatively few differences between humans and Neanderthals. One particularly surprising result was the discovery of *Streptococcus* present in both modern *Homo sapiens* and Neanderthals, which suggests that starchy foods were an important part of the human diet well before the appearance of modern humans.

While diet clearly influences the members of the oral microbiome, several studies suggest that your microbiome can influence what foods you crave! The microbes in your mouth form biofilms on your tongue, which act as physical barriers limiting access of taste receptors to foods. The metabolism of these foods also influences the activation of our taste receptors and taste sensitivity. Perhaps it is comforting to know that some of our repeated failures to eat what we know are more healthful foods may be due to our microbes' influence, rather than our own lack of self-control.

Oral Microbiome Dysbiosis

In what will emerge as a common theme, a well-functioning microbiome that is working in concert with the immune system is able to limit or inhibit certain diseases. In the case of the oral microbiome, one critical role is in the prevention of **dental caries** (**cavities**) and **periodontal** (**gum**) **disease**, which are caused by imbalances in the microbiome, rather than the presence of a single pathogenic species. When this balance is disrupted (dysbiosis), pathogens can gain a foothold. The shift from a balanced community with mostly aerobic bacteria to an imbalanced one results in the propagation of anaerobic bacteria. This includes the growth of *Streptococcus mutans*, most closely associated with tooth decay, and *Porphyromonas gingivalis*, associated with gum inflammation and disease.

Box 3.4 provides an overview of the complex ecological interactions involved in the development of gum disease. As plaque accumulates below the gum line, dysbiosis occurs and the microbiome transitions into a diseased state. At the heart of this dysbiosis is the **keystone species**, *Porphyromonas gingivalis*. The "keystone" label reflects its outsized impact, relative to its general rarity in the oral microbiome, in producing chronic inflammation of the gums, which, over time, can damage connective tissue and the bones that ensure teeth remain in place. Some microbes even secrete enzymes that break down the host tissue, resulting in more inflammation and more acute disease. Once in place, gum disease becomes irreversible and difficult to treat, which is why it is so critical to maintain good oral health.

Cavities are formed from the interaction between complex carbohydrates we eat and acid-producing bacteria in our mouth. The metabolism of carbohydrates results in the generation of acid, which can result in a thinner layer of tooth enamel, which protects your teeth from cavities. Further, as the microbial biofilms on teeth mature, acid-producing microbes accumulate, resulting in a lower pH of the mouth. The resulting slightly acidic environment helps minerals diffuse out of our teeth, a process called demineralization. This creates weaker teeth that are even more prone to cavities.

The influence of the oral microbiota goes far beyond the mouth. Microbes that cause cavities have been found in the small intestines, placenta, lungs, and brain. Other oral microbes have been associated with heart disease and increased blood pressure (LaMonte et al., 2022). It isn't yet clear why such connections exist, but several studies have shown that microbes that cause cavities influence immune responses well beyond the mouth (Sedghi et al., 2021).

Contrary to what has been taught since the first dental school opened in 1840 (Baltimore College of Dental Surgery), the goal of good oral hygiene is no longer

BOX 3.4. THE GUM MICROBIOME AND GUM DISEASE

Periodontitis is a serious gum infection that damages the soft tissue in the mouth and can eventually destroy the bone that supports your teeth. It is caused by an increase in the density of oral microbes capable of thriving in an inflamed microenvironment (**Figure 3.15**). *Porphyromonas gingivalis* may be the keystone pathogen of periodontal disease because, even in low abundance, it has profound effects on the oral microbial community structure. *P. gingivalis* possesses an array of virulence factors that can subvert host innate immunity without blocking the inflammatory pathways; this is referred to as "nonproductive inflammation." This chronic inflammation supplies nutrients to *P. gingivalis* and provides a selective advantage for co-colonizing microbes able to tolerate and thrive in an inflammatory environment.

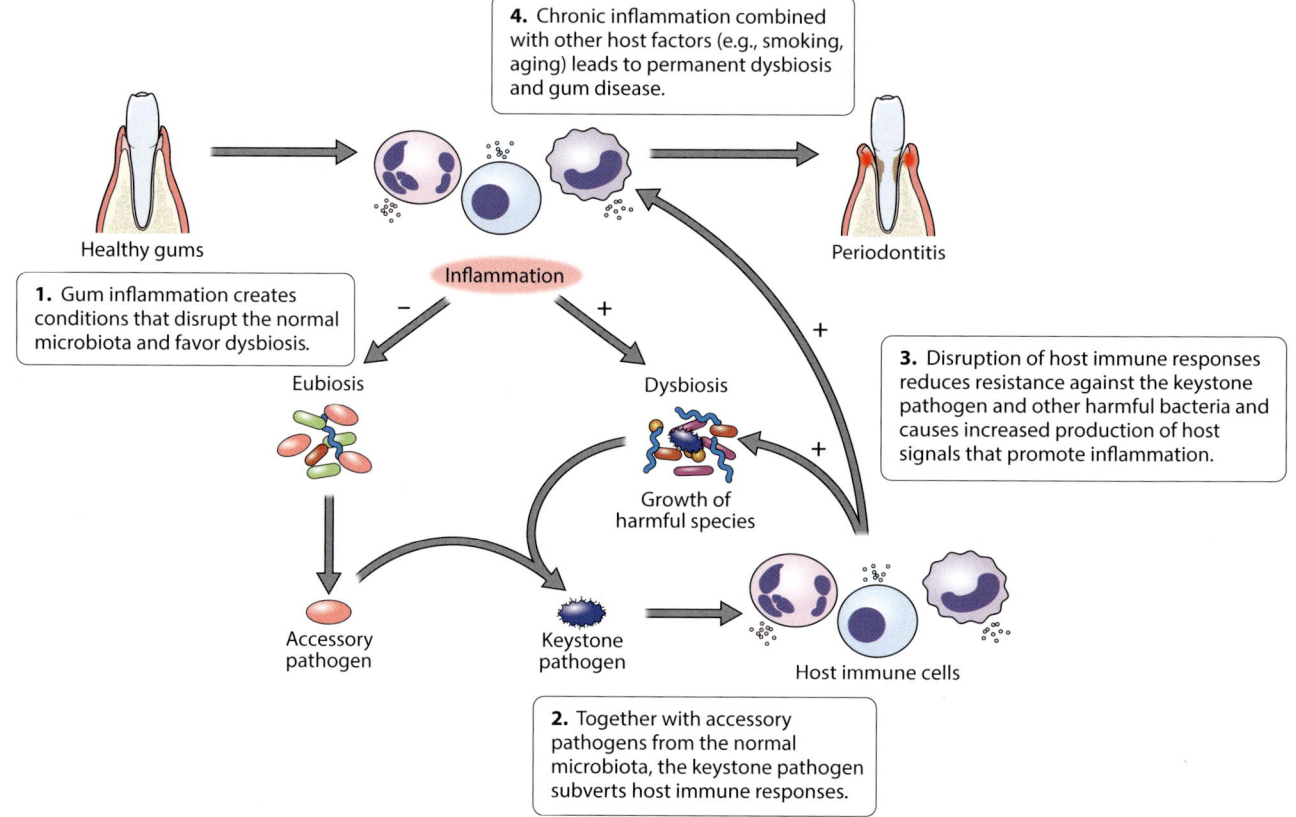

Figure 3.15 The Ecology of Gum Disease The progression of dysbiosis that leads to gum disease involves several bacterial pathogens. The keystone species, *Porphyromonas gingivalis*, triggers production of inflammatory cytokines, which result in pathogen overgrowth, further inflammation, and ultimately gum disease. (After Lamont et al., 2018.)

aimed at killing bad bacteria, but rather aimed at ensuring that a healthy oral microbiome is preserved. According to Gerry Curatola, "an unbalanced or unhealthy oral microbiome is like a garden overgrown with weeds" (Curatola & Reverand, 2017). Dr. Curatola is a dentist who has developed a new approach to dentistry he calls the mouth-body connection, in which he proposes that the health of your mouth impacts the health of your entire body. Oral disease results in chronic inflammation throughout your body, which can result in cardiovascular disease, obesity, and diabetes. His oral health program focuses on restoring the natural ecology of the mouth, and he argues that keeping these microbial communities in balance is the key to well-being.

3.5 THE MICROBES ON OUR SKIN

Let's move our focus from the mouth to the skin, which is the largest organ of the human body. There is over 1.8 m^2 (19 ft^2) of skin enveloping an average adult, which includes the skin surface, sweat glands, pores, and hair shafts. Much of the skin is relatively dry, salty, and acidic, except for the more nutrient-rich areas around hair follicles. Our skin harbors a discrete collection of millions of bacteria, fungi, and viruses whose composition varies across the body's surface. The relatively dry and cool areas, like the arms and legs, have fewer microbes than moist and warm areas, such as the armpits or nose. The forehead has very high densities of microbes, but relatively few types, whereas your palm has a lot of variety, but relatively low numbers.

There are approximately 700 microbial species that can take up residence on our skin. The most abundant are bacteria and include the same four key bacterial phyla (Actinobacteria, Firmicutes, Bacteroidetes, and Proteobacteria) found in our gut and oral microbiomes. Key members include the oil-loving Actinobacteria (e.g., Corynebacterineae, *Propionibacterium*, and *Cutibacterium*) that reside near **sebaceous glands**, which secrete sebum, an oily substance that lubricates the hair and skin. In contrast, dry areas, such as the forelegs, are enriched with Proteobacteria. *Staphylococcus* and *Corynebacterium* grow well in moist areas, such as the crease of the nose. *Cutibacterium acnes*, the bacterium that can cause acne, is also one of the primary members of the skin microbiome—up to 80% of the microbes on the palm of your hand can be of this species alone. Another common skin microbe is *Staphylococcus epidermidis*, which is a harmless relative of the potentially dangerous pathogen *S. aureus*.

The nonbacterial constituents of the skin microbiome are quite limited. Fungal members include *Malassezia* and *Candida*, both of which are yeast. *Malassezia* is a **lipophilic** (fat-loving) yeast that feeds on fat secreted by sebaceous glands. Although considered a normal commensal in the skin microbiome, *Malassezia* can metabolize the lipid found on the skin surface and produce by-products that can potentially create harmful interactions with the skin. For example, it has been proposed that *Malassezia* is involved in causing skin cancer, because some of the breakdown products of fat metabolism can activate pathways in the skin related to tumor promotion, similar to those induced by sunlight (Velegraki et al., 2015). Archaeans are rarely found on the skin, and there does not appear to be a core set of viruses, aside from those bacteriophages associated with *Cutibacterium* and *Staphylococcus*.

A Nutritional Desert

Compared with our gut and mouth, our skin has far fewer nutrients for microbes to snack on. To survive, they feed on the outer layer of skin cells as well as sweat and sebum. For example, *C. acnes* consumes arginine found in skin cell proteins and degrades sebum lipids. *Staphylococci* are **halotolerant**, which means they can survive in salty environments and can metabolize the urea present in sweat. Several species of *Staphylococci* can also produce **adherens**, which are sticky proteins that help them attach to the skin, and proteases that break apart bonds in proteins, which are then consumed as nutrients. One species, *Staphylococcus hominis*, which lives in the human armpit, is responsible for the $74 billion deodorant market. These bacteria feed on odorless chemicals released in sweat, which they break down into thioalcohols, which are responsible for the often-offending smell of body odor.

The skin microbiota is distributed in a highly predictable manner. **Figure 3.16** shows the key taxa found in various skin sites. The skin microbiome was once thought to exist solely on the skin's surface. However, we now know that it also thrives in the deeper layers, such as the subcutaneous fat layer, and it turns out that some of the most intimate communication between the microbiome and our immune system occurs here (Nakatsuji et al., 2013).

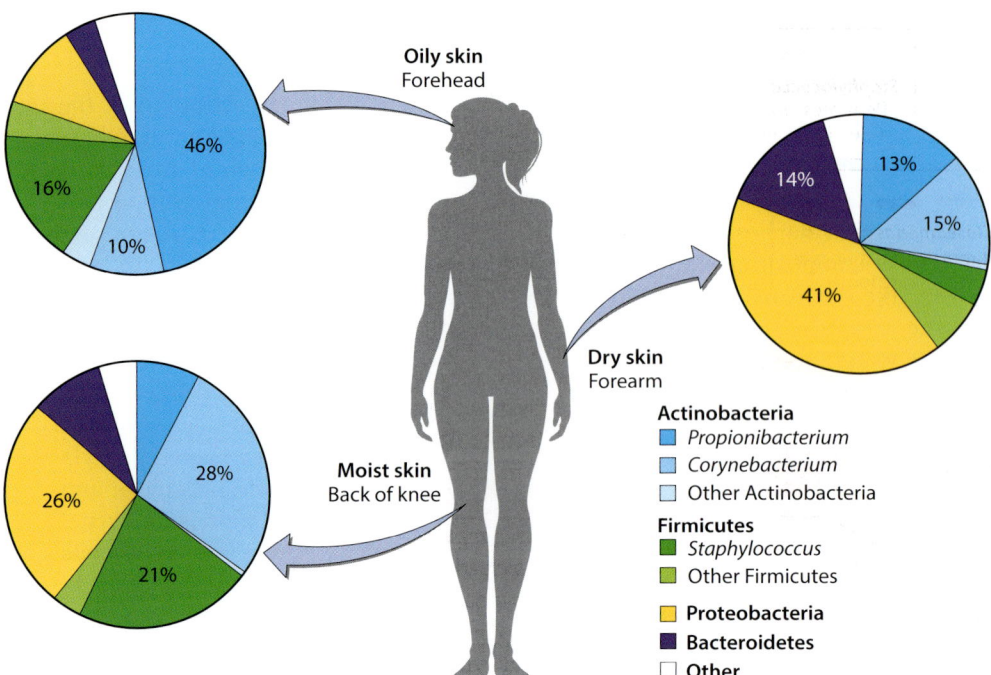

Figure 3.16 Composition of Our Various Skin Microbiomes Understanding the community distribution of an organ's microbiome can help us to gain key insight into its functions. Make note of how different the communities are on different parts of your skin! Much as in the large intestine, different areas of the same system can have completely different microbiomes. (After Grice et al., 2009.)

Primary Functions of the Skin Microbiome

The skin microbiome plays a variety of health-related roles, including wound healing, limiting exposure to allergens, and retaining moisture (**Figure 3.17**). For example, *S. epidermidis* aides in wound healing by recruiting immune cells to the site (Leonel et al., 2019). It may also protect us from harmful UV rays. One study found that when a mouse's skin was colonized with *S. epidermidis*, it developed significantly fewer tumors when exposed to UV rays (Patra et al., 2016). *Malassezia* spp. and *Roseomonas mucosa* stimulate production of keratinocytes, which are the dominant cell type in the epidermis, playing multiple roles essential for skin repair.

A healthy skin microbiome protects against infection in much the same way a good gut microbiome does: by occupying the niches required by potential pathogens. The skin microbiome prefers a slightly acidic environment, with a pH around 5.0, which is also effective at inhibiting the growth of many pathogens. The methanogenic archaeans present are likely involved in nitrogen metabolism, in particular the oxidation of ammonia from sweat. Removal of nitrogen results in an even lower skin pH, which further prevents pathogen invasion. Commensal skin bacteria also produce toxic metabolites and antibiotics that mediate interactions between the commensal flora and invading pathogens. Some skin bacteria simply compete for resources better, such as binding sites and nutrients, which blocks colonization by pathogens. For example, *S. epidermidis* binds to certain skin cell receptors, which excludes the binding of its close relative, *S. aureus*. It can both stimulate and suppress inflammation that contributes to stability of the microbial ecosystem and skin integrity. Other species such as *Roseomonas mucosa*, *Malassezia* spp., or *Corynebacterium accolens* can fine-tune host immune responses. Last, even the often-problematic *C. acnes* has beneficial interactions with the host as it metabolizes sebum secretions, which in turn aids in maintaining an acidic skin pH, making the skin suitable for only select organisms.

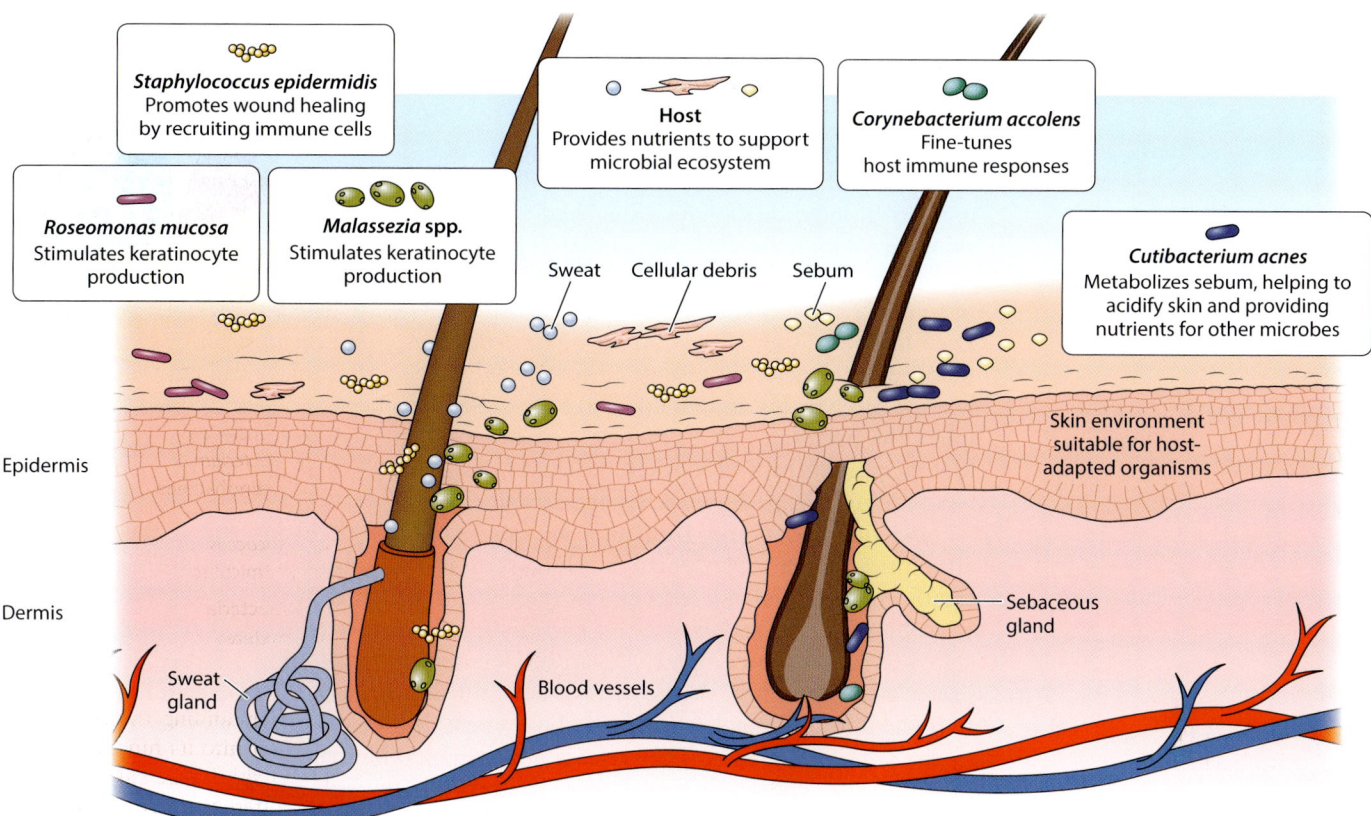

Figure 3.17 Functions of the Skin Microbiome Members of your skin microbiome play an active role in maintaining skin health. Several species stimulate keratinocyte production, such as *Roseomonas* mucosa and *Malassezia* spp., which helps to create a tight barrier that prevents foreign substances from entering the body. Others are involved in recruitment of immune cells during wound healing, such as *Staphylococcus epidermidis*, or the fine-tuning of host immune responses, such as *Corynebacterium accolens*. Still others provide nutrients for other members of the microbiome, such as *Cutibacterium acnes*. (After Eisenstein, 2020.)

Skin Microbiome Dysbiosis

Dysbiosis of the skin microbiota can result in several common diseases. One example is eczema, or **atopic dermatitis**, which results from an increase in *S. aureus* and a concomitant decrease in overall bacterial diversity on the skin. People with eczema have dry and easily irritated skin. Further, *S. aureus* releases **delta toxin**, which causes cells in the skin to release tiny granules that cause inflammation, which makes you want to scratch. Scratching makes your skin itchier, and so a vicious cycle is established.

Acne and Your Skin Microbes

One of the most common skin diseases is acne, affecting over 50 million people in the US. It has a long history with humans, as ancient writings describe Egyptians using sulfur to treat it. Acne is a chronic disease resulting in inflammation and the development of blackheads, whiteheads, and other types of pimples. The areas of the body most affected by acne are the face, neck, chest, shoulders, and back, where sebaceous glands are common. These glands produce sebum—an oily substance that serves to lubricate the hair and skin. When too much sebum is produced, it can clog the pores and create a perfect environment for some bacteria to thrive, which then leads to inflammation and acne development. *C. acnes* uses sebum as a food source, and as its numbers increase, it elicits an immune response that results in inflammation. However, *C. acnes* is abundant in the pores of all individuals, those with and without acne, which indicates that it is not simply the presence of this bacterium that drives the skin condition.

Dr. Huiying Li and her colleagues at the UCLA School of Medicine sought to learn more about the role of bacteria in acne development. They sampled the skin of a similar number of individuals who either had or did not have acne, and they determined the skin microbiota for each individual. The *C. acnes* strains taken from diseased skin looked very different from the strains taken from healthy skin, with two unique strains of *C. acnes* appearing in one out of five volunteers with acne but rarely in healthy skin. Even more exciting, a new strain of *C. acnes* was common in healthy skin yet rarely found when acne was present. Li notes, "We suspect that this strain contains a natural defense mechanism that enables it to recognize attackers and destroy them before they infect the bacterial cell." Offering new hope to acne sufferers, the researchers believe that increasing the body's friendly strain of *C. acnes* through the use of a simple cream or lotion may help eliminate spotty complexions.

Individuals suffering from acne are taught to clean their skin and slather on topical antibiotics. We now recognize that long-term use of antibiotics to address acne carries significant potential side effects, not the least of which is dysbiosis of the normal skin microbiome. This results in acne becoming ever harder to treat. This recognition resulted in a start-up company, Xycrobe, focused on developing genetically engineered bacteria that can limit the growth of harmful skin bacteria. The bacteria are designed to travel into the deep layers of the skin, where they consume some of the oil produced by the sebaceous glands. They release antioxidants that help to prevent the overgrowth of pathogens. The result is that good bacteria can flourish and help the skin heal. The CEO and founder of Xycrobe, geneticist Thomas Hitchcock, notes, "People were going overboard and actually killing off whole communities of bacteria, including the good ones, ending up with more organisms that were resistant to treatment or that had built up a tolerance. . . . We want to change the way we design acne treatments and understand how the body interacts with the skin microbiome to come up with more effective solutions" (Puniewska, 2017).

3.6 SAMPLING YOUR OWN MICROBIOME

Sometimes seeing really is believing. If you have never had the opportunity to view microbes under a microscope or on a petri dish, you may still be wondering what all the fuss is about. In this final section you will learn how to sample your very own skin microbiome. This activity may seem silly at first, but it will give you the chance to "see" some of your own bacteria and will teach you how to design and carry out a scientific experiment. There is nothing like producing your own data to open your eyes to the power of science!

Identifying a Testable Hypothesis

The very first step in designing an experiment is to determine what you wish to explore. You learned earlier that different locations on your skin harbor different types of microbes. Perhaps you wonder how different the environment of your inner elbow could be from your hand. Let's generate a **testable hypothesis** to assess whether different areas of the skin host different microbes. A **hypothesis** is a proposed explanation, based on limited evidence, that serves as a starting point for further investigation. A testable hypothesis is one that can be proved or disproved because of experimentation. Testable hypothesis in hand, we now design our experiment, which is quite simple in this instance and requires only three steps: (1) sampling the skin microbiome from your hands and inner elbow, (2) growing the microbes in these samples, and (3) comparing the results, that is, what grows in our samples.

BOX 3.5. BACTERIAL GROWTH MEDIA PROTOCOL

We will need a sterile growth medium on which we can grow numerous types of microbes. There are two options for you to consider. You can purchase pre-made nutrient agar petri dishes online, or you can make your own. If you prefer to make your own growth medium, gather the following ingredients:

- 25 g (12.5 teaspoons) of agar powder or flakes (can be found in the specialty aisle in grocery stores or purchased online), which contains nutrients for growth and agar for solidifying the growth medium
- 625 ml (2 2/3 cup) of water
- 5 clean ramekins or sterile empty petri dishes
- Roll of plastic wrap
- Spoon or fork
- Clean microwave-safe medium-sized glass bowl

Pour 625 ml of water into a clean microwave-safe medium-sized glass bowl. Add 25 g of agar (with nutrients) into the same bowl and stir with a clean spoon or fork until completely dissolved. Put the agar solution into the microwave and set the timer for 4 minutes. Watch the solution to ensure it does not boil over the top of the bowl. Let the solution cool for 1 minute before removing from the microwave. Be careful, the glass and agar solution will be hot! Next pour the solution into your five empty petri dishes or ramekins, and cover with the plastic lid or plastic wrap. After one hour, the solution will have solidified and be ready for use (**Figure 3.18**).

Figure 3.18 Nutrient-Filled Petri Dishes (Photo from iStock.com/tonaquatic)

Preparing Nutrient Media

Before we begin sampling, we need to prepare a food source so that if our samples have microbes present, those that can grow aerobically and utilize the food source we provide will do so. **Box 3.5** provides information on how to purchase or make the required growth medium and create sterile growth surfaces. A sterile medium is one which is free of all life forms. It is usually sterilized by heating to a temperature at which all contaminating microorganisms are destroyed. By following the instructions, you will end up with petri plates, or the equivalent, filled with sterile growth medium. We are now almost ready to sample our skin microbiome.

Experimental Controls Are Key

But before we jump into our microbiome sampling experiment we need to identify and employ **experimental controls** to ensure that the results of our experiment are valid. Controls allow the experimenter to minimize the effects of factors other than the one being tested. It's how we know an experiment is testing what it claims to be testing. This goes beyond science—controls are necessary for any sort of experimental testing, no matter the subject area. Say you wanted to test a new treatment for the common cold. If you gave patients a drug and found that half of them felt better after three days, how would you know whether your treatment helped or if half of the people just got better on their own? You'd need a control group of people not given the drug, to show that your treatment made a difference and led to a faster recovery time. Take a moment to consider what you think is important to control for in our experiment. Hint—we have created agar plates that we are assuming are sterile. What if they aren't?

When you are ready, examine **Box 3.6** to learn the two most critical controls required for our experiment. The first is designed to ensure that our growth medium is sterile. We mark one of our petri plates as Control 1 (or sterility control) and set it aside without ever opening the lid of the plate. If we are correct and the medium is sterile, then no microbes should grow on the plate's agar surface.

The second control is focused on the cotton swabs we use to sample our skin microbiome. We are assuming that they are also sterile. To assess this factor, we label a second petri plate as Control 2 (or swab control) and take a fresh swab and use it to swab the control plate's surface—without ever letting the swab touch our skin or any other surface. If we are correct in our assumption that our swabs are sterile, then no microbes should grow on the Control 2 plate. During an experiment, if your controls don't work as expected, then the results of the experiment aren't valid. If you find that the control plates aren't sterile, you have no way to know whether the bacteria growing on treatment plates are from the skin microbiome or are contaminants from the nutrient agar or swabs. One control we will not employ is a test to determine if our nutrient agar supports the growth of bacteria, that is, did we add the right food source? What if you conduct this experiment and nothing grows? It could mean that you are the one person on the planet to harbor no skin microbes or, more likely, that you forgot to add something to the growth medium and the microbes present have nothing to eat. If nothing grows on your plates, consider that possible interpretation.

Sampling Your Skin

You are now ready to carry out your sampling. Box 3.6 provides detailed instructions on this phase of the experiment. After the samples have had time to grow, examine them carefully. What will you see? That will depend upon what samples you took. **Figure 3.19** shows the colony morphologies of several of the more common microbial

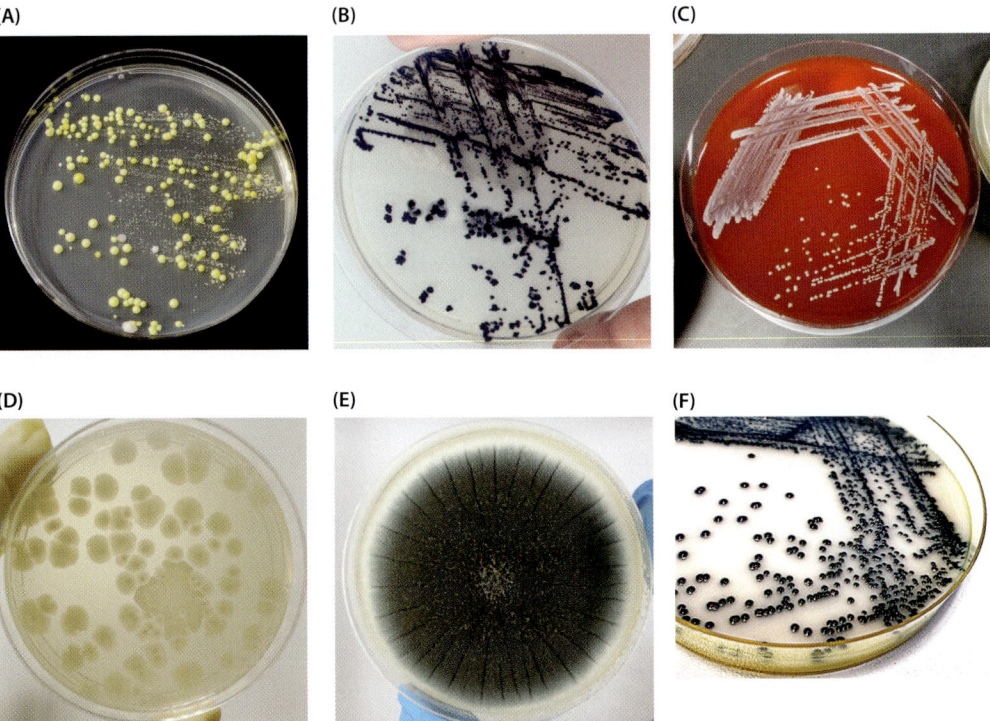

Figure 3.19 Bacteria Colony Morphology Shown are some of the more common members of the skin microbiome: (A) *Micrococcus*, (B) Clostridiales, (C) *Staphylococcus*, (D) *Bacillus*, (E) *Aspergillus*, (F) *Candida*. (Photos from iStock.com/Dr_Microbe [A]; iStock.com/Scharvik [B]; Sun14916, CC BY-SA 3.0, via Wikimedia Commons [C]; Kateryna Kon/Shutterstock [D]; TachiNui/Shutterstock [E]; iStock.com/Scharvik [F])

BOX 3.6. CULTURING YOUR SKIN MICROBIOME

1. Label the bottom of your plates with a marker.
2. Take a clean swab and use it to streak back and forth on the agar surface of the Control 2 plate (**Figure 3.20**).
3. Toss out that swab.
4. Take a fresh swab and rub it back and forth several times against one of the skin locations you wish to sample.
5. Now streak back and forth on the agar surface of the skin sample 1 plate.
6. Toss out that swab.
7. Repeat steps 4 and 5 for as many skin locations you wish to sample.
8. Leave the plates at room temperature (about 25°C) for 7 days, taking note of any growth that occurs each day.

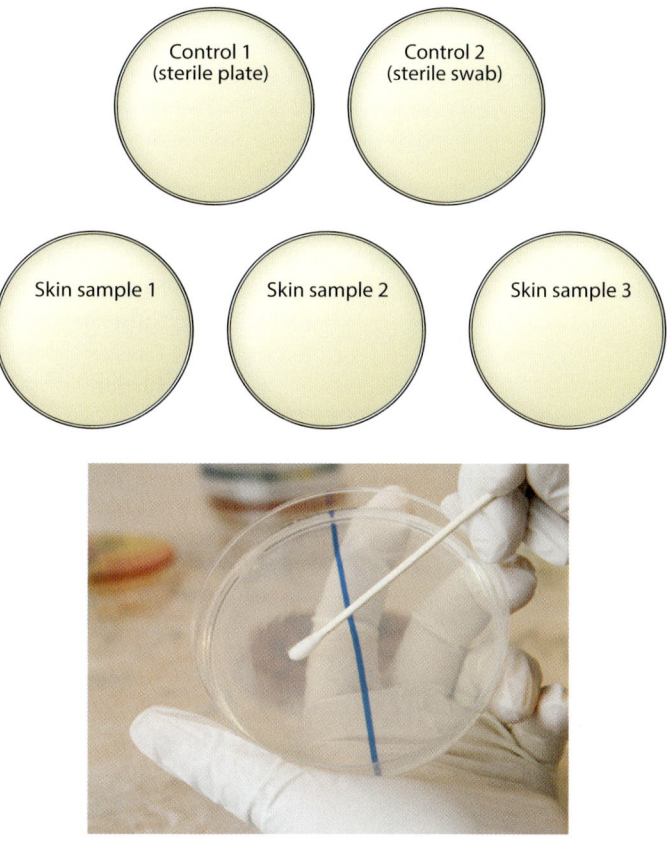

Figure 3.20 Control and Treatment Design To ensure the plates and swabs are sterile, two controls are used. Two or more treatment plates are used, which have been swabbed with samples from the skin microbiome. Significant microbial growth is expected to be seen on each treatment plate. (Photo from iStock.com /Zaharia_Bogdan)

skin residents. Note the different colony sizes, shapes, colors, and textures. Although we often think of bacteria as these tiny blobs of cells, they produce extraordinarily beautiful morphological diversity. You are likely to see striking variations in colony morphologies from your skin microbiome samples! If you want to try to identify some of the microbes you found, Figure 3.20 may help in that endeavor, which turns out to be a challenging task, and one beyond the scope of this exercise.

Congratulations on performing a simple, yet well-designed scientific experiment! If all went well, you should now have an image in your mind of what some of the microbes that contribute to your skin microbiome look like and how the microbial composition of one body site differs from another. Our next task is to learn how scientists sample the microbiome in a far more rigorous manner, what the resulting data looks like, and how we engage in its analysis. While we are at it, you will learn how to read and understand a scientific research article on the microbiome. Chapters 4 and 5 provide a rigorous introduction to the experimental methods and modes of analysis employed in most microbiome studies.

CHECK YOUR UNDERSTANDING

1. What is the term used to describe an ecological unit of a host and the species living on or in the host, such as the human microbiome?
   a. Community
   b. Holobiont
   c. Parasite
   d. Environment

2. What term describes the average microbiome of individuals who don't have an obvious disease?
   a. Healthy microbiome
   b. Population-wide microbiome
   c. Holobiont microbiome
   d. Normal microbiome

3. What makes up most of the cellular diversity in our microbiome?
   a. Viruses
   b. Archaeans
   c. Bacteria
   d. Parasites

4. Humans have about 22,000 genes. How many genes does an average human's microbiota have?
   a. 5,000
   b. 250,000
   c. 9 million
   d. 7 billion

5. What genus is *Staphylococcus aureus* a member of?
   a. Aureus
   b. Staphylococcus
   c. Epidermis
   d. Bacteria

6. What term describes a subgroup within a bacteria species, such as *E. coli* O157:H7?
   a. Strain
   b. Subtype
   c. Class
   d. Order

7. Which phylum is not a core member of the normal human microbiome?
   a. Bacteroidetes
   b. Proteobacteria
   c. Firmicutes
   d. Chordata

8. Rather than understanding the microbiome using the abundance of distinct taxa, which can vary over time and between individuals, some scientists use microbial functions to characterize the microbiome.
   a. True
   b. False

9. A stable microbiome does not change over time.
   a. True
   b. False

10. How is the gut able to host a high concentration of microbiota?
    a. Large volume
    b. High oxygen content
    c. High surface area
    d. Low temperature

11. Gut microbiota help us maintain a stable state known as _____ by digesting indigestible components of food,

producing vitamins, preventing pathogen colonization, training the immune system, and more.
   a. resilience
   b. dysbiosis
   c. homeostasis
   d. consistency

12. What part of the GI tract has the lowest density of microbiota? The highest?
    a. Stomach; large intestine
    b. Small intestine; large intestine
    c. Trachea; small intestine
    d. Stomach; kidney

13. Which carbohydrate cannot be digested by humans but can be digested by microbiota?
    a. Glucose
    b. Sucrose
    c. Galactose
    d. Cellulose

14. What important product is produced when bacteria digest resistant starch?
    a. Short-chain fatty acids
    b. Oxygen
    c. Antigens
    d. Adenosine triphosphate

15. Actinobacteria and Firmicutes outcompeting pathogenic *Clostridioides difficile* for food and space and thus preventing it from infecting the host is an example of
    a. host immune response.
    b. resilience.
    c. colonization resistance.
    d. antimicrobial compounds.

16. Viruses that attack bacteria are known as
    a. bacteriophages.
    b. bacterial parasites.
    c. coronaviruses.
    d. cholera.

17. Antibiotics, lifestyle changes, and illness may cause a state of imbalance in the gut microbiota called
    a. instability.
    b. homeostasis.
    c. misregulation.
    d. dysbiosis.

18. What disorder is caused by the host immune system attacking gut commensals and altering the intestinal epithelium?
    a. Recurrent *C. difficile* infection
    b. Inflammatory bowel disease
    c. Food allergy
    d. Peptic ulcers

19. Which of the following is a mechanism that has been proposed to explain how gut microbiota contribute to colorectal cancer?
    a. By recruiting pathogenic bacteria
    b. By damaging the intestinal epithelium
    c. By producing toxins that damage host DNA
    d. By infecting and killing healthy human cells

20. One key difference between the gut and oral microbiota is that the gut contains primarily _____ bacteria and the oral microbiome contains primarily _____ bacteria.
    a. toxic; nontoxic
    b. anaerobic; aerobic
    c. commensal; pathogenic
    d. gram negative; gram positive

21. Oral microbiota form a _____-based community on teeth that is commonly known as dental plaque.
    a. holobiont
    b. biofilm
    c. infection
    d. oral commensal

22. Oral microbiota produce nitrite, which becomes nitric oxide in the bloodstream. What does nitric oxide do?
    a. It reduces blood pressure and prevents heart disease.
    b. It plays a role in blood clotting.
    c. It prevents infections in the bloodstream.
    d. It decreases blood flow.

23. What keystone species is involved in gum disease?
    a. *Escherichia coli*
    b. *Acinetobacter baumannii*
    c. *Porphyromonas gingivalis*
    d. *Streptococcus pyogenes*

24. How can oral microbiota contribute to the development of cavities?
    a. By using the nutrients needed for tooth health
    b. By decreasing oxygen levels in the mouth
    c. By blocking calcium
    d. By producing acid

25. Oral microbes can colonize other parts of the body, such as the gut, where they are associated with the creation of certain diseases.
    a. True
    b. False

26. The abundance and composition of microbiota on the skin is the same throughout all locations on the body.
    a. True
    b. False

27. Which is not true about microbiota and acne?
    a. Different strains of *Cutibacterium acnes* may play different roles in acne development.
    b. *C. acnes* is significantly more abundant in the pores of people with acne than without acne, showing that its presence alone contributes to acne.
    c. Adults with acne had *C. acnes* with more toxin-producing genes.
    d. Long-term antibiotic use for acne treatment may have side effects.

28. Why are experimental controls important?
    a. To prevent contamination
    b. To minimize the likelihood of falsely reporting a significant difference when none exists
    c. To ensure that any differences observed were caused by your experimental treatment, and not other factors
    d. To keep conditions the same for all subjects in the study

29. What is one key limitation to our skin microbiome sampling study?
    a. Not all skin microbes will be able to grow on the medium we used.
    b. Skin microbes will vary from location to location.
    c. Our skin doesn't host many microbes.
    d. Different people host different skin microbes.

30. If we follow the experimental protocol provided to sample our skin microbiome, we are unlikely to observe any microbes growing on our nutrient medium.
    a. True
    b. False

Answers: 1B, 2D, 3C, 4C, 5B, 6A, 7D, 8A, 9B, 10C, 11C, 12A, 13D, 14A, 15C, 16A, 17D, 18B, 19C, 20B, 21B, 22A, 23C, 24D, 25A, 26B, 27B, 28C, 29A, 30B

DIVING DEEPER

1. What's the difference between a normal microbiome and a healthy microbiome, and why are scientists currently only able to characterize the normal microbiome?

2. What are the four major phyla that make up the human microbiome?

3. What are some factors that may cause a person's microbiome to change over time?

4. Why might we describe the core microbiome based on the functions encoded, rather than using a list of species?

5. What are the three characteristics of a healthy microbiome?

6. What are the possible outcomes after a person's microbiome is perturbed?

7. What characteristic of the GI tract allows such a rich community of microbes to thrive?

8. Why does metabolism of carbohydrates like cellulose by the gut microbiome benefit humans?

9. Why are SCFAs important, and how are they produced?

10. How are T cells trained, and what role does each type of T cell play in preventing and destroying infections?

11. How is dysbiosis of the gut microbiome involved in the development of IBD and of colorectal cancer?

12. What are adherens and how do they affect the geographical diversity of the oral microbiome?

13. List and briefly describe the functions of the oral microbiome.

14. What role does your microbiome play in preventing or causing cavities?

15. Why do you think scientists previously considered the deeper layers of the skin to be sterile? (Think back to previous chapters.)

DISCUSSING AND REFLECTING

1. In this chapter we discuss how several microbiomes perform essential functions for a healthy human body. What do you think these communities look like in different species? What functions may other species require their microbiomes to perform that we don't? For example, consider the microbiome of an orange tree—would you expect to find community members similar to those in the human microbiome? Why or why not?

2. Many companies and research institutes are beginning to explore microbiome-based therapies targeting all different kinds of conditions and diseases. Which microbiome do you think has the best potential to lead to a major health breakthrough and why? Which potential therapeutic are you most excited about?

3. Reflection. Since you were introduced to your microbiome in chapter 2, and we're now diving into the specific habitats that exist in and on your body, take a few minutes to digest how this information makes you feel. Are you surprised to learn that different parts of your body are host to entirely different microbial communities? The next time you get sick, will you be thinking of your potential microbiome dysbiosis? How do you now feel about the use of antibiotics to treat patients?

RECOMMENDED READINGS

Popular Science Review
Gorman, C. (2012). Explore the Human Microbiome [Interactive]. *Scientific American*. https://www.scientificamerican.com/article/microbiome-graphic-explore-human-microbiome/

Popular Science Book
Yong, E. (2016). *I Contain Multitudes: The Microbes within Us and a Grander View of Life* (First US edition). Ecco, an imprint of Harper Collins.

Scientific Review
Gilbert, J. A., Blaser, M. J., Caporaso, J. G., Jansson, J. K., Lynch, S. V., & Knight, R. (2018). Current Understanding of the Human Microbiome. *Nature Medicine*, 24(4), Article 4. https://doi.org/10.1038/nm.4517

Generating Microbiome Data

4

CHAPTER CONTENTS

- **4.1** An Opportunity and Many Challenges
- **4.2** Designing Our Study
- **4.3** Entering the Experimental Phase
- **4.4** The "Omics"

Who is responsible for one of the most commonly occurring sexually transmitted diseases? That would be me, *Neisseria gonorrhoeae*. My moniker comes from the fellow who first described me, Albert Neisser, in 1879, coupled with the disease he was working on at the time, gonorrhea. This disease is a devil to deal with and it has been afflicting humans for over 3000 years! The *Ebers Papyrus* from circa 1550 BCE mentions it and it has become more common the longer I have been co-evolving with humans. I am now called a super bug, which sounds awesome to me, although I doubt that title was meant in any sort of honorific manner. Yes, I am now able to resist just about any antibiotic you toss my way. Here is one easy way to remember me:

There once was a bug called *Neisseria*,
Whose resistance grew quite superior,
Antibiotics would fail, It would laugh and prevail,
Now treating it's causing hysteria!

(Photo from iStock.com/Gilnature)

When a scientist is choosing a research question, they are trying to find the boundary between knowledge and ignorance and extend it by advancing our knowledge base. That is the way that most science is done: inch by inch we move the boundary and reveal the knowledge we find. The goal of this chapter is to provide a primer on how to design and carry out a microbiome study so that you can move that boundary as well. We will begin by learning how to develop a compelling research question and design a study that will permit us to meaningfully address our question. We will then explore some of the more common microbiome research techniques and learn about their strengths and limitations. Along the way we will learn how to read a research article and gain practical experience as we apply our newfound knowledge in the design of our very own microbiome study.

> "Developing a solid research question requires knowing where the boundary between current knowledge and ignorance lies."
>
> —R. Brian Haynes (Haynes, 2006)

4.1 AN OPPORTUNITY AND MANY CHALLENGES

Imagine that you have the unique opportunity to create your very own microbiome study. You are granted all the funds you require from the NIH (National Institutes of Health) (lucky you!), and you run a world-class microbiome laboratory filled with the most powerful sequencing machines and plenty of well-trained, brilliant colleagues. However, it is up to you to identify the question you wish to address, design the study, and, once the data are generated, analyze and interpret those data. How simple is that?

Training to Become a Microbiome Scientist

Well, it really isn't simple at all. Your average microbiome scientist has spent 4 years slogging through an undergraduate science major seeking a general understanding of biochemistry, statistics, and microbiology. They attend chemistry laboratories, learn physics, and tackle calculus. Their reward for all this hard work, focus, and drive is to immerse themselves in an even more rigorous graduate program for 5 to 6 years, a task that in many ways will make their undergraduate efforts seem like a gentle warm-up. As graduate students they will focus intensely on one very narrow slice of microbiome research and become an expert on that subject. At the end of their PhD program, they will write a dissertation explaining their research findings. If they succeed and earn their doctorate, they often engage in another 2 to 4 years of even more grueling research, this time as a research apprentice, otherwise known as a postdoctoral fellow. Sometimes, instead of coming from a PhD program, a postdoc has earned an MD at a medical school and has chosen to pursue research in addition to clinical work. In the postdoc position they have a bit more autonomy and are expected to possess the skills required to carry out research that will add knowledge to our ever-expanding understanding of the microbiome. They will also learn how to raise the research funds required to support their work, while pushing themselves to publish, publish, publish (or perish!), and will give seminars to spread their story. And, oh yes, don't forget that they now must look for their first independent, professional position, perhaps in academia, medicine, the microbiome industry, or a combination of all three!

Designing a Microbiome Study

Let's dispense with all that slogging and jump right to the fun bit, where we get to decide what questions to ask. Perhaps you come from a large family and were the only member whose birth required a C-section. You are also the only sibling with asthma, and you read somewhere that C-section delivery, which can impact the diversity of the gut microbiome, is **correlated**, or positively associated, with an increased incidence of asthma. You start to wonder what impact that single, isolated event, your delivery by C-section, might have had on your health. This is how most research starts: you make an observation and then design an experiment to attempt to explain what you observed.

Your question—Did C-section delivery cause me to have an altered microbiome that predisposed me to experience asthma—sounds simple, right? Well, as you consider how to test your idea, you quickly realize that it isn't very simple at all. First of all, your participants will be pregnant mothers and their newborns. You propose to focus on one treatment imposed on one or more groups that is expected to have an impact on some **outcome**. In this case your groups are the mothers, and the *treatment* is delivery mode, while the *outcome* is the level of gut microbiome diversity in the newborns and, thus, incidence of asthma. I imagine that you can already see some of the challenges you will face. In fact, the number of factors to consider are almost unimaginable. First, the occurrence of asthma, even if partially due to the impact of delivery mode, can also be due to a host of other factors, such as a viral infection, diet, and many other confounding variables. Was the health of your birth parent the same during your and your siblings' births? Were you and your siblings breastfed

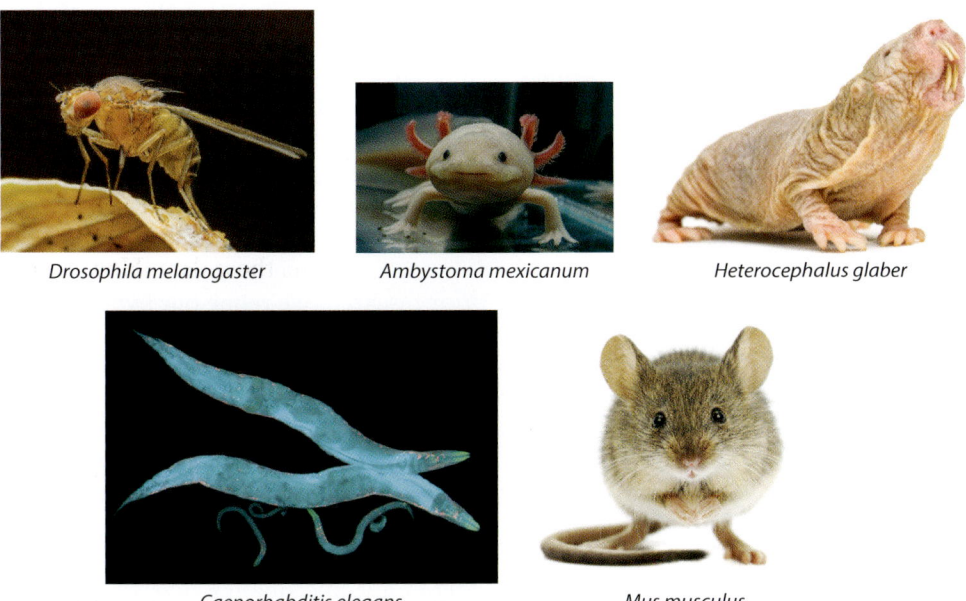

Figure 4.1 **Model Organisms** Using model organisms is a practice as old as biomedical research itself. Because we know so much about the biology of model organisms, we can more easily determine whether changes we observe are due to some treatment we imposed. This figure identifies several of the more common model organisms: fruit fly (*Drosophila melanogaster*), axolotl (*Ambystoma mexicanum*), naked mole rat (*Heterocephalus glaber*), nematode (*Caenorhabditis elegans*), and mouse (*Mus musculus*). (Left-Right: photos from nechaevkon/Shutterstock; istock/Iva Dimova; Eric Isselee/Shutterstock; Heiti Paves/Shutterstock; Szasz-Fabian Jozsef/Shutterstock)

and, if so, for the same amount of time? Was the birth parent given antibiotics prior to your or your siblings' births? We now know that the presence of pets and the degree of social interactions with the newborn can impact their health in general and their tendency towards asthma specifically. There are also logistical challenges to consider: At what age did you first develop asthma? Will we have to track participants for years to determine whether they experience asthma?

Using Model Organisms When Practical

Maybe now you can appreciate the power we gain by conducting microbiome studies with **model organisms** such as mice as our subjects, so we have more control over these factors and the hundred others we haven't yet mentioned or even identified. Figure 4.1 illustrates some of the more commonly employed model organisms, including mice, worms, and fruit flies. The advantages of focusing on a model organism are immense. We have detailed knowledge about their genetics, development, physiology, and the like. We can dictate the environmental conditions they experience, provide precisely prescribed diets, maintain large sample sizes, and as we learned in chapter 2, we can even eliminate their microbiomes. Recall the germ-free mice that we discussed in chapter 2 and how critical they were in establishing the importance of our microbiome. One of the drawbacks is the obvious one: they aren't humans, and what we learn from them will not inform us about the human condition specifically, which, in this case, is our interest.

A Testable Hypothesis Is Key

Let's go back to our initial idea and transform it into a **testable hypothesis**, which is an explanation for an observation that is backed up by limited evidence and which serves as a starting point for further investigation. One easy way to write out our

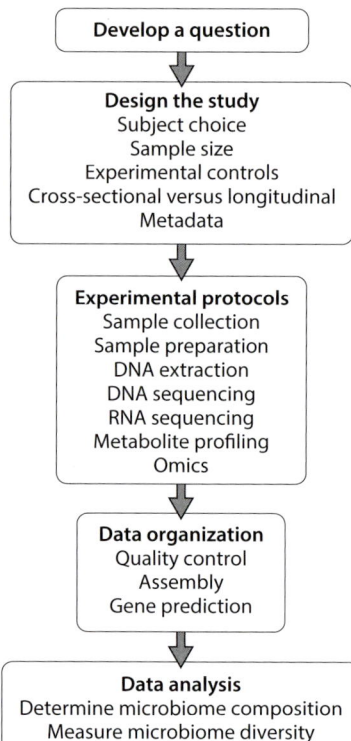

Figure 4.2 The Sequential Steps of Conducting a Microbiome Study

hypothesis is using a conventional **if/then** format. For example, *if* a baby is delivered via C-section, *then* their microbiome diversity will be altered, and they will have a higher incidence of early-onset asthma compared with a baby delivered vaginally. That's a start, but how would we create an experimental protocol to test such a hypothesis? What do we mean by *altered microbiome*, and how do we measure that? This is where our many years of training help us. Imagine we spent 5 years in graduate school focused on mice and the impact of C-section deliveries on the earliest wave of gut microbiome colonization. We learned from those studies that birth mode has a long-lasting impact on a newborn mouse's gut microbiome for well over 6 months and sometimes much longer. Based on this prior research, we refine our hypothesis to this: *if* a baby is delivered via C-section, *then* the diversity of their gut microbiome will be altered over the first 6 months of life. Super, we can test that well-defined hypothesis. However, where does the asthma bit come in?

It turns out we focused our postdoctoral research on the impact of gut microbiome diversity on the incidence of asthma in infants, and we learned that alterations to a newborn's gut microbiome can result in early-onset asthma. Further, it turns out that about 50% of babies who develop early-onset asthma during their first 6 months of life will go on to develop adult asthma later in life. Great, we have a proposed link between the fact that C-sections result in altered gut microbiome diversity and the fact that changes in gut microbiome diversity can result in early-onset asthma. We also have a reasonable time frame, 6 months, over which we can observe the newborn's microbiomes and record asthmatic events. We are now prepared to design a study that ties these facts together. Our final hypothesis is this: *if* a baby is delivered via C-section, *then* the diversity of their gut microbiome will be altered, and the incidence of early-onset asthma will be increased during the first 6 months of life.

In this chapter and the next two, we will address each of the challenges we so quickly swept away above and will learn how to think critically about the design of our experiment, how to generate the data we seek, and how to analyze those data to make sense of it all. **Figure 4.2** provides an overview of the sequential steps of conducting an average microbiome study. In this chapter we will address the first three steps: developing a question, designing the study, and identifying the appropriate experimental protocols. We will address the final two steps in chapter 5: data organization and data analysis. In chapter 6 we will apply all that we have learned by examining one published study in detail using an online data analysis tool, thus providing hands-on experience with manipulating microbiome data.

We have already made great headway in thinking through the question we wish to address, but there is still more for us to consider. Let's, yet again, restate our hypothesis, this time a bit more formally. We will transform our idea into a **null hypothesis**, which proposes that no significant difference exists in a set of observations (**Figure 4.3**). In other words, the null hypothesis assumes the difference between test variants is *null* or amounts to nothing at all. In our study, we will assume that our treatment (birth mode) has no impact, giving us our specific null hypothesis: *Birth mode has no impact on the colonization pattern and taxonomic diversity of a newborn's gut microbiome, and there is no detectable impact of birth mode on the incidence of early-onset asthma over the first 6 months of life.*

Why did we do that? It seems like we are complicating our story by proposing the opposite of what we expect. However, this transformation provides us with an easily tested expected outcome. We propose that delivery mode does not measurably impact the diversity of a newborn's microbiome or result in an increased incidence of asthma (even though we think it does). If correct, then we should see no difference in gut microbiome diversity or incidences of asthma (regardless of birth mode) over the first 6 months of life in our study participants. As we will see, we have statistical methods that permit us to easily test for any differences that might appear in our data. As the concept of a null hypothesis can be confusing, there are two additional examples of the transformation of a hypothesis into a null hypothesis provided in Figure 4.3. Take a moment to ensure that this key step in experimental design is clear to you.

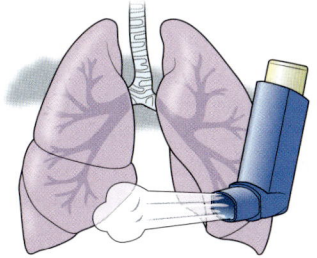 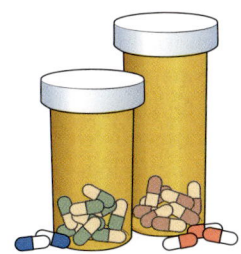 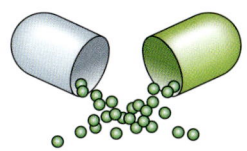

Hypothesis: If a baby is delivered via C-section, then the diversity of their gut microbiome will be altered, and the incidence of early onset asthma will be increased during the first 6 months of life.

Null hypothesis: Mode of birth has no impact on the colonization pattern and taxonomic diversity of a newborn's gut microbiome and there is no detectable impact of birth mode on the incidence of early onset asthma, over the first six months of life.

Independent variable: Birth mode

Dependent variables: Gut microbiome diversity and incidence of asthma

Hypothesis: If a person is given antibiotics, then the diversity of their gut microbiome will be lower after the course of antibiotics.

Null hypothesis: Gut microbiome diversity is the same before and after taking a course of antibiotics.

Independent variable: Antibiotics

Dependent variables: Gut microbiome diversity

Hypothesis: If a patient with IBD is given a daily probiotic, then their gut microbiome diversity will increase and their symptoms of IBD will become less severe.

Null hypothesis: A probiotic does not affect a person with IBD's gut microbiome diversity or symptoms.

Independent variable: Probiotic

Dependent variables: Gut microbiome diversity and IBD symptoms

Figure 4.3 The Null Hypothesis and Variables for Three Example Studies A null hypothesis is usually the opposite of the experimental hypothesis and assumes that the treatment of interest has no effect. For example, if the hypothesis is that C-section babies will have lower gut microbiome diversity, then the null hypothesis is that there will not be a difference in gut microbiome diversity between babies born vaginally and babies born by C-section. IBD = inflammatory bowel disease.

Experimental Variables

Let's spend some time dissecting what we mean when we talk about an experimental design. In research, **variables** are any characteristics that can take on different values, such as gut microbiome diversity or asthma incidence. Researchers often manipulate or measure independent and dependent variables in studies to test cause-and-effect relationships (see Figure 4.3). The **independent variable** is the cause. It's called "independent" because it's not influenced by any other variables in the study. In our case, the independent variable is birth mode. The **dependent variable** is the affected variable, which changes because of manipulation of the independent variable. It's called "dependent" because it is influenced by another variable. In our study, the dependent variables are gut microbiome diversity and incidence of asthma. These are the outcomes you're interested in measuring, and their values *depend* on your independent variable. If a baby is born by C-section (independent variable), we expect that treatment to impact the levels of microbiome diversity (dependent variable) and result in a higher incidence of asthma (dependent variable). Reexamine the three examples provided in Figure 4.3 to ensure that you understand the distinction between an independent and dependent variable.

Once we have arrived at what we believe is a testable null hypothesis and we have identified our variables, we are finally ready to design our study. The importance of a robust study design simply cannot be overstated. Many microbiome studies have produced inconclusive data, which a stronger experimental design could have prevented. The goal in designing a study is to eliminate (or limit) the occurrence of confounding factors that might impact our ability to draw meaningful conclusions.

Exploring the Primary Literature

Our next step is to engage in a deep exploration of the relevant literature to learn the precise state of the field and what approaches have worked (or not) for prior researchers. We could simply type our search terms into Google, such as "C-sections" and "microbiome diversity." We would get millions of hits, and we might read fascinating things, some of which might even be true. To avoid the challenge of sifting through fact and fiction, we turn to the **primary literature**, that is, publications of original research findings. These findings are found in **peer-reviewed** journals, which engage journal editors and other experts to critically assess the quality and scientific merit of the research presented. Articles that pass this form of professional evaluation are published in the primary literature. Peer-reviewed journals include the research of scholars who have collected their own data or of those who have performed novel analyses of existing data. The primary literature focused on the microbiome is rapidly growing. A paltry 200 or so publications in 2000 had exploded to well over 10,000 by 2020!

There are several tools we might use to find this literature, such as **Google Scholar** (https://scholar.google.com/) or **PubMed** (https://pubmed.ncbi.nlm.nih.gov/). The trick is to try a variety of different search terms to cast your search net as wide or as narrow as you wish. We may also wish to use reviews, which summarize recent research on a topic. Although reviews don't present original findings, they may help us gain a broad understanding of the topic we're interested in and help us to identify key articles in the primary literature to read.

Performing a Literature Search

Let's dive in and carry out our own Google Scholar search of the terms *birth mode* and *microbiome* and see what we find. Take a moment and carry out your own literature search using these precise search terms. Note how many hits you get and read through the titles of some of the articles to get a feel for the relevant scientific literature. Imagine that when we do this search, we find one article that stands out as the very first publication on the impact of birth mode on the microbial colonization of newborns (Dominguez-Bello et al., 2010). This study, which we will work through in detail in chapter 6, informs us that newborns do, indeed, have very different gut microbiomes when delivered via C-section. From our many other search hits we find articles that discuss the incidence of asthma in newborns, the frequency of C-sections, and their impact on a newborn's health. We also learn about the pitfalls of employing human participants in microbiome studies, which predicts a high dropout rate for the participants. Good to know! The primary literature is rich with studies that may have some bearing on our own research—so it is well worth our time to read as many articles as time permits before committing ourselves to a particular experimental design.

4.2 DESIGNING OUR STUDY

Now that we've taken the time to learn from the existing literature, we are ready to think about how to properly design our study. There are the obvious technical challenges, such as how to obtain and store samples, how to properly extract DNA from those samples, etc. However, there are even more unpredictable challenges, especially since we are focusing on humans. Will we get enough pregnant women to enroll in the study? Will the parents keep the records they promise to keep? Will the babies (or pregnant moms) require antibiotics that might impact our results? The number of relevant variables to track is considerable and will require us to keep reams of information, so-called metadata, on each person in the study and the techniques, both experimental and analytical, that we employ. **Metadata** is a set of data that describes and gives information about other data. Sounds confusing, right? In our study, the metadata is information about the study methods, the participants, their health status, anything that we feel might be relevant to how we interpret the results we obtain. Members of the professional microbiome community have developed a set of principles for generating metadata, the so-called FAIR principles; *FAIR* stands for *findable* (structured in a hierarchical way), *accessible* (compatible with existing microbiome data repositories), *interoperable* (linked to other metadata), and *reusable* (follows community standards).

Choosing Our Subjects

We made the decision early on to sample humans—pregnant women, and their newborns, in particular. However, as you learned above, by choosing to work on humans, we have made our study considerably more challenging, and the results may be far more difficult to interpret because of the confounding variables that come into play. Imagine that, instead, we had chosen to work with mice. We could have ensured that the only difference between our treatment and control mice was birth mode. We could have fed all the mice identical diets, provided them with the same number of social interactions, and controlled all those other variables that we now know come into play in developing a healthy, diverse microbiome. There is even a breed of mice that exhibit asthma, as we expect to find in some of our newborns (Aun et al., 2017). An animal study is far easier to design and control, but, of course, we lose the direct link to human health. There are pros and cons to any subject you choose. The bottom line—think this part through very carefully.

Statistical Power Is Key

Choosing the sample size is, perhaps, one of the most important decisions you will make in designing a microbiome study, or any kind of scientific study. The choice requires a consideration of **statistical power**. Before your eyes glaze over, understand that you are not required to become proficient in statistics to understand the information presented in this textbook. However, microbiome studies are based upon comparisons of enormous datasets. Statistics isn't just an occasionally useful tool in these studies; it's a foundational requirement for all microbiome research. The dictionary defines *statistics* as "the practice of collecting and analyzing numerical data in large quantities, especially for the purpose of inferring proportions in a whole from those in a representative sample" (Merriam-Webster, 2022). Our study fits that bill precisely. We wish to generate and then compare large microbiome datasets for babies born vaginally or by C-section and then use those data to infer the impact of birth mode on the data we collect. If the experiment is designed well, the conclusions we draw from our necessarily small sample size will be applicable to a much wider group of newborns.

The good news is that there are powerful and elegant computer programs that will take care of the required statistical analysis for us. However, even with those programs in our tool kit, it is essential that when we are designing our study, we either engage a statistician or become fluent in the statistical factors that will be important to the success of our study. In this textbook, we will provide a simple overview of the statistics required—so no worries for now.

Let's focus on one statistics-related term, **power**, which is the probability that a test of significance will pick up on an effect that is present. By **significant** we mean that there is a solid chance that the relationship we observe between a treatment and outcome is not just a chance occurrence. In our study, we want a sample size with enough power to determine whether or not our data support our null hypothesis, which is that microbiome diversity measures and incidence of asthma will *not* differ between newborns with different birth modes (even though we think they will). We chose this null hypothesis for a reason. If we employ a large enough sample size, we will have the power to test such a hypothesis. The larger the sample size, the more precise the estimates from our study will be and the greater power we will gain to detect small effects.

Testing for Statistical Significance

Statistical significance is the probability of finding a given deviation from the null hypothesis in a sample. In our case, it is the probability that babies born with C-sections will have higher or lower levels of gut microbiome diversity than babies born vaginally. Our null hypothesis is that there will be no difference. However, if we do find a difference, we want to know whether that difference is significant. In research papers, statistical significance is often referred to as the **p-value** (or simply p), short

Figure 4.4 Coin Toss Example of Probability The graph in (A) shows the distribution of the expected probability of heads and tails in a coin toss conducted 10 times. In (B) you see the far narrower distribution obtained from 1,000 coin tosses.

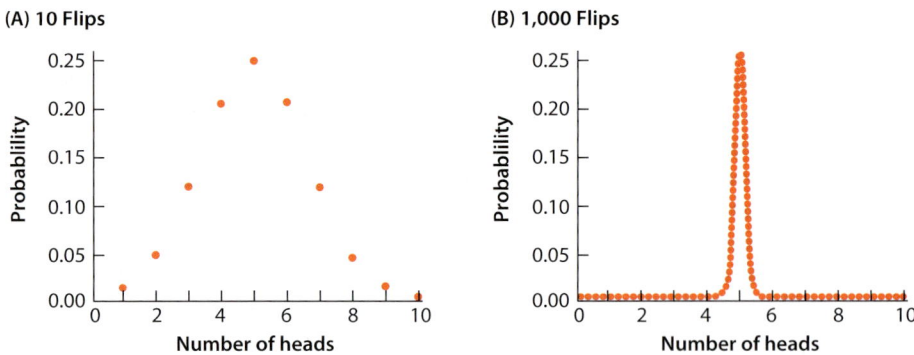

for "probability value." A small p-value basically means that your data are unlikely under some null hypothesis. A somewhat arbitrary convention is to reject the null hypothesis if $p < 0.05$.

Let's illustrate this concept with an example of tossing a coin. Let's say that I believe I have a balanced coin and so my null hypothesis is that, since it is balanced, there is a 50:50 chance of the coin landing heads up. I flip my coin 10 times and get 3 heads and 7 tails. The probability for this outcome—assuming my coin is really balanced—is shown in the distribution in **Figure 4.4A**. This figure tells us that it is possible to get 3 heads and 7 tails with a balanced coin about 12% of the time. If we were to look up the p-value for this outcome (which is beyond the scope of this introduction to statistics), we would find that it is reasonably likely that we obtain this outcome, with a p value of 0.34. We only reject our null hypothesis if the p value were ≤ 0.05, so in this case we would *not* reject our null hypothesis that the coin is fair. Now, imagine that I toss my coin 1,000 times (distribution in **Figure 4.4B**). Then the probability of getting 300 heads and 700 tails becomes far lower, in this case it is extremely low and has a p-value less than less than 1×10^{-37}. In this case, we would certainly reject our null hypothesis and conclude the coin was unbalanced.

Returning to our study, if we sample the gut microbiomes of only 10 babies, versus 1,000, we face the same issue of potentially accepting the null hypothesis and being wrong—just like with the 10-coin-toss experiment above. In **Figure 4.5** we have plotted microbiome diversity and used error bars to reveal how variable the measurements we took are. If every sample had the exact same level of diversity, there would be no error bars. If samples varied from high to low microbiome diversity levels, within either treatment group (vaginal birth or C-section), then the error bars would be very long, reflecting how different any two samples can be. The more variable our measurements, the larger the sample size we need to detect significant differences between our treatment groups. Data are significant when a meaningful difference can be detected between the treatment groups that is unlikely to be the result of random chance. The datasets for both case A and case B show what appears to be a

Figure 4.5 The Importance of the Mean and Standard Deviation As the difference between group means increases or the difference between standard deviations decreases (case A vs. case B), the more likely it is that a test can detect an effect (i.e., a statistically significant result).

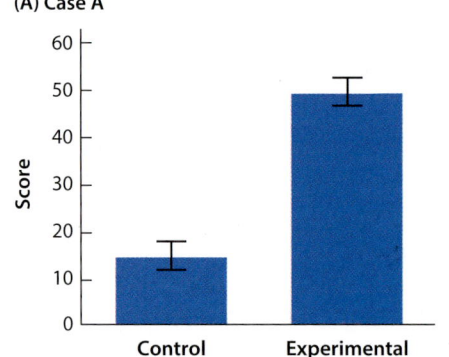
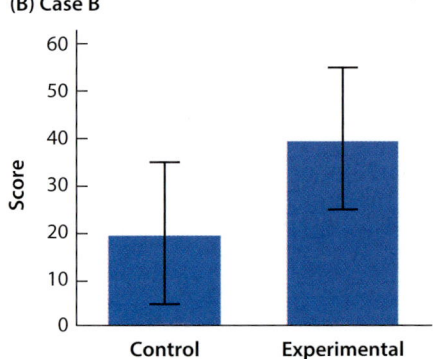

difference in the measurements taken from the control versus experimental samples. However, if you examine the error bars (the thin lines superimposed over the score bars), you will see that the error bars in Figure 4.5A are small and do not overlap each other. In contrast, the error bars in Figure 4.5B are large and do overlap each other. The length of an error bar helps reveal the uncertainty of a data point: a short error bar shows that values are concentrated, signaling that the average value is more likely, while a long error bar indicates that the values are more spread out and less reliable. Using the error bars as our reference, we can conclude that the control and experimental samples are significantly different in case A but not in case B. It is often the case that having a larger sample size will result in smaller error bars and an improved ability to detect significance in your data.

So why not just sample a million newborns? There are several reasons. Having too many samples can be costly, and in some cases it requires ethical considerations, and so there is often a trade-off involving power, cost, and ethics. For example, if we were to use mice in our study, we would be able to include a far larger sample size than is possible with humans. This increase in sample size would provide increased power in terms of our ability to reject or accept our null hypothesis. However, even mice are costly to keep, so the larger the sample of mice, the more costly the experiment. Further, we would have to euthanize the animals at the end of the study, and it's important for this to be taken into consideration. Scientists must strike a balance between enough replicates for powerful data and no superfluous replicates which would cause needless cruelty. However, underpowered studies are one of the main causes of **irreproducibility** in microbiome studies because they lead to high false-positive rates, that is, large numbers of studies in which researchers incorrectly accept or reject the null hypothesis. *Irreproducible* in microbiome studies refers to experiments that, when repeated, result in different outcomes.

Sample size calculation is not an exact science, since it relies on numerous assumptions about the underlying **distribution** of our data. What is a distribution? Imagine that we are given the gut microbiome diversity measures for all babies born in 2023. That is a lot of newborns, roughly 140 million. We plot these diversity measures in a **histogram** as seen in **Figure 4.6A**. Note that the outline of our graph resembles the shape of a bell, in which most values cluster around a central region, with values tapering off as they go farther away from the center. In other words, most babies have a moderate level of diversity in their gut microbiomes, while some have very high or very low levels. This shape is so often found when we take measurements in nature that it is called a **normal distribution**.

Now compare the normal, also known as Gaussian, distribution with the one shown in **Figure 4.6B**. This second distribution is called **bimodal** because it has two

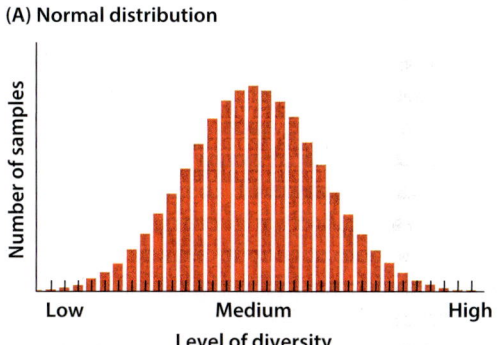
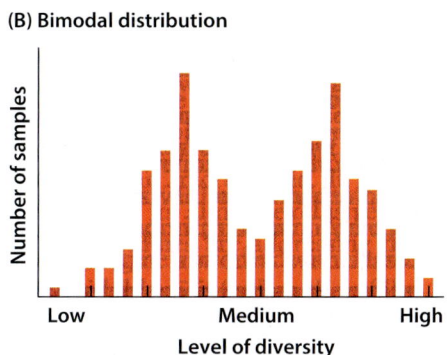

Figure 4.6 Normal and Bimodal Distributions If we plot the gut microbiome diversity of all babies born in a year, we may observe a normal distribution (A) or some other distribution, such as a bimodal distribution (B). In the case of a bimodal distribution, we would want to determine why the data are split in this way. In the normal distribution the variability in our measurements is evenly distributed around the mean, or average value.

peaks rather than one. If our data had a distribution that looked like this, we would want to know why—what is causing this split such that about half of the babies have very high or very low levels of diversity. Now imagine that we learn that the microbiome diversity measures of all babies born via C-section fall into the far-left distribution, while those in the far-right distribution are from babies delivered vaginally. In this contrived scenario we don't need statistics to see that the clusters are significantly different, but this isn't often the case with complex, multidimensional datasets.

What we need to know before we start our study is how many samples are required to determine whether our plotted data resemble a normal distribution or not. As you think about power, note that in Figure 4.6 we plotted about 140 million actual diversity measurements—the entire universe of values for one year of newborns. However, when we carry out our study, we will sample a minuscule portion of these newborns. What if our tiny sample shows something other than a normal distribution, not because of something biologically interesting, but simply because our samples just happened to have unusually high or low levels of microbiome diversity? In other words, just by bad luck we happened to pick an outlier set of newborns to include in our study. That is what our power calculation is all about. It helps you determine the smallest sample size needed to give you an acceptable chance of meeting your study aims.

In our study, we decide that we want a far more robust sample size. We ask our colleagues about the **retention rates** in these types of studies (on average how many parents/babies will remain for the study duration), about the challenges of getting the parents to perform the required fecal sampling from the newborns, and about numerous other experimental details. We also inquire about the current costs of the technical aspects of the experiment, such as the cost of sequencing, usually provided as cost per sequenced base pair. We realize that the signal we hope to detect, the impact of delivery mode on gut microbiome diversity and health, may be weak, relative to all the factors that impact the colonization of a microbiome during the first few months of life.

The Power of Our Sample Size

Finally, we consider a **power analysis**, which employs mathematical equations and simulations of existing data obtained from prior studies to provide an estimate of the number of samples required to ensure that we would reject our null hypothesis if it were not correct. Power analysis requires a level of mathematics beyond the purview of this book. Just be aware that as our knowledge of microbiome diversity expands, we gain the ability to predict more accurately how many samples are required to be confident that we will have the power to either reject or accept our null hypothesis.

Based upon all these factors, we propose to track 500 newborns, of whom 50% will be delivered via C-section. We assume a low retention rate—50%. This percentage includes pregnant women, or their newborns, who may have to be excluded from the study, which we will discuss in section 4.3 Entering the Experimental Phase. Thus, even with a highly conservative retention rate of 50%, we should still have a robust number of samples to work with. Many studies requiring human participants offer a financial incentive to both recruit and retain participants. Most often this is structured with initial and recurring payments to encourage participants to continue with the study. Like our other research expenses, this money will come from our NIH funding.

Figure 4.7 shows the sample size we propose for our study, which includes 500 pregnant mothers and their ~500 newborns, assuming no twins are delivered. We also determine that we need a fecal and vaginal sample from each pregnant woman just prior to delivery, which amounts to 1,000 samples. These samples will provide a baseline for use in understanding whether the mother has a particularly high or low diversity of her microbiome, some of which gets shared with her newborn. If we happen to sample mothers who have unusual vaginal or gut microbiome diversity measures, we can take that into account in our downstream, or subsequent, data analysis. We then determine that, although we would prefer to have weekly samplings

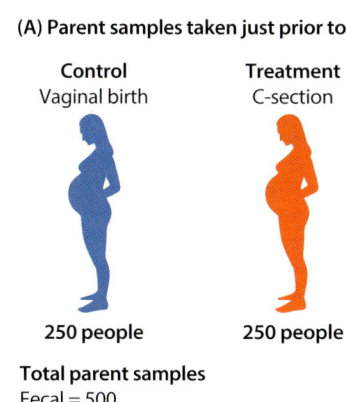
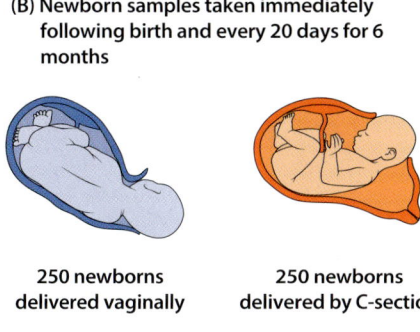

Figure 4.7 Our Study Design (A) 500 participants are enrolled in the study, half who give birth vaginally and half who give birth via C-section. (B) Fecal and vaginal samples are collected from each birth parent just prior to giving birth, and fecal samples are collected from the newborns each month for the first 6 months of their lives.

from our newborns over the proposed 6-month period, we are willing to settle for a monthly fecal sample from each baby, for a total of 3,000 samples. We realize that it simply isn't feasible to ask parents of newborns to provide more samples—their lives are already hectic enough without having to send us fecal samples each week. So, for each pregnant mother, we have two samples from just prior to delivery and 6 samples from her baby.

From this discussion you can see that it is not trivial to determine sample size, and yet it may become one of the most important factors in the end. An excessively large sample size will make your study unnecessarily time-consuming and expensive. With too small a sample, you might just see the hint of a signal but be unable to draw a solid conclusion after several years of very hard work.

Taking Control of Our Experiment

Experimental controls are employed to minimize the effects of extraneous factors (outside of the researcher's control) on the outcome. Choosing the appropriate experimental controls is critical in all scientific investigations, and particularly so in microbiome studies. A controlled experiment generally involves two or more treatments: one producing observations without interference (in our case, vaginal delivery), while another receives the interference (in our case, C-section delivery). Controls help to identify whether a signal is real or just a spurious result. In our case, the biological signal we seek might be a significant shift in gut microbiota diversity measures or an elevated number of asthma attacks in newborns delivered by C-section.

It is challenging, if not impossible, to obtain fully controlled samples in most human microbiome studies, where microbial composition is affected by so many extraneous factors, such as gender, age, genotype, ethnicity, and numerous other lifestyle factors. In animal studies, many factors that are difficult to manipulate in human studies, like diet, can be controlled. However, there are still factors that could impact the results, such as the animal housing facilities, handling protocols, and animal breeds. We have an even stranger confounding variable when we work with mice. They engage in **coprophagy**, which refers to the consumption of feces. Given that the feces are chock full of microbes, that habit could impact our results, and we must ensure the mice don't have access to each other's feces. Who would have imagined that we would need to be concerned with fecal consumption in an experimental design? More subtle issues constantly arise; for example, genetically identical mice raised in different locations within a lab or in separate labs can have different levels of microbiome diversity. It is important to try to control and document as many factors as possible so we can isolate the effect of our independent variable to the greatest extent possible. These records, which become part of our metadata, will be used in downstream statistical analyses to help account for confounding factors.

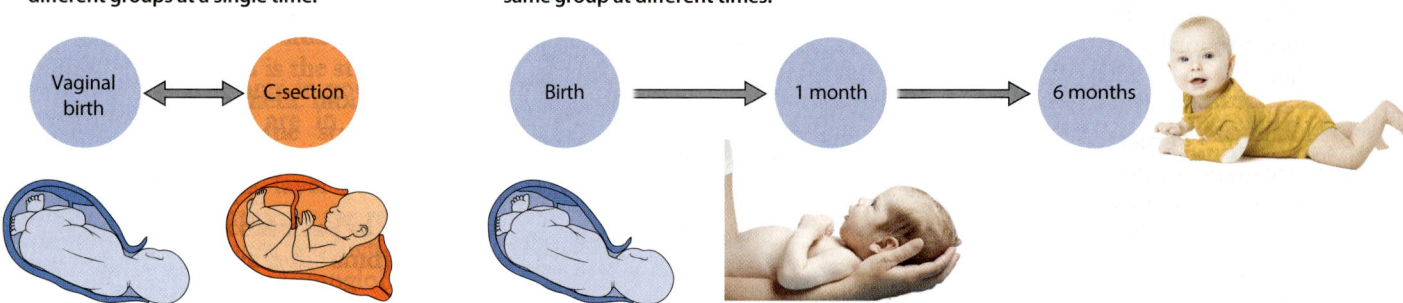

Figure 4.8 Longitudinal and Cross-Sectional Study Designs (A) A cross-sectional component comparing two groups: vaginal versus C-section babies. (B) A longitudinal component that follows newborns for 6 months. (B photos from Ventura/Shutterstock [1-month-old baby]; iStock/inarik [6-month-old baby])

Let's think a bit more about what might go wrong with our study. You might be asking yourself, Why did we propose to sample the pregnant woman's gut and vaginal microbiomes just prior to birth? Imagine that those who require C-sections just happen to have very low (or high) gut microbiome diversity. That could easily translate into newborns with much lower (or higher) gut microbiome diversity that is not due to birth mode at all! Finally, there is a different set of controls required to ensure that our experimental protocols are producing what we want. These controls are specific to each technology and will be discussed as we learn more about microbiome-related technologies below.

Cross-Sectional versus Longitudinal Study Design

A **cross-sectional study** is designed to compare two groups, whereas a **longitudinal study** is designed to track changes in an individual or group over time (**Figure 4.8**). Our study has a cross-sectional and longitudinal component to it. We are comparing measures taken for babies in the vaginal and C-section groups, and we are tracking measures in each newborn over a 6-month period. We will, ultimately, pool the data from the babies under the two treatments at each time point and then track how their combined microbiome diversity measures change over time. This pooling of data helps ensure that a single infant does not skew our results. Yes, sample size comes back into play! We want to see how long the impact of birth mode lasts, if we find one, and whether this impact is correlated with increased incidence of asthma. Thus, we have included a longitudinal aspect to our study, by sampling each newborn monthly over 6 months.

Experimental Data versus Metadata

Let's distinguish between experimental data and metadata. In our study, the experimental data we wish to produce will be DNA sequences. To be precise, we will obtain sequences of the 16S rRNA genes in our samples to identify which microbial taxa are present and at what densities. Just as we learned about Carl Woese employing 16S rRNA sequences to distinguish archaeal from bacterial species (chapter 1), we can use the 16S rRNA sequences we generate from the DNA in our samples to identify which taxa are present. In addition, we will be collecting whole-genome information for those same taxa so that we can explore what genes are encoded in each species, which tells us quite a bit about what the taxa do: what they use for an energy source, what metabolites they produce, and much, much more.

In contrast to our experimental data (DNA sequences), metadata is information about our experimental data. It is the who, what, when, where, and why of these data. Metadata includes information about samples: when, where, and how they were collected and information about experimental methods used. However, the personal information attached to these samples, such as the person's name, address, or any other piece of data that could link the samples they provided to who they are, is excluded.

Generation of metadata is critical to our downstream analysis. Generally, participants are given a questionnaire to collect some of the relevant information, which is tabulated in a spreadsheet, such as Excel. There are programs, such as **QIIMP** (pronounced like *chimp*), which stands for Quick and Intuitive Interactive Metadata Portal (https://metadata-wizard-tutorial.readthedocs.io/en/latest/), that provide a standardized approach to creating metadata files. We will return to a discussion of metadata in chapter 5, when we learn how to analyze our experimental data. For now, you should understand that metadata is critical to the success of your study, and you want to think hard about what information you should collect before you begin your study.

4.3 ENTERING THE EXPERIMENTAL PHASE

We are finally ready to enter the experimental phase of our study. This next section will walk us through the more common protocols employed in microbiome studies, which are also listed in Figure 4.2. To begin we require our microbiome samples. In our study, we already know what sample types we'll use: vaginal and stool samples from pregnant mothers and stool samples from their newborns. The vaginal and fecal samples from the mother are compared with the fecal samples of newborns, under the assumption that babies born vaginally will have microbes similar to those of their moms, while babies born via C-section will not.

Obtaining Our Samples

The good news is that both types of sampling are quite simple to do. However, if we choose to sample stool, we must accept the fact that stool does not capture all the microbes in the gut—in particular, those microbes that hold fast to the mucosal lining of the large intestine and many microbes in the small intestine. In addition, stool sits in the rectum for varying periods of time, where dehydration takes place, and there is the selection for microbes that are not common in other parts of the lumen. As a result, the microbes found in stool will not perfectly represent those found in the gut. Because we are handling all our participants similarly, we will not be concerned with these issues, and we will use fecal microbial diversity as a proxy for measuring actual gut microbiome diversity.

The standard protocol for collecting fecal samples is to collect the whole stool, immediately **homogenize** it (or thoroughly mix with a blender, for example), then flash freeze all but a small portion of the homogenate. This small portion, known as an **aliquot**, is preserved in glycerol for future **culturing** (or growth) if required (Allaband et al., 2019). Right! Picture that equipment next to the diaper changing table in our participants' homes. We can follow standard procedures when sampling the pregnant parents, as this will occur immediately before birth and generally in a hospital setting. However, since the parents will be obtaining the newborns' fecal samples, we will follow a somewhat simpler protocol for those. Before the study begins, we will provide the parents with six small ice chests, prepackaged in mailing boxes with shipping labels affixed, complete with six sets of sterile gloves, tubes, sampling spatulas, and Parafilm strips (the lab version of plastic wrap).

An email reminder will be sent the night before each sampling date, and the parents will obtain the first fresh sample the baby produces that day. It is critical that they not use the feces deposited in the diaper during the night, as microbes will have had time to either overgrow or die. We want a fresh sample, within 1 hour of production. The parents will take off the diaper, put on their gloves, and use the spatula we provide to fill the sample tube. They will then screw on the cap, wrap Parafilm around the cap to ensure it is secure, deposit the sample in the ice chest, add some ice, and call FedEx for a pickup. We will maintain close contact with each of the 500 parents to ensure that the sampling is done in a timely manner and shipped immediately. We will also hold an online training session for parents to learn how to gather the samples. Imagine the staff we need to oversee this one key step in sampling.

At room temperatures, the microbial composition of our samples is altered as some microbes grow while others die. Immediate freezing will minimize these effects by slowing microbial growth. However, that may be difficult in some scenarios, such as our study, in which we have parents in charge of the sample procurement. Even though simply storing samples briefly on ice can impact the composition of the microbes, these effects are generally small compared with the variation found in individual microbiomes. The most important consideration is that samples are handled as consistently as possible in each study.

We will also require vaginal samples from the pregnant individuals just prior to birth. That sampling is invasive, which means that it requires inserting equipment into the body. It is done by a nurse who inserts a sterile swab 5 cm (2 in.) into the vagina and gently rotates it for 30 seconds before removing it and placing it in a sterile receptacle. The swab then holds a sample of the pregnant woman's vaginal microbiome.

So far, we have focused on two of the more easily obtained microbiome samples. What happens when your study is focused on even more internal sites, such as the stomach, or the lungs? Clearly these samplings require highly invasive procedures. Regardless of your body site target, there will be published protocols that you can use to obtain the most representative, uncontaminated, and robust sample possible. In the field of microbiome research, it is well established that how one obtains samples can play a significant role in what one finds in the study.

There are some general principles to consider in your sampling protocol. First, you want to avoid freeze-thaw cycles, which can impact the quality of the DNA you will obtain. In fact, you want to avoid all temperature fluctuations, if possible, as these can create stress on the microbes, which can kill some taxa. When practical, you want to minimize the transportation time so that undesired microbial overgrowth or decline can be minimized. Finally, the most common problem one encounters in sampling is variability in the quantity of microbial DNA present at different body sites. For example, the microbial biomass of skin is far lower than that found in feces. Having a large enough sample for sequencing remains a critical factor to consider when you determine the type and size of sample to obtain.

We finally have our samples stored in the freezer and can now pause and take a moment to breathe, or perhaps a short vacation? Numerous prior studies have assured us that freezing a sample can maintain a stable representation of the original microbial community for up to 2 years (Shaw et al., 2016). We just need to ensure that the power doesn't go off or to have a backup generator in place, just in case. The next step is to obtain our desired materials from the samples, which is usually DNA but might also be RNA, protein, or metabolites. In our case, we are interested in obtaining microbiome DNA, which will be used to determine the 16S rRNA gene sequences present so that we can identify the taxa present. This same DNA will also be used to determine whole genome sequences, which provide valuable information about the microbial functions encoded. As the cost of sequencing has dropped, more researchers opt for obtaining whole-genome information. In our study, we will do both. In either case, we require high-quality, pure DNA from our samples.

DNA Chemistry

Before we start to extract the DNA from our samples, let's pause for a moment and learn about the chemistry of DNA, which will help us understand the extraction, amplification, and sequencing steps that follow. **Figure 4.9** shows the structure of DNA, which is made of two linked strands of **nucleotides**, which are compounds

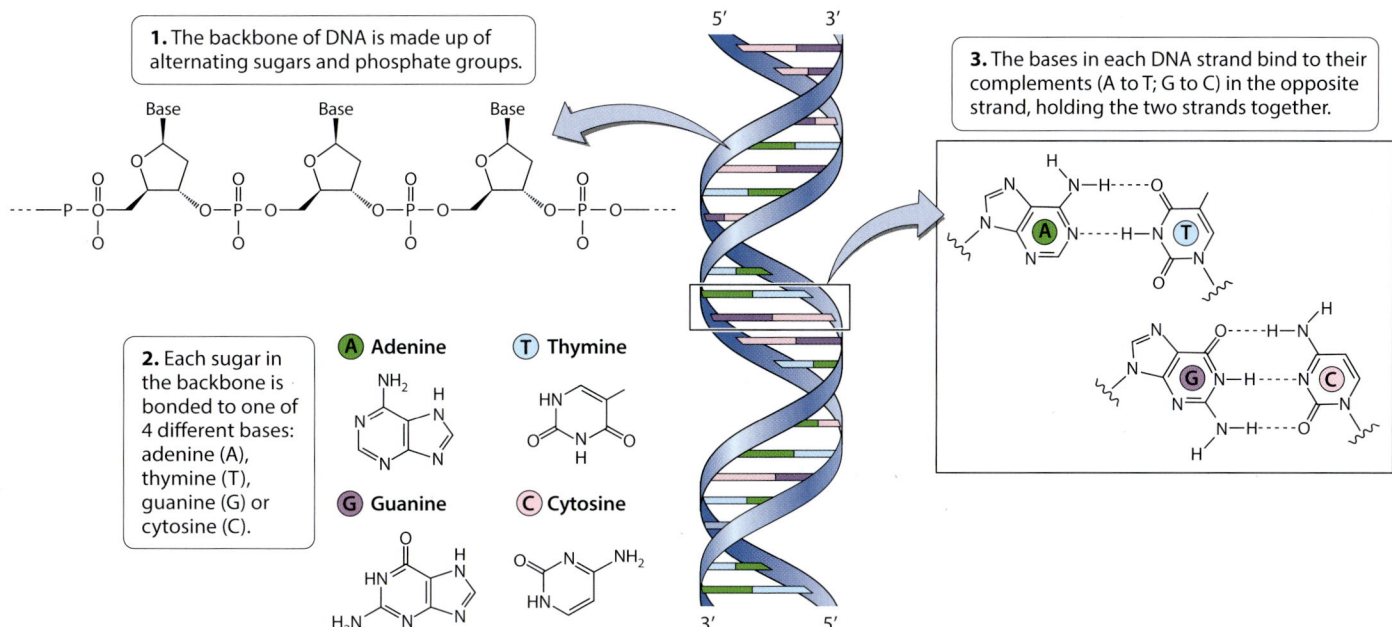

Figure 4.9 The Chemical Structure of DNA Understanding the basic chemistry that makes up our DNA is essential for designing sequencing experiments. If we didn't know which nucleotides would form bonds together, our current sequencing technology would be impossible!

each consisting of a **nucleoside** (a sugar plus a base) linked to a **phosphate group**. These are the building blocks of DNA that wind around each other to form that famous **double helix**. Each strand of DNA has a backbone of alternating sugar (**deoxyribose**) and phosphate groups. Each sugar is attached to one of four **bases**: adenine (A), **cytosine** (C), **guanine** (G), or **thymine** (T). Hydrogen bonds form between the bases, and adenine (A) always forms hydrogen bonds with thymine (T), while guanine (G) always bonds with cytosine (C). The human genome has 3×10^9 base pairs, or 3,000 megabases, while your average microbes have significantly smaller genomes, from 0.6 to 8 megabases. When you look at the sugar-phosphate backbone in Figure 4.9, you will see that one end of the strand is labeled **5'** (**five prime**) and the other is labeled **3'** (**three prime**). DNA is "read" in a specific direction (5' to 3'), just as words in the English language are read from left to right. The 5' and 3' designations refer to the number of the carbon atom in the deoxyribose sugar molecule to which the phosphate group bonds.

When a cell divides, its DNA must be copied so that the resulting daughter cells will each acquire a complete set of genetic information. **Figure 4.10** illustrates the process of DNA replication, which results in two copies of the DNA. Each copy is made up of half old and half new DNA; that is, it is a **semiconservative replication** product.

Extracting Metagenomic DNA

There are three steps involved in extracting DNA from our microbiome samples: sample homogenization, cell lysis, and DNA purification (separation from non-DNA substances), as shown in **Figure 4.11**. Step 1 is **homogenization** (see Figure 4.11A). We homogenized our samples when they first arrived, but as the sample may have separated during the freezing and thawing, we will repeat that step. We need to ensure that if we take a subsample from the fecal matter that it is an accurate representation of the starting material. Once the material is thawed, we can readily homogenize the sample using a **vortex**, which is a desktop machine that shakes the material in our tube using rapid back-and-forth motions.

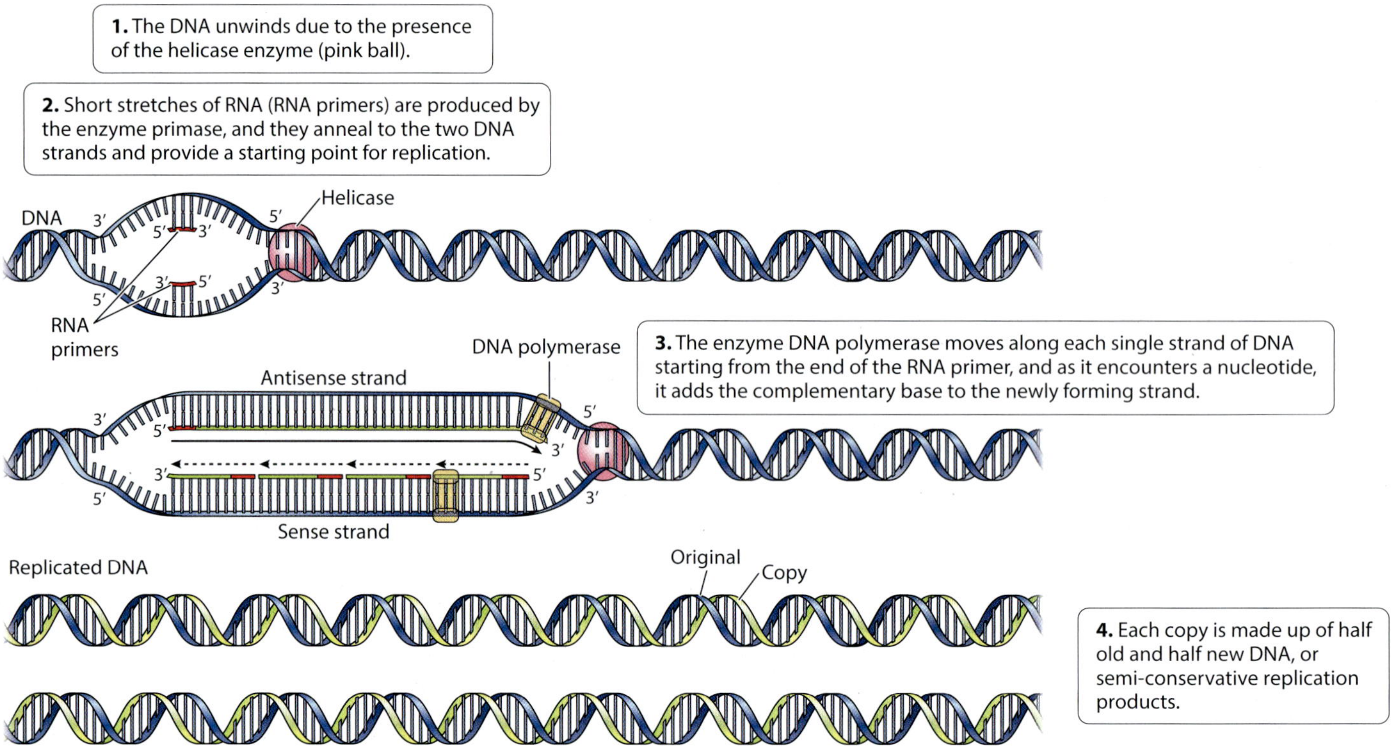

Figure 4.10 DNA Replication The specific bonding pattern of DNA nucleotides is what allows this important molecule to store information. Imagine if nucleotides bonded randomly; it would be impossible for DNA replication to be consistent! Fortunately, consistent base pairing allows for identical DNA molecules to be produced in the cell according to the semiconservative model of DNA replication.

We then liberate the microbial DNA by breaking open the cells, called **lysis** (see Figure 4.11B). There are chemical and mechanical lysis protocols we can employ, or a mixture of both, and the method we choose is based upon factors such as whether we expect to have primarily bacteria and archaeans or microbial eukaryotes in the sample. Figure 4.11B provides an illustration of mechanical (left) and chemical (right) lysis. In the case of mechanical lysis, we are employing a hard substrate, such as glass beads, to break down the cell walls and membranes. The glass beads are added to the

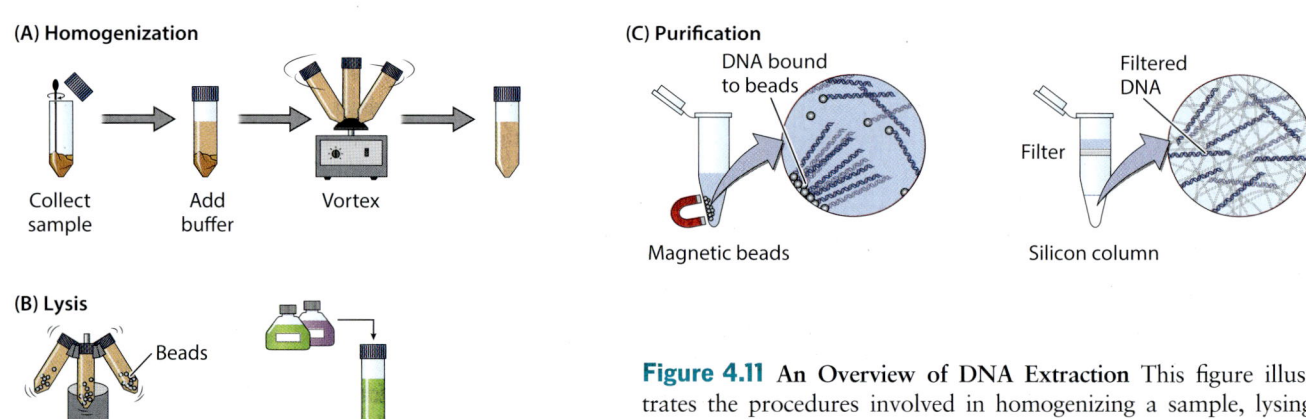

Figure 4.11 An Overview of DNA Extraction This figure illustrates the procedures involved in homogenizing a sample, lysing open the cells, and purifying the DNA.

feces, and the mixture is vortexed, after which the glass pellets will be removed during the purification step. If we employ magnetic beads, we can very easily separate out the beads with a magnet. How cool is that! In the case of chemical lysis, various chemicals will be used that will result in pH changes that break apart the cell walls and membranes.

We now have a sample in which the microbial cells are lysed open, and the genetic material has been released. Our next task is to extract the DNA, or RNA or proteins or metabolites, from what is now cellular debris (see Figure 4.11C). The more commonly employed DNA extraction protocols use magnetic beads or silica columns. The negatively charged DNA molecules will bind to the beads, while contaminants stay in solution. Once bound, the beads are washed and then transferred to a solution that has a pH that causes the DNA to be released from the beads. DNA separation by adsorption to silica is based on the same principle; DNA binds to the silica surface under certain pH conditions. Once bound, the DNA is washed and then released from the silica by a simple change of the pH of the solution. The result for all these procedures is concentrated, pure metagenomic DNA that is ready for **amplification** (generation of multiple copies) and **sequencing** (determining the precise order of nucleotides).

Quite often the DNA extraction procedure is accomplished with one of the numerous commercially available DNA isolation kits. For example, Dominguez-Bello and colleagues (2010), whose work motivated our own study used a bead beating tube that homogenizes the sample followed by a combined mechanical and chemical cell lysis procedure. The choice of kit generally depends upon the nature of your sample and what sort of environment it was obtained from.

Amplifying Metagenomic DNA

In microbiome studies we want to enrich the microbial DNA present and eliminate the contaminating human host DNA. We do that by making lots of copies of the microbial DNA using a procedure known as **polymerase chain reaction**, or **PCR**. This amplification method mimics how DNA is copied in nature, which we learned about above. **PCR amplification**, sometimes called "molecular photocopying," is a rapid, inexpensive method used to "amplify," or make many copies of, small segments of DNA. Significant amounts of DNA are required for most molecular analyses, and PCR provides large quantities of DNA quickly. PCR is considered one of the most important inventions in molecular biology, and its creator, Kary Mullis, was awarded the Nobel Prize in Chemistry in 1993.

PCR is at the heart of one commonly used method for microbial identification known as **target gene amplicon sequencing**. The goal is to amplify the target gene so that it can be sequenced. In many microbiome studies, the target gene is the **16S rRNA gene** we have already mentioned frequently. **Figure 4.12** shows the process of gene amplification. Each cycle of amplification involves three steps: denaturing, annealing, and extension. Denaturing involves separating the two strands of the double helix, in this case using heat rather than the helicase enzyme used in DNA replication. Short DNA sequences, called **primers**, are added that will serve as starting points of the synthesis of new strands. Since DNA is double stranded, two types of primers are needed. They are known as **forward and reverse primers**, which correspond to which strand of the gene they anneal to. Forward primers anneal with the antisense DNA strand, and reverse primers anneal with the sense, or coding, strand. The coding strand refers to the DNA strand that encodes a gene product.

There are two primer pairs that are commonly employed in amplifying 16S rRNA. Their names (515F and 806R) refer to the locations in the 16S rRNA gene where these primers anneal and whether they anneal to the forward (F) or reverse (R) strand of the gene. Note that not only do these primers have a region that is complementary to the target gene (**primer pad region**), but the forward primer also includes a **barcode**. Each sample gets a different set of primers, which contain a unique barcode for that sample. If we have 10 samples, then we need 10 pairs of primers that differ only in their barcode. The adapter region, identified as Illumina 5′ and 3′ adapters

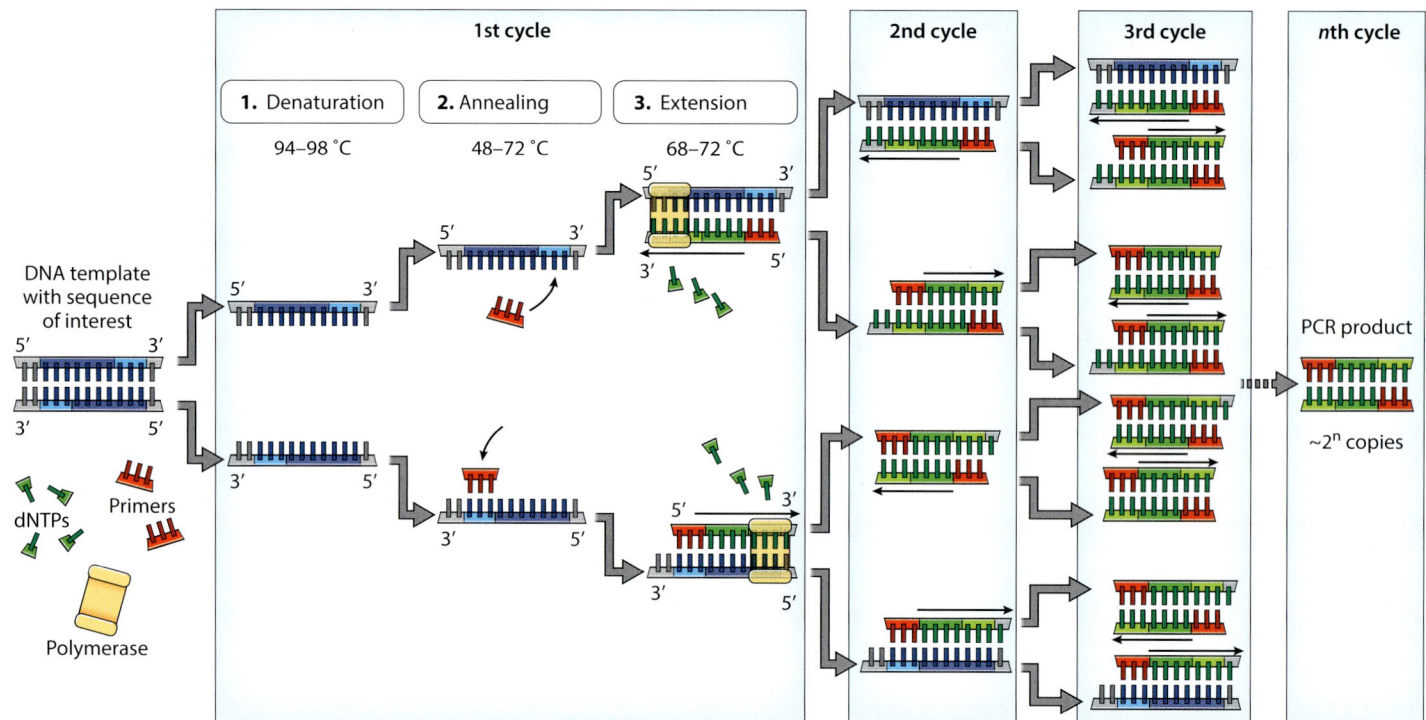

Figure 4.12 Polymerase Chain Reaction Overview Many of our most powerful molecular biology tools were developed using preexisting biological machinery, and PCR is no different. Leveraging the chemistry of DNA and DNA polymerase from other organisms, we're able to replicate specific genes from a sample during an experiment.

in **Figure 4.13**, are so named because Illumina is one of the leading companies in DNA sequencing, and the adapters are designed based upon its sequencing platform. We will discuss the roles of these adapters below, as they play a role in the sequencing of our fragment after amplification is complete.

Let's return to Figure 4.12 and the third step in the first amplification cycle. The primers have annealed to the 5′ and 3′ ends of the target gene. The next step is the addition of an enzyme, known as **Taq polymerase**, which is a specially engineered polymerase, originally derived from bacteria that live in hot springs. Through natural selection, this enzyme has evolved to be insensitive to a wide range of temperatures that would normally inhibit most enzymes. If you live in a hot spring, you adapt or die! The result is a polymerase that can withstand the range of temperatures required for PCR. The polymerase starts at the 5′ end of the primer sequence and, as it reads the target gene sequence, adds the complementary nucleotides to the 3′ end of the primer. This is the extension phase of PCR, and the result is two copies of the target gene. This cycle of denaturing, annealing, and extension is repeated up to 40 times, resulting in more than a billion copies of the original DNA sequence. That's right, more than 1 billion copies are produced in very short order! The PCR process is automated and can be completed in just a few hours using a **thermocycler**, which is a machine programmed to alter the temperature of the reaction every few minutes to allow DNA denaturing at one temperature, and then primer annealing and strand extension at different temperatures.

Billions of Amplification Products

The product of our amplification effort is billions of copies of the 16S rRNA gene, which, as you have previously learned (chapter 1), encodes the small subunit of the ribosomal complex that serves in protein synthesis. This gene is highly conserved,

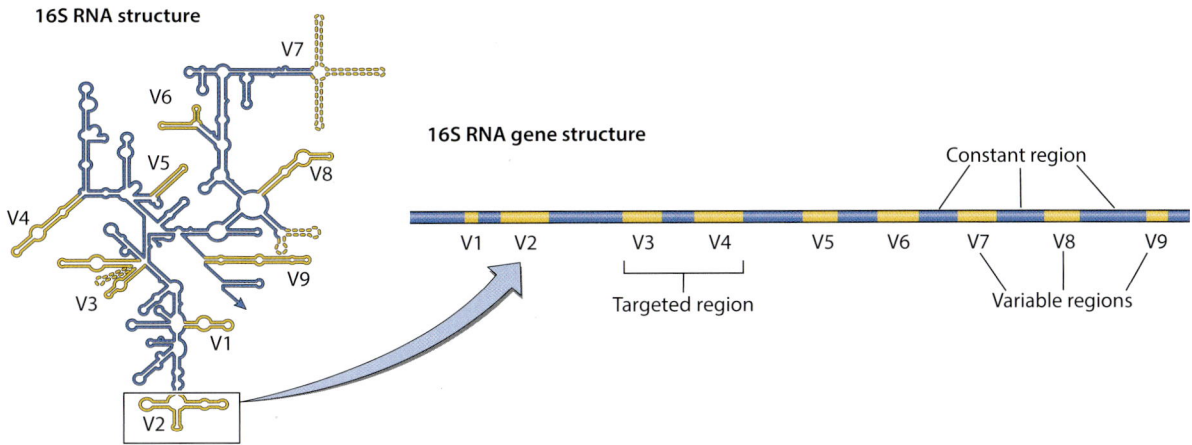

Figure 4.13 **The Variable Regions in the Prokaryotic 16S rRNA Genes** This figure shows the 16S rRNA target gene, with the highly variable regions indicated. The encoded 16S molecule is shown to the left as a 2-dimensional structure. You can see that the variable regions (marked with V and a number) usually correspond with regions of the molecule where the actual base does not matter, since there is no base pairing in these regions, called loops. A box is placed over one of these variable regions, labeled V2. Notice how small this region is, within an already small gene!

meaning that over long stretches of evolutionary time, there have been relatively few changes in the DNA sequences of the constant regions indicated in Figure 4.13. In fact, those constant regions are so constant that you can compare 16S gene sequences from organisms that span billions of years of evolutionary divergence, as is the case for Archaea and Bacteria. This high level of conservation argues that the molecule plays a crucial role in cellular function and survival, and thus we can be confident that when we compare 16S gene sequences, we are comparing **homologs**, which can be traced back to a single common ancestor. All the 16S genes we examine will have shared a common evolutionary ancestor.

To ensure that we also identify microbial eukaryotes in our screen, we need to expand our amplification target to include the **18S rRNA gene**, which is homologous to the 16S rRNA gene and encodes the small eukaryotic ribosomal subunit. When the eukaryotes diverged from their archaeal ancestors, so too did their rRNA genes begin to diverge. What had been an ancestral 16S rRNA gene in the earliest of eukaryotes became different enough in sequence and function over time to warrant a new name, in this case 18S rRNA. Just as seen for the 16S gene, the 18S gene has constant and variable regions, and the variable regions are the target when our interest is in identifying organisms at the genus and species level. The variable regions will have accumulated enough differences to permit us to distinguish between relatively closely related taxa. With the 16S and 18S target genes, we can identify all cellular life forms, which all have one of these two homologous ribosomal RNA genes.

We are at a loss if we wish to compare a homologous gene in viruses, since they don't encode ribosomal genes. In fact, viruses do not share any genes that are found in all of Bacteria and Archaea, or in microbial eukaryotes. That is one reason why we still know relatively little about the viral components of microbiomes. However, rapid progress is being made in cataloging viral genomes. Camarillo-Guerrero et al. (2021) published the Gut Phage Database, which is a collection of ~142,000 viral genomes available for use as a reference database.

One final, but key, feature of PCR amplification is that it will produce copies of the gene in numbers that are representative of its abundance in the original sample. In other words, if 95% of the microbial cells in a sample are one species and 5% another, the amplification products will be produced in the same ratio: 95:5. In

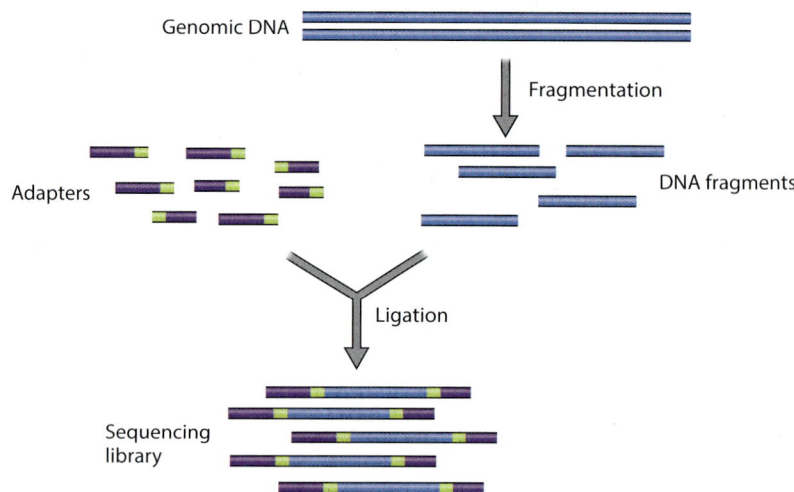

Figure 4.14 Tagmentation While many researchers believe that the experiment ends after DNA extraction, there are several more steps known as library preparation that are needed to get ready for sequencing.

principle, you simply divide the number of a specific 16S rRNA sequence (representing one taxon) by the total number of 16S sequences produced from that sample. This ratio provides an estimate of the relative abundance of each sequence type, and thus each taxon.

If our goal is to reveal the entire genome of each microbe present, rather than just a single target gene, we need a slightly different approach to DNA amplification. We start by randomly fragmenting all the microbial DNA extracted from our sample, which we call the **metagenome**, into small pieces, roughly 200 to 500 base pairs in length. One way to create these fragments is to apply ultrasonic sound waves, which result in DNA shearing. There are also physical methods, such as beating the DNA with glass beads, and enzymatic methods as well. The resulting fragments are then tagged by adding short DNA adapters onto their ends, a procedure called "tagmentation," which is short for "tagging and fragmentation" (**Figure 4.14**). These adapters contain segments that will be used in the amplification and sequencing step we describe below. The final product is a **metagenomic sequence library** that contains fragments from every organism present in the sampled microbiome.

High-Throughput DNA Sequencing

The standard method to determine the DNA sequences in our sample DNA is called high-throughput sequencing (**Figure 4.15**). DNA is first fragmented into smaller input-sized fragments by enzymes or by sonication. The ends of these fragments are repaired, and specific adapters are ligated to the ends of the fragments, allowing hybridization to a flow cell to occur. A bridge amplification step is performed to create a cluster of fragments with the same sequence. One strand of the DNA is removed, and fluorescently labeled nucleotides are added. From the resulting fluorescent image of the flow cell, a computer records which nucleotide was incorporated at each cluster's coordinates (cycle 1). The fluorescent label is then cleaved and a second round of fluorescently labeled nucleotides is passed by each cluster and the nucleotide incorporated at each coordinate is again recorded (cycle 2). Repeated cycles produce a sequence for each fragment (a read). These reads are then aligned to a reference genome. By assembling reads (merging short reads together), it is therefore possible to reconstruct the unfragmented original sequence. It is more complicated to describe these methods than it is to view them—so take a close look at Figure 4.15 to be sure you have a general idea of the steps involved in high-throughput sequencing.

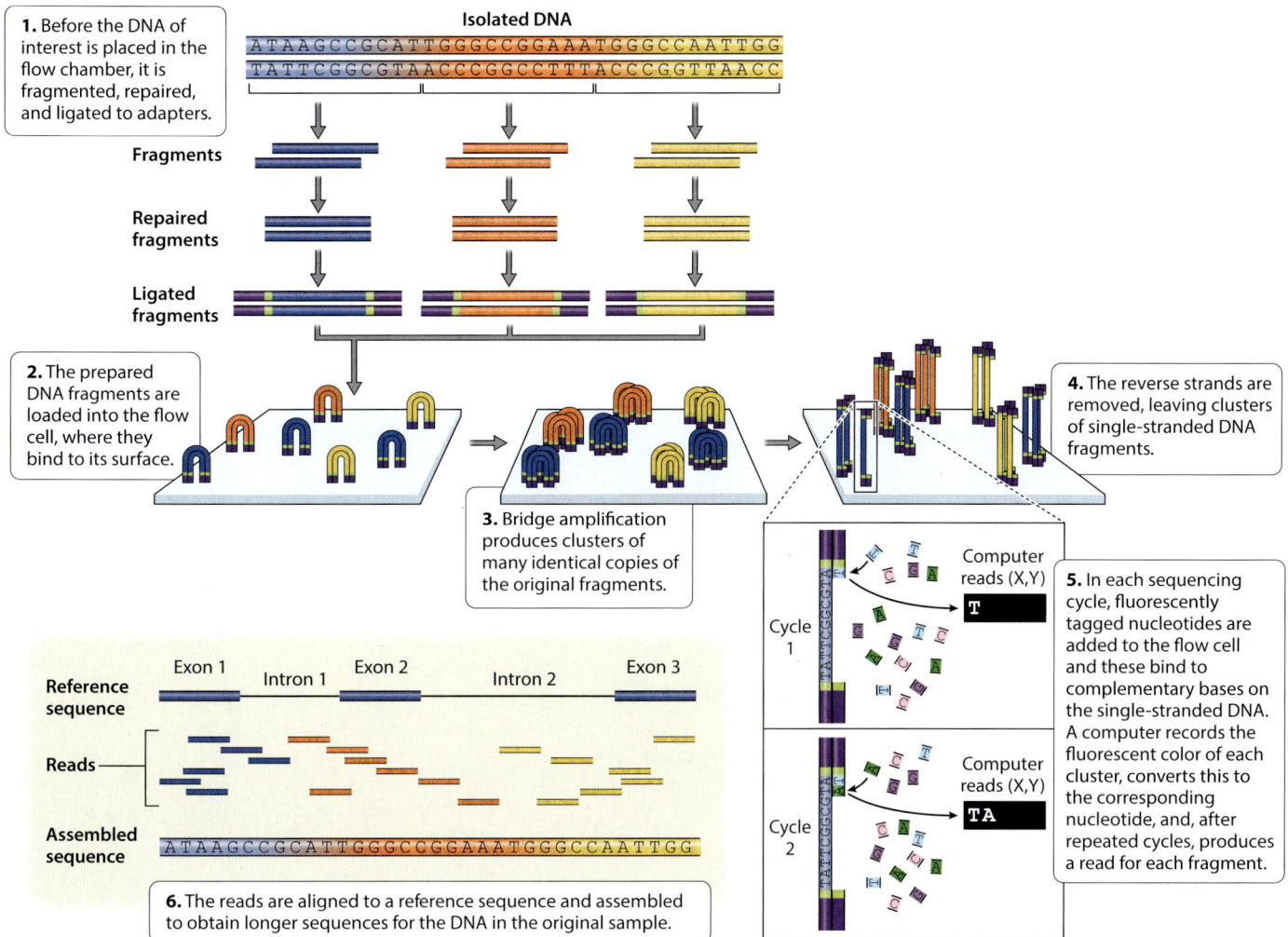

Figure 4.15 A Close-Up Look at High-Throughput Sequencing Understanding the basics of how sequencers work is important for designing experiments and interpreting the data. However, because the machines and reagents are so expensive, most researchers use third-party companies to perform their sequencing. (After Churko et al., 2013.)

Whether we added metagenomic fragments or target gene amplicons to the flow cell, the result is that each fragment of DNA is sequenced. The adapters we added identify the sample from which the sequence was obtained and uniquely label each of the metagenomic fragments or gene amplicons. When we move on to the analysis section in the next chapter, we will learn how we translate these signals into our final sequence dataset.

Target gene amplification and metagenomic fragment sequencing are both valuable ways to understand the composition of a microbiome sample. Each method has its benefits and drawbacks, and when designing a study, it's important to consider which method best fits the needs of the study.

TARGET GENE AMPLIFICATION Target gene amplification is a quick and relatively inexpensive method to identify the microbial taxa in a sample. To ensure that even the rarer members of a microbiome are identified requires 50,000 sequenced reads for each sample (Peterson et al., 2021). Each sequence read is the result of the sequencing procedure mentioned above and refers to the corresponding set of laser images for one amplification product. There are some drawbacks to this approach. Due to the highly conserved nature of the target gene, coupled with the short length of sequence reads produced (less than 1,500 base pairs), this method has limited taxonomic resolution, meaning that we are not always able to identify a microbe to the species level. Further, the 16S rRNA gene may allow us to identify the organisms present but does not inform us about the functional capacity of those microbes. We miss useful information, like metabolic pathways. Finally, amplification can introduce sequence artifacts (PCR errors) and result in a bias in the taxa identified in our sample. However, this technology has been utilized for over 25 years, and there are numerous experimental controls that permit us to identify potential amplification biases or error-prone regions of DNA and omit them from our analysis.

METAGENOMIC (SHOTGUN) SEQUENCING   Metagenomic sequencing, often called **shotgun sequencing**, amplifies essentially all the DNA in a sample. The name comes from the indiscriminately fragmented metagenome, from which an incredible number of short, asymmetrical segments are sequenced. This increased complexity requires more sequenced reads, from 5 to 200 million per sample, depending on community complexity and project aims (Zaheer et al., 2018). This extra coverage results in significantly higher costs but yields much more information, as it provides the whole genome sequence of every species present in the sample. The increased amount of information substantially increases our ability to identify organisms to the species level (Ranjan et al., 2016). Metagenomic sequencing has the additional benefit of providing direct evidence of gene function in the microbiome. However, analyzing metagenomic data typically relies heavily on comparisons to published microbial genomes, and it can be challenging to identify novel species or functions. Additionally, these large-scale assemblies are much more computational-resource intensive, where programs may take hours or days to run, depending on the facilities available.

The Earth Microbiome Project (EMP) is a collaborative effort to characterize microbial life on Earth. This community has created standardized protocols for data collection, curation, and analysis, which has gone a long way towards ensuring that such complex datasets produced in different laboratories will be comparable. Although we focus solely on the human microbiome in this book, the EMP has a much larger goal, which includes sampling microbiomes across the planet.

4.4 THE "OMICS"

We now know the essential experimental steps required to perform sequencing of either a target gene or metagenomic fragments. Depending upon the questions we wish to address, the resulting DNA sequence information may be all we require. With the target gene amplicon data, we are empowered to identify the taxa in our samples, sometimes to the species level, and quantify their relative abundances. With the metagenomic sequence data, we have added information about the genes encoded in their genomes. This permits us to predict the functions each species might carry out, such as digesting a particular food source or producing key metabolites. In short, metagenomics offers a powerful lens for viewing the microbial world through their genetic data.

However, we can go well beyond simply sequencing the metagenome of our samples. We have techniques that permit us to determine which of the genes in the microbiome are transcribed into messenger RNA (**metatranscriptomics**), what proteins are produced from the messenger RNA transcripts (**metaproteomics**), and what metabolites result from the actions of the microbes present (**metabolomics**). Let's spend some time learning a bit about each of these "omic" technologies, which are illustrated in **Table 4.1**.

Metatranscriptomics

As we have learned, metagenomics focuses on the genomic content present within a sample. One of the main limitations is that it does not distinguish active from inactive genes in the metagenome. Only some of the genes in a genome are transcribed, or

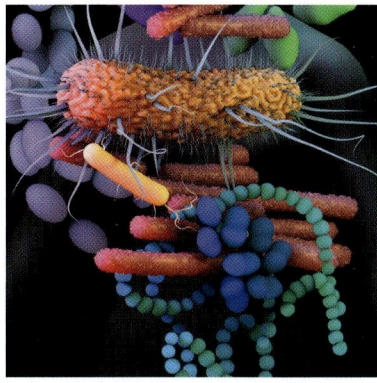

(Photo from iStock.com/Design Cells)

Table 4.1 What "Omics" Technology Can Do for Microbiome Studies

OMIC TECHNOLOGY	ALLOWS US TO EXPLORE
Metagenomics	The microbial taxa present
	Their relative abundance
	The genes they encode
Metatranscriptomics	The microbial RNA transcripts present
	The taxa producing them
	Their relative abundance
	The functions they encode
Metaproteomics	The microbial proteins present
	The taxa producing them
	Their abundance
	The functions they encode
Metabolomics	The microbial metabolites present
	The taxa producing them
	Their abundance
	The functions they provide

copied, into messenger RNA, which is the first step in using the instructions encoded in DNA to make a protein. RNA is very similar to DNA but is different in one nucleotide base—uracil (U) in place of thymine (T)—and it has a larger and more reactive hydroxyl group on each nucleotide, which allows it to be a more flexible molecule. Your DNA is like the blueprint for a house; it contains all the information you need for building, but also some information you may not need. RNA then performs the functions of the builders, using the information from the blueprint to lay the bricks (proteins) and start the construction.

The technology required to sequence RNA transcripts and measure their abundance has been in place for many years and is like those used for shotgun metagenomic sequencing, so we won't dive into the details here. Metatranscriptomics indicates which genes are expressed in the metagenome and can inform us about which organisms are producing those transcripts. We can explore how the **metatranscriptome**, which refers to all the RNA transcripts in a sample, changes under states of disease or when a treatment is applied. The advantage of metatranscriptomics is that it can tell us about what the microbial community is doing, what functions it is performing. This approach provides a more accurate snapshot of the genes that are being used by the members of the microbiome.

There remain numerous technical challenges in carrying out metatranscriptomic studies. It is not always possible to capture the entire metatranscriptome. This can be due to the complexity of the microbial community and the speed with which RNA degrades. Despite those challenges, metatranscriptomics is a powerful tool in the microbiome toolbox. This approach to probing the functions of a microbiome holds great promise in helping us identify and understand the biologically active members of these microbial communities.

Metaproteomics

Metaproteomics is an umbrella term for experimental approaches, such as **mass spectrometry**, designed to study all proteins in a microbiome. Mass spectrometry is a powerful tool that measures the ratio of the mass (or weight) to the charge of substances

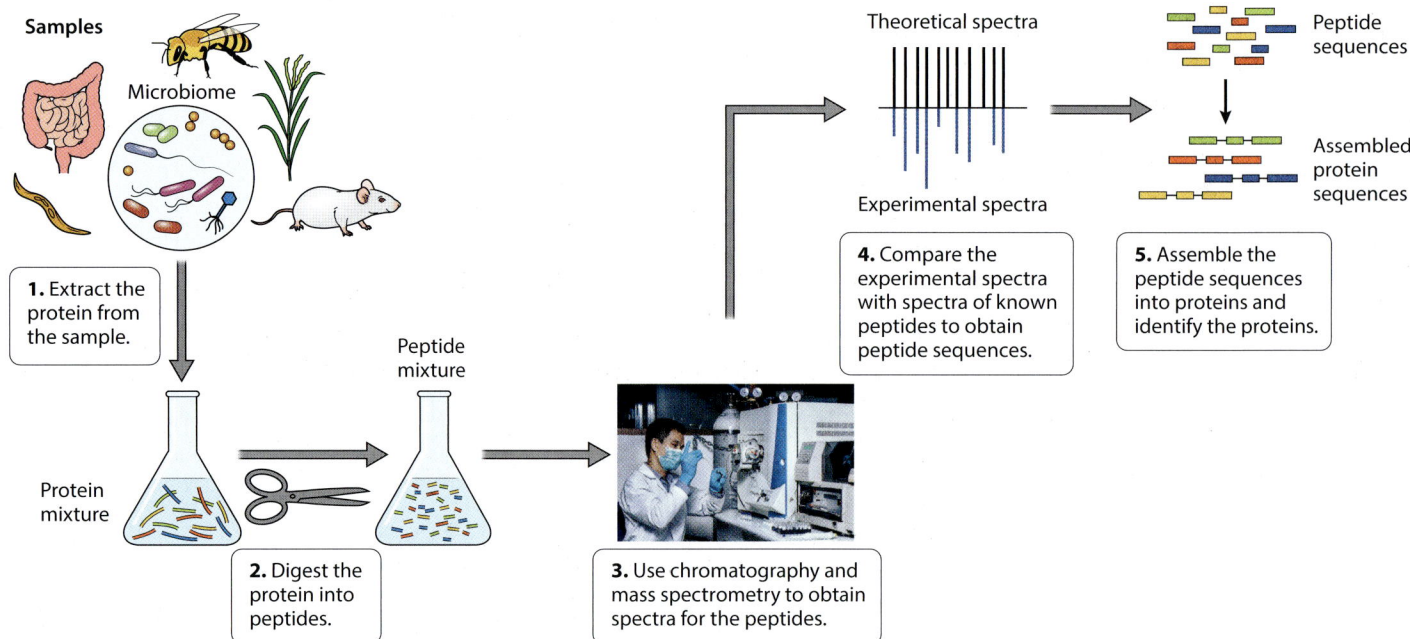

Figure 4.16 An Overview of Proteomics Unlike sequencing the genome or transcriptome of a sample, proteomics requires different instruments capable of quantifying proteins. Proteomics experiments can be incredibly expensive and require special reagents and equipment, keeping many labs from pursuing this approach. (Photo of mass spectrometer from S. Singha/Shutterstock)

in a sample. This ratio is key to identifying proteins and metabolites. **Figure 4.16** provides an overview of the steps involved. First, the microbial cells are separated by centrifugation from the rest of the sample and are lysed open to release all the proteins, which can range in size from several to thousands of amino acids. Then the proteins are digested into short fragments using proteolytic enzymes that break apart the bonds holding amino acids together. The result is a mixture of protein fragments, which are then injected into an HPLC (high-pressure liquid chromatography) column. The column is packed with irregularly or spherically shaped particles, and the protein fragments will move through the column at different rates, depending on their size, charge, and other physical chemistry, which allows them to be separated into fractions.

The fractions are then given a positive or negative charge by attaching different **ions**. An ion is an atom that has gained or lost an electron, resulting in a net non-neutral charge. Imagine that you are spray-painting a car, but instead of paint you spray ions, and instead of a car you coat protein fragments. Each possible combination of amino acids in a peptide has a unique mass-to-charge ratio; therefore, the mass-to-charge ratio of an unknown peptide can be used to infer its amino acid composition. These charged, fractionated proteins are then fed into a mass spectrometer. Although there are different kinds of mass spectrometers, they all employ electromagnetic fields to record the motion of ions in a sample, such as our ionized protein fragments, and determine their mass-to-charge ratio. The resulting mass spectrum helps to determine the mass of the substance and its chemical structure.

In the past, metaproteomics was rarely employed in gut microbiome studies, at least in part due to the lack of efficient bioinformatic tools. Fortunately, the recent development of metaproteomic data-processing tools has improved our ability to analyze metaproteomic data. This has enabled deep characterization of microbiome protein compositions, with some reports quantifying >50,000 unique microbial proteins in a single study.

Metabolomics

We have one final omics technology to explore, metabolomics. Metabolites are the intermediate and end products of reactions catalyzed by various enzymes that naturally occur within cells. One example of a class of metabolite is the short-chain fatty acids produced by the gut microbiome. The **metabolome** is the metabolites present in a microbiome. These metabolites are identified using analytical techniques like those employed for proteomic studies, such as mass spectrometry.

Metabolomics presents a significant technical challenge because, unlike genomic and proteomic methods, its goal is to identify molecules (metabolites) that have disparate physical properties. They possess a wide range of polarity ranging from water-soluble organic acids to nonpolar lipids. Metabolomic technology platforms typically take the strategy of dividing the metabolome into subsets of metabolites—often based on compound polarity, common functional groups, or structural similarity—and devise specific sample preparation and analytical procedures optimized for each. They also divide metabolites into those that are already known and those that are unique. Targeted metabolomics employs a predetermined list of metabolites, whose mass-to-charge ratios are already known, which comprise our reference standards. This method is quite sensitive and provides a more rigorous quantification of the metabolites present as well. However, because the metabolites are targeted to what we already know, we can't discover new metabolites using this approach. In fact, most metabolites produced by a microbiome are unknown and won't be picked up using targeted methods. Untargeted metabolomics will detect many more metabolites. However, it is not always clear what to do with the substances detected, since they don't have names or known functions. We don't yet have the information to annotate many of the molecules identified. **Sequence annotation** is the process of marking specific features in a DNA, RNA, or protein sequence with descriptive information about structure or function. However, there are bioinformatics methods that permit us to link these novel spectra to known metabolites through a technique called **molecular networking**. This is a new area of research, and the data generated at this point should be considered preliminary. It is so new that we don't even know what we don't know about it, if that makes sense.

The metabolome is therefore measured as a patchwork of results from different analytical methods. Metabolomics is an emerging field, and the methods used in it continue to improve. However, a consequence of metabolomics laboratories using multiple procedures that are potentially subject to frequent refinement is that individual laboratories tend to have unique methods, and there are comparatively few standard operating procedures. This means that it is challenging to compare data generated in different laboratories. The **Metabolomics Standards Initiative** served to develop guidelines for data reporting, control tests to evaluate the capabilities of different methods and laboratories to obtain similar results, and open-access repositories for metabolomics, such as the Metabolomics Workbench (http://www.metabolomicsworkbench.org) funded by the National Institutes of Health's Common Fund in the United States.

Spatial Omics

The potential of **multi-omics** technologies (genomic, metagenomic, proteomic, etc.) is enormous and rapidly increasing. One of the most exciting new methods is referred to as **spatial transcriptomics**. By combining advances in imaging technologies and sequencing, spatial transcriptomics can map where a particular RNA transcript exists. According to one of the leaders in this field, George Church, a professor of genetics at Harvard Medical School, "*Medicine is moving from very blunt instruments to molecules that are the finest scalpel you could ever have.*" (https://www.genengnews.com/insights/spatial-the-next-omics-frontier/). These new spatial technologies will provide a deeper understanding of what and where things are happening in a microbiome.

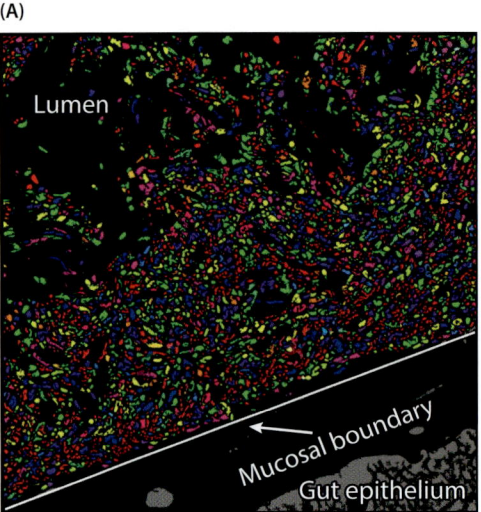
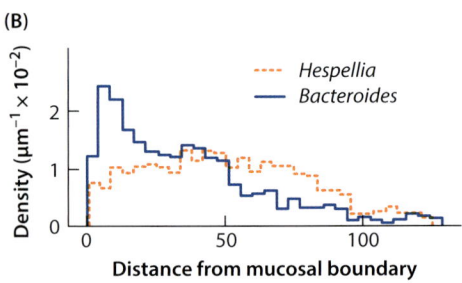

Figure 4.17 The Striking Clarity of Mouse Gut Spatial Omics (A) A section of mouse colon with the microbiota fluorescently labeled. (B) The relationship between bacterial density and distance to the mucosal barrier is plotted for two genera. *Bacteroides* is enriched near the mucosal boundary, while *Hespellia* cells are more evenly distributed away from the mucosal boundary. (A from Shi, H., Shi, Q., Grodner, B. et al. Highly multiplexed spatial mapping of microbial communities. *Nature* 588, 676–681 (2020), with permission from Springer Nature; B after Shi et al., 2020.)

One example of the application of **spatial omics** is provided by a group interested in the precise location of members of a gut microbial community (Shi et al., 2020). The researchers targeted 16S rRNA and created fluorescent labels for each novel 16S rRNA gene in their sample. They then caused the fluorescent markers to glow and used a microscope to see the colors (**Figure 4.17**). Starting with a palette of 10 colors, which generated 1,023 color combinations, each labeled with a unique barcode, this method produced a stunning image! "The imaging itself leads to very beautiful, rich images with all bacterial cells in different colors," says Iwijn De Vlaminck, one of the researchers (Nutt, 2020). By providing useful information about not just what species are present but where they are present, spatial mapping will help us study diseases caused by microbes, such as infection or inflammatory bowel disease. We can apply this approach to mapping the location of RNA transcripts as well—but that field is truly in its infancy.

In this chapter, we have discussed how to generate a hypothesis and design a study to test it. We learned that careful consideration must be given to determine an appropriate sample size with sufficient power and to implement controls to eliminate any confounding factors. We then discussed how to collect and process samples through sequencing and other omics methodologies. One of the greatest challenges encountered with all the omics technologies is amassing, analyzing, and interpreting this highly complex, dynamic information. We will tackle these challenges in the next chapter, where we learn how to analyze microbiome data.

CHECK YOUR UNDERSTANDING

1. The "treatment" in an experiment is also known as the
   a. dependent variable.
   b. independent variable.
   c. hypothesis.
   d. control.

2. All of the following are examples of model organisms except
   a. mice.
   b. fruit flies.
   c. humans.
   d. worms.

3. The use of model organisms allows us to control
   a. the environment they grow and develop in.
   b. their diet.
   c. their genetic background.
   d. all of the above.

4. A null hypothesis
   a. proposes that no significant difference exists for a set of given observations.
   b. is the exact opposite of an experiment's hypothesis.
   c. proposes that your experiment will produce no data.
   d. is a boring hypothesis.

5. The variable that is affected by a study is called the
   a. independent variable.
   b. null hypothesis.
   c. dependent variable.
   d. control variable.

6. Experimental results may be inconclusive due to
   a. small sample size.
   b. too many variables affected.
   c. participant dropout.
   d. all of the above.

7. Primary literature
   a. is peer-reviewed research published in journals.
   b. doesn't need formal review to be published.
   c. is research written up before being published.
   d. can often be found in magazines and newspapers.

8. Metadata can be defined as
   a. information about the participants in our study.
   b. a collection of protocols employed in our study.
   c. background data about a set of research subjects.
   d. all of the above.

9. Ensuring that your experiment has enough statistical power to answer your question requires you to specify which of the following?
   a. How many variables you're changing
   b. The type of research subject used in your experiments
   c. How many participants to include and/or samples to take
   d. How many statisticians to employ

10. Power analyses are based on
    a. subject retention rate.
    b. results from previous experiments.
    c. type of subject.
    d. all of the above.

11. All of the following are true about experimental controls *except* that
    a. they remain unchanged between different experimental conditions.
    b. controlling more variables in the experiment produces more reliable the data.
    c. researchers can fully limit extraneous factors.
    d. using model organisms helps us to maintain experimental controls.

12. The data generated for our proposed experiment will be
    a. information about pregnant and newborn microbiomes.
    b. 16s rRNA sequences.
    c. the impact of microbiome diversity on asthma.
    d. microbiome community composition.

13. Stool samples are
    a. a useful proxy for understanding gut microbiome diversity.
    b. helpful for comparing oral microbiome composition.
    c. a perfect representation of gut microbiome communities.
    d. rarely used in microbiome research.

14. By the time newborns' stool samples have reached the lab,
    a. their microbiome composition will have slightly changed.
    b. the community profile of the microbes will be the same as when the sample was taken.
    c. being held on ice will have radically changed the microbiome composition.
    d. all the bacteria will have died.

15. All of the following are nucleotides found in DNA *except*
    a. adenine.
    b. guanine.
    c. cytosine.
    d. uracil.

16. DNA replication is
    a. completely conservative.
    b. semiconservative.
    c. error-free.
    d. not a prerequisite for cell division.

17. DNA is read from
    a. left to right.
    b. right to left.
    c. 5′ to 3′.
    d. 3′ to 5′.

18. DNA polymerase
    a. unwinds the DNA helix to allow replication.
    b. adds complementary nucleotide bases to an existing DNA strand.
    c. separates DNA into sense and antisense strands.
    d. is the same in all organisms.

19. Bead and column-based DNA extraction protocols rely on which molecular property of the DNA?
    a. Net charge
    b. Handedness
    c. Complementary base pairing
    d. Double-stranded helix shape

20. PCR experiments
    a. synthesize additional copies of target genes if they are present in the sample.
    b. copy all of the DNA present in the reaction.
    c. are often used to copy whole genomes of organisms.
    d. are outdated technology.

21. The 18S rRNA gene is
    a. found exclusively in bacteria.
    b. a eukaryotic gene comparable to 16S rRNA.
    c. upstream of the 16S rRNA gene.
    d. downstream of the 16S rRNA gene.

22. The metagenome is
    a. metadata about an organism's genome.
    b. information about DNA methylation across an organism's genome.
    c. a collection of the total DNA from a sample.
    d. a library of genes most relevant to our experiment.

23. Shotgun sequencing
    a. got its name from the use of metal beads.
    b. amplifies only our target genes of interest.
    c. is used to produce long reads representing the entire genome.
    d. produces short fragments allowing for the sequencing of the whole genome.

24. Which of the following is true about DNA but not RNA?
    a. Uracil is one of the bases.
    b. The sugar is deoxyribose.
    c. It is often used as a messenger molecule.
    d. It is usually single stranded.

25. Mass spectrometry uses electromagnetism to record
    a. the mass-to-charge ratio of proteins.
    b. whole transcriptome data.
    c. DNA sequences of interest.
    d. chemical composition of metabolites in a sample.

Answers: 1B, 2C, 3D, 4A, 5C, 6D, 7A, 8D, 9C, 10D, 11C, 12B, 13A, 14A, 15D, 16B, 17C, 18B, 19A, 20A, 21B, 22C, 23D, 24B, 25A

DIVING DEEPER

1. Say you wanted to study the effect of antibiotics on the microbiome. What is one hypothesis you could test?
2. What is a null hypothesis used for? Can you generate one for the hypothesis you wrote in the previous question?
3. How is an independent variable different from a dependent variable?
4. Why is it useful to explore relevant literature before beginning a study?
5. What are the benefits and drawbacks of choosing humans as study participants?
6. What does it mean if our data is statistically significant?
7. Why is it important to choose a sample size that is large enough but not too large?
8. What are experimental controls, and why are they necessary?

9. What is the difference between a cross-sectional study and a longitudinal study? Can you think of an example of each?
10. What are the four types of metadata we can collect?
11. What is the purpose of collecting metadata?
12. How might you obtain a sample of your gut microbiome that is more accurate than a stool sample?
13. Not only is it important for scientists to store backup samples, but they also back up their data. What are some ways you could make backup copies of your data?
14. Why might researchers combine mechanical and chemical lysis methods?
15. Why is it important that we know which nucleotides bond with each other?
16. Why has the development of PCR frequently been described as one of the most important scientific advances in molecular biology?
17. What is a metagenome? Why is it important?
18. What role does DNA polymerase play in high throughput sequencing?
19. When would it be helpful to take a target gene amplicon approach to a sequencing experiment instead of metagenomics?
20. What types of questions can shotgun sequencing approaches answer that target gene amplicon experiments can't?
21. How is sequencing a community transcriptome different from sequencing a genome? What information is gained or lost based on your approach?
22. Metaproteomics experiments can provide increased insight into the proteins produced and used by your microbiome. What information can't we learn with this technique?
23. Think back to previous chapters—why would learning about a person's metabolome be important?
24. Pairing transcripts with spatial data provides much more information to researchers. What types of research questions can this information be used to ask/answer?

DISCUSSING AND REFLECTING

1. Early in the chapter we discussed using model organisms to conduct your microbiome experiments, using mice as an example. Take some time to read about other model organisms in use by researchers and consider what other species might suit this type of research. What are the benefits and drawbacks of the species you considered?
2. If you could use any of the "omics" that we discussed at the end of the chapter to analyze your own microbiome, which would you choose and why? What questions would you be able to ask and answer with the data?
3. Reflection. Reflect on some of the experimental tools and techniques that we've discussed in this chapter. How do they align with your preconceived image of a researcher? Were you surprised by the amount of lab work required for even a "simple" experiment? Did you consider the computational resources needed for sequencing experiments?

RECOMMENDED READINGS

Popular Science Book

DeSalle, R., & Perkins, S. L. (2016). *Welcome to the Microbiome*. Yale University Press.

Scientific Reviews

Gotschlich, C. G., Colbert, R. A., & Gill, T. (2019). Methods in Microbiome Research: Past, Present and Future. *Best Practice & Research: Clinical Rheumatology*, 33(6).

Zhu, Q., et al. (2022). Phylogeny-Aware Analysis of Metagenome Community Ecology Based on Matched Reference Genomes while Bypassing Taxonomy. *mSystems*, 7, e00167-22. https://doi.org/10.1128/msystems.00167-22.

Analyzing Microbiome Data

5

CHAPTER CONTENTS

5.1 The Data Analysis Pipeline

5.2 Reconstruction of Genes and Genomes

5.3 Microbiome Data Analysis

5.4 Species Diversity Measures

5.5 Visualizing Diversity Estimates

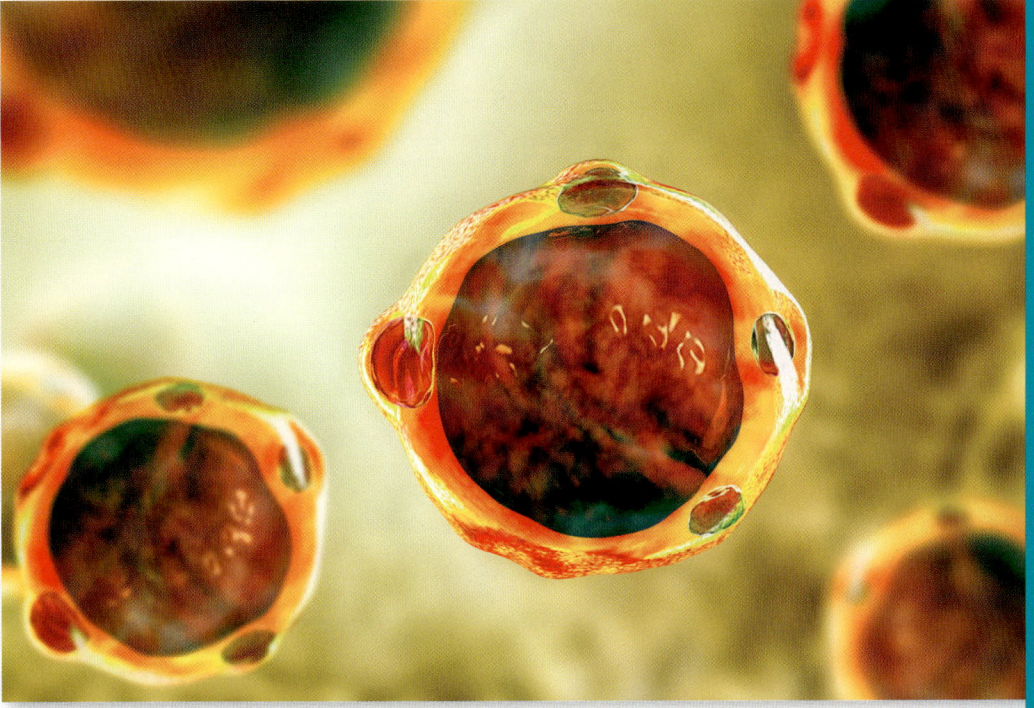

Hi there, human hosts. I am *Blastocystis hominis* and some consider me the party crasher of the human microbiome! I often show up uninvited and am the kind of guest who can be quiet in the corner, barely noticed, or—when the mood strikes me—stir up some unexpected "gut feelings." While scientists debate whether I am harmless, helpful, or a bit of a troublemaker, I have a knack for leaving my hosts guessing. So, if your gut's got an unpredictable vibe, don't blame the tacos just yet; it might just be me adding a little spice to the mix! (Photo from KATERYNA KON / Science Source)

This quote aptly summarizes the goal of this chapter. We have learned how to generate billions of DNA sequences, and we wish to turn those data into information about our microbiome. The more we learn, the sooner we will gain insight into some of the great questions before us, such as how the microbiome trains our immune system, regulates our metabolism, or influences our emotions. We will begin with the raw data, the sequence reads, and learn how to use those data to identify the taxa in our samples and determine their relative abundance, to predict genes encoded in their genomes, and to infer the functional roles of the different taxa. The remainder of the chapter will then explore how we can compare our samples and determine whether any differences we may find in their microbiomes are significant. Finally, we will test our knowledge by diving deeper into our case study as we apply what we have learned.

> "The goal is to turn data into information, and information into insight."
> —Carly Fiorina (Hämäläinen, 2021)

5.1 THE DATA ANALYSIS PIPELINE

Congratulations! We have completed the experimental phase of our microbiome study and have generated an enormous amount of raw data, perhaps tens of thousands of 16S rRNA gene amplicon sequences and billions of metagenomic fragment

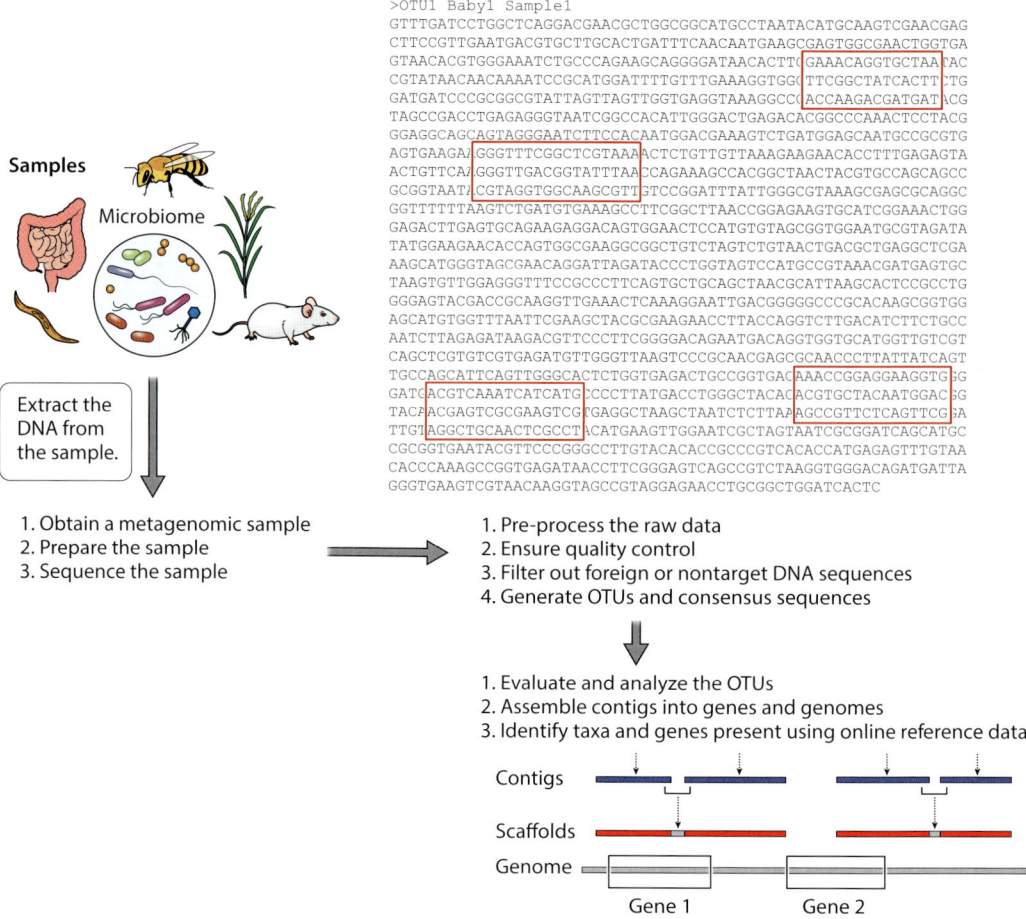

Figure 5.1 A Flowchart of Metagenomic Sequence Data Generation and Analysis An overview of the data collection process, from obtaining a sample to analyzing the resulting sequencing data is shown. The first phase involves obtaining a metagenomic sample, extracting the DNA, and sequencing it. The second phase involves filtering out contaminating sequences, applying quality control standards to the raw data, and generating operational taxonomic units (OTUs). The third phase involves analysis of the OTU data, which results in the identification of taxa and genes in the metagenomic sample.

sequences, not to mention the RNA transcripts, proteins, and metabolites we may have chosen to determine. The next phase in our research is data analysis, where we seek to make sense of all that we have generated. A common problem among inexperienced scientists is the tendency to focus more on the creation of data than on its analysis. From my perspective, if you are not spending more time thinking about your data than generating it, there is something wrong!

The analytical phase of our study answers the following three questions: (1) How do we transform our raw data (such as DNA sequence reads or protein/metabolite mass-to-charge ratios) into a summary of how many of each species (or gene, protein, and metabolite) are present in the samples? (2) How do we link these summary data to relevant clinical variables (i.e., the metadata) to predict their effect on our data or vice versa? (3) How do we determine the significance of what we find; do we have sufficient evidence to support or reject our null hypothesis? We address the first two questions in this chapter. The third question is the focus of all of the subsequent chapters.

Let's now dive even deeper into the microbiome data we generated from our C-section study as we apply the most common analytical methods for both gene amplicon and metagenomic fragment reads. **Figure 5.1** provides an overview of this data analysis process for DNA sequencing, which includes transforming fluorescent signals

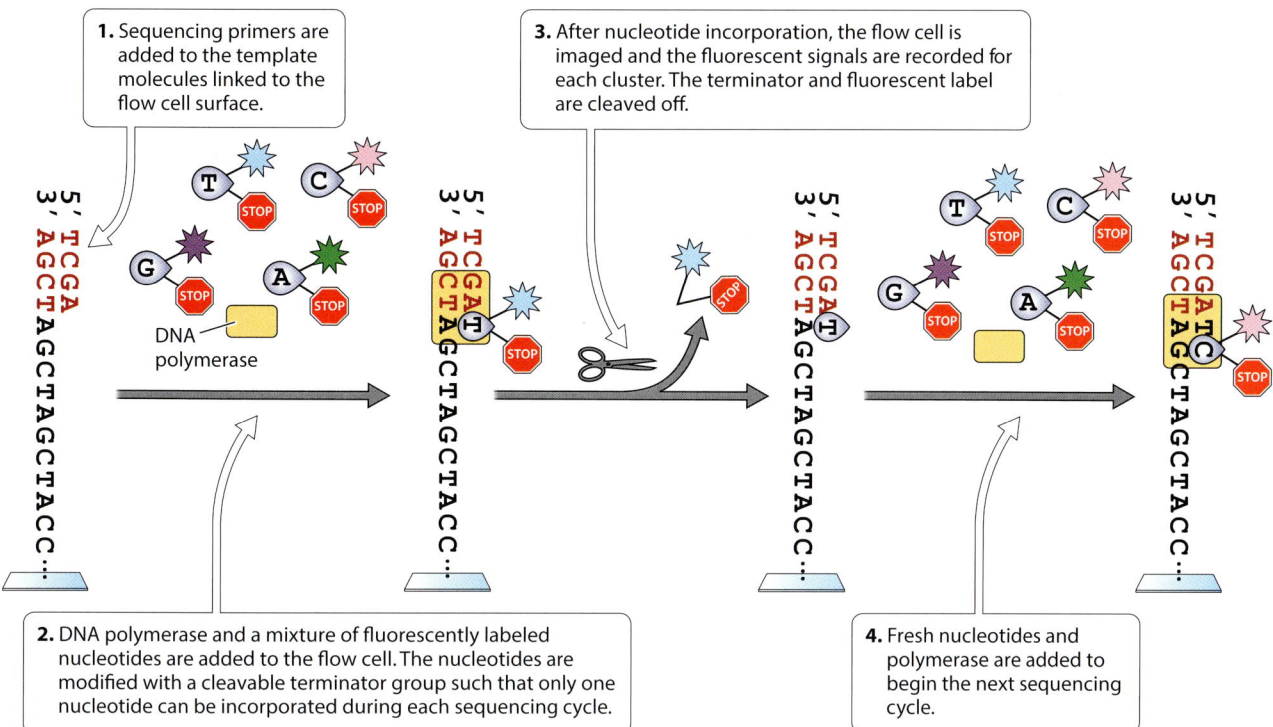

Figure 5.2 Generating a Sequence Read The steps involved in generating DNA sequences from the metagenomic samples is illustrated. (After Anderson et al., 2010.)

into DNA sequence reads, analyzing the reads, determining their quality, processing the resulting sequence data, and applying it to address biologically interesting questions.

The Raw Data

Let's first consider what raw sequence data looks like. As described in chapter 4, the sequencing method we employed, sequencing by synthesis, involves a flow cell with single-stranded DNA adapters physically attached to its wells (see Figure 4.15). DNA amplicons and metagenomic fragments are added to the flow cell, and they anneal to their complementary adapter sequences on the flow cell surface. In sequencing by synthesis, sequencing primers are annealed to the gene amplicons and metagenomic fragments (**Figure 5.2**). Then, DNA polymerase and fluorescently labeled nucleotides are added. Each nucleotide has one end blocked so that only a single nucleotide can be incorporated, after which the base is imaged and the fluorescent signal recorded. The fluorescent label is removed, and fresh nucleotides and polymerase are added, and a new cycle ensues.

There are four fluorescent colors employed in sequencing by synthesis, one corresponding to each of the four nucleotides (A, T, G, C). These visual signals are transformed into the corresponding sequence of nucleotides, which is called a sequence read. One read is generated per gene amplification product or metagenomic fragment attached to the flow cell. The fluorescent signals are captured for each nucleotide in each of the millions of strands annealed to the flow cell, which are sequenced simultaneously, producing millions to billions of sequence reads. Each unique sequence produced will be flanked by the PCR primers employed in gene amplification, the barcodes that were added after amplification to identify the sample origin, and the adapters added to ensure the amplicon anneals to the flow cell. Now recall that all the DNA amplicons and fragments from a study are combined, or multiplexed, which saves time and significantly lowers the cost of sequencing. For example, for our C-section study, all 500 parental fecal samples were combined before sequencing.

Before we can tackle this massive dataset, the raw sequence reads must first be **demultiplexed**, or sorted based on their sample origin, which is accomplished using their barcode identifiers. We now have raw sequence reads of either the target gene amplicons or random fragments from the various microbial genomes present in our samples.

Garbage In, Garbage Out

The expression *garbage in, garbage out* originated in computer science to illustrate that the quality of the information generated by a computer program depends on the quality of the inputted information. However, the phrase is equally appropriate in microbiome science. If we don't bother to check the quality of our data, or curate our sequences, it really isn't worth using them in an analysis. This critical **quality control** step involves inspecting our sequence reads for potential contamination or low-quality sequences. Microbiome samples are almost always contaminated with extraneous DNA. For example, if we sample a human microbiome, our reads will include human sequences—through no one's fault, it is simply impossible to prevent contamination of samples with host DNA. Samples from the gut may also be contaminated with DNA from the foods recently eaten or even from random DNA introduced during our manipulations in the lab. Sophisticated programs exist to screen our sequence reads against the human genome (or any other host genomes). The screening involves literally comparing each sequence read against the entire host genome sequence to identify similarity. Sequence reads that show significant similarity with a nonmicrobial genome are removed from further consideration.

During sequencing by synthesis, also known as next-generation sequencing, process errors can occur. Recall that each base that is added has a blocker agent so that only one end can be active, which ensures that only one base is added per cycle. The blocker is then released with a chemical agent, and the next base is added. One of the more common sequencing errors is a failure of the blocker agent release; then in the next round of sequencing, no new base is added, and the prior base is thus detected a second time. If a sequence is TAT, an error might result in the A blocker not being removed, resulting in TAAT. Once that happens, that growing DNA fragment is now one cycle behind the rest, or out of phase. Alternatively, two nucleotides might get added during a single cycle, and this fragment will then be one cycle ahead of the others—again, out of phase. These errors occur with a low probability. But over time, and with increasing read length, multiple errors can occur in the same read, causing the signal to get more and more asynchronous.

We can identify and remove sequence reads that contain an unacceptable number of errors by using **quality scores** (**QS**), which measure the probability that a base is identified incorrectly. Each base in a read is assigned a QS, based upon factors such as the signal intensity and the similarity in sequences across members of a cluster. A quality score of 10 means that for that read, it is predicted that 1 in 10 bases were read incorrectly. More generally, a quality score of x means that 1 in $10^{x/10}$ bases may contain an error, so a read with a QS of 40 means 1 in 10,000 bases likely contain an error. Generally, reads that have a quality score of 20 or below, which corresponds to 1 error for every 100 or fewer bases, are removed from the analysis. The result is a large set of high-quality raw sequence reads, which are predicted to be at least 99% accurate.

5.2 RECONSTRUCTION OF GENES AND GENOMES

We have generated millions, or even billions, of sequence reads from each sample. Some are simply target gene reads, such as 16S rRNA gene sequences. However, if we chose to sequence the fragmented metagenomic DNA, then we now must figure out

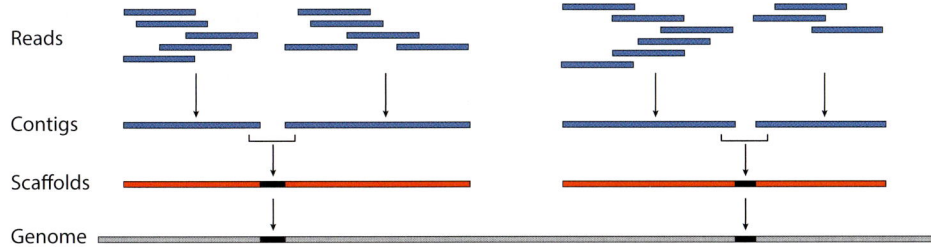

Figure 5.3 Contig Assembly of Metagenomic Fragments Once sequencing reads are generated from a metagenomic sample, they must be aligned to form contigs. These contiguous segments can then be sorted into groups based on high similarity in a process called binning. Contigs are then linked together to form genomic scaffolds, which are then used to identify which genomes are present in the sample.

how those fragments are stitched back together to reform the microbial genomes they were originally derived from. We have essentially taken numerous books and put them through a shredder, and now we must sort through the shredded pages to determine which books they came from and how the pages within each book are organized.

As the sequence fragments are aligned, regions of overlap between fragments emerge, and these overlapping sequences form a **contig**, or contiguous sequence. Members of a contig may disagree at a given base due to errors or to sequence variants present in the original sample. A contig sequence is created based upon a consensus of all the bases found at a site. That means that at each base, the consensus base is chosen to be whatever is the most common base at that site in a contig. If one sequence has an A at a site and four other sequences have a T, then the consensus will be T. As we accumulate contigs, they are placed into what is called a genome scaffold, which is a series of contigs with gaps for the missing data (**Figure 5.3**). The goal is to end up with one contig for each genome present in our sample. Believe me, that is no small feat!

There are many challenges encountered in whole-genome reconstruction. Recall that we may have billions of sequence reads, which correspond to hundreds of species. There may be issues such as uneven species abundance in the original samples and sequencing errors to contend with. Assembling a whole genome from scratch (**de novo**) is a computationally intensive task. However, in many cases, we will be sequencing microorganisms with known genome reference sequences available online. Comparing our reads and contigs to these reference genomes can speed up genome assembly by providing scaffolds against which we can map our contigs. Additionally, bacteria, and prokaryotes in general, tend to have small genomes, usually on the order of 1 million base pairs, or a megabase. A typical eukaryote has a genome 1,000 times larger!

Locating Open Reading Frames

With our gene amplification reads, we know from the start that the sequence is from the target gene, often the 16S rRNA gene. In contrast, our metagenomic fragment reads will include thousands of genes, as well as **noncoding regions**, which we must identify. Noncoding DNA corresponds to the portions of an organism's genome that do not code for amino acids, the building blocks of proteins. The goal is to locate ORFs, or **open reading frames**, that signal the likely presence of a functional gene, which encodes a protein. An ORF is a sequence that has a length divisible by 3 (as a set of three nucleotides, or **codon**, corresponds to an encoded amino acid), begins with a translation start codon (ATG), and ends at a stop codon. To locate ORFs, we turn to a genetic dictionary that will help us identify the meaning of each codon in

Figure 5.4 Genetic Code If we know the sequence of a putative gene, we can determine the amino acid sequence of its corresponding protein using a codon chart. Each codon in a gene encodes a start codon, an amino acid, or a stop codon. For example, AAA encodes the amino acid lysine. (Photo from iStock.com/Rujirat Boonyong)

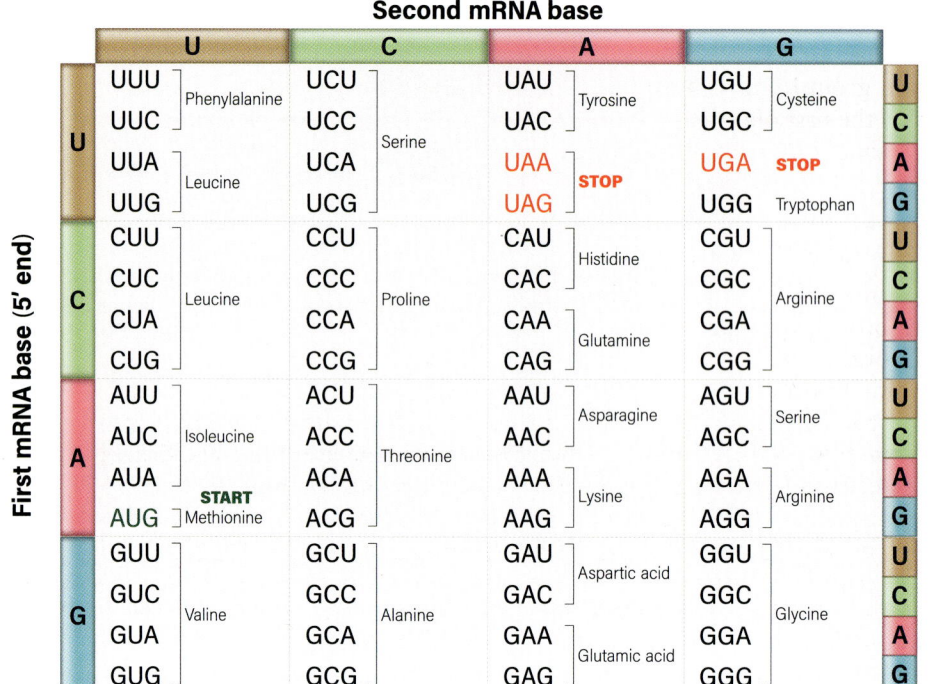

our sequence. **Figure 5.4** shows the dictionary of the genetic code, which identifies the amino acids that correspond with codons, or triplets of nucleotides, in functional genes. Sixty of the triplets encode amino acids, one encodes a translation start site, and three encode translation stop sites. Which codons are used to encode an amino acid varies between species, and an examination of the patterns of codon usage can help us identify ORFs. Notice that four codons (CUU, CUC, CUA, CUG) each direct the addition of a leucine during the translation of an RNA message into a protein. Some species will use CUU frequently to encode leucine, while others may use CUA more often. This preference is known as **codon bias**, and it helps scientists more accurately predict the presence of a functioning ORF.

Predicting Sequence Functions

Although knowing a gene's sequence is a solid first step, our goal is to then identify the function that gene encodes. The most common method for gene function prediction is similarity based, which identifies genes by searching for similar sequences, just as we did when we searched for contaminating reads. The **Basic Local Alignment Search Tool** (**BLAST**) (Altschul et al., 1990) is used to search for similarities between an ORF and genes found in sequence repositories. Many genetic databases are available online. One of these, **GenBank**, is the NIH genetic sequence database that contains an annotated collection of all publicly available DNA sequences (Benson et al., 2013). Once a putative gene sequence is identified in a BLAST search, the encoded protein can be inferred and used in a second round of similarity searching. In this case we turn to protein databases, such as **UniProt**, which is a complete compendium of all known protein sequence data linked to functional information about known proteins (UniProt Consortium et al., 2021). Similarity-searching algorithms have been around for over 40 years, and we have success rates of over 97% in identifying already-known genes and proteins. The most significant limitation of similarity-based methods is that you can't identify what you don't know. If a gene encodes an unidentified protein, which is common in microbiome studies, you will have located what appears to be an ORF but will not be able to attach a gene name or function to it.

5.3 MICROBIOME DATA ANALYSIS

Finally, we are ready to start the analysis we have been waiting for. The main questions microbiome researchers usually have include: How does one group differ from a second group? and Is a sample indicative of a particular disease? Our study is focused on the first question: Do babies born by C-section (one group) differ from those born vaginally (second group) in terms of gut microbiome diversity and incidence of asthma?

Online Data Analysis Package Programs

Most of the analyses of microbiome data occur within one or more of the numerous online software packages. One of the very first such resources was **mothur** (https://mothur.org/), which was released in early 2009 and has become one of the most cited bioinformatics tools for analyzing 16S rRNA gene sequences (Schloss et al., 2009). The most widely used microbiome analysis package today is **QIIME 2** (pronounced like *chime two*), which stands for **Quantitative Insights into Microbial Ecology** (Caporaso et al., 2010). These, and numerous other such data analysis resources, allow you to submit your raw DNA sequence reads and choose the types of analysis you wish to perform, and then the algorithms do most of the hard work for you. They even perform the quality control, contig assembly, and gene prediction methods we discussed previously.

Instruction in how to run your data through such programs is beyond the scope of this book. Further, it is no trivial task to learn how to take advantage of these powerful computer programs. However, there are numerous online tutorials if you wish to dive in further on your own. Some of these resources provide sample datasets and instruct you on how to engage in the analysis process, step by step. I highly recommend you give one of these a shot—as there is no substitute for learning by doing!

Operational Taxonomic Units (OTUs)

We now have high-quality sequence reads that are 16S rRNA gene amplicon sequences or whole or partial genome sequences or some combination of those. What do we do next? Let's focus on the 16S amplicon reads first. We will have perhaps 50,000 or more such sequence reads, each of which includes a portion of the 16S rRNA target gene. Our goal is to use these sequences to identify which microbes are present in our samples. **Figure 5.5** shows the procedures we employ. First, the sequence reads are aggregated into clusters of similar reads. These clusters are called **operational taxonomic units** (**OTUs**), which are defined as sequence reads that share sequence identity above a given threshold, often set to 97%. Clustering our sequences results in a reduction of the size of the raw dataset and also decreases the computational requirements for analysis. For example, if each OTU contains an average of 100 reads, we reduce our sequence read complexity from 100 to 1 by creating a consensus. Clustering our reads also limits the impact of sequencing errors, since the consensus generated from the cluster will use the most common nucleotide at each site, as illustrated in Figure 5.5.

Assigning Taxonomic Identities

Once we have generated our 16S rRNA consensus OTUs, we want to identify which taxons they represent, preferably species names. To do this we turn to reference data. There are numerous online options, such as Greengenes and SILVA. **SILVA** provides comprehensive, quality-checked, and regularly updated aligned ribosomal RNA gene sequences for all three domains of life (Quast et al., 2012). The SILVA home page can be found online at https://www.arb-silva.de/. At last count there were well over 11 million named ribosomal sequences in SILVA, as well as a phylogenetic tree that shows the evolutionary relationships among these sequences.

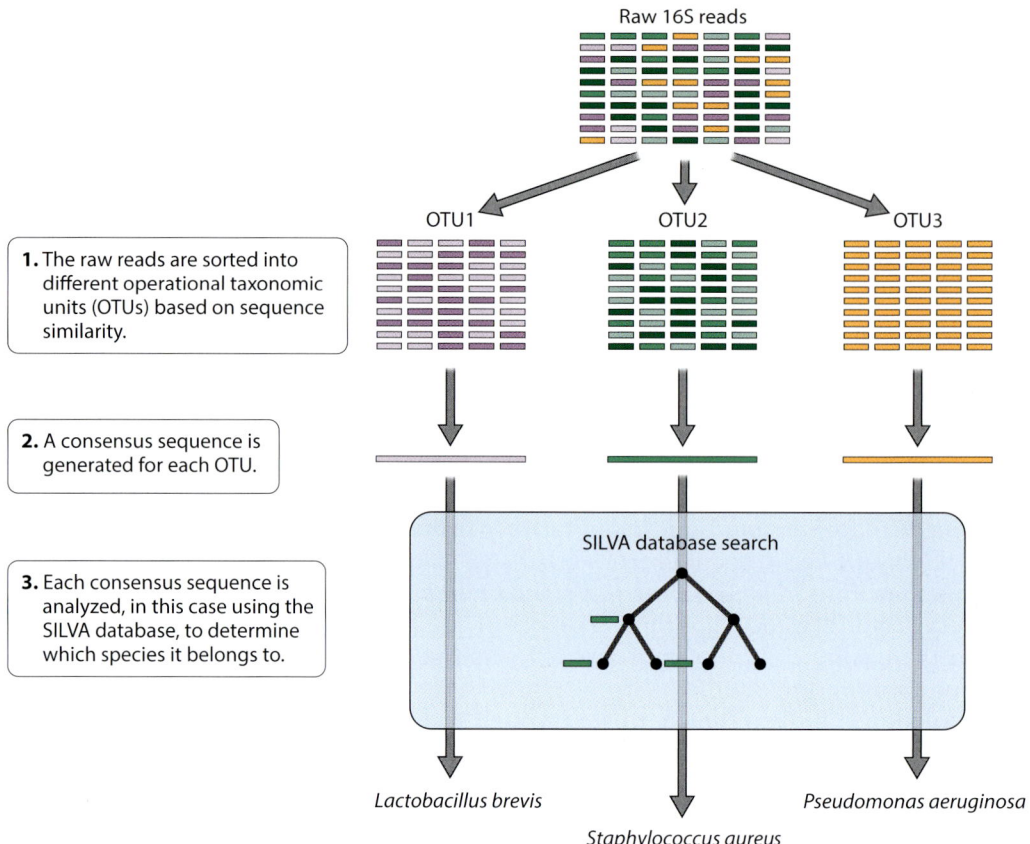

Figure 5.5 Identifying Genera and Species from 16S rRNA Sequence Reads Highly similar sequence reads can be grouped together into OTUs. A consensus sequence from an OTU can be input into a database of 16S rRNA sequences, such as SILVA, which can identify the taxon, often to the genus or species level, that this OTU represents. (After Zeng et al., 2017.)

Let's employ the pilot data we generated in chapter 4 and the SILVA database to examine how an OTU is named. In **Figure 5.6A** you will see the first OTU (OTU1) consensus sequence generated in our targeted gene amplification pilot study. OTU1 is in the form of a **FASTA** file, which is a format for representing DNA sequences, in which bases are given in a single-letter code [A, C, G, T, N] where N stands for a base that couldn't be identified accurately. The FASTA format begins with a **single-line identifier**, which is distinguished from the sequence data by a "greater than" symbol (>) in the first column. The text immediately following the symbol is the identifier of the sequence, which is a name given to this piece of data, while the remainder of that line may provide an optional description separated from the identifier by a space. The sequence data starts on the next line and ends if another line starting with a ">" appears.

To identify the taxonomic identity of each OTU, we paste the contents of a FASTA file into the SILVA search engine as shown in **Figure 5.6B** and click on the search button. SILVA quickly returns the information that OTU1 has a 100% match with the 16S rRNA gene identified from the bacterial species *Lactobacillus brevis* and provides an alignment between OTU1 and the *L. brevis* 16S rRNA sequences (**Figure 5.7**). How cool is that! We have attached a species name to our first piece of real data. In some cases, we may only be able to identify an OTU to the level of genus, or even family. However, if our OTU sequence is 100% identical to a sequence previously identified to the species level, SILVA will return that species identification.

In Figure 5.7 our sequence (OTU1) is found as the bottom line in the **sequence alignment**, which is a way of arranging nucleotide or amino acid sequences to identify regions of similarity. The assumption is that the similarity reflects functional, struc-

(A)

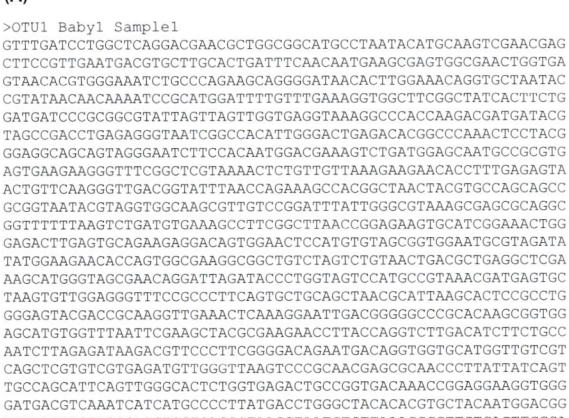

(B)

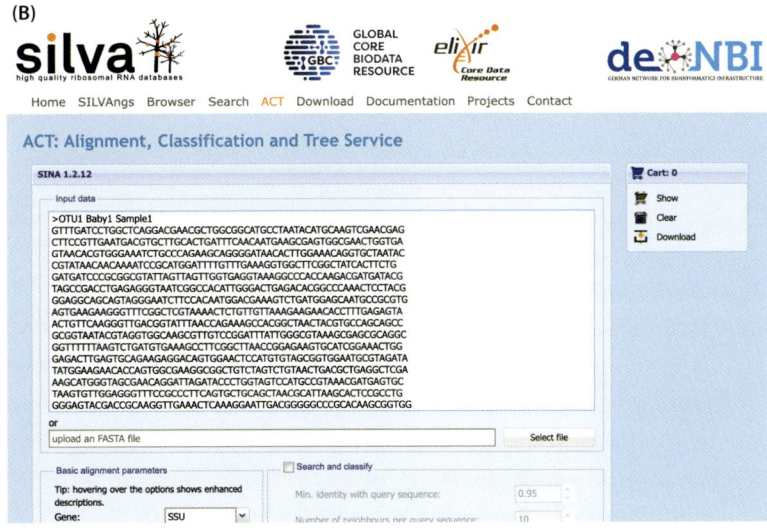

Figure 5.6 SILVA Database Search (A) The DNA sequence for OTU1 is shown in the FASTA file format. (B) This sequence is pasted into the sequence search browser in the SILVA database. (B SILVA Alignment, Classification and Tree (ACT) Service, release 138, CC-BY 4.0, https://creativecommons.org/licenses/by/4.0/)

tural, or evolutionary relationships between the sequences. In other words, more-similar sequences generally come from more closely related bacterial species and encode similar, if not identical, functions. Alignment algorithms usually seek to maximize the similarity between sequences. Gaps are inserted between the nucleotides or amino acids so that identical or similar characters are aligned. These gaps may represent insertion or deletion mutations that have occurred over time.

In our alignment we notice that OTU1 is identical to the 10 other *Lactobacillus brevis* 16S rRNA sequences previously submitted to SILVA. The color coding makes it easy to see that, starting at the first base, all the sequences have a U (highlighted in yellow) followed by an A (highlighted in blue) and then another U, etc. This extremely high level of sequence identity is not due to chance, and we are safe in assuming that these sequences all share a close evolutionary relationship. You might be wondering how our sequence acquired Us in place of the original Ts we submitted to SILVA. Since SILVA is devoted to ribosomal RNA, rather than RNA genes, it presents the data as an RNA sequence, inserting Us in the place of Ts in the original DNA sequence. In other words, SILVA automatically transcribes our DNA into RNA sequences.

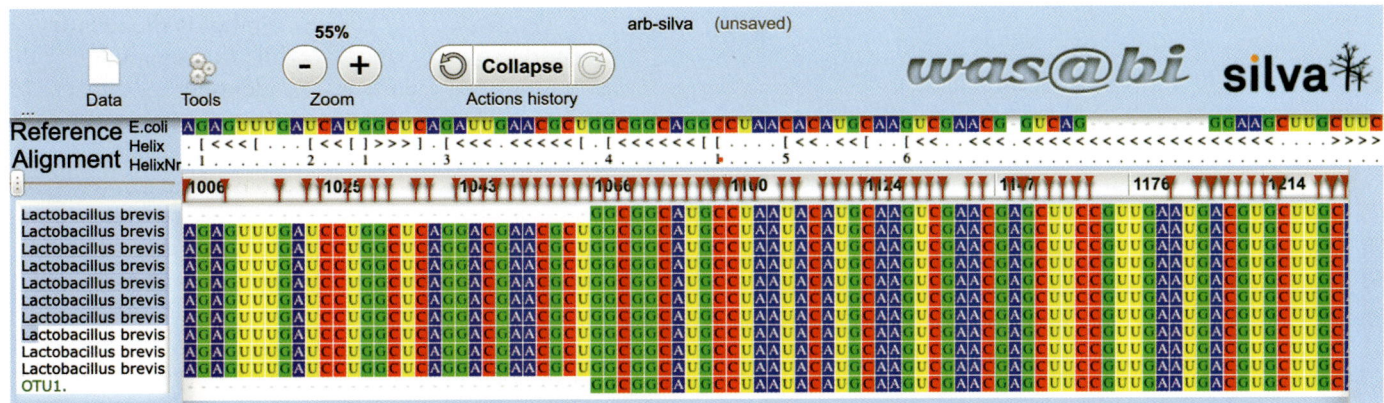

Figure 5.7 Alignment of OTU1 and SILVA Sequences from Our Pilot Study An alignment of *Lactobacillus brevis* 16S rRNA sequences in the SILVA database (first 10 rows) and our OTU (last row) is shown. All of these sequences are identical. (SILVA Alignment, Classification and Tree (ACT) Service, release 138, CC-BY 4.0, https://creativecommons.org/licenses/by/4.0/)

Table 5.1 OTUs and Their Assigned Species Names

OTU	SPECIES	PHYLUM	ABUNDANCE
OTU1	Lactobacillus brevis	Firmicutes	190
OTU2	Bifidobacterium bifidum	Actinobacteria	71
OTU3	Bacteroides fragilis	Bacteroidetes	81
OTU4	Staphylococcus aureus	Firmicutes	154
OTU5	Clostridium beijerinckii	Firmicutes	27
OTU6	Prevotella albensis	Bacteroidetes	12
OTU7	Enterobacter aerogenes	Pseudomonadota	74
OTU8	Propionibacterium avidum	Actinomycetota	129
OTU9	Corynebacterium durum	Actinomycetota	54

There are software packages, such as QIIME 2, that handle all the manipulations we just described in SILVA and incorporate our sequences into existing rRNA reference alignments. Based upon these alignments, each OTU is paired with the closest match in the reference database. **Table 5.1** shows the taxonomic identity of the first 9 OTUs in our C-section study. These species names are based upon a 100% match with one or more SILVA sequences. If you examine the phyla the species belong to, you might notice all are bacterial, which is not surprising since bacteria make up a very large fraction of our gut microbiota. Finally, there is a column for abundance which provides the total number of reads with that identical sequence in our study.

Species Accumulation Curves

When we describe a microbial community with sequence data, we are only taking a sample of that community. Not every piece of DNA present in the original community is sequenced. Not every microbe ends up in our sample. This can impact our description of the community. Recall that we sampled 1,000 babies in our C-section study and identified the OTUs present. **Table 5.2** shows a subsample of these data—just the first 6 babies sampled, 3 from natural delivery and 3 from C-sections. There are 9 OTUs in total (column 1), each identified to the species level (column 2). For

Table 5.2 Distribution of OTUs among Babies Born Vaginally or by C-section

			VAGINAL BIRTH				C-SECTION BIRTH				
OTU	SPECIES	PHYLUM	BABY 1	BABY 2	BABY 3	TOTAL	BABY 4	BABY 5	BABY 6	TOTAL	GRAND TOTAL
OTU1	Lactobacillus brevis	Firmicutes	55	77	51	183	3	1	3	7	190
OTU2	Bifidobacterium bifidum	Actinobacteria	14	9	33	56	7	3	5	15	71
OTU3	Bacteroides fragilis	Bacteroidetes	27	15	21	63	15	2	1	18	81
OTU4	Staphylococcus aureus	Firmicutes	5	4	3	12	41	57	44	142	154
OTU5	Clostridium beijerinckii	Firmicutes	1	2	1	4	7	11	5	23	27
OTU6	Prevotella albensis	Bacteroidetes	3	1	3	7	2	1	2	5	12
OTU7	Enterobacter aerogenes	Pseudomonadota	2	3	1	6	11	21	36	68	74
OTU8	Propionibacterium avidum	Actinomycetota	1	2	1	4	36	41	48	125	129
OTU9	Corynebacterium durum	Actinomycetota	3	1	4	8	22	15	9	46	54
			111	114	118	343	144	152	153	449	792

Note: Species richness = 9 for all 6 samples

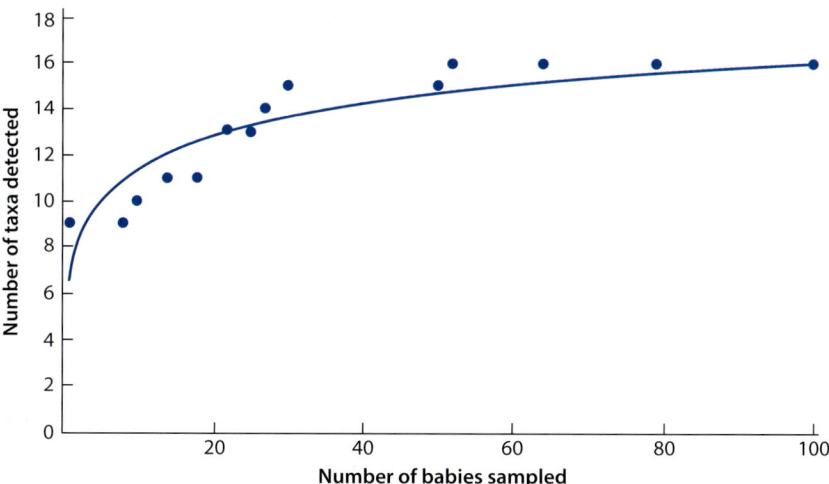

Figure 5.8 Species Accumulation Curve for Our Pilot Study A species accumulation curve shows the number of OTUs identified for each sample taken. The maximum number of OTUs identified here is 16, and only 55 samples are needed to capture this amount of diversity. A species accumulation curve tells us at what point taking additional samples no longer provides a meaningful benefit to our study.

ease of identification, column 3 provides the phylum-level rank. The subsequent columns provide various summaries of the OTU data. Look first at baby 1, the very first baby we sampled. If you scan the OTU abundance column (under "Baby 1"), you find that the fecal sample from this baby contains at least one representative of all the 9 unique OTUs. Also note that some of the OTUs are common in baby 1 (such as OTU1) and some are not (such as OTU8). Table 5.2 provides this same information for each of the first 6 babies in our pilot study.

Clearly sampling 3 babies from the treatment group and 3 from the control group is not sufficient to accurately reflect the microbiome diversity in the two groups. But what number would be enough? Let's take the first 100 babies sampled in our study and pool the OTU abundances to generate a **species accumulation curve**, also known as a **rarefaction** curve, which can tell us how many taxa we can expect to detect from a given sample size (**Figure 5.8**). This curve, which plots the number of babies sampled (*x*-axis) versus the number of taxa detected (*y*-axis), shows us how much more information we obtain with increasing sample sizes. With 1 baby we identified 9 OTUs. With 22 babies we identified slightly more, a total of 13 unique OTUs. With 64 babies we identified 16 OTUs, which is the maximum number of taxa identified in the entire set of 100 babies. This tells us critical information, which is that a sample size of about 55 babies would have captured all the OTU diversity present in our pilot study.

Another way to think about this is, how much effort are we willing to put into our study to ensure that we capture all or most of the actual species present in these samples? As we increase our effort, we will repeatedly find the same abundant OTUs, as well as some rarer ones. In other words, the benefits of increasing sample size will level off eventually or reach a plateau. Sampling more babies will not greatly increase the number of unique OTUs we can identify and will add significant costs to our study. If we had chosen to carry out a pilot study with just 50 samples from the treatment group and 50 from the control group, we could have used this information to significantly reduce the sample size in our real study, from 500 to 100 from each group. Although our rarefaction curve tells us we could reduce our sample size even more, to a mere 55 samples per treatment or control, without losing significant information, we still expect as much as a 50% dropout rate, so we revise our study design and propose a total sample size of 100 per group. What a savings! We had originally proposed 500 babies per group, so we have eliminated 800 newborns in our study! That is an enormous savings in time and effort on everyone's part!

5.4 SPECIES DIVERSITY MEASURES

Now that we have identified the species (and their relative abundance) present in the samples, we are ready to dive into a comparison of their microbial diversity. **Biodiversity**, which is the variety of life in a particular habitat or ecosystem, can be measured at several scales, two of which you will encounter repeatedly in microbiome studies: alpha and beta diversity. **Alpha diversity** describes the species diversity *within* a sample, while **beta diversity** describes the species diversity *between* samples. **Figure 5.9** provides a colorful illustration of four microbiome samples. Each sample is shown as a box containing some number of colored shapes, where each shape is an individual microbe, and each color is a species (or OTU). There are in total seven species (shapes) in this set of 4 samples (boxes). The number of species in each box represents its alpha diversity. A box harboring only a single species of microbe has no alpha diversity, as there is only one species. The more species in a box, the more diversity in that sample and the higher the measures of alpha diversity. In contrast, beta diversity refers to the difference in the number of species when 2 samples (boxes) are compared. If 2 samples share identical species, their beta diversity is zero. The more difference there is in the types of species found in the 2 samples, the higher the beta diversity. Since alpha and beta diversity are extremely important and commonly used measurements in microbiome studies, we will now describe in greater detail exactly how these values are calculated and what information they provide about our samples.

Alpha Diversity

Alpha diversity is a measure of the **species richness** (number of species), **evenness** (abundance of each species), or both in a sample. It is a measure that informs us about how diverse a sample is and provides a means of comparing diversity among samples.

SPECIES RICHNESS One of the first questions you might ask is simply, How many species are present, or what is the species richness? Let's return to Figure 5.9 and

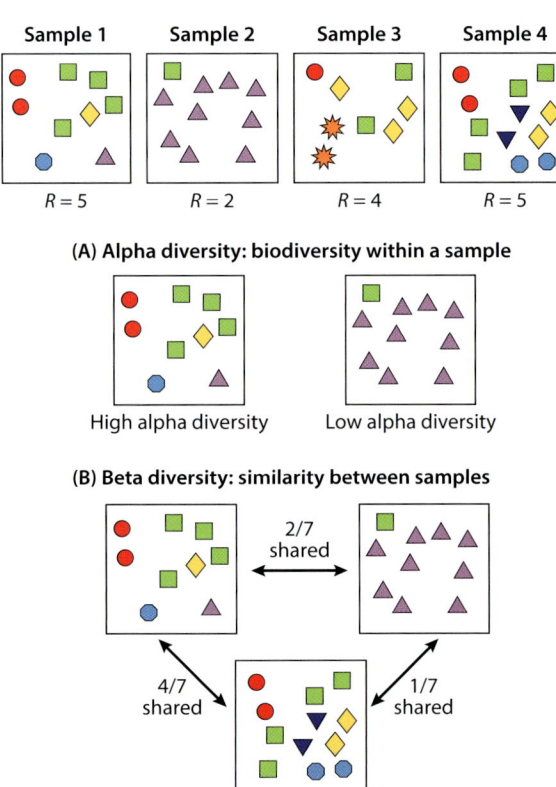

Figure 5.9 Biodiversity Estimates Applied to Samples of Microbial Species Each box represents a microbiome sample, and each colored dot represents a species. The two main ways of measuring the biodiversity of microbiomes are alpha (A) and beta (B) diversity. Alpha diversity measures the richness and evenness of one sample. In this example, we simply count up the number of colors (species) in each sample, which gives us a measure of species richness (R) for each. Here, sample 1 has higher alpha diversity than sample 2. Beta diversity measures how similar or dissimilar multiple samples are. In this example, we add up the number of colors that are shared by each pair of samples and divide that number by the total number of colors over all of the samples being compared (7 in this case). This calculation shows us that samples 5 and 2 have greater beta diversity than samples 5 and 1.

Table 5.3 Numbers of Each Species in Our Colored-Dot Example

SPECIES	SAMPLE 1 ABUNDANCE	FREQUENCY	SAMPLE 2 ABUNDANCE	FREQUENCY	SAMPLE 3 ABUNDANCE	FREQUENCY	SAMPLE 4 ABUNDANCE	FREQUENCY
● (red)	2.00	0.22	0.00	0.00	1.00	0.13	2.00	0.17
■ (green)	4.00	0.44	1.00	0.10	2.00	0.25	4.00	0.33
◆ (yellow)	1.00	0.11	0.00	0.00	3.00	0.38	2.00	0.17
● (blue)	1.00	0.11	0.00	0.00	0.00	0.00	2.00	0.17
▲ (purple)	1.00	0.11	9.00	0.90	0.00	0.00	0.00	0.00
✹	0.00	0.00	0.00	0.00	2.00	0.25	0.00	0.00
▼	0.00	0.00	0.00	0.00	0.00	0.00	2.00	0.17
Total	9.00		10.00		8.00		12.00	

Source: Shi, Baochen. Workshop 11. Metagenomics Analysis. CNSI 4338, UCLA. Provided by Qiime website

estimate the species richness of these 4 samples. Species richness, or R, is simply the species count per sample, which ranges from 2 to 5 in these 4 samples. This measure tells us that some of the samples have few species, such as sample 2 with a species richness of 2, while others have more species, such as sample 4 with a species richness of 5. With this measure alone, we can easily distinguish between samples with low and high species richness.

Now let's apply that same measure to a slightly more complicated dataset, the OTU data from the first 6 babies sampled in our pilot study, which is found in Table 5.2. We note that every sample, regardless of birth mode, hosts a total of nine species, which is a species richness of 9. If species richness were the only tool we had to analyze our microbiome data with, we might come to the erroneous conclusion that the samples are identical. However, we can clearly see that these samples do not appear to be very similar to each other at all.

SPECIES ABUNDANCE A second, and perhaps more useful, descriptor of our community is species evenness or abundance, which refers to the number of times a particular species is found in a sample. Let's return to Figure 5.9. **Table 5.3** shows the frequency of each color (species) in each sample. In sample 1 we note that there are nine shapes (microbes): two red, four green, and one each of yellow, blue, and purple. Their frequencies are then $2/9 = 0.22$, $4/9 = 0.44$, and $1/9 = 0.11$. Sometimes researchers convert these frequencies into percentages by multiplying each number by 100, but we will stick with frequencies here.

As we look across Table 5.3, we see differences in the species numbers in the 4 samples. However, it is challenging to make sense of even this relatively small dataset. So, we decide to plot the frequency of each species in a sample using a **stack plot**, which you see in **Figure 5.10**. Each column represents 1 sample, and it is divided into sections that are color coded to represent the species, with the height of each colored section corresponding to the frequency of that species in the sample. Turn to the first column and you see that the red species has a frequency of 0.22 while the green species has a frequency of 0.44, etc.

In a stack plot the relative contribution of each species jumps out at us. This format permits us to follow one species (or color) to see how it varies in frequency between samples. We can see that the red species has about the same frequency in samples 1, 3, and 4 but is absent in sample 2. We also see that some of the samples have a fairly even distribution of species (such as in sample 4), while others have highly skewed frequencies (such as in sample 2). An image is truly worth 1,000 rows in a spreadsheet! You will often see stack plots like this in microbiome research articles.

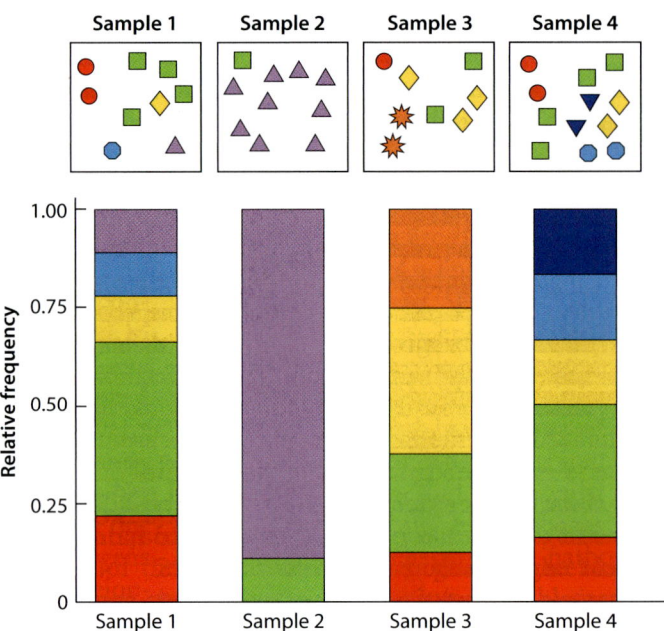

Figure 5.10 Relative Species Frequencies in the Colored-Microbe Example Comparing relative frequencies is a great way to get a sense of what our data looks like. Look at the 4 different samples, and you can immediately see that they're different, but we can't do that with bacteria! Instead, we use the 16S rRNA data to compare the abundance of species in a stack plot like this. Each column refers to one of our microbiome samples, and the various species present are indicated by different colors in the column, with their relative frequencies provided on the y-axis.

Let's return to our study data and perform the same type of species frequency analysis. **Table 5.4** shows the species frequencies in the first 6 babies, 3 born vaginally and 3 born via C-section. The first column lists the species in order from most to least frequent for the vaginal birth samples, and the fifth column does the same for the C-section samples. We can now readily see that the babies in the treatment group (C-section) versus the control group (vaginal birth) possess quite different species frequencies. The babies born vaginally have gut microbiomes dominated by *Lactobacillus* and *Bacteroides*, while those born by C-section have microbiomes dominated by *Staphylococcus* and *Propionibacterium*. Although both groups have the same species richness ($R = 9$), they have dramatically different species frequencies. That is why, when considering alpha diversity, we cannot look at species richness or species frequency alone. To get a full picture of the data, we must think of alpha diversity as a combination of the two.

Table 5.4 Species Frequencies of 6 Infant Gut Microbiomes

VAGINAL BIRTH				C-SECTION BIRTH			
SPECIES	**OTU**	**COUNT**	**FREQUENCY**	**SPECIES**	**OTU**	**COUNT**	**FREQUENCY**
Lactobacillus brevis	OTU1	183	0.53	*Staphylococcus aureus*	OTU4	142	0.32
Bacteroides fragilis	OTU3	63	0.18	*Propionibacterium avidum*	OTU8	125	0.28
Bifidobacterium bifidum	OTU2	56	0.16	*Enterobacter aerogenes*	OTU7	68	0.15
Staphylococcus aureus	OTU4	12	0.03	*Corynebacterium durum*	OTU9	46	0.10
Corynebacterium durum	OTU9	8	0.02	*Clostridium beijerinckii*	OTU5	23	0.05
Prevotella albensis	OTU6	7	0.02	*Bacteroides fragilis*	OTU3	18	0.04
Enterobacter aerogenes	OTU7	6	0.02	*Bifidobacterium bifidum*	OTU2	15	0.03
Clostridium beijerinckii	OTU5	4	0.01	*Lactobacillus brevis*	OTU1	7	0.02
Propionibacterium avidum	OTU8	4	0.01	*Prevotella albensis*	OTU6	5	0.01
		343				449	

Source: Dominguez-Bello et al., 2010.

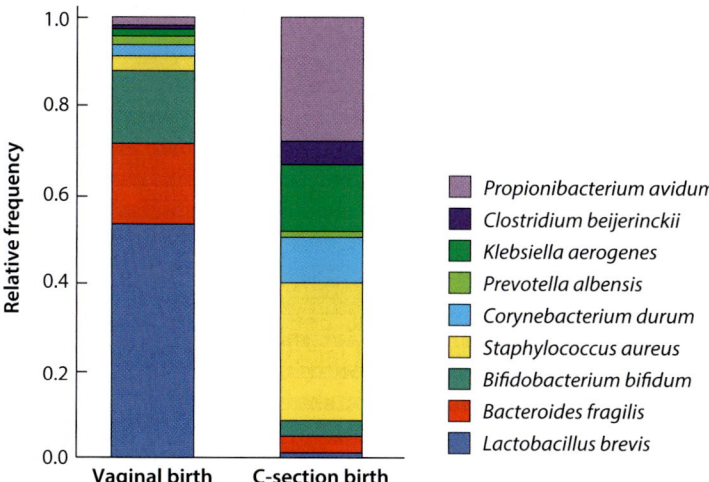

Figure 5.11 Comparing Relative Species Frequencies in Our Pilot Study Plotting the relative species frequencies is a great way to get a quick look at how our samples may be similar or different. It's not an incredibly robust measurement (more on that later), but for a brief comparison it's sufficient.

Diving more deeply into the details, we immediately notice that the samples from babies born vaginally have large numbers of *Lactobacillus brevis*, whereas the babies born via C-section have very few of that species. Vaginal microbiomes are usually dominated by lactobacilli, and so it makes sense that when a baby is born vaginally, it should get covered with the most abundant bacteria in the mother's vagina. Voila, we have identified one key difference between babies born vaginally and those born by C-section! That might seem like a trivial difference, but it is a relief to see that not only did our study produce meaningful data, but these data match the results from prior studies. This outcome helps us gain confidence in our experimental methods.

Figure 5.11 plots the relative species frequencies for our pilot data in the form of a stack plot. We can now clearly see that the two groups, C-section versus vaginal birth, have different species distributions. The vaginal samples are dominated by *Lactobacillus, Bacteroides*, and *Bifidobacterium*, while the C-section samples are dominated by *Staphylococcus* and *Propionibacterium*. We have successfully measured the species richness and the species frequencies in our samples. Both measures of alpha diversity provide informative descriptions of the species diversity in our samples. However, to obtain an even richer measure of diversity, we will turn to a slightly more complicated metric that incorporates both species richness and frequency, as well as phylogenetic relationships.

Alpha Diversity Indices

A **diversity index** is a mathematical measure of species diversity in a sample. It's a way for us to understand the diversity of a sample using a single number, making it much easier to compare large datasets. One of the most common measures is called the Shannon diversity index, and it is a measure of the species diversity in a community. Denoted as H, this index is calculated as follows:

$$H = -\Sigma p_i * \ln(p_i)$$

where

Σ = a Greek symbol that means "sum"

p_i = the proportion of the entire community made up of species i

$p_i = n_i/N$ = the number of species i divided by the total number of individuals

ln = the natural log

Table 5.5 Calculating Shannon Diversity for the Colored-Dot Example

SPECIES	SAMPLE 1			SAMPLE 2			SAMPLE 3			SAMPLE 4		
	FREQUENCY	$\ln p_i$	$p_i (\ln p_i)$	FREQUENCY	$\ln p_i$	$p_i (\ln p_i)$	FREQUENCY	$\ln p_i$	$p_i (\ln p_i)$	FREQUENCY	$\ln p_i$	$p_i (\ln p_i)$
●	0.22	–1.51	–0.33	0.00		0.00	0.13	–2.04	–0.27	0.17	–1.79	–0.30
■	0.44	–0.82	–0.36	0.10	–2.30	–0.23	0.25	–1.39	–0.35	0.33	–1.10	–0.37
◆	0.11	–2.21	–0.24	0.00		0.00	0.38	–0.97	–0.37	0.17	–1.79	–0.30
●	0.11	–2.21	–0.24	0.00		0.00	0.00		0.00	0.17	–1.79	–0.30
▲	0.11	–2.21	–0.24	0.9	–0.11	–0.09	0.00		0.00	0.00		0.00
✹	0.00		0.00	0.00		0.00	0.25	–1.39	–0.35	0.00		0.00
▼	0.00		0.00	0.00		0.00	0.00		0.00	0.17	–1.79	–0.30
		Sum	–1.41			–0.32			–1.32			–1.56
		H (–[Sum])	1.42			0.33			1.33			1.56

Source: Shi, Baochen. Workshop 11. Metagenomics Analysis. CNSI 4338, UCLA. Provided by Qiime website.

This is our first equation, so let's take time to ensure we understand what it is telling us. To read this equation, you simply follow the symbols across the page from left to right. We first find H, which is the symbol used to denote the Shannon diversity measure. H is equal to the negative value (–) of the sum (Σ) of each species' frequency (p_i) times (*) the natural log (ln) of that frequency (p_i). Wow—what a mouthful, and yet we made it through that equation with ease! However, there is one term, **natural log**, that may be new to you. In its simplest form, a logarithm (log) answers the question: How many of one number multiply together to make another number? For example, how many 2s multiply together to make 8? The answer: $2 \times 2 \times 2 = 8$, so we had to multiply three of the 2s to get 8, so the logarithm is 3. The natural log uses a special number and is one of the most useful functions in mathematics, with applications throughout the physical and biological sciences. It is beyond the scope of this book to explain the mathematics behind it, but one very simple definition is as follows: It is how many times we need to use e (Euler's number, equal to, roughly, 2.71828) in a multiplication to get our desired number.

To calculate H, say for our colored-dot example, we will employ a new spreadsheet. We already determined the frequency of each species (see Table 5.3), which we will carry over to a new spreadsheet (**Table 5.5**). We then add, for each sample, a column in which we take the natural log of each species frequency. The final column we require for each sample is the product of multiplying the frequency of each species by the natural log of that frequency. We now have all the numbers we require to calculate H. For each species in a sample, we calculate its frequency times the natural log of that frequency, then calculate the sum of those values and take its negative.

The resulting metric, H, tells us if a community has high or low levels of species diversity. H will usually vary between 0 and 3.5, although there is no theoretical limit to diversity measures. The higher the value of H, the greater the diversity of species in a particular community. The lower the value of H, the lesser the diversity. With $H = 0$, there is only one species in a sample. Take a moment and compare the H values of the 4 samples. Which sample has the highest H and which the lowest? We can create a graph of these H values (**Figure 5.12**), which makes it very simple to compare the diversity for the colored-dot example. Sample 2 has a dramatically lower diversity than any other sample, while the remaining 3 samples have quite similar levels of diversity.

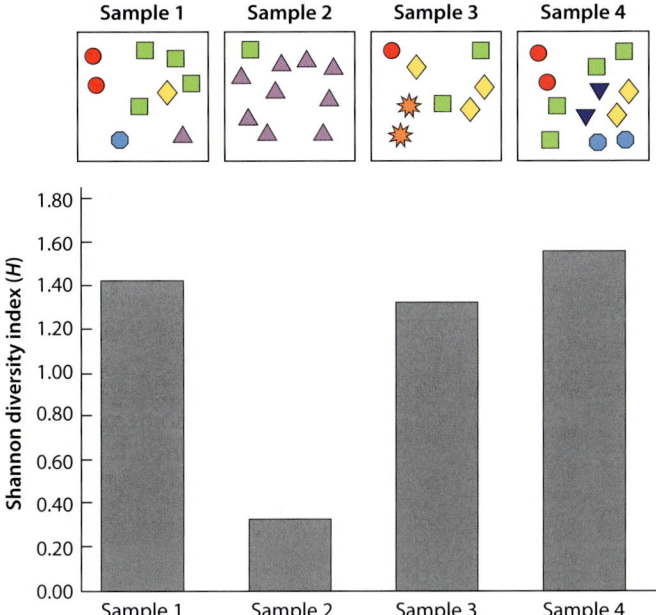

Figure 5.12 A Comparison of Shannon Diversity for the Colored-Microbe Example Reducing alpha diversity into a single value helps us to better visualize and compare our data. Graphs like this make excellent figures for our future publications!

Now let's apply these indices to the first 6 samples from our study. **Table 5.6** shows these same calculations separately for the pooled vaginal babies and **Table 5.7** shows the pooled calculations for the pooled C-section babies. We find that our two groups have slightly different H values, with the vaginal group showing less diversity ($H = 1.4$) and the C-section group showing higher diversity ($H = 1.7$). It is beyond the scope of this chapter to discuss how one might determine whether our H values are significantly different. If we wished to do so, we would need to perform a statistical test, but that would require far larger sample sizes than our 6 babies and numerous

Table 5.6 Calculating the Shannon Diversity Index for the Vaginal Samples in Our Pilot Study

SPECIES	OTU	VAGINAL SAMPLES ABUNDANCE	p_i	$\ln p_i$	$p_i (\ln p_i)$
Lactobacillus brevis	OTU1	183	0.53	−0.63	−0.3352
Bacteroides fragilis	OTU3	63	0.18	−1.69	−0.3352
Bifidobacterium bifidum	OTU2	56	0.16	−1.81	−0.3113
Staphylococcus aureus	OTU4	12	0.03	−3.35	−0.2959
Corynebacterium durum	OTU9	8	0.02	−3.76	−0.1173
Prevotella albensis	OTU6	7	0.02	−3.89	−0.0794
Enterobacter aerogenes	OTU7	6	0.02	−4.05	−0.0708
Clostridium beijerinckii	OTU5	4	0.01	−4.45	−0.0519
Propionibacterium avidum	OTU8	4	0.01	−4.45	−0.0519
		343	1		−1.4013
				×(−1)	1.4013
					$H = 1.40$

Source: Shi, Baochen. Workshop 11. Metagenomics Analysis. CNSI 4338, UCLA. Provided by Qiime website.

Table 5.7 Calculating the Shannon Diversity Index for the C-section Samples in Our Pilot Study

SPECIES	OTU	C-SECTION SAMPLES			
		ABUNDANCE	p_i	$\ln p_i$	$p_i (\ln p_i)$
Staphylococcus aureus	OTU4	142	0.32	−1.15	−0.3641
Propionibacterium avidum	OTU8	125	0.28	−1.28	−0.356
Enterobacter aerogenes	OTU7	68	0.15	−1.89	−0.2859
Corynebacterium durum	OTU9	46	0.10	−2.28	−0.2334
Clostridium beijerinckii	OTU5	23	0.05	−2.97	−0.1522
Bacteroides fragilis	OTU3	18	0.04	−3.22	−0.129
Bifidobacterium bifidum	OTU2	15	0.03	−3.40	−0.1136
Lactobacillus brevis	OTU1	7	0.02	−4.16	−0.0649
Prevotella albensis	OTU6	5	0.01	−4.50	−0.0501
		449	1		−1.749
				×(−1)	1.75
					H = 1.75

assumptions about the distribution of our data. Let's not worry about whether the difference in *H* we observe is significant for now and just note that they differ slightly.

Let's make things a bit more complicated, but also more realistic. You might imagine that 2 samples could have similar measures of richness and evenness but contain different species. Say one sample had all closely related species, whereas the other sample had more distantly related species. Species richness, species abundance, and even the Shannon diversity index would suggest that these samples are equally diverse, when we know that the latter sample incorporates a greater breadth of evolutionary distance and is therefore more diverse. We make a reasonable assumption that the greater the evolutionary distance between our species is, the more likely they are to have different metabolic functions and the like. How do we take that critical difference in evolutionary distance into account? One approach is to incorporate a phylogeny into our diversity indices. Faith (1992) first introduced the idea that you could employ the branch lengths of a phylogenetic tree to "weigh" species differently. Let's see how that is done.

Figure 5.13 provides an example of 2 new samples (red and blue) mapped onto a phylogeny. Our interest is in determining and then comparing diversity measures for these 2 samples while incorporating information we gain from their phylogeny. If we only considered species richness, we would conclude that the blue sample, which has six species, has a higher diversity than the red sample, with only five species. Now let's consider the phylogenetic relationships of the species.

Faith's Phylogenetic Diversity Metric

Faith's **phylogenetic diversity** (**PD**) involves measuring the branch lengths in the phylogeny. You may recall from chapter 1 that when Carl Woese was generating a 16S rRNA–based tree of life, he noted that the more differences in 16S fragments observed between two taxa, the greater their evolutionary distance. Regardless of whether we are comparing morphology, 16S fragments, or DNA sequences, in general, the more differences between two taxa, the greater their evolutionary distance. Let's apply this to PD measurements in Figure 5.13. This phylogeny shows the relationships among species identified in 2 samples (red and blue) together with several close relatives not found in either sample (yellow). The branches connecting the red

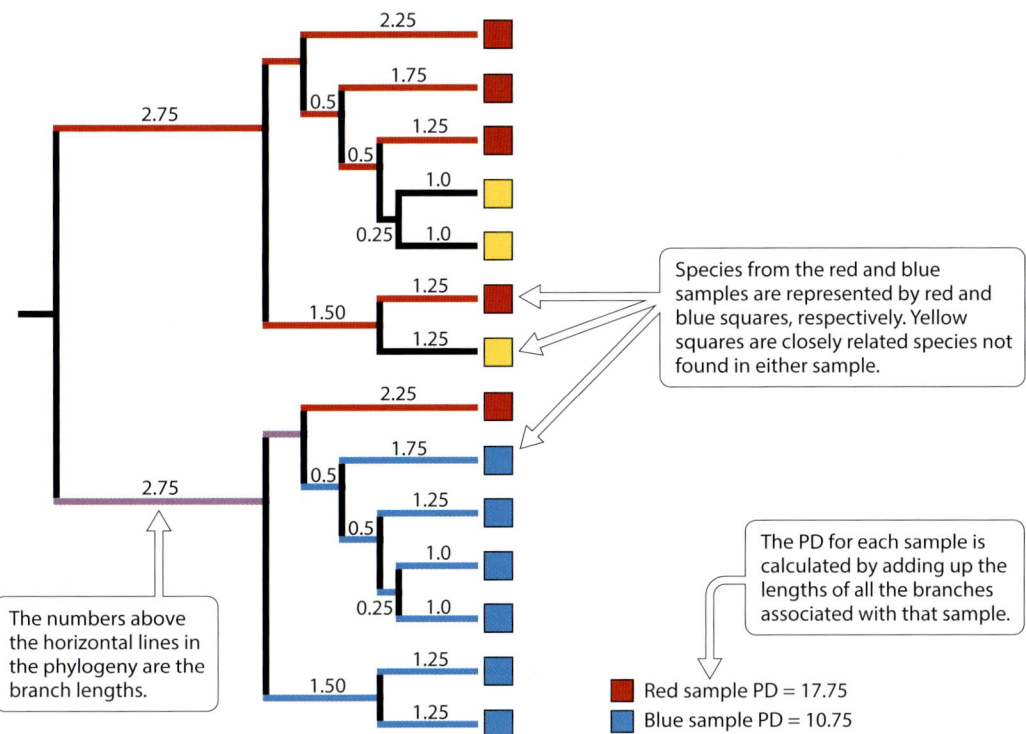

Figure 5.13 Incorporating Phylogenetic Distance in Diversity Measures An example of how one can incorporate the phylogenetic distance represented in a sample into a diversity measure is shown. In this example, the blue taxa were obtained from one sample and the red taxa from a separate sample. The yellow boxes represent taxa not found in either sample. A phylogenetic tree is inferred and a comparison made between the branch lengths connecting the blue boxes and those connecting the red boxes. By incorporating these branch lengths into our diversity measure, we can more readily see that the red sample includes far more diversity of taxa than the blue sample. PD = phylogenetic diversity

or blue taxa are indicated with red or blue shading. The purple branches are common to both samples and are included in calculating branch length sums. The sum of all the red and purple branch lengths is 17.75, which is this sample's PD. The vertical branches are not included in the total because they have no evolutionary significance—they simply indicate the connections between our taxa. Now add up the branch lengths for all of the blue-shaded and purple-shaded branches, and you obtain a PD of 10.75 for the blue taxa. When we consider the PD metric, we realize that although the red sample has fewer species, those taxa cover a greater evolutionary distance, which suggests that their genomes may encode more-diverse functions.

Figure 5.14 incorporates a phylogeny into our colored-microbe example. Since we weren't given a phylogeny, we have had to create one, so we aligned 16S rDNA sequences that correspond to the species, which provides information about how closely related any of the taxa are to each other. The more differences there are between a pair of sequences, the greater the evolutionary distance between them. In the resulting phylogeny (Figure 5.14), we can see that in this example the yellow and red species are each other's closest relatives (the shortest total branch lengths separate them relative to any other species), while the green and yellow species are quite distantly related (longer branch lengths separating them relative to any other species).

We can use the branch lengths in this tree to calculate the PD for each sample. For example, the distance from the base of this tree to the green taxon is 6. If we include the branch lengths to the purple taxon, the distance increases to 7, and so on. If we add together all possible branch lengths, the total distance represented in this tree is 16. Now let's look only at the branches that connect dots from sample 1—so

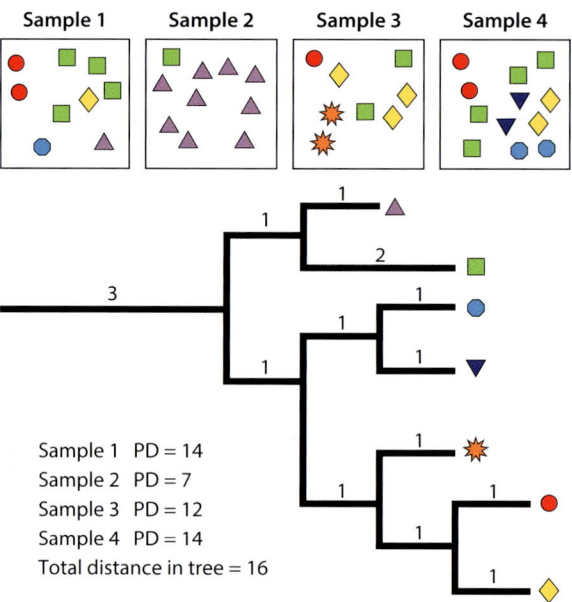

Figure 5.14 Incorporating Phylogeny Distance in Diversity Measures of the Colored-Microbe Example The details involved in incorporating phylogenetic distance into diversity measures for the colored-microbe example is shown. A phylogenetic tree is inferred based upon the DNA sequences obtained for each of the colored-microbe "species." The total branch lengths represented in each sample are then computed. In this example, the second sample includes only two taxa and thus has a far shorter branch length and therefore less phylogenetic diversity, while the taxa in samples 1 and 4 encompass the greatest branch lengths and thus those samples share the highest phylogenetic diversity.

only the branches that connect the purple, green, blue, red, and yellow taxa. Count the branch lengths that connect this subset, and you find the total distance is 14. Do the same for each of the samples, and you arrive at the numbers shown in the PD table provided in Figure 5.14. You can see that the species present in sample 4 span 88% of the full phylogenetic diversity represented in this tree (14/16 PD), while the species in sample 2 span only 44% of the diversity (7/16 PD). The use of a phylogenetic tree can be important because, in general, more-similar species can have more-similar ecological roles. A sample that is truly diverse will not just have more species or a more balanced composition of species, but will also have a range of different species with unique functions and niches.

A common way to visualize diversity measures is a **box and whiskers plot**, such as is shown in **Figure 5.15** for the 6 babies. The plots provide a way to display the level of variation in your data. To create a box and whiskers plot, you need to determine five statistics from your relative species abundance data: the **minimum value** (the smallest value in the dataset), the **second quartile** (the value below which the lower 25% of the data are contained), the **median value** (the middle number in a range of numbers), the **third quartile** (the value above which the upper 25% of the data are contained), and the **maximum value** (the largest value in the dataset). Once you have those values, your spreadsheet program can create a plot such as we see in Figure 5.15 for our pilot data.

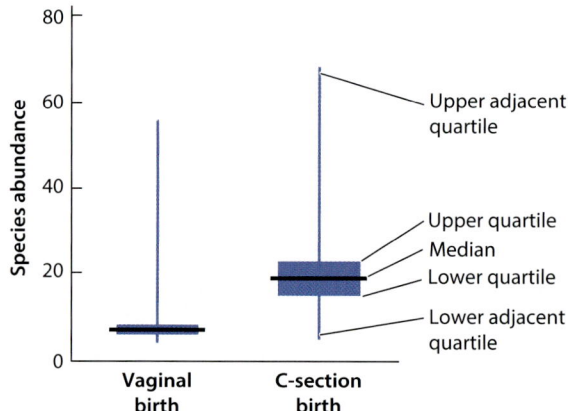

Figure 5.15 Bar and Whisker Plot of Pilot Data We can compare our alpha diversity measures between samples more readily when we plot our data in a bar and whiskers plot. In this example we have plotted the pilot data for the two birth modes separately.

The key information you glean from the type of plot shown in Figure 5.15 is whether the aggregate samples from each group (in our case vaginal versus C-section babies) are different, and if so, how? The blue boxes in the figure represent the **interquartile range**, or the middle half of the values in each group. If two boxes do *not* overlap with one another, say one is completely above or below the other, then there *is* a difference between the two groups. In Figure 5.15 we see that the two boxes showing the interquartile ranges do not overlap—the C-section box is higher than, and outside the distribution of, the box for the vaginal data. That observation tells us that our two datasets are indeed different! Although we cannot provide a level of significance to the difference observed, we now feel confident that the variability in alpha diversity *within* our sample of C-section or vaginal microbiomes is smaller than the variability *between* them. If you think back to our null hypothesis from chapter 4, recall that we stated that the microbiome diversity of our two treatments would not differ. We have now proven that they do differ, so we can reject our null hypothesis and state with confidence that the gut microbiomes of babies born vaginally are different from those of babies born via C-section. Congratulations, you have rejected your first null hypothesis!

Let's imagine for a moment that the two boxes did overlap in distribution. Could we still determine whether the distributions were different? To do so we would have to consider the lines inside the boxes. The middle lines are the medians, or the "middle" values of each group. If the value of the median of one box lies outside another box, there is *likely* to be a difference between the two groups. However, if both median lines lie within the overlap between two boxes, then more analysis is required. The whiskers are the lines projecting out from each box. They extend from the largest to the smallest values of each group. Larger ranges indicate a wider distribution, that is, more scattered data. That's something to look for when comparing box plots, especially when the medians are similar. This set of box and whisker plots reveal that the C-section and vaginally born babies have different distributions of their alpha diversity and that there is considerable variation in the data, shown by the long whiskers projecting out of both boxes.

Beta Diversity

A second key measure used in microbiome studies is beta diversity, which measures the difference in species diversity between samples. Beta diversity is often represented as a measure of dissimilarity, rather than similarity. Thus, a high beta diversity index indicates that 2 samples have a very different species composition from each other, while a low beta diversity index shows a high level of similarity. One very commonly employed beta diversity index is the Bray-Curtis (BC) dissimilarity measure, which is based on the differences in species counts and abundances between 2 samples (Bray & Curtis, 1957).

Bray-Curtis Dissimilarity Metric

The **Bray-Curtis dissimilarity** (**BC**) metric is a bit more complicated to calculate than the Shannon index. The standard formula is this:

$$BC_{ij} = 1 - ((2 * C_{ij}) / (S_i + S_j))$$

where

C_{ij} = the sum of the lesser values for the number of species found in each sample

S_i = the total number of species found in sample i

S_j = the total number of species found in sample j

Let's read this equation from left to right. The BC value for a comparison of 2 samples (i and j) is BC_{ij}. We calculate BC_{ij} by taking two times the sum of the lesser values for the number of species found in a sample ($2 * C_{ij}$) and dividing it by the sum of the total number of species found in each of the 2 samples. Finally, we subtract this number from 1. When calculating the quotient, taking the lesser value for any shared species means that only the number of species that is shared by both samples is counted as similar.

Yuck! What in the world does that mean, and why are we doing it? Let's walk through an example to see if that makes things clearer. Imagine that you have 2 samples and sample A has 5 members of a species, and sample B has 8 members of that same species. In this example only 5 units are counted because that is the abundance that is shared by both samples. Sample A does not have the extra 3 members of this species that sample B has, so those 3 extra members would not count towards the samples' similarity. You can think of this quotient as a measure of the species abundance the samples share divided by the total species abundance the 2 samples have, which gives the fraction of their species abundance that is the same. By subtracting this measure of similarity from 1, the overall dissimilarity is calculated.

The Bray-Curtis dissimilarity metric ranges from 0 to 1 where

- 0 indicates that two sites have zero dissimilarity—that is, they share the exact same species, and
- 1 indicates that two sites have complete dissimilarity—they share none of the same species.

Let's work through this equation with our colored-dot data. **Table 5.8** provides the values for all the variables for the colored dots, and the bottom of the table gives the BC for each pair of samples. Let's start with a comparison between samples 1 and 2. The first step is to identify which species the 2 samples have in common, which are green and purple. We then identify which of the 2 samples has the smallest number of each species. In this example, sample 1 has 4 greens and sample 2 has 1 green, while sample 1 has 1 purple and sample 2 has 9 purples. We add the numbers 1 (the number of greens in sample 2) and 1 (the number of purples in sample 1). This means that $C_{1,2}$ is 2 for this pair of samples. We also note that sample 1 has a total of 9 bacteria ($S_1 = 9$), while sample 2 has a total of 10 bacteria ($S_2 = 10$). We can plug these numbers into the BC formula:

$$BC_{ij} = 1 - (2 * C_{ij}) / (S_i + S_j)$$

with $C_{1,2} = 2$

$S_1 = 9$

$S_2 = 10$

$BC = 1 - (2 * 2) / (9 + 10) = 0.79$

Table 5.8 Calculating the Bray-Curtis Dissimilarity Measure

SPECIES	SAMPLE 1 ABUNDANCE	SAMPLE 1 FREQUENCY	SAMPLE 2 ABUNDANCE	SAMPLE 2 FREQUENCY	SAMPLE 3 ABUNDANCE	SAMPLE 3 FREQUENCY	SAMPLE 4 ABUNDANCE	SAMPLE 4 FREQUENCY
●	2.00	0.22	0.00	0.00	1.00	0.13	2.00	0.17
■	4.00	0.44	1.00	0.10	2.00	0.25	4.00	0.33
◆	1.00	0.11	0.00	0.00	3.00	0.38	2.00	0.17
●	1.00	0.11	0.00	0.00	0.00	0.00	2.00	0.17
▲	1.00	0.11	9.00	0.90	0.00	0.00	0.00	0.00
✹	0.00	0.00	0.00	0.00	2.00	0.25	0.00	0.00
▼	0.00	0.00	0.00	0.00	0.00	0.00	2.00	0.17
Total	9.00		10.00		8.00		12.00	

Source: Shi, Baochen. Workshop 11. Metagenomics Analysis. CNSI 4338, UCLA. Provided by Qiime website

Table 5.9 BC Values Taxa Pairs

SAMPLE 1 vs 2		SAMPLE 1 vs 3		SAMPLE 1 vs 4		SAMPLE 2 vs 3		SAMPLE 2 vs 4		SAMPLE 3 vs 4	
C_{12}	2.00	C_{13}	4.00	C_{14}	8.00	C_{23}	1.00	C_{24}	1.00	C_{34}	5.00
S_1	9.00	S_1	9.00	S_1	9.00	S_2	10.00	S_2	10.00	S_3	8.00
S_2	10.00	S_3	8.00	S_4	12.00	S_3	8.00	S_4	12.00	S_4	12.00
BC_{12}	0.79	BC_{13}	0.53	BC_{14}	0.24	BC_{23}	0.89	BC_{24}	0.91	BC_{34}	0.50

Source: Shi, Baochen. Workshop 11. Metagenomics Analysis. CNSI 4338, UCLA. Provided by Qiime website.

We obtain a BC value of 0.79 for this pair of samples. We then continue with our calculations, only now we employ a spreadsheet to keep track of our data. Table 5.9 shows the BC values for each pair of taxa. We see that BC values range from 0.24 (samples 1 versus 4) to 0.91 (samples 2 versus 4). When we plot these data (Figure 5.16), we can readily see that samples 2 and 4 are the most dissimilar (or least similar) and samples 1 and 4 are the least dissimilar (or most similar). Wow, that was a lot of work to find out how similar, or not, 2 samples are. And, just as we noted above, these measures are fine if our 2 samples have species with roughly the same evolutionary distance in them. However, when they don't, we need to take that into account.

Unique Fraction Metric

Just as phylogenetic information can render alpha diversity metrics more informative, beta diversity metrics also can be improved by incorporating phylogenetic information. The most common phylogenetic beta diversity metric is the unweighted UniFrac. The **unique fraction** (**UniFrac**) metric was initially introduced by Lozupone and Knight (2005) and has been widely applied in microbial ecology since. The purpose of this metric is to incorporate information about the phylogenetic relationships of the taxa within a sample. If one sample has numerous species but they are all closely related, that sample will have less beta diversity when compared with one that has the same number of species that are more distantly related.

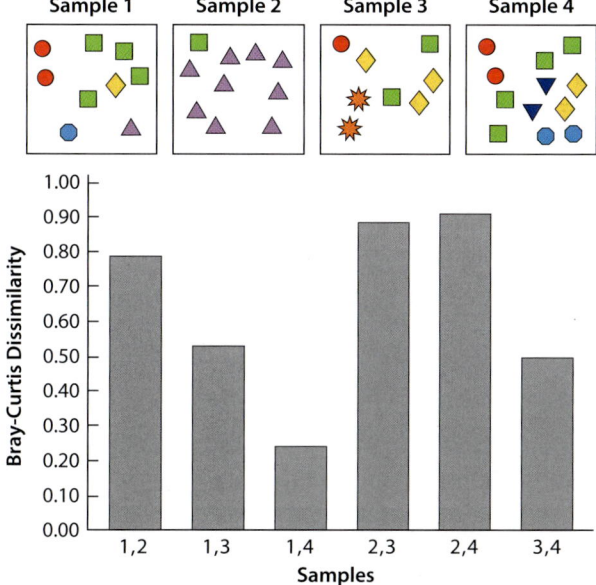

Figure 5.16. Plotting Bray-Curtis Dissimilarity for the Colored-Microbe Example This figure provides a plot of the Bray-Curtis dissimilarity measure for each of the colored-microbe samples. Samples 1 and 4 are the most similar, while samples 2 and 4 show the greatest dissimilarity.

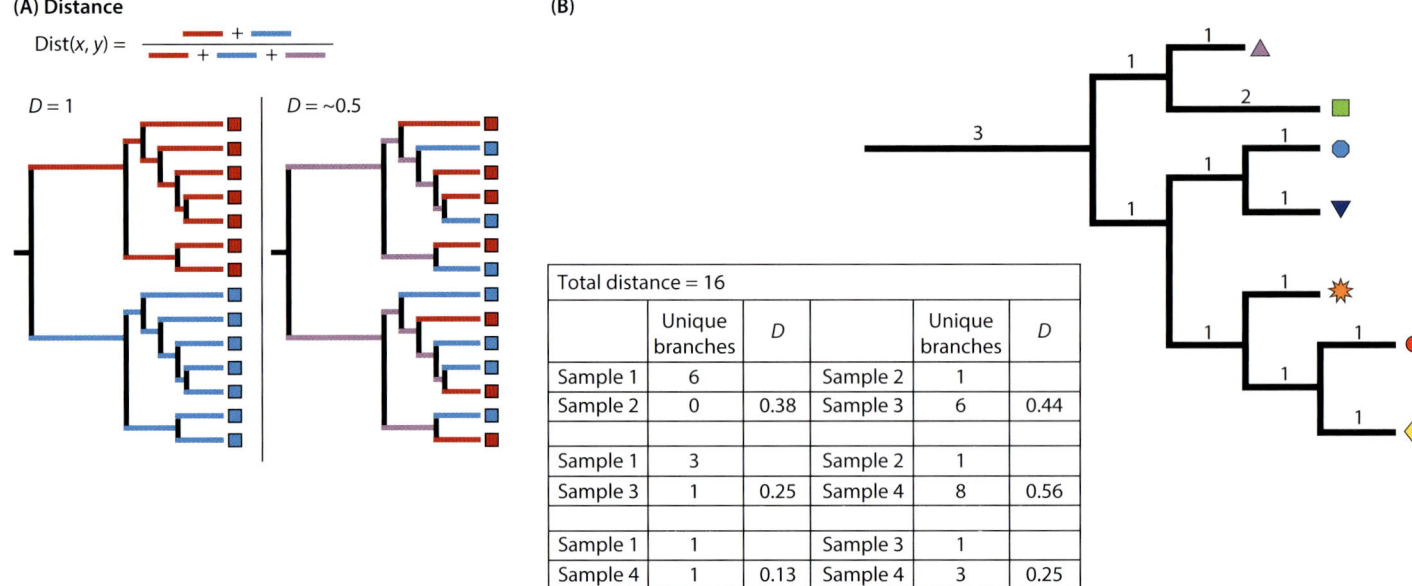

Figure 5.17 A Visual Comparison of UniFrac Trees Unlike some beta diversity calculations, UniFrac considers the phylogenies of the species identified in a given sample. (A) These two phylogenies include one in which branches that are red or blue are unique to that sample, and another in which purple branches are not unique. Notice how the trees on the left and right are the same shape, but changing which samples the species are from changes our UniFrac calculation drastically! (B) This phylogeny is for the colored-microbe example. The table contains colored-microbe UniFrac data.

The unweighted UniFrac distance between a pair of samples (A and B) is a branch-length-based qualitative beta diversity measure defined as follows:

$$\text{UniFrac distance } (D) = U_{AB} = \text{unique/observed}$$

where

- *unique* refers to the unique branch length, or branch length that only leads to OTU(s) observed in either sample A or sample B, and
- *observed* refers to the total branch length observed in either sample A or sample B.

This equation should be relatively easy for you to understand. You start with the UniFrac distance, which is given the symbol D and is equal to U_{AB}, which is simply the length of unique branches in either sample divided by the total branch length of the tree. **Figure 5.17A** shows an example phylogeny. The red branches are unique to the red sample, the blue branches are unique to the blue sample, and the purple branches are shared by both. The tree on the left *illustrates* the situation in which the 2 samples share none of their branches and the entire tree is composed of unique branches. In this case, the UniFrac distance between the samples is 1 (number of unique branches divided by the total number of branches). The tree on the right shows a situation in which the 2 samples share about half of their branches, so their unique branches represent only about half of the total branches in the tree. In this case, the UniFrac distance is ~0.5.

Let's apply this metric to our colored-dot example. **Figure 5.17B** shows a phylogenetic tree for the species in these samples. The spreadsheet in the figure shows the UniFrac calculations. We first determine the total length of the branches in the tree, which is 16. We then calculate the unique branch lengths for each pair of samples. For example, when comparing samples 1 and 2, the red, blue, and yellow species are unique (see Figure 5.16), so we add up those branch lengths and arrive at 6 units. The second sample has no unique taxa, or 0 units. The sum of the unique branch lengths

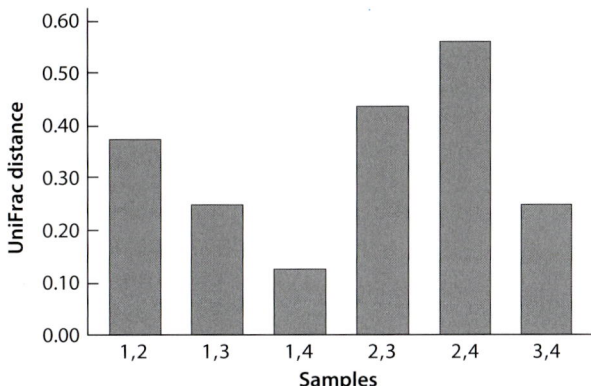

Figure 5.18 Using UniFrac Distance to Visualize Diversity in the Colored-Microbe Example Using a plot of UniFrac distances to visualize the diversity of our colored microbes, we can easily see which samples are most similar (such as 1 and 4) and which are the least similar (such as 2 and 4).

is thus 6 + 0, which we divide by the total of all the branch lengths in the tree (16) and arrive at a UniFrac measure of 0.375 units. We apply this same calculation for each pair of samples and find UniFrac distances that range from 0.13 to 0.56. When we plot these distances (**Figure 5.18**), the differences between the samples jump out at us. Samples 1 and 4 are the most similar, with the lowest value of D, while samples 2 and 4 are the least similar, with the highest value of D. The UniFrac measure we just employed only considered the presence or absence of lineages. It is beyond the scope of this section, but one could factor in the abundance of those shared and unique lineages in the UniFrac distance calculation, in which case it is a weighted UniFrac.

If we wished to determine the significance of any of the differences we observed between our UniFrac measures, we would need to go a further step in the analysis. In short, we would return to our null hypothesis, which you may recall was that the two birth modes would not result in different taxa or different taxa abundances in our samples. Another way to say that is that the within-group (within birth mode) UniFrac distances would not be significantly smaller than the between-group (between birth mode) distances. We would then employ one of several statistical tests to determine whether the null hypothesis would be accepted or rejected. For example, we could employ a t-test to look for a significant difference between the average UniFrac measures of two groups. Determining which statistical test is appropriate for our data is beyond the scope of this discussion. The point is that UniFrac measures can be employed to determine whether the differences we observe between treatments are significant or not.

5.5 VISUALIZING DIVERSITY ESTIMATES

Our beta diversity measures result in large pairwise distance matrices. Imagine that we had 100 samples in which we identified 50 species. The species abundance data would require a 50×100 table to show the relative abundance of each species in each sample. The beta diversity measures would require a 100×100 distance matrix, or 1,000 cells, as each sample would have to be compared against all others. How would you make sense of all that data? How do you know which samples are similar to one another, and which ones are different? How do you know how similar a group of samples is overall, and what subgroups of similar samples may be present? How do you know which taxa are responsible for such similarities or differences? What a mess of data!

Principal Component Analysis

Fortunately, there are several approaches to visualizing these types of data. We will discuss one of the more commonly employed measures in microbiome studies, **principal component analysis** (**PCA**). PCA is a dimension-reducing method that can reduce the size of large datasets. It does so by transforming a large set of variables into a smaller set that still contains most of the information in the larger set. Sounds like magic, eh? Our 100×100 distance matrix would have 100 dimensions if we were to

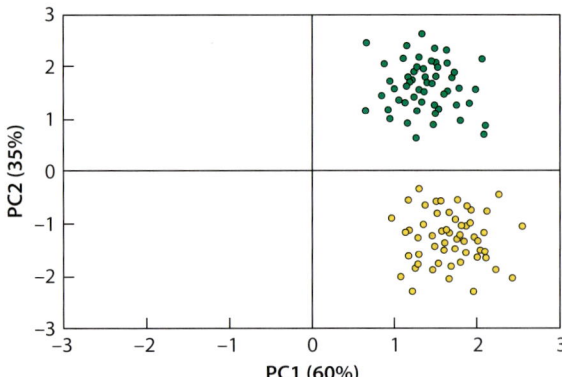

Figure 5.19 PCA Plot of Beta Diversity for Our Pilot Study This figure provides a PCA plot for the data generated in our pilot study, with data from C-section babies indicated in yellow and data from babies born vaginally in green. It is immediately clear that these two birth modes result in babies with quite different levels of beta diversity. The *x*-axis is the first principal component, which explains 60% of the variance in our data. The *y*-axis is the second principal component, which explains 35% of our data. Although there are numerous other principal components, these first two explain 95% of the variation in our 2 samples!

attempt to plot the data, which would be impossible. With a PCA plot, we can see a summary of that information condensed into a mere 2 or 3 dimensions, those which contain the greatest variation in our data. Of course, reducing the number of variables in a dataset comes at the cost of accuracy, but we are okay with that trade-off because the resulting smaller datasets are easier to analyze and make sense of. We don't need to know every aspect of how each pair of samples is similar or different; we just need an overall summary of their similarities and differences. So, to sum up, the idea of PCA is to create simplicity—reduce the number of variables in a dataset—while preserving as much information as possible.

The process of creating a PCA plot is not very simple at all. In fact, it is quite complicated. There are numerous software programs that can do this for us, including mothur and QIIME 2, so we won't get into the details of the mathematics involved here. However, a PCA plot generally tries to find ways to separate out the data points as much as possible, which is called maximizing the variance. Software programs will try to find variables or combinations of variables (such as abundance of species A plus abundance of species B) that can be used to distinguish between the samples in your plot. These combinations of variables are called principal components, and a PCA plot can be created by graphing the samples using the principal components that maximize the variance between samples most effectively. In other words, PCA analysis finds the most variable measures and focuses on those. Finally, it is important to note that when you examine a PCA plot, the patterns of clustering do not signify a statistical difference. There are methods that would permit you to draw confidence intervals around the clusters and, using those confidence intervals, interpret whether the clusters are significantly different, but those methods are beyond the scope of this introduction.

Let's employ PCA to plot the beta diversity of our pilot data for the vaginal versus C-section babies. The result might look something like the plot in **Figure 5.19**. The *x*-axis is the first principal component, which explains 60% of the variance in our data. The *y*-axis is the second principal component, which explains 35% of our data. There are, in fact, numerous principal components (as there are numerous species detected in our samples). However, we will ignore most of them as we have already captured most of the variation (60% and 35%) in our data with just the first two. That is the beauty of PCA. We take a very large set of data and transform it into something that is far less messy.

In Figure 5.19 we can easily distinguish two subsets of data points. To see how birth mode is distributed among these points, let's color the data from C-section babies yellow and the data from vaginally born babies green. We can now easily see that the C-section babies cluster together and well away from the vaginally delivered babies. There is certainly variability within the data for each delivery mode (not all the yellow points or green points are on top of each other), but that level of variability is far less than the variability between any pair of yellow and green points. You will often see the results of PCA presented in microbiome studies, and although it is beyond the scope of this chapter to explain the algebra that underlies this analysis, you should have enough information to begin to understand its use. Note that we have simply plotted the most informative components of the variance; no statistical test was conducted. Using a PCA plot like this, we're able to condense a vast amount of data into a two-dimensional graph. While maybe this type of analysis is overkill for our pilot study, it's a necessity for larger datasets that would be impossible to visualize otherwise.

We have covered a fair bit of information in chapters 4 and 5. However, the experimental methods and analytical tools you have learned should permit you to tackle the increasingly complicated microbiome studies we will encounter in future chapters. You should spend time familiarizing yourself with the figures and tables in the present chapter, as you will likely find yourself returning to them frequently as we move on. The good news is that, although there is much about microbiome data generation and analysis we did not cover, you should have a solid enough knowledge base to be able to work your way through many of the microbiome studies you may wish to read on your own.

With respect to our proposed study, we have learned so much. We now understand the importance of a careful consideration of sample size. We conducted a pilot study of 100 babies, which told us that we could reduce our proposed sample sizes from 500 to 50 babies per treatment and not lose much information in terms of species diversity estimates. We are now far more confident that our proposed sample size is large enough to provide the statistical power we require. Finally, we have the analytical methods in place to fully exploit the data we generate. We are comfortable determining alpha and beta diversity measures, and we understand the power of a PCA to visualize the enormous number of UniFrac measures of beta diversity we will generate. If we are lucky (yes, there is also a degree of luck in science), our data will be powerful enough to help us understand whether babies born by C-section tend to have higher levels of asthma, which would be a fascinating result.

As we make our way through the subsequent chapters in this book, we will repeatedly turn to the primary literature to discuss some of the more compelling, and provocative, results that are being reported from microbiome studies. In the next chapter we will start with one research article and employ what you have learned from reading chapters 4 and 5 to understand the science presented. The tools that you have learned in this and the prior chapter will ensure that you can follow these discussions with ease. Congratulations on making it through the densest material in this book. Now, we get to learn about some of the exciting discoveries being made in microbiome science every day!

CHECK YOUR UNDERSTANDING

1. The 16S rRNA gene
   a. codes for a subunit of the bacterial ribosome
   b. is most often used for eukaryotic phylogeny analysis.
   c. used to be used for bacterial phylogenetic analysis but has since been replaced by 18S rRNA sequencing.
   d. is a highly variable region of the genome.

2. "Multiplexing" a sample means to
   a. combine multiple samples that are all genetically identical.
   b. pool differently bar-coded samples and sequence them together.
   c. sequence using multiple "plex" flow cells.
   d. run each sample from an experiment on its own flow cell.

3. Sequencing outputs from human microbiome studies often contain genes from
   a. bacteria present in the microbiome.
   b. the human the sample was collected from.
   c. food or other biological contaminants present in the microbiome.
   d. all of the above.

4. Studying an organism with a published genome
   a. requires fewer computational resources needed for genome assembly.
   b. requires a de novo assembly.
   c. is often more difficult than studying an organism without reference information.
   d. limits you to a small amount of bacterial species.

5. Each codon
   a. corresponds to a triplet of noncoding nucleotides.
   b. translates into a single amino acid, a "start translation" message, or a "stop translation" message.
   c. codes for a gene within the coding region of the genome.
   d. is made up of six coding nucleotides.

6. A novel protein
   a. is easily found in databases such as UniProt.
   b. represents 97% of the proteins searched for in sequence similarity databases.
   c. is an unidentified protein, and not recognized in databases.
   d. is a group of proteins less than 100 amino acids in length.

7. What is the raw sequencing data produced through sequencing by synthesis?
   a. Gel electrophoresis lengths
   b. Fluorescent signals
   c. Microscopy images
   d. Electronic current signals

8. Which of the following is not an example of quality control performed on sequencing data?
   a. Removing contamination from sample processing
   b. Removing low-quality sequencing data
   c. Removing data from the human host
   d. Removing ultralong reads

9. A quality score of 30 represents what error frequency?
   a. 1 in 10 bases
   b. 1 in 100 bases
   c. 1 in 1,000 bases
   d. 1 in 10,000 bases

10. When an assembly is being built, overlapping reads are combined into
    a. contigs.
    b. a consensus base.
    c. combined read segments.
    d. compressed fragments.

11. We can use 16S rRNA sequencing data to identify protein functions.
    a. True
    b. False

12. What segments of the genome potentially encode a protein?
    a. Noncoding regions
    b. Codons
    c. Open reading frames
    d. Promoters

13. Highly similar 16S rRNA sequences are combined into
    a. open reading frames.
    b. assemblies.
    c. operational taxonomic units.
    d. contigs.

14. A species accumulation curve
    a. can be used to estimate how many individuals should be sampled to more fully capture the diversity in the population.
    b. will likely increase the cost of a study but provide higher-quality data.
    c. is a plot of the number of sequencing reads versus the number of taxa identified.
    d. must be a part of any microbiome study.

15. Which of the following is not true about alpha diversity?
    a. Alpha diversity measures species richness, evenness, or a combination of the two.
    b. Alpha diversity can be measured by Faith's phylogenetic distance.
    c. A Shannon diversity index of 1 represents a sample with only 1 species.
    d. Alpha diversity is commonly visualized for a group of samples in a box and whiskers plot.

16. What is a benefit of using Faith's phylogenetic diversity (PD) index instead of the Shannon diversity index?
    a. PD includes more bacterial species in its measurement of diversity.
    b. PD accounts for evolutionary distance.
    c. PD requires less computational power to calculate.
    d. PD excludes contaminants.

17. The Bray-Curtis dissimilarity
    a. measures the levels of similarity between individuals with disease.
    b. may be higher for a healthy individual than an individual with disease.
    c. is between 0 and 1.
    d. is higher in samples with more species in common.

18. Why do scientists use PCA plots to analyze data?
    a. To decrease computational complexity
    b. To visualize variation in alpha diversity across treatment groups
    c. To summarize the quality of sequencing data
    d. To simplify and summarize many components of a complex dataset

19. The first principal component
    a. is the component that explains the most variation in the data.
    b. represents a single variable.
    c. can only be used to analyze species information, and not omics data.
    d. is a combination of every other variable used in the analysis.

20. How is the UniFrac metric similar to Faith's phylogenetic diversity index?
    a. It measures diversity in a single sample.
    b. Higher values represent less diversity.
    c. It incorporates evolutionary distance.
    d. It is the sum of each branch length.

21. Which of the following is not a challenge of de novo gene assembly?
    a. Sequencing errors
    b. Uneven coverage of the genome
    c. The small proportion of the species in the sample
    d. Very large prokaryotic genomes

22. What file type contains the bases found in a sequence and is uploaded to online tools to identify OTUs?
    a. Illumina file
    b. FASTA file
    c. CSV file
    d. SEQ file

23. When OTUs are being identified, two nucleic acid sequences are compared using a
    a. Contig.
    b. Sequence alignment.
    c. Gene scaffold.
    d. Reference genome.

Answers: 1A, 2B, 3D, 4A, 5B, 6C, 7B, 8D, 9C, 10A, 11B, 12C, 13C, 14A, 15C, 16B, 17C, 18D, 19A, 20C, 21D, 22B, 23B

DIVING DEEPER

1. What is the raw data produced during sequencing? How is this data transformed into a sequencing read?
2. What are the two main ways sequence reads are filtered for quality?
3. What is a QS score? What frequency of errors does a QS score of 20 represent?
4. What is a contig, and how are contigs used to assemble a whole-genome sequence?
5. If overlapping reads disagree, how is the correct sequence determined, and what is this process called?
6. What tools can we use to identify the function of a gene, and what is one limitation of these tools?
7. What is an OTU? How can we determine the species an OTU represents?
8. What are species accumulation curves used for?
9. What are the two main ways we can measure diversity, and how are they different?
10. Why might "species abundance" be more important than "species richness" to researchers?

11. Why is calculating a diversity index instead of specific values important?
12. How is Faith's diversity index different from the Shannon diversity index?
13. What type of information could you gain from calculating beta diversity?
14. What does it mean to reduce the dimension, and why is it important?

DISCUSSING AND REFLECTING

1. If you were conducting our hypothetical experiment trying to determine the impact of birth delivery mode on the microbiome but could only calculate one of the diversity measures we discussed in this chapter, which would you choose and why? What information could you get from those data, and what would you miss?
2. The use of PCA permits us to visualize very large datasets. However, when we apply a PCA to our data, we choose to eliminate most of the between-sample comparisons. Why is this loss of information a reasonable decision? In other words, what do we gain (or lose) by reducing the complexity of our dataset?
3. Reflection. Although it was not the focus of this chapter, we learned that birth by C-section results in a newborn's gut microbiome dominated not by mother's vaginal bacteria, but by bacteria more often found on mother's skin (or the hospital environment). What does this understanding make you think about the choice to use a C-section when not medically necessary?

RECOMMENDED READINGS

Popular Science Review

Rusting, R. (2021). *The Microbiome: Your Inner Ecosystem*. Scientific American audiobook. Blackstone.

Popular Science Book

Blaser, M. J. (2014). *Missing Microbes: How the Overuse of Antibiotics Is Fueling Our Modern Plagues* (first edition). Henry Holt and Company.

Scientific Review

Hayes, W., & Sahu, S. (2020). The Human Microbiome: History and Future: Microbiome. *Journal of Pharmacy & Pharmaceutical Sciences*, 23, 406–411. https://doi.org/10.18433/jpps31525

Applying Microbiome Analysis

6

CHAPTER CONTENTS

- 6.1 The Scientific Primary Literature
- 6.2 Dissecting a Research Article
- 6.3 Introduction to Qiita
- 6.4 Beyond Qiita

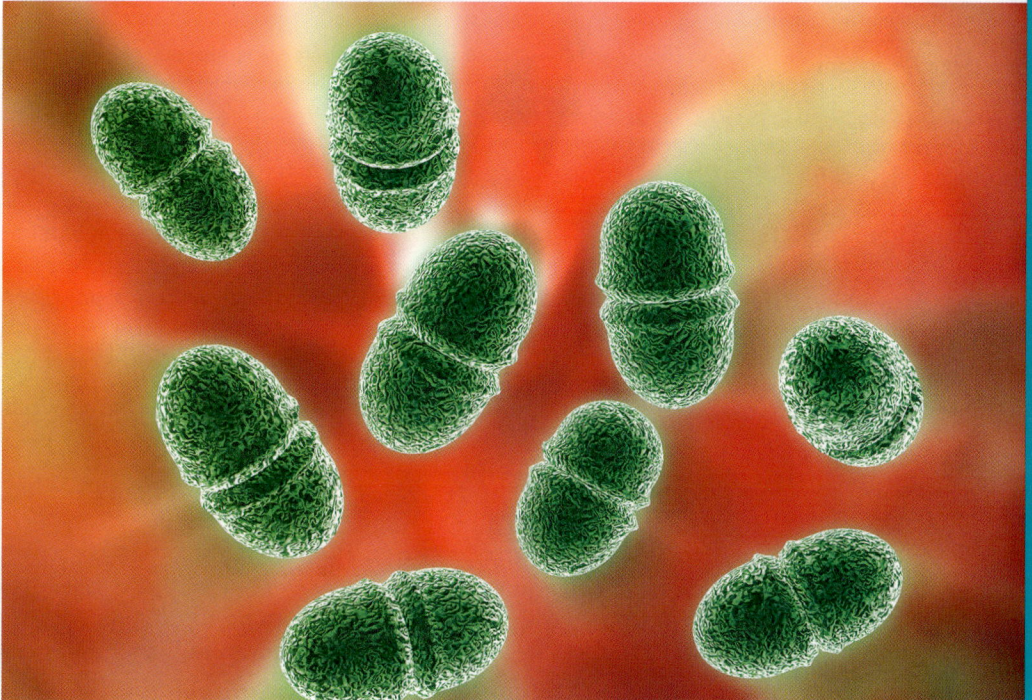

Ah, there you are! I've been waiting. My name is *Enterococcus faecalis*, but you can call me 'E. Faec.' Yeah, I know, fancy name for a gut microbe, right? Some of us have to keep it classy. I'm a proud member of your microbiome—the digestive diva, the bacterial big shot, the unsung hero of your gut! You may not know this, but I'm one of the old guys around here. I've been living in human guts for, oh, forever! I hang out in your intestines, just minding my own business—helping break down food, keeping those harmful bugs in check. You know, typical gut microbe stuff. I'm super good at tolerating stress, so whether you've had too much spicy food or a surprise run-in with antibiotics, I'm still standing, cool as a cucumber! So, I am a helpful guy, most of the time, I'm your buddy, a hard worker. But, uh . . . I have a bit of a rebellious side. If I wander off—like, say, into your bloodstream or urinary tract—things can get dicey. I've been known to cause infections, especially if you're feeling a little weak. Hey, everyone makes mistakes, right? (Photo from Pasieka / Science Source)

You have worked hard to slog through chapters 4 and 5 which are chock-full of experimental techniques and analytical methods. The good news is that you now have a solid knowledge base with which to delve further in microbiome research. Even more exciting, you are now ready to tackle the primary literature itself—where numerous, novel microbiome studies are published every day! In this chapter we will test our understanding of the materials presented in chapters 4 and 5 and apply this knowledge as we work through a microbiome research publication. We will dissect the content of the article as we learn some tips and tricks on how to read scientific articles so that we don't get bogged down in the fine detail and can still see the forest for the trees. Finally, we will apply some of our newly developed analytical skills as we explore the sequence-based dataset produced in this case study. To this end, we will learn how to use one of the most user-friendly microbiome analysis software packages available, known as Qiita (pronounced like *cheetah*). Through

> "A research journal serves that narrow borderland which separates the known from the unknown."
> —Prasanta Mahalanobis (Mahalanobis, 1933)

Table 6.1 Types of Literature

	PRIMARY	SECONDARY	TERTIARY
Description	Original research and/or new scientific discoveries. Immediate results of research activities. Often includes analysis of data collected in the field or laboratory	Summarizes and synthesizes primary literature. Usually broader and less current than primary literature	Summaries or condensed versions of materials. Usually with references to primary or secondary sources. Good place to look up facts or get a general overview of a subject
Examples	Original research published as articles in peer-reviewed journals. Dissertations. Technical reports. Conference proceedings	Literature review articles. Books	Textbooks. Dictionaries. Encyclopedias. Handbooks

Source: Ohio Northern University Library, adapted from University of San Diego Library: http://ucsd.libguides.com/MCWP/sources

these activities, we will be trained and ready to tackle the many fascinating research topics ahead of us.

6.1 THE SCIENTIFIC PRIMARY LITERATURE

Before we proceed further, let's take a minor detour to learn how to read a scientific article. This may seem like a strange transition, but for the remaining chapters of this book, we will repeatedly dive into what we call the **primary literature**, which refers to publications that present original research. There are several levels of scientific publications that serve as the entry points into the world of science. They are known as the primary, secondary, and tertiary literature (**Table 6.1**). The primary literature is the focus of this chapter and is rich with articles that present original research findings that have been reviewed by scientific peers. The **secondary literature** is composed of review articles, which may also be peer-reviewed, and specialty books, written with scientists in mind. The **tertiary literature** is composed of textbooks (such as this one) and includes dictionaries, encyclopedias, and the like, written with the general public in mind.

What makes the primary literature and review articles so critical to the progress of science is the **peer review** process (**Figure 6.1**). Peer review ensures that scientific articles meet certain minimum standards for quality. This process typically involves the following sequence of events: Scientists complete a study and write it up in the form of a research article, which at this point is called a manuscript. They submit the manuscript to a relevant scientific journal to have it considered for publication. The journal's editors determine whether the manuscript presents findings that are relevant to the journal's mission and readership. If so, they have several scientists who work in the same field (peers) read the article, which is what is meant by *peer review*. The reviewers assess the quality of the science and the manuscript, provide feedback for the author, and tell the editor whether they think the study is of high enough quality to be published. Depending upon these reviews, the manuscript may be rejected, or the authors may be asked to revise the article and resubmit it for another round of peer review. Less frequently, the manuscript is simply accepted as is for publication. The peer review process is lengthy and highly competitive. For example, the premier scientific journals—*Science*, *Nature*, and the *New England Journal of Medicine*—accept fewer than 8% of the articles that are submitted to them.

Peer-reviewed articles are the gold standard of scientific communication, as they create a sense of trust among the scientific community. Because peer-reviewed articles meet the basic standards of scientific quality, researchers can trust the results. Since scientific discoveries build on prior work, this trust is critical to the scientific enter-

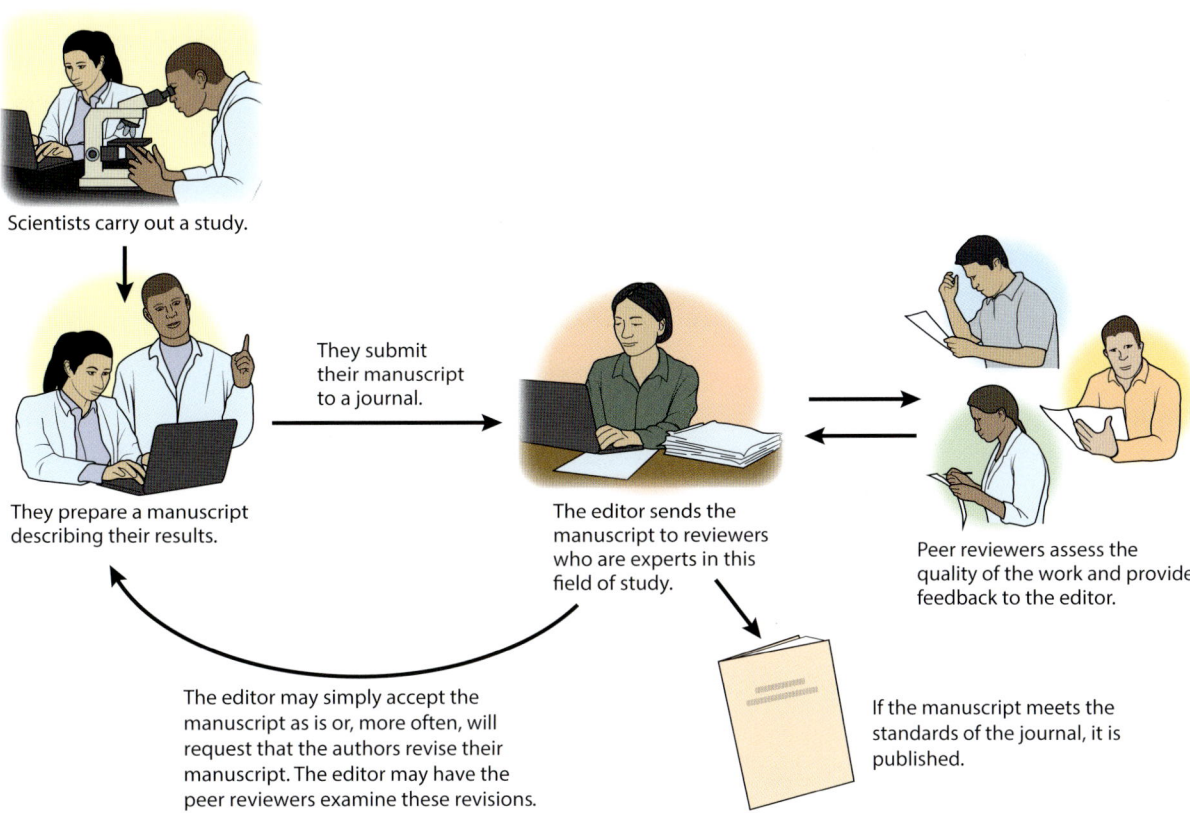

Figure 6.1 The Peer Review Process

prise. To be clear, peer-reviewed articles aren't necessarily correct or conclusive in their interpretation of the data, but they do meet the standards of scientific research. That means that once a manuscript is peer-reviewed and published, it stands in the scientific literature forever. It either becomes incorporated into the body of knowledge in a field or is further investigated to determine whether its conclusions are wrong, primarily through attempts to replicate its results. That is what makes peer review so critical to the advancement of science. If poor science is published, most likely in the so-called **predatory journals**, it confuses everyone and slows the advancement of the field. Predatory journals accept articles for publication without performing quality checks for issues such as plagiarism, ethical approvals, or scientific rigor. Though they are tricky to identify, you may recognize them from some common traits—they generally promise rapid publication, design their homepage language to target authors rather than readers, and do not have clear policies on retractions, corrections, or plagiarism.

We call journals that publish peer-reviewed articles the primary literature. These articles have a typical format, which usually includes the following components: title, abstract, introduction, methods, results, discussion, and references (**Figure 6.2**). The **title** provides a simple, direct, informative, and interesting description of the research presented. An **abstract** summarizes the main contents of the article, and it is often used by a reader to determine whether the article is relevant. The **introduction** describes the question under study while leading the reader from the general subject area to a specific topic of inquiry, which often involves proposing a testable hypothesis. The **methods** section explains how the data were generated or collected and how they were analyzed. It describes the research actions taken in enough detail for the reader to critically evaluate the validity of the study's approach and the reliability of

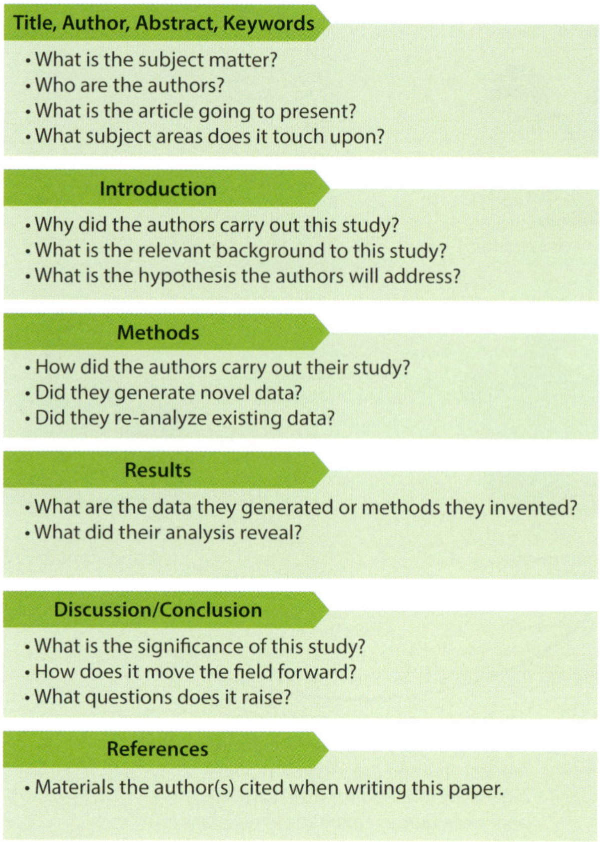

Figure 6.2 Components and Functions of a Scientific Article

Title, Author, Abstract, Keywords
- What is the subject matter?
- Who are the authors?
- What is the article going to present?
- What subject areas does it touch upon?

Introduction
- Why did the authors carry out this study?
- What is the relevant background to this study?
- What is the hypothesis the authors will address?

Methods
- How did the authors carry out their study?
- Did they generate novel data?
- Did they re-analyze existing data?

Results
- What are the data they generated or methods they invented?
- What did their analysis reveal?

Discussion/Conclusion
- What is the significance of this study?
- How does it move the field forward?
- What questions does it raise?

References
- Materials the author(s) cited when writing this paper.

the data, and to replicate the study if they wish. The **results** section is where the findings of the study are provided. It presents the data and analyses, without interpretation. The **discussion** section helps the reader understand why the results should matter to them. It synthesizes the main points, places the results in the context of the field, and may point out areas for future research. The **references** are the sources used during the preparation of the manuscript. If you are unsure whether an article is from the primary literature, you should start by reading the abstract, in which you should find a statement that the authors engaged in actual research activities. If you remain unsure, read a portion of the methods section, which should reveal the details of the experiment(s) performed.

6.2 DISSECTING A RESEARCH ARTICLE

We will now focus our attention on a microbiome research article obtained from the primary literature. This article was published in the journal *Proceedings of the National Academy of Sciences*, or *PNAS*, which provides free user access. Once published, a research article is assigned a **DOI**, or **digital object identifier**, which allows readers to search for the publication quickly and accurately. If you have the DOI for the article shown in Figure 6.3, you can type it into the address bar of a search engine (e.g., Google) to quickly access the article online. We will work our way through this article as we practice some of the analytical methods discussed in chapter 5.

Figure 6.3 provides the first page of the Dominguez-Bello et al. (2010) study and introduces you to some components of a research article. The title is "Delivery Mode Shapes the Acquisition and Structure of the Initial Microbiota across Multiple Body Habitats in Newborns," and we reasonably deduce that the study focused on assessing the impact of birth mode on a newborn's microbiome. Below the title we find a list of the authors and their affiliations. That list usually identifies a corresponding

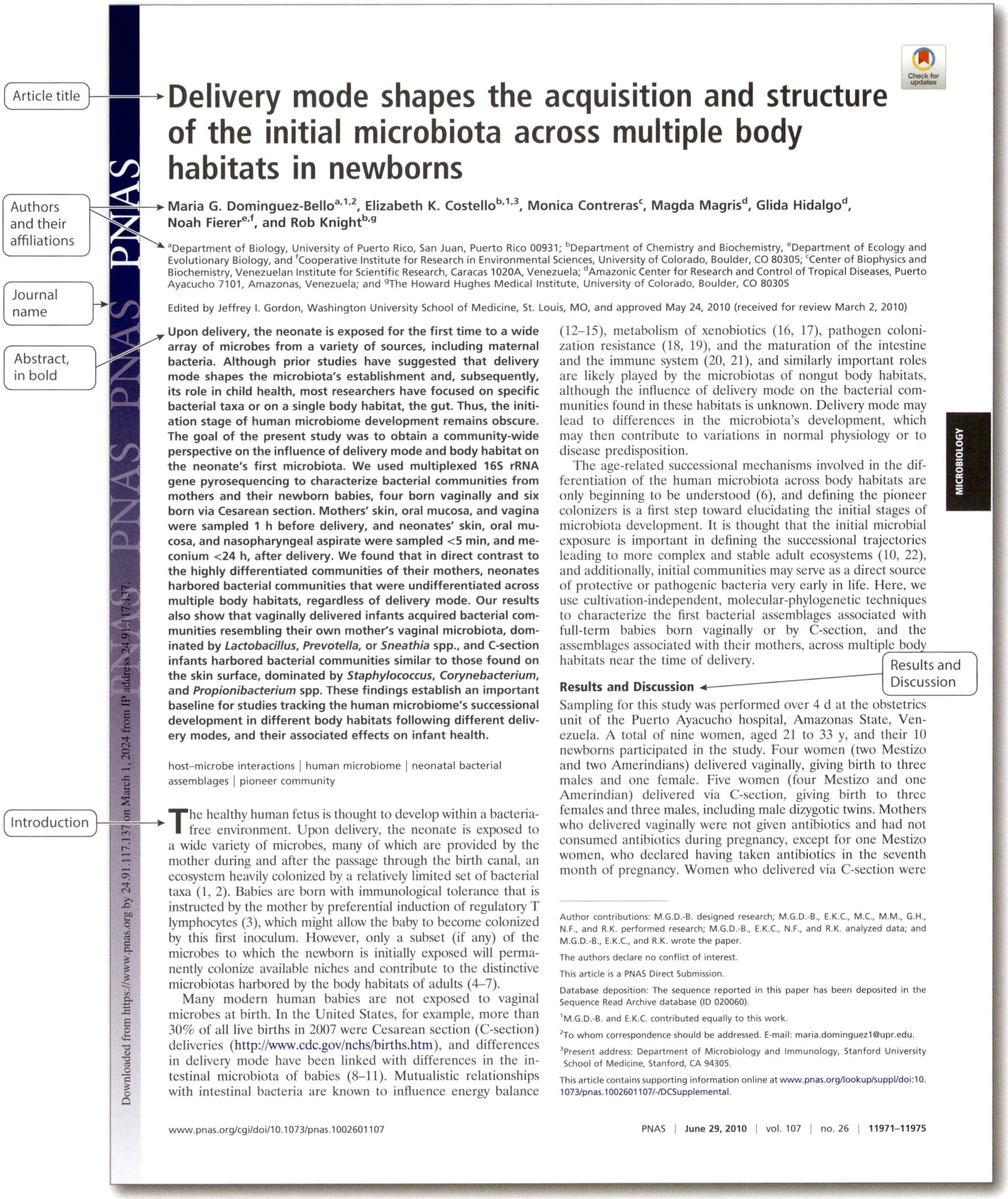

Delivery mode shapes the acquisition and structure of the initial microbiota across multiple body habitats in newborns

Maria G. Dominguez-Bello[a,1,2], Elizabeth K. Costello[b,1,3], Monica Contreras[c], Magda Magris[d], Glida Hidalgo[d], Noah Fierer[e,f], and Rob Knight[b,g]

[a]Department of Biology, University of Puerto Rico, San Juan, Puerto Rico 00931; [b]Department of Chemistry and Biochemistry, [e]Department of Ecology and Evolutionary Biology, and [f]Cooperative Institute for Research in Environmental Sciences, University of Colorado, Boulder, CO 80305; [c]Center of Biophysics and Biochemistry, Venezuelan Institute for Scientific Research, Caracas 1020A, Venezuela; [d]Amazonic Center for Research and Control of Tropical Diseases, Puerto Ayacucho 7101, Amazonas, Venezuela; and [g]The Howard Hughes Medical Institute, University of Colorado, Boulder, CO 80305

Edited by Jeffrey I. Gordon, Washington University School of Medicine, St. Louis, MO, and approved May 24, 2010 (received for review March 2, 2010)

Upon delivery, the neonate is exposed for the first time to a wide array of microbes from a variety of sources, including maternal bacteria. Although prior studies have suggested that delivery mode shapes the microbiota's establishment and, subsequently, its role in child health, most researchers have focused on specific bacterial taxa or on a single body habitat, the gut. Thus, the initiation stage of human microbiome development remains obscure. The goal of the present study was to obtain a community-wide perspective on the influence of delivery mode and body habitat on the neonate's first microbiota. We used multiplexed 16S rRNA gene pyrosequencing to characterize bacterial communities from mothers and their newborn babies, four born vaginally and six born via Cesarean section. Mothers' skin, oral mucosa, and vagina were sampled 1 h before delivery, and neonates' skin, oral mucosa, and nasopharyngeal aspirate were sampled <5 min, and meconium <24 h, after delivery. We found that in direct contrast to the highly differentiated communities of their mothers, neonates harbored bacterial communities that were undifferentiated across multiple body habitats, regardless of delivery mode. Our results also show that vaginally delivered infants acquired bacterial communities resembling their own mother's vaginal microbiota, dominated by *Lactobacillus*, *Prevotella*, or *Sneathia* spp., and C-section infants harbored bacterial communities similar to those found on the skin surface, dominated by *Staphylococcus*, *Corynebacterium*, and *Propionibacterium* spp. These findings establish an important baseline for studies tracking the human microbiome's successional development in different body habitats following different delivery modes, and their associated effects on infant health.

host–microbe interactions | human microbiome | neonatal bacterial assemblages | pioneer community

The healthy human fetus is thought to develop within a bacteria-free environment. Upon delivery, the neonate is exposed to a wide variety of microbes, many of which are provided by the mother during and after the passage through the birth canal, an ecosystem heavily colonized by a relatively limited set of bacterial taxa (1, 2). Babies are born with immunological tolerance that is instructed by the mother by preferential induction of regulatory T lymphocytes (3), which might allow the baby to become colonized by this first inoculum. However, only a subset (if any) of the microbes to which the newborn is initially exposed will permanently colonize available niches and contribute to the distinctive microbiotas harbored by the body habitats of adults (4–7).

Many modern human babies are not exposed to vaginal microbes at birth. In the United States, for example, more than 30% of all live births in 2007 were Cesarean section (C-section) deliveries (http://www.cdc.gov/nchs/births.htm), and differences in delivery mode have been linked with differences in the intestinal microbiota of babies (8–11). Mutualistic relationships with intestinal bacteria are known to influence energy balance (12–15), metabolism of xenobiotics (16, 17), pathogen colonization resistance (18, 19), and the maturation of the intestine and the immune system (20, 21), and similarly important roles are likely played by the microbiotas of nongut body habitats, although the influence of delivery mode on the bacterial communities found in these habitats is unknown. Delivery mode may lead to differences in the microbiota's development, which may then contribute to variations in normal physiology or to disease predisposition.

The age-related successional mechanisms involved in the differentiation of the human microbiota across body habitats are only beginning to be understood (6), and defining the pioneer colonizers is a first step toward elucidating the initial stages of microbiota development. It is thought that the initial microbial exposure is important in defining the successional trajectories leading to more complex and stable adult ecosystems (10, 22), and additionally, initial communities may serve as a direct source of protective or pathogenic bacteria very early in life. Here, we use cultivation-independent, molecular-phylogenetic techniques to characterize the first bacterial assemblages associated with full-term babies born vaginally or by C-section, and the assemblages associated with their mothers, across multiple body habitats near the time of delivery.

Results and Discussion

Sampling for this study was performed over 4 d at the obstetrics unit of the Puerto Ayacucho hospital, Amazonas State, Venezuela. A total of nine women, aged 21 to 33 y, and their 10 newborns participated in the study. Four women (two Mestizo and two Amerindians) delivered vaginally, giving birth to three males and one female. Five women (four Mestizo and one Amerindian) delivered via C-section, giving birth to three females and three males, including male dizygotic twins. Mothers who delivered vaginally were not given antibiotics and had not consumed antibiotics during pregnancy, except for one Mestizo women, who declared having taken antibiotics in the seventh month of pregnancy. Women who delivered via C-section were

Author contributions: M.G.D.-B. designed research; M.G.D.-B., E.K.C., M.C., M.M., G.H., N.F., and R.K. performed research; M.G.D.-B., E.K.C., N.F., and R.K. analyzed data; and M.G.D.-B., E.K.C., and R.K. wrote the paper.

The authors declare no conflict of interest.

This article is a PNAS Direct Submission.

Database deposition: The sequence reported in this paper has been deposited in the Sequence Read Archive database (ID 020060).

[1]M.G.D.-B. and E.K.C. contributed equally to this work.

[2]To whom correspondence should be addressed. E-mail: maria.dominguez1@upr.edu.

[3]Present address: Department of Microbiology and Immunology, Stanford University School of Medicine, Stanford, CA 94305.

This article contains supporting information online at www.pnas.org/lookup/suppl/doi:10.1073/pnas.1002601107/-/DCSupplemental.

www.pnas.org/cgi/doi/10.1073/pnas.1002601107

PNAS | June 29, 2010 | vol. 107 | no. 26 | 11971–11975

Figure 6.3 The Components of Dominguez-Bello et al. (2010) This article is composed of a title, list of authors, abstract, introduction, and results-plus-discussion section. What is not shown here is a methods section, acknowledgements, and references. We will return to this article several times in this chapter as it will serve as a case study for assessing our ability to understand and interpret a microbiome study. (After Dominguez-Bello et al., 2010. *Proc Natl Acad Sci USA* 107[26]: 11971–11975.)

author, the individual who, when working on a paper with multiple authors, takes primary responsibility for communicating with the journal they intend to publish in. There follows an abstract, which is a short, succinct summary of the study, as well as an introduction. A section for results and discussion and a methods section follow. Finally, there is an acknowledgments section and a list of references that are cited in the article.

Abstract

Look at the online version of Dominguez-Bello et al. (2010) and find the abstract, which is a condensed summary of what is presented in the body of the article. An abstract will include information about why the study was done, what methods were used, and what findings were made. Let's read the abstract and discuss what we learn.

The first three sentences are as follows: "Upon delivery, the neonate is exposed for the first time to a wide array of microbes from a variety of sources, including maternal bacteria. Although prior studies have suggested that delivery mode shapes the microbiota's establishment and, subsequently, its role in child health, most researchers have focused on specific bacterial taxa or on a single body habitat, the gut. Thus, the initiation stage of human microbiome development remains obscure." These sentences inform us that the focus of this article is on how the human microbiome is colonized at birth, and the impact of birth mode on that process. Excellent, that is precisely what we had hoped for when we chose this article in chapter 4. The third sentence explains a gap in the knowledge in the field—we don't have a good understanding of how the microbiome is established.

The goal of the study is then stated: "The goal of the present study was to obtain a community-wide perspective on the influence of delivery mode and body habitat on the neonate's first microbiota." This is clearly what we are interested in, so reading this article is great preparation for the study we designed in chapters 4 and 5. Understanding the state of the field through review of primary literature is an essential first step before designing a series of experiments. The next sentences describe what the authors did: "We used multiplexed 16S rRNA gene pyrosequencing to characterize bacterial communities from mothers and their newborn babies, four born vaginally and six born via Cesarean section. Mothers' skin, oral mucosa, and vagina were sampled 1 h before delivery, and neonates' skin, oral mucosa, and nasopharyngeal aspirate were sampled <5 min, and meconium <24 h, after delivery." So, we now know that they employed target gene amplification of the 16S rRNA gene, and then amplicon sequencing to obtain estimates of the microbiome diversity in the mothers and their babies at several body locations. So far, so good.

Figure 6.4 illustrates what the abstract of the article told us about the experimental design of the study, which included a total of nine mothers, who gave birth to 10 newborns, including one pair of twins. Three microbiome samples (skin, mouth, and vagina) were taken from each of the birth parents, and four (skin, mouth, nose, and first bowel movement) from each of the newborns. The first thing we note is that this study has a tiny sample size. That may seem like a concerning factor at first. However, in the early days of microbiome studies, the driving factor in study design was the cost of sequencing (which was prohibitive). It turns out that over the past 13 years, these costs have plummeted by over sixfold, and it is no longer prohibitive to include a robust number of samples in our microbiome studies. Further, this study took place prior to 2010 and tough decisions were required by Dominguez-Bello and her colleagues to determine how large a sample size they could afford, rather than how large a sample they needed in order to accept or reject their null hypothesis. In fact, one of the authors, Rob Knight, and his colleagues subsequently invented methods (such as attaching barcodes to each amplicon fragment) that helped to significantly reduce these costs for future studies (Hamady et al., 2008).

In these early days of microbiome research, the idea of actually acquiring microbiome samples was still quite foreign, and a variety of **ethical considerations** were

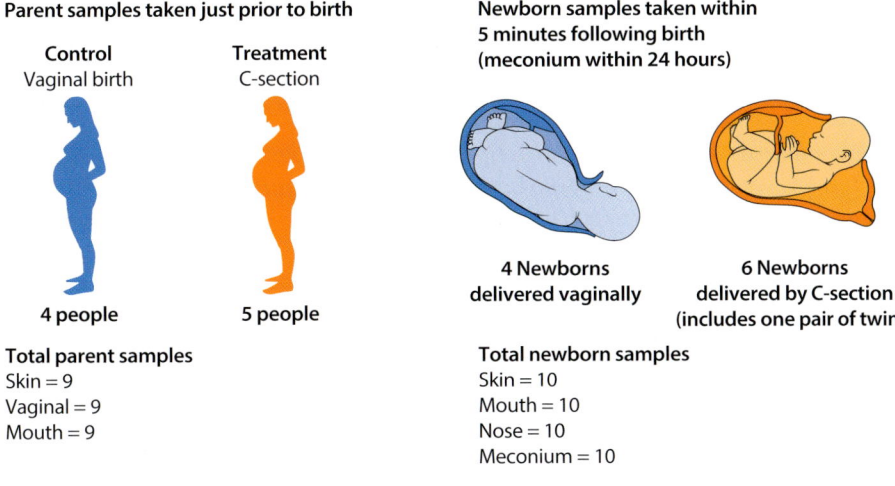

Figure 6.4 The Dominguez-Bello et al. (2010) Study Design This study included nine pregnant participants: four who gave birth vaginally and five who gave birth via C-section. Samples from the parents' skin, vagina, and mouth were collected, as well as samples from the newborns' skin, mouth, nose, and first bowel movement (meconium).

required, such as whether it is appropriate to sample a newborn's microbiome without their knowledge. We note that the sampling employed did not require an invasive procedure and did not impact the newborn in either a positive or negative manner, so those ethical considerations are, essentially, moot if the data we generate is **de-identified**, which means that identifying information, like a participant's name, has been removed. The point is that when Dominguez-Bello and her team carried out their study, it simply wasn't possible to design one based upon power considerations, as the limiting factors were cost and ethical considerations. In this case, the novelty outweighs having a robust research design.

The next section of the abstract describes what they found: "We found that in direct contrast to the highly differentiated communities of their mothers, neonates harbored bacterial communities that were undifferentiated across multiple body habitats, regardless of delivery mode. Our results also show that vaginally delivered infants acquired bacterial communities resembling their own mother's vaginal microbiota, dominated by *Lactobacillus*, *Prevotella*, or *Sneathia* spp., and C-section infants harbored bacterial communities similar to those found on the skin surface, dominated by *Staphylococcus*, *Corynebacterium*, and *Propionibacterium* spp." We learn that the mothers' skin, oral, and vaginal microbiomes are distinct in terms of microbiota, which doesn't surprise us after reading chapter 3. In contrast, the newborns possessed the same microbes at each site sampled: skin, mouth, nose, and first stool. Further, we learn that babies born vaginally harbor microbiota that resemble their mother's vaginal microbiome, while babies born via C-section harbor microbes found on the mother's skin. That finding is significant to us, as we are interested in the impact of C-section on a newborn's microbiome.

The final portion of the abstract describes how the authors interpret their findings: "These findings establish an important baseline for studies tracking the human microbiome's successional development in different body habitats following different delivery modes, and their associated effects on infant health." The authors conclude that their study provides a baseline for future studies that will track how a baby's distinct microbiomes develop into their unique adult composition. In other words, this result is highly novel and will direct future studies that will seek to replicate and explore even further the impact of delivery mode on microbiome establishment.

Congratulations for those of you who have made it through your first scientific abstract, and you didn't encounter anything that you couldn't understand! That won't always be the case, but if you take it one sentence at a time, you will be surprised how well you can follow the author's meaning. As you can see, the abstract provides a useful summary of the study's main points and can help you decide whether the paper is relevant to you. We agree that this article is relevant to us, so we move on to the introduction.

Introduction

The introduction is located just after the abstract and provides a more detailed description of the state of the field prior to the present study. We won't read each sentence together as we did for the abstract. Instead, take some time to read the first paragraph on your own, and see whether you can answer the following questions: Does a fetus have a microbiome? Are the microbes acquired at birth the same microbes the baby will host over a lifetime? Why doesn't the newborn's immune system kill the microbes acquired during birth?

How did you do? We have covered enough of this type of material that it should be possible for you to read this introduction and understand most of the relevant background material. One thing you might have noticed is that many of the sentences in the introduction have numbers in parentheses. Each number corresponds to a research article or review that presents information relevant to this study and that was used by the authors when writing this article. These citations provide the information you need to locate the articles mentioned by the authors so that you can learn more about the topic, or double-check a claim made by the authors. Click on one of the numbers in the online version of our target article and see that it opens a sidebar in which appears a numbered list of references. If you are interested in reading any of these articles, simply click on the link provided for each reference. In some cases, such as reference number 1, clicking on the Google Scholar link will take you directly to the article, which you are then able to read for free. **Open-access** journals use a publishing model that makes research material freely available to readers, unlike the traditional subscription model, where readers gain access to scholarly information by paying a subscription fee. In contrast, if you click on the Google Scholar link for reference 8, you will find that unless you have a subscription to that journal, you will not be able to read that article.

Continue reading the introduction, and make sure that you identify the following key points:

- 30% of babies in the United States were born via C-section in 2007;
- differences in mode of delivery have been linked to differences in gut microbiota;
- the gut microbiota has been linked to energy balance, metabolism of xenobiotics, resistance to pathogen colonization, and the maturation of both immune system and intestines;
- "delivery mode may lead to differences in the microbiota's development, which may then contribute to variations in normal physiology or to disease predisposition";
- defining the pioneer colonizers across body habitats is crucial information that we currently lack; and
- "the initial microbial exposure is important in defining the successional trajectories" that lead to the stable, adult microbiomes.

You have read the entire introduction and identified all the key points expressed by the authors. You should feel very proud of this accomplishment! Reading and understanding an introduction to a research article is no easy thing. The prior two chapters have helped you gain enough background knowledge so that you could relatively easily comprehend the author's thoughts. However, while this article is relatively short and concisely written, that will not be the case for all articles. It may take hours to work through some introductions. That is why scientists spend so much time reading, rather than exclusively doing experiments. Grasping a field of research requires an enormous investment in reading the primary literature. If that is something that you wish to do, and if you are willing to dissect the introduction one sentence at a time, you are very likely to succeed and develop an understanding of the author's intentions and results.

Results

In this journal (*PNAS*), the results and discussion sections are merged. In part, this is a way to reduce the article's length so that more studies can be published within each volume of the journal. The impact is that in this article you do not have a clean separation of the scientific facts and the author's interpretation of their meaning. Normally a results section presents just the facts (the data) and the data analysis, and the discussion section is where the author is free to interpret the analysis as they wish. In this article the facts and the interpretation are presented together, per the requirements of this journal.

Let's now work our way through the results portion of the article. To begin, the authors lay out the experimental design, which we have already discussed (see Figure 6.4). The methods section, which we are skipping for now, can be found at the end of the paper, and it is the place in which the precise details of the research are provided. However, in some cases, researchers briefly explain their methods at the start of the results section, to help clarify the intent of the study. Recall that this study consists of nine mothers and their 10 newborns (one pair of twins), of which 4 were delivered vaginally and 6 delivered by C-section. We understand that the reason for including such a small sample of newborns is related to both the cost of undertaking this research at that time and ethical considerations about sampling a newborn's microbiome.

Next the authors describe the steps involved in data generation and analysis. **Figure 6.5** provides a flowchart of the methods employed. First, the authors remind us that samples of the mother's microbiomes were taken just prior to birth and included the skin, mouth, and vagina. Samples of the newborns' microbiomes were taken immediately after birth in the case of skin, mouth, and nose, or within 24 hours in the case of meconium. These samples were subjected to 16S rRNA gene amplification and next-generation sequencing. We are already aware of those methods, so all good so far! The samples resulted in the generation of 157,915 16S rRNA amplicon sequences. Imagine trying to handle a dataset of that size! And yet, current studies involve 10 to 1,000 times more data—we have come such a very long way in such a very short time.

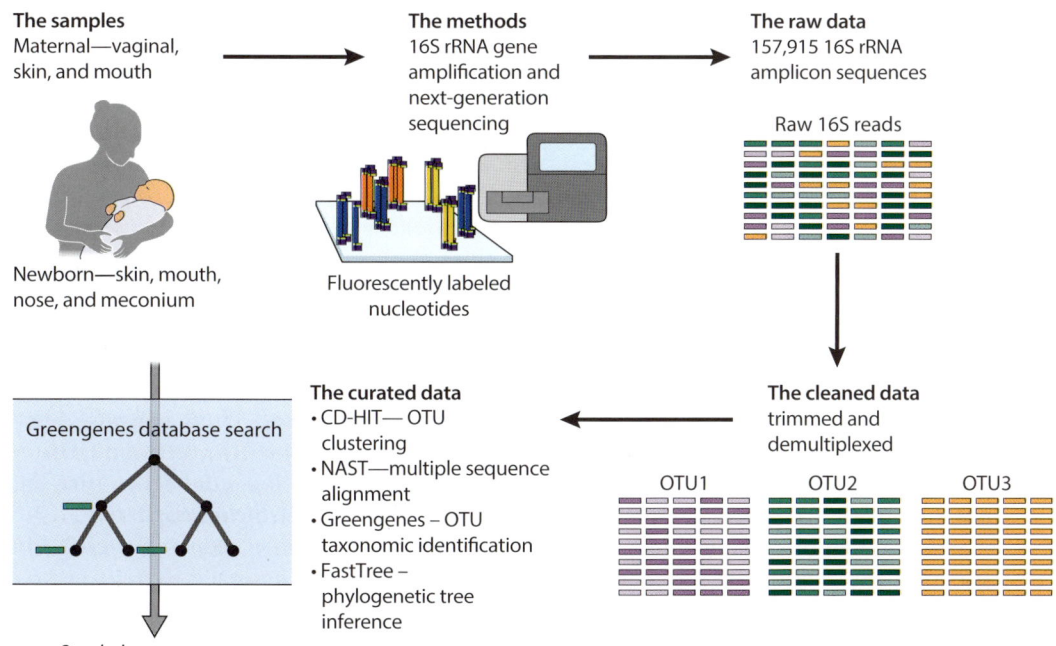

Figure 6.5 Flowchart of a Typical Microbiome Study This flowchart shows the major stages of a typical microbiome study. For each stage, the methods employed by Dominguez-Bello et al. (2010) are provided.

After the sequence reads from this study were cleaned up (trimmed) and demultiplexed, the analysis was carried out using a series of microbiome analysis programs, such as **cluster** (Li & Godzik, 2006), which clusters the operational taxonomic units (OTUs) generated; **NAST** (deSantis et al., 2006), which is a multiple sequence alignment server; **Greengenes**, which identifies taxonomic origin; and **FastTree** (Price et al., 2009), a program that generates a phylogenetic tree of 16S rRNA sequences. Since it is critical that we understand the analytical methods they employed in a bit more detail, we are going to pause our discussion of the results section and turn to the methods to ensure we understand precisely how these data were generated.

Methods

The methods section of the article is found after the results and discussion section and provides the specifics of how each of the experimental and analytical steps was accomplished. Since you are now well informed about most of the experimental methods the authors employed and as they employed quite standard methods, we will only touch on a few of the analytical techniques. According to the authors, "similar sequences were clustered into OTUs using CD-HIT with a minimum coverage of 97% and a minimum identity of 97%." CD-HIT is a widely used program for clustering and comparing protein or nucleotide sequences (Huang et al., 2010). The researchers then determined which taxa were represented by the OTUs they identified: "A representative sequence was chosen from each OTU by selecting the longest sequence that had the largest number of hits to other sequences in the OTU. Representative sequences were aligned using NAST and the Greengenes database, with a minimum alignment length of 150 and a minimum identity of 75%." This means that the authors accepted the inferred species identity of an OTU only if the representative sequence aligned with at least 150 consecutive bases of the reference sequence with at least 75% of the bases being identical. The Greengenes database is like SILVA, which we learned about in chapter 5, in that it provides a highly curated database of rRNA sequences. However, it includes only 16S (bacterial and archaeal) sequences, and not 18S (eukaryotic). The authors then used representative sequences to infer a phylogeny of the taxa they identified. To this end, the authors chose one representative sequence from each OTU, rather than creating a consensus sequence, and used that to identify its taxonomic classification. We will incorporate more of their analytical methods as we dive deeper into the results and discussion section.

Data Analysis

Return to the results and discussion session and find their Figure 1B. This figure is recreated in our **Figure 6.6**, which shows a stack plot of the OTUs obtained from the birth parents and the newborns, identified to the taxonomic level of genus. Let's focus first on the columns for mothers and note that each site sampled is displayed separately (Mom-Oral, Mom-Vagina, Mom-Skin). You can readily see that each of the birth parents' body sites has a different composition of genera. The authors note that "the mothers' aggregate bacterial communities were dominated by taxa typical of these habitats; for example, *Streptococcus* spp. in the oral cavity, *Staphylococcus, Corynebacterium*, or *Propionibacterium* spp. on the skin, and *Lactobacillus* or *Prevotella* spp. in the vagina." They state that, as one would expect, the different body sites showed the anticipated site-specific taxa, just as you learned in chapter 3. This increases our confidence that the authors' methods have accurately determined the microbiota of their samples.

If you examine Figure 6.6 very carefully, you will also note that for each body site, there are significant numbers of OTUs that could not be assigned to a genus and are listed as "other." The Greengenes database identified the OTUs as belonging to the domain Bacteria but, for this subset, was unable to provide a genus-level designation. This study was published in 2010, and since then there have been thousands of additional human microbiome studies. The list of taxa identified continues to grow, and the "other" category is beginning to shrink.

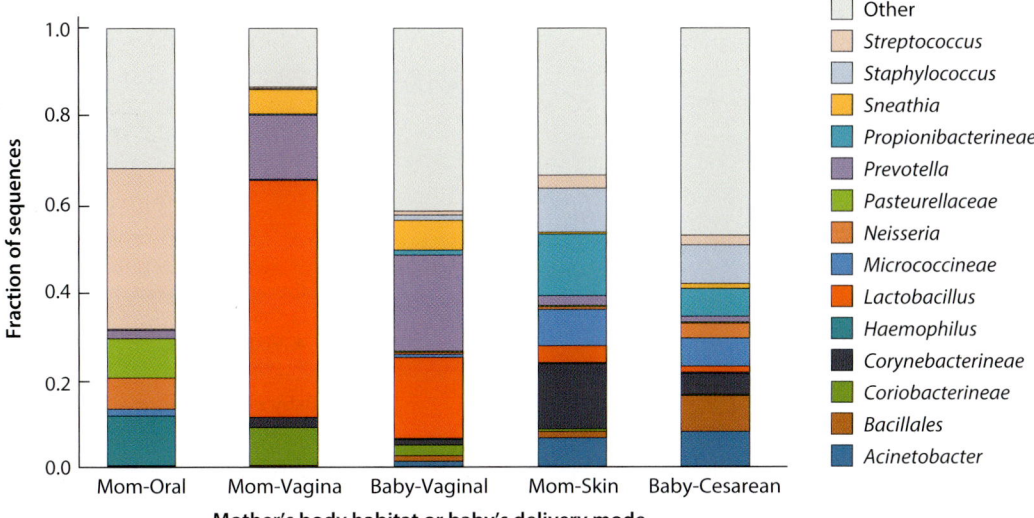

Figure 6.6 Species Abundance in the Mother's Body Habitat or Baby's Delivery Mode Graphs like this are commonly used in microbiome literature, so it's important to have a clear understanding of what they mean. Each color represents a taxonomic identification, while the sizes of the bars indicate the fraction of that taxon in a sample. Notice how different the species composition of each body site is, and think back to chapter 3; do these data match your expectations? (After Dominguez-Bello et al., 2010. *Proc Natl Acad Sci USA* 107[26]: 11971–11975.)

Let's return to Figure 6.6 and, this time, focus solely on the newborns, whose data are separated into those delivered vaginally (Baby-Vaginal) or by C-section (Baby-Cesarean). The first thing we notice is that the two groups of newborns have very different microbiome compositions. According to the authors, "the primary determinant of a newborn's bacterial community composition was his or her mode of delivery. Vaginally delivered infants harbored bacterial communities (in all body habitats) that were most similar in composition to the vaginal communities of the mothers; as expected, C-section babies lacked bacteria from the vaginal community. On the other hand, infants delivered via C-section harbored bacterial communities (across all body habitats) that were most similar to the skin communities of the mothers." By comparing the mothers' skin and vaginal samples to the newborns' samples, the researchers were able to identify taxa transmitted from mother to baby, depending on the mode of delivery. They wrote, "Accordingly, the dominant taxa found in infant communities were reflective of delivery mode: *Lactobacillus, Prevotella, Atopobium,* or *Sneathia* spp. were abundant in aggregate samples from vaginally delivered babies, and typical skin taxa, including *Staphylococcus* spp., appeared in samples from C-section infants."

We have just learned something about birth mode that is highly relevant to our interests. The babies born vaginally had microbiome samples that most closely resembled those of the mothers' vaginal microbiome, whether the baby's sample was taken from the mouth, skin, or feces. In contrast, the babies born by C-section had microbiome samples that most closely resembled their mothers' skin microbiome, again across all the body sites sampled. That confirms what we have been assuming all along, that mode of delivery significantly impacts the acquisition of a newborn's microbiome! As you look closely at Figure 6.6, note that even with a very small sample of babies, one can readily see a difference in the compositions when the data are viewed in a stack plot. This type of plot is frequently used in microbiome studies, and now you know why—it is a powerful way to visualize these types of data. You will also note that the placement of the babies' columns next to particular "Mom" samples is deliberate. The placement allows you to clearly see that the C-section babies harbor microbiomes that most closely resemble the skin microbiome of the birth

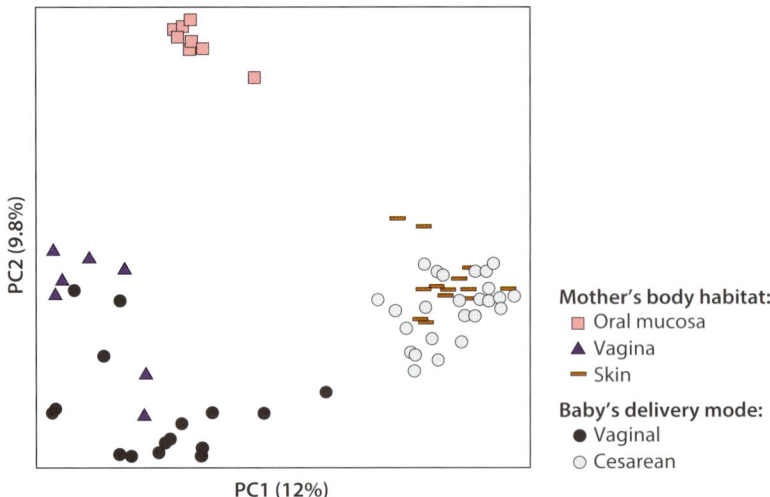

Figure 6.7 PCoA Plot of Beta Diversity PCoA plots permit you to summarize a vast amount of data in just two or three dimensions. Further, they inform you about how much of the variability in your data the plot captures. In this plot, the first principal component (PC1) captures 12% of the variance while the second (PC2) captures 9.8%. (After Dominguez-Bello et al., 2010. *Proc Natl Acad Sci USA* 107[26]: 11971–11975.)

parents. In contrast, the vaginally born babies have microbiome compositions that resemble the vaginal samples from the moms.

Turn to Figure 1A in Dominguez-Bello et al. (2010), which has been re-created as our **Figure 6.7**. The data were employed in a principal coordinate analysis (PCoA). You may recall from chapter 5 that we introduced a related procedure, principal component analysis, PCA, which is a dimension-reducing method that is often used to make large datasets easier to interpret. In this study we have an enormous dataset, ~157,000 16S rRNA sequences. Can you imagine trying to comprehend this dataset by eye? It is simply not possible. So, the authors correctly employed analytical methods that allowed them to reduce the complexity of their data without losing essential patterns that might emerge. Rather than use PCA, they chose a slightly different method to reduce the complexity of their data—PCoA. This method starts by creating a distance matrix, which involves computing the distances between every pair of data points (which is a very large set of distances) and then plotting those distances.

In Figure 6.7 the markers represent samples taken from either mothers or babies. The squares, triangles, and bars represent the oral, vaginal, and skin samples, respectively, from the mothers, while each black dot represents a composite of the samples taken from the vaginally delivered babies and each gray dot represents a composite of the samples taken from the babies delivered by C-section. Note that the markers cluster based upon the body site the sample was taken from, not based on the individual. In other words, the same sites from different mothers were more similar than different sites from the same individual! That is precisely what we learned in chapter 3. If the different body sites shared taxa, then each mother's square, triangle, and bar would cluster. They do not. There is variation in the taxa present within each cluster of samples taken from a single body site, which explains why the squares, triangles, and bars are not superimposed on one another. However, the spread between markers of the same type is far less than the spread between markers of different types. Thus, this figure reinforces what we learned in chapter 3, which is that microbiomes adapt to the specific habitat in which they reside, such that samples obtained from the skin of several individuals are more similar than samples from the skin, mouth, and vagina from the same individual.

Now let's turn to the PCoA of the baby's data. We observe that the black circles (vaginally born babies) cluster near to the birth parents' vaginal microbiome samples, while the gray circles (Cesarean-delivered babies) cluster near to the mothers' skin samples. In other words, the babies born vaginally have microbiomes that are most similar to their mothers' vaginal microbiomes, while babies born by C-section have microbiomes that are most similar to their mothers' skin microbiomes. This elegant representation converts an overwhelming amount of data into a visually clear representation of the relationships between birth mode and a newborn's microbiome. Fantastic—we have tackled most of the challenging information provided in the Dominguez-Bello et al. (2010) article, and it all makes sense!

Discussion

Throughout the results and discussion session, the authors relate their findings to relevant prior studies. Let's focus on two of their key discussion points. Immediately after their Figure 2, the authors state: "Our results demonstrate that the mother's vaginal microbiota provides a natural first microbial exposure to newborn body habitats. In C-section babies, the lack of a vaginal exposure leads to first microbial communities resembling the human skin microbiota, with an abundance of *Staphylococcus* spp." This statement gets to the heart of why the authors carried out this study in the first place and provides compelling evidence that C-section deliveries impact a newborn's microbiome. The analysis does not suggest that the impact is either good or bad with respect to the newborn's health. It merely shows that the different birth modes result in different initial microbiome compositions.

The role of a discussion section is to permit the authors to relate their findings to prior observations or questions that have been raised. In this article, the authors relate their findings to the frequencies of certain bacterial infections in newborns. The authors note: "This finding may, in part, explain why susceptibility to certain pathogens is often higher in C-section than in vaginally delivered infants. For example, 64 to 82% of reported cases of methicillin-resistant *Staphylococcus aureus* (MRSA) skin infections in newborns occurred in Cesarean-delivered infants." The direct transmission of the vaginal microbiota to the baby may serve a defensive role, occupying niches and reducing colonization by MRSA and other pathogens as site-specific communities develop." The authors bring their data to bear on the observation of higher levels of skin infection in babies delivered by C-section and propose that this health outcome may be attributable to a protective function of the mother's vaginal microbiome being passed on to the newborn.

The final paragraph of this section provides suggestions for the direction of future work. For example, the authors note: "Finally, as we have found that the neonatal microbiota is essentially undifferentiated across body habitats, it will be important to determine the timeline over which the distinctive microbiotas found in adult body habitats establish. Our findings emphasize the need to design prospective studies tracking the successional development of the baby's microbiome in different body habitats and after different modes of delivery, and the effects that any associated microbial community shifts may have on infant health." They propose that it is now important to determine the time course over which the baby's microbiome becomes differentiated by site, as is found in adults.

We chose this article not only because it focuses on a topic we are interested in, but also because it is a relatively small, manageable study (although not considered small at the time!), and it employed very standard methods of analysis. Chapters 4 and 5 provided enough background information that you should be able to read and comprehend most of this study on your own. That will not always be the case, but if you find yourself drowning in data, just take things one sentence at a time and you may very well be able to forge through to the end of the article. You may often encounter the use of novel methods or means of analysis, and you will then have to spend time reading enough background material to familiarize yourself with those methods. However, you do not need to become an expert in those methods, as you may make the assumption that the methods are valid due to the peer review process employed by the journal. So, try not to get bogged down by new technologies applied to generate the data or by new analytical methods employed. As long as the article has been peer-reviewed, you may feel confident that experts in that area have done the serious work of evaluating whether these methods are appropriate. Your job is to understand enough so that you understand how the data were generated. Good luck with your investigations into the primary literature. It is enormous, rich, and filled with exciting discoveries, most of which never make it to the public discourse.

6.3 INTRODUCTION TO QIITA

The Basics

We will now turn our attention to a more in-depth investigation into the methods of analysis employed by Dominguez-Bello et al. (2010). You learned above that the authors carried out a series of independent analyses, using programs and databases such as CD-HIT, NAST, and Greengenes. As the field grew, software developers created packages that incorporated all these functions into a sort of one-stop shopping center. One of the more popular packages is QIIME 2 (Bolyen et al., 2019). Instead of taking the raw data and feeding it into a series of software packages to trim it, identify the taxonomic identities of an OTU, and so forth, as was the case in Dominguez-Bello et al. (2010), researchers now have the option of feeding their raw data into QIIME 2 and other similar software packages, which carry out all of the required data manipulations in one place.

QIIME 2 is a particularly appealing option, as it is free, open-source, and generated by the community of scientists using this resource. Unfortunately for us, most microbiome analysis packages require a significant investment in learning how to code and manipulate complex datasets and the associated metadata. That precludes us from diving into these highly powerful packages and directly manipulating the microbiome data, which is, perhaps, the best way to learn about the microbiome. However, there is one software package, Qiita, that is also free and open-source but was designed for those with interest in manipulating microbiome data but without experience in coding (Gonzalez et al., 2018). Qiita was developed to broaden participation in the field of microbiome investigations by permitting those with limited skills using computers to engage in complex analyses of microbiome data.

We will employ Qiita to reinvestigate the Dominguez-Bello et al. (2010) data, that is, the 16S rRNA gene amplicon sequences generated. If you have access to the internet, you will be able to follow the instructions below and explore these data yourself from within Qiita. If you do not have access to the internet or don't wish to invest that time, you can use our figures that show screenshots of some of the key steps in the procedure and follow along as if you were doing the analysis yourself. For those who choose to engage in the online analysis, which I highly recommend, you will be instructed to take an action in each of the numbered steps below. Taking an action means that you will enter text, click on a tab, or choose an option from a drop-down menu. Those instructions are presented in **BOLD BLUE CAPS** to clearly indicate that you are meant to take an action. When you are instructed to look for a signal or text in the Qiita window, the signal or text will be presented in italics below.

Let's start by finding the software package and creating a user account. Go to the following website: https://qiita.ucsd.edu/. The homepage of Qiita should appear. Take a moment to read the overview provided, and you will learn that Qiita is an open-source microbiome study management platform. The term **open-source software** refers to computer code that is designed to be publicly accessible, which means that anyone can see, modify, and distribute the code as they see fit. Qiita further provides access to microbiome databases and computational resources, along with fairly simple instructions, so that individuals without significant experience in writing computer code can still explore these rich datasets. Get ready, because we are going to do precisely that right now.

On the Qiita homepage you will find a sign-up button in the top right corner:

1. **CLICK** on the light-blue *Sign up* button in the top right corner of the homepage.

2. A new window appears and you must **TYPE** your email address and a password and then **CLICK** on the green *Create User* tab.

Once your account is available, return to the homepage and log in to Qiita.

3. **CLICK** on the green *Sign In* button on the top right corner of the homepage. You will be instructed to enter your username and password.

4. **CLICK** on the blue term *documentation* that you find partway down the homepage text.

This link will take you to a more detailed description of Qiita's inner workings. There is an enormous amount of information about how to use Qiita provided here, and it may be useful to spend some time learning some of these details. However, you do not need to do so in order to follow the remainder of this exercise.

Return (go back) to the Qiita homepage, and in the black bar at the top of the page, that is, the toolbar, you will find the following terms: *Analysis, Study, redbiom, Help, Software,* and *More Info*, as well as a *Welcome* to you and a *Log Out* button.

5. **HOVER** your pointer over the *Study* tool, and a drop-down menu will appear.

6. **CLICK** on the subheading *View Studies*, and a new page will appear (**Figure 6.8**).

7. **SCROLL** down the page to the *Public Studies* section.

You will find a table with column headers such as *Expand for analysis, Title, Study ID*, etc. At the time this book went to print, there were 700 public studies available.

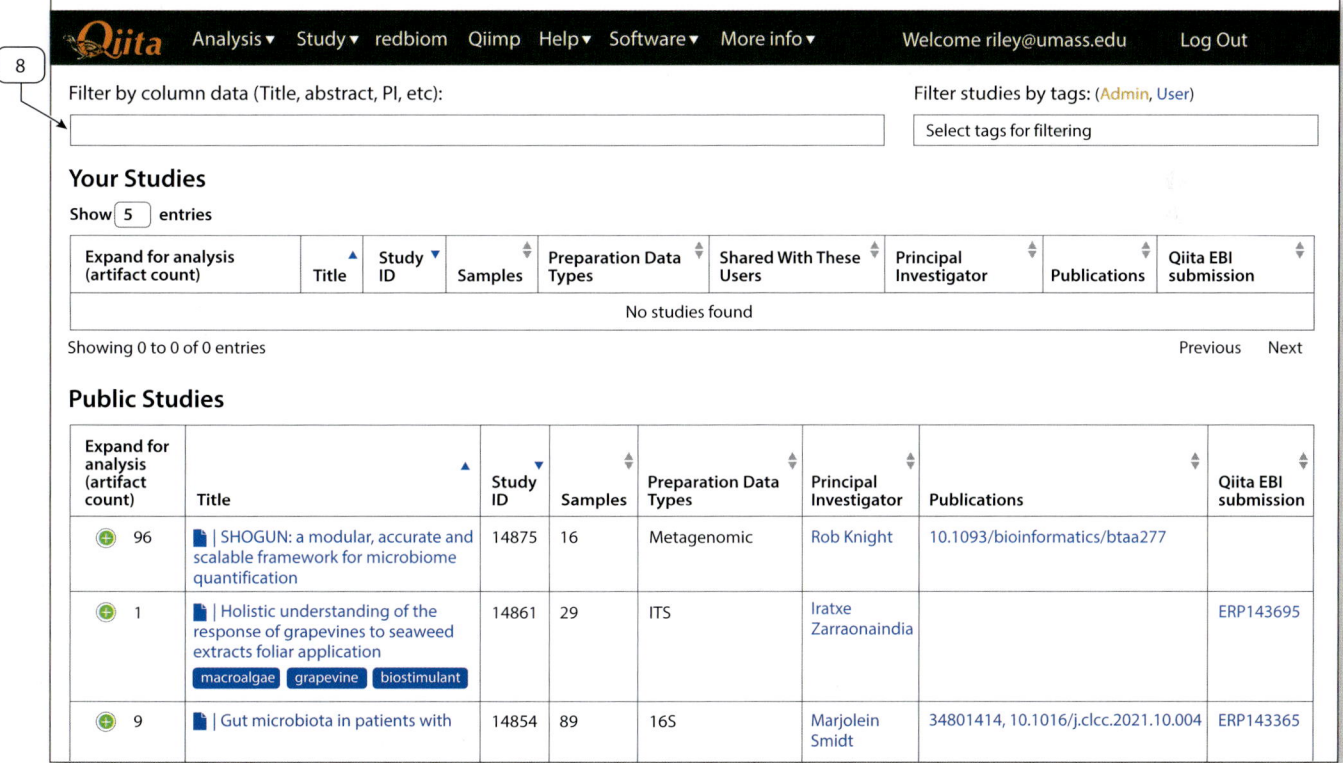

Figure 6.8 Qiita View Studies Page This view is the default when you request to view the studies available in Qiita. The public studies interest us as we will not be generating our own microbiome data in this course. You can scroll through to see the diversity of studies available for you. Note that the page returned may differ slightly from this figure when you click on Qiita, as new studies will be added over time. (After Gonzalez et al., 2018.)

You can find this number at the bottom of the current web page. Now, let's find the study data we are interested in: Dominguez-Bello et al. (2010). Scroll back to the top of that page and take the following action:

8. **TYPE** *Dominguez-Bello* into the *Filter by column data* tool near the top of the page and **HIT** the return (or enter) key (see Figure 6.8, where the 8 at top left refers to step 8).

A new page appears. Examine the datasets that are provided in the *Public Studies* section at the bottom of that page. Each should have *Dominquez-Bello* in the column labeled *Principal Investigator*. At the time this book went to print, there were seven such studies provided.

9. Look at the bottom of the first page of returns and **CLICK** on the *2* or *Next* button to reach the second page of Dominguez-Bello studies.

The top of **Figure 6.9** shows the resulting window, where you will see, under *Title*, *Delivery mode effects on newborn microbiota*. That is the study we have been reading. Note that for each study, in addition to showing *Title* and *Principal Investigator* as mentioned above, Qiita shows the *Study ID*, number of *Samples*, *Preparation Data Types*, and any *Publications* the data were used for. You should see that the study we will focus on has a study ID of *395*, consists of *80* samples, from which *16S* rRNA gene amplicons were obtained.

10. **CLICK** the icon in the first column of our focal study to expand the row.

As soon as you click on it, it will turn from green to red, while four rows of *Artifacts* will appear below the table (see Figure 6.9). A Qiita **artifact** is a collection of files and

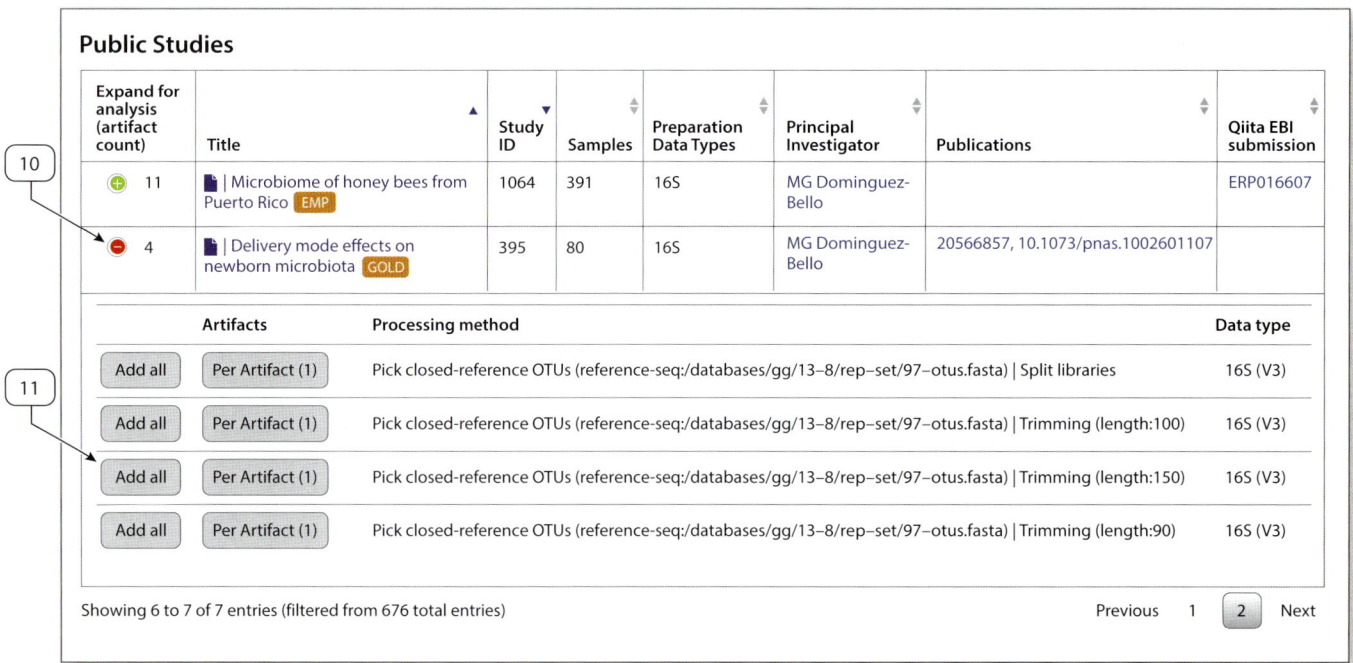

Figure 6.9 List of Dominguez-Bello Microbiome Studies This image is the second page of microbiome studies entered by Dominguez-Bello. Here you see that we have clicked on what was the green button to the left of the study entitled "*Delivery mode effects on newborn microbiota*." The green button turned to red, and the artifacts available for this study were revealed. We can choose to load all or any number of these artifacts for our analysis. We will choose to load the third artifact, which is home to the OTU sequences of a particular length (150) identified in this study. (After Gonzalez et al., 2018.)

their metadata that represent the input or output of a processing or analytical command. In this case, the four artifacts are different versions of the Dominguez-Bello 16S rRNA data following the cleaning process, which involves trimming to remove poor-quality sequence, binning of sequences, and the like, which we learned about in the data analysis section of chapter 5. We will focus here on only one artifact, the third from the top, in which the sequences were trimmed to a length of 150, which is noted in the description provided in that row.

11. **CLICK** the *Add all* button for the third artifact, which temporarily saves this dataset to your Qiita clipboard.

12. Return to the black toolbar at the top of the page (see Figure 6.8), and from the *Analysis* drop-down menu, **SELECT** *Create From Selected Samples* to build your Qiita workspace.

13. **CLICK** on the green *Create Analysis* button.

A new window will open (not shown here) in which you will follow steps 14 and 15.

14. **TYPE** your own text into *Analysis Name* and *Short Description*.

15. Then **CLICK** on the gray *Create analysis* button at the bottom of this window.

Now, you must be patient. Depending upon the time of day and the size of your task, the program may take a bit of time (perhaps up to 15 minutes) to gather the data you have selected and create a new workspace for you. Another benefit of using Qiita is that the computational resources required are hosted by the University of California, San Diego. This means that your computer isn't actually doing any of the work to perform these analyses. However, computational resources are sometimes limited, so the time to process a job will depend on global usage. At some point, the small window in which you created your analysis name and description will disappear, and you will be returned to the Qiita homepage.

A message will appear—*Hang tight, we are processing your request*—as Qiita is creating your workspace. While you wait for the program to create your workspace, notice the name of your request for analysis, which should be displayed on the first line in the current window. Make sure you see the name you created for your analysis and note the ID provided, which will differ for each new analysis you engage in.

Within several minutes your **network workspace** will appear, which is a box in which you will find arrows to expand or minimize as well as one prominent triangle labeled *dflt_name (BIOM)* (**Figure 6.10**). This workspace is where you will follow the stages of analysis you request. The prominent black-sided triangle represents the dataset you chose in step 11, that is, the 16S rRNA sequences that were trimmed to a length of 150 nucleotides. But before we begin the analysis, let's take a moment to examine the data represented by the prominent triangle.

16. **CLICK** on the triangle and then **SCROLL** down to find an overview of the sample statistics, as well as the location where we will perform our own analyses.

Figure 6.10A shows the top portion of this page. To ensure you are looking at the correct dataset, find the *Available files* section and note that there are four clickable boxes. The title of the first box ends with the number *150.biom (biom)*. If you find a different number, return to step 11 and make sure you have chosen the correct artifact (which has sequence lengths of 150). Now look farther down to find *Table summary* (see Figure 6.10B), and you will note that this dataset is composed of *80 microbiome* samples and *3,467* unique sequences (OTUs), which appear in the 80 samples for a combined total frequency of *157,371*.

17. **CLICK** on the term *Feature Detail* found to the right of the *Overview* tab (see Figure 6.10B).

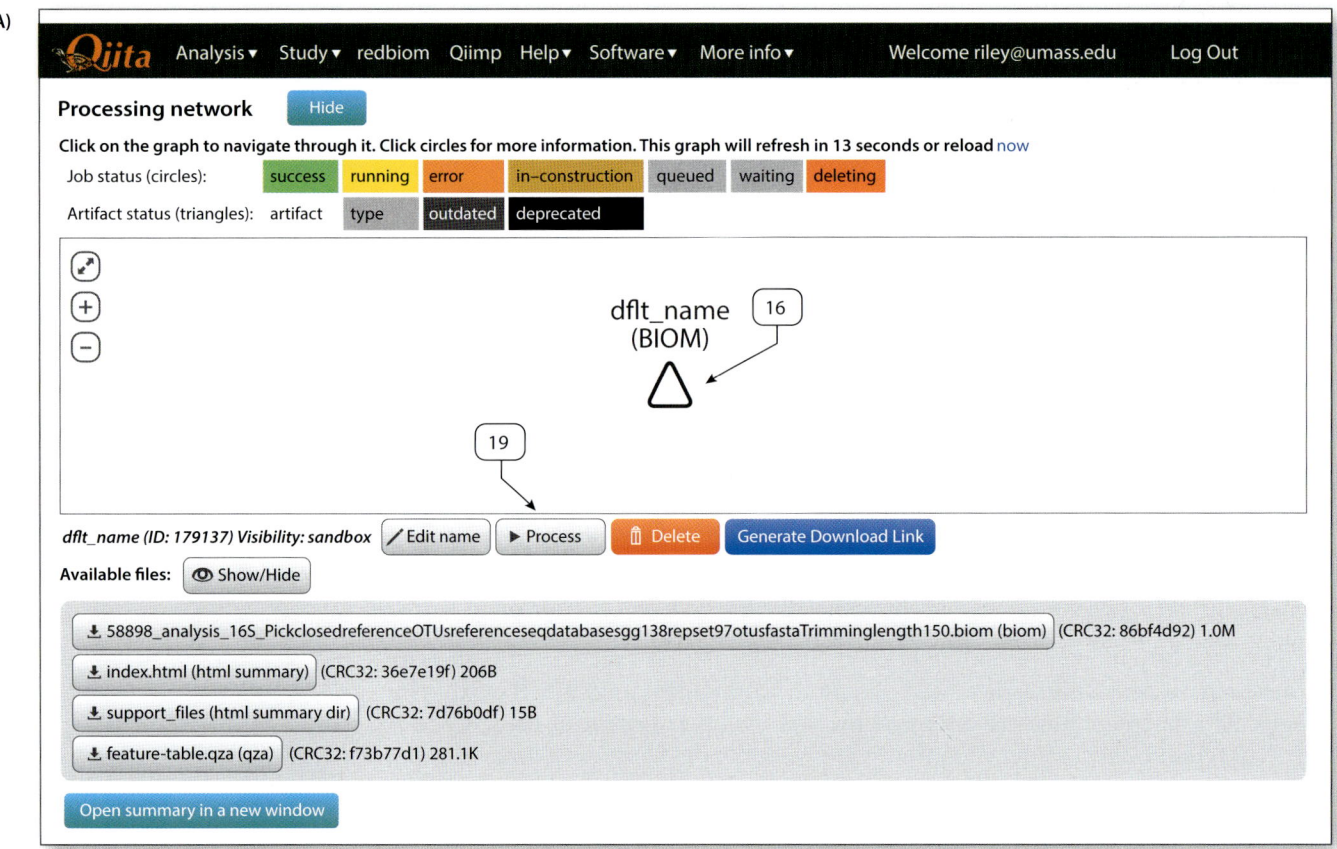

Figure 6.10 Qiita Workspace Populated with an Artifact In this figure you see the workspace created within Qiita to track the analyses you engage in. In the screenshot you see a sole clear triangle labeled *dflt_name (BIOM)*, which is the dataset we chose in the prior step. That file contains the 16S rRNA sequences along with their metadata, such as birth mode, sex, and time of sampling. As we begin to request actions of the program, the job status legend will display the status of our requests as symbols projecting out from this triangle and colored according to the status of the job. (After Gonzalez et al., 2018.)

A very long table will appear (of which we show just three entries in **Figure 6.11**) that lists each sequence ID (OTU) in the first column, followed by the number of times that ID was identified and the number of samples it was found in. For example, at the top of this table you find item sequence ID *235591*, which was observed *17,426* times in the total dataset and occurred in *57* of the 80 microbiome samples. If you scroll to the bottom of this very, very long table, you will find sequence ID *936354*, which was present once in the total dataset and therefore occurred in only 1 of the 80 microbiome samples. These are the heart of the matter when we discuss microbiome data. Spend some time further examining the data summary if you wish.

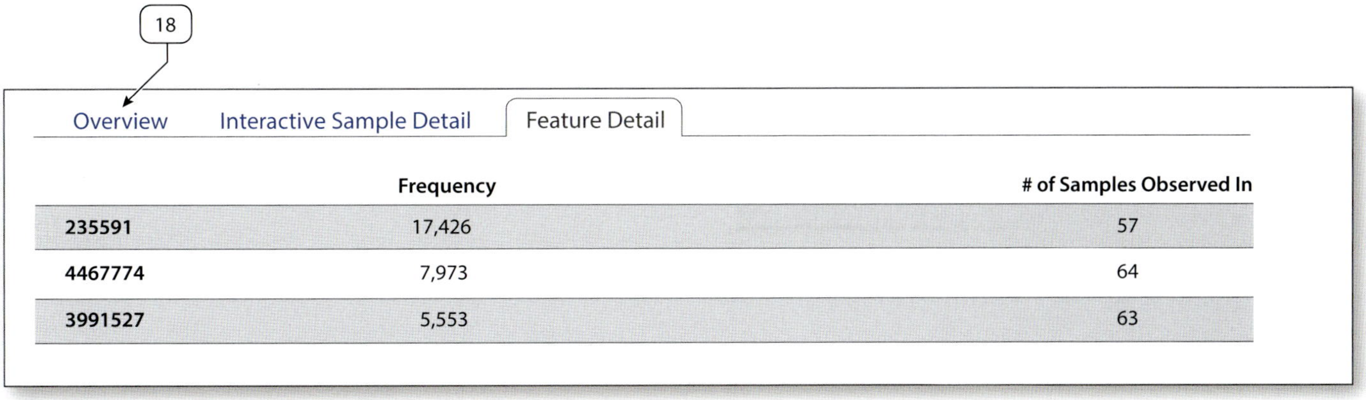

Overview	Interactive Sample Detail	Feature Detail	
	Frequency		# of Samples Observed In
235591	17,426		57
4467774	7,973		64
3991527	5,553		63

Figure 6.11 Qiita OTU Frequency Table (After Gonzalez et al., 2018.)

18. **CLICK** on the *Overview* tab and **SCROLL** down the page, you will see a number of data summaries.

We will not go into these features, but feel free to explore them on your own. If you have questions, don't forget that Qiita has a detailed help section you may choose to explore.

Diving Deeper into Qiita

We will now engage in our first analysis of the data, **data transformation**, which is the process of converting, cleansing, and structuring data into a usable format that can be analyzed to support decision-making processes. Our first analysis will be to visualize the taxonomic distribution of the bacteria represented by these data. What that means is that the program is going to compare the 16S rRNA sequences generated in this study with a database, in this case Greengenes, to identify which microbial species possess those 16S rRNA sequences in their genomes. You learned in chapter 5 that Greengenes is a repository for 16S rRNA sequence data with their taxonomic identities. This transformation allows us to put names on our OTUs, and with those names comes additional, critical information, such as how closely or distantly related the OTUs are.

Let's return to Figure 6.10A.

19. **CLICK** on the box labeled *Process*, which you find just below the graph in which the prominent triangle, which is called an artifact triangle, is located. This will allow us to transform our data.

20. From the *Command* drop-down menu (not shown), **SELECT** *Visualize taxonomy with an interactive bar plot [barplot]*, and then **CLICK** *Add command*.

You will be directed back to the network workspace view, which shows your original artifact triangle, with two symbols added (**Figure 6.12**). The initial artifact triangle in the network space (or data) has an arrow that leads to a dark-orange circle entitled *Visualize taxonomy with an interactive bar plot [barplot]*. A second arrow then leads to a new, gray triangle entitled *visualization (q2_visualization)*. These arrows show you that you have requested an analysis, which involves identifying the OTUs in terms of their taxonomy and summarizing the resulting taxa in a bar plot. By examining the color of the symbol, you can tell whether the action requested is waiting to run (gray), is running (light orange), or has successfully finished running (green), according to the *Job status* legend immediately above the workspace. It is also possible that a run could fail, in which case other colors would appear, as well as error messages to help sort out the problem.

21. **CLICK** on the green *Run* button that appears above your network space.

The circle will turn from dark orange to gray, which means it is queued to run. Be patient, as the job may take some time to run, at which point the gray circle will turn

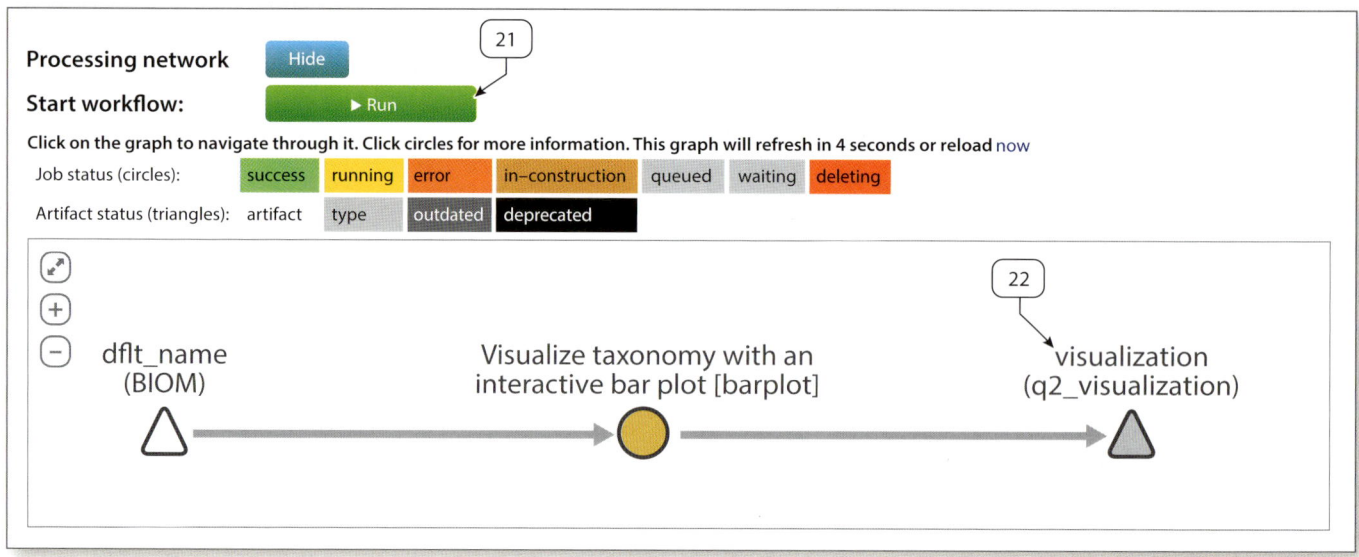

Figure 6.12 Command Processing in the Workspace In this view you can see the original data file (clear triangle) with an arrow pointing to a new symbol, an orange circle. If you look above the workspace, you will see a legend defining each of the colors you may see in the workspace. At this point in the command process, the command we gave, indicated by the circle, is orange, which means that our command is under construction. The color of the symbols in the workspace will change as the program initiates (gray), is running (light orange), or successfully completes the task (green). If there are problems with the data or the command, different colors will be observed according to the job status legend above the activity block that holds the triangles and circles. (After Gonzalez et al., 2018.)

orange until the job is complete, at which point the circle will turn green. If you pick a busy time to do this exercise, it may take many minutes to hours for the job to be complete. You must be patient; there is no way to speed it up—one of the drawbacks of using a free service!

22. Once you see that the circle is green, **CLICK** on the triangle to the right of it, which should open up a visualization (bar plot) of the data transformation you requested (**Figure 6.13**).

23. **SCROLL** down the page to view the results of the analysis indicated in a bar plot.

The bar plot shows relative frequency on the *y*-axis for the 67 microbiome samples arranged as green bars along the *x*-axis. We are currently in the default taxonomic level (level 1), which is the domain level. We learn from this plot that the OTUs from each of the 67 samples shown contain only members of the domain Bacteria.

Qiita-Based Taxonomic Distribution Analysis

Determining the taxonomic distributions in our microbiome samples is a critical component of our analysis. Let's dive into this analysis in more detail. Many studies will focus on the distribution of species in an analysis. Let's do this here.

24. Under the heading *Taxonomic level* **CHOOSE** a number from the drop-down menu that ranges from *1* to *7*.

25. **SELECT** different numbers to see what the bar plots of taxa look like for different levels: *1* for domain, *4* for family, and *7* for species.

For each selection, look at the resulting bar plot of the taxonomic level distribution. Figure 6.13 shows the bar plot for the taxonomic distribution at the genus-and-species

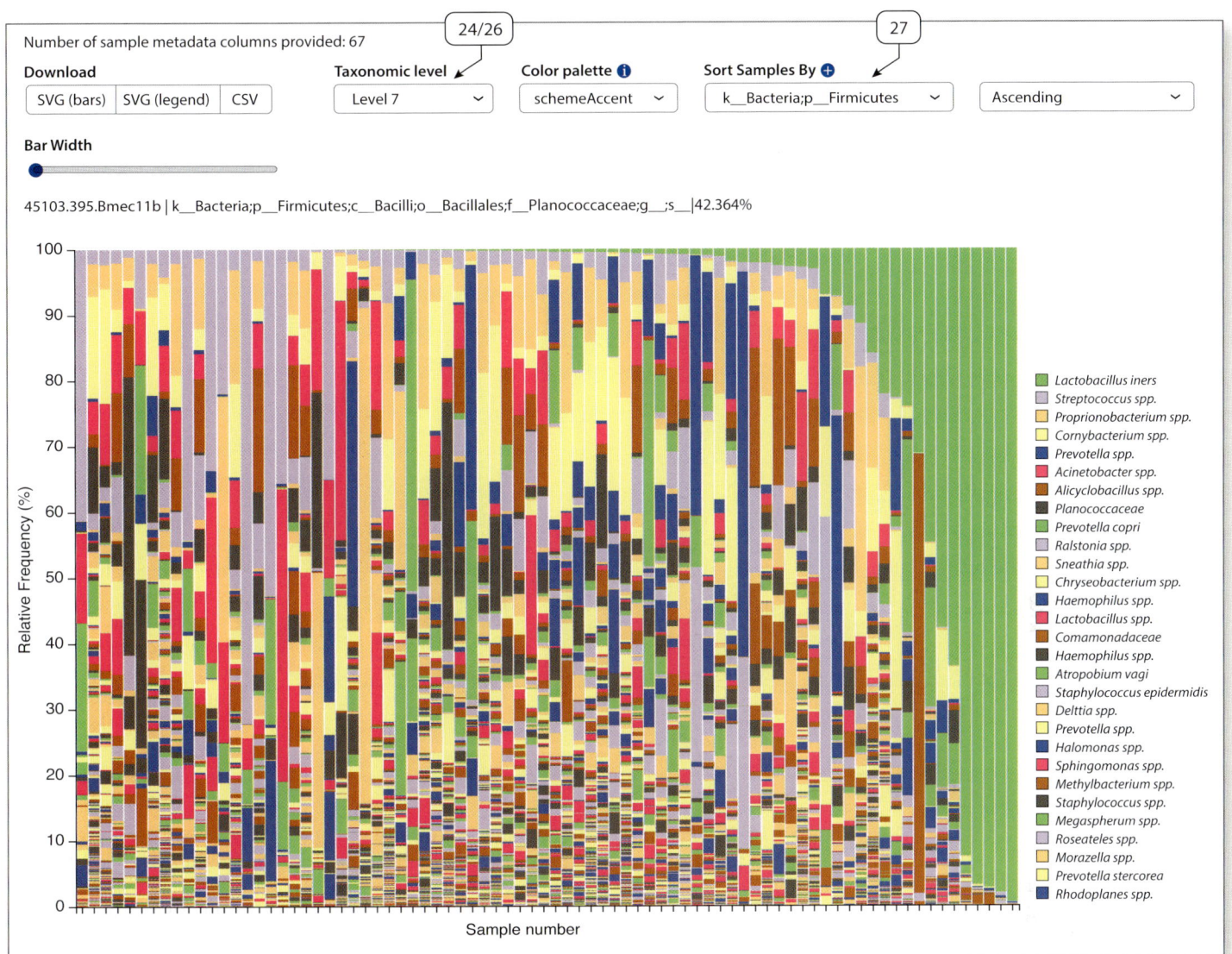

Figure 6.13 Species-Level Frequency Distribution The stack plot shows that for each of the 67 microbiome samples, there is a different assemblage of species present. The key to the right of the plot identifies each color as close to the species level as was possible with the Greengenes dataset. (After Gonzalez et al., 2018.)

level (level 7). Note that you may have to **SCROLL** the chart back and forth or up and down to see all 67 samples and view the entire species name legend on the right.

One pattern emerges in this complex bar plot. Look at the samples shown in the far right of the plot, and you will see a high frequency of one particular species, *Lactobacillus iners*. Given the extraordinary diversity of species represented on this plot, it seems surprising that several samples appear to be composed of primarily this one species. We will return to this observation when we consider Figure 6.16.

Let us now learn how to sort these data based upon various other criteria from the metadata. In addition to sorting by taxon level, we can also sort by different features of our data, for example, by delivery mode, age, or sex. Let's sort by delivery mode.

26. Under the heading *Taxonomic level* **CHOOSE** level *4* (genus).

27. Under the heading *Sort Samples By* **SELECT** *Age*, then **CLICK** on the plus sign above *Sort Samples By* to add another sorting criterion. **SELECT** *Delivery mode*.

The resulting distribution is shown in **Figure 6.14**. Look at the bar plot labels found on the *x*-axis, and you can see that the newborns are plotted together on the left side

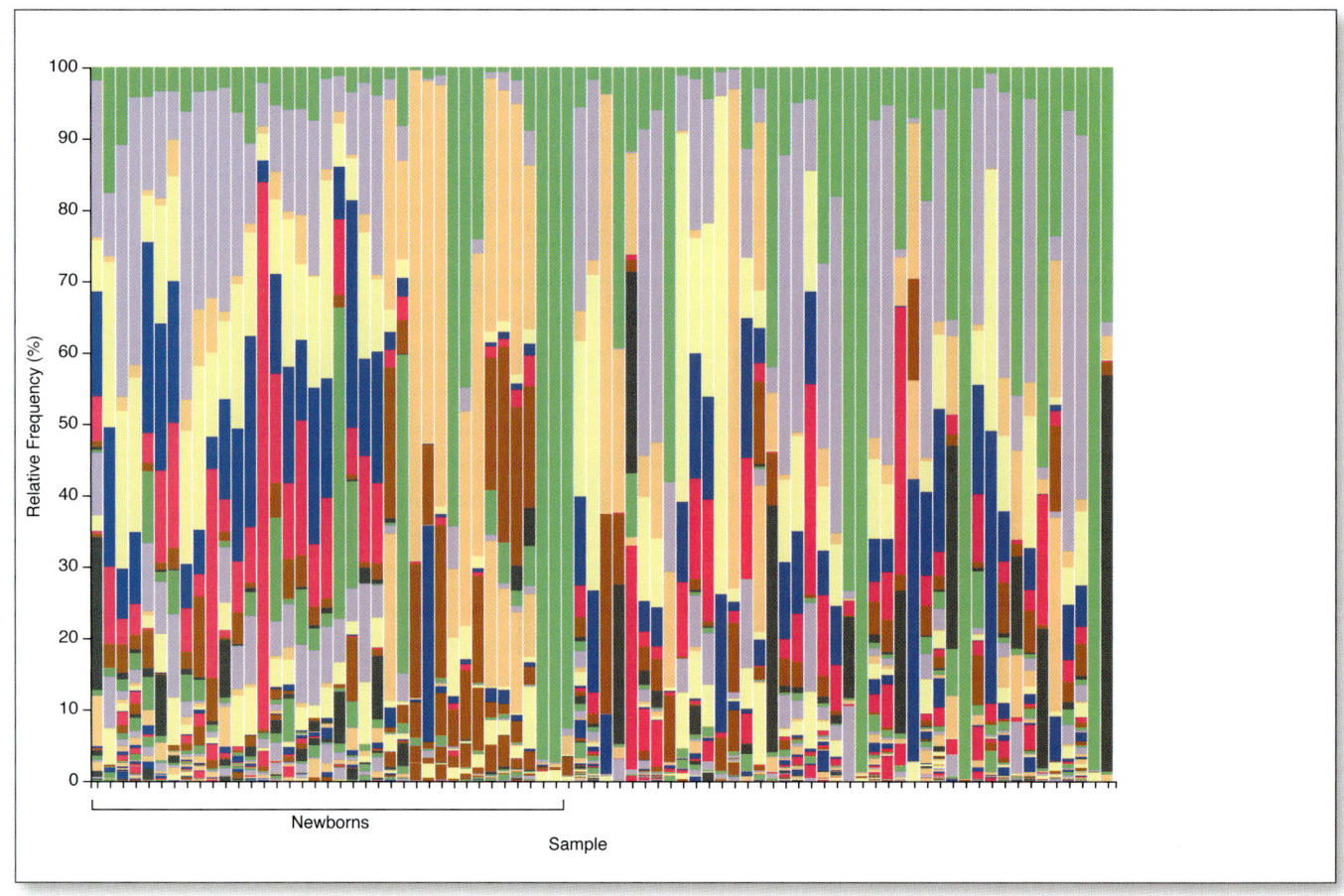

Figure 6.14 Bar Plot Sorted by Age and Delivery Mode Once you have assigned taxonomic identities to the samples, you can search for patterns by sorting the data in a number of ways. This chart shows the data sorted first by age and then by delivery mode. The newborn samples are indicated and appear on the left portion of the *x*-axis. (After Gonzalez et al., 2018.)

of the plot. Within this section they are sorted based upon their delivery mode. You can easily see patterns emerge in the distributions of genera. The C-section babies have a different composition of genera in their microbiomes than the vaginally delivered babies, just as we learned by reading the original article. You can use this analysis feature to dive into numerous aspects of the data, such as delivery mode, age, sex, and body site. The authors used this type of analysis in their publication, which is shown in Figure 6.6. The authors did not use Qitta, so we can't precisely re-create their figures here, but you can see how they grouped the data to look for patterns in the genera diversity for the different birth modes and by comparing babies versus parents.

Qiita-Based Alpha Diversity Analysis

Our next analysis will employ an alpha diversity metric, which you may recall from chapter 5 is used to describe the within-sample diversity. It's a measure of how diverse a single sample is, usually taking into account the number of different species observed. Alpha diversity metrics are also often weighted by the abundances at which the individual microbes are observed. We will employ the Shannon diversity index, which we also learned about in chapter 5. Do you recall how the Shannon index is calculated? Let's apply this index to our samples.

Return to the workspace portion of the current Qiita page and click on the original triangle symbol to refresh us to the original dataset. In the new window (**Figure 6.15**), look for the box labeled *Process* under the workspace.

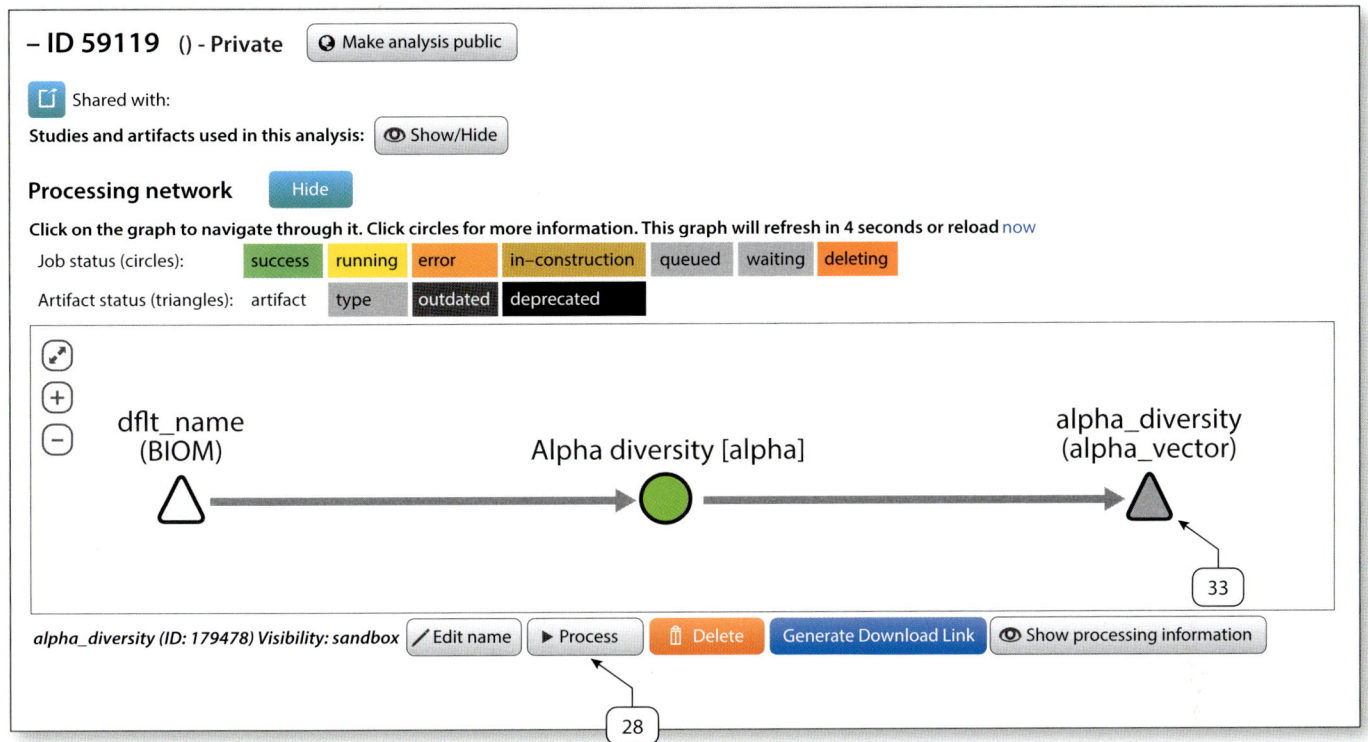

Figure 6.15 Workplace for Alpha Diversity In this view of the workplace, you can see the original data file (clear triangle) with an arrow pointing to the alpha diversity action circle, which has completed its analysis, hence it is green. The process key can be seen next to the step 28 label. (After Gonzalez et al., 2018.)

28. **CLICK** on the box labeled *Process*.

You will see a drop-down menu that provides access to a long list of alpha diversity measures, and you can choose different measures depending upon your analytical needs and the type of data you have (not shown). You should be able to follow the next four steps without referring to a figure.

29. From the drop-down menu, **CHOOSE** *Alpha diversity*. Then look below at the parameter settings.

30. Under *optional parameters*, **CHOOSE** *Shannon index* in the drop-down menu.

31. **CLICK** on the blue *Add command* box that appears.

Note that in the refreshed workspace a new set of arrows and figures appears. There is a new circle corresponding to the alpha diversity task you have asked Qiita to perform and a new triangle that will provide the resulting output.

32. **CLICK** on the green *Run* button at the top of the screen.

Wait for the program to run, which may take several minutes or longer, and when the job is done, the circle will turn green.

33. **CLICK** on the gray-outlined triangle to the right of the alpha diversity circle to view the visualization of the alpha diversity analysis.

Scroll down to view the alpha diversity visualization for these data. You have the option to sort the data in the same manner as for our prior analysis. **Figure 6.16A** shows the alpha diversity measures sorted based upon delivery mode. We see a box and whiskers plot, which we learned about in chapter 5. Test your knowledge to see whether you can interpret this graph. When you are ready, read on.

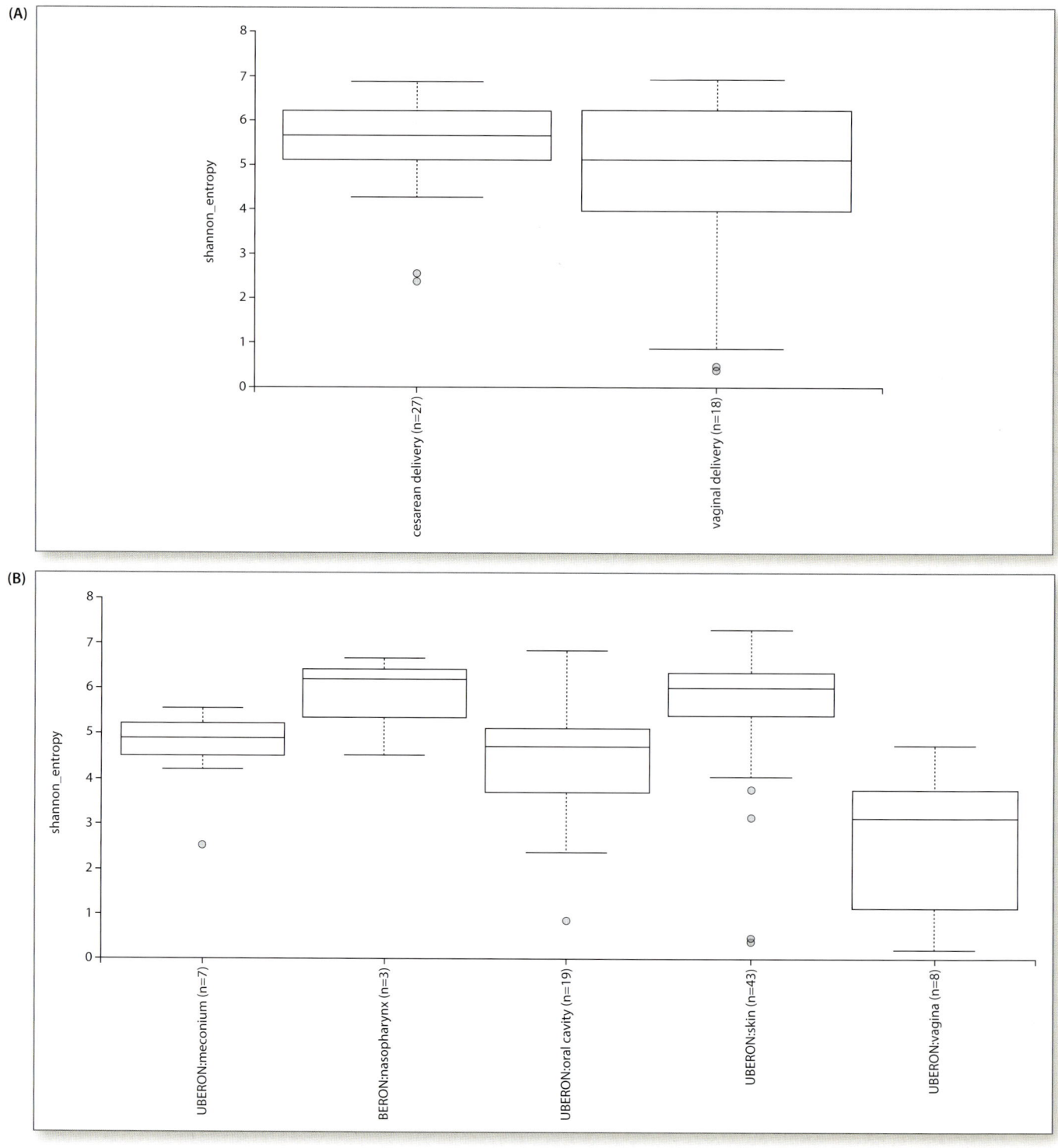

Figure 6.16 Alpha Diversity Measures (A) In this plot, the alpha diversity measures are provided separately for the newborns based upon delivery mode. The babies born vaginally show a wider range of alpha diversity and a slightly lower mean value than the babies born by C-section. However, these differences are not pronounced in this plotting of the data. The gray dots are outlier points in the analysis. (B) In this plot the samples have been separated by body site sampled. The various body sites harbor significantly different levels of alpha diversity as measured by the Shannon index. The gray dots are outlier points in the analysis. (After Gonzalez et al., 2018.)

The Shannon diversity index values for the babies born vaginally show a wider range of values (longer whiskers). However, the median values of the two groups of babies are similar, approximately 5.2 for vaginal babies versus 5.6 for C-section babies. Note that the median value for vaginal babies is even with the lowest value within the box plot for the C-section babies, suggesting that the difference between the medians is nearing the level of significance. Recall from chapter 5 that two box plots are different if the median value of one box plot falls outside the range denoted by the box for the other, as is almost the case here.

Interpreting these results, we might state that the vaginal samples were slightly less diverse in terms of the species present than were the C-section samples, that is, have a lower median Shannon diversity value. That result might seem strange at first, since we noted in chapter 3 that a vaginal birth results in a more diverse microbiome in babies. However, during the first few minutes after birth, the baby has primarily been exposed to the mother's vaginal microbiome, which is dominated by *Lactobacillus*. If you return to Figure 6.14 and search for the color green in the stack plot, you will see that the vaginal birth newborns have far more green (the color for *Lactobacillus*) in their taxonomic distributions than do the C-section newborns. Later in time, even a few days after birth, the vaginally born babies will begin to show higher levels of diversity.

Figure 6.16B shows a chart of the alpha diversity measures provided separately for the different body sites sampled. To produce this plot, you simply choose to sort based upon body sites in the same manner you did above, by choosing *Body sites* from the drop-down menu labeled *Sort Sample By* to the right of the screen (not shown). The alpha diversities of the body sites examined differ, as measured by the Shannon index. You may wish to try some of the other alpha diversity measures to see whether they make a difference in the conclusions we have reached so far, which agree with those presented in our target article.

Qiita-Based PCA Analysis

For our final Qiita-based exploration of these data, we will carry out a principal component analysis (PCA), similar to the PCoA carried out by the investigators and reported in their publication (see Figure 6.7). To complete this analysis, we must employ the following command: *Core diversity metrics (non-phylogenetic) [core_metrics]*. This will generate multiple artifacts, one of which will be the PCA we desire.

34. **CLICK** on the original black-sided triangle (on the left of the workspace) to return to the command view (not shown).

35. **CLICK** on the process key just below the workspace box (not shown).

36. From the command drop-down menu, **SELECT** *Core diversity metrics (non-phylogenetic) [core_metrics]* (not shown).

You will be prompted to enter several parameter values, only one of which we will fill in.

37. **SCROLL** down to the *parameters* section, and **TYPE** "500" where it asks *the frequency that each sample should be rarefied to prior to computing diversity metrics* (not shown).

We introduced the concept of rarefaction in chapter 5, where we learned that a rarefaction curve answers the question: What would be the species richness in a community if we obtained fewer samples? In general, the higher the number, the more confidence you will have in the diversity estimates produced. In this case, since we have 164,300 sequences in our dataset, choosing a number of 500 or greater is appropriate, particularly since there is a trade-off between higher numbers and program run times.

38. **CLICK** on the blue *Add command* button (not shown).

The network view will change again, this time adding a circle for the *Core diversity metrics* we requested, and there will be arrows connected to a suite of artifacts, including an *evenness vector*, *shannon vector*, and *jaccard_pcoa_results* (**Figure 6.17A**).

39. **CLICK** on the green *Run* button (not shown).

This analysis may take some time, so be patient. When it's complete, you may wish to take some time to explore some of the analyses provided. We will focus our attention here on the PCA analysis.

40. **CLICK** on the *jaccard_pcoa_results* artifact (see Figure 6.17A).

Scroll down and you will see a PCA plot (not shown). From the drop-down menus to the right, we can choose how to present the PCA results. We decide to re-create the color sets from the Dominguez-Bello study and choose black circles for vaginal and gray circles for C-section using the color drop-down menu on the right. We also choose to invert the second axis using the axis drop-down menu on the right, which results in an orientation of the data similar to that in the figure in the Dominguez-Bellow article. The result (**Figure 6.17B**) looks very much like the PCoA plot provided in the publication (see Figure 6.7). The two delivery modes appear as well-separated clusters—of black and gray circles in Figure 6.7 and as black circles and gray squares in Figure 6.17B.

We have investigated only two of the many types of analyses that Qiita provides. For those who wish to pursue this topic, there are several options available. You could spend time diving even deeper into Qiita, reading the manual and testing out one or more of the 676 studies available. You might also choose to take one of the numerous tutorials available on the Qiita website, which can be found through the help command. Finally, you can keep up to date with improvements to Qiita, by visiting the website of the Center for Microbiome Innovation (https://cmi.ucsd.edu/news-events/), where Qiita was developed.

6.4 BEYOND QIITA

There are several other freely available microbiome analysis programs designed for those without extensive coding skills. One example is called **EzMAP**, for **Easy Microbiome Analysis Platform** (Shanmugam et al., 2021), which provides a user-friendly microbiome analysis pipeline that provides access to a large suite of functions, including metadata profiling, read pre-processing, sequence processing and classification, OTU clustering, taxonomy assignment, and visualization. A second example is **VAMPS**, for **Visualization and Analysis of Microbial Population Structures** (https://vamps2.mbl.edu/), which offers a range of visualizations and analyses for the exploration of microbial communities (Huse et al., 2014). **MicrobiomeAnalyst** (https://www.microbiomeanalyst.ca/) provides abundance profiles and taxonomic signatures studies (Chong et al., 2020). Finally, there is **Mian** (https://miandata.org/), which contains a rich set of visualization and machine learning tools (Jin, 2018).

Qiita was chosen for use here because, although all of the programs listed above have interesting and unique features, Qiita is the most user-friendly for those without extensive coding knowledge or experience. Qiita was designed to transfer the power of microbiome data analysis from the hands of bioinformaticians to anyone interested in exploring this exciting new field. The complex datasets generated in microbiome studies are now readily available to relative novices who wish to gain an appreciation for the richness and complexity of these data. You can learn so much by manipulating and analyzing the data yourself—there is simply nothing better than getting your hands dirty by diving into the original data to learn more about the field of microbiome studies!

We are now trained in the basic experimental methods and modes of analysis that underlie most microbiome studies. We have also gained experience in reading the primary literature and manipulating microbiome data ourselves. We shall now begin a new section of this book, an exploration of how our microbiomes contribute to our health—and there are so very many ways indeed!

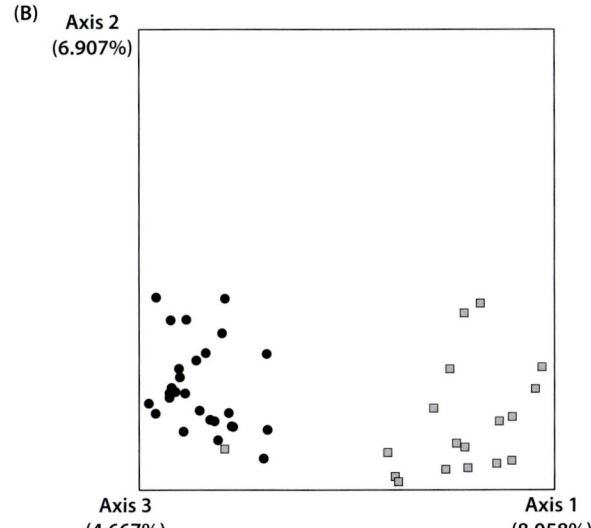

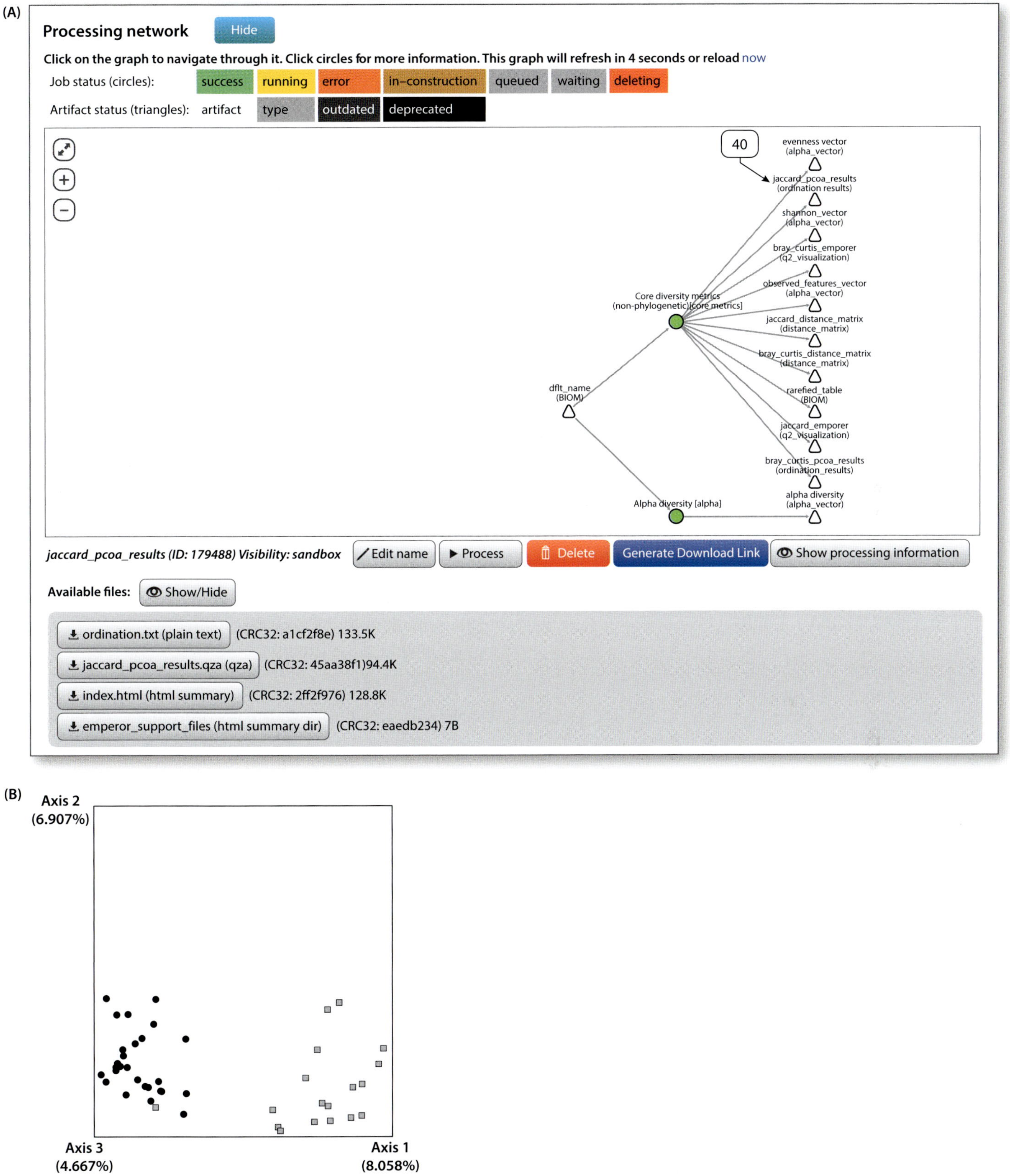

Figure 6.17 Principal Component Analysis of Impact of Birth Mode (A) The command screen for the PCA results. (B) The Qiita-generated PCA of the Dominguez-Bello data colored by birth mode, with black circles for vaginally born babies and gray squares for C-section babies. (After Gonzalez et al., 2018.)

CHECK YOUR UNDERSTANDING

1. Primary literature is
   a. peer-reviewed.
   b. written for the general public.
   c. often easy to quickly read and understand.
   d. usually written by a single author.

2. Peer review includes all of the following except
   a. editors.
   b. scientists.
   c. feedback.
   d. a "general public" reader.

3. Peer-reviewed scientific articles are always correct.
   a. True
   b. False

4. Outlining reproducible protocols in the methods section of your manuscript is important because
   a. it allows other scientists to verify your findings and conclusions.
   b. the methods section includes the results from your study.
   c. peer reviewers often repeat each experiment in a reviewed manuscript before approval.
   d. you may need to go back to your paper for a refresher on what you did.

5. In a manuscript, references to other literature are cited most often in which section?
   a. Methods
   b. Discussion
   c. Results
   d. Introduction

6. The results section always contains only the results of the experiments, and never discussion about their interpretation.
   a. True
   b. False

7. If you were going to read only one section of a manuscript and wanted to get an overview of the study and its findings, which would you read?
   a. Results
   b. Discussion
   c. Introduction
   d. Abstract

8. The article we dissected in this chapter was published in which of the following journals?
   a. *New England Journal of Medicine*
   b. *Microbiome Monthly*
   c. *Proceedings of the National Academy of Sciences*
   d. *Astrophysics for Dummies*

9. A DOI can be used to
   a. find a specific article after publication.
   b. contact the authors of a manuscript.
   c. provide the citation of an article.
   d. both find an article and provide its citation.

10. The goal of a study is usually directly stated in which section of the article?
    a. Abstract
    b. Discussion
    c. Results
    d. Introduction

11. Dominguez-Bello and colleagues used target gene amplification and sequencing of the 16S rRNA gene to
    a. identify eukaryotes in the neonate and maternal microbiomes.
    b. compare bacterial communities across neonatal sample sites.
    c. identify bacterial species in our sites of interest.
    d. quantify microbiome community members.

12. One limiting factor to the Dominguez-Bello study published in 2010 was
    a. sequencing costs.
    b. availability of pregnant people.
    c. newborns birthed in hospitals.
    d. understanding of microbial communities.

13. Which of the following has contributed the most to decreasing the cost of sequencing since 2010?
    a. Inflation in the United States
    b. Development of new sequencing technologies
    c. Cheaper computational resources
    d. Lower salaries for researchers

14. C-section deliveries have no impact on the newborn's gut microbiome colonization.
    a. True
    b. False

15. If a newborn is delivered vaginally, their colonizing skin microbiome is likely closest in composition to their birthing parent's
    a. skin microbiome.
    b. oral microbiome.
    c. vaginal microbiome.
    d. gastrointestinal microbiome.

16. What percentage of babies born in the US are delivered via C-section?
    a. 75%
    b. 60%
    c. 30%
    d. 15%

17. Modern microbiome composition studies generate approximately 10^5 to 10^6 16S rRNA amplicon sequences.
    a. True
    b. False

18. After analysis of 16S rRNA sequencing data, reads classified as "other" represent
    a. bacteria species yet to be identified.
    b. contaminating bacterial or eukaryotic DNA.
    c. species that are normally located at a different body site than the one tested.
    d. bacterial taxa from other kingdoms.

19. When multiple samples from the same person cluster together on a PCA chart, it indicates that
    a. the microbiota in the different samples are very different.
    b. the microbiota in the different samples are very similar.
    c. these individuals likely have low alpha diversity.
    d. the statistical power of your experiment is too low.

20. After sampling the skin, oral, and vaginal microbiomes of 5 individuals, we would expect PCA clusters to form based on the
    a. individual.
    b. alpha diversity of each site.
    c. time of sample collection.
    d. body site sampled.

21. Qiita, developed at the University of California, San Diego,
    a. is a coding language for microbiome studies.
    b. is a platform for people to use established microbiome analysis software without needing to code.
    c. needs a computer with lots of processing power to be used effectively.
    d. is closed-source software.

22. To access the data from Dominguez-Bello et al. (2010), which resource would you use?
    a. Email of the first author
    b. The SILVA database
    c. QIIME 2
    d. Qiita

23. To understand the taxonomy represented by the 16S rRNA genes in our sample, we'll use all of the following *except*
    a. Unix.
    b. Greengenes.
    c. QIIME 2.
    d. Qiita.

24. All of the following are microbiome analysis programs *except*
    a. EzMAP.
    b. VAMPS.
    c. MicroProfiler.
    d. QIIME 2.

Answers: 1A, 2D, 3B, 4A, 5D, 6B, 7D, 8C, 9D, 10A, 11C, 12A, 13B, 14B, 15C, 16C, 17A, 18A, 19B, 20D, 21B, 22D, 23A, 24C

DIVING DEEPER

1. What are the major sections of a scientific paper, and what information does each section contain?

2. Can you summarize the abstract of Dominguez-Bello et al. (2010)?

3. Why is Qiita more user-friendly than other microbiome analysis software?

4. Describe the differences between primary, secondary, and tertiary literature.

5. What are the steps of the peer review process, and how do they contribute to rigorous science?

6. Based on the outline of the peer review process you just described, how would it be possible for incorrect conclusions to be published?

7. If you wanted to design another experiment based on the Dominguez-Bello et al. (2010) article, which section would likely be most helpful?

8. Why is it important that a DOI is assigned to each article upon publication?

9. Which section of a manuscript would you read to learn more about the knowledge gap that researchers are trying to address?

10. How do the experiments described in Dominguez-Bello et al. (2010) address the goal of the study as stated in the abstract?

11. Why do you think researchers chose mouth, skin, and vagina as the three sample sites from the mothers in the study?

12. What ethical concerns did researchers have at the outset of this study, and how were they resolved?

13. If 30% of babies in the US are delivered via C-section, why is it so difficult for researchers to find large cohorts of study participants?

14. What are the benefits and drawbacks of combining the results and discussion in a manuscript?
15. Which four microbiome samples did Dominguez-Bello and colleagues collect from the newborns in this study, and why?
16. List the steps taken in this study to sequence and analyze the samples collected.
17. This study had low statistical power, which we discussed in chapter 5, due to the small sample size. Why was it published anyway?
18. Why are PCAs often used in microbiome studies, and how do they help us to visualize large datasets?
19. Briefly summarize the steps needed to reinterpret the Dominguez-Bello et al. (2010) dataset using Qiita.
20. Which body sites had the highest alpha diversity in this cohort? Would using different alpha diversity measures change this?

DISCUSSING AND REFLECTING

1. The Dominguez-Bello et al. (2010) paper that we used as our case study is over 10 years old! Consider some of the technological and software advances that we discussed during this chapter. If we performed the same study today, even with the same small sample size, do you think we would find different data?
2. The article that we used as a template for this chapter was already familiar to you, and we've discussed most of the relevant terminology and techniques. Keeping this in mind, are there any sections of the manuscript that you still found difficult to understand? Which parts did you find particularly clear?
3. Reflection. For a portion of this chapter, we read through a scientific publication, one of many that are freely available to the public (or, if not, that can still be found online!). Was this your first time reading this type of primary literature? Reflect on the language and format that the researchers used to communicate their findings. Did you find the paper easy to read? Difficult? Why do you think a peer reviewed article, such as this one, is the standard for publishing research?

RECOMMENDED READINGS

Scientific Reviews

Gonzalez, A., Navas-Molina, J. A., Kosciolek, T., McDonald, D., Vázquez-Baeza, Y., Ackermann, G., DeReus, J., Janssen, S., Swafford, A. D., Orchanian, S. B., Sanders, J. G., Shorenstein, J., Holste, H., Petrus, S., Robbins-Pianka, A., Brislawn, C. J., Wang, M., Rideout, J. R., Bolyen, E., . . . Knight, R. (2018). Qiita: Rapid, Web-Enabled Microbiome Meta-Analysis. *Nature Methods*, 15(10), Article 10. https://doi.org/10.1038/s41592-018-0141-9

Knight, R., Vrbanac, A., Taylor, B. C., Aksenov, A., Callewaert, C., Debelius, J., Gonzalez, A., Kosciolek, T., McCall, L.-I., McDonald, D., Melnik, A. V., Morton, J. T., Navas, J., Quinn, R. A., Sanders, J. G., Swafford, A. D., Thompson, L. R., Tripathi, A., Xu, Z. Z., . . . Dorrestein, P. C. (2018). Best Practices for Analyzing Microbiomes. *Nature Reviews Microbiology*, 16(7), Article 7. https://doi.org/10.1038/s41579-018-0029-9

Mother's First Gift

7

CHAPTER CONTENTS

- **7.1** The Microbiome of the Female Reproductive Tract
- **7.2** The Mother's Microbiome during Pregnancy
- **7.3** The Birthing Process and the Newborn Microbiome
- **7.4** The Infant's Core Microbiome
- **7.5** Beyond the Gut Microbiome
- **7.6** The Wonder of Mother's Milk
- **7.7** Transitioning to Solid Foods
- **7.8** Environmental Impacts on the Infant's Microbiome
- **7.9** Health Impacts of a Newborn's Dysbiotic Microbiome
- **7.10** Microbiome-Based Therapies

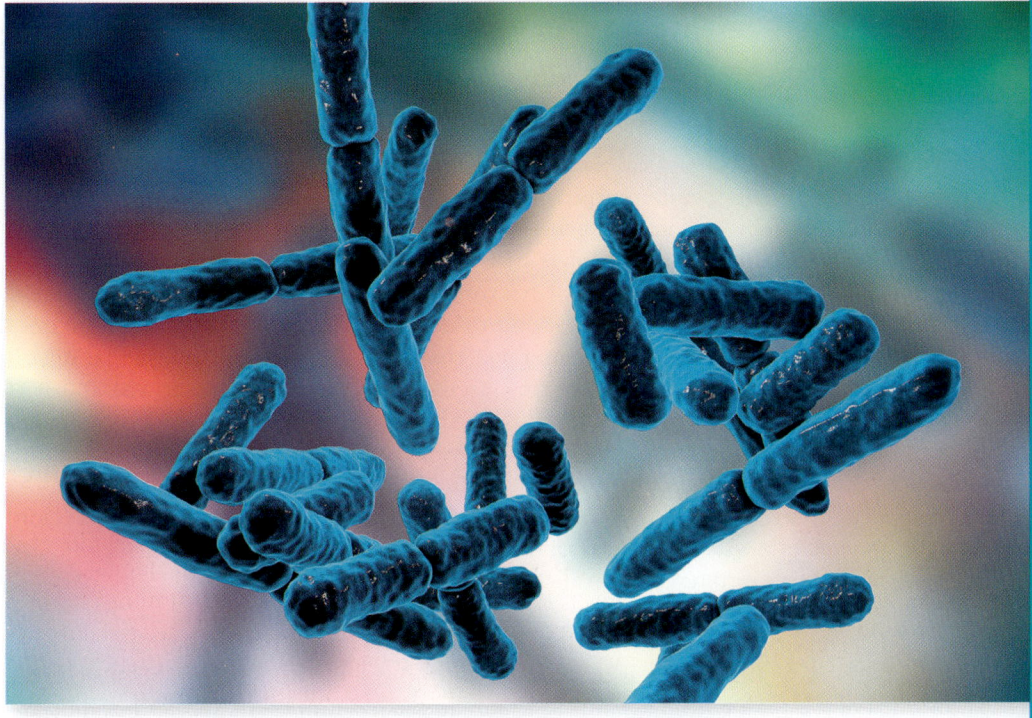

Hello, I'm *Bifidobacterium infantis*, your gut's tiny yet mighty friend! Pleased to meet you, human! You may not see me, but I'm hard at work helping digestion, boosting your immune system, and making sure you're feeling as 'regular' as clockwork. Remember, a happy gut means a happy life—and that's what I'm here for! Now, let's get this fermentation party started! (Photo from Kateryna Kon/Shutterstock)

We have come so far in our journey into the microbiome. We took a quick peek into the composition of the major human microbiomes, such as the gut, skin, vagina, and mouth, and focused even more intently on how we, as scientists, developed the tools and knowledge to explore them. This chapter marks a transition, where we'll start to look at specific microbiomes, their community members, and their impact on human health in greater detail. In the last two chapters, we employed an example that focused on the impact that birth mode has on a newborn's microbiome composition—but what about the 9 months before birth? The human microbiome plays such an important role in all other aspects of human health, it only makes sense that it will have a significant impact on a pregnancy. The focus of this chapter is precisely that. When and how do newborns acquire their microbiomes, what roles do their microbiomes play in their growth and development, and what impacts can a **dysbiotic** microbiome, or one that has an altered microbiota, have on a baby's health? As this chapter's opening quote notes, bacteria are not inherently good or bad. If a newborn acquires the right combination of certain bacteria, their

> "There's no such thing as good and bad bacteria or fungi. It's not good and bad. It's just whether there's too much of it or too little of it and things are out of balance."
>
> —Nigel Palmer (Palmer, 2020)

microbiome can influence their health in amazing and positive ways. If, in contrast, the newborn has a different combination, the results can be both devastating and long-lived with respect to health. Let's explore several studies that have revealed the very earliest moment of our microbiome acquisition. However, before we dive in, we will spend some time learning about a normal microbiome of the female reproductive tract and then assess the changes that occur when a person becomes pregnant.

7.1 THE MICROBIOME OF THE FEMALE REPRODUCTIVE TRACT

Let's apply some of the analytical tools we acquired from chapters 5 and 6 as we explore the microbiome of the female reproductive tract. **Figure 7.1A**, a pie chart indicating the number and frequency of bacterial genera in the upper and lower regions of the female reproductive tract, should look familiar to you by now, as this is a standard approach to illustrating the diversity found in a microbiome (Chen et al., 2017). The different slices of the pies are colored to correspond to different bacterial genera, whose names we will not worry about for now. From these charts we can see that the vaginal region exhibits low genera richness (as estimated by the number of genera shown) and it is dominated by a single genus, *Lactobacillus*. In sharp contrast, we see significant genera diversity in the upper regions of the reproductive tract, where *Lactobacillus* abundance is reduced and 10 or more bacterial genera are found as well. If we were to plot the phylum level of diversity, we would see that Firmicutes dominates the lower region, while Proteobacteria, Actinobacteria, and Bacteroidetes populate the upper region.

We decide we are interested in directly comparing the bacterial diversity of these regions, and we recall that beta diversity measures are designed to do just that. Chen et al. (2017) employed one of the measures discussed in chapter 5, called UniFrac. Recall that the purpose of this metric is to incorporate information about the phylogenetic relationships of the taxa within a sample. If one sample has numerous species but they are all closely related, that sample will have less beta diversity than one that has the same number of species that are more distantly related. The authors do not show us the UniFrac data; however, they do provide a **principal coordinate analysis** (**PCoA**) of these measures (**Figure 7.1B**). We learned in chapter 5 about PCA (principal component analysis), which summarizes multiple variables in the minimum number of components so that each component explains the most variance. In the present study, the authors used the slightly different method PCoA, which we discussed in chapter 6 and which focuses on distances and extracts the dimensions that account for the maximum distances between components. We aren't concerned here with how to conduct a PCA or PCoA analysis. However, they are visually quite rewarding and help us to understand our data better. Figure 7.1B shows a PCoA of the UniFrac measures for the upper and lower regions of the reproductive tract. We can readily see that the upper and lower regions separate in the PCoA. Clearly these microbes appear to experience quite different ecological pressures in the two regions of the reproductive tract. Why might that be?

Lactobacillus is considered a **keystone** genus in the lower portions of the reproductive tract, which refers to the fact that numerous other species in an ecosystem depend upon its presence and if it were removed, the ecosystem would change drastically. In the lower regions of the female reproductive tract, which is highly acidic and anaerobic, lactobacilli flourish. As their name implies, these bacteria make lactic acid as a product of fermenting lactose, glucose, and other sugars, contributing to the low pH in this region. The growth of numerous species of *Lactobacillus* is further regulated by the host through estrogen production. Higher levels of estrogen correlate with increases in vaginal glycogen and more lactobacilli. Decreases in estrogen, for example during part of the menstrual cycle, are associated with lower levels of glycogen and thus decreased lactic acid production and a higher pH—selecting against lactobacilli. Changes in hormone levels, such as those that occur during pregnancy, sexual activity, and stress, can all cause fluctuations in the vaginal microbiome.

The lower regions of the reproductive tract are far more likely to be exposed to introduced bacteria, through sexual activity or invasion of fecal microbes, etc. It

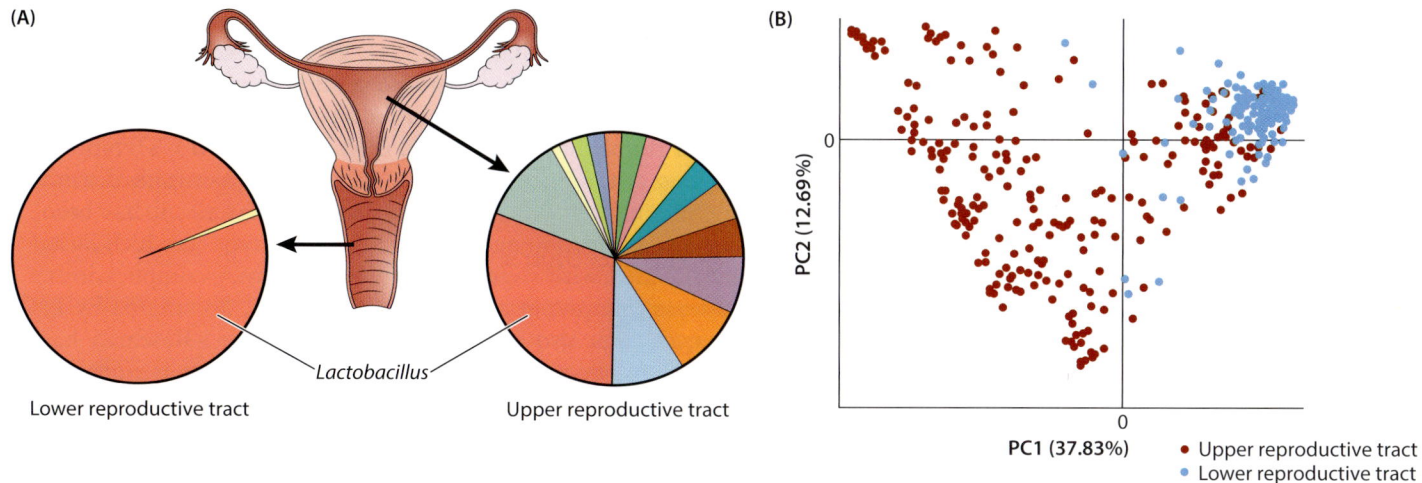

Figure 7.1 Composition of the Microbiome in the Upper and Lower Reproductive Tract (A) These two pie charts provide a snapshot of the relative abundance of major bacterial taxa in the upper and lower regions of the reproductive tract. Clearly, the upper and lower regions harbor quite different levels of bacterial diversity. (B) UniFrac distances were employed in a PCoA to further examine the microbiota differences seen in the pie charts. (After Chen et al., 2017.)

makes sense that this region would be dominated by bacteria that produce potent antimicrobials, such as lactic acid and a bacteriocin called nisin, that can eliminate contaminants before they invade and cause infection. Let's now dive into the female reproductive tract microbiome diversity in more detail.

Vaginal Community State Types

While every vaginal microbiome is unique, most fall into one of five types, called **community state types** (**CSTs**), characterized by the major bacterial species present and their diversity (**Table 7.1**) (Cassano, 2022). At the time that CSTs were introduced, most vaginal samples were obtained from the lower vaginal regions. What we have learned since is that the upper and lower regions of the reproductive tract in females harbor significantly different levels of diversity (see Figure 7.1), and thus the scientific community will need to reconsider this generalization.

Nonetheless, CSTs continue to be useful in describing vaginal microbiome diversity. The main difference between types is in which species of *Lactobacillus* are present. Types 1–3 and 5 are dominated by *L. crispatus*, *L. gasseri*, *L. iners*, and *L. jensenii*, respectively. Type 4 is characterized by a lack of *Lactobacillus* and an increase in overall microbial diversity. Age, pregnancy, sexual activity, menstruation, race, and ethnicity are some of the factors that determine a woman's CST (Chen et al., 2021), as well as host genetic factors, such as immune system, ligands on the surface of epithelial cells, and the quantity and components of vaginal discharge.

CST type 1 is one of the healthiest vaginal community states, as it is the most effective in preventing the invasion of pathogens that cause **urinary tract infections**

Table 7.1 Community State Types of Vaginal Microbiome

	DOMINANT BACTERIUM	EFFECT
Type 1	*Lactobacillus crispatus*	Typically, protective
Type 2	*Lactobacillus gasseri*	Typically, protective
Type 3	*Lactobacillus iners*	Neutral, may be protective or disruptive
Type 4	Diverse bacteria, no *Lactobacillus* dominance	Typically, disruptive
Type 5	*Lactobacillus jensenii*	Typically, protective

(**UTIs**) and **sexually transmitted infections** (**STIs**). This type has also been shown to have the lowest risk for infertility, pelvic inflammatory disease, and toxic shock syndrome. Type 2 is also considered a healthy CST, as it appears to result in a similarly low incidence of pathogen infection and is associated with a lower risk of urogenital disease. In contrast, CST type 3, which is dominated by *L. iners*, is less stable than CST types 1 and 2. *L. iners* can cause itching and unusual discharge if it is found alongside disruptive bacteria. However, in combination with other protective lactobacilli, it has been shown to provide a high level of protection against infection. CST type 4, which lacks lactobacilli, is most often associated with vaginal dysbiosis, pregnancy complications, and pelvic inflammatory disease (Sharma et al., 2014). CST type 5, dominated by *L. jensenii*, is quite rare. However, it is also one of the healthiest types. Research has shown CST type 5 to be the most effective in preventing infections such as STIs and UTIs, and its presence is linked with a very low risk for urogenital disease.

Primary Functions of the Vaginal Microbiome Types

The host provides the nutrients needed to support bacterial growth, which include sloughed cells and glandular secretions. The microbes, on the other hand, serve several roles in maintaining a healthy vaginal environment (**Figure 7.2**). Most critically, they aid in preventing pathogen colonization that can cause bacterial vaginosis, sexually transmitted diseases, and urinary tract infections. Nearly 90% of the sugar that *Lactobacillus* ferments is converted to lactic acid, which results in a low vaginal pH, less than 4.5. This more acidic environment inhibits the growth of invading pathogens. *Lactobacillus* also produces hydrogen peroxide and bacteriocins, which are both potent antimicrobials, which prevent the proliferation of invading pathogens. Furthermore, *Lactobacillus* can stick to vaginal cells and successfully compete with other microbes for cell surface binding sites. This is accomplished either

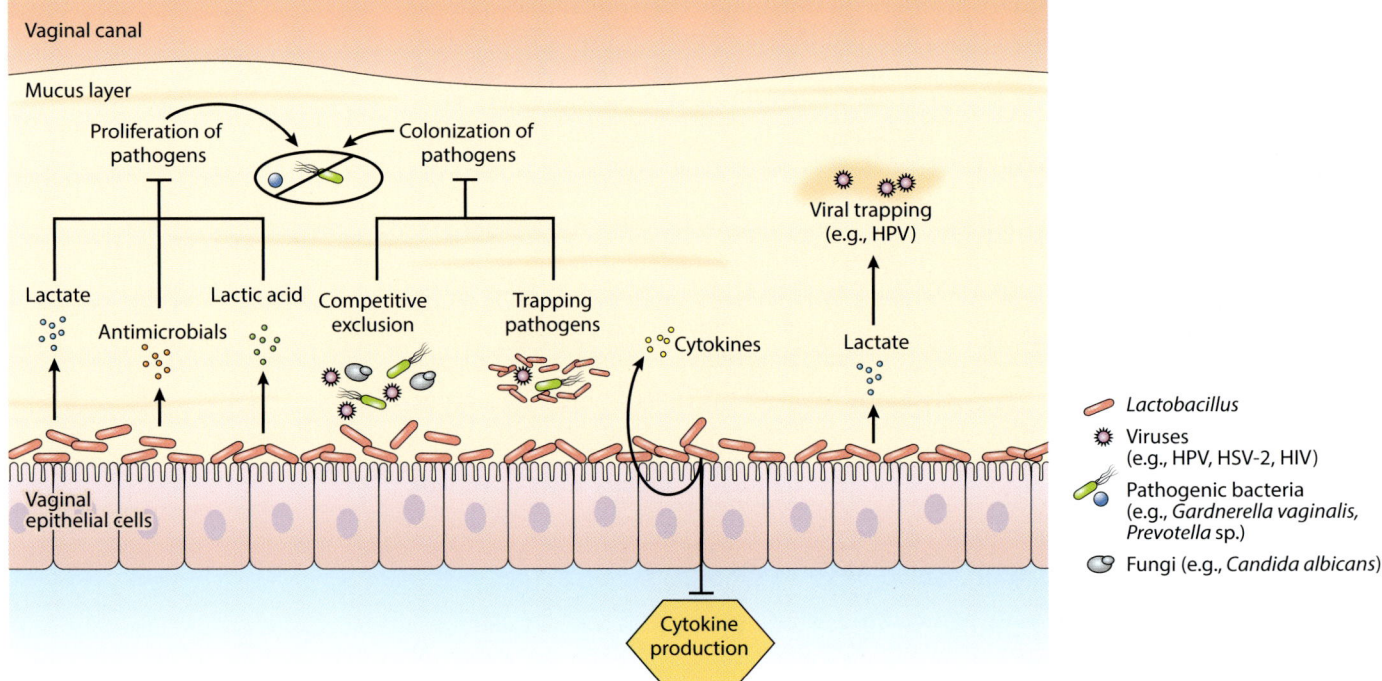

Figure 7.2 Primary Functions of the Vaginal Microbiomes Not only do microbes that reside inside us play key roles in regulating our body systems, as is clearly the case with our vaginal microbes, our exterior microbes do too! Having *Lactobacillus* bacteria on or near the external surface of the vaginal canal helps to prevent pathogenic bacteria and viruses from infecting this tissue. (After Kaur et al., 2021.)

through **competitive exclusion**, in which pathogens simply can't compete against the resident *Lactobacillus*, or through **pathogen trapping**, in which pathogens are physically trapped inside a cluster of *Lactobacillus*. These key microbes also impact the inflammatory response by decreasing production of cytokines, which are compounds secreted by your immune cells. Finally, *Lactobacillus* produces lactate, which increases the viscosity of mucus, helping to trap viruses and other pathogens.

Dysbiosis of the Vaginal Microbiome Types

Bacterial vaginosis (**BV**), the most common vaginal infection, affects nearly 30% of women globally and causes discomfort, itching, and odor. It results from bacterial overgrowth, which can happen when the vaginal microbiome gets out of balance. Doctors don't know what causes the outgrowth of bacteria that results in BV, but risk factors include having sex without a condom, having multiple partners, and having recently changed partners.

BV is characterized by a drop in the numbers of *Lactobacillus* in the vagina, corresponding with a 100-fold or more increase in certain anaerobic microbes, such as *Gardnerella* (Actinobacteria) and *Prevotella* (Bacteroidetes). Researchers are developing the tools to map out precisely which bacteria are present during a BV infection and, equally important, what genes are being expressed over the course of the infection. One recent study showed that just prior to the onset of symptoms, one species, *Lactobacillus iners*, increased its gene expression such that it was responsible for 20% of the metabolic products produced by the vaginal microbiome (Ng et al., 2021). Another group showed that members of Veillonellaceae appear to play a foundational role in BV by altering levels of inflammation and metabolism in the vagina through the production of certain fat molecules (Salliss et al., 2021). These same bacteria impact the production of lactic acid and appear to be involved in the production of the odor typically associated with BV. These studies are just beginning to help us understand the complexity of this polymicrobial vaginal disease.

A second vaginal disease that results from dysbiosis is candidiasis. *Candida albicans* is the fungus most associated with this disease, and when the normal defenses are disrupted in a dysbiotic microbiome, it is able to proliferate, resulting in the common symptoms of a yeast infection, such as vaginal itching and discharge (Dekaboruah et al., 2020). Although the acid produced by *Lactobacillus* generally prevents fungal infection, when yeast numbers explode, the balance of *Lactobacillus* can be tipped and vaginal infection can result.

7.2 THE MOTHER'S MICROBIOME DURING PREGNANCY

From the moment of conception, when an egg and sperm fuse to create an embryo, through 9 months of development in a mother's womb, her microbiome is hard at work directing many aspects of fetal development. The fetus is protected within a seemingly sterile amniotic sac, which blocks the mother's microbes from gaining access (**Figure 7.3**). There is considerable debate about whether a fetus is sterile or not. However, since there are no clear answers yet, we will continue with the most accepted hypothesis, which is that a newborn acquires its microbiome at birth. So, assuming that is correct, and microbes can't physically reach the fetus, how can they impact fetal development? This is a 9-month-long story that involves extraordinary changes in the mother's physiology, while she works overtime sustaining the fetus in her womb.

During pregnancy, the mother's body undergoes numerous changes, not just the obvious increase in girth. The gut microbiome composition shifts, the host and microbiome metabolisms change, and the cells lining the gut become more permeable. The maternal microbiota, and the factors that shape it, such as genetics, diet, exposure to antibiotics, infections, and stress, influence which metabolites are produced and transferred to the fetus through the placenta, where they can directly influence fetal development.

Figure 7.3 A Barrier between Parent and Child We know that the amniotic sac provides a space within the mother's body for an embryo to grow and develop into a fetus. For years, researchers have assumed this protective environment is sterile—but how do we know for sure? Think back to chapters 2 and 4–5; how could we test this hypothesis?

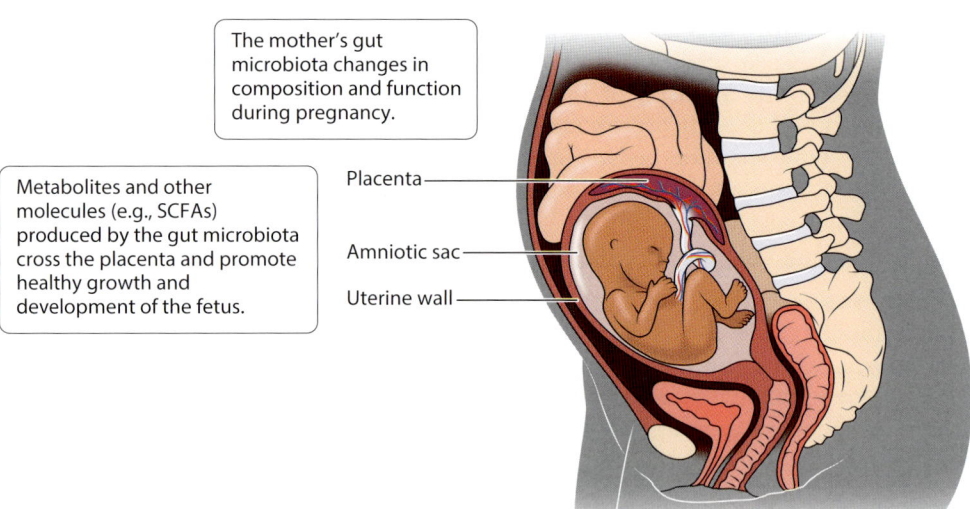

The mother's gut microbiota changes in composition and function during pregnancy.

Metabolites and other molecules (e.g., SCFAs) produced by the gut microbiota cross the placenta and promote healthy growth and development of the fetus.

Placenta
Amniotic sac
Uterine wall

What We Can Learn from the Mouse about a Mother's Microbiome during Pregnancy

Much of what we have learned about a mother's microbiome during pregnancy has been obtained through studies in mice. **Figure 7.4** provides an overview of some of the more critical changes that occur. These studies revealed that complex interactions occur between the fetus and the maternal microbiome. In our discussion of the role of birth mode in infant microbiome composition in the previous chapters, you may have wondered how the commensal microbes from the parent, which enter the newborn's body during the first wave of colonization at birth, don't cause harm. Shouldn't the baby's immune system attack these colonizing microbes?

We learned from mouse studies that the maternal gut lining becomes more permeable during pregnancy, increasing the level of communication between her gut microbiome and immune system. This cross talk produces commensal-recognizing antibodies, which are passed to the fetus through the umbilical cord. These antibodies allow the newborn's immune system to recognize commensal microbes, and not target them for destruction. Finally, and most surprising of all, we learned that maternal microbiome metabolites, produced primarily in the maternal gut, move across the placenta and that the transfer is dependent upon maternal antibodies chaperoning these molecules. If a pregnant mouse lacks all antibodies, due to a deficient or dysbiotic gut microbiome, the effect of the maternal microbiota on the fetus is largely absent. These maternal metabolites play a particularly critical role in educating the developing fetal immune system, helping to protect the fetus from infections and preventing a host of immune-related illnesses that appear later in life, such as allergic asthma and type 2 diabetes (Macpherson et al., 2017).

Immune Interactions between the Developing Fetus and the Maternal Microbiome

Interactions between the fetus and mom's gut microbiome are critical for the maturation of the fetal immune system, which is designed to recognize and dispatch foreign invaders, such as pathogens, or foreign objects, like splinters. Over 75% of a human's immune cells are in the gut—that's right, in the gut! Specifically, they are found inside the cells that line the intestines, the gut epithelium. The intestine is long, and its lining is heavily folded, resulting in the body's largest and most important interface with the outside. (Yes, topologically speaking, your gut lumen is outside your body!) It plays a key role in defending the host from harmful substances and organisms.

There are two major components of the immune system, **innate** and **acquired**. The body's first line of defense is its **innate immune system**, which nonspecifically prevents

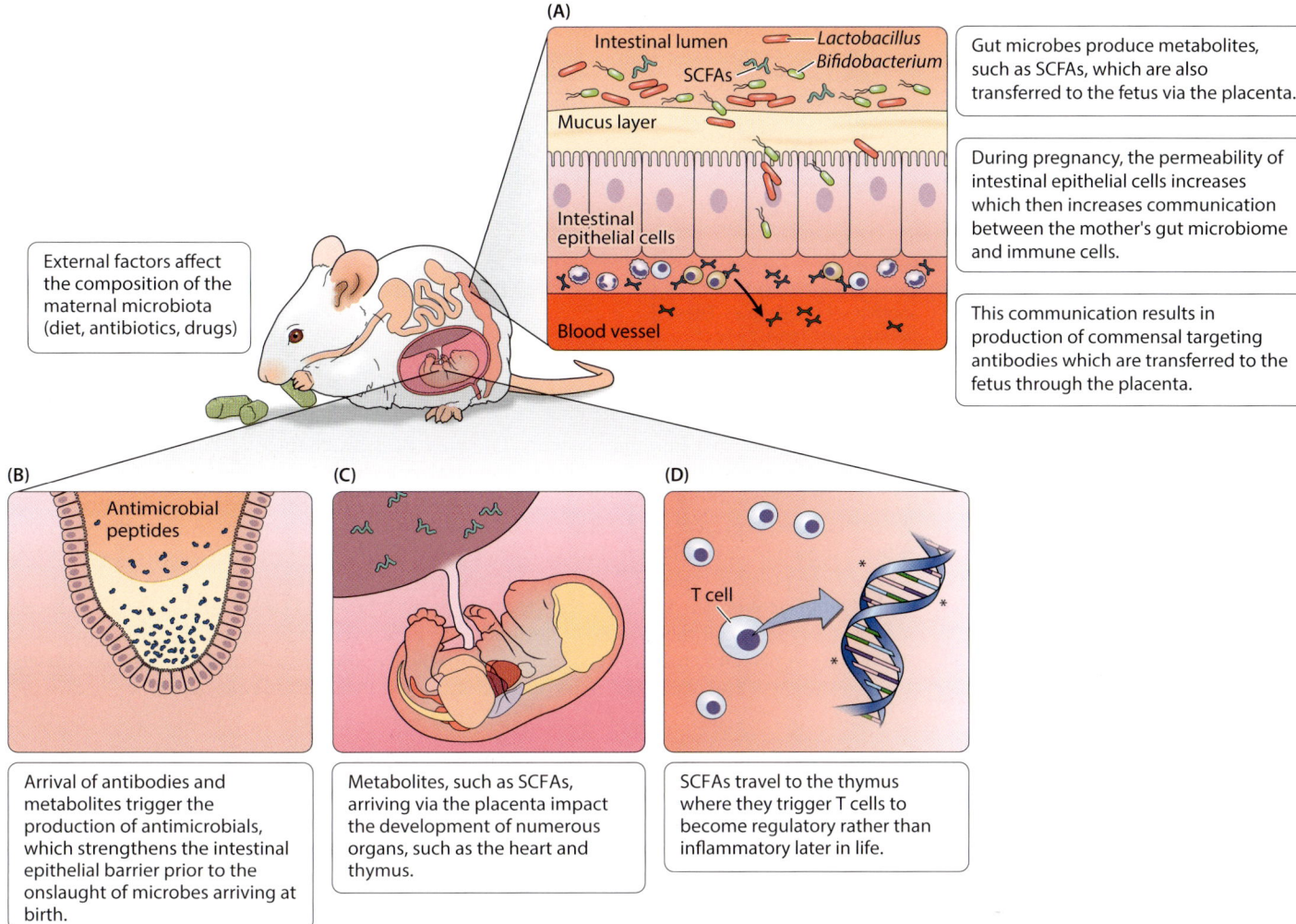

Figure 7.4 What Mice Can Tell Us about the Impact of Mom's Microbiome on the Fetus We have learned so much from mouse studies that focus on the maternal microbiome's impact on the fetus. In this figure, the several impacts identified are shown based upon the locations of their impact in the mouse or her fetuses, and then defined below. (After Thomson and McCoy, 2021.)

pathogen invasion; for example, tears and mucus help block invaders from entering the body. *Numerous* innate immune cells are produced, such as **macrophages** and **neutrophils**, which engulf foreign entities in a process called **phagocytosis**. There are also dendritic cells, which are named for their dendritic (i.e., branching like a tree) shapes and are responsible for initiating an adaptive immune response and thus function as "sentinels" of the immune system. Natural killer T cells are best known for killing virally infected cells and detecting and controlling early signs of cancer. As well as protecting against disease, specialized natural killer cells are also found in the placenta and may play an important role in pregnancy. Finally, there are eosinophils, which play a role in host defense against nematodes and other parasitic infections and are active participants in many immune responses, as well as basophils, a type of bone-marrow-derived circulating leukocyte that are recruited to sites of inflammation. Newborns rely primarily on innate immunity when born, and these cells can detect a wide range of foreign objects and respond to their presence rapidly, within minutes to hours. This is one of the reasons that the commensal-targeting antibodies received through the umbilical cord are so important. Without them, the newborn's innate immune system would go into overdrive trying to eliminate their own microbiomes, doing far more harm than good.

Unlike the innate immune system, the **adaptive (acquired) immune system** develops over time as it learns to respond to specific pathogens. The adaptive immune

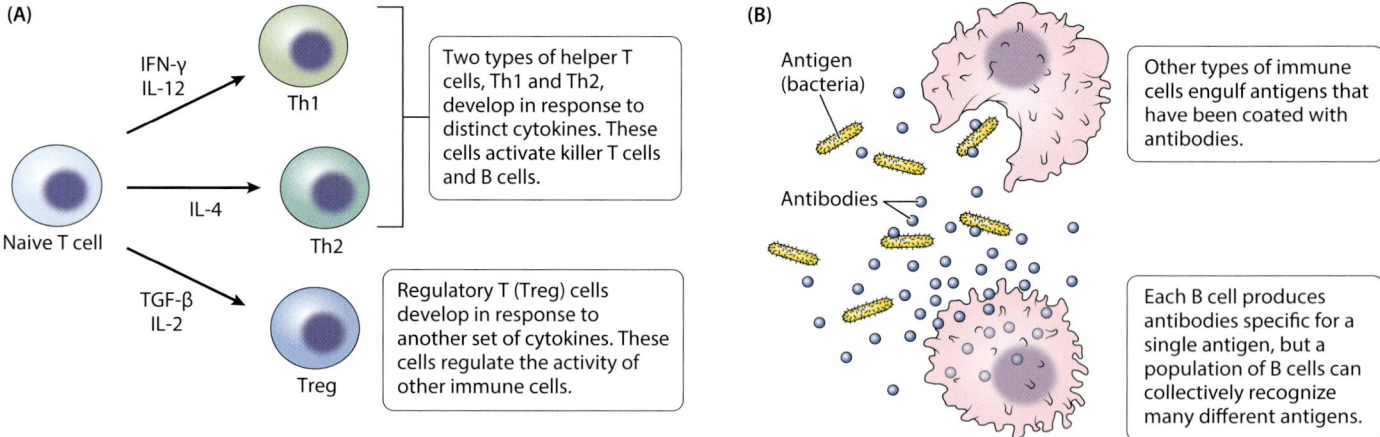

Figure 7.5 Your Two Immune Systems Thinking back to chapter 3, both your innate and adaptive immune systems play unique, essential roles in maintaining health. While many people have heard about T and B cells before, they may not realize that these specifically targeted cell types represent only a fraction of a possible immune response. (A) The naive T cells, when exposed to particular cytokines, develop into one of several types of mature T cells. Only some of the possible types of T cells are shown here. The cytokines shown here are interferon (IFN)-γ, interleukin (IL)-12, IL-4, transforming growth factor (TGF)-β, and IL-2. (B) A B cell, which is a type of cell that produces antibodies to fight bacteria and viruses. These antibodies are Y-shaped proteins that are specific to each pathogen and are able to lock onto the surface of an invading cell or virus and mark it for destruction by other immune cells.

system is composed of **B cells** and **T cells**, which can be considered the special forces of the immune system. They use past behaviors and interactions to learn to recognize specific threats and attack them when they reappear. Naive T cells are formed in the thymus. Under certain conditions, these naive cells differentiate into several types, including **helper T cells** and **killer T cells** (**KTCs**) (**Figure 7.5A**). One major function of helper T cells is to activate KTCs, which functionally are part of the innate immune system and are therefore shown in an overlap between the adaptive and innate immune systems (see Figure 7.5A). The KTCs bind to infected cells and secrete cytotoxins, which induce apoptosis (cell suicide) in the infected cells, and perforins, which cause perforations in the infected cells. B cells produce and secrete **antibodies** (**Figure 7.5B**). These Y-shaped proteins are pathogen specific and interface with a microbial cell surface to mark it for destruction by other immune cells. The main difference between T cells and B cells is that T cells recognize an infected host cell, while B cells recognize the foreign object itself.

The Maternal Impact on Development of the Fetal Immune System

The newborn's adaptive immune system is underdeveloped at birth, and the baby must rely on its innate immune system and the temporary adaptive immunity provided by the mother (**Figure 7.6**). The maternal microbiome produces a wide variety of metabolites that have the ability to modulate immune functions. Figure 7.6 illustrates the interface of the mother's microbiome with the fetus, showing how at least some maternal antibodies and gut metabolites contribute to the education of the fetal immune system.

The types and amounts of short-chain fatty acids (SCFAs) produced in the mother's gut depend on mom's microbiome and diet. When a pregnant woman consumes a diet rich in fiber, those microbes that can ferment fiber thrive and produce SCFAs that are then transferred to the fetus through the placenta. These compounds can impact the development of the fetal immune system and promote the production of a third type of T cell, **regulatory T cells** (**Tregs**) (see Figure 7.5A). Tregs act to suppress immune responses, ensuring that the host is tolerant of its own cells, a characteristic that we call **self-tolerance**. **Autoimmunity** is the condition in which the im-

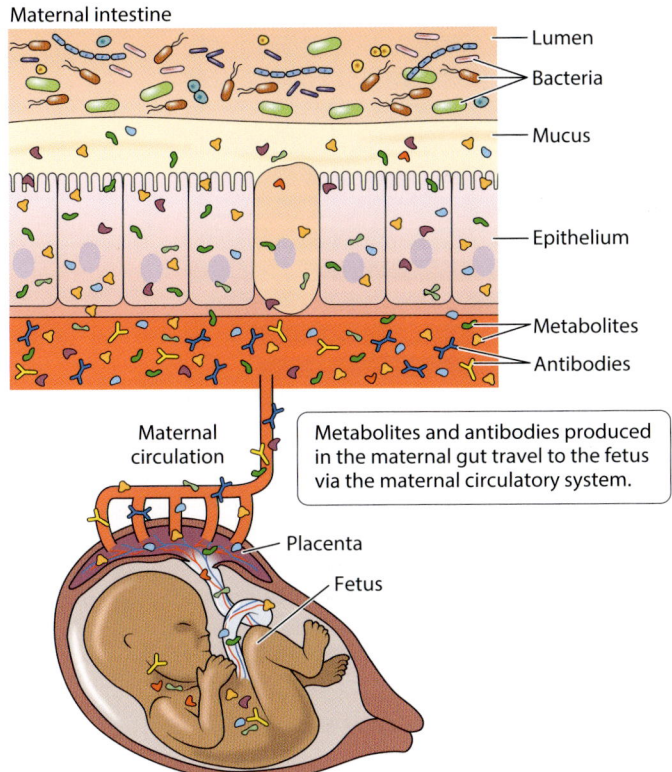

Figure 7.6 Education of the Fetal Immune System The adaptive immune system responds to previously identified threats, so how does that work for a newborn who's never encountered foreign microbes before? Well, the metabolites produced by the pregnant mom are passed to the fetus through the placenta and umbilical cord, and they play essential roles in configuring this prenatal immune system.

mune system recognizes and attempts to eliminate its own healthy cells and tissues. Tregs protect against autoimmunity. They are long-lived and remain throughout the life of the host. If the maternal microbiota impacts the development of the Treg cells, there could be significant impacts on the health of the offspring.

Let's take a moment to examine some of the research that underpins our understanding of the importance of the maternal diet, the production of maternal SCFAs, and their impact on the fetus and newborn. Critical research in self-tolerance was carried out by Akihito Nakajima and colleagues at Juntendo University in Tokyo (Nakajima et al., 2017). They took pregnant mice and fed them either a high-fiber diet (HFD) or a no-fiber diet (NFD) and examined the levels of short-chain fatty acids in the blood of the moms and their pups. The researchers reported that mouse pups had more Tregs if their mothers were fed an HFD, and they suggested that SCFAs produced by the maternal microbiota may act remotely to influence T cell education in the developing fetus. They then performed 16S rDNA–based microbiome analysis of the maternal and pup feces. Although the entire experimental protocol they employed is a bit too complicated to tackle here, you should be proud that you can easily interpret one of their more important figures, shown in **Figure 7.7**. Let's explore this figure. First, you will see a column of numbers from 1 to 21, which refer to the individual mice and pups used in this study. Second, you see that the NFD and HFD mice are presented separately, as are their offspring. Finally, note that the bacterial phyla identified in the gut microbiome samples are presented in a bar graph similar to those we learned about in chapter 5. Take a moment and see whether you can interpret what these data tell us. Once you are ready, read on.

The phylum-level abundance of Bacteroidetes (dark purple part of the bar) in NFD mice was clearly lower, on average, than in HFD mice. Firmicutes (blue) were found at higher levels in the NFD mice. Thus, diet has a clear impact on the gut microbiomes of the pregnant mice. The pups have gut microbiomes that are completely different from their moms. There were almost no Bacteroidetes present, and the dominant phylum was Firmicutes in all offspring, with no detectable difference in gut flora between the pups born from the two sets of moms. We know that *Bacteroides* are the key producers of SCFAs, so how do we explain the fact that the pups born from

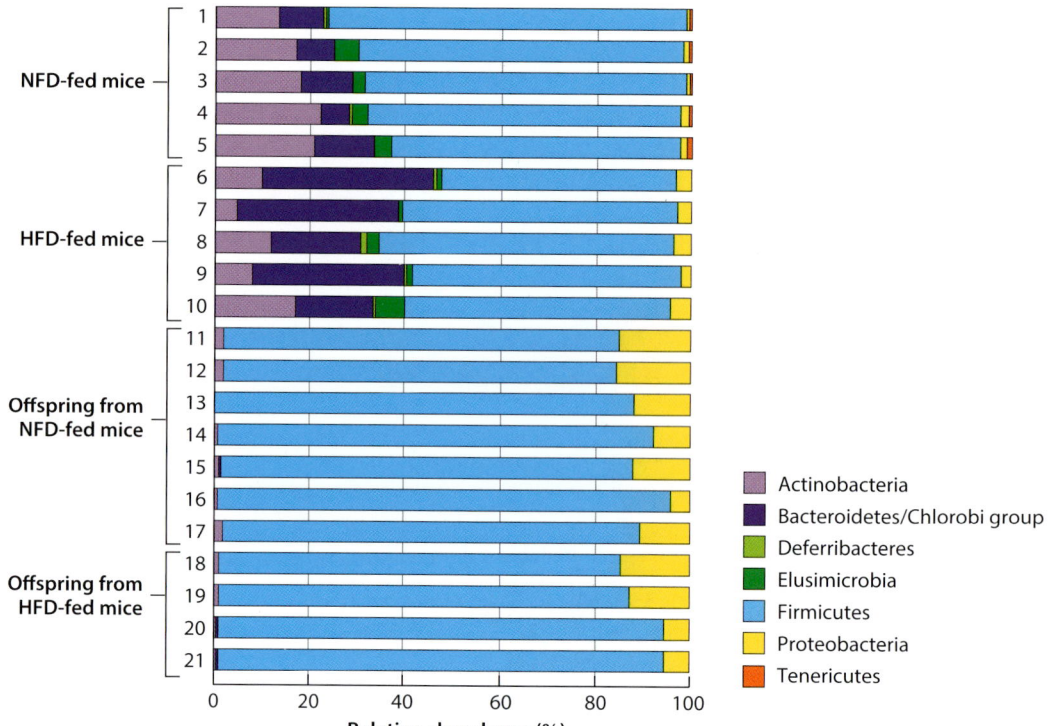

Figure 7.7 **Maternal Mouse and Newborn Pup Gut Microbiome Diversity** This figure is taken from Nakajima et al. (2017) and presents a microbiome analysis at the phylum level based on 16S rRNA amplicon sequencing of feces from mice fed a no-fiber diet (NFD) (1–5), mice fed a high-fiber diet (HFD) (6–10), offspring from NFD-fed mice (11–17), and offspring from HFD-fed mice (18–21). (After Nakajima et al., 2017.)

moms on a HFD diet have higher levels of SCFAs and yet have no *Bacteroides*? The authors suggest that the babies obtained their SCFAs from the moms through the placenta and, after birth, through lactation. This is an elegant example of how there is a coevolution between the colonization of the newborn's gut microbiome, which acquires vaginal microbes, and the ability of the mother to provide the SCFAs the newborn requires before their own microbiomes are equipped to supply them.

Meanwhile, another research team revealed that immune education is more efficient when a specific SCFA (acetate) is provided by the mother through the placenta (Thorburn et al., 2015). Also using mice, this study showed that maternally produced acetate permanently alters T cells in the pups, skewing T cell differentiation toward a Treg and thereby protecting pups from developing asthma and other autoimmune conditions that would be caused by this inflammation. These researchers discovered a direct link between the SCFAs produced by pregnant mice and the levels of self-tolerance in their offspring—yet another reason we love our moms so much! The SCFAs they provide ensure that we don't start out fighting off those very microbes we require for health.

As we are learning, multiple factors influence the fetal immune maturation process. Traditionally, it was believed that the maternal antibodies provided through the placenta were those that recognized pathogens. It is now clear that these antibodies can also recognize commensal bacteria, which ensures that these commensal bacteria don't cross the newborn's gut epithelial cells during the first wave of colonization that occurs during birth. This immune training further protects the newborn from some inflammatory diseases later in life. We will return to the immune system of the newborn a bit later in the chapter. But first, we will turn our attention to one other key impact the pregnant parent's microbiome has on the developing fetus.

Much of the research into the effects of the maternal microbiota on fetal development has focused on immune education. However, there is increasing interest in how it impacts fetal brain function and behavior as well. We now recognize the importance of the maternal gut microbiota for normal fetal brain development in mice and that this effect is driven by microbiome-produced metabolites.

One role played by mom's metabolites was revealed in a study by Morgane Thion and colleagues from the PSL University in Paris (Thion et al., 2018). They found that changes in the maternal microbiome can result in differences in the abundance of the cells that serve as a form of innate immunity in the **central nervous system**. These **microglial** cells are a specialized form of macrophage that removes damaged neurons. They also target pathogens—one of the reasons that your brain tends not to get infected is that these cells are trained by the microbiome and serve as the front line of immune defense in the central nervous system. Upon sensing signals of infection, microglia transition from a surveillance state to an activated state, facilitating antimicrobial production programs that restore homeostasis. Thion and colleagues compared germ-free pregnant mice to conventional pregnant mice and found that in the germ-free mice, which don't produce the short-chain fatty acid acetate, the microglial cells fail to develop. Further, we now understand that microglia contribute to the pathogenesis of several neurodegenerative and neurodevelopmental disorders, such as Parkinson's disease and autism. So, the mind-microbiome connection is increasingly recognized as critical in ensuring neural health.

Another example of the impact of the maternal microbiome on fetal neural development is provided by the work of Vuong et al. (2020). An **axon**, also called a **nerve fiber**, is the portion of a nerve cell that carries nerve impulses away from the nerve cell body. In short, axons allow neurons to send information, which enables us to think, move, and do everything else we need to do to survive. Vuong and colleagues studied mouse neural development and showed that depletion of the maternal microbiome altered the expression of numerous genes in the fetal mouse brain, including many involved in the generation of axons, which resulted in certain behavioral abnormalities (**Figure 7.8**).

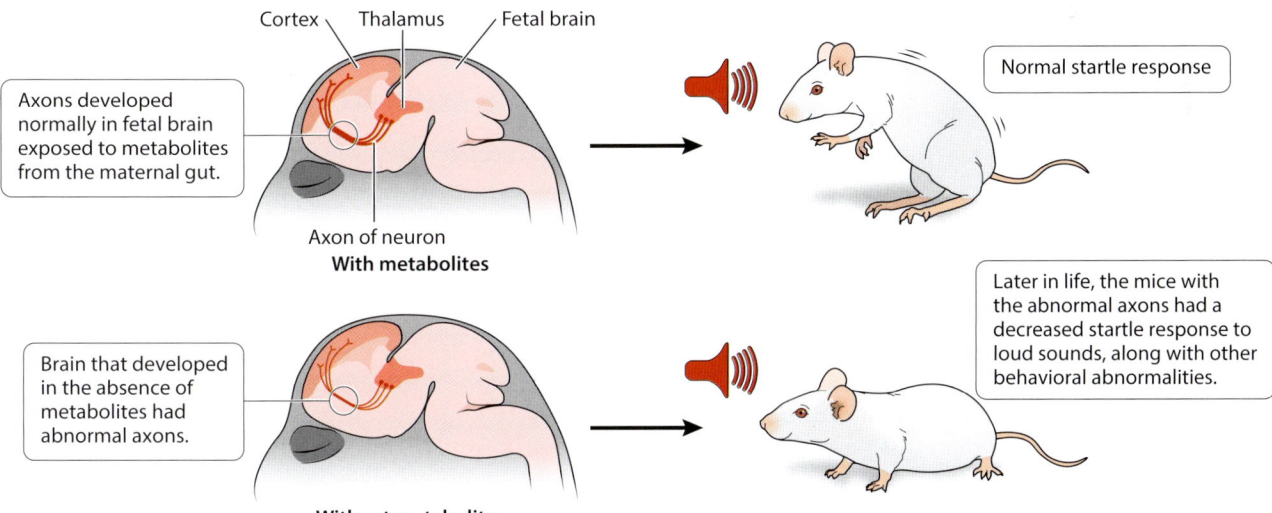

Figure 7.8 Metabolites from Maternal Microbiome Influence Brain Development Understanding how the embryonic brain develops has been a hot topic in research for decades. It is only more recently, however, that the role microbes play in this process has been explored. While developing in utero, mouse embryos receive metabolites from maternal gut microorganisms. These metabolites aid normal wiring by neuronal projections, called axons, that connect brain regions called the thalamus and the cortex. Such connections are needed for sensory processing. If pregnant mice lacked their usual gut microbes and their fetuses did not receive maternal microbial metabolites, then neurons forming thalamocortical projections had axonal defects, including thinner-than-normal axons. These animals displayed behavioral abnormalities when tested later in life. (After Meckel and Kiraly, 2020.)

These studies suggest that a mother's microbiota may impact fetal neurodevelopment and may even protect the fetus from neurological disorders later in life. This area of research is still quite young, and we have much to learn about the influence of mom's microbiome on fetal development. We will return to the brain-microbiome connection for a more in-depth exploration of what is known in the next chapter.

7.3 THE BIRTHING PROCESS AND THE NEWBORN MICROBIOME

From the moment we are born, an entire microbial army is poised to protect us against the perils of the world around us. Babies born vaginally first encounter microbes that reside in their mothers' birth canal. Those microbes seed the first of many waves of colonization. The resulting microbial communities develop into robust microbiomes that support metabolism, immunity, and many other aspects of health. Let's explore this initial colonization process in more detail.

Vaginal Delivery

To get out of the womb, most babies travel through the birth canal. This journey is particularly important because the birth canal, unlike the womb, is bursting with bacteria. The newborn gets slathered with and swallows these vaginal microbes and is literally bathed in friendly bacteria such as *Lactobacillus*, *Bacteroides*, and *Prevotella*. Then, as the baby is greeted by a midwife, doctor, or Uber driver, they also meet members of the others' skin microbiomes, as well as mom's gut microbiome. We won't go into the details of how this happens, but let's just say that birthing can be a messy process. It shouldn't be surprising, then, to learn that a newborn's microbiome looks very much like a trimmed-down version of mom's vaginal and gut microbiomes (**Figure 7.9A**). Some have referred to this first wave of microbial colonization as mother's first gift—and what a gift it is! These microbes direct the next phase of development, the maturation of the newborn's immune and digestive system functions, which helps prevent infection and further influences brain development. But before we explore the many important functions of the baby's microbiomes, let's discuss what happens when newborns don't get mom's first gift.

Cesarean Section

Nearly one-third of the babies in the United States are born through a Cesarean section (C-section) surgical procedure. This mode of birth results in a newborn missing that key first wave of colonization by mom's vaginal microbiome. As we learned in chapter 6, C-section newborns tend to have microbiomes that resemble their mothers' skin microbes (Dominguez-Bello et al., 2010). Since the pioneering study by Dominguez-Bello and colleagues, numerous other investigations into the impact of C-sections on a newborn's microbiome have been published. One such study was led by Dr. Trevor Lawley, a microbiologist at the Wellcome Sanger Institute in the UK, who examined the microbiomes of nearly 600 newborns (a far more robust sample than was possible for the Dominguez-Bello study) and noted that C-section babies are missing bacterial species found in vaginally born babies (Shao et al., 2019). Because babies born via C-section aren't exposed to the vaginal microbes, they are more likely to harbor microbes found in the hospital environment and on hospital staff. Since hospitals are full of people with infections, they are a breeding ground for pathogens, which can easily be transmitted between patients. Rather than being dominated by vaginal microbes, these newborns' first wave of colonization is dominated by potentially harmful pathogenic microbes (**Figure 7.9B**). Ewen Callaway, who wrote a companion piece for Nature notes: "The level of colonization by health-care pathogens is shocking in these children. When I first saw the data, I couldn't believe it" (Callaway, 2019). Lawley's team also identified antibiotic resistance genes encoded by bacteria in the microbiomes of C-section babies and confirmed that these bacteria were related

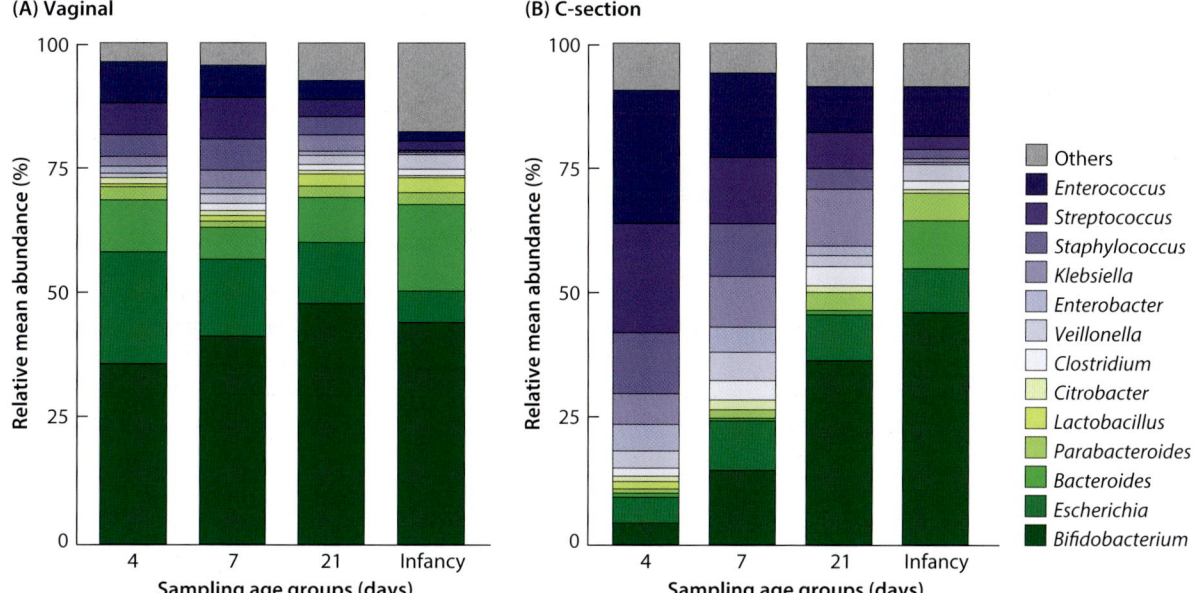

Figure 7.9 Microbiome Acquisition during Birth As we discussed in the two prior chapters, birth mode has a profound influence on an infant's microbiome. This figure provides a series of stack plots of the relative abundances of microbial taxa for newborns at different ages. The graph in (A) is from babies delivered vaginally, whereas the graph in (B) is from babies delivered via C-section. We'll learn in this chapter the health outcomes that these infants may experience due to their microbiomes. (After Shao et al., 2019.)

to hospital pathogens. In fact, the differences were so dramatic that Lawley noted, "I could take a sample from a child and tell you with a high-level certainty how they were born" (Callaway, 2019).

Over the next 9 months the microbiota of infants born by C-section will grow more like the microbiota of infants born vaginally, with one critical exception: the continued absence of *Bacteroides*. One year later, these babies' guts remain deficient in *Bacteroides*. As we learned earlier in this chapter, members of the genus *Bacteroides* are critical in producing metabolites (SCFAs) that train the immune system and reduce inflammation (Wu & Wu, 2012). Even though babies born by C-section can recover a mostly healthy microbiome, not having a diverse community during the critical first few months of life can have lasting effects. C-section birth is associated with an increased risk for a variety of immune-related disorders, such as asthma, later in life—but more on those topics later in this chapter.

7.4 THE INFANT'S CORE MICROBIOME

After birth, the microbial floodgates open. The first year is a dynamic phase of microbiome development, with repeated waves of colonization. As newborns transition at 1 month into infants, they begin to acquire microbes from everyone and everything around them, including mothers, fathers, siblings, the family pets, caregivers, and what they encounter on the floor or in the soil of potted plants and pretty much everywhere they wander. They also ingest microbes in their food—the more diverse the diet, the more diverse the microbial exposure (**Figure 7.10**).

Structuring the Infant's Core Microbiome

Despite their contact with the baby, most microbial visitors never become residents of the baby's microbiome; they can't live in the habitats provided by the human body. A key concept in this process is **selection**, in which those microbes that can adapt to

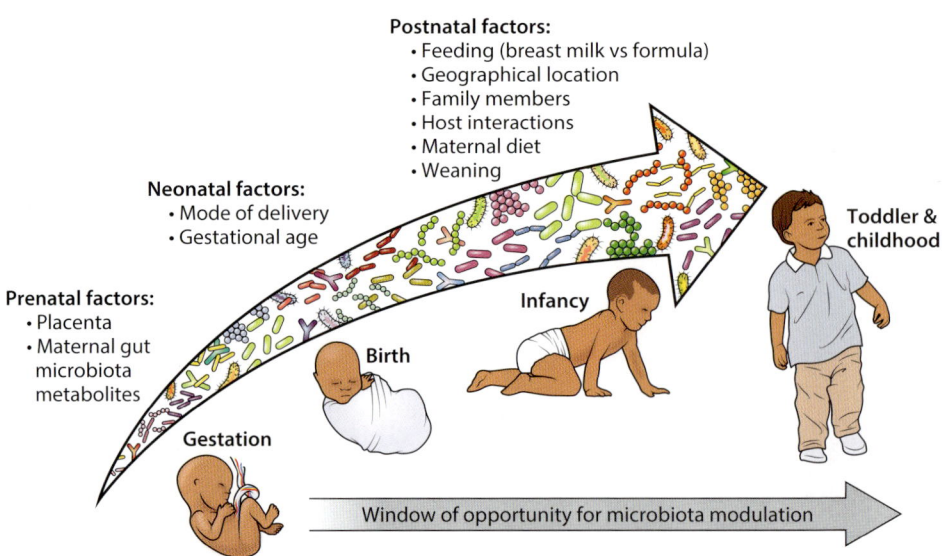

Figure 7.10 The Dynamic Gut Microbiome The newborn's microbiome is like an unpainted canvas, just waiting for diverse microbes to take up residence. This image highlights some of the more important factors that mold the process of microbiome establishment through life. (After Milani et al., 2017.)

their new home survive and reproduce while others fail to establish. Selection is not limited to the microbes and their ability to survive; the human body itself is selecting for microbes that provide critical services or metabolites, such as vitamins, and those that do no harm. Newcomers must compete with those microbes already established, leading to a second key concept, **priority**, the enormous advantage held by those microbes that arrive first. It is hard to displace a resident microbe without the use of a strong selection pressure like antibiotics.

The baby's genetics also play a role in structuring their microbiome. Bacteria interact with specific markers or receptors found on the surfaces of the host cells. These cues inform many microbes about where to grow. The mechanisms governing this process of selection remain unknown; however, there is clearly a great deal of chemical communication, as well as physical cues such as temperature and humidity. The collection of microbes that make up a mature microbiome is far from random. There is a mutual selection process that occurs between the baby and the microbes they encounter, especially during the first 2 years of life.

Relative to the gut microbiota of older children or adults, the infant gut microbiota is unstable and highly dynamic. Those bacteria that dominate the infant gut are also found in adults, just at differing densities. **Figure 7.11** shows the dominant taxa in the infant's gut core microbiome, as well as their oral, skin, and nose microbiomes. Immediately after birth, the gut is dominated by aerobic bacteria, such as members of the Proteobacteria (e.g., *Escherichia*) and Firmicutes (e.g., *Staphylococcus*). However, there is a rapid transition to a gut dominated by Actinobacteria, particularly *Bifidobacterium*, which metabolizes glycans and breast milk nutrients. Mucin, a host-produced glycan, creates a barrier that protects the gut lining. *B. bifidum* is one of the few microbes that can use mucin as a carbon source. It gets even more complicated than that. *Bifidobacterium* communities establish **trophic interactions** with each other, which are, essentially, food chains. These interactions may lead to cooperative sharing of nutrients. As these cross-feeding interactions increase, more species are able to survive in the developing gut microbiome, and these species provide new metabolic functions. Cross-feeding interactions can therefore promote an expansion of the gut microbiome's ability to consume carbohydrates. Such cross-feeding interactions in the gut are common. *Bifidobacterium* species hydrolyze certain complex carbohydrates found in milk, such as glycan, and provide the resulting mono- and oligosaccharides to other microbes. In addition, fermentation of these carbohydrates by *Bifidobacterium* generates metabolites that are food for the next level in the food chain. So, level by level, the species present in the infant gut microbiome and the functions their genes encode diversify.

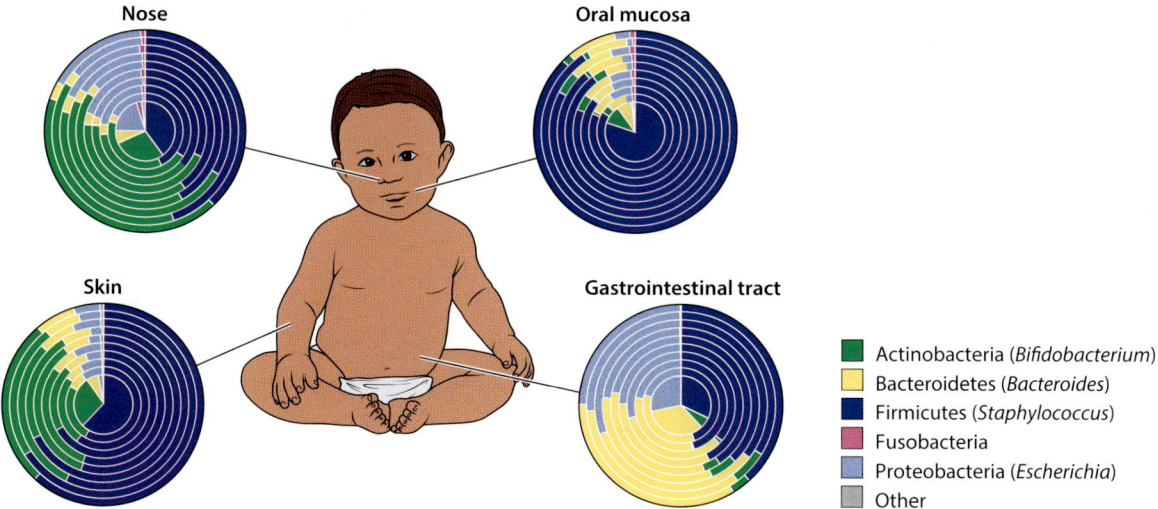

Figure 7.11 The Infant's Core Microbiomes The infant's microbiomes are rapidly established after birth, with each acquiring its own specific constellation of taxa, such as shown here for the nose, skin, oral mucosa and gut. (After Milani et al., 2017.)

Members of the *Bacteroides* genus are another early colonizer of the infant gut. They can metabolize glycans produced by the host, such as breast milk nutrients and mucins, as well as complex polysaccharides, such as starch and cellulose, found in plant matter. *B. fragilis* produces a critical metabolite, polysaccharide A, which plays a role in mediating host-microbe cross talk and immune modulation.

One of the least studied members of the newborn's gut are the viruses, which are abundant members of the microbiota throughout life. Guerin and Hill (2020) revealed that as the infant gut microbiota diversifies, there is a decrease in the abundance of bacteriophages, also called phages, which are viruses that target bacteria. They suggest that this contraction facilitates bacterial diversification (**Figure 7.12**). In contrast, viruses that target eukaryotic cells are present in low numbers at birth, and their diversity and abundance increase over the first 2 years (Matijašić et al., 2020).

Wave after Wave of Microbial Colonization

Over the first few years of life, wave after wave of microbes enter the infant's gut and then engage in intense biological warfare to occupy every available niche. **Figure 7.13** provides an illustration of some of the factors that serve to impact the development of this young microbiome. For example, the microbiota found on mom's skin and in her milk play a key role in how well the newborn is able to digest mother's milk. A newborn's gut is aerobic, meaning that it has far higher levels of oxygen than the

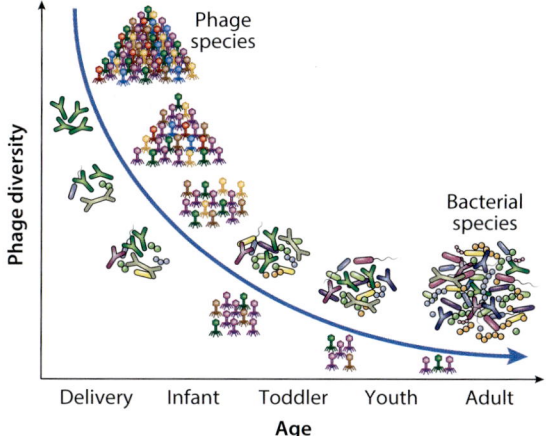

Figure 7.12 Contribution of Bacteriophages to the Maturation of the Gut Microbiome The contribution of phages to the gut microbiota development over time is shown. As phage diversity and densities decrease, the diversity of the microbiome increases. (After Milani et al., 2017.)

Figure 7.13 Factors That Affect the Development of a Newborn's Microbiome The newborn starts with a microbiome that resembles the mother's vaginal or skin (depending on birthing mode) microbiome. Over time, and depending on numerous factors, such as mode of feeding and environmental exposures, the baby will acquire distinct microbes across their body that begin to resemble the normal adult body site-specific microbiomes we discussed in chapter 3.

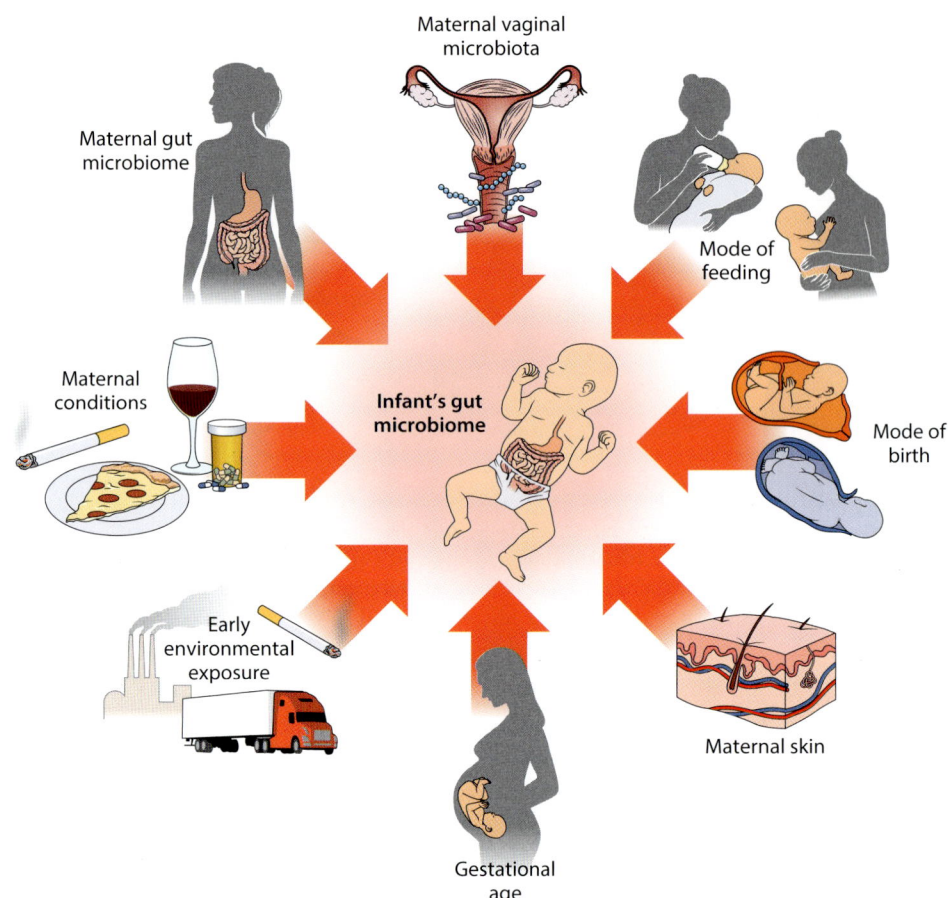

adult gut. This oxygen-rich environment selects for oxygen-tolerant microbes. Proteobacteria and Actinobacteria are perfectly happy growing in the presence of oxygen. Firmicutes and *Bacteroides*, not so much. As the early inhabitants use up the available oxygen, they create a more anaerobic environment that favors a new wave of anaerobic microbes, such as *Lactobacillus*. If we jump ahead 6 months, the baby's microbiome is now dominated by *Bifidobacterium*, which is responsible for breaking down breast milk and coating the intestinal surface to prevent the attachment of pathogens. Members of *Bifidobacterium* encode some or all of the enzymes required to break down mother's milk and produce an array of metabolites, such as SCFAs, that are critical for the newborn's continued microbiome-mediated immune system training.

7.5 BEYOND THE GUT MICROBIOME

We tend to focus on a newborn's or infant's gut microbiome, as it plays such a prominent role in the newborn's development. But all of a newborn's microbiomes, including those found in the mouth, on the skin, and in the respiratory and urogenital tracts, are also rapidly colonized at birth and with primarily the same microbes found in the baby's gut. At this early time point, the distinct microbiomes of the body have not yet developed. Let's take a short detour and examine the newborn's skin microbiome.

The Newborn's Skin Microbiome

At birth, a newborn's skin leaves an aqueous, sterile womb and enters an aerobic, dry environment chock-full of microbes. This new habitat results in a reduction in the water content of the skin and a decrease in skin pH. These changes create an environment that invites colonization of some microbes and prohibits invasion by others.

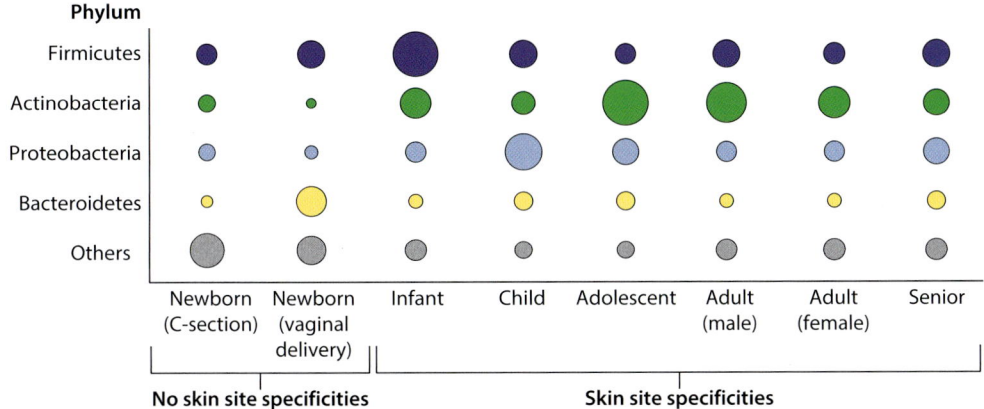

Figure 7.14 Acquiring an Adult Skin Microbiome A newborn's skin is coated with the mom's vaginal microbiome at birth. However, over the first year of life, a series of colonizations result in skin-specific, adult-like microbiomes. By the time the infant becomes a child, their microbiome has achieved an adult composition. The relative abundances of different bacterial phyla are represented here by colored dots of different sizes: the larger the dot, the greater the relative abundance. (After Luna, 2020.)

While the adult skin microbiome is relatively stable over time, the newborn's experiences an explosion of microbial diversity (Manus et al., 2020).

During birth, the skin is colonized by surrounding microorganisms, which differ depending on the mode of delivery (**Figure 7.14**). Babies born vaginally have skin microbiomes resembling their birthing parents' vaginal microbiomes, as we discussed above and in chapter 6 (Dominguez-Bello et al., 2010). The next wave of colonization is dominated by Firmicutes, followed by the gradual transition to site-specific skin microbiomes that eventually resemble those in the adult. Establishment of a healthy skin microbiome during this early phase might play a crucial role in preventing pathogens from gaining a hold and potentially impacting the composition and stability of the adult microbiome. The skin microbiome modulates inflammatory responses of the skin and helps maintain homeostasis. Thus, early microbial colonization is critically important in the development of the skin immune function.

7.6 THE WONDER OF MOTHER'S MILK

The next major source of microbes for the newborn is breast milk. Perfected by thousands of years of evolution, human breast milk contains exactly what a baby needs to grow. The microbes present in breast milk—yes, breast milk is not sterile—will help to seed the newborn's gut microbiota with the species required to metabolize human milk. It also provides additional passive immune components from the breastfeeding parent, such as SCFAs, which serve to increase the generation of Tregs in the baby.

The Composition of Breast Milk

The bacteria introduced to the baby during breastfeeding, along with others already in the baby's gut microbiome, jump-start the baby's digestive system, help prevent infection, and even affect brain development. Breast milk is composed of unique and complex oligosaccharides, which function as **prebiotics**, or food sources that are not digested by the host but which feed beneficial gut microbes, resulting in increased health and diversity of the gut microbiome. The result of breastfeeding is a gut microbiome quickly dominated by *Bifidobacterium*.

Human milk has highly abundant and very special ingredients, **human milk oligosaccharides** (**HMOs**), which are a diverse class of complex carbohydrate molecules synthesized by the mammary gland. With approximately 200 different types, HMOs are the third most abundant ingredient in breast milk, after lactose and fat. HMOs reach the colon intact, where they shape the microbiota. Paul György, a biochemist and pediatrician from the University of Pennsylvania, unknowingly referred to HMOs in the mid-1900s when he proposed the existence of a "bifidus factor," which was a substance unique to breast milk that fed *Bifidobacterium* (György et al., 1954). The bifidus factor turned out to be HMOs. Although humans are unable to digest

Figure 7.15 The Milk-Oriented Microbiome Digestion of HMOs by MOM *Bifidobacterium* results in the production of lactate and the short-chain fatty acid acetate, which are secreted into the gut lumen. These molecules lower the pH in the intestinal milieu, which improves their transport into the epithelium for use by the host and creates an undesirable environment for potential pathogens. (After Smilowitz and Hazard Taft, 2020.)

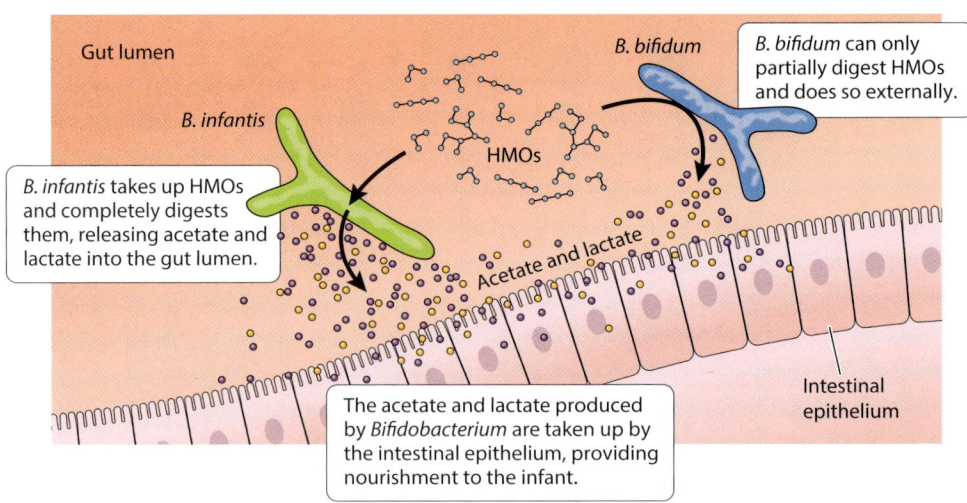

HMOs, *Bifidobacterium* can, especially *B. infantis*, which uses HMOs as its primary energy source.

The community of bacteria in the infant gut capable of consuming HMOs is called the **milk-oriented microbiome** (**MOM**) (**Figure 7.15**). *B. infantis* is the most efficient HMO consumer in the MOM. It possesses all the genes necessary to degrade HMOs. It secretes several proteins that attach to HMOs in the gut lumen and produces others that direct HMOs into its cells. Once inside *B. infantis* cells, the HMOs are fermented into monosaccharides that are then fermented into lactate and the short-chain fatty acid acetate, both of which are secreted back into the lumen (Chichlowski et al., 2020). These metabolites result in a lower gut pH, which supports transport of the metabolites into the cells lining the intestine.

B. infantis is the only species that can completely digest the HMOs. Other species, such as *B. breve* and *B. bifidum*, can consume some of the HMOs present in milk, but only externally, using enzymes attached to their cell membranes and relying on transporter enzymes to uptake some of the products of HMO digestion. This extracellular digestion provides free simple carbohydrates that can be used by other, perhaps less beneficial, microbes. This cross-feeding is common among microbes and results in a more highly diversified gut microbiome. Although this diversity is beneficial in adults, it may not be in infants. When milk is the primary food source, having lower microbial diversity appears to be just fine in newborns and infants.

One fascinating feature of HMOs is that their composition differs depending on the diet, genetics, and other factors of the mother producing breast milk. Infants receive a form of personalized nutrition, or combinations of prebiotics, through breast milk, that change over the period of breastfeeding. Mothers' milk provides an essential form of nutrition, which has long-term consequences on the health of the infant. However, for various reasons, not all infants are breastfed, so infants who are not breastfed are given formula instead. HMOs are complex and expensive to manufacture, so infant formula has only a fraction of the diversity and levels of HMOs found in breast milk. This impacts the abundance of some of the highly beneficial gut microbes that are capable of preventing pathogen invasion.

Figure 7.16 illustrates the differences between breast milk and formula with respect to their impact on the infant microbiome. Breast milk contains lactoferrin, oligosaccharides, immunoglobulins, and human milk microbiota, which aid in modulating a healthy infant gut. In contrast, infant formula lacks the immunoglobulins and the human milk microbiota and often has additional supplements, such as pre- and probiotics, added to attempt to mimic human milk. The impact on the infant's gut microbiome is significant. Infants who are breastfed have higher levels of *Lactobacillus* and *Bifidobacterium* in their gut microbiomes than their formula-fed counterparts, who

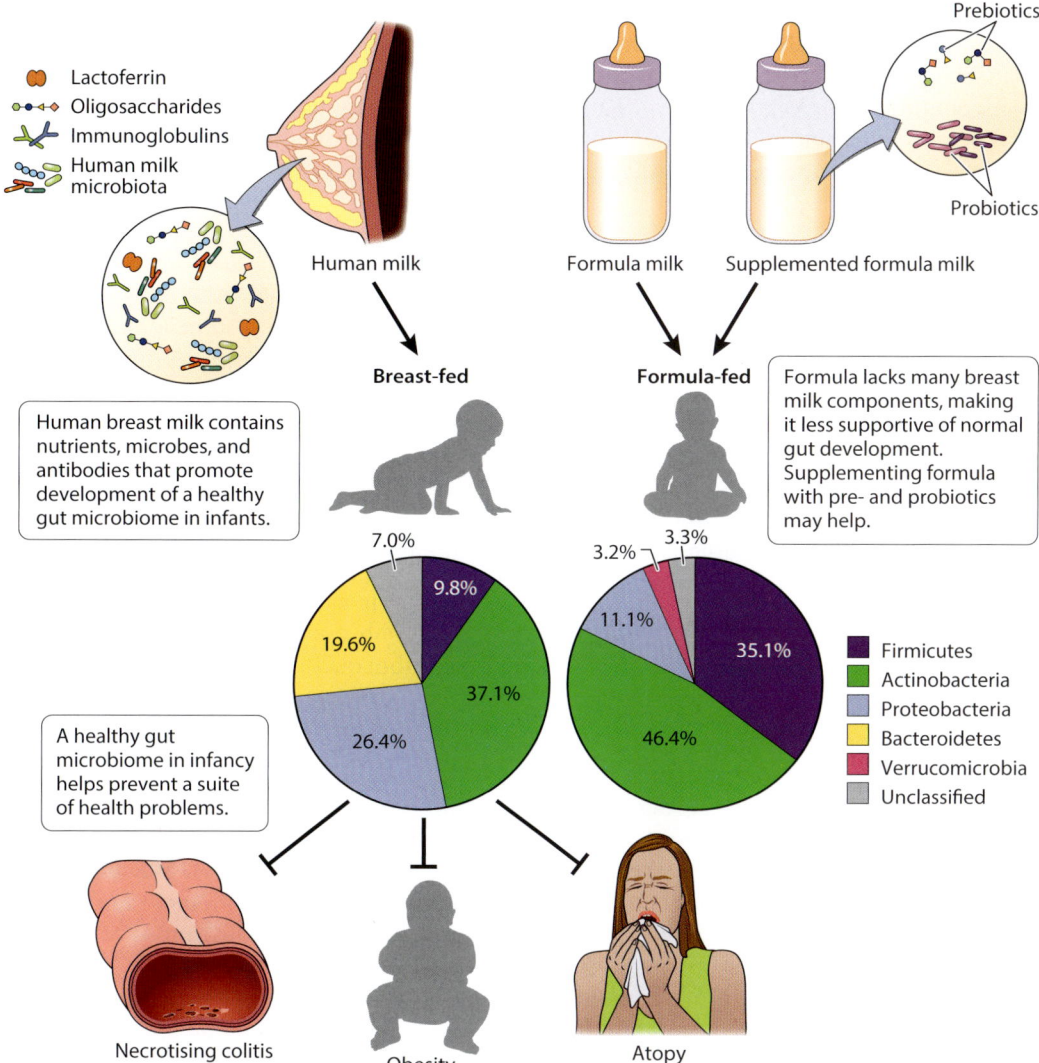

Figure 7.16 The Breastfeeding Versus Bottle-Feeding Debate Illustration of how the different feeding modes modulate the infant gut microbiome. Human milk naturally contains ingredients that support a healthy microbiome and aid in the prevention of obesity, necrotizing colitis, and atopy. Infant formula, which often has additional supplements such as probiotics and prebiotics, results in a significantly different gut microbiota, and one that is not as protective against this suite of diseases. (After Baumann-Dudenhoeffer et al., 2018.)

have higher levels of diversity in their microbiomes and a dominance of *Staphylococcus*, *Bacteroides*, and Enterobacteriaceae. Because of these differences in microbiota, the SCFAs are produced in different ratios, with higher levels of propionate and butyrate in formula-fed babies (Ho et al., 2018), along with fewer pathways related to lipid and vitamin metabolism. We have previously learned that microbiome diversity is a good thing. Here is an example where the specific taxa, rather than generalized diversity, is key. Even though formula-fed babies have a more diverse microbiome, the species composition isn't as beneficial. For example, formula-fed babies have less vitamin metabolism due to the changes in species composition, which can impact their health.

Researchers are attempting to produce formula that more closely resembles human milk, and there has been some success using probiotics to supplement formula (Moya-Pérez et al., 2017). The infants who were fed the bacterial-supplemented formulas had gut microbiota that were more like the gut microbiomes of breastfed babies. However, we don't know whether the *Bifidobacterium* species that babies get from mom through breastfeeding have other, yet-unidentified benefits. Infants get microbes from the breast milk and the skin around the nipple, and these bacteria, which first colonize the newborns' microbiomes, will often be major members of the community for years to come. Knowing what we know now about the impacts of a healthy microbiome on an individual, gaining an understanding of what compounds are required to feed a neonatal microbiome is exceptionally important.

7.7 TRANSITIONING TO SOLID FOODS

The introduction of solid foods to an infant's diet coincides with some major changes to the gut microbiota. The abundance of *Bifidobacterium*, *Clostridium*, and *Bacteroides* increases, and the microbiota of breastfed children versus formula-fed become more similar. This latter wave of colonization results in the establishment of an adult-like gut microbiome dominated by Bacteroidetes and Firmicutes. As the infant consumes more-complex food, the functions related to carbohydrate metabolism increase, as do those required for the biosynthesis of amino acids and vitamins. A mere 5 days after breastfeeding is ceased, there is an increase in the relative abundances of the *Bacteroides* and several members of the Firmicutes, and a decrease in *Bifidobacterium*, *Lactobacillus*, and Enterobacteriaceae (Laursen, 2021). These community transitions continue until 3 years of age, when a relatively stable microbiome is established.

The now adult-like gut microbiota of the infant harbors more functional diversity, required to metabolize more plant-derived polysaccharides and increased protein intake. New members of the Firmicutes appear, such as Lachnospiraceae, and members of the Bacteroidetes, such as Prevotellaceae. Two bacterial species, *Faecalibacterium prausnitzii* (Firmicutes) and *Akkermansia muciniphila* (Verrucomicrobia), increase in abundance over the next 2 years (Guo et al., 2020). *A. muciniphila* is considered by some to be the wonder child of the infant's microbiome, thanks to its action in degrading mucin in the intestinal mucosa and its production of SCFAs. *A. muciniphila* modulates intestinal permeability and affects inflammation in the digestive tract, as well as in the liver and blood. It also appears to play a key role in the metabolism of carbohydrates and fats, as low levels of *A. muciniphila* are often found in patients with obesity and diabetes. It is one of the most abundant species within the adult gut, and its abundance rises quickly during the first years of life. Its presence results in a more robust and rapid development of anti-inflammatory properties of the newborn's microbiome.

7.8 ENVIRONMENTAL IMPACTS ON THE INFANT'S MICROBIOME

Some of the most influential environmental factors that shape an infant's maturing microbiome include family members and pets. If you are an animal lover, you already know that having a pet can provide all sorts of benefits. But we now know that pets can also improve a newborn's health (Marrs et al., 2019) (**Figure 7.17**). For example, interactions between a newborn and either cats or dogs are associated with lower levels of asthma and allergies. A study led by Jasmine Alsukhon (2020) found that infants who were raised with a dog were 90% less likely to develop food allergies. Further, this study revealed that a mother living with a dog while pregnant showed a positive impact on the developing infant's immune system (Alsukhon et al., 2020). Further, exposure to pets may result in increased levels of *Oscillospira* and *Ruminococcus* in the infant's gut microbiome, which may be protective against obesity.

One highly intriguing finding is that **immunoglobulin E** (**IgE**) levels were far lower in children whose mothers had prenatal exposure to a pet. IgE binds to allergens and helps create an allergic response. The higher the levels of IgE in a person's blood, the higher the sensitivity they display to common allergens. Further, they also exhibit

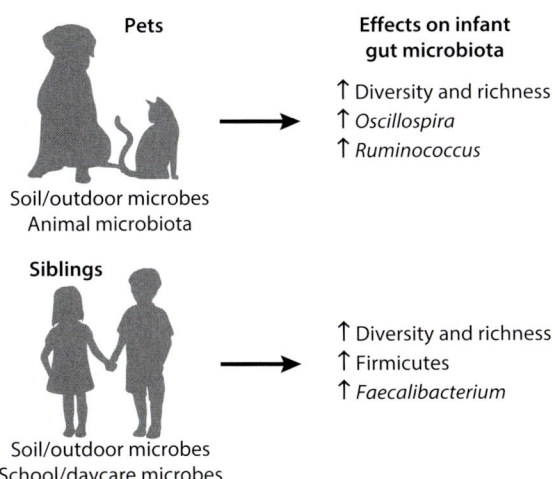

Figure 7.17 Pets and Siblings Impact Your Microbiome Research suggests a pet—especially a dog—and siblings may improve a newborn's health. Having a cat or dog or brothers and sisters present early in life is associated with lower levels of asthma and allergies. (After Azad et al., 2013.)

higher levels of food allergies and asthma. Owning a dog, but not a cat, results in lower IgE levels. It isn't clear why, but one hypothesis is that dogs tend to spend more time outdoors than cats and thus carry more diverse microbes on their paws and in their fur. Figure 7.17 shows the impact of the pet microbiome on the composition and biodiversity of the infant gut microbiome. This increased exposure to foreign microbial diversity drives the immune system to mature earlier and helps to prevent the development of allergies later in life.

The impact of animals on the infant's microbiome can start even earlier (see chapter 12). Where a pregnant person goes, who they meet, all the experiences they have during pregnancy can impact their microbiome and, in turn, the developing fetus. One of the more fascinating observations is that exposure of a pregnant person to farm animals can result in long-term, positive impacts on their child's health. In fact, children of parents who live on farms have less asthma, hay fever, and eczema later in life (Fall et al., 2015). These early microbial exposures create a long-lasting imprint on the immune system. However, it is critical that these exposures occur in utero and during the newborn's first year of life. Moving to a farm later in life does not provide the same reduction in allergies. In response to these and other studies, new treatments are being explored that leverage exposure to pets to lower levels of asthma, obesity, and even depression (Salas Garcia et al., 2020).

Family members and close relatives also impact the colonization process of the infant gut microbiome (see Figure 7.17). Newborns with older siblings have increased levels of *Bifidobacterium* in their gut and a more diverse microbiome than infants without siblings (Vacca et al., 2020). And the more siblings you have, the more species of Bacteroidetes and Firmicutes you have—likely due to the fact that more siblings handle the baby right after birth (Vacca et al., 2020). This study also suggest that parents normally change how they care for a baby as more children are born into the family. Younger siblings may be exposed to a more diverse sample of microbes because more family members handle the baby and because family members are less stringent about hygiene practices when handling them. Finally, the researchers noted that several key species, such as *Faecalibacterium prausnitzii*, increase in frequency more quickly with older siblings present.

A very powerful dataset is being developed in the Netherlands that will permit future dissection of numerous environmental impacts on newborns, infants, and adults. The KOALA Birth Cohort Study has engaged more than 2,500 children born between the years 2001 and 2003, who are being followed from gestation into adulthood, together with their families. A unique feature of this study is the inclusion of both social science and biomedical approaches in collecting data from the families, including biomarker and genetic studies, food choices, spirituality, and more.

7.9 HEALTH IMPACTS OF A NEWBORN'S DYSBIOTIC MICROBIOME

The microbiome and host are engaged in a continuous dialogue that, if disrupted, may result in long-lasting health disorders. The **fetal programming hypothesis** suggests that there is a critical window, the period between conception and the first 1,000 days after birth, when factors such as birth mode, use of antibiotics, diet, and presence of pets and family members may lead to long-term impacts on immune system programming (Forgie et al., 2020) (**Figure 7.18**). Establishing a healthy microbiome during this critical window results in a mutualistic relationship between host and microbes that can reduce disease susceptibility, whereas disruptions to the microbiome can have long-lasting consequences. Factors that alter this critical phase of microbial colonization in the gut are associated with a variety of illnesses later in life, such as asthma, allergies, diabetes, and obesity. Let's explore several of the health impacts of a dysbiotic microbiome in the newborn and infant.

Antibiotics

Newborns are frequently prescribed antibiotics at birth, particularly those born premature or via C-section. They are also exposed to the antibiotics taken by their mothers, via breast milk. These exposures can result in a significant loss of gut microbiome diversity. Levels of *Bifidobacterium* and *Bacteroides* drop. As we learned above, *Bifidobacterium species* are key players in the digestion of human breast milk and support the immune defense against pathogen invasion. In their place, we find an increase in pathogenic bacteria and the presence of antibiotic resistance genes (Reyman et al., 2022). Debby Bogaert at the University of Edinburgh led a study, described in **Box 7.1**, that showed that following antibiotic administration, newborns and infants had a significant decrease in the alpha diversity of their gut microbiomes. Since the infants did not already have an established microbiome, it is harder for them to reach a resilient microbiome following the treatment. As Bogaert notes, "We were surprised with the magnitude and duration of the effects of broad-spectrum antibiotics on the infants' microbiome. . . . This is likely because the antibiotic treatment is

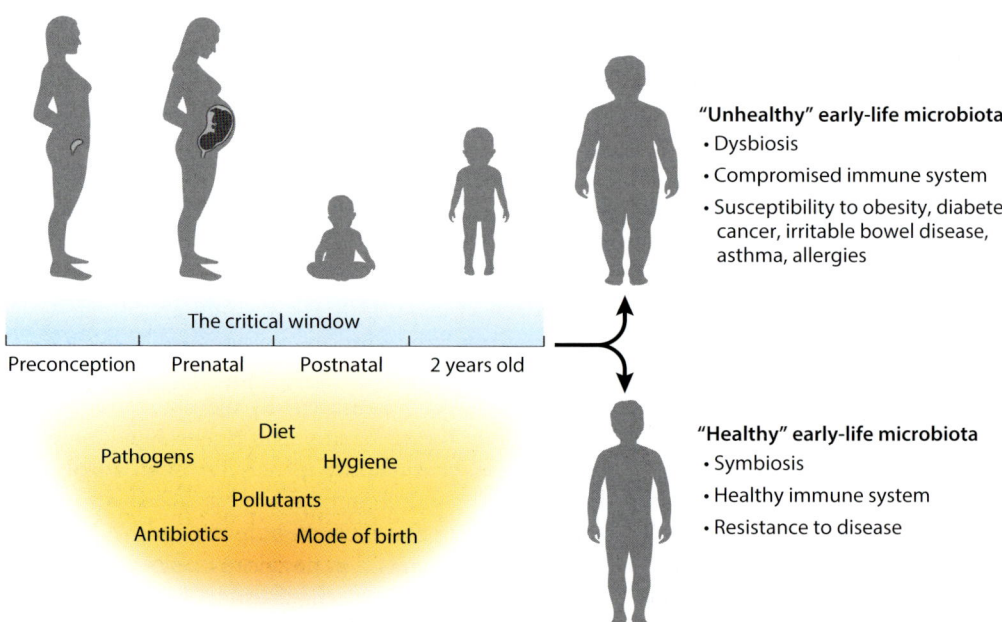

Figure 7.18 Infant Health and Their Microbiome The fetal programming hypothesis states that there is a critical window when factors such as birth mode, use of antibiotics, diet, and presence of pets and family members may lead to long-term impacts on the programming of the immune system.

BOX 7.1. RESEARCH IN ACTION
Antibiotics and the Newborn Gut—Long-Term Impacts on Microbiome Development

- ❖ **Hypothesis.** Broad-spectrum antibiotics administered in the first week of life may have pronounced effects on gut microbiome development and selection of antimicrobial resistance when administered in the first week of life.

- ❖ **Methods.** 147 infants each received one of three commonly prescribed intravenous antibiotic combinations. A subset of 80 non-antibiotic-treated infants served as controls. Rectal swabs were collected before and immediately after treatment, and at 1, 4, and 12 months of life. Microbiota were characterized by 16S rRNA-based sequencing.

- ❖ **Results.** Starting with the control individuals in which antibiotics were not administered, we see a rapid increase in *Bifidobacterium* by the end of week 1, while the treatment individuals show no such increase until 12 months after treatment (**Figure 7.19**). Further, in lieu of the increase in *Bifidobacterium*, there is an increase in Proteobacteria (such as *Klebsiella*) and several potential pathogens (such as *Enterococcus* and *Streptococcus*). By month 12, the antibiotic-treated groups had acquired microbiomes that appear similar in composition to the control group.

- ❖ **Conclusions.** These data suggest that the use of antibiotics early in life can have an enormous impact on a newborn's gut microbiome development. Although the treatment groups' microbiomes approach the control group's microbiome composition by month 12, this prolonged dysbiosis during a critical window of development may be responsible for increased levels of asthma, obesity, and other diseases in children who are routinely treated with antibiotics.

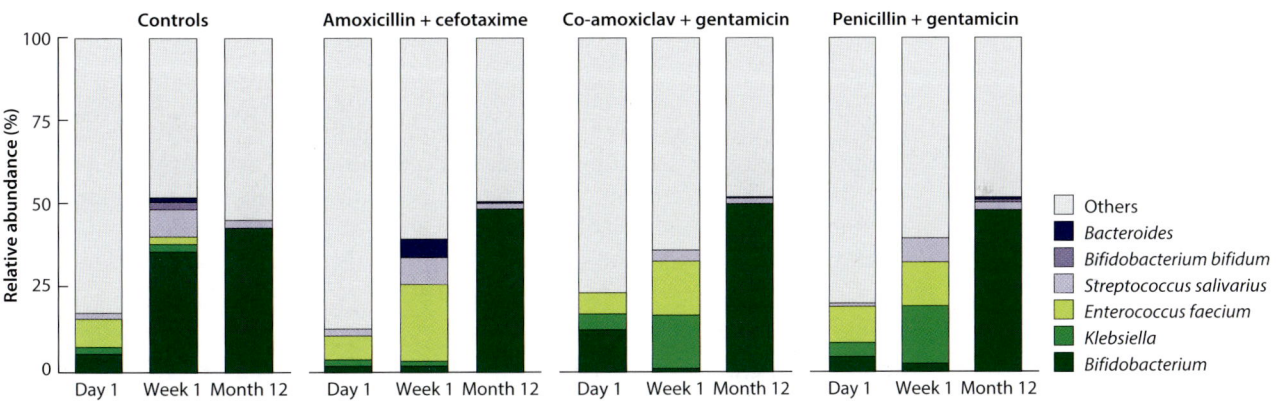

Figure 7.19 Antibiotic Treatment and Impact on the Microbiome Stack plot of relative abundance across groups of infants who received one of three antibiotic treatments. Shown are the relative abundance of operational taxonomic units (OTUs) prior to antibiotic treatment or 1 week after treatment or 12 months after treatment. (After Reyman et al., 2022.)

given at a time that infants have just received their first microbes from their mother and have not yet developed a resilient microbiome" (Reyman et al., 2022).

Malnutrition

Childhood malnutrition results from several factors, which include maternal malnutrition, antibiotic use, stress, and lack of hygiene, which can result in an increase in the density of microbial pathogens. Not surprisingly, the microbiomes of malnourished children are different from those of well-nourished children (**Figure 7.20**) (Iddrisu et al., 2021). Dysbiotic microbiomes are more likely to be infected by pathogens that cause diarrhea, which is a major contributor to malnutrition. This increased rate of intestinal infection suggests that the child's microbiome is not exerting its protective function. Diarrhea results in an infant not absorbing the essential nutrients and

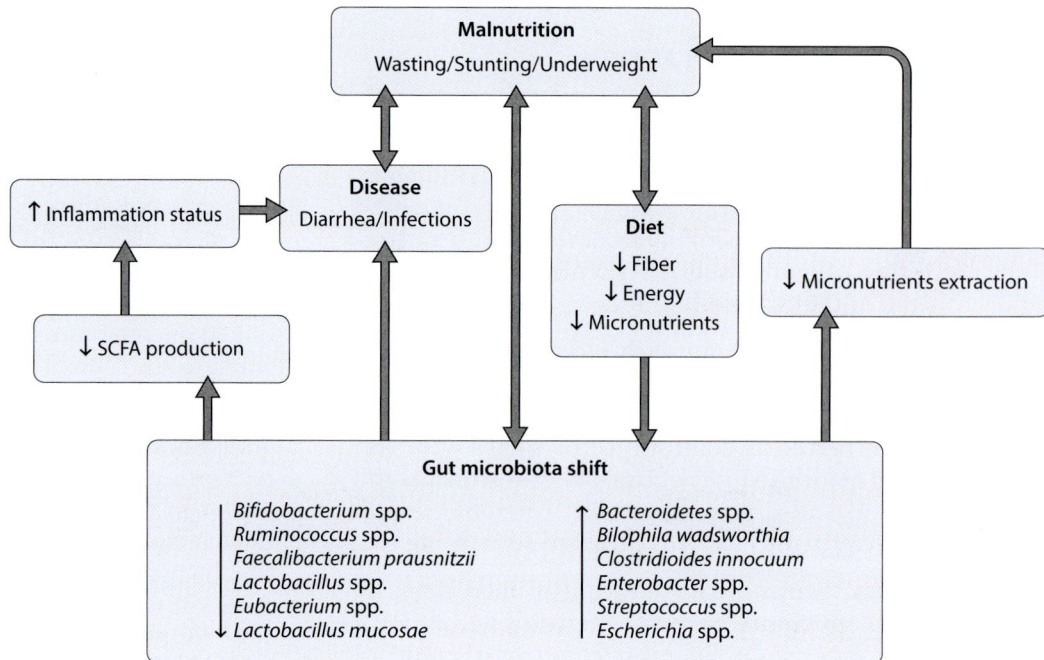

Figure 7.20 Malnutrition Results in a Shift in the Gut Microbiota Malnutrition has a profound impact on the members of an infant gut microbiome, which results in an increase in a variety of disease-related outcomes, such as a decrease in the ability to extract micronutrients, a decrease in short-chain fatty acid production, and an increase in overall inflammation. (After Iddrisu et al., 2021.)

calories required for normal growth and in losing weight. Loss of water due to diarrhea is also a leading cause of dehydration and death in countries lacking access to healthcare resources. Malnourished children show increased levels of Proteobacteria and decreased *Bacteroides*. Transplanting the microbiomes of malnourished infants into germ-free animals results in underweight animals (Schröder et al., 2020), suggesting a firm link between gut microbiome health and malnutrition.

Allergic Diseases

Atopy is the tendency to develop allergic diseases such as asthma and atopic dermatitis (eczema), all of which are associated with a more sensitive immune response to common allergens. More than 32 million people have at least one food allergy, a clear signal that we are experiencing a virtual epidemic of allergic responses. Dysbiotic gut microbiomes are now recognized as a significant factor fueling this epidemic. More than a decade ago, a pioneering study reported that infants who suffered from a dysbiotic gut microbiome subsequently developed atopic diseases, in particular **eczema** (Fouhy et al., 2012), which is a condition that causes the skin to become dry and itchy. It is caused by a combination of immune system hypersensitivity, genetics, environmental triggers, and stress. Studies have shown that infants with eczema have gut dysbiosis, characterized by increased levels of Firmicutes, such as Clostridiaceae, and a reduction in *Bifidobacterium* (Zheng et al., 2016) (**Figure 7.21**). In fact, it has been proposed that the ratio of *Bacteroides* to *Faecalibacterium* may provide a biomarker that predicts the risk of eczema in infants (Alsharairi, 2020). Recall that *Bifidobacterium* plays a key role in the production of SCFAs, which are integral to training the immune system. If levels of *Bifidobacterium* are low, then the immune system will not respond optimally to allergens. This connection is so tight that the ratio of two bacteria in the gut microbiome can predict whether a child will develop eczema. This matter is of keen interest to many, so we will return to it in chapter 12, when we devote the entire chapter to your microbiome and allergies.

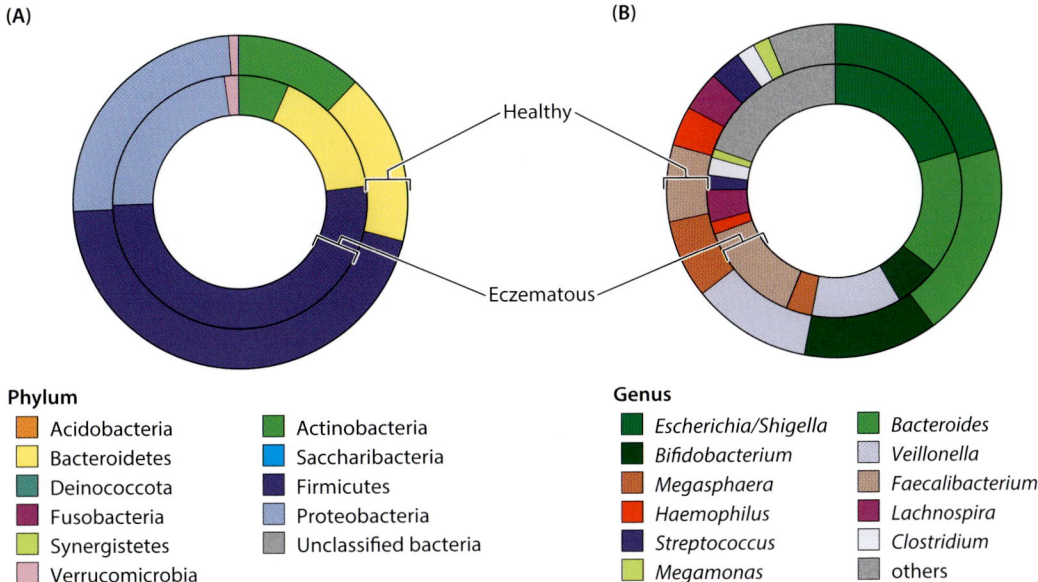

Figure 7.21 Altered Gut Microbiota Composition Associated with Eczema in Infants Relative abundance of bacteria at the phylum (A) and genus (B) levels. The outer circle represents the gut microbiota of healthy infants, and the inner circle represents the gut microbiota of eczematous infants. (After Zheng et al., 2016.)

Obesity

Obesity in infants has reached alarming levels, and the gut microbiota is now included in the wide spectrum of factors implicated in this shift. Recent studies suggest that the gut microbiota composition as well as functions encoded by it are associated with obesity. They may even be the cause of it, which we briefly introduced in chapter 2. The gut microbiota dictates the host energy flow by contributing metabolites and controlling nutrient absorption in the intestine (Thursby & Juge, 2017), which can promote energy storage or favor leanness (Meliț et al., 2022).

Some of the most revealing research into microbiome-induced obesity was carried out in germ-free mice. We described this research in chapter 2, but it bears repeating here. Transplantation of the gut microbiomes from human twins where one twin was lean and the other obese resulted in lean or obese mice, respectively (Ridaura et al., 2013). The weight and fat composition differences were correlated with differing capacities of the gut microbiome to ferment nutrients. Microbiota in lean mice produced more SCFAs, while those in obese mice produced more branched-chain amino acids—a difference that impacts the metabolic health of the host. Feeding the obese mice a different diet, one high in fruits and vegetables, resulted in decreased obesity, and caging lean and obese mice together led to the obese mice becoming leaner and with a gut microbiome more similar to the lean mice (Ridaura et al., 2013). So, maybe there is light at the end of this obesity-related tunnel. Imagine if all you needed to do was hang out with lean friends and eat what they eat! Put more elegantly by Nancy Mure, a New York–based, holistic nutrition and natural healing practitioner, "If there's one thing to know about the human body; it's this: the human body has a ringmaster. This ringmaster controls your digestion, your immunity, your brain, your weight, your health and even your happiness. This ringmaster is the gut."

A comparison of many obese and normal weight infants revealed that obesity was associated with elevated levels of Firmicutes and depleted levels of Bacteroidetes. In fact, it was shown that the abundance of one species, *Bacteroides fragilis*, can serve as a biomarker for an elevated **body mass index** (**BMI**), which is a measure of body fat based upon weight and height. This single measure might serve as an indication that the infant's gut microbiome is dysbiotic and thus requires attention. Obese children also have more-dense microbiomes. More microbes mean more metabolism, and higher levels of metabolites, suggesting elevated substrate utilization (Moszak et al., 2020). In short, obesity appears to involve a dysbiotic microbiota that simply extracts more energy from the food a child eats.

You are now well aware that an infant's gut is generally dominated by *Bifidobacterium*. Babies fed formula, as well as those born by C-section, have lower levels of this key bacterial genus. Such babies experience higher levels of colic and necrotizing enterocolitis within the first 2 weeks of life. Decreased levels of *Bifidobacterium* can also lead to the development of obesity and autoimmune diseases, as we just learned above. *Bifidobacterium* appears to impact immune stimulation and, by producing SCFAs, create a more acidic intestinal environment. These data suggest an important role for *Bifidobacterium* in establishing infant health.

Diabetes

Type 1 diabetes is an autoimmune disease characterized by the destruction of beta cells, which are found in the pancreas and produce and release insulin in response to blood glucose levels. Several factors may affect the development of type 1 diabetes, including gut microbiota dysbiosis. Individuals with type 1 diabetes have lower gut microbiome diversity and altered ratios of Firmicutes and Bacteroidetes. These individuals also have a decreased abundance of *Faecalibacterium prausnitzii*. In fact, *F. prausnitzii* has been proposed as a key player in reducing the risk of type 1 diabetes. **Figure 7.22** provides an overview of the proposed benefits of *F. prausnitzii* in inflammation due to diabetes and obesity. It is a key producer of butyrate, and its growth alters the gut environment, which then limits the growth of Proteobacteria and, thus, production of lipopolysaccharides (LPS) and free fatty acids (FFAs). *F. prausnitzii* is now considered a standard treatment for type 1 diabetes (Ganesan et al., 2018). The prior observations tell us, as succinctly stated by Sherry Rogers, MD, "The road to health is paved with good intestines!" (Rogers, 2002).

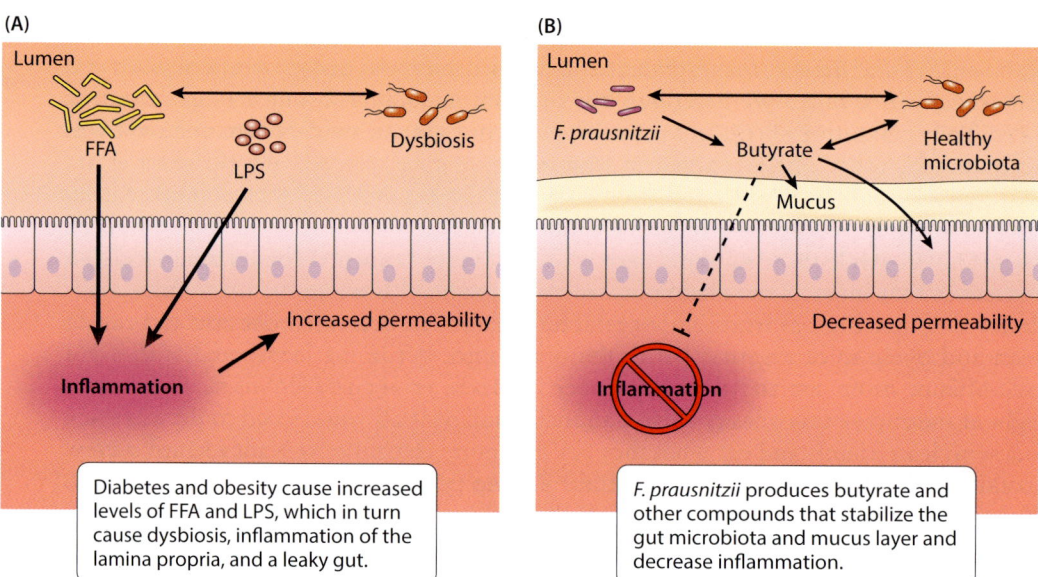

Figure 7.22 Novel Strategy for Diabetes Prevention Diabetes is due, in part, to the hyperglycemia resulting from gut microbiome dysbiosis, resulting in increased levels of free fatty acids (FFAs) and lipopolysaccharide (LPS), and resulting disruptions in intestinal permeability. The probiotic treatment with *Faecalibacterium prausnitzii* can increase the butyrate concentration, stabilize the microbiota and mucus layer, and decrease the activation of inflammatory pathways. (After Maioli et al., 2021.)

7.10 MICROBIOME-BASED THERAPIES

The microbial makeup of a newborn's gut microbiome has changed dramatically over the past hundred years, primarily due to changes in our diet and the increased use of antibiotics. There has been a loss of *Bifidobacterium*, an increase in potential pathogens, and increase in the pH from 5.0 to 6.5. This loss is likely due to an increase in C-section deliveries, the use of infant formula, and antibiotics. The absence of *Bifidobacterium* results in an increased presence of proinflammatory microbes during the critical period of immune system development, which may increase the risk for immune disease later in life. To address these and other significant health challenges posed by our ever-changing maternal and newborn microbiomes, several novel therapeutic approaches are being developed. It is still early days for this area of research, and many such studies have conflicting outcomes, but let's look at some of the more promising approaches.

Probiotics

During pregnancy, *Lactobacillus* increasingly dominates the pregnant woman's vaginal microbiome and results in higher levels of lactic acid production and, ultimately, a lower vaginal pH. At the same time, a woman's general ability to mount a robust immune response is diminished, which is thought to be beneficial for the fetus, as it results in less rejection of fetal antigens by the mother's immune system. Several studies have sought to modulate dysbiotic maternal microbiomes with the application of **probiotic** bacteria, which are live microorganisms that are intended to have health benefits when consumed or applied to the body (**Figure 7.23**). In one study, pregnant women were treated with *Lactobacillus rhamnosus*, which helped to maintain a lower vaginal pH and prevented pathogenic microorganisms from invading (Superti & De Seta, 2020). The same species of probiotic was shown to colonize the intestines of the newborns, even when only administered to the pregnant mother, and the presence of the probiotic was correlated with an increase in the abundance of *Bifidobacterium* (Navarro-Tapia et al., 2020). Although these results are compelling, other studies found no impact of the probiotic on the maternal vaginal microbiome at all (Husain et al., 2019).

Probiotics have also been explored to reduce some of the common negative effects of pregnancy, such as nausea and vomiting. A group of researchers explored whether administration of a probiotic could improve gastrointestinal function during pregnancy. The researchers found that the probiotic significantly reduced nausea (by 16%) and vomiting (by 33%). It also resulted in less fatigue and improved appetite. Dr. Albert Liu, a professor at UC Davis School of Medicine and a coauthor of the study, notes, "Over the years, I've observed that probiotics can reduce nausea and vomiting and ease constipation. It's very encouraging that the study proved this to be

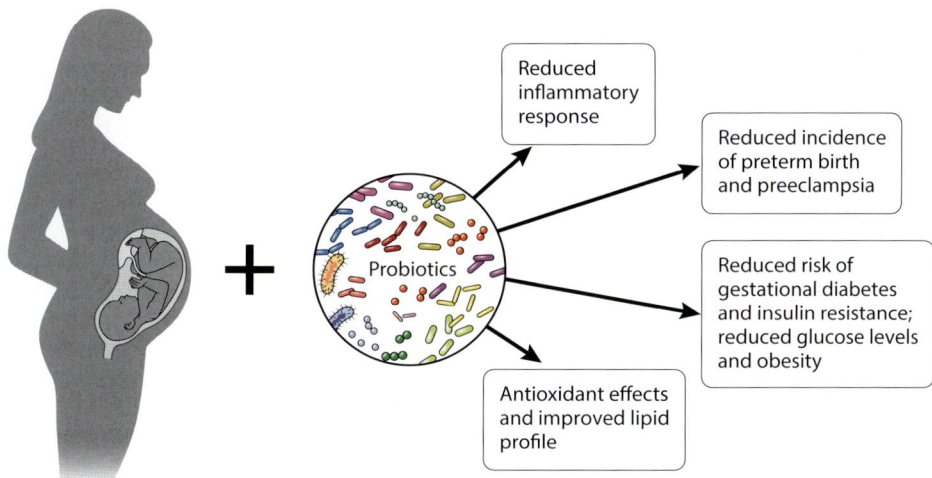

Figure 7.23 Potential Impact of Probiotic Therapy during Pregnancy The impact of probiotics during pregnancy may be associated with increased health of mother and fetus.

true. . . . Probiotics have also benefited many of my other patients who weren't in the study" (University of California, Davis Health, 2021).

Another area of interest is in the use of probiotics to decrease the risk of preterm labor. The limited studies available so far have mixed results. One analysis of over 4,000 pregnant women, which employed a randomized controlled experimental design, found no impact of the use of probiotics on the risk of preterm labor (Jarde et al., 2018). A second study that engaged 5 times the number of participants found a link between ingesting probiotic foods and a decrease in preterm labor (Myhre et al., 2011). A third study found that drinking milk laced with probiotics lowered the risk of preterm labor (Nordqvist et al., 2018). These latter two studies were observational studies, which aren't considered as reliable as a randomized controlled trial. With observational studies, an unknown underlying factor could be influencing patient outcomes, so it's impossible to prove cause and effect. In the third study, for example, women who chose to drink probiotic milk may have had overall healthier diets or more access to healthcare, and these factors could have been what lowered the risk of preterm labor, not the probiotic milk. There's no way to know whether it was the milk or not, only that drinking the milk was associated with a lower risk of preterm labor. Overall, the use of probiotics to influence the vaginal microbiome remains controversial. A major review of this area of research concluded that there is some promise that the use of probiotics can modulate the vaginal microbiome, even though most studies do not result in the probiotic strain colonizing the vagina, or even being detectable, after the dosing period (van de Wijgert & Verwijs, 2020).

Vaginal Microbiome Transplant

Vaginal microbiome transplant (**VMT**) is being explored as a novel method for restoring a dysbiotic vaginal microbiome. VMTs involve the transfer of one person's cervicovaginal secretions to another, much like the fecal microbiota transplants we discussed in chapter 2 Gardner and Dukes (1955) engaged in one of the earliest VMTs, in which they inoculated the vaginas of healthy women with *Gardnerella vaginalis* sourced from the vaginas of women with bacterial vaginosis. The healthy women developed BV after transfer. A far more recent, albeit exploratory, study was done to evaluate the use of VMT from healthy donors in patients with BV. Most of those patients showed full remission of the BV (Lev-Sagie et al., 2019). However, this therapeutic strategy is in its infancy, and numerous randomized, placebo-controlled studies are required before progress will be made (Junca et al., 2022).

Fecal Microbiota Transplant

A modification of FMT, fecal microbiota transplant, is being explored as a means to ensure that babies born via C-section are exposed to their birthing parent's vaginal microbiome. The procedure involves inserting sterile gauze into the pregnant person's vagina, then swabbing the baby with the gauze immediately after birth. We described in Chapter 4 the very first newborn VMT study, which was carried out by Dominguez-Bello and colleagues in 2015, and showed that the microbiota of the gut, skin, and mouth of C-section newborns who were swabbed were dominated by vaginal bacteria, as is the case for babies born vaginally (Dominguez-Bello et al., 2016). Dominguez-Bello notes that the maternal vagina may provide the pioneer colonizers for many different infant organs thanks to what she and her fellow authors refer to as "the pluripotent nature of the perinatal vaginal microbiome" (Song et al., 2021).

Some scientists argue against this approach. Their concerns include the fact that we aren't sure what we are trying to achieve, since the newborns don't have disease to begin with and since there is no evidence about the potential long-term impacts of the swabbing. Dominguez-Bello feels that the goal is to simply restore something that the use of surgical birthing prevents, which is the intimate exposure to the mother's vaginal microbiome. She argues that the only way we will know is by carrying out controlled trials (Reardon, 2019). There are three clinical trials in the works at this time—and I am sure we will learn more about this therapeutic application in the near future.

Oral Probiotics

Babies who are born early (preterm) are far more likely to develop intestinal infections, due to their underdeveloped immune status, the more frequent application of antibiotics, and delayed gut microbiome colonization. Several studies have shown that the application of oral probiotics in preterm babies can reduce the risk of necrotizing enterocolitis (Morgan et al., 2020). There is also an apparent decrease in several other conditions, such as allergies and obesity, and in the occurrence of colic. A team led by Arun Wadwah in a New Delhi clinic evaluated the impact of using a specific strain of *Lactobacillus reuteri* on babies with colic (Wadhwa et al., 2022). The amount of time spent crying was significantly less in infants who received the probiotic. There is high variability in the outcome of these types of probiotics studies. However, we are learning how to create better study designs, and it is possible that we will develop therapeutic approaches to prevention or alleviate the symptoms of colic.

When designing a probiotic, researchers must choose species that are able to survive in the gut, so that the probiotic strains can establish themselves in the gut and have the intended effect. One group of researchers revealed that feeding breastfed babies a probiotic that included a specific strain of *Bifidobacterium longum* (EVC001) resulted in a significant increase in levels of *B. infantis* in the feces (Chichlowski et al., 2020), an impact that persisted for a month after the probiotic was consumed. The researchers wondered what ecological niche EVC001 filled such that it could persist in the gut. One key factor in the persistence of bacterial strains is the presence of their food sources. For a probiotic to survive, the food source it uses must match what's available and must not be consumed more readily by other bacteria. When these same researchers supplemented infants' diets with the probiotic *B. infantis* EVC001, those babies exclusively breastfed had guts dominated by the *Bifidobacterium*. Adding this one specific strain resulted in a drastically different overall gut microbiome composition. Many infants never acquire *B. infantis*, and there is interest in promoting the combination of breastfeeding and probiotic supplementation with EVC001.

Oral Prebiotics

Prebiotics are forms of dietary fiber that you can't digest but that members of your gut microbiome can. Unlike probiotics, prebiotics don't contain live bacteria, but they do contain food for the gut microbiota. A baby's gut microbiome naturally obtains prebiotics from breast milk, and this selects for an abundance of *Bifidobacterium*. Given the critical importance of breast milk in promoting gut microbiome health, there has been a push to develop prebiotic additives that can serve the same purpose in babies who receive formula rather than mother's milk. **Table 7.2** shows the potential health benefits of adding prebiotics to formula, and several studies have shown

Table 7.2 Prebiotics & Health

MECHANISM OF ACTION	POTENTIAL HEALTH BENEFITS
Regulates intestinal barrier	Reduces colic in babies
Optimizes pH	Reduces atrophy
Increases mucin production	Minimizes infections
Stimulates microbiota	Reduces mucosal inflammation
Suppresses pathogens	Stimulates peristalsis
Enhances epithelial development	Reduces abdominal gas production
Orchestrates immune pathways	Improves stool frequency and consistency
Improves lipid and sugar homeostasis	
Influences brain functions	

Source: Miqdady et al., 2020

that at least some of these benefits are achieved. Infant formulas supplemented with prebiotics have been studied in randomized clinical studies and have been shown to reduce the risk of infection. Their consumption also results in a gut microbiota like that of breastfed infants. Preterm newborns given prebiotics in formula had both an increase in *Bifidobacterium* numbers and a significant reduction in *Escherichia coli*.

Prebiotics, by serving as a fuel source for health-promoting microorganisms, protect against pathogen invasion, improve the barrier function of the intestinal wall, orchestrate immune pathways, and impact brain function. SCFAs are the primary products of selective fermentation of prebiotic food sources, and this provides an energy source to the gut epithelial cells. The SCFAs also serve to improve gut motility and immune functions, increase absorption of metabolites, and maintain sugar and lipid homeostasis. Finally, the SCFAs result in increased production of mucin that can prevent bacterial movement across the gut barrier.

We are just beginning to understand the crucial interactions that occur between a newborn and its microbiome and the potential impact these interactions have on immediate and future health. Much as we want to act on this knowledge and do something to change our babies' microbiomes, in a good way, most scientists are reluctant to make practical recommendations. They are rightly concerned about feeding a newborn billions of bacteria that have not been rigorously tested. Even so, probiotics are already being hyped commercially as the new wonder drug. As we learned above, there is intriguing research suggesting that certain probiotics may be effective in modulating the immune system, reducing allergic response; and improving the function of the gut epithelium. However, because the probiotic marketplace is unregulated, it's impossible for a consumer to know what, if anything, you're getting when you buy a probiotic. Investigations into numerous commercially available probiotics have shown that few even contain the microbial species listed on the label. So, for now, buyer beware, and perhaps the best advice is to eat as if your health depends upon it, because it does!

CHECK YOUR UNDERSTANDING

1. A PCoA plot
   a. is the same as a PCA plot.
   b. provides more-detailed information than a PCA plot by not compressing multiple variables.
   c. identifies dimensions that are best able to differentiate between different groups.
   d. usually uses alpha diversity measures.

2. What is the keystone genus in the vaginal microbiome?
   a. *Staphylococcus*
   b. *Lactobacillus*
   c. *Gardnerella*
   d. *Prevotella*

3. Which is true about vaginal community state types (CSTs)?
   a. They are based on the microbiomes in an individual's upper and lower reproductive system.
   b. A person's CST is based primarily on the amount of alpha diversity present.
   c. Genetics play a large role in determining a person's CST.
   d. The dominant species present is the main determinant of a person's CST.

4. An unhealthy vaginal microbiome
   a. has high alpha diversity.
   b. is determined primarily by genetics.
   c. contains an overabundance of *Lactobacillus*.
   d. has a low alpha diversity.

5. The least healthy CST is
   a. type 1.
   b. type 2.
   c. type 3.
   d. type 4.

6. Which of the following is not a role the vaginal microbiome plays in maintaining human health?
   a. Preventing infections through metabolic functions that result in a low pH
   b. Preventing colonization of pathogens through competitive inhibition
   c. Upregulating the immune response through the activation of cytokines
   d. Producing compounds that build a mucus layer, which traps pathogens

7. Bacterial vaginosis is
   a. caused by a single pathogen infection.
   b. the result of an overgrowth of non-*Lactobacillus* species.
   c. impacted by the species present but not by gene expression.
   d. likely caused primarily by hormones.

8. It has been clearly proven that the fetus exists in a sterile environment during pregnancy.
   a. True
   b. False

9. Which of the following is *not* a way the maternal gut microbiome influences the fetus's development?
   a. The fetus receives antibodies that allow them to identify and coexist with commensal microbes.
   b. Microbial metabolites can be transferred through the placenta.
   c. The maternal microbiome influences the likelihood of the fetus developing autoimmune disease later in life.
   d. It prepares the fetus's immune system to attack any microbe that it encounters.

10. Which is an example of the acquired immune system?
    a. Avoiding getting the flu because you had it last month
    b. A macrophage consuming a pathogen
    c. The mucus barrier in the nose trapping a virus
    d. Tears removing a pathogen from the eyes

11. Short-chain fatty acids produced by a pregnant person's microbiome
    a. can only impact the fetus indirectly.
    b. promote the production of killer T cells.
    c. are most strongly influenced by the birthing parent's genes.
    d. can promote self-tolerance and decrease autoimmunity.

12. A study in pregnant mice found that
    a. the amount of fat in the diet does not influence SCFA production.
    b. the effect on the mouse pups' Treg levels is limited to the time they are in utero.
    c. mouse pups can receive SCFAs through lactation, regardless of the ability of their own microbiomes to produce SCFAs.
    d. pregnant mice fed a high-fat diet had pups with high levels of Bacteroidetes.

13. The primary purpose of maternal antibodies passed to the fetus is to protect against pathogens, which kill 3 million neonates each year.
    a. True
    b. False

14. Maternal antibodies are able to recognize commensal bacteria, preventing them from crossing the newborn's intestinal tract epithelial cells.
    a. True
    b. False

15. Microglial cells are specialized
    a. T cells.
    b. B cells.
    c. macrophages.
    d. lymphocytes.

16. Microglial cells in the central nervous system signal for the production of _____ when activated?
    a. antimicrobials
    b. anti-inflammatory signals
    c. growth hormones
    d. microbiome-oriented bacterial growth factors

17. The microbiome of a newborn delivered via C-section is most similar to the birthing parent's
    a. gut microbiome.
    b. microbiome.
    c. vaginal microbiome.
    d. skin microbiome.

18. By looking at a newborn's microbial composition, scientists could likely determine the
    a. baby's age.
    b. hospital the baby was born in.
    c. method of delivery.
    d. birthing parent's identity.

19. Within about 9 months, the microbiomes of babies born via C-section come to resemble the microbiomes of babies born vaginally
    a. except with an overabundance of *Lactobacillus*.
    b. except with a lack of *Bacteroides*.
    c. except with a lack of Firmicutes.
    d. with no noticeable differences.

20. One method proposed to restore the microbiome of babies delivered via C-section is to
    a. swab newborns with the birthing parent's vaginal fluids.
    b. do a fecal microbiota transplant from the birthing parent.
    c. house babies born vaginally and babies born via C-section in the same hospital room.
    d. have hospital workers interact with the baby to encourage the transfer of skin flora.

21. Only microbes that can survive in the gut and also outcompete other gut microbiota will become long-term residents of the microbiome. These concepts are known as
    a. fitness and competitive inhibition.
    b. evolution and pathogenicity.
    c. selection and priority.
    d. stability and resilience.

22. In the months after birth, the baby's gut microbiome changes from one dominated by aerobic bacteria to one dominated by anaerobes, primarily
    a. *Staphylococcus*.
    b. Proteobacteria.
    c. *Lactobacillus*.
    d. *Bifidobacterium*.

23. Breast milk is sterile.
    a. True
    b. False

24. The "bifidus factor" in breast milk that feeds *Bifidobacterium* is
    a. short-chain fatty acids.
    b. antibodies.
    c. regulatory T cells.
    d. human milk oligosaccharides.

25. The only *Bifidobacterium* species capable of digesting HMOs completely is
    a. *B. breve*.
    b. *B. bifidum*.
    c. *B. infantis*.
    d. none of the above.

26. A difference in the microbiota of formula-fed vs breastfed babies is
    a. breastfed babies have a more diverse gut microbiome.
    b. breastfed babies produce more butyrate and propionate.
    c. formula-fed babies have greater levels of vitamin metabolism.
    d. breastfed babies have higher levels of *Bifidobacterium* and *Lactobacillus*.

27. A stable, adult-like microbiome is typically established at
    a. 9 months old.
    b. 3 years old.
    c. 6 years old.
    d. 10–12 years old.

28. What is the important species that digests mucin, produces SCFAs, prevents inflammation, regulates gut permeability, and more?
    a. *Akkermansia muciniphila*
    b. *Bifidobacterium infantis*
    c. *Faecalibacterium prausnitzii*
    d. *Lactobacillus iners*

29. Having a pet as a child decreases your risk of allergic disease, likely through the regulation of
    a. Tregs.
    b. IgE.
    c. LPS.
    d. SCFAs.

30. The theory that if the microbiome is disrupted during the first 1,000 days of life, there is a greater risk of long-term health impacts is known as the
    a. hygiene hypothesis.
    b. neonate microbiota hypothesis.
    c. fetal programming hypothesis.
    d. gut microbiota–immune axis hypothesis.

31. In addition to having allergic diseases later in life, infants with dysbiotic microbiomes are more likely to
    a. be underweight.
    b. have a disrupted circadian rhythm.
    c. have eczema.
    d. have diabetes.

32. There is not strong, consistent evidence that prescribing a probiotic supplement during pregnancy changes the birthing parent's gut microbiome.
    a. True
    b. False

33. Which of the following is true about supplements to infant formula?
    a. Most infant formulas contain all or almost all of the HMOs found in breast milk.
    b. Prebiotics have been shown to help formula-fed babies' microbiota become more like breastfed babies' microbiota.
    c. Prebiotic supplements in infant formula may decrease SCFA production, but they improve the permeability of the gut barrier.
    d. Most probiotics tested on newborns are effective because they are able to survive in the gut for the long term.

Answers: 1C, 2B, 3D, 4A, 5D, 6C, 7B, 8B, 9D, 10A, 11D, 12C, 13B, 14A, 15C, 16A, 17D, 18C, 19B, 20A, 21C, 22D, 23B, 24D, 25C, 26D, 27B, 28A, 29B, 30C, 31C, 32A, 33B

DIVING DEEPER

1. How are antibodies from the pregnant parent transferred to the fetus?
2. Think back to chapter 3; why does your gut host the densest microbiome? How does this play a role in communicating with the immune system?
3. Compare and contrast the innate and acquired immune systems.
4. What role does your birthing parent's microbiome play in the development of your central nervous system?
5. In chapters 4 and 5 we learned about the differences in infant microbiome composition based on birth mode. What are some of the biological factors that cause these differences?
6. What compounds are found in breast milk that play a role in microbiome composition but may not be found in formula?
7. How does an infant's microbiome composition change as they transition to solid foods?
8. How is the infant's core microbiome different from that of an adult?
9. Why does it make sense that germ-free mice will have a weaker immune system than their wild-type counterparts?
10. How do antibiotics affect the microbiome, in both adults and infants?
11. Think back to the section about the gut microbiome in chapter 3; how might a dysbiotic microbiome affect nutrition and nutrient absorption?
12. Using the germ-free mice that we introduced in chapter 3 as a model, how does microbiome composition affect obesity?
13. Compare and contrast vaginal microbiome transplants with fecal transplants.
14. Why is the presence of antimicrobial genes in the infantile microbiome a concern?
15. Prebiotics don't directly add microbes to your microbiome—so how do they work to change your community composition?

DISCUSSING AND REFLECTING

1. Many of the studies we've discussed only focus on the community profile of infant microbiomes for a few months after birth. What would be the benefits of extending these studies to 1–2 years? How many participants would you need to make a claim that any observed differences are due to birth mode?

2. Think back to chapter 3, where we briefly introduced many of the microbiomes found in and on your body. Using your new knowledge from this chapter, which microbiomes are most heavily impacted by your birth conditions and the lifestyle of your birthing parent? Which are the least impacted? Why?

3. Reflection. In this chapter we learned about how the 9 months before your birth played a significant role in developing your first microbiome and may still be impacting you. What do you think about this information? Do you view your own development differently? Can you connect any of your own health experiences to your birth mode or conditions of your birthing parent while pregnant?

RECOMMENDED READINGS

Popular Science Review

Callaway, E. (2019, September 18). C-Section Babies Are Missing Key Microbes. *Nature*. https://www.scientificamerican.com/article/c-section-babies-are-missing-key-microbes/

Popular Science Book

Finlay, B. B., & Arrieta, M.-C. (2016). *Let Them Eat Dirt: Saving Your Child from an Oversanitized World* (first edition). Algonquin Books of Chapel Hill.

Medical Profession Review

Valentine, G., Prince, A., & Aagaard, K. M. (2019). The Neonatal Microbiome and Metagenomics: What Do We Know and What Is the Future? *NeoReviews*, 20(5), e258–e271. https://doi.org/10.1542/neo.20-5-e258

Scientific Reviews

Yao, Y., Cai, X., Ye, Y., Wang, F., Chen, F., & Zheng, C. (2021). The Role of Microbiota in Infant Health: From Early Life to Adulthood. *Frontiers in Immunology*, 12, 708472. https://doi.org/10.3389/fimmu.2021.708472

Consales, A., Cerasani, J., Sorrentino, G., Morniroli, D., Colombo, L., Mosca, F., & Giannì, M. L. (2022). The Hidden Universe of Human Milk Microbiome: Origin, Composition, Determinants, Role, and Future Perspectives. *European Journal of Pediatrics*, 181(5), 1811–1820. https://doi.org/10.1007/s00431-022-04383-1

The Microbiome and the Brain

8

CHAPTER CONTENTS

- 8.1 The Nervous System
- 8.2 The Maternal Microbiome and Neural Development
- 8.3 The Microbiota-Gut-Brain Axis
- 8.4 Gut Microbiota and Neuropsychiatric Disorders
- 8.5 Gut Microbiome-Based Therapies
- 8.6 The Evolved Dependence of Our Microbiota

Hello! I'm *Escherichia coli* (a.k.a. *E. coli*), one of the gut microbiome's busiest and most versatile members. I'm a microscopic multitasker and love to break down undigested food, produce essential vitamins like K and B12, and help train your immune system to recognize friend from foe. I also play a key role in maintaining gut homeostasis by competing with less friendly bacteria for resources and space. But like that one friend who sometimes takes things a bit too far at the party, I have my mischievous moments, such as when I cause food poisoning or infections, reminding you that even the best team players can have a bad day. Most of the time, though, I'm the gut's unsung hero, quietly keeping everything running smoothly while dreaming of microbiome Olympic glory! (Photo from iStock.com/peterschreiber.media)

M ost people think of their brains and bowels as quite distinct. Gray matter is where higher-order thinking happens, while the gut is where food is digested. But some of the most exciting microbiome research today is revealing the intimate connection between our gut, its microbes, and our brain, through the **microbiota-gut-brain axis** (**MGBA**). It turns out that our brain is constantly talking with our gastrointestinal (GI) tract and its microbiome, and vice versa. So how does this work, what do they say to each other, and what does it mean for our mental health? Following a quick refresher on the nervous system and an introduction to the MGBA itself, we will focus on how the gut influences our brain's development and activity in healthy individuals. We will then turn to an exploration of how dysbiosis of our gut microbiome is associated with, and may even cause, some of the more common neurological disorders: depression, autism spectrum disorder, and Parkinson's disease. Finally, we will learn about some of the more novel therapeutic approaches targeting the gut microbiome that may help alleviate some of these devastating diseases.

"Brain chemistry essentially determines how we feel and respond to our environment, and evidence is building that chemicals derived from gut microbes play a major role."
—Frank Schroeder (Langille, 2019)

8.1 THE NERVOUS SYSTEM

But first, let's have a refresher on our nervous system, which is responsible for all mental activity, such as learning and memory, as well as keeping us in touch with our bodies, inside and out. Our nervous system is made up of billions of cells, many of which host receptors that, in concert with our endocrine system, keep our bodies in homeostasis. The nervous system has two major divisions: the **central nervous system** (**CNS**), which includes the brain and spinal cord, and the **peripheral nervous system** (**PNS**), which includes sensory neurons and clusters of neurons called ganglia (**Figure 8.1**). The CNS controls most of what we do and think. In contrast, the PNS, which is a network of nerves, contains sensory receptors that help the brain to sense changes in our environment. You can think of the CNS and PNS working together like a person driving a car. The driver (our CNS stand-in) both receives information from the car (PNS), such as speed and direction, and uses this information to direct the car's movement. The CNS and PNS work intimately together. In fact, nerves from the PNS enter and merge into the CNS. The PNS also carries out two main types of functions: the **somatic** and **autonomic**. The somatic, or voluntary, nervous system has control of skin, bones, joints, and skeletal muscle. When you decide to jog across the road, your somatic system is at work. The autonomic nervous system controls our internal organs, a process that occurs without our conscious thought or direction. When you listen to your heart beating, it is your autonomic system that keeps that muscle contracting.

There are three major and overlapping activities of the nervous system: sensory, integrative, and motor. Imagine the millions of **sensory receptors** located across our bodies that detect changes both inside our bodies and in our local environment. Sensory

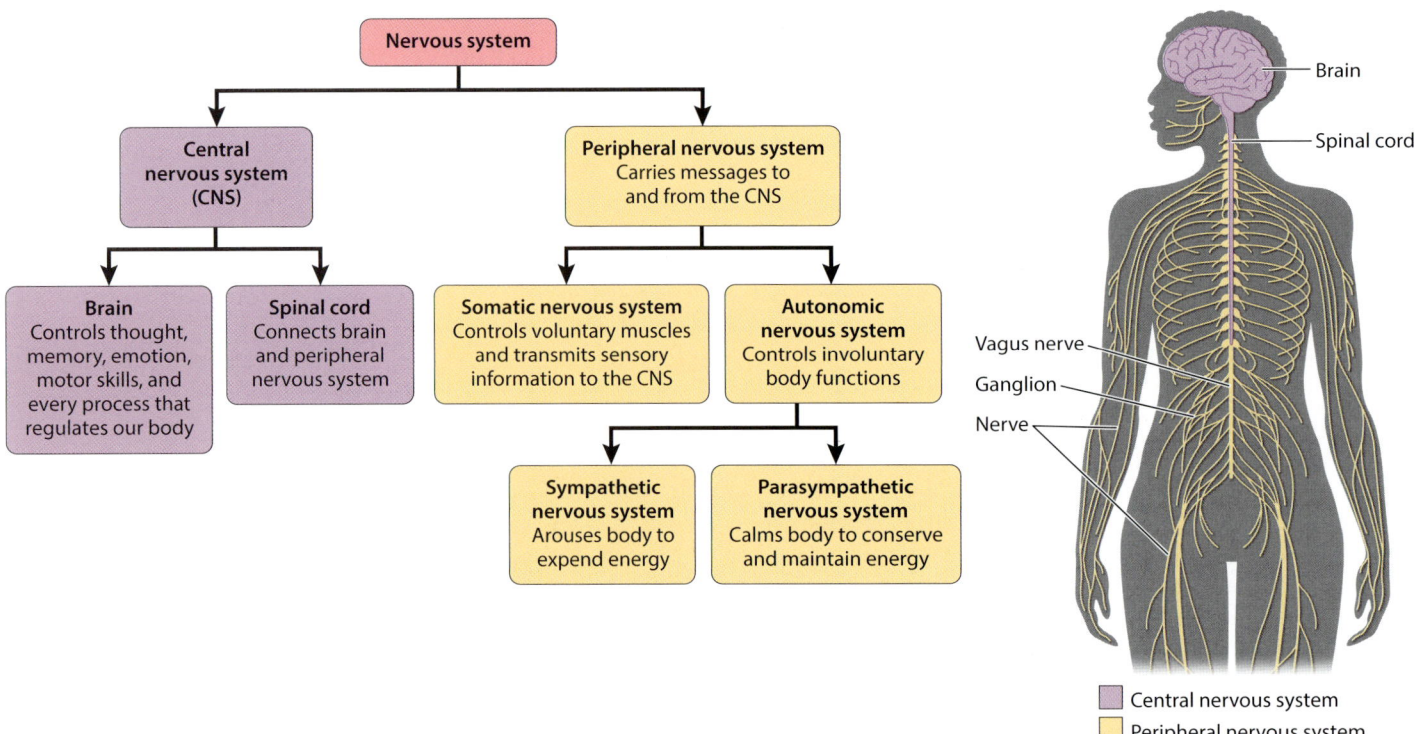

Figure 8.1 The Parts of the Human Nervous System The nervous system is composed of two distinct but interrelated systems. The central nervous system contains the brain and spinal cord, while the peripheral nervous system includes nerves and ganglia. Within the peripheral nervous system, we can also distinguish a somatic portion, which directs voluntary movements such as your choosing to read this figure legend, and an autonomic portion, which oversees involuntary actions such as your heart beating.

receptors can be found in massive clusters that form sensory organs, such as the eye, or they can be scattered across the body, as in the skin. They monitor temperature, light, sound, pressure, changes in pH and carbon dioxide, and the levels of electrolytes in the body. The sensory input is converted by our receptors into **nerve impulses**, which are simply electrical signals that are sent to the brain, where they create sensations, produce thoughts, or add to our memory. The brain makes decisions constantly as sensory input is received. This is the process of **integration**. The nervous system takes in the sensory stimuli, integrates the information, and sends signals to the body, which causes a response, such as muscle contraction or glandular secretion. This response to the brain's signals is called an effect, and the organs that respond are called **effectors** because they create a response based on input from the brain. This is what is meant by the **motor function of the CNS**.

In addition to sensing and responding to environmental stimuli, the nervous system regulates the breakdown of food we consume. This key component of the nervous system, called the **enteric nervous system** (**ENS**), is composed of a vast web of some 500 million neurons, which are embedded among the epithelial cells that line the GI tract (**Figure 8.2**). To put this number in perspective, the spinal cord has a similar number of neurons. These neural circuits detect the condition of the gut, integrate that information, and provide outputs that control the gut's actions. Many ENS circuits do not require input from the CNS, and because of this large level of autonomy, the ENS is referred to as a **second brain**. However, the ENS has extensive neural connections with the CNS, many mediated by the **vagus nerve**, which connects the brain to our body organs, including the GI tract, and carries signals between the brain and the rest of the body. This link between the CNS and ENS is called the **gut-brain axis** (**GBA**), and it is critical in connecting the brain with peripheral intestinal functions. The GBA involves not only links between the ENS, CNS, and intestine; the neural links also tie into endocrine, humoral, metabolic, and immune routes of communication.

In order to fully understand how microbes communicate with the GBA, we need to learn about one more neural component, the **neuroendocrine system**. The endocrine system consists of glands located throughout the body that produce hormones.

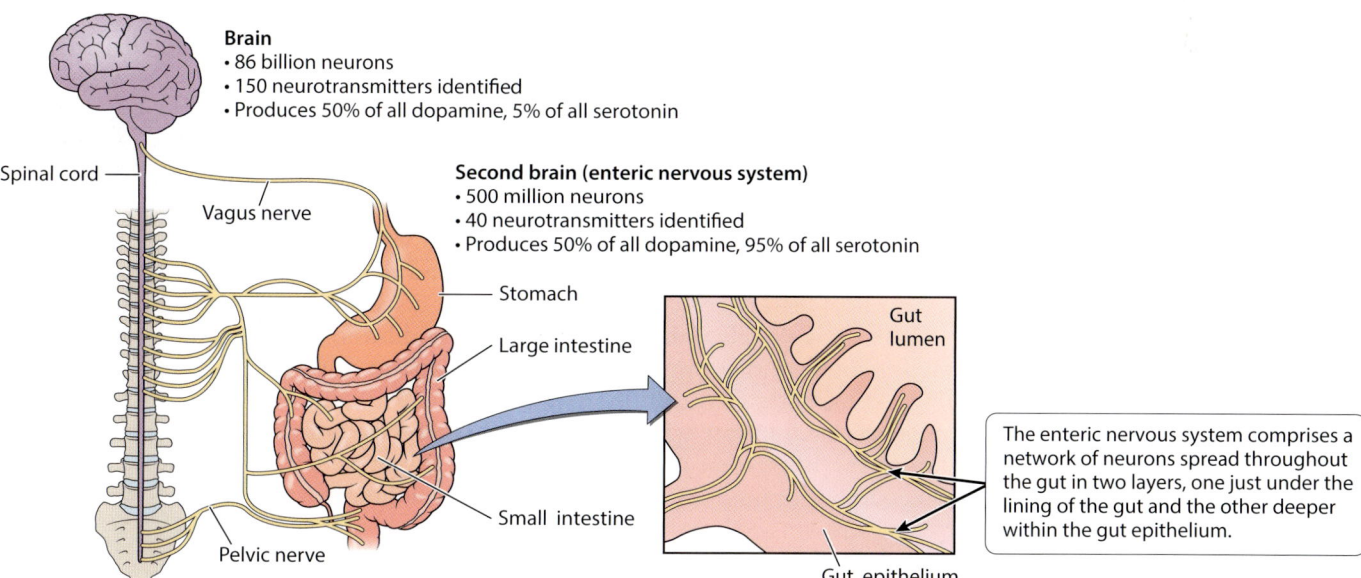

Figure 8.2 The Enteric Nervous System Your enteric nervous system, sometimes referred to as your "second brain," is located in the tissues of the GI tract. Some nervous pathways project outwards from the gut. Other pathways send signals from the CNS to the ENS via the vagal nerve, sympathetic nervous system, and pelvic nerves.

Figure 8.3 The Hypothalamic-Pituitary-Adrenal Axis *HPA axis* is a term used to represent the interaction between the hypothalamus, pituitary gland, and adrenal glands; it plays an important role in the body's response to stress. The pathway of the axis results in the production of cortisol, a hormone with several functions throughout your body, such as reducing inflammation and suppressing digestion.

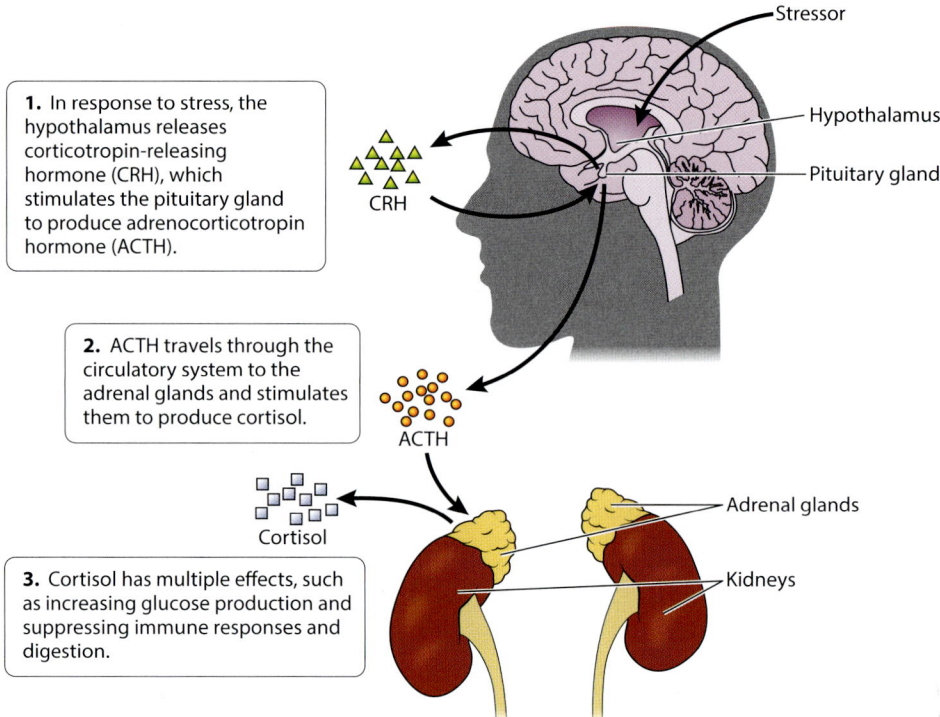

You may have heard of some of these hormones, such as insulin, which acts to keep blood sugar levels normal, or melatonin, which helps regulate sleep and circadian rhythms. Hormones also interact with the gut through the hypothalamus, the pituitary gland, and the adrenal glands—a set of organs called the **hypothalamic-pituitary-adrenal** (**HPA**) axis (**Figure 8.3**). When a stressor is detected, the hypothalamus, the region of the brain that coordinates the autonomic nervous system, is activated to produce and release corticotropin-releasing hormone (CRH). CRH binds to receptors in the nearby pituitary gland, which controls hunger and thirst. The binding of CRH triggers the release of the hormone adrenocorticotropin (ACTH). ACTH travels through the bloodstream and binds to receptors on the adrenal glands, which are found on top of the kidneys. The adrenal glands then produce the stress hormone cortisol. Cortisol has numerous effects throughout the body, from the brain to the immune and reproductive systems. It increases glucose production in the liver, which provides important energy to fight or flee, but it can also lead to increased blood sugar levels. We will return to how and why the HPA axis and the gut microbiome interact.

One final feature critical in our subsequent discussions about microbiota impacts on the brain is the **blood-brain barrier** (**BBB**), a diffusion barrier established early in fetal development that provides a critical checkpoint to protect the brain from circulating toxins, pathogens, and more (**Figure 8.4**). While it's important for the brain to be able to communicate with the rest of the body, the brain is a very sensitive organ. Without the BBB, even the healthy constituents of your microbiome could wreak havoc on your brain both during and after development.

8.2 THE MATERNAL MICROBIOME AND NEURAL DEVELOPMENT

To fully understand how the gut microbiome shapes our nervous system, we must first learn how its development is impacted by the bacteria in our GI tract (**Figure 8.5**). Relative to the diminutive nature of its members, the gut microbiota has an oversized role in our neural development. It is involved in neurogenesis, creation of the BBB,

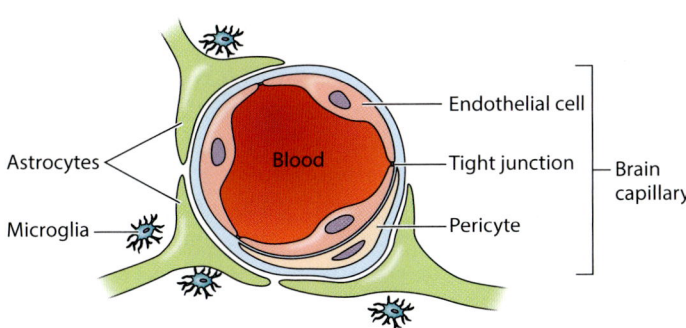

Figure 8.4 Schematic Diagram of the Blood-Brain Barrier (BBB) The BBB is a critical and complex structural feature at the intersection between the CNS and the circulatory system. It serves as both a structural and functional roadblock to microorganisms, such as bacteria, fungi, viruses, or parasites, that may be circulating in the bloodstream. It is composed of endothelial cells, pericytes, and tight junctions in the brain capillaries. The endothelial cells are interspersed with pericytes, which are cells that help mediated immune cell entry into the CNS, and tight junctions, which provide a continuous intercellular barrier between endothelial cells. Two types of non-neuronal cells in the CNS, astrocytes and microglia, also play a role in mediating communication between the CNS and the capillary.

myelination of nerve axons, and the maturation of microglia, which you may recall are the brain's version of an immune system. Let's walk through some of the more impressive contributions microbes from a pregnant woman make to fetal brain and nervous system development.

Just three short weeks after conception, neural development has already begun. Neural cells are gathering into what will emerge as the brain. These cells are instructed in where to travel by **neurotrophic signals**, which are proteins that function in mediating the development and functions of neurons. The pregnant woman's microbiome provides some of these signals as metabolites, such as short-chain fatty acids (SCFAs) and neuropeptides, which are small enough to cross the placenta. If the mother's microbiome is disrupted, such as through stress or drug use, the production of these key metabolites is altered, which can impact gene expression patterns in the fetal brain.

One of the best-characterized examples of microbiome and CNS cross talk involves microglial development. We learned in chapter 7 that microglia are the first

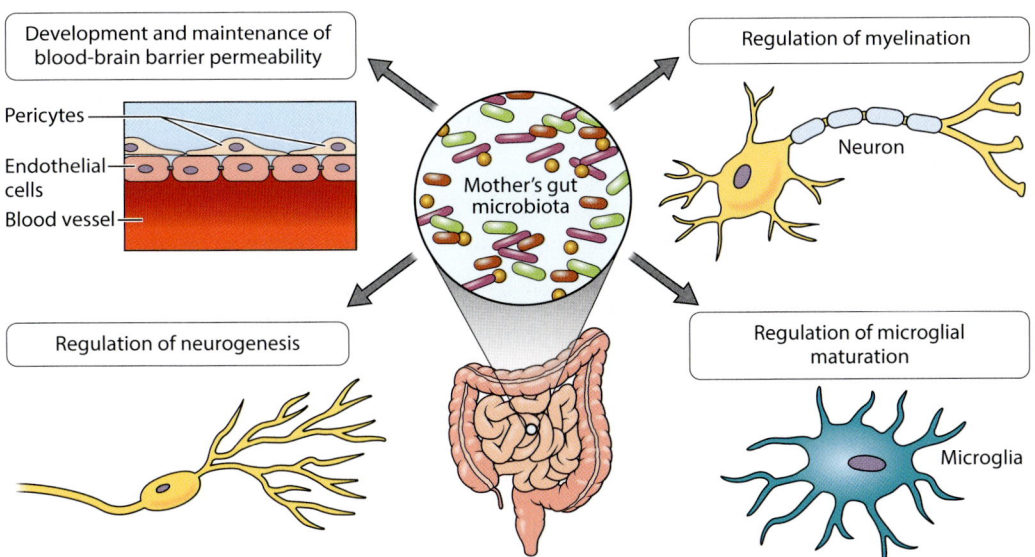

Figure 8.5 Role of a Pregnant Woman's Gut Microbiome in Brain Development During pregnancy, the woman's gut microbiome plays a key role in fetal neural development. The state of the parent's microbiome affects nerve cell differentiation and maturation, the HPA axis, the blood-brain barrier, myelination, and immune cells in the brain. A dysbiotic parental microbiome can have lasting impacts and is associated with various neurological disorders in the offspring.

line of immune defense in the CNS. They are also the most abundant innate immune cells of the CNS. Their job is to scan the CNS for potential intruders, such as pathogens, and destroy them. Your central nervous system is extremely sensitive to invasion, so having a robust detection system is essential for survival. SCFAs have been shown to contribute to the maturation and proper functioning of microglia, although the precise mechanism through which they act is not yet known. For example, oral application of SCFAs to germ-free mice was shown to drive microglial activation, which would normally be lacking (Silva et al., 2020). An interaction also occurs between gut microbes and **astrocytes**, which are also glial cells, but ones that play a dominant role in neuroinflammation. Microbial metabolites can reduce inflammation by regulating the signaling of astrocytes. One mechanism involves the metabolism of dietary tryptophan into aryl hydrocarbon receptor agonists, which are substances that initiate gene expression when they attach to a receptor, resulting in activation of astrocytes to limit CNS inflammation (Liu et al., 2022).

One group of researchers provided exquisite insight into how the parental microbiota impacts prenatal brain development in mice. In chapter 7 we learned about the research of Vuong et al. (2020), which showed that metabolites produced by the mother's microbiota aid in the creation of the wiring that connects the thalamus and cortex in the fetal brain, which is required for sensory processing (Vuong et al., 2020) (see Figure 7.8). The authors raised pregnant mice with either a normal gut microbiome or one depleted by antibiotics. During normal development, axons merge to form a thick bundle that creates what is called the internal capsule. The internal capsule is an information superhighway connecting the brainstem and cortex. Without it, communication between the brain and the rest of the CNS would be impossible. In pregnant mice with a depleted gut microbiome, whose fetuses did not receive maternal microbial metabolites, the connections between the offspring's thalamus and brain cortex are defective (see Figure 7.12).

Vuong and colleagues sought a mechanistic understanding of the defective capsule development. **Box 8.1** provides the details of two experiments that compared the brains of fetuses developing in either normal mice or germ-free mice. The experiments measured the differences in gene expression and in circulating metabolites. As revealed in Figure 8.6A, the differences in gene expression help explain the lack of axon formation that connects the thalamus and cortex in the fetal brain.

Vuong's team next explored the long-term consequences of these disrupted neural connections. They examined the adult offspring using behavioral tests that reveal sensorimotor deficits (see Figure 7.8. One common experiment used to understand sensorimotor impairment is to expose mice to a sudden, loud sound. Mice in the control group quickly sense the stimulus and startle in response. This method is helpful because a "normal" startle response requires the mice to sense, integrate, and respond to a stimulus, demonstrating a healthy sensorimotor system. Mice delivered by mothers with a dysbiotic gut microbiome showed altered response to sound.

To determine whether the maternal microbiome was responsible for the behavioral changes observed in their adult offspring, Vuong's team introduced specific bacteria into germ-free mice. They hypothesized that if the presence of these specific bacteria alone allow the offspring to develop normally, then their metabolites likely play a significant role in neurodevelopment. Among the many bacteria tried in this series of experiments, the most interesting result occurred when the germ-free pregnant mice were inoculated with *Clostridium*. The offspring from these mice displayed no abnormalities in brain development or behavior, suggesting that metabolites from this genus play an important role in fetal neurodevelopment (see Box 8.1; Figure 8.6B and C). You learned about metabolomics in chapter 4. These researchers engaged in discovery metabolomics, in which one identifies the metabolites produced in a microbiome and track where they end up in the body. In this study, the researchers revealed that the gut microbiota was the source of numerous metabolites in the maternal blood and in the fetal brain tissue. They also showed which metabolites were impacted by the presence of *Clostridium* in the maternal microbiome.

BOX 8.1. RESEARCH IN ACTION
Maternal Microbiota—Shaping the Fetal Brain's Biochemistry

- **Hypothesis.** The maternal microbiome influences fetal development, particularly axonal development, by regulating circulating metabolites and thus conditioning metabolic profiles in the fetus.

- **Methods.** Conventional and germ-free pregnant mice (dams) were studied, with a focus on fetal gene expression patterns and serum brain metabolomes in the dams and fetuses. Principal component analysis was used to identify whether metabolome profiles differ between the control and experimental mice.

- **Results.** Levels of gene expression differed between fetuses of dams with a microbiome and fetuses of dams without, with many of the 333 differentially expressed genes involved in axon formation. The heat map in **Figure 8.6A** reveals a snapshot of these variations in gene expression. Further, as shown in **Figure 8.6B**, serum metabolomic profiles differ between offspring of conventional and germ-free dams. **Figure 8.6C** provides examples of two such metabolites whose relative concentrations were more similar in the fetal brains from dams with normal microbiomes (SPF and Sp in the figure) compared to those from germ-free dams (ABX and GF in the figure).

- **Conclusions.** These findings reveal that the maternal microbiota regulates biochemical profiles and select metabolites in the fetal brain. The findings also offer support for the hypothesis that the maternal microbiome is required to support formation of axon connections.

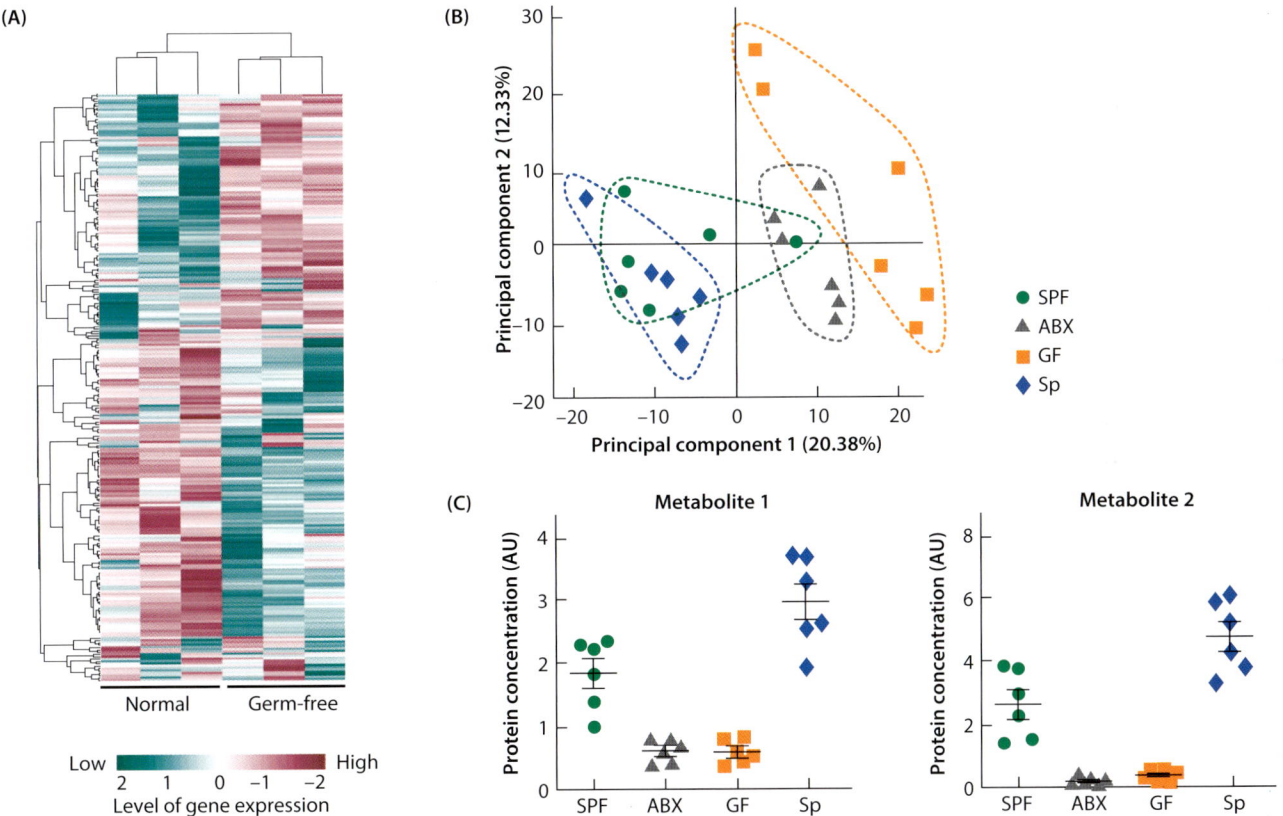

Figure 8.6 Depletion of the Mouse Maternal Microbiome Impairs Fetal Axonogenesis (A) This seemingly complex illustration is called a heat map, which is used to plot differentially expressed genes. Each row of the map represents a different gene. Each column represents a different fetal brain sample from either the conventional (normal) or microbiome-depleted (germ-free) parental mice. The colors of the boxes, ranging from greenish-blue to purple, show the relative level of expression of each gene represented in the sample, with greenish-blue and purple indicating lower or higher levels of expression. (B) Principal component analysis of maternal circulating (serum) metabolomes from dams with normal microbiomes (SPF), microbiomes with *Clostridium* (Sp), and depleted microbiomes (ABX, GF). Each data point represents a single dam. (C) Examples of metabolites that are significantly decreased in fetal brains from embryos of ABX and GF dams versus SPF and Sp dams. Metabolite 1 is imidazole propionate; metabolite 2 is N,N,N-trimethyl-5-aminovalerate. Protein concentrations are given in absorbance units (AU). (A from Vuong, H.E., Pronovost, G.N., Williams, D.W. et al. The maternal microbiome modulates fetal neurodevelopment in mice. *Nature* 586, 281–286 (2020), with permission from Springer Nature, B,C after Vuong et al., 2020.)

Altogether, these findings reveal that the maternal gut microbiome promotes connections between the fetal thalamus and cortical region of the brain. It is important to remind ourselves that these key discoveries were found in mice, not humans. However, they demonstrate the close connection between the microbiome and neurodevelopment, which can help us to better understand how the same process might occur in humans. Further, these experiments provide an excellent example of how one can proceed from observations (that mice born from germ-free mothers have altered startle response) to an understanding that specific microbes produce metabolites that are involved in the development of neural connections that promote the normal startle response.

The implications for promoting healthy brain development are considerable. According to Vuong, "The gut microbiota has the incredible capability to regulate many biochemicals not only in the pregnant mother but also in the developing fetus and fetal brains" (Vuong et al., 2020). Mouse models are a long way from clinical applications, but not so far off that they're not worth considering. Currently, we administer folic acid during pregnancy to help prevent neural tube defects in developing fetuses. Someday, we may promote supplementing the diet of pregnant women with microbial metabolites to promote healthy fetal brain development!

Formation of the Blood-Brain Barrier

The presence of a normal maternal gut microbiota and appropriate levels of their metabolites are essential in regulating the formation of the BBB. This is done through a variety of pathways, including the vagus and sympathetic nerves and immune and endocrine systems (**Figure 8.7**). Maternal inflammation during pregnancy, which can

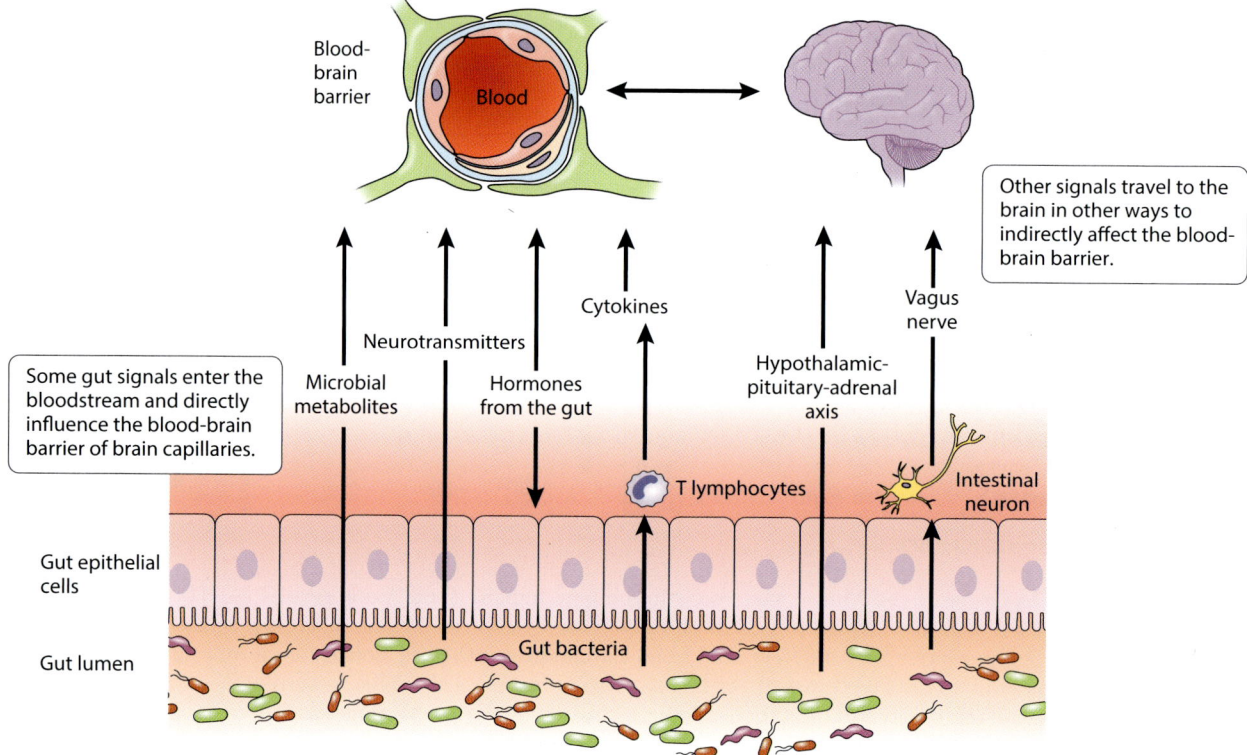

Figure 8.7 Gut Microbiota and the Blood-Brain Barrier In addition to affecting CNS function through the HPA axis and vagus nerve, the gut microbiome can also alter neurological activity through the BBB. Metabolites, neurotransmitters, hormones, and cytokines produced or influenced by microbial species can affect the development of the BBB, which is essential for protecting the brain.

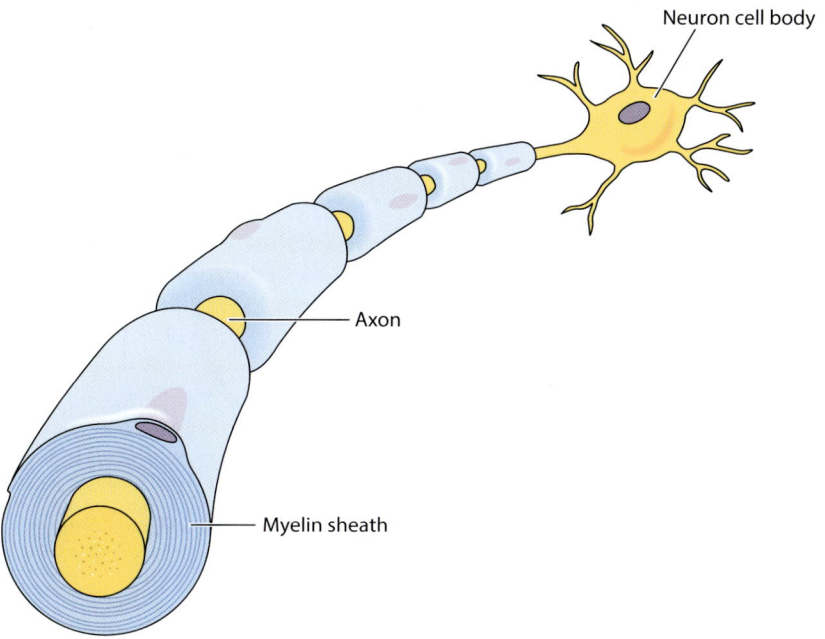

Figure 8.8 Myelin Sheath Neurons are coated by a protective layer known as a myelin sheath. Myelination occurs within the first few years of life and is essential for proper neural signaling.

result from a dysbiotic gut microbiome, perturbs the formation of the BBB, leading to abnormal brain function and chronic brain inflammation in the baby.

The pregnant woman's microbiome is also involved in ensuring that the axons of nerve cells are myelinated. Myelination is the process by which nerves are coated in a protective layer of lipids and proteins known as myelin (**Figure 8.8**). Not only does this layer protect the nerves from damage, but insulating them from the rest of the body allows more-rapid transmission of neural information, just as the rubber coating of a copper wire you may find in your house does. Myelination of maturing axons occurs shortly after birth. Any disruption in this process can lead to long-lasting CNS defects. The constituents of a normal microbiome are directly involved in myelination, due to their ability to regulate myelination-related gene expression.

Finally, the pregnant woman's gut microbiome plays an important role in the activation of fetal immunity. We first heard about the immune activation functions of the maternal gut microbiome in chapter 7 For example, prior to birth, gut bacteria produce **cytokines** that cause certain immune cells to elicit an inflammatory response. Cytokines are small proteins involved in controlling other immune and blood cells, directing both their growth and their activity. When cytokines are released, they signal the immune system to protect the host and fend off pathogens. Fetal stem cells can perceive these metabolites, which impacts the production and function of immune cells and plays a vital role in their "training."

8.3 THE MICROBIOTA-GUT-BRAIN AXIS

The microbiome continues its communication with the nervous system after birth and involves the brain, our "second brain" (ENS), and the neuroendocrine system (HPA axis). The brain is the captain: it takes charge of thinking, movement, speech, memory, vision, and certain vital functions like breathing and blood circulation. It speaks directly to the gut microbiome via the HPA axis and the autonomic nervous system. When you are frightened, your fight-or-flight response is triggered, and your stomach and bowels respond. Stress can rapidly influence the type of bacteria inhabiting the gut, resulting in less diversity. It can also increase inflammation in the bowel. Conversely, the gut communicates with the CNS by producing a variety of metabolites, neuroactive substances, and hormones that travel via the circulatory system, the ENS,

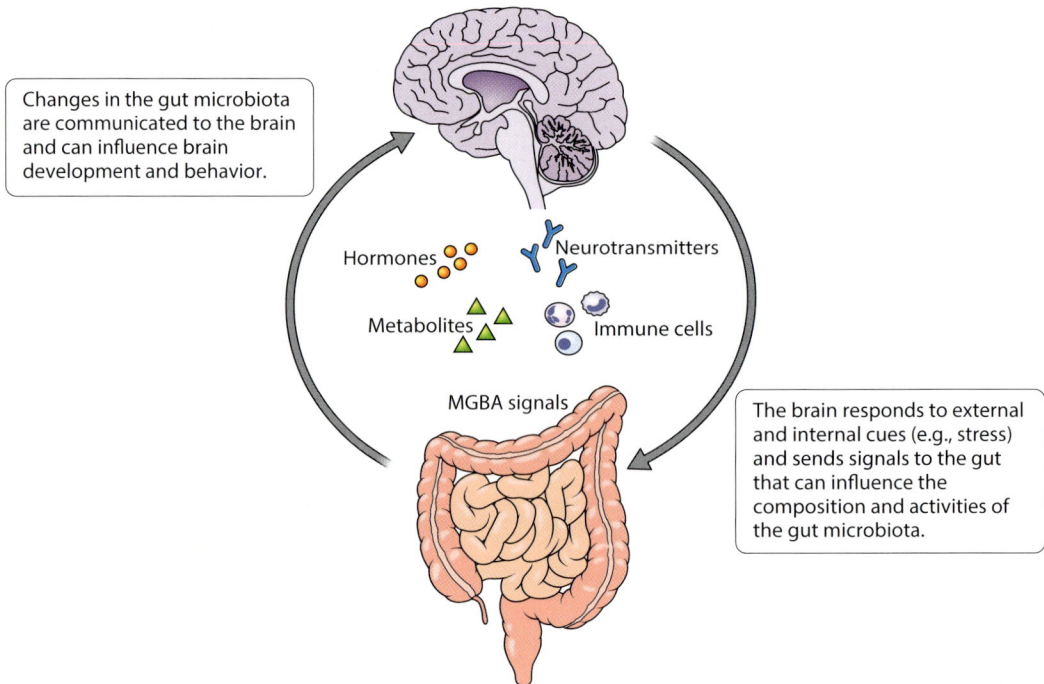

Figure 8.9 Microbiota-Gut-Brain Axis The microbiota-gut-brain axis involves a multidirectional channel for communication mediated by hormones, metabolites, neurotransmitters, and immune cells. (After de la Fuente-Nunez et al., 2018.)

the vagus nerve, or the immune system to reach the brain. These pathways are all part of the microbiota-gut-brain axis (MGBA) (**Figure 8.9**) (Liu et al., 2022).

MGBA Communication via the Endocrine Pathway

Let's focus first on what the microbiome has to say. The gut microbiota communicates with the brain through three main pathways (Chakrabarti et al., 2022) (**Figure 8.10**). The **endocrine pathway**, which involves circulating hormones, provides a

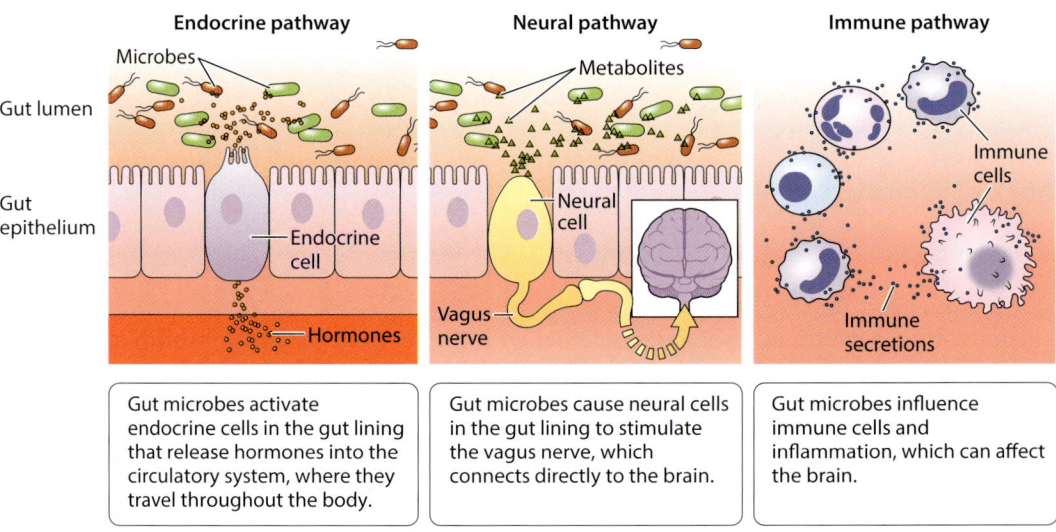

Figure 8.10 The Gut Talks to the Brain Bacterial residents of the intestines may influence neurons and the brain through several routes. (After Pennisi, 2020.)

relatively slow and indirect channel for movement of microbial metabolites to all parts of the body. The **neural pathway** provides the fastest and most direct route for the microbiota to influence the brain. The neurons in this pathway connect the soft tissue organs, such as the liver, lungs, and heart, to the brain. Finally, there is the **immune pathway** in which cytokines produced in the gut result in intestinal and neural inflammation as well as changes in microglia functions. Let's learn a bit more about each of these three main MGBA pathways.

The endocrine pathway (**Figure 8.11**) involves the movement of metabolites produced by microbes in the gut, such as SCFAs, into the circulatory system, which can deliver them directly to the brain, passing through the BBB. Once in the brain, SCFAs can influence **neurogenesis,** which is the formation of neural cells, and neuroinflammation. SCFAs can compete for attachment sites on brain receptors and thus alter the release of signaling molecules from brain cells. Numerous microbes, such as *Bifidobacterium* and *Lactobacillus*, produce these metabolites.

These same metabolites can act indirectly by stimulating the ENS to send signals to the brain, by way of the vagus nerve, that play a role in memory and learning processes. More specifically, the intestinal epithelium has enteroendocrine cells (EECs), which directly interact with intestinal neurons and other epithelial cells through projections. Some EECs directly synapse with the vagus nerve and provide for more instantaneous communication with the brain. Most EECs function through the endocrine pathway, as they manage digestion, regulate hunger signaling, and sense ingested toxins. EECs comprise the largest endocrine system of the body, producing numerous hormones and neurotransmitters, which are directly impacted by the gut microbiota. A recurring theme you may have noticed by now is that many of your body's systems that are essential for its healthy functioning also interface directly with your microbiome. This is not by coincidence, and we'll discuss the coevolution of us, *Homo sapiens*, and our microbes in chapter 13.

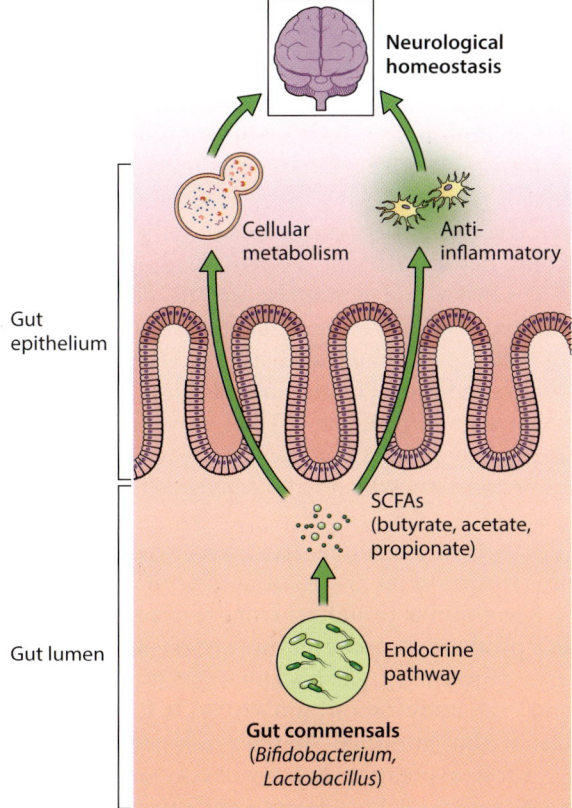

Figure 8.11 MGBA Communication via the Endocrine Pathway The endocrine pathway involves the movement of microbial metabolites, such as SCFAs, through the circulatory system and their ultimate delivery to the brain. This pathway provides a slow and indirect channel of microbial metabolites to all parts of the body. (After Shoubridge et al., 2022.)

Figure 8.12 MGBA Communication via the Neural Pathway The neural pathway provides the fastest way for the microbiota to influence the brain. The microbiota signal neural cells to produce neurotransmitters and neurotrophic factors, which are then delivered directly to the brain. (After Shoubridge et al., 2022.)

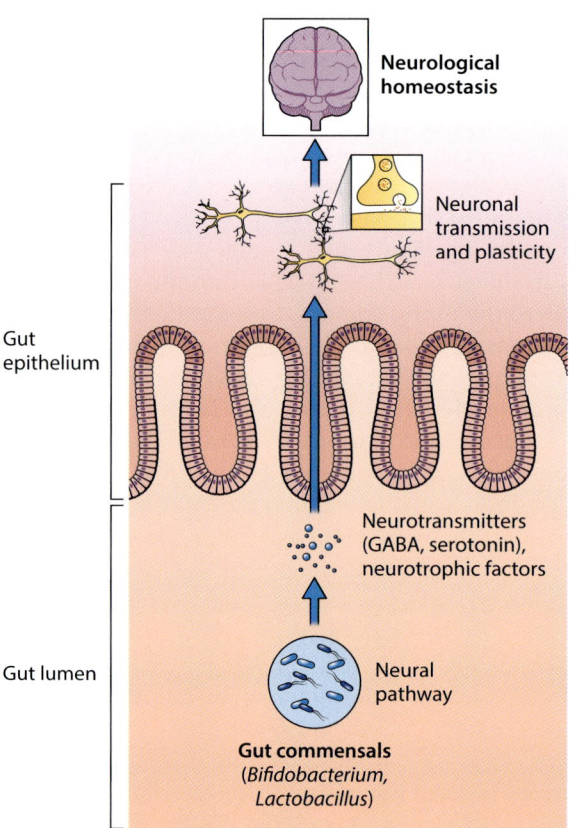

MGBA Communication via the Neural Pathway

The neural pathway (Figure 8.12) involves the movement of microbially produced **neurotransmitters**, which carry signals from nerve cells to muscle, gland, or other nerve cells. These messages are responsible for moving a hand when you wish, keeping your heart beating whether you think about it or not, and taking in all kinds of information around you and in you. **Gamma-aminobutyric acid** (**GABA**) is one of the key inhibitory neurotransmitters. It regulates brain activities involved in anxiety, irritability, and depression. When GABA is released, it results in feelings of calmness and reduced anxiety. The excitatory neurotransmitter **glutamate** performs the opposite function and stimulates the nervous system, leading to increased signaling. Appropriate amounts of GABA and glutamate are needed for the brain to function properly, and unbalanced levels of these neurotransmitters are associated with anxiety, depression, and other psychological disorders. GABA can be produced in the brain, but it can also be synthesized by the microbiome. *Lactobacillus* and *Bifidobacterium* metabolize glutamate to produce GABA, which can directly affect our mood. Studies have found that use of *Lactobacillus* as a probiotic in mice led to dramatic changes in GABA brain activity, which led to changes in how the mice responded to stress (Bravo et al., 2011).

Serotonin is a key neurotransmitter that helps regulate anxiety, pain, sexuality, and more. Imbalances in serotonin are associated with depression, anxiety, and several other disorders. Serotonin is produced from tryptophan, which is an amino acid that our bodies don't produce. Tryptophan must instead be obtained from the food we eat, or from the metabolism of gut microbes, such as *Candida*, *Escherichia*, and *Enterococcus*. Most of the body's serotonin (90%) is found in the ENS, where it directs the movements of our intestinal muscles. The remainder is produced in the brain, where it plays a role in creating feelings of contentment and well-being. Drugs that affect levels of serotonin, such as many antidepressant medications, often have marked impacts on the gut homeostasis. For example, individuals with irritable

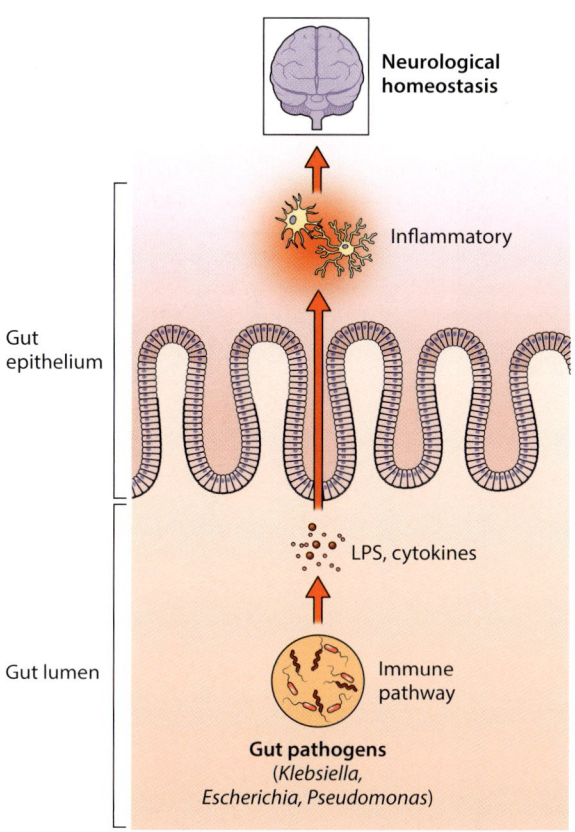

Figure 8.13 MGBA Communication via the Immune Pathway The microbiome influences immunity through the interaction of microbial products with receptors found on intestinal epithelial and immune cells. These interactions trigger the production and circulation of proinflammatory cells and cytokines, which then interact directly with the brain. (After Shoubridge et al., 2022.)

bowel syndrome (IBS) have 10 times higher levels of serotonin in the gut than people without IBS. Thus, serotonin is a messenger that provides cross talk between the microbiome and the brain, and when serotonin levels in the gut are too high, brain function is impacted.

Neurotrophic factors, which are a family of biomolecules that support neurons, provide yet another means of gut-brain communication. One of the most important of these is **brain-derived neurotrophic factor** (**BDNF**), which plays a key role in regulating the growth and essential functions of neurons, such as growing to where they are needed and becoming stronger. Several studies have demonstrated that gut bacteria have a profound impact on BDNF levels in mice. For example, removing *Bifidobacterium* from otherwise healthy mice results in substantially less BDNF expression and exaggerated stress responses (Zhu et al., 2021). Several mental disorders are associated with low levels of BDNF, including depression, memory loss, and cognitive decline.

MGBA Communication via the Immune Pathway

The microbiome influences immunity through the interaction of microbial products with receptors found on intestinal epithelial and immune cells (**Figure 8.13**). These interactions trigger the circulation of proinflammatory cells and cytokines, which then interact directly with the brain. SCFAs can even cross the BBB to influence the innate immune cells of the brain, including microglia and astrocytes, the importance of which we discussed earlier in the chapter. **Lipopolysaccharides** (**LPS**) make up another metabolite. Found in the cell walls of certain bacteria, including *Pseudomonas aeruginosa*, *Klebsiella pneumoniae*, *Escherichia coli*, and several other human pathogens, they can attach to certain receptors in brain cells to stimulate microglia. These metabolite and neural receptor interactions are involved in the development of several neurodegenerative conditions, including multiple sclerosis, Alzheimer's disease, and Parkinson's disease.

Figure 8.14 MGBA Communication via Autophagy Autophagy is a relatively newly described process for communication between the gut microbiome and the brain. The gut microbiome can trigger autophagy through the production of certain metabolites. (After Shoubridge et al., 2022.)

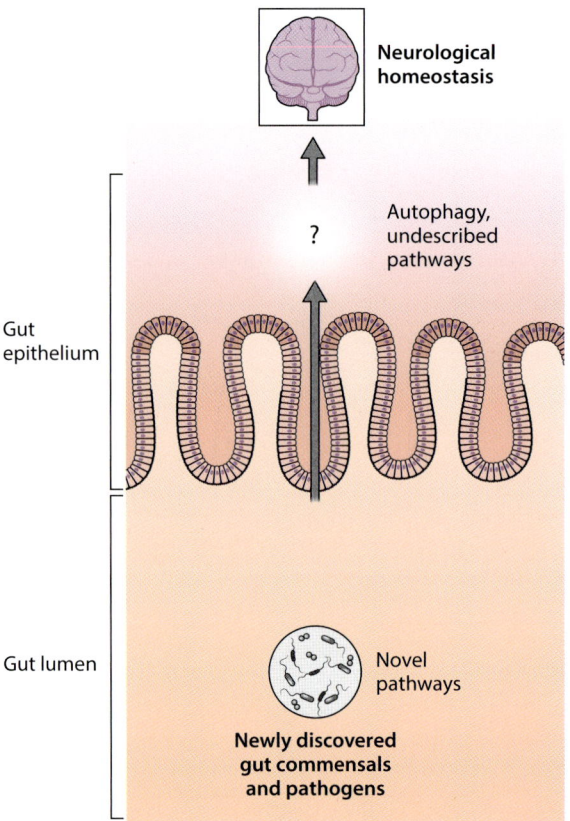

MGBA Communication via Autophagy

A novel form of communication was discovered to play a role in maintaining mental health (**Figure 8.14**). **Autophagy** is a process that prevents cellular damage and helps to maintain cell function by clearing damaged material from cells and breaking it down into reusable biomolecules. This damage control prevents toxin accumulation and inhibits neuronal inflammation, which are cellular characteristics associated with dementia. Autophagy is triggered by numerous factors, such as infection, nutritional state, and oxygen levels, and is mediated, at least in part, by the gut microbiome. One example of this triggering is the suppression of certain gene regulatory functions by the production of SCFAs (Gassen & Rein, 2019). Both SCFA biosynthesis and the presence of LPS increase levels of autophagy.

Your gut microbiota can influence autophagy activity, but autophagy also plays an important role in shaping your microbiome. Autophagy allows your human cells to remove and digest pathogenic invaders from your body. The pathogens are enclosed in a vacuole, which then merges with a lysosome (think of a vacuole filled with digestive enzymes), and the microbe is destroyed (**Figure 8.15**). While there are untold numbers of helpful bacterial species in your body, you still need mechanisms to remove the ones that can make you sick, a concept first introduced in chapter 3.

8.4 GUT MICROBIOTA AND NEUROPSYCHIATRIC DISORDERS

We have identified several critical roles played by gut microbes in communicating with the brain. It is hardly surprising to learn that when the gut microbiome is perturbed, it can play a role in causing a number of neuropsychiatric diseases. Most of the early work in this area was conducted in mice and rats. However, there is a mounting body of evidence from human studies that confirm many of these earlier findings. It was a rocky start for those pioneering investigators who proposed a direct

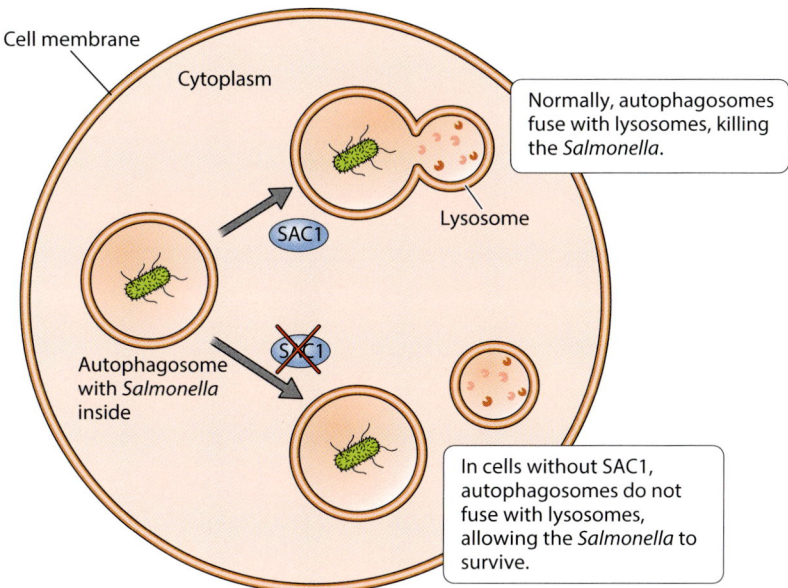

Figure 8.15 Demonstrating the Importance of Autophagy as an Immune Response The *SAC1* gene allows for mammalian cells to break down *Salmonella* pathogens through a process known as autophagy. Without this gene, cells can capture the *Salmonella* bacteria but are unable to break them down, resulting in a population increase.

link between the microbiome and neurological disorders. Much like what Lynn Margulis endured when she proposed the endosymbiotic theory for the origin of eukaryotes (chapter 1), the early microbiological neurologists were frequently dismissed, and it required spunk, determination, and patience to persevere. One scientist who recognized early on the potential link between the gut microbiota and neurological disease was Jane Foster at McMaster University in Canada. She observed that germ-free mice behaved differently than conventional mice. They seemed less anxious, and they would choose open paths in a maze challenge, as opposed to covered or walled-in paths. She noted that bacteria in the mouse gut seemed to be influencing the mouse's behavior. She attempted to publish her findings, but her manuscripts were repeatedly rejected. It took 3 years and seven submissions to see one through to publication (Neufeld & Foster, 2009). According to Dr. Foster, "People didn't buy it. They thought it was an artifact" (Willyard, 2021).

Establishing a causal role of the microbiome with respect to neurological disease remains challenging. The most common approach is to transplant gut microbiota from individuals with a disorder into germ-free mice to see whether disease attributes appear. Although this strategy has numerous limitations—for instance, mouse studies don't always reflect what happens in humans—it provides an excellent direct experimental route to investigate the impact of the gut microbiome on conditions such as neurodegenerative diseases and depression. As discussed in chapter 4, this kind of controlled experiment is necessary to show that alterations to the microbiome are a factor causing a disorder, rather than a dysbiotic microbiome resulting from that disorder. Additionally, since there are no germ-free people to volunteer for these clinical trials, model organisms that share key human characteristics are our best way to build experimental evidence. Next, we highlight recent advances in our understanding of how the gut microbiome plays a role in neuropsychiatric disorders.

Depression

Let us start with depression, but not your Monday morning blues type of depression. We're talking about the kind of depression that causes a persistent, intense feeling of sadness and a loss of interest in most things. This form of depression has a clinical name, **major depressive disorder** (**MDD**), and it affects how your brain creates feelings, thoughts, and behaviors. MDD can result in numerous physical and emotional challenges. It affects more than 300 million people globally and is one of the leading causes of psychiatric disability.

Table 8.1 Establishing a Connection between Depression and the Gut Microbiota

KOCH'S POSTULATES	EVALUATION AGAINST CRITERIA
1. The microorganism must be found in patients with the disease.	Isolation of *Mycobacterium neoaurum* from fecal samples
2. The microorganism can be isolated and cultured from a diseased individual.	Culture of *M. neoaurum* in vitro
3. Inoculation with the isolated microorganism from a diseased individual should reproduce the symptoms of the disease in a healthy individual.	Fecal microbiota transplantation experiments Inoculation with isolated *M. neoaurum* Construction and inoculation with recombinant *E. coli* expressing 3β-HSD
4. The microorganism causing the symptoms should be reisolated from the inoculated diseased individual.	??

Source: Li et al., 2022

Before we dive into depression, let's think back to something we learned in chapter 2, which was that in 1890 Robert Koch established a series of postulates to identify whether a microorganism causes a particular disease (see Figure 2.6). In a simple yet elegant study, Dr. Li and colleagues from Wuhan University employed Koch's postulates to identify a microbe and an enzyme it produces that play a crucial role in some types of depression (Li et al., 2022) (**Table 8.1**). First, they established that men with depression have a particular bacterial species present in their gut microbiomes, *Mycobacterium neoaurum*. Next, they showed that *M. neoaurum* produces 3β-HSD, an enzyme that can degrade testosterone. Further, they showed that introducing *M. neoaurum* into rats resulted in lower levels of testosterone and more severe symptoms of depression. Finally, a large screen of men revealed that the enzyme 3β-HSD is higher in men with depression. Taken together, these data show that this bacterial species, or at least an enzyme it produces, has a direct impact on levels of depression experienced by some men. Perhaps even more exciting, their application of Koch's postulates reveals the value of moving beyond simply assessing which microbes are present, to identifying the precise molecular mechanisms at work. In this case, it is the production of an enzyme that is key, and presumably it won't matter which bacterium produces that enzyme, as was noted by the authors of this study. As quoted by researchers Doolittle and Booth, "It's the song, not the singer" (Doolittle & Booth, 2017).

A second research team has also succeeded in reaching a mechanistic level in their studies of the impact of microbiota on depression. Let's start with some things we knew before their work. We knew that when certain foodborne pathogens, such as *Citrobacter rodentium* or particular strains of *E. coli*, are introduced into a mouse gut, stress circuits in the brain are activated through triggers from the vagus nerve. We also knew that one outcome of this triggering is an increase in the expression of the c-FOS enzyme in the brain, which is a generalized marker of neural activity.

A recent multi-omics study analyzed the levels of amino acids in the diet, blood plasma, and fecal samples of patients with no depression, mild depression, and major depression (**Figure 8.16A**) (Mayneris-Perxachs et al., 2022). A positive correlation was observed between high levels of proline and more severe depression. Proline is an amino acid that can be metabolized to GABA, which, as you may recall from earlier, contributes to feelings of well-being. However, when the levels of proline increase too much, its presence can disrupt the interaction of GABA and its neural receptors during prolonged stimulation. Depressed individuals also tend to have higher levels of plasma proline, which isn't metabolized properly.

The researchers used a mouse model to establish a cause-and-effect relationship between excess proline and depression. They found that administering a proline supplement to mice exacerbated depression symptoms. Further, when they took fecal samples from depressed humans and transplanted them into mice, the mice showed more behaviors associated with depression (**Figure 8.16B**).

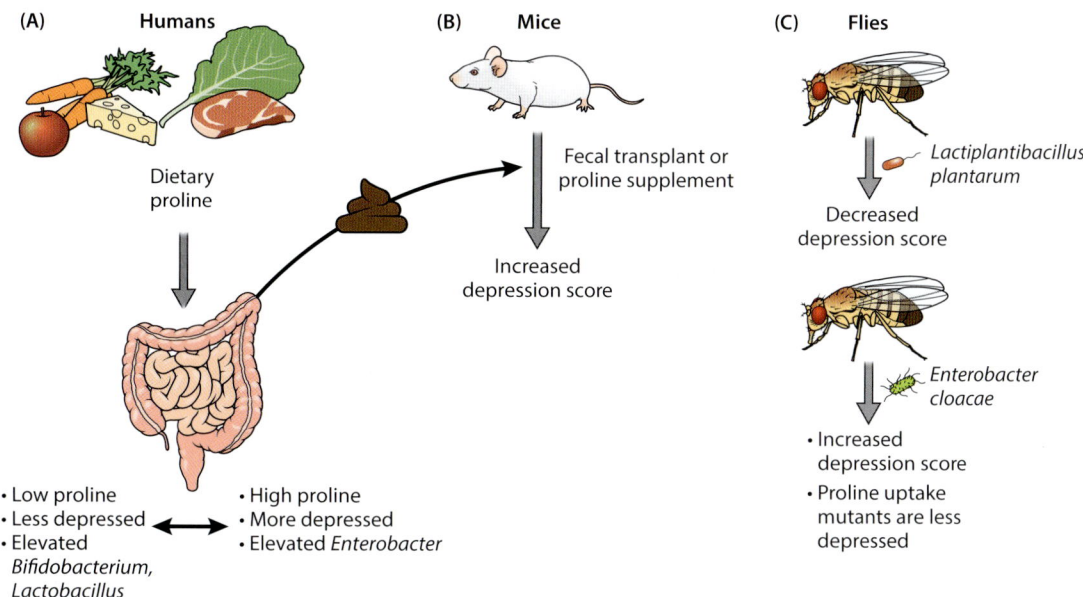

Figure 8.16 Relationships between Microbiome-Based Metabolism of Proline and Depression (After Mayneris-Perxachs et al., 2022.)

The next stage of the research involved determining which microbes affect proline levels. Mayneris-Perxachs and colleagues (2022) found that participants experiencing milder depression had higher levels of *Bifidobacterium* and *Lactobacillus*, while more-depressed participants had higher levels of *Enterobacter*. They then fed fruit flies food laced with *Lactobacillus* or *Enterobacter* and noted that flies fed *Lactobacillus* ate more and were more active in general than those given the *Enterobacter* (**Figure 8.16C**). In their final experiment, they tested the impact of mutations in the flies that inhibited transport of proline to the brain and found that those mutated flies were highly resilient to depression, suggesting that proline's effect on GABA production in the brain can contribute to depression. These studies not only help us to understand a second mechanism involved in depression, but also offer new opportunities to develop novel approaches to treat depression. Not to mention that this team gets the prize for the most amazing combination of in vivo assays: mice, men, and fruit flies!

As you can see from these two studies, the causes of depression are incredibly complex and involve many neurotransmitters, metabolites, and biochemical interactions. Although there is much still to learn about MDD at a molecular level, it is clear that the gut microbiome is a central hub in many of the chemical pathways that are disrupted in depression. Most studies of depression, or any neurological impairment, have not yet reached this level of mechanistic detail and instead rely on associations between particular microorganisms and disease states. Two bacterial genera in particular, *Morganella* and *Klebsiella*, seem to play a causal role in depression. One group of scientists were interested in a possible link between depression and inflammation and reported that individuals with depression also displayed an enhanced immune response to metabolites produced by *Morganella* and other related bacteria, such as *Klebsiella* and *Enterobacter*, in the gut. Yang et al. (2020) compared fecal metagenomes between MDD and healthy control volunteers. They identified a microbial signature of MDD that included 3 bacteriophages, 47 bacterial species, and 50 fecal metabolites. The most significant differences in the microbiomes of individuals with MDD were an increased abundance of members of the genus *Bacteroides* and fewer *Blautia* and *Eubacterium*. The gut microbiomes of individuals with MDD were enriched for 18 bacterial species, including 10 species of *Bacteroides*, and were depleted in 29 species, including 5 species of *Blautia*, 5 of *Eubacterium*, and 3 of *Clostridium*. Examination of metagenomics, transcriptomics, and proteomics revealed an MDD signature of gene expression disturbances. One of the most fascinating results of this study was the fact that most of these changes were associated with amino acid metabolism, in particular the metabolism of GABA, phenylalanine, and tryptophan. Imagine if MDD could be solved as simply as by taking amino acid supplements. Of course, biology is never that simple!

Autism Spectrum Disorder

Neurodivergent is a nontechnical term that describes individuals who experience and interact with our world differently due to developmental or functional differences in their brain. Psychological and psychiatric research suggests that neurodivergence is incredibly common. Autism spectrum disorder (ASD) is one such condition with a diverse spectrum of symptoms. Individuals with autism may have difficulty focusing, have a tendency for repetitive behavior, or communicate differently. About 2.8% of the children in the United States have been diagnosed with ASD.

The mechanisms underlying ASD remain unknown; however, there appears to be a strong link between infection during pregnancy and the chance of a child being diagnosed with ASD. A study of a cohort of nearly 1.8 million Swedes revealed that mothers who had an infection during pregnancy had a remarkable increase (79%) in the chance of their child being diagnosed with ASD (Lee et al., 2015). To study this association, researchers employed a mouse model and showed that when pregnant mice were injected with double-stranded RNA, the mice were tricked into sensing an invading virus (Careaga et al., 2017). The pups of these mice exhibit ASD-like behaviors, such as repetitive behaviors.

One group focused on the role of cytokines in ASD (**Figure 8.17**) (Sgritta et al., 2019). When the researchers mimicked an infection in pregnant mice, T helper 17 cells (a type of immune cell) became overactive and produced high levels of the cytokine interleukin-17. This molecule made its way into the CNS of the pups and attached to brain receptors, which resulted in autism-like behaviors in the adult offspring. The researchers then examined filamentous bacteria that were known to promote formation of T helper 17 cells. Following treatment with antibiotics, the pups of pregnant mice showed no behavioral differences. This suggests that the T helper 17 cells, and not other aspects of the immune response, contribute to ASD-like behaviors in mice. As in previous experiments we've discussed, it's important to understand the limits of our model system. Symptoms of ASD in mice are generally described as changes in social interaction and communication, and increased repetitive behavior. However, because neurodivergent conditions present with a spectrum of symptoms, it's impossible to tell what signs of ASD a human being might display given the same circumstances.

That same team has identified bacteria that may be useful in treating ASD (Sgritta et al., 2019). Working with mice whose offspring have autism-like symptoms, they noticed that when those mice were co-caged with conventional mice, their ASD-like behaviors disappeared. The researchers proposed a possible explanation: the mice had engaged in coprophagy—they ate each other's feces—and thus shared their gut microbiomes. The researchers also determined that mice displaying ASD-like behaviors lacked *Lactobacillus reuteri*. When this bacterium was introduced into the mice, some of the ASD-like behaviors disappeared. It is not yet known what signal is being sent by this bacterium and why only some strains of *L. reuteri* can reverse the ASD-like behaviors. The team is currently determining which bacterial genes are involved.

These promising studies have garnered the interest of the pharmaceutical industry. Axial Therapeutics is a clinical-stage biopharmaceutical company developing gut-targeted, small molecule therapeutics. They are currently engaged in clinical studies that involve a first-in-class molecular therapeutic (AB-2004) that targets the

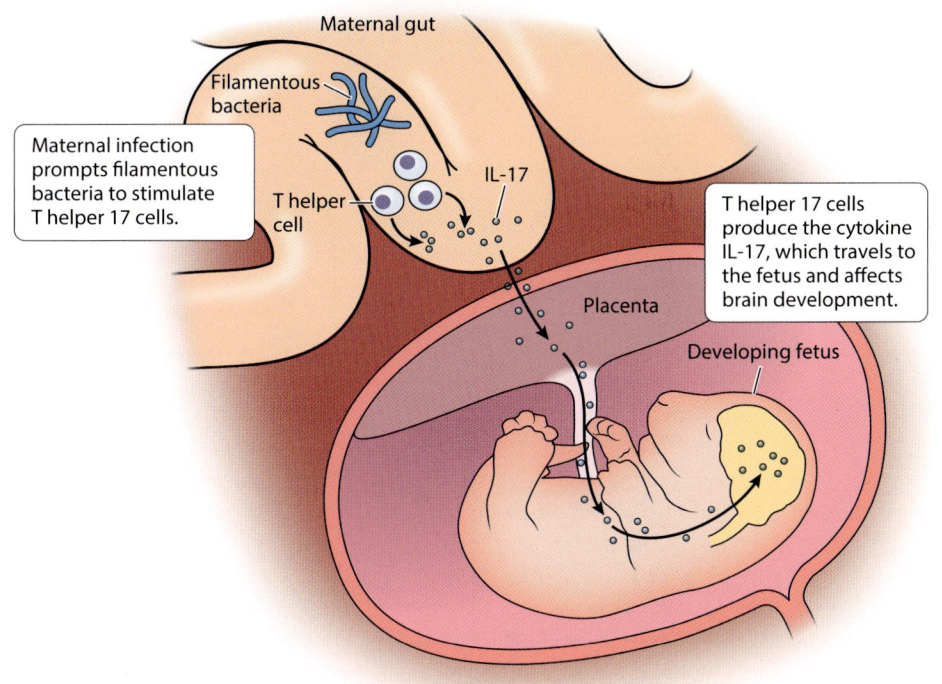

Figure 8.17 Autism Spectrum Disorder and the Microbiome Studies in mice suggest an infection in pregnancy can set off a cascade of activity. In the mother's gut, segmented filamentous bacteria stimulate T helper cells. They produce immune molecules that travel to the fetus's brain and provoke autism-like behavior. (After Willyard, 2021.)

microbiota-gut-brain axis and its role in autism. John Cryan, a neuroscientist at University College Cork in Ireland notes that microbiome-based therapies are showing enormous potential and help us view some of these disorders from a different perspective: "Unlike your genome, which you can't do much about except blame your parents and grandparents, your microbiome is potentially modifiable. And that gives great agency to patients. That's really exciting" (Willyard, 2021).

Many people who have been diagnosed with autism spectrum disorder or other neurodivergent conditions don't believe in the necessity of a "cure." Just because neurodivergent individuals interact with the world differently, doesn't mean there's anything wrong with them. However, the ability to "treat" conditions such as ASD would give people the choice and would open opportunities for individuals who otherwise might never be able to have full independence.

Parkinson's Disease

Parkinson's disease (PD) was first described in 1817 by an English surgeon, James Parkinson. He described what he referred to as "shaking palsy" that is now recognized as PD. He also noted that in one of his patients, the symptoms were accompanied by a swollen abdomen. Parkinson administered a laxative, and after the patient's bowels were emptied, his symptoms disappeared. It is now known that many people who develop PD experience intestinal issues prior to the appearance of PD-like symptoms.

PD is a neurodegenerative disorder that initially presents as tremors, stiffness, and slowness of movement. These involuntary symptoms are due to the death of neurons responsible for coordinating motion. Why the neurons die is still unknown; however, the protein α-synuclein, which is involved in signaling between neurons, appears to play a key role. In people with PD, this protein misfolds, creating a domino effect, or more misfolding, until clumps of misfolded proteins create what are known as Lewy bodies that accumulate in the brain. This area of research is in its infancy, and this explanation remains a working hypothesis and not proven fact.

Figure 8.18 The Microbiome and Parkinson's Disease Changes to the gut microbiome can result in abnormal aggregation of misfolded α-synuclein, which is then transported from the intestine into the central nervous system via the vagus nerve. The α-synuclein accumulations in the CNS result in the formation of Lewy bodies in the cytoplasm and mitochondrial dysfunction in certain neurons.

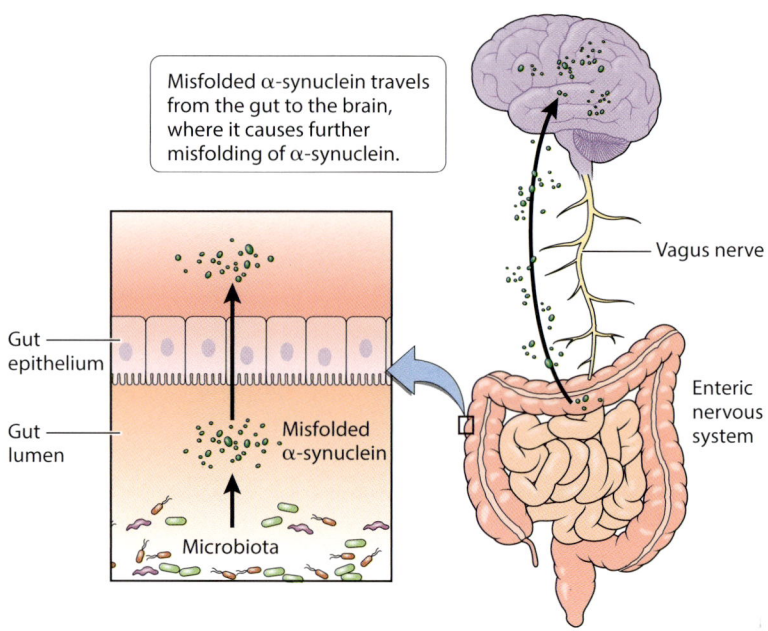

A neurologist, Robert Friedland, came up with a hypothesis to test why the misfolding occurs. He knew that certain bacteria in the gut produce proteins similar to the misshapen α-synuclein proteins, and he hypothesized that these bacterial proteins provided a misfolding template (Friedland, 2015). When rats were fed a bacterium that produces these proteins, higher levels of α-synuclein were detected in the animals' brains (Chen et al., 2016). We aren't sure how the gut bacterial protein is serving as a template and thus signaling the brain, but the vagus nerve is one strong possibility (**Figure 8.18**). In a study using mice, it was found that if misfolded α-synuclein was injected into the gut, then misfolded proteins were found in the brain. However, if the vagus nerve was removed before the protein was injected, the misfolding didn't occur in the brain. These results suggest that the vagus nerve plays a key role in the propagation of this misfolding. Curiously, the α-synuclein injected into the gut seems to remain there and initiates a misfolding domino chain in which misfolded proteins transmit the folding error up the vagus nerve until it reaches the brain, and misfolded α-synuclein is then detected in the brain (Kim et al., 2019). Protein misfolding appears to be associated with other neurodegenerative diseases, such as Alzheimer's disease.

8.5 GUT MICROBIOME-BASED THERAPIES

The interactions that occur in your gut microbiome, unlike your genes and aging, have the potential to be modified relatively easily. Several approaches to impacting the gut microbiome are being explored in this regard. There is the obvious one: simply change the composition of the gut microbiome, or introduce new microbes that result in changes to the metabolic output. One could also propose to target how the host responds to microbial metabolites (Shoubridge et al., 2022). Most prior research has focused on the first two options, employing **probiotic** (adding specific bacteria) or **prebiotic** (changing conditions to promote the growth of specific bacteria) strategies.

In 2013, Dinan and colleagues first coined the term **psychobiotic**, which refers to a living organism, such as a bacterium, that when ingested, produces a health benefit in individuals with psychiatric illness (Dinan et al., 2013). The term is now used more broadly to include other external means of influencing the brain by altering the microbiome, including probiotics, prebiotics, and fermented foods.

Probiotic Therapies

Numerous probiotic-based therapies are now receiving rigorous testing through clinical trials, the gold standard for proving the efficacy of a therapeutic treatment. Clinical trials require that researchers demonstrate, in a scientifically rigorous way, that their therapeutic interventions treat a disease or its symptoms, and that we understand any potential side effects. One recent clinical study explored whether high-dose probiotic supplements can limit the symptoms of depression. Study participants received a probiotic supplement containing *Lactobacillus helveticus* R0052 and *Bifidobacterium longum* R0175 for a period of 8 weeks (Wallace & Milev, 2021). After 4 weeks, many participants showed significant changes in their clinical symptoms and no side effects. Another pilot clinical study employed Vivomixx, a widely available brand of probiotics, which has eight different bacterial strains: *Streptococcus thermophilus, Bifidobacterium breve, B. longum, B. infantis, Lactobacillus acidophilus, L. plantarum, L. paracasei,* and *L. delbrueckii* subsp. *bulgaricus*. Study participants received either a probiotic supplement or placebo for 31 days, in addition to their usual treatments. Participants completed the Hamilton Rating Scale for Depression before and after intervention, as well as a brain-imaging and gut microbiota analysis. Compared with the placebo group, participants who received the probiotic saw improvements in their depressive symptoms. The probiotic group also experienced an increase in gut levels of *Lactobacillus*, as well as neural changes documented by imaging. It is not yet known which species were behind the improved symptoms, but future work is aimed at determining just that.

Prebiotic Therapies

Prebiotics, which include nutrients that are degraded by gut microbiota, may provide an alternative way to alter the gut microbiome for mental illness prevention and treatment strategies. The literature on this mode of treatment is still quite limited. One early clinical study focused on the impact of trans-galactooligosaccharide, which is a nondigestable carbohydrate, on the gut microbiota and, subsequently, on the symptoms of irritable bowel syndrome (IBS) (Silk et al., 2009). Application of this carbohydrate stimulated the growth of bifidobacteria and resulted in reduced symptoms in patients with IBS. A more recent study in rats and humans has revealed that prebiotic applications and the subsequent increases in levels of SCFAs produced by the gut microbiota can directly impact intestinal health, as well as stimulate the immune system, and that they have properties that are antagonistic to detrimental gut bacteria (Chidzick et al., 2021). Moreover, several studies showed a significant decrease in the symptoms of depression following prebiotic applications (Maguire & Maguire, 2019).

8.6 THE EVOLVED DEPENDENCE OF OUR MICROBIOTA

We are well along in developing a robust understanding of how the MGBA functions. However, if we think about how easily this chemical and electrical communication channel can be disrupted, we might reasonably wonder why it exists in the first place. One team applied evolutionary theory to this question and proposed several possible explanations (Johnson & Foster, 2018). The simplest of these is **evolved dependence**, the concept that host physiology, and ultimately fitness, is compromised when a microbe is absent. When a host and a symbiont, such as a member of our gut microbiota, coevolve, the host may become dependent on that microbe for certain functions.

One example of evolved dependence is provided by our reliance on certain microbes to digest foods that we are unable to digest on our own, such as resistant starches. This dependence can reach such an extreme that when the microbe is not present, and its particular physiological or neurological impact is missing, the fitness of the host is reduced. Selection originally acts upon the microbes' ability to digest a particular starch to benefit itself, and not because of any advantage it may also provide the host. However, over time, the host becomes reliant on the microbe for the benefits it inadvertently provides, in this case access to increased nutrients from plant matter. It gets even more complicated because our gut microbiota can be functionally redundant, and thus numerous microbes may be able to serve that functional role to fulfill any such evolved dependence.

Johnson and Foster predict that microbial metabolites, such as neurotransmitters or SCFAs, that impact the host physiology likely evolved first as a by-product of selection on the microbes' fitness. For example, numerous microbes produce quorum sensing molecules that allow individual bacteria within communities to coordinate and carry out community-wide functions such as sporulation, bioluminescence, virulence, conjugation, competence, and biofilm formation. The production of these molecules provides a clear benefit to the bacterium: they permit members of the species to communicate and alter their lifestyle when environments change. However, we are learning that these molecules can also impact the physiological status of the host, in both positive and negative ways. For example, quorum sensing in pathogenic bacteria activates host immune signaling and prolongs host survival, by limiting the bacterial intake of nutrients. In this way, quorum sensing allows a commensal interaction between host and pathogenic bacteria (Judger et al., 2022). Bacteria produce vast numbers and types of metabolites and other compounds whose effects on gut physiology are currently unknown. Metabolites and neurotransmitters immediately come to mind, but microbes also produce toxins that inhibit other microorganisms, as well as molecules that play a role in communication.

In the next chapter we will turn to an exploration of how our gut microbiome influences the development of yet another key component of our body, the immune system. The immune system is dependent upon a eubiotic, or normal, gut microbiome both for normal development and for the proper education of immune cells in distinguishing friend from foe. It is also similarly impacted by a dysbiotic gut microbiome, which can result in a wide array of immunological diseases.

CHECK YOUR UNDERSTANDING

1. Sensory neurons are a part of the
   a. central nervous system.
   b. peripheral nervous system.
   c. spinal column.
   d. gastrointestinal tract.

2. The somatic nervous system is also known as the
   a. voluntary nervous system.
   b. involuntary nervous system.
   c. autonomic nervous system.
   d. MGBA.

3. Effectors in the human body
   a. are small, secreted proteins.
   b. receive sensory input from the PNS.
   c. are organs that generate a response based on information from the brain.
   d. integrate sensory receptor input into the central nervous system.

4. Which of the following is a function the PNS handles?
   a. Memory
   b. Breathing
   c. Vision
   d. Speaking

5. Which of the following is an autonomic nervous system function?
   a. Doing homework
   b. Processing auditory information
   c. Digestion
   d. Walking

6. The enteric nervous system
   a. communicates with the CNS via the vagus nerve.
   b. has about 3 times as many neurons as the spinal cord.
   c. requires CNS input to conduct all of its functions.
   d. only carries out somatic functions.

7. Which of the following is not a major component of hormone regulation?
   a. Hypothalamus
   b. Adrenal gland
   c. Pituitary gland
   d. Amygdala

8. Microbiota are involved in neural development by
   a. producing SCFAs that downregulate microglia.
   b. increasing inflammation.
   c. producing tryptophan, which alters gene expression.
   d. microbial metabolites being blocked from passing through the placenta.

9. Without communication between microbial metabolites and the fetal nervous system, the _____ cannot properly develop, leading to a dysfunctional thalamus-cortex connection.
   a. internal capsule
   b. adrenal gland
   c. vagus nerve
   d. enteric nervous system

10. The thalamus-cortex connection is important for
    a. autonomic processes.
    b. memory.
    c. movement.
    d. processing sensorimotor stimuli.

11. It is important for neural development and function that some microbial metabolites can pass through the blood-brain barrier.
    a. True
    b. False

12. The brain is protected from harmful substances by
    a. the thalamus.
    b. cytokines.
    c. the blood brain barrier
    d. myelination.

13. A study found that an overgrowth of _____ was associated with a greater risk of brain damage in premature babies.
    a. certain pathogenic bacteria
    b. beneficial gut bacteria
    c. *Clostridium*
    d. *Acinetobacter*

14. Which of the following is not one of the three main ways the gut and brain communicate?
    a. Circulatory pathway
    b. Endocrine pathway
    c. Immune pathway
    d. Neural pathway

15. The vagus nerve connects enteroendocrine cells and the brain, providing signals regarding
    a. motor functions.
    b. vision.
    c. digestion.
    d. immune functions.

16. GABA
    a. is a stimulatory neurotransmitter.
    b. produces a calm feeling.
    c. can only be synthesized in the brain.
    d. performs a function similar to that of glutamate.

17. What percentage of serotonin is found in the ENS?
    a. 0%
    b. 25%
    c. 60%
    d. 95%

18. What microbial metabolite can bind to certain neural receptors and stimulate microglia?
    a. Butyrate
    b. GABA
    c. BDNF
    d. LPS

19. Which is true about studies of the gut-brain connection?
    a. Discovering an association between an alteration to the microbiome and disease is sufficient to prove causality.
    b. All mouse studies translate directly to humans.
    c. Transplanting microbiota from humans with a disease into germ-free mice is a common way to study the impact of microbiota on disease progression.
    d. Controlled human studies of causality are easy to conduct.

20. Which is a proposed link between *Mycobacterium neoaurum* and major depressive disorder in some men?
    a. *M. neoaurum* produces important SCFAs that help reduce feelings of depression.
    b. *M. neoaurum* downregulates GABA production.
    c. *M. neoaurum* produces an enzyme that degrades testosterone.
    d. There is no link between *M. neoaurum* and MDD.

21. Excessive production of proline
    a. is the result of higher levels of *Enterobacter*.
    b. is associated with mild depression.
    c. can decrease GABA production.
    d. can lead to increased feelings of well-being.

22. Which of the following is not true about ASD?
    a. Infection during pregnancy is associated with an increase in the likelihood of the child being diagnosed with ASD.
    b. A decrease in *Lactobacillus reuteri* decreased ASD-like symptoms in mice.
    c. Pregnant mice injected with double-stranded RNA (mimicking an infection) were more likely to have pups with ASD-like symptoms.
    d. T helper 17 cells in particular are likely to contribute to ASD-like behaviors.

23. Gut microbiota may produce proteins that provide a template that leads to misfolded α-synuclein and an increased risk of Parkinson's disease.
    a. True
    b. False

24. A living organism administered to confer psychiatric health benefits is called a
    a. probiotic.
    b. prebiotic.
    c. psychobiotic.
    d. medibiotic.

25. Abnormality in human behavior when normal microbiota are absent is referred to as
    a. host-microbiota integration.
    b. evolution.
    c. symbiosis.
    d. evolved dependence.

26. Microbiota may produce compounds that alter human behavior in ways that are beneficial to bacterial fitness and survival.
    a. True
    b. False

Answers: 1B, 2A, 3C, 4B, 5C, 6A, 7D, 8C, 9A, 10D, 11B, 12C, 13A, 14A, 15C, 16B, 17D, 18D, 19C, 20C, 21A, 22B, 23A, 24C, 25D, 26A

DIVING DEEPER

1. What is the name for the connection between the microbiome and the central nervous system?

2. Can you think of examples of your somatic and autonomic nervous systems in action that were not mentioned in this chapter?

3. What is the "second brain" in your gut called, and what are the two main ways it communicates with your brain?

4. How does the birthing parent's gut microbiome affect sensory processing?

5. What are three aspects of brain development that are affected by the birthing parent's microbiome?

6. What protective cells in the brain are affected by the microbiota-gut-brain axis?

7. Why might it be important for your microbiome to have so many ways to communicate with your nervous system? When would it be important for signals to travel quickly?

8. What connection between proline and depression did Mayneris-Perxachs et al. observe?

9. What is the connection observed by Lee et al. between infection during pregnancy and diagnosis of autism spectrum disorder during the child's life?
10. How might the vagus nerve be connected to the development of Parkinson's disease?
11. Why is changing your gut microbiome composition easier to accomplish than some other types of therapeutic interventions?
12. List some examples of how we are dependent on the microbes in our gut microbiome.
13. Before looking ahead to chapter 13, how do you think our gut microbiome composition compares with that of our evolutionary relatives?

DISCUSSING AND REFLECTING

1. In the past 10 years you've probably heard news about the development of CRISPR-based technology and therapeutics to alter patients' genomes. Why might clinical researchers believe microbiome-based therapies could be easier to develop and implement? What do you think these therapeutic interventions may look like in the future?
2. This and the previous chapter focused on how the microbiome of a pregnant parent can directly influence the health outcomes of their children. This is a relatively new field of study, with major clinical trials either in development or expected far in the future. How do you think this research will change how we think about prenatal care someday?
3. Reflection. Consider the last section of this chapter, which focused on how your microbiome can influence neurodivergence and neurological disorders. Would you have thought these conditions could have roots in your microbiota? Does it change your perspective on mental health to think that so much may rely on your microbiome?

RECOMMENDED READINGS

Popular Science Review

Willyard, C. (2021). How Gut Microbes Could Drive Brain Disorders. *Nature*. https://www.nature.com/articles/d41586-021-00260-3

Popular Science Book

Perlmutter, D. (Ed.). (2020). *The Microbiome and the Brain* (first edition). CRC Press.

Scientific Review

Dash, S., Syed, Y. A., & Khan, M. R. (2022). Understanding the Role of the Gut Microbiome in Brain Development and Its Association with Neurodevelopmental Psychiatric Disorders. *Frontiers in Cell and Developmental Biology*, *10*, 880544. https://doi.org/10.3389/fcell.2022.880544

The Microbiome and Immunity

9

CHAPTER CONTENTS

9.1 Components and Activation of the Immune System

9.2 The Gut Microbiome and Immune Response

9.3 The Mucosal Firewall Influences Immune Function

9.4 Immune System and Extra-Intestinal Microbiome Cross Talk

9.5 Influence of the Maternal Microbiota during Fetal Development

9.6 Influence of the Maternal and Fetal Microbiota during the Infant Years

9.7 Immune Diseases and Dysbiosis of the Gut Microbiome

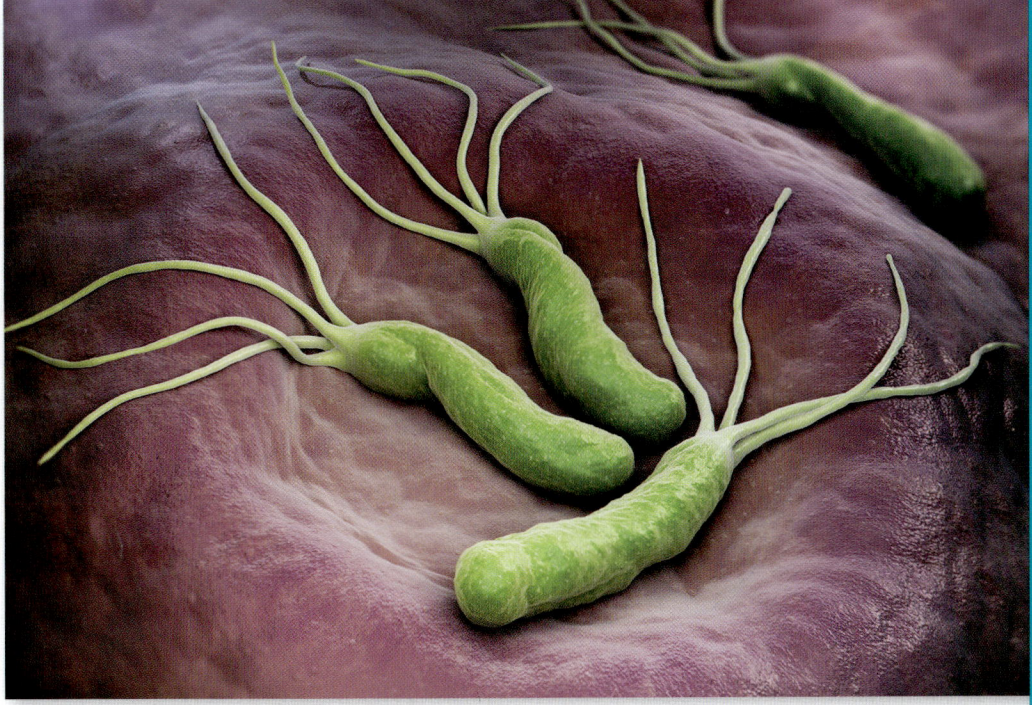

Well, hello there. I am *Helicobacter pylori*, a proud member of your stomach microbiome—those few of us who can withstand a continuous acid bath. And I am nasty. I mean truly nasty. I cause serious disease. Not your minor acne breakout or a cold sore on the lip kinda thing. No, I mean *real* disease, we're talkin' stomach cancer, ulcers, and the like. I am so good at my job that you all didn't even know you had me living in your tummy—until a mere 20 years ago. Talk about successful subterfuge. Ha-ha. And just when you decided you should try to get rid of me with those nasty chemical antibiotics you throw at us, you learn that I am also key to keeping other types of cancer from occurring. We have a sorta "can't live with me and can't live without me" situation. So, deal with it, meaning me, and get on with your reading. (Photo from iStock.com/iLexx)

"Simply put, our immune system is mostly in our gut."
—Anonymous

T he opening quote relates to the fact that most of our immune cells are in our large intestine and their functions are informed by the resident gut microbiota. Based upon this education, immune cells identify microbes as either friend or foe. We learned a bit about the immune system in several prior chapters. Here we will fill in the details as we dive a bit deeper into the components of our adult immune system and then explore how our gut microbiome and immune system communicate. We will then turn to a pregnant woman's gut microbiome and learn how it impacts the maturation of the fetal immune system, which is the first phase of the neonatal "window of opportunity," where lifelong host-microbial interactions are defined and immune homeostasis is established. Finally, we will explore how dysbiosis of the gut microbiome during these two critical phases of immune development can wreak havoc on our immune system and potentially trigger a variety of immune-associated diseases.

Figure 9.1 The Immune System Our immune system is a complex network of organs, bone marrow, and white blood cells.

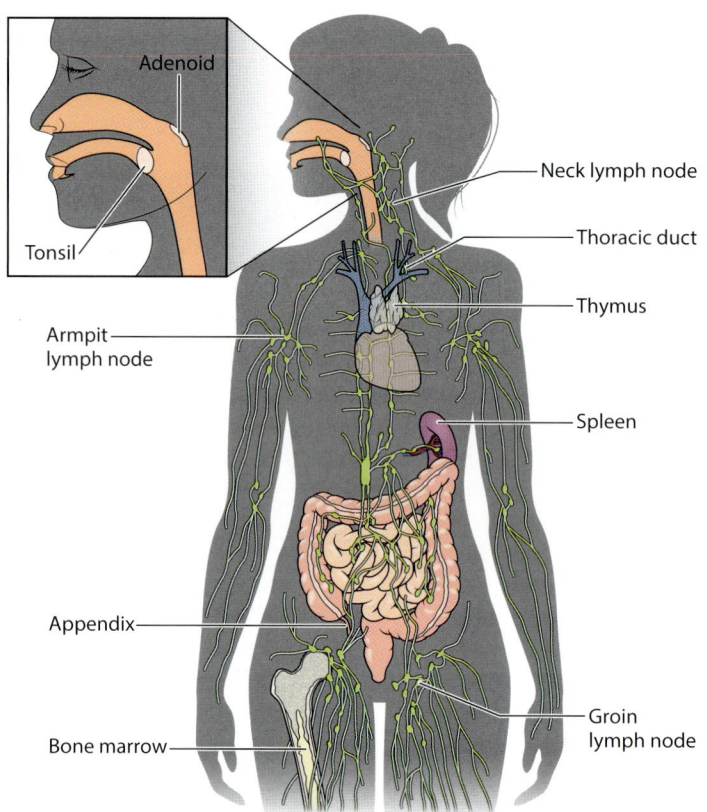

9.1 COMPONENTS AND ACTIVATION OF THE IMMUNE SYSTEM

Our immune system is one of our body's primary defenses against infectious agents and cancerous cells. It is made up of a complex network of organs, such as the spleen, adenoids, and tonsils, as well as our lymph system, bone marrow, white blood cells, and proteins, all of which serve to protect us from pathogens, cancerous cell changes, and harmful substances. When your immune system is working properly, you aren't conscious of its presence. But when things go wrong, perhaps because it doesn't recognize something as harmful, you won't be able to ignore its presence, because you will feel ill. Imagine life without an immune system—we would die younger and have far more bouts of infectious disease, and cancers would spread more rapidly.

Primary Components of the Immune System

Figure 9.1 provides an illustration of the immune system's primary components. All key immune cells are first formed in the **bone marrow**, including those that end up in the blood (**myeloid** cells) and those that end up in the lymph nodes (**lymphatic** cells). Like the circulatory system that channels blood throughout the body, the lymphatic system is connected by numerous thin tubes, and these link the lymph nodes. It plays a key role in our immune response as it fights against the invasion of infectious agents, such as bacteria and viruses. It is also responsible for maintaining fluid levels in the body, removing waste, and transporting immune cells. Perhaps you have had a parent or physician check to see whether your lymph nodes were swollen, which is potentially a sign of infection.

   Once immune cells are generated in the bone marrow, they can remain there and mature or they can be transported to the thymus, a small gland located under your breastbone. Where they mature determines which type of immune cell they will become. If they remain in the bone marrow, they become **B lymphocytes** (**B cells**). If they are transported to the thymus, they become **T lymphocytes** (**T cells**), also called thymus-dependent cells. Both B and T cells help recognize disease and infection, but they do so in different ways, which we will explore later in this chapter.

Activating the Immune System

The immune system is activated when it encounters an **antigen**, which is anything the immune system doesn't recognize as "self." Antigens are often foreign entities, such as bacterial cells, viral particles, pollen, or even snake venom. They can also be produced by our bodies, in which case they are called **autoantigens**. Immune cells have highly specialized receptors that identify specific antigens. When antigens attach to immune cell receptors, they trigger a series of responses. The first encounter of an antigen provides a memory, such that if the immune system sees that antigen again, it can more rapidly mount an immune response in the future. If the immune system detects an autoantigen, it may mistake it for a foreign antigen and mount an attack. This is known as an **autoimmune response**, in which your immune system targets your own body's cells.

Let's review what we learned about the immune system in chapter 7, where we distinguished between innate and adaptive immunity. These two immune components work together whenever an antigen triggers an immune response. The **innate immune system** can be described as nonspecific, meaning that it defends our body against anything that is not self. Due to this lack of specificity, the innate immune system can detect a wide range of foreign objects and respond to their presence rapidly, within minutes to hours. Our skin is part of the innate immune system, and it serves as a physical barrier against foreign objects and life forms. Innate immune cells include a variety of **white blood cells**, such as **neutrophils** and **eosinophils** (**Box 9.1**), as well as **natural killer cells** and phagocytes, which literally eat the targeted items. We also learned that newborns rely primarily on this form of immunity, since their bodies have not yet learned to identify specific invaders.

In contrast, the **adaptive immune system** is highly specific. It makes antibodies, which target foreign antigens that the body has previously experienced. This is called a learned immune response. The adaptive immune system screens everything it encounters. When it recognizes an antigen, it then directs special immune cells to attack. Once an adaptive immunity cell contacts an antigen, it will remember it, and future interactions will result in more-rapid responses. The adaptive immune system is composed of two types of cells: T cells and B cells (see Figure 9.2B).

Humoral immune responses target pathogens found outside of our cells, such as bacteria and fungi. Helper T cells recognize antigens, for example proteins found on a bacterial or fungal cell wall, and then instruct B cells to create antibodies that are specific to those proteins. Antibodies will recognize their specific antigens and mark them as non-self. The marked foreign substances are then destroyed by other immune cells. **Cell-mediated immunity**, in contrast, targets pathogens such as viruses that are located inside host cells. Infected cells send signals, and certain T cells recognize these signals and then trigger the infected cells to self-destruct. The main difference between T and B lymphocytes is that T cells are trained to recognize an infected host cell, while B cells recognize the foreign object itself.

The **innate lymphoid cells** (**ILCs**) represent a newly discovered player on the innate immunity team. They are somewhat like T cells, in that they secrete cytokines and regulate the functions of other innate and adaptive immune cells. However, they do so in a fundamentally different way. ILCs do not possess the antigen-specific receptors that are present on T cells. They do have receptors that detect cytokines released from tissue damage and various microbial products. Thus, these immune cells appear to monitor various signals that suggest damage or infections, and they release cytokines in response. ILCs are most abundant at mucosal barriers, and it is here that they are exposed to allergens and potential pathogens. There are three types of ILCs: ILC1–ILC3, each of which releases a particular set of cytokines when triggered.

BOX 9.1. COMPONENTS OF THE INNATE AND ADAPTIVE IMMUNE SYSTEMS

Hematopoietic stem cells in the bone marrow give rise to red blood cells and platelets, as well as the myeloid innate immune cells (Figure 9.2A), including the following:

Neutrophils. When an injury occurs or a foreign agent invades, signaling molecules called cytokines are secreted that instruct certain cells to express adhesion molecules. Neutrophils then recognize and adhere to the foreign agent, ultimately destroying the pathogen through phagocytosis.

Eosinophils. These cells respond to the presence of parasites and mediate allergic reactions.

Basophils and Mast Cells. These cells produce cytokines and histamine in response to immunoglobulin E (IgE) antibody activities, resulting in an inflammatory response.

Monocytes. These cells travel to foreign cells and tissues and transform into macrophages or dendritic cells. Macrophages surround and kill microbes, ingest foreign material, and boost immune responses. Dendritic cells initiate an adaptive immune response.

Adaptive immune cells are also derived from hematopoietic stem cells, which produce lymphoid progenitor cells (Figure 9.2B). These cells then produce T and B lymphocytes and natural killer cells, which are involved in the adaptive or antigen-specific immune response. These cells are the only immune cells able to recognize and respond to specific antigens:

T Cells. These cells have receptors that recognize specific antigens. They communicate with other cells to coordinate an immune response. There are three types of T cells: helper, regulatory, and cytotoxic. We will learn more about T cells later in this chapter.

B Cells. B lymphocytes are transformed into plasma cells when activated. Plasma cells then synthesize antibodies to specific antigens.

Natural Killer (NK) Cells. These cells attack other cells that have markers identifying them as defective. All human cells are marked as "self" with major histocompatibility complex (MHC) antigens that identify us as unique individuals. If the MHC antigens are either missing or altered, perhaps due to transformation into cancerous cells or viral infection, NK cells recognize them as foreign and destroy them.

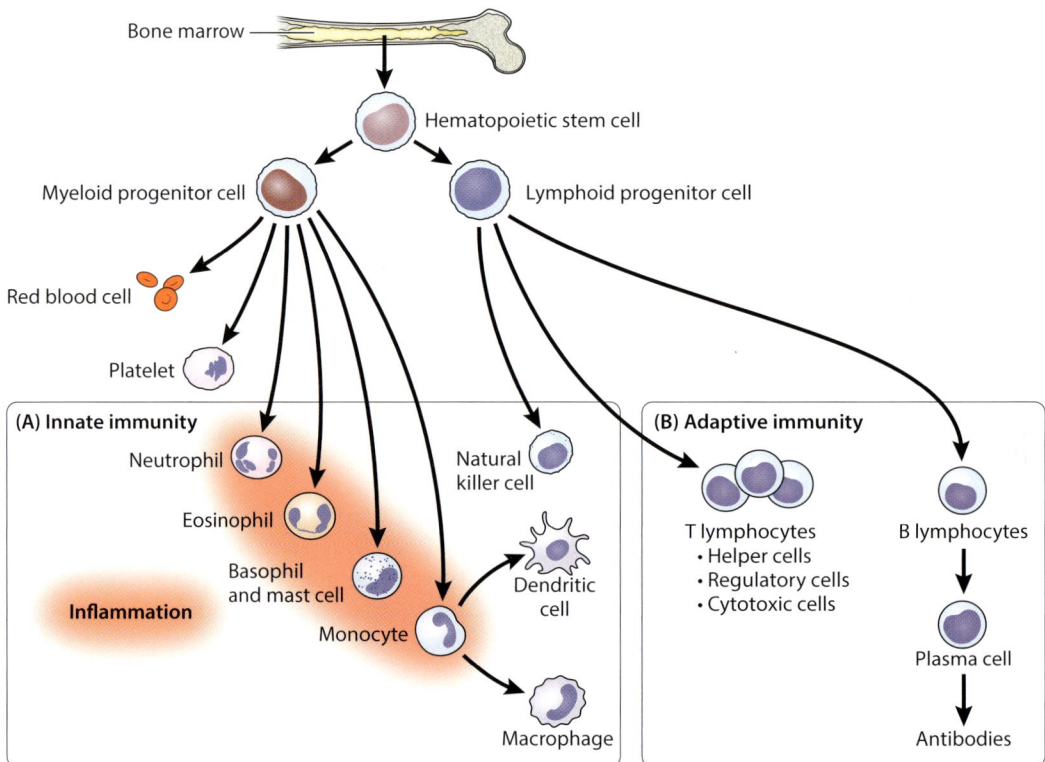

Figure 9.2 Developmental Origins of Immune Cells Our diverse immune cells are derived from a single ancestral source—hematopoietic stem cells, which give rise to both the innate and adaptive immune systems.

9.2 THE GUT MICROBIOME AND IMMUNE RESPONSE

The highly folded epithelial lining of your intestinal tract is an internal extension of your skin, providing the body's largest and most important interface with the external world. The skin, both external and internal in the GI tract, serves as the first line of defense against invasion by pathogens or other harmful substances. In the intestinal tract, this physical barrier consists of the epithelial cells themselves and the layer of mucin that lines the internal intestinal wall. This internal epithelial layer also hosts an enormous number of immune cells. In fact, over 70% of your immune cells are located within the epithelium lining of your large and small intestines.

As we've discussed previously, the immune system faces a massive challenge of consistently and accurately distinguishing between the body's own cells and invading pathogens. However, the gut microbiota is part of an interesting third category, as it is both "self" and "non-self." These microbes are decorated with cell surface proteins and they produce metabolites, all of which incite the host immune system to respond. The immune system must learn to tolerate members of the gut microbiota while still regulating them, while still being able to recognize and kill any pathogens that appear. However, the gut microbiota doesn't just sit passively in the gut lumen and hope the host immune system doesn't attack. Gut microbes produce numerous metabolites, some of which have bioactive functions that interface with host immunity. A prime example, and one that we are already quite familiar with from chapter 7, is the synthesis of short-chain fatty acids (SCFAs), which are the products of microbial fiber fermentation. Numerous gut commensals produce SCFAs that can reach high concentrations locally. These metabolites serve as an energy source for intestinal epithelial cells, stimulate mucin production, trigger inflammatory responses, and influence the differentiation and function of various immune cells.

9.3 THE MUCOSAL FIREWALL INFLUENCES IMMUNE FUNCTION

The intestinal immune system is programmed to tolerate a diverse and constantly changing commensal gut microbiome while at the same time responding to infection by pathogens. In a healthy state, a host's intestinal immune response is concentrated at the mucosal surface (**Figure 9.3**). As we have mentioned previously, a dense mucus layer separates the intestinal epithelial cells from microbes present in the gut lumen. This mucosal firewall not only limits interactions between microbes in the gut and these cells lining the gut, but also directs dendritic cells towards an anti-inflammatory state. Finally, the mucus is rich with antimicrobial peptides produced by the underlying epithelial cells.

If cells make it past these defenses, dendritic cells (DCs) screen for pathogen presence by recognizing antigens. Once an antigen is identified, DCs then present this information to components of the adaptive immune system, such as T cells, and this triggers an immune cascade of events resulting in inflammation, constricted blood vessels, increased production of mucin, and several other allergic responses. To ensure that DCs do not go overboard and signal pathogen presence unnecessarily, they are governed by numerous factors, including cytokines, growth factors, and transcriptional programming, that suppress DC responses.

9.4 IMMUNE SYSTEM AND EXTRA-INTESTINAL MICROBIOME CROSS TALK

The immune responses that occur at the interface between the intestinal microbes and the layer of mucus lining the gut epithelial cells have been the focus of most research on how the microbiome and immune system interact. However, it is becoming clear that there are numerous interactions that occur outside of the gut that mediate organ-specific interactions. We will explore several of these types of interactions, including those that occur on the skin and in the lungs and liver.

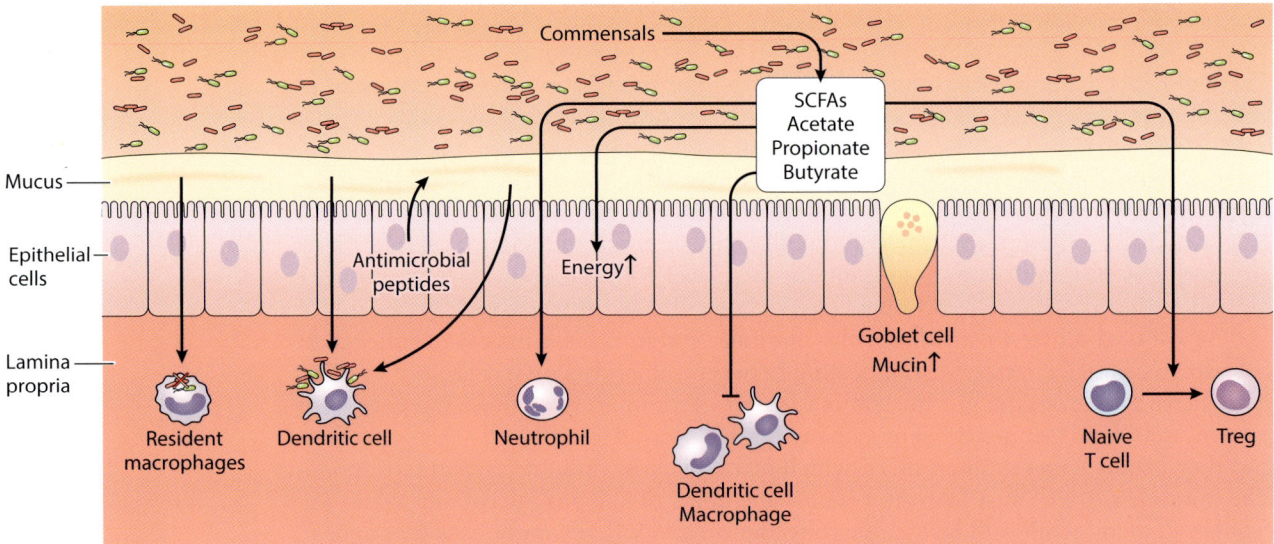

Figure 9.3 **The Mucosal Firewall** The intestine provides a series of barriers to limit the ability of microbes to cross the epithelial barrier: (1) Mucus is the first line of defense, and it is produced by epithelial cells, which also produce (2) antimicrobial peptides. Some microbes may evade these defenses and cross the epithelial layer, but they will be eliminated by (3) macrophages or captured by (4) dendritic cells. The mucus layer, the antimicrobial peptides of the epithelial cells, the macrophages, and the DCs and T cells comprise the "mucosal firewall." (After Belkaid and Hand, 2014 and Blacher et al., 2017.)

Skin

As we learned in chapter 3, the skin harbors its own complement of commensal microorganisms, which provide both a protective role and one involving immune education, or the process by which commensal microbes train the immune system to recognize them as friends (**Figure 9.4**). Colonization of the skin with commensal microbes at birth provides a robust mechanism for colonization resistance, which limits the ability of potential pathogens to take up residence on the skin. These commensals also trigger the rapid accumulation of regulatory T cells as well as the production of a diverse array of antimicrobial proteins (AMPs) by epithelial cells, which aid in wound healing. One common skin commensal, *Staphylococcus epidermidis*, has evolved a mechanism to employ a human-derived AMP to eliminate one of its closest competitors from the skin (Pastar et al., 2020). It does so by stimulating production of the perforin-2 AMP, which increases the ability of skin cells to kill intracellular *S. aureus*. *S. epidermidis* benefits from the lack of one of its closest competitors, while the host benefits from the elimination of a pathogenic microbe.

Dysbiosis of the skin microbiome is associated with several inflammatory skin disorders, such as atopic dermatitis (or eczema) and psoriasis, in which skin cells build up and create itchy dry patches. It is not yet clear whether these disorders are the cause or consequence of dysbiosis of the skin microbiome. However, there are suggestions that both genetics and impaired skin barriers may contribute to these diseases. For example, colonization of the skin with the bacterial pathogen *S. aureus* promotes an allergic response on the skin of mice through mast cell activation (Nakamura et al., 2013).

Lung

The lungs, which were long considered to be sterile, are now known to host their own specific, low-density, resident pulmonary microbiota. These microbes are tolerated by the host immune system and serve to prevent colonization from lung pathogens, primarily through the production of various antimicrobial molecules. The lung tissue and the pulmonary microbiome communicate through the production of a variety of

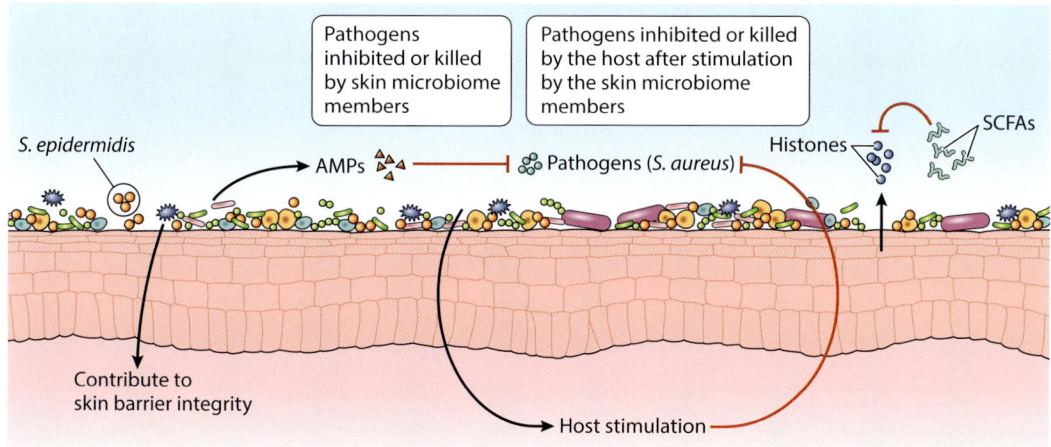

Figure 9.4 Microbiome Metabolite-Mediated Interactions in the Skin A variety of chemically diverse metabolites are produced by the skin microbiota. Benefits such as pathogen inhibition, immune education, and homeostasis are produced by this diversity of metabolites, yet an inflammatory response can develop under the right circumstances. (After Canchy et al., 2023.)

proteins and metabolites. Commensal microbes influence the immune system by producing numerous metabolites, including short-chain fatty acids and lipopolysaccharides. The immune system simultaneously influences the lung microbiome by secreting immunoglobulin A (IgA), producing antimicrobial peptides, and recognizing the commensal members as "self." IgA is the most abundant antibody in the body; it serves to protect the mucosal tissues from microbial invasion and maintain immune homeostasis with the microbiota. AMPs, which were discussed in relation to the skin microbiome above, are also protective against microbial invasion. When there is pulmonary microbiome dysbiosis, immune homeostasis can be impacted, resulting in the potential development of respiratory inflammation and subsequent disease.

Scientists from University Medical Center Göttingen demonstrated a direct link between the pulmonary microbiome and CNS autoimmune disease in rats (**Figure 9.5**) (Hosang et al., 2022). Some of the main immune players in this disease are microglial cells—members of the innate immune system that you can think of as trash collectors, clearing damaged or dead cells from brain tissue. The researchers hypothesized that the pulmonary microbiome may impact the activity of these microglial cells and sought to test this by administering an antibiotic (neomycin) to one set of mice while leaving a control set untreated. They found that when the pulmonary microbiome was altered by antibiotic treatment, the severity of the autoimmune disease changed. More specifically, the application of neomycin resulted in increased levels of *Bacteroides* in the pulmonary microbiome. *Bacteroides* coat their outer membrane with lipopolysaccharides (LPS). As the levels of LPS in the lungs increased, the level of microglial activity in the brain decreased, resulting in an alleviation of the CNS autoimmunity. As team leader Flügel explains, "there seems to be a continuous relation between lung bacteria and brain microglia. The microglia cells continuously sense these signals coming from the microbiome of the lung" (Robitzki, 2022).

Liver

The liver is constantly exposed to products derived from the gut microbiota, including metabolites and toxins. The gut and liver engage in constant communication through the portal vein, as well as the bile ducts and systemic circulation. This cross talk is called the **gut-liver axis**. The liver is home to a wealth of innate immune cells, of which macrophages are the dominant members. These cells serve as the primary line of defense against invading microorganisms by either phagocytosing pathogens themselves or by serving as **antigen-presenting cells** (**APCs**). APCs recognize LPS and other proteins produced by bacterial pathogens and present these toxins to T cells,

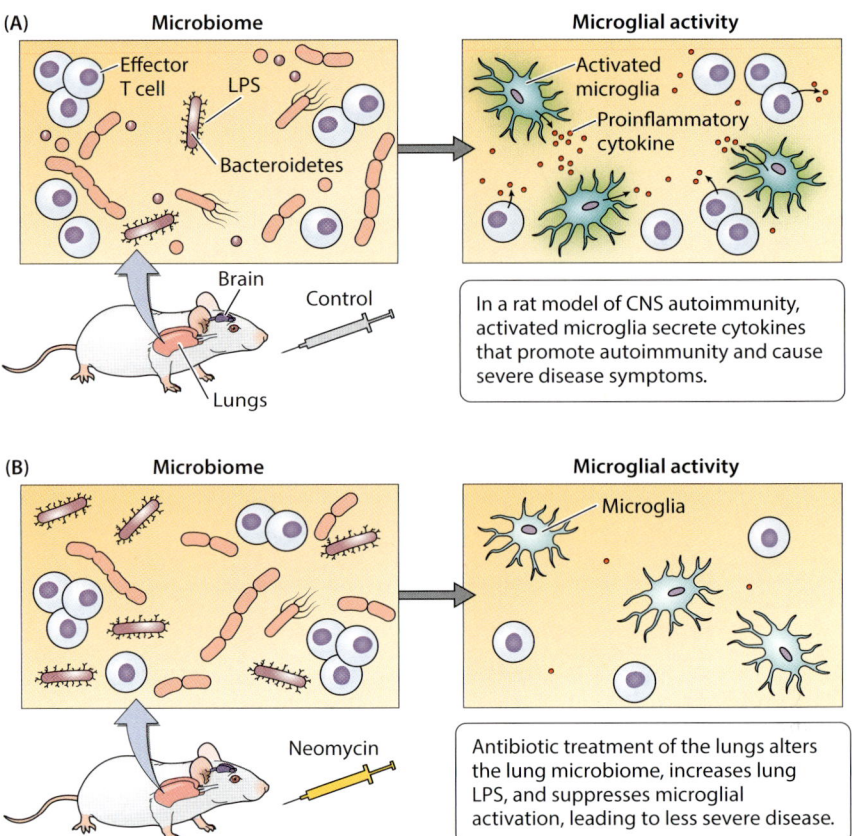

Figure 9.5 Lung Bacteria Modulate Autoimmunity in the CNS (A) In this in vivo model of CNS autoimmunity, researchers have shown how the rat pulmonary microbiome can incite autoimmunity. The rat lung hosts a diverse pulmonary microbiome, including members of the phylum Bacteroidetes which pepper their outer surfaces with lipopolysaccharides (LPS). The LPS incite T cells that then travel to the central nervous system and release cytokines, which results in inflammation and may lead to CNS autoimmune disease. (B) When rats are treated with an antibiotic, their pulmonary microbiome is altered, resulting in even higher levels of LPS. These rats show significant impacts in the brain, including fewer T cells and thus lower cytokine levels. Most important, these antibiotic-treated animals are healthier than the control rats. (After Schonhoff and Mazmanian, 2022.)

which then mount an immune response. The resulting increased inflammation in the liver can result in cirrhosis, or scarring, of liver tissue and may lead to liver failure.

9.5 INFLUENCE OF THE MATERNAL MICROBIOTA DURING FETAL DEVELOPMENT

A healthy gut microbiome is required to develop a resilient and well-tuned immune system. Beginning at conception and continuing for the next 1,000 days, babies develop and maintain a balance between the members of the intestinal microbiome and the components of the immune system, a homeostasis that is essential for lifelong health. Once a baby is born, the gut is populated with microbes that play a critical role in training the immune system to recognize the good guys. This crucial task, which was initiated during pregnancy with input from the maternal gut microbiome, continues as the neonate's microbiome matures, further exposing the immune system to new friends and foes. The newborn, particularly in the first few days and weeks after birth, is more susceptible to infections, as their immune system must rely on the generalized and relatively slow responses of the innate immune system provided by their mother through the placenta. Over time, the baby's adaptive immune system learns to defend against invaders and to tolerate harmless friends, such as most members of the gut microbiome.

Immune Cells in Fetal Development

We have known for some time that maternal antibodies cross the placenta to protect the fetus from infection. We also suspected these antibodies could influence immune system development. We also knew that pathogenic microbes triggered the development of antibodies in mom that were passed on to the fetus through the umbilical cord. It is now clear that maternal microbial metabolites are transferred and that these products initiate the process of exposing the fetus to the microbial world well before birth! Thus, both bacteria-produced molecules and maternally derived antibodies play a significant role in driving immune development in utero.

By week 4 of pregnancy, the fetus is the size of a poppy seed. But even in this diminutive state, its immune system is already developing. All immune cells come from a semispecialized type of stem cells called **hematopoietic stem cells** (see Figure 9.2). Hematopoietic stem cells, also called blood stem cells, can develop into all types of myeloid cells, including platelets and red blood cells, as well as lymphoid cells, including T and B lymphocytes. You can think of these stem cells as college students who are undeclared STEM majors—they can become any type of scientist or engineer, but anyone can become an author and anyone can become a business executive, regardless of college education. Hematopoietic stem cells and their immune cell descendants, myeloid and lymphoid cells, ultimately give rise to our innate and adaptive immune systems, respectively. Hematopoietic cells are delivered by the circulatory system to the liver, spleen, and thymus. By week 11, the fetus has started to produce macrophages that travel to the intestine and rapidly increase in number through the 5th week of pregnancy. By week 13, B and T cells appear in the intestine, and by week 19 they can be found in Peyer's patches, which are specialized lymph nodes in the intestine.

Although immune cells are now present in the fetus, there are no pathogens for them to attack. The fetus is protected in the sterile amniotic sac, and its only exposure to microbes is through microbial factors, such as metabolites, supplied by mom through the placenta. Maternal antibodies, primarily **immunoglobulin G** (**IgG**), start to cross the placenta and impact fetal development by the 13th week of gestation. However, most cross much later in pregnancy, so there will be plenty available to fight any infections that might occur at birth. Because most antibodies arrive later in a pregnancy, babies born prematurely, which lack many of these late-arriving antibodies, are more susceptible to infections. Traditionally, it was believed that the maternal antibodies were designed to recognize pathogens in the fetus or newborn. However, it is now clear that commensal bacteria are also targeted, and this ensures these commensals don't cross the newborn's gut epithelial cells during the first wave of microbiome colonization that occurs during birth.

Distinguishing Self from Non-Self

Babies are not genetically identical to their mothers, which raises the questions: Why doesn't a pregnant person's immune system attack this "non-self" entity, the fetus, and why doesn't the developing fetal immune system attack the cells being transferred from the mother? Simply put, we don't yet know. However, there are clues that may help us to understand what is happening. First, the fetal immune system is highly suppressed. Second, by week 13 of gestation a fetus does make dendritic cells (DCs), which are responsible for initiating all adaptive immune responses. However, they respond differently to proteins delivered through the umbilical cord. Rather than mark mom's proteins as foreign, the DCs are more likely to activate regulatory T cells, whose role is to suppress, rather than initiate, an immune response. In other words, after birth, foreign particles will be destroyed by the immune system, but before birth, in the relative safety of the womb, the fetal immune system learns which "foreign" materials are safe and should not be destroyed, such as maternal compounds. Thus, the fetal immune system is not simply an immature version of an adult immune system, but rather one that has its own distinct functions, primarily focused on learning what to tolerate versus what to attack.

Maternal Inflammation and Infection and the Fetus

Another way in which the maternal immune system shapes fetal immune development is through inflammation and infection of the mother. Although precisely how a maternal infection will affect the fetus is not yet known, it is predicted to lead to hypersensitivity disorders as well as increased resistance to certain infections as newborns. This second mode of immune transfer, one that is driven by inflammation in the mom, turns out to have long-term impacts on the child's immune function.

When a pregnant woman eats a high-fiber diet, increased amounts of short-chain fatty acids are produced by her gut microbiome and transferred to the developing fetus. These compounds play a role in fetal immune system maturation by triggering the development of regulatory T cells (Tregs), which help reduce inflammation. Tregs not only protect us from runaway inflammation, but also train our immune system to tolerate food and commensal bacteria. The fact that Tregs are long-lived means that they, or their progeny, will be present through life and thus will have long-term consequences for the offspring's health. A study with mice revealed that if a pregnant mouse is fed a high-fiber diet, she will have increased levels of SCFAs in her feces, and her pups will have higher levels of SCFAs and more Tregs (Nakajima et al., 2017). The authors suggest that the maternal SCFAs are passed through the umbilical cord and influence fetal T cell development, which would account for the higher levels of Tregs observed in the pups.

9.6 INFLUENCE OF THE MATERNAL AND FETAL MICROBIOTA DURING THE INFANT YEARS

Just after birth, the newborn's immune system faces several challenges. As noted above, the fetal immune system is suppressed, and it must be activated at birth. Further, the fetal immune system has never been directly exposed to pathogens and is thus "antigenically inexperienced." It needs to be trained to respond, and while that is happening, the maternal immune system is there to help keep pathogens at bay.

We learned in chapter 7 that the newborn's microbiome is rapidly colonized by an enormous diversity of microbes, with sources ranging from the mother's vagina, gut, and skin to the hospital, family members, and pets. The gut and the immune system then develop simultaneously to support one another to promote health. One well-known example of this coordination is seen in the interaction between the gut microbiome and T cells. When the gut harbors a diversity of microbes, T cells are trained to tolerate harmless bacteria while destroying potential pathogens. When the microbiome and the immune system are working in concert, the body responds to pathogens and tolerates the commensal bacteria. When the gut microbiome diversity is depleted, perhaps through the application of antibiotics or malnutrition, this reduced diversity can quickly result in a reduction of immune response and disease. But how does this synergy between the immune system and gut microbiota happen?

The composition of the gut microbiome is at its most variable immediately following birth and through the first 1,000 days of life. During this period a newborn's microbiome is constantly fluctuating until a more stable composition is reached by about 3 years. The critical importance of these early phases of microbial colonization was shown through studies on germ-free mice, which demonstrated that an absence of commensal microbes is associated with significant defects in immune function (Zheng et al., 2020). Germ free mice have low levels of IgA, but if microbes are introduced into their intestines, those levels are rapidly normalized. The production of IgA is under control of plasma cells. These cells are delivered to the intestine through the blood, where they cross into the intestinal lumen and bind to bacterial antigens and prevent infection. The formation of plasma cells from their B cell precursors is also induced when B cells sense microbe-derived antigens. Without microbial exposure, B cells are not induced to produce IgA.

Breast Milk

Breast milk, in addition to serving as an ideal nutritional source for infants, helps defend against infections. Immediately following birth, when the baby is most vulnerable to infection, breast milk provides a starter kit of sorts to give the fetal immune system what it needs to begin development. It contains antibodies, macrophages, and other immune-related factors, such as **cytokines**, which are signals that initiate an immune response. The first few days of breastfeeding may be the most important, as the **colostrum**, which is the very first milk produced after birth, is packed full of macrophages, over 3 million per milliliter of milk. As the infant's own immune system becomes more active, the components of breast milk change. It begins to serve primarily a nutritional, rather than immune, role, and the levels of mom's immune cells plummet just as the infant's own immune system kicks into gear.

Immunoglobulin A

The immune cells that cross the placenta during gestation are IgG. After birth, the IgA provided via breast milk becomes the primary immune contribution from mom. IgA protects against microbial pathogens, just as IgG does. However, IgA has another key role, which is to protect mucosal surfaces, such as those found in the mouth, intestine, and respiratory tract. One example of this role is found in the large intestine. Maternal antibodies supplied in breast milk coat the newborn's intestinal epithelium and protect against viruses. This type of protection, through antibodies provided by the mother, is called **passive immunity**, which is a short-term solution to protecting against infection as the newborn continues to develop their own immune responses.

9.7 IMMUNE DISEASES AND DYSBIOSIS OF THE GUT MICROBIOME

If something goes wrong during the neonatal window of opportunity, or if the microbiome is disrupted later, this can lead to dysbiosis. This is now known to influence numerous inflammatory conditions, including type 1 diabetes, which is a chronic disease impacting the production of insulin by cells of the pancreas; celiac disease, which results from an immune response to eating gluten; and asthma. One suggestion to account for increases in these and other autoimmune diseases is the **hygiene hypothesis** (Figure 9.6). This hypothesis argues that a series of factors involved in Westernization, including the extensive use of antibiotics and improved hygiene, have resulted in an absence of some of the key bacteria in our gut microbiome that we historically have been exposed to. These missing microbes are sometimes referred to as our "old friends." Further, this absence is proposed to result in a dysregulation of our immune responses. In other words, if we no longer encounter the relatively mild infections experienced by our ancestors, our immune system does not get trained properly to recognize friend versus foe and thus more often attacks healthy cells or commensal microbes. Although initially focused on the absence of pathogens, the hypothesis now embraces the importance of a lack of beneficial gut commensal flora as well.

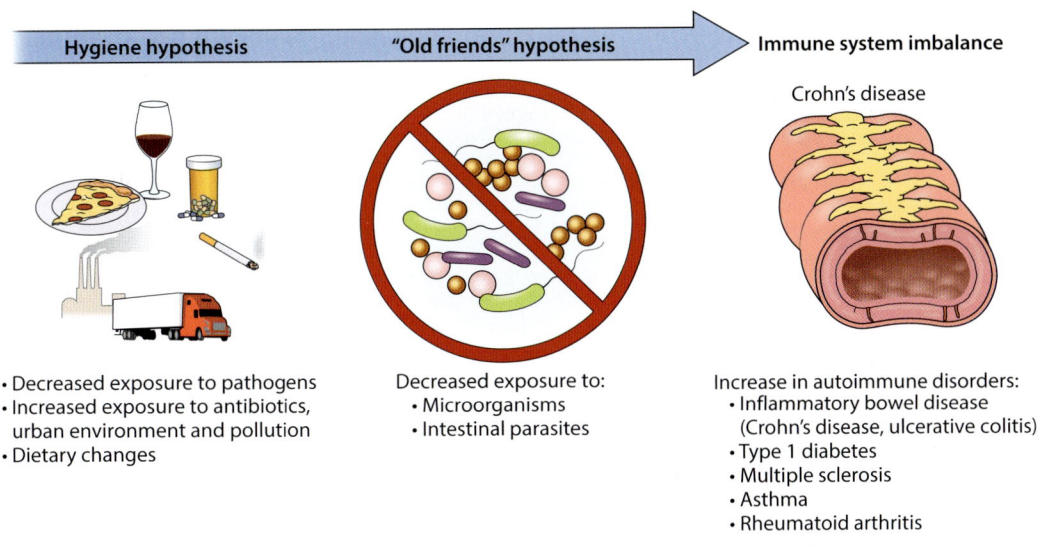

Figure 9.6 The Hygiene Hypothesis, Old Friends Hypothesis, and Autoimmune Diseases

When the gut microbiome is in homeostasis, it produces key metabolites, such as SCFAs and polysaccharides, that play a role in the regulation of immune responses. Dysbiosis, which results in the production of a different population of metabolites and toxins, can lead to changes in gut epithelial integrity and immune tolerances (see Figure 9.6). A Westernized lifestyle has led to this dysbiosis on a grand scale, resulting in skyrocketing levels of autoimmune disorders. The most extensively studied associations between gut dysbiosis and immune-associated diseases include inflammatory bowel disease (IBD), which will be discussed in chapter 10; systemic autoimmune diseases, such as rheumatoid arthritis; cardiometabolic diseases; and cancer.

Rheumatoid Arthritis

Rheumatoid arthritis (RA) is a **systemic** (not localized) autoimmune disease that results in inflammation in the joints. It is estimated that 14 million people worldwide have RA, and although it is more common in older individuals, it can develop at any age. The gut microbiota of individuals with RA exhibit decreasing microbial diversity as the disease progresses, with a simultaneous expansion of formerly rare lineages like the bacterial genera *Eggerthella* and *Collinsella*. The presence of *Collinsella* in a mouse model of RA results in increased levels of several metabolites, including beta-alanine, alpha-aminoadipic acid, and asparagine, and increased gut permeability. It also reduces the levels of tight junction proteins, which help form the barrier that separates epithelial cells from other tissue spaces (Chen et al., 2016). An inflammatory signal is triggered, which can signal immune responses elsewhere, such as the joints, causing cartilage and bone damage (**Figure 9.7**). The presence of *Collinsella*

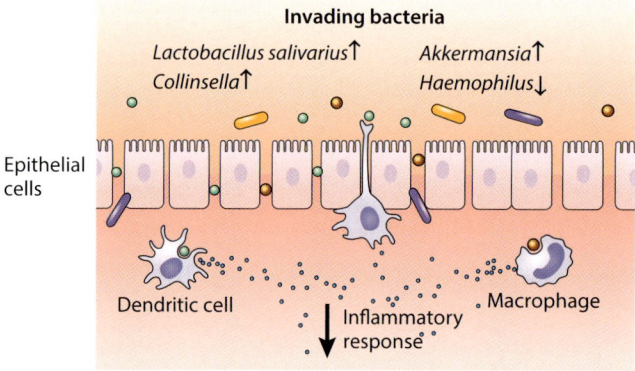
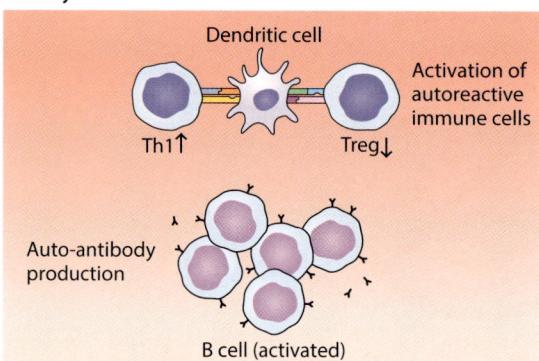
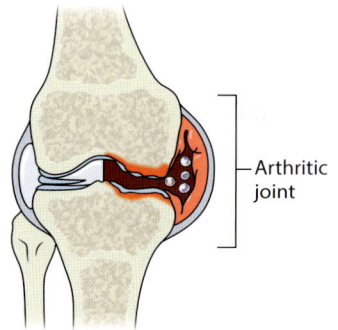

Figure 9.7 The Gut Microbiota and Rheumatoid Arthritis (A) Changes in the gut microbiota may trigger an inflammatory response, which can signal immune responses elsewhere, such as the joints. (B) These signals result in the activation of key immune cells, resulting in a shift in the ratio of regulatory T cells (Tregs) and helper T cells (Th1) and activation of B cells. These autoreactive cells migrate from the lymph to the joints, causing cartilage and bone damage.

also induces production of a proinflammatory cytokine, interleukin-17 (IL-17), which is produced by T cells and mediates innate immunity to pathogens. When in excess, it promotes inflammation and contributes to the development of inflammatory diseases, such as rheumatoid arthritis.

Another group of researchers explored the use of probiotics in the treatment of mice with RA. The administration of *Prevotella histicola* led to decrease in the frequency and severity of RA symptoms in the mice (Marietta et al., 2017). It appears that *P. histicola* helped to lower gut permeability and increase the expression of tight junction proteins. Essentially, *P. histicola* allowed the mouse GI system to properly seal the gut, preventing harmful compounds and microbes from leaking into the bloodstream, while at the same time producing fewer of the side effects that accompany traditional medications used to treat RA.

Cardiometabolic Disease

Cardiovascular and metabolic diseases, also known as **cardiometabolic diseases** (**CMDs**), are the top cause of death globally. This heterogeneous group of conditions includes type 2 diabetes, obesity, atherosclerosis, chronic heart disease, and kidney failure. CMDs primarily result from unhealthy lifestyles that include tobacco smoking, poor diet, and a lack of physical activity. However, genetics, age, sex, and environmental exposures also play a major role in CMDs. Given that these same factors influence the composition of the gut microbiome, it is not surprising that the microbiome appears to play a role in CMD progression.

CORONARY HEART DISEASE Researchers sought to establish signatures within the gut microbiota and metabolome that correspond to the onset and progression of one class of CMD, coronary heart disease (CHD) (Zhong et al., 2022). Participants were grouped into cohorts based upon disease severity and their current treatment status. There were (1) healthy controls (HC), (2) individuals with obesity or diabetes that were not diagnosed with CHD (i.e., metabolically matched controls, MMC), and (3) individuals diagnosed with coronary heart disease (CHD) (**Figure 9.8A**). Microbiome data, including levels of gut microbial diversity and measures of microbial metabolites, were collected from over 1,000 individuals. A subset of these data are presented using a violin plot, which is simply a box plot on steroids! In addition to the summary statistics provided by the box plot, the violin plot shows the entire distribution of the data, which is particularly important when the data is multimodal, that is, there is more than one peak. In this example (**Figure 9.8B**), we are comparing the species identified in fecal samples from subsets of the three groups. What we observe is that there is a significant difference in the species compositions of HC samples relative to either CHD or MMC samples. In fact, the MMC samples are more distinct relative to HC than are the CHD. The CHD and MMC samples are not significantly different in species composition. A more detailed examination of the data reveals a significant shift in the *Bacteroides* and *Ruminococcus* communities in the healthy controls to one enriched in a different subset of *Bacteroides* species in the CHD and MMC samples. The MMC samples also had significantly lower levels of *Faecalibacterium* and a far lower overall bacterial cell count. In other words, disease progression was closely associated with loss of microbial density and richness. Both the taxonomy and functional outputs of the gut microbiomes were significantly different between the healthy control group and those with CHD, showing that CHD is associated with an altered microbiome.

A whopping 121 bacterial species emerged as markers of metabolic dysfunction, and 85% of these were depleted in the CHD cohorts. Of these, 23 unique species were identified as markers specific to CHD, some of which were depleted and others enriched (Zhong et al., 2022). Specifically, depleted levels of *Acinetobacter*, *Turcimonas*, and *Acetobacter* were tightly linked to CHD. Low levels of one species, an uncharacterized member of the *Ruminococcus* genus, identified it as a marker of disease escalation. *Ruminococcus* include butyrogenic bacteria (i.e., butyrate producers). Thus,

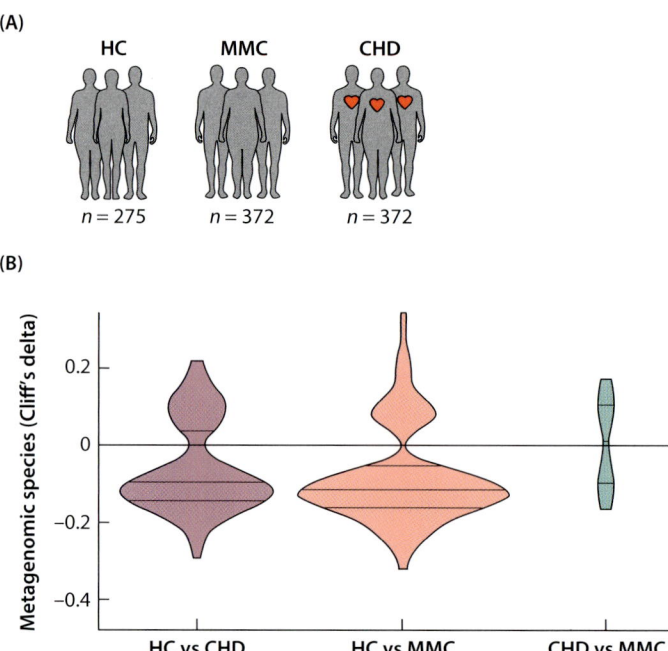

Figure 9.8 The Gut Microbiota Contributes to Coronary Heart Disease (A) The study design of a comparison of gut microbiota between individuals with coronary heart disease (CHD), with metabolic syndrome but not diagnosed with CHD (MMC), and healthy controls (HC). (B) A violin plot shows a significant difference in the microbial species identified in the three groups of participants. (After Fromentin et al., 2022.)

depletion of these bacteria is associated with insufficient production of short-chain fatty acids—molecules that we have seen are essential for gut energy metabolism, gut barrier integrity, and immune regulation.

Cancer

Cancer is a heterogeneous group of diseases that all share the feature that cells proliferate in an uncontrolled manner. The role of microbe-induced inflammation in cancer development is well established. First, dysbiosis can lead to systemic inflammation and an increased likelihood of developing cancer. Multiple chronic inflammatory conditions, such as inflammatory bowel disease (IBD) and ulcerative colitis, are associated with an increase in cancer incidence. Second, specific pathogens may play a role in cancer. Two of the earliest specific examples of infection leading to inflammation and then cancer are seen with *Helicobacter pylori*, leading to gastric cancer, and *Schistosoma haematobium*, resulting in bladder cancer.

The best-known bacterial cause of cancer is *H. pylori*, which is a spiral-shaped bacterium that survives in the mucus layer that coats the inside of the human stomach. Immune cells, which cannot survive in the acidic stomach environment, are unable to reach *H. pylori*, rendering this infection invisible to immune-mediated protection. Although we have known for some time that infection with *H. pylori* can promote gastric cancer, the molecular mechanisms underlying this outcome have only recently been revealed (**Figure 9.9**). There is a transmembrane protein, semaphorin 5, which was first known for its ability to direct the development of neural networks by instructing cells where to go, when to branch, and when to terminate. More recently it has been determined that interactions between *H. pylori* and semaphorin 5 result in the activation of a signaling pathway (ERK/MMP-9), resulting in increased expression of the kinase ERK and a metalloproteinase (MMP-9). These two proteins promote tissue remodeling, and increased levels are associated with carcinoma. A bacterial signal results in an increase in proteins involved in directing a cell to become cancerous!

In addition to the microbiome affecting cancer through systemic inflammation, we now know that tumors host their own microbiomes, whose members can directly affect the growth and progression of cancer. One example is provided by

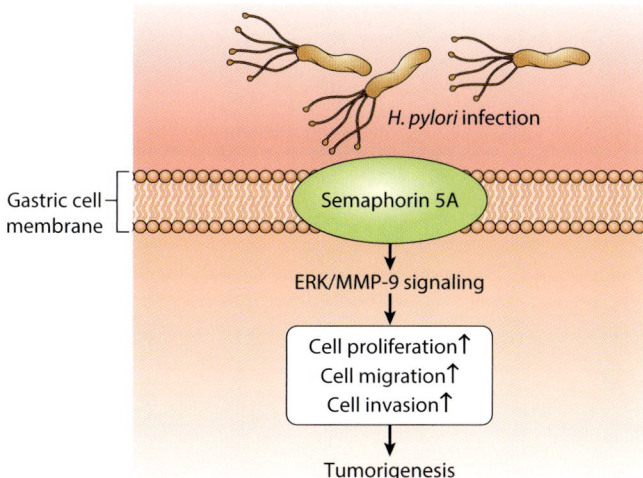

Figure 9.9 *Helicobacter pylori* **and Gastric Cancer** *H. pylori* infection can promote gastric cancer by activating gastric cancer cells to proliferate, migrate, and invade tissues. The mechanism involves interactions between *H. pylori* and a member of the semaphorin protein family (5A). Semaphorins are transmembrane proteins that mediate immune cell regulation, vascular growth, and neural genesis. The bacterium recognizes the protein and impacts a signaling pathway (ERK/MMP-9). (After Guoqing et al., 2021.)

colorectal cancer in which *Fusobacterium* species, which are found inside the tumors, can promote tumor growth. They do so by several mechanisms, including inhibiting T cell and macrophage activity, as well as natural killer cell cytotoxicity. This means that *Fusobacterium* is actively undermining the body's ability to fight cancer by blocking the immune response, allowing colon cancer to progress more rapidly. Figure 9.10 provides a schematic of the proposed interactions between tumor, immune, and microbial cells. This area of study is in its infancy, and there is great excitement about how this knowledge will impact our ability to treat a diversity of human cancers.

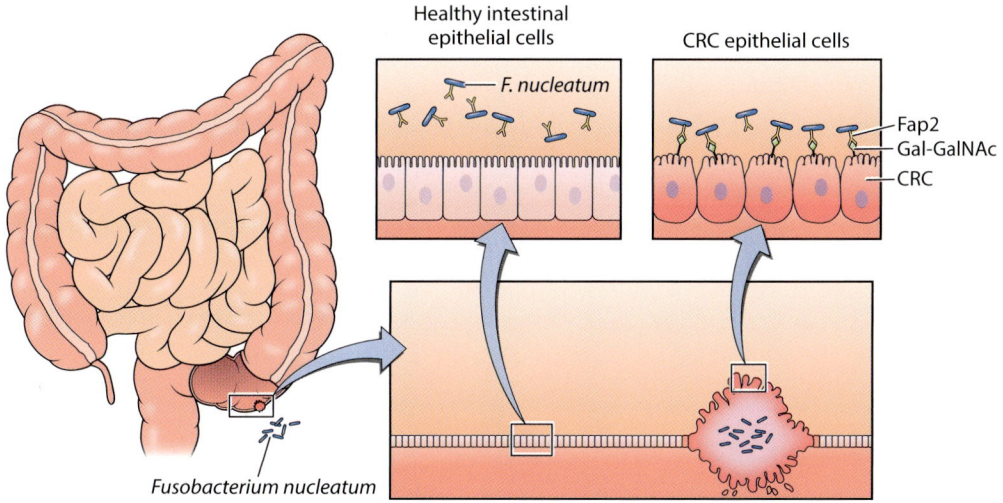

Figure 9.10 Human Tumor Microbiome Tumors have their own distinct microbiomes, and the microbiota may reside inside or outside the tumor and associated immune cells. The precise roles of these microbes remain unclear but are almost certain to impact therapeutic interventions. In this image we find a blow up of healthy and cancerous (colorectal cancer [CRC]) epithelial cells in the colon showing the location of *Fusobacterium nucleatum* in the healthy gut lumen on the left or producing Fap2 protein as a cell surface receptor on the right, bound to a sugar derivate of galactose (N-Acetylgalactosamine [GalNAc]), which is overexpressed in CRC cells. (After Abed et al., 2016.)

Figure 9.11 Microbes Used in Treating Cancer Several species of bacteria have shown potent activity in the destruction of cancerous tumors. (After Mills et al., 2022.)

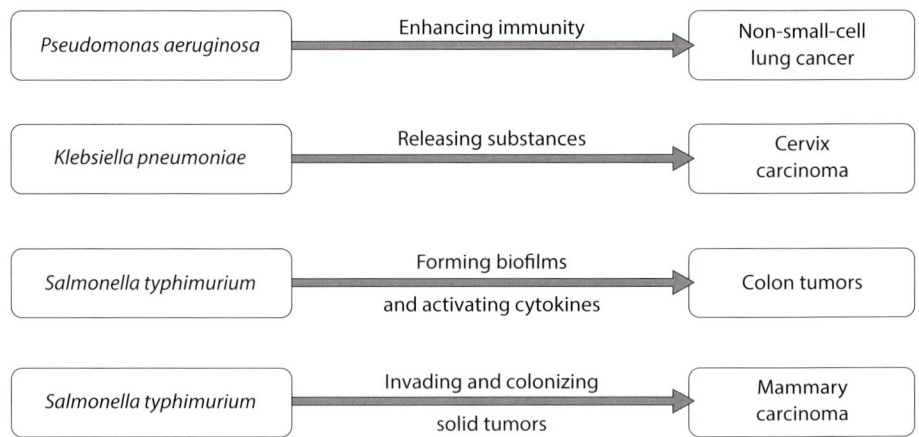

Several studies have suggested that microbes could be developed for use as cancer therapies (Sedghi et al., 2021). For example, direct injection of *Streptococcus* into tumors was shown to result in lower rates of tumor growth, which it accomplishes by activating certain cytokines that help to destroy tumors (Phan et al., 2015). **Figure 9.11** shows examples of three bacterial species that can impact cancerous tumor growth (Mills et al., 2022). A variety of mechanisms are involved, including the invasion of the tumor directly, the use of nutrients required for tumor growth, and the signaling of immune responses that then target tumor cells.

The future of microbiome-related cancer therapy appears bright as we recognize the many ways in which we may be able to harness the power of the microbiome in our war against cancer. We are developing ways to alter the microbiome by adding beneficial microbes, inactivating harmful microbial metabolites, and even genetically engineering bacteria and viruses to target cancer cells or incite appropriate immune responses against growing tumors. The future of microbiome-mediated cancer interventions is truly bright indeed!

The prior sections have revealed how critically intertwined are the microbiome and immune systems. The microbes present in the gut of a newborn determine how our immune cells are educated, or trained, to respond to future antigen presentations. This training process is required for our body to achieve the delicate balance required of our immune cells, between recognizing and eliminating pathogens and also permitting our commensal bacteria to remain and provide the appropriate signals that help to reduce inflammation and the attendant disease.

CHECK YOUR UNDERSTANDING

1. Immune cells are first generated in the
   a. bone marrow.
   b. spleen.
   c. appendix.
   d. liver.

2. Your skin is considered to be part of your immune system.
   a. True
   b. False

3. Most of your immune cells are located in the
   a. GI tract.
   b. blood.
   c. bone marrow.
   d. appendix.

4. Which antibody is most commonly transferred from parent to fetus across the placenta?
   a. IgA
   b. IL-17
   c. IL-21
   d. IgG

5. Which immune cell type is able to suppress immune responses?
   a. Mature B cells
   b. Regulatory T cells
   c. Macrophages
   d. Lymphocytes

6. Most of the antibodies found in breast milk are
   a. IgG.
   b. IL-17.
   c. hemoglobin.
   d. IgA.

7. Newborns have an increased risk of infection because their
   a. adaptive immune system is still developing.
   b. body is very small.
   c. innate immune system is still developing.
   d. microbiome composition is still variable.

8. How long does it take for a stable adult-like microbiome to be achieved in newborns?
   a. 1 month
   b. 3 months
   c. 1 year
   d. 3 years

9. All of the following can be sources of colonizing bacteria for an infant's microbiome *except*
   a. the hospital environment.
   b. pets.
   c. the birth parent's urogenital, skin, and gut microbiomes.
   d. the womb.

10. Newborns have fewer _____ than adults do.
    a. cytotoxic T cells
    b. bones
    c. B cells
    d. macrophages

11. Pulmonary microbes influence the host immune system through
    a. secretion of proteins and metabolites.
    b. production of antibiotics.
    c. emission of carbon dioxide.
    d. cytokine secretion.

12. Microglial cells
    a. are members of the innate immune system.
    b. are found in the bone marrow.
    c. are found exclusively in the gut.
    d. help train the acquired immune system.

13. The hygiene hypothesis posits that dysregulated immune responses may be due to
    a. Decreased exposure to farm animals.
    b. widespread antibiotic use.
    c. Increased personal hygiene.
    d. all of the above.

14. Rheumatoid arthritis
    a. is inflammation localized to only one part of the body.
    b. is a systemic autoimmune disease.
    c. Is an autoimmune disease that effects the brain
    d. only develops in old age.

15. IL-17
    a. is a proinflammatory cytokine.
    b. helps to reduce inflammation.
    c. is produced by B cells.
    d. has no role in a healthy immune system.

Answers: 1A, 2A, 3A, 4D, 5B, 6D, 7A, 8D, 9D, 10A, 11A, 12A, 13D, 14B, 15A

DIVING DEEPER

1. List the components that make up your immune system.
2. What is an antigen?
3. Describe the difference between your adaptive and innate immune systems.
4. What are stem cells, and why are they important?
5. Describe the differences between fetal and adult immune systems.
6. Briefly summarize the hygiene hypothesis.
7. How can bacteria from your gut microbiome reach other parts of your body?
8. What are autoimmune diseases, and how do they relate to our study of the microbiome?
9. What can cause a decrease in gut microbiome diversity?
10. Describe the role of antimicrobial proteins in our immune system.
11. What disorders are associated with skin microbiome dysbiosis?
12. How is homeostasis maintained in the lung microbiome?
13. Briefly summarize the findings by Flügel and his team on the connection between lung microbiota and autoimmune disease.
14. What is the role of macrophages in the immune system?
15. How can the microbiome contribute to tumorigenesis?

DISCUSSING AND REFLECTING

1. When most people think of the immune system, they picture an apparatus meant to fight off pathogenic invaders. However, we know from previous chapters that bacteria are found everywhere in our body. If our immune systems were constantly fighting these commensal critters, it would cause a major problem! Describe how your body is able to recognize members of your microbiome and why this is important.
2. While many parents begin breastfeeding their newborns, economic, social, and health factors often lead them to stop within 2 months after delivery. Formula provides newborns with an alternative food source containing all of the nutrients needed for healthy development. Using the information you learned in this chapter, what factors that are found in breast milk are lacking in manufactured formula? How might this impact newborn development?
3. Reflection. The immune system is one of the most complex networks in our body, and yet—when working correctly—it goes entirely unnoticed. Understanding how we're able to differentiate good from bad bacteria is an essential goal for developing microbiome-based therapeutics in the future. This chapter spends a lot of time describing how your immune system identifies "self" versus "non-self" in the context of commensal microbiota. Take a few minutes to reflect on how your own thinking of "self" versus "non-self" has changed since you began reading this textbook.

RECOMMENDED READINGS

Popular Science Review

Zeliadt, N. (2010). When Good Germs Go Bad: "Friendly" Gut Bacteria Can Trigger Rheumatoid Arthritis in Mice. *Scientific American*. https://www.scientificamerican.com/article/gut-bacteria-can-trigger-rheumatoid-arthritis-in-mice/

Popular Science Book

Mayer, E. A., & Casey, N. (2021). The Gut-Immune Connection: How Understanding the Connection between Food and Immunity Can Help Us Regain Our Health (first edition). HarperWave, an imprint of HarperCollins.

Scientific Reviews

Zheng, D., Liwinski, T., & Elinav, E. (2020). Interaction between Microbiota and Immunity in Health and Disease. *Cell Research*, 30(6), 492–506. https://doi.org/10.1038/s41422-020-0332-7

Sharma, P., Jain, T., Sethi, V., Iyer, S., & Dudeja, V. (2020). Gut Microbiome: The Third Musketeer in the Cancer-Immune System Cross-Talk. *Journal of Pancreatology*, 3(4), 181–187. https://doi.org/10.1097/JP9.0000000000000057

Microbiome Dysbiosis

10

CHAPTER CONTENTS

10.1 Defining Eubiosis and Dysbiosis

10.2 An Ecological Perspective on the Microbiome-Host Relationships

10.3 A Framework for Assessing Microbiota-Disease Associations

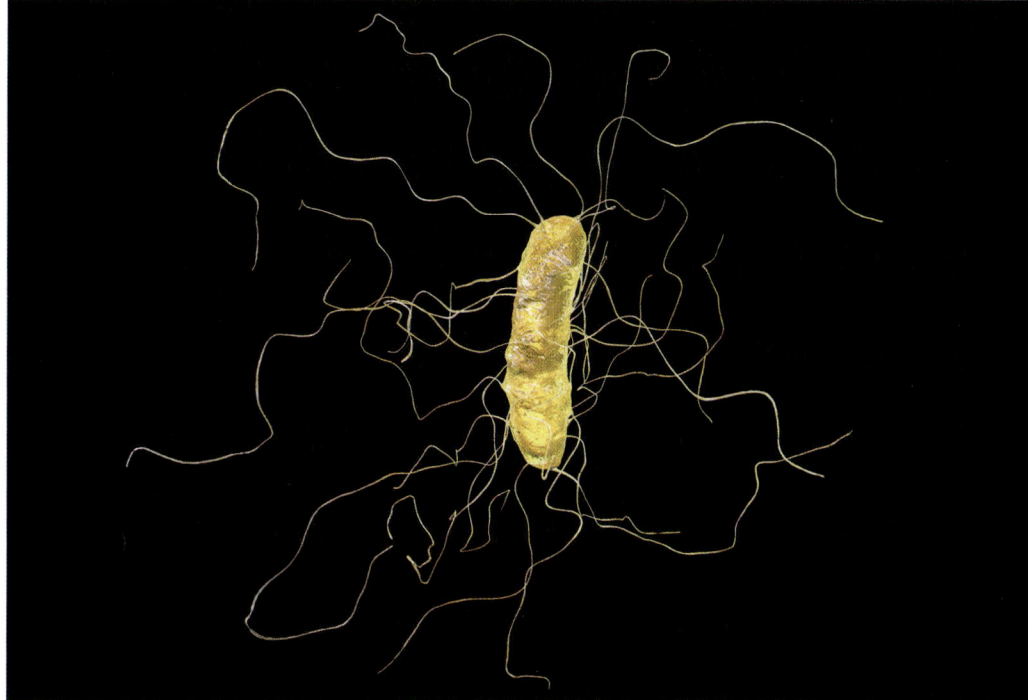

Have you ever wondered what prompted the first physician to consider taking feces from one person and inserting it into another? Well, that would be me, *Clostridioides difficile* (*C. diff* for short). It was 1958 and I was causing some very tough forms of recurrent diarrhea. One ingenious physician took an idea first proposed in the 4th century and transferred feces from a healthy person into a sick patient. Not only did that transfer work, in over 95% of transfers since, the cure has been quick and permanent. Yes, that's right—I wasn't able to compete against the so-called "normal" gut microbes, and I surrendered. So, the moral of this story is that I may cause disease (OK, I do cause disease), but I inspired ingenuity in return! (Photo from iStock.com/Dr_Microbe)

We have developed a font of knowledge with respect to our microbiome, its members, and their contributions to human health. Pioneering work has provided rich catalogs of the species present and their frequencies, and there are clear associations linking imbalances of the microbiome with various diseases. The hope is that with investments in this relatively young field of research, new therapeutic solutions for human disease may be uncovered. However, it has also become clear that merely identifying what is in a microbiome is simply not good enough. We must move beyond considering microbiomes as discrete communities housed within us, to a more nuanced perspective in which the microbiome is assessed as an integral component of its host and focus directly on host-microbiome interactions. Only then do we stand a chance of understanding what constitutes a "healthy" microbiome. As the chapter's opening quote makes clear, we won't be able to resolve dysbiosis if we don't solidly understand what a healthy microbiome is first.

> "We can't talk about dysbiosis . . . if you don't know what a healthy microbiome is."
> —Hans Verstraelen
> (Scott et al., 2019)

10.1 DEFINING EUBIOSIS AND DYSBIOSIS

The dominant perspective until recently has been that a microbiome is a discrete entity—which some consider to be the equivalent of an organ—that includes the microbiota, the encoded genes, and gene products. This microbe-centric view suggests that if we could just come up with a complete list of what is present, their genomes and metabolites, we would eventually come to a complete understanding of the microbiome. Indeed, the first phase of the NIH-funded **Human Microbiome Project** (**HMP**), introduced in chapter 2, had just such a goal: identifying taxa that are commonly present in a healthy microbiome. The scientists involved characterized the gut microbial communities from 242 healthy individuals but failed to define a core "healthy" gut microbiome. This led to calls for yet more data and resulted in the NIH funding additional studies that resulted in the production of over 42 terabytes of multi-omics data. Unfortunately, this massive dataset did not bring us substantially closer to revealing what constitutes a healthy microbiome.

We introduced the term *holobiont* in chapter 3 and defined it as a host and the species living in or on it, which form a discrete ecological unit. This term perfectly describes the relationship between humans and their microbiomes. From this perspective, the microbiome is far greater than simply the sum of its parts, such as genes, transcripts, and metabolites. Indeed, it is becoming clear that our microbiomes are part and parcel of the human holobiont, and early efforts to separate out the microbiome as akin to an organ in the human body may be futile.

The Eubiotic Microbiome

Eubiosis is a term used to describe a state in which the host and microbiome work in synergy resulting in host (and microbiota) health. However, a more precise definition of this state is required if we wish to employ a healthy microbiome as an experimental control, that is, to identify "unhealthy" microbiomes present in an individual with disease. If we can't define eubiosis, how can we detect differences in microbiomes that result in disease? The original approach to defining eubiosis involved producing extremely large sample sizes so a more complete picture of the microbial diversity present in the human gut could be captured. For example, a population study carried out in Belgium employed 1,106 individuals of average health (Falony et al., 2016). The resulting microbial census revealed that 14 of the 664 microbial species detected were common to all participants. Using questionnaires, the investigators also revealed 69 factors, such as stool consistency and birth mode, that were associated with the variation in microbiome composition detected. However, these factors explained no more than 15% of the variation in taxon abundance, leaving 85% of the differences between healthy individuals' gut microbiome compositions unexplained. This leaves us with a fairly unsatisfying definition of eubiosis.

The Dysbiotic Microbiome

The flip side of the coin is that if we can't define a microbiome in a state of health, it is even more challenging to define one associated with disease, which we refer to as **dysbiosis**. The dysbiotic state is variably described as one in which taxonomic and/or functional changes in a microbiome occur in individuals with disease that do not occur in healthy individuals. These changes might include the loss of commensal microbes, the presence of pathogens, and/or an overall loss of microbial diversity. However, since it remains unclear what constitutes a eubiotic microbiome, it is even less clear how to define an impaired one associated with a disease state.

Olesen and Alm, two biological engineers at the Massachusetts Institute of Technology, aptly noted, "Dysbiosis is not an answer" and rejected the concept as having no explanatory benefit (Olesen & Alm, 2016). They went so far as to compare it to the humoral theory of disease from the Middle Ages, which argued that the cause of disease was exposure to "bad air" (chapter 2). It is no wonder that they disparaged the concept of dysbiosis when we learn that the most common definition of dysbiosis is an imbalance or loss of homeostasis of the microbiota. This vague definition clearly lacks scientific value. Another group (Levy et al., 2017) have suggested an exciting, albeit challenging, solution: the application of Koch's postulates to identify the microbiome or its constituent members as the cause of disease. Recall from chapter 2 that **Koch's postulates** are the four criteria designed to assess whether a microorganism causes a disease (refer back to Figure 2.6). To apply these criteria in evaluating the potential impact of members of a microbiome in causing disease would require that investigators disentangle cause from effect, which, at this moment, is a difficult path to follow in human studies. We will return to that thought later in this chapter.

More Is Not Necessarily Better

Most definitions of dysbiosis include some reference to the microbial diversity present, such as the alpha and beta diversity measures we learned how to estimate in chapter 5, and the loss of such in a dysbiotic state. The assumption is that a more diverse microbiome is associated with better health outcomes, and it is true that high species diversity is associated with the microbiomes of healthy individuals, in general. Ecological theory supports the notion that increased diversity within a community is expected to confer overall functional efficacy. In the same way that a forest with many different plants and animals is more resilient to change, a diverse microbiome has a greater variety of functions and abilities.

However, we must be careful with the direct application of the ecological diversity argument to the microbiome. For instance, babies who are breastfed have far lower levels of gut microbiome diversity than their bottle-fed counterparts, and yet they experience lower levels of numerous diseases later in life, such as asthma and allergies. Another example is provided by the vaginal microbiome, which usually has very low levels of diversity, and in some individuals increased microbial diversity is associated with detrimental effects, such as bacterial vaginosis. Thus, the concept of diversity, alone, has limited value in our efforts to distinguish healthy from disease-related microbiomes, primarily because it requires a contextual interpretation. Diversity alone is not necessarily a benefit. To determine whether there is a benefit, we must identify how that diversity, in that specific setting, confers a benefit. In the world of the microbiome, more is not necessarily better!

10.2 AN ECOLOGICAL PERSPECTIVE ON THE MICROBIOME-HOST RELATIONSHIPS

It has become clear that scientists must take into account the ecological and evolutionary factors that underpin microbiome-host relationships (Gilbert & Lynch, 2019). The critical role served by the host in structuring internal microbial ecosystems through nutrient flow, immune response, and niche generation is taking center stage in the design of microbiome research studies. Thus, there is a growing appreciation for the incorporation of ecology as a theoretical framework and for the recognition that a microbiome is meaningless without the context of its host. This new perspective requires that we consider not only how the microbiome impacts its host, but how the host shapes the environment inhabited by the microbiota, and thus shapes the microbiome itself (Tiffany & Bäumler, 2019).

Ecological Interactions between the Gut and Its Microbiome

Let's apply this perspective to our gut microbiome. Which microbes are present and what functions they serve are under the joint control of both the host and the microbiome (**Figure 10.1**). The host provides selective pressures that support some, and eliminate other, microbes. These pressures include the production of glycoproteins

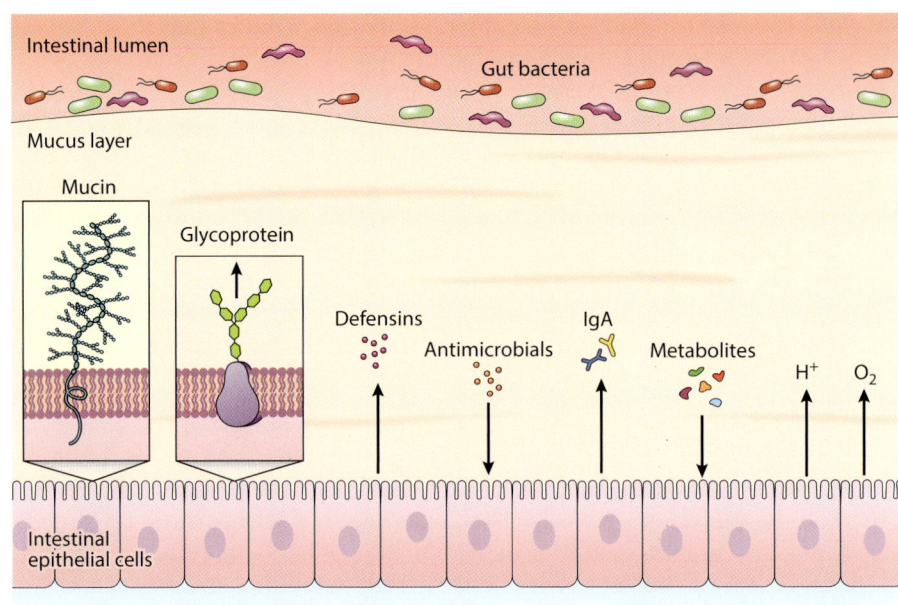

Figure 10.1 Host-Microbiome Interactions in the Mucin Layer of the Gut The host dictates which microbes can survive and thrive in their gut. For example, the host secretes nutrients into the large intestine, such as the glycoproteins that create the mucin layer that lines the intestinal epithelial cells. Some microbes consume mucin as food. The host also releases immunoglobulin A (IgA), which limits which microbes can reside on and feed from the mucin layer. Antimicrobial proteins, such as defensins, as well as levels of oxygen and hydrogen also impact the mucin layer microbial community. The microbes that thrive in this niche produce metabolites, such as short-chain fatty acids, that nourish the host gut epithelial cells. They also produce antimicrobial proteins that further limit who can reside in the mucin layer.

that form the basis of the mucin layer, which serves as an energy source for numerous microbes, and the release of IgA and defensins, which detect and eliminate select microbes. Meanwhile, the microbiome is also selecting its members through the production of antimicrobials while it supports its host through the production of a variety of metabolites, such as short-chain fatty acids (SCFAs). All of these selective pressures act on mechanisms that regulate microbial community assembly, diversity, and stability (Foster et al., 2017). As Kevin Foster and his colleagues at Oxford University so aptly state, the microbiome can be regarded as an ecosystem on a leash. The leash refers to the host functions that attempt to maintain the microbiome within certain boundaries, perhaps those that support a healthy host. The leash can be tightened, which will significantly reduce the microbial species that are able to enter the host, invade, and colonize a particular microbiome. But what, precisely, is a microbial leash?

The Microbial Leash Metaphor

Let's start with one of the more striking examples of the host controlling its microbiomes. For the majority of reproductive-aged women, most (>97%) of the vaginal microbiota belong to the genus *Lactobacillus* (**Figure 10.2A**). The mechanism driving this selection process is very likely host-produced estrogen, which coordinates the breakdown of glycogen in vaginal mucosal fluid, thus nourishing *Lactobacillus*. Why would the host want a *Lactobacillus*-dominated vaginal microbiome? The answer seems to be that these bacteria create a highly effective barrier against vaginal infection by lowering the pH in the vagina and producing a plethora of antimicrobial compounds. In this case, the vaginal environment selects for a limited number of bacterial species that can survive in that specific environment—the so-called leash effect. These bacteria then create a suite of controls to limit the possibility of sexually transmitted disease from taking hold.

Another example of host control at play is in the microbiomes of breastfed infants. Human milk is composed of complex carbohydrates, many of which infants cannot digest on their own. We learned in chapter 7 that these complex carbohydrates reach the large intestine intact where *Bifidobacterium* are able to ferment them, resulting in proliferation of these bacteria in the infant's feces (**Figure 10.2B**). The dominant species (*B. longum*) produces all of the enzymes required to ferment milk oligosaccharides, and their fermentation efforts result in the production of SCFAs (refer back to Figure 7.16). The presence of key fermentation products,

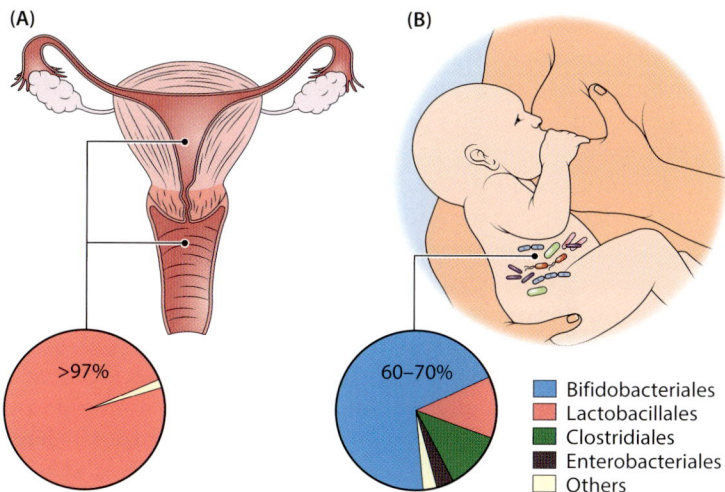

Figure 10.2 Hosts Steer Microbiome Composition (A) Hosts direct the composition of the vaginal microbiota towards a dominance of *Lactobacillus*. (B) Breast milk drives a dominance of *Bifidobacterium* in the infant gut microbiota. (After Byndloss, Pernitzsch, and Bäumler, 2018.)

particularly SCFAs, are critical in training the infant's underdeveloped immune system to both tolerate the SCFA producers and to inhibit other bacteria from invading this niche. Thus, the host, by consuming breast milk, shapes the composition of the gut microbiome during a very critical period when the immune system is still developing, and the microbiota trains the host to recognize commensals as part of self. In this example, the leash is the food provided by the host (breast milk), which selects for specific bacteria with the gene content that permits them to break down breast milk for their own nourishment. In return the bacteria occupy this niche and produce SCFAs that not only feed the host's gut epithelial cells, but also help train the newborn's developing immune system.

Once the infant is weaned from breast milk, the microbial gut community diversifies and becomes dominated by obligate anaerobic bacteria. At this point, it appears that the host no longer selects for specific microbes, but rather dictates their overall density and selects for production of key metabolites. For example, in the large intestine, airways, and reproductive tract, **goblet cells** secrete mucin to form a densely packed inner mucus layer that excludes bacteria. In the airways there is the added presence of a **mucus escalator**, composed of mucus and cilia, which moves the mucus up and out of the lungs. In that process the escalator catches and removes over 100,000 bacteria per day before they reach the delicate lung tissues. Gastric acid, produced in the stomach, prevents most microbes from flourishing there. In other niches, such as the small intestine, there is the production of antimicrobial peptides by **Paneth cells** that limit microbial growth. The leashes in these examples are quite different from the one that selects for specific bacteria. Rather than supporting one species over another, the leashes in these situations ensure bacterial invasion is prevented.

Host Control of Microbiomes in Small versus Large Intestines

Let's compare how the host controls the microbiomes of the small versus the large intestine. Paneth cells in the small intestine can mediate the microbiota present. We learned in chapter 3 that these specialized epithelial cells are located in the invaginations (**crypts**) found at the bases of the epithelial villi that line the small intestines (refer back to Figure 3.12). Crypts are the site of production and secretion of granules filled with antimicrobial peptides and immunomodulating proteins, such as defensins, that help to regulate the gut microflora. These secretions, in turn, play a role in regulating host injury and repair mechanisms of the intestinal epithelial cells and, subsequently, the levels of intestinal inflammation. This intensive secretory activity makes Paneth cells vulnerable to stress.

Crohn's disease, a form of inflammatory bowel disease, is associated with disruption in the functioning of Paneth cells (see Figure 3.13). One risk factor for this disease is childhood exposure to certain dietary emulsifiers, which are food additives that help immiscible components combine. For example, typical flavored yogurt, without an emulsifier, would separate out into a layer of mostly fat sitting above a soupy, water-based liquid containing the flavorings. These emulsifiers trigger inflammation in the intestinal epithelium and impact Paneth cell performance. The result in individuals with Crohn's disease is an expansion of *E. coli* in the gut, which settle along the intestinal mucosal surface and consume mucin. The antimicrobial peptides (AMPs) produced by Paneth cells normally serve to protect the epithelial surface from the grazing activities of *E. coli*, whereas impaired AMP synthesis might result in enhanced colonization of this niche, degradation of the mucin layer, and increased permeability of the intestinal lining, resulting in disease.

In the small intestine, the host controls not only the specific taxa that reside there, but also their density. Facultative anaerobic bacteria dominate the microbiome of the small intestine, with densities ranging from 10^2 to 10^6 bacteria/ml of lumen content. This community synthesizes vitamin B_{12}, which is absorbed by epithelial cells. Although B_{12} is a critical metabolite, the density of the bacteria producing it is controlled by the host through a combination of gastric acid, secretion of immunoglobulin, food transit time in the gut, and antibacterial secretions from the pancreas and bile system. This control system keeps all microbes in check. The host provides a wealth of nutrients that would support immense populations of microbes. The host must keep these numbers in check to prevent malnutrition of the host.

Host control of the microbiome in the large intestine is quite different. First, there is a much higher density of microbes, reaching 10^{12} bacteria/ml of lumen content. As we first learned in chapter 3, this dense community performs numerous vital functions, in particular breaking down complex carbohydrates that the host cannot digest. Obligate anaerobic bacteria ferment these carbohydrates and produce key metabolites, such as SCFAs, which the host absorbs. In support of these obligate anaerobes, the host maintains low levels of oxygen in the large intestine. Hosts do this by way of a complex and cyclical monitoring system. The bacteria produce SCFAs, which activate a fatty acid sensor in the epithelial layer of the large intestine. Once fatty acid production is sensed, the epithelial cells trigger an oxygen-consuming pathway of energy metabolism called **β-oxidation**. The result is that oxygen levels remain low in the lumen. Low oxygen levels select for more obligate anaerobic bacteria that continue to break down fiber into SCFAs, and a positive feedback loop is established. This cycle provides stability and very likely contributes to microbiota resilience in the colon.

However, if levels of oxygen increase in the colon, the anaerobes are overrun by facultative anaerobic bacteria that can survive better in the presence of oxygen and that produce carbon dioxide, rather than butyrate, as a waste product, which interferes with host nutrition. The lack of butyrate triggers the epithelial cells to use a different energy production pathway, anaerobic glycolysis, resulting in higher levels of oxygen in the lumen and selecting for even more facultative anaerobes. This expansion of facultative anaerobes is a well-known microbial signature of dysbiosis in the colon. As these bacteria lack the ability to break down complex carbohydrates, their proliferation negatively impacts the host's ability to digest key food sources and results in the production of harmful metabolites, such as lipopolysaccharides. The picture emerging is that the host has metabolic mechanisms to maintain **hypoxia** in the colon, which maintains homeostasis. The gut epithelial cells literally function as a control switch that governs a shift between eubiotic and dysbiotic microbiota outcomes.

Host Control of a Microbiome-Based Disease

Let's examine an example of the host control of a microbiome-based disease. Ulcerative colitis (UC) is a type of inflammatory bowel disease where there is a predictable shift in the dominant members of the gut microbiome, from the normal obligate anaerobes to facultative anaerobes (refer back to Figure 3.13). Presence of Enterobacteriaceae, which can grow in the presence or absence of oxygen, is a signature of UC. These observations suggest that reestablishing an anaerobic environment in the large intestine might serve to remediate the resulting gut dysbiosis. In fact, the drug mesalazine, used to treat UC for over 40 years, very likely acts by signaling activation of the β-oxidation energy pathway, thus resulting in lower levels of oxygen in the gut (Byndloss et al., 2018). In this example we find that during homeostasis the host controls the gut microbiome by maintaining anaerobiosis, thereby fostering the growth of beneficial anaerobic bacteria. A loss of host control results in a shift in oxygen levels and the selection of facultative anaerobes. These epithelial signaling mechanisms provide a promising therapeutic target for restoring host control in colitis.

Microbial Control of the Gut Microbiota

Members of the gut microbiota can impact how the host controls its own microbiome, sometimes resulting in disease. For example, invasion of the gut by *Salmonella typhimurium* results in acute gastroenteritis, a disease characterized by inflammation of the stomach, small intestine, or large intestine, leading to a combination of abdominal pain, nausea, vomiting, and diarrhea. Normally members of the gut microbiome work with the host to prevent proliferation of *S. typhimurium*, and other pathogens. However, these control systems can be subverted by *Salmonella* to promote disease. **Figure 10.3** illustrates the process by which *S. typhimurium* is able to impact host control in such a way as to cause modifications in the niche. These modifications then make it easier for the pathogen to invade and cause disease.

Members of the microbiome provide **colonization resistance**, first discussed in chapter 3, which refers to a microbiome's ability to repel microbial invasion. This characteristic is critical to homeostasis, enabling the microbiome to remain unchanged regardless of environmental perturbations. To coexist, different members of a community utilize different resources, with each species adapted to best utilize that resource. Once established, it is difficult for latecomers to displace these now-resident strains.

Several ecological mechanisms are at work. One mechanism is **niche preemption**, which is the process in which a resident strain occupies a niche and prevents access to that niche by other microbes that come later (**Figure 10.4**). For example, if a pathogen such as *Salmonella* relies on a certain nutrient as a food source but resident microbes (here, Clostridia) already consume that food source, then the pathogen cannot invade. A second mechanism is **niche modification**, which involves organisms

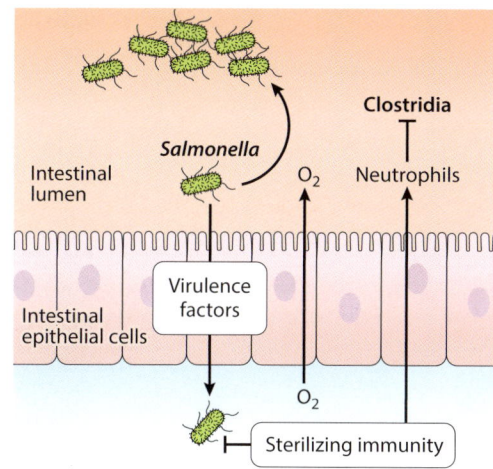

Figure 10.3 Niche Modification by a Pathogen *Salmonella typhimurium* is depicted as triggering inflammation by the production of virulence factors that induce an immune response resulting in the production of neutrophils, which results in "sterilizing immunity," as indicated by the negative impact on the commensal Clostridia. *S. typhimurium* also triggers epithelial cells to shift to anaerobic glycolysis, which results in increased levels of oxygen in the lumen, which enhances the growth of itself and other facultative anaerobes that are associated with disease. (After Litvak and Bäumler, 2019.)

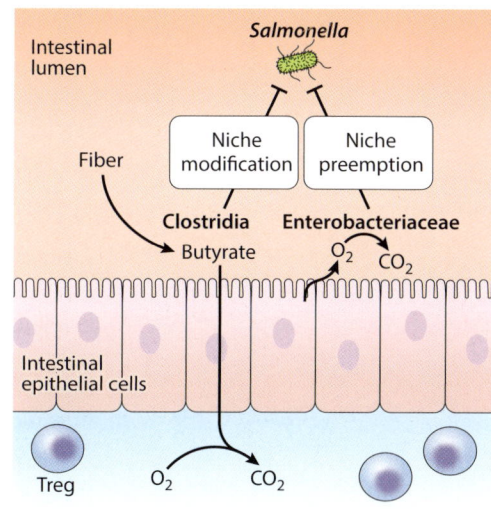

Figure 10.4 Niche Modification and Preemption by a Commensal Fiber-fermenting gut microbiota fuel intestinal epithelial cells with a key metabolite, butyrate. The mitochondria in these cells are then triggered to engage in aerobic glycolysis of the butyrate. This process consumes oxygen, resulting in depressed levels of oxygen at the epithelial surface. Low levels of oxygen select for obligate anaerobic bacteria to convert fiber into more butyrate. In this illustration, Clostridia are depicted as the anaerobic cells producing butyrate and engaging in niche modification. The presence of SCFAs, such as butyrate, also triggers the production of regulator T cells (Tregs), which result in decreased levels of inflammation in the gut. Niche preemption results from these modifications, as potential pathogens, such as *Salmonella* illustrated here, cannot invade. (After Litvak and Bäumler, 2019.)

actively modifying their local environment. For example, the activities of Clostridia keep oxygen levels very low in a healthy gut—so low that pathogenic facultative anaerobes such as *Salmonella* cannot thrive.

Dysbiosis, the Cause or Consequence of Disease?

The goal in understanding dysbiosis is to develop a clear understanding of the ecological interactions occurring between the host and its microbiome and between different members of the microbiome. Only then are we able to identify the key physical or physiological forces in play that may serve as the focus for therapeutic interventions. One remaining ambiguity is that our emerging framework does not distinguish between dysbiosis as a cause of disease or as a result of it. For example, when antibiotics are used to treat an infection, one outcome can be gut and oral microbiome dysbiosis. This altered microbiome can then cause new disease. For instance, antibiotic use can result in opportunistic *Clostridioides difficile* infections. The *C. difficile* overgrowth can then trigger new disease states. One reason that fecal microbiota transplantation is so effective in this situation is because the therapy actually targets the underlying cause, a dysbiotic gut microbiome. Restoring eubiosis results in the elimination of the *C. difficile* overgrowth. However, there are numerous diseases in which some function of the host is impacted, such as the production of mucus or a weakened immune system, which then results in dysbiosis. In these cases, dysbiosis is a result of the disease, rather than the cause of it.

Between these two extremes are numerous more-complex interactions, such as when dysbiosis is caused by an illness, but the resulting change in the microbiome exacerbates the illness. What this leaves us with is the challenging requirement to determine, experimentally, whether dysbiosis is a cause or consequence in every disease state we examine. Even more challenging is the fact that only rarely will there be a single species that causes or is impacted by dysbiosis.

10.3 A FRAMEWORK FOR ASSESSING MICROBIOTA-DISEASE ASSOCIATIONS

Studies frequently find associations between changes to the microbiota and disease state. However, we must assess these associations in the context of the host to understand which associations are strong and merit further investment, and which may be the result of normal variation and not necessarily related to disease. Several structures for evaluating findings and discerning the strength and potential future applications of microbiota-disease associations have been proposed. Gilbert and colleagues (2016) employed such a framework when they identified three types of microbiome-disease associations. They proposed that there are predictable associations, as observed for irritable bowel syndrome (IBS); intriguing associations, as observed in obesity; and surprising associations, such as what is being found in major depression. These degrees of association may, ultimately, actually reflect the likelihood of microorganisms causing disease. For instance, the mechanisms behind the predictable association between the gut microbiome and IBS result from a comprehensive understanding of the underlying mechanisms of host and microbiota interaction in this disease. Specific changes in the gut microbiota result in increased levels of inflammation that trigger a cascade of outcomes that often result in IBS. In contrast, the potential mechanisms that underly the intriguing association between obesity and gut microbiome dysbiosis are far less clear. We are just beginning to untangle strands of the web of interactions involved, which include genetics, lifestyle, microbiome, and more. Surprising associations are those that we might never have predicted, such as what we are now learning from studies of the relationship between members of the gut microbiome and the central nervous system. For example,

Table 10.1. Categories of Microbiome-Disease Associations

CATEGORIES	DISEASE TRIGGER	THERAPEUTIC STRATEGY	EXAMPLE
Limited pathogens	Small number of potential pathogens	Narrow-spectrum antibiotics	Colorectal cancer caused by bacteria such as *Porphyromonas* and *Enterobacter*
Loss of health-associated bacteria	Lacking key bacteria	Probiotics	IBD caused by lack of Ruminococcaceae and Lachnospiraceae
Gross changes of microbiota diversity	Major changes in composition	Fecal microbiota transfer	*C. difficile*–associated diarrhea caused by overgrowth of *C. difficile*

the observation that there is a correlation between certain members of your gut microbiome and levels of depression was simply not something we envisioned a mere 15 years ago.

Developing a framework for understanding associations between microbiota and disease can also help us understand the role microbiota play in disease and may help identify potential treatments. A group led by Claire Duvallet of MIT (Duvallet et al., 2017) employed a meta-analysis of a large sample of microbiome studies across different disease types and made some fascinating observations. They identified several categories of microbiome-disease associations (**Table 10.1**). The first is the single or limited pathogens category, which is characterized by a small number of potential pathogens. In this category, one or a small number of species or genera cause or contribute to a disease. One example of this type of disease is colorectal cancer and the frequent presence of *Fusobacterium, Porphyromonas, Peptostreptococcus, Parvimonas,* and *Enterobacter*. Based upon this categorization, they propose that narrow-spectrum antibiotics, such as bacteriophages or bacteriocins, may be an effective preventative. Eliminate or minimize the potential impact of one or a few taxa, and, perhaps, you avoid the disease. A second category comprises diseases associated with the loss of health-associated bacteria. For example, patients with inflammatory bowel disease (IBD) are consistently lacking select genera of Ruminococcaceae and Lachnospiraceae. For this category, the authors suggest that probiotics may be able to supplement the missing bacteria. A final category is composed of diseases characterized by gross changes of microbiota composition. In other words, major alterations to the composition of the microbiota are observed across many genera. Perhaps the clearest example of this category is *C. difficile*–associated diarrhea. In this instance, although one bacterium triggers the transition in microbiota, it is the subsequent increase in a suite of pathogens that results in disease. As has been repeatedly confirmed, fecal transplantation is effective in ameliorating this dysbiotic state. These categories are particularly useful because they predict the type of therapy one might want to pursue.

Figure 10.5 provides an illustration of how gut dysbiosis, in this case an increase in abundance of a limited number of facultative anaerobes, such as Proteobacteria, is associated with a plethora of disease states, including colitis, IBS, and colorectal cancer. The dysbiotic signature of the gut microbiome in these instances is an overabundance of facultative anaerobes. This dysbiosis is also observed following antibiotic usage, chronic alcohol use, and genetically induced colitis. In this case, which might be described as a gross change of microbiota composition, Duvallet et al. might suggest the use of fecal microbiota transplant therapy.

Polymicrobial Synergy and Dysbiosis

The gross changes of microbiota composition category introduced above predicts that major alterations in microbiome composition may be the cause of a disease state. One possible example of such a microbiome-wide induced disease state is periodontal, or gum, disease. Wang (2015) and his team invoke a new model, **polymicrobial synergy**

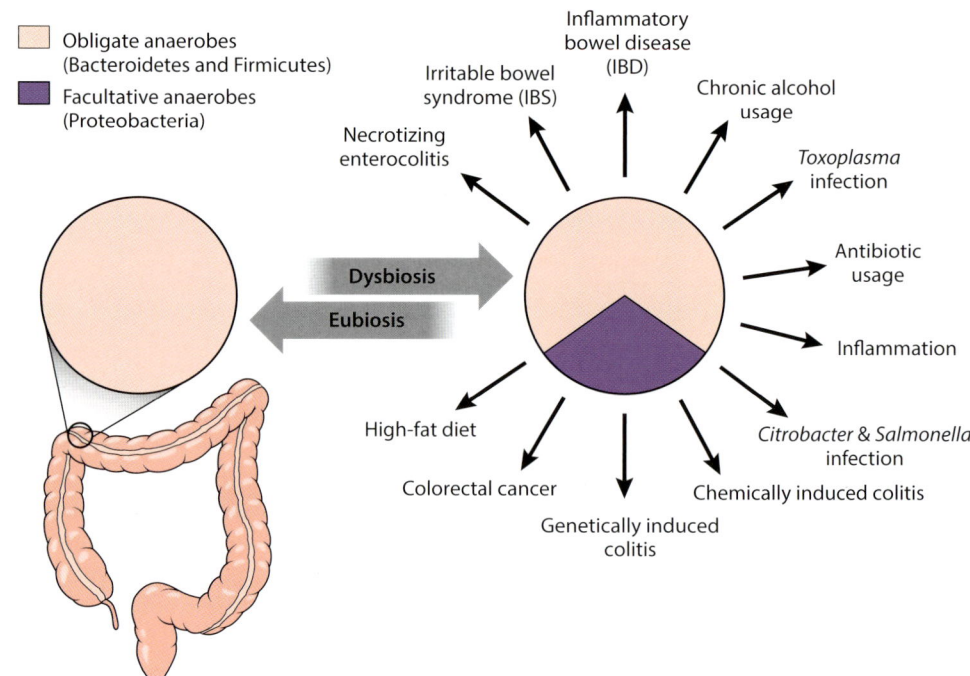

Figure 10.5 A Microbial Signature of Dysbiosis in the Gut Microbiome A eubiotic gut microbiome is dominated by obligate anaerobic bacteria, such as Bacteroidetes and Firmicutes. When the gut microbiome is disturbed, there is often an expansion of Proteobacteria, which are facultative anaerobes, and an attendant increase in numerous disease states, such as colorectal cancer and irritable bowel syndrome. (After Tiffany and Bäumler, 2019.)

and dysbiosis (**PSD**), to explain our current understanding of gum disease (**Box 10.1**). This model suggests that disease is triggered by a dysbiotic oral microbiome, rather than by any specific pathogens. A synergy among the members of this dysbiotic environment alters oral homeostasis and facilitates its transition to a chronic inflammatory state. In other words, the entire microbial community, rather than specific pathogens, drives gum disease progression. Thus, PSD may be considered an example of Duvallet and colleagues' third category of microbiome-disease associations, where widespread alterations to the gut microbiota, rather than the presence of a pathogen or lack of some commensals, are associated with a disease state.

Polymicrobial Synergy in Gum Disease

We discussed in chapters 2 and 3 how multiple members of the oral microbiome are involved in plaque and periodontal disease. In fact, three bacterial species, *Porphyromonas gingivalis*, *Treponema denticola*, and *Tannerella forsythia*, are considered the primary cause of periodontitis, or gum disease. More recently, however, this species-specific concept of gum disease has evolved into a far more complex process, in which several species trigger the transition to a dysbiotic oral microbiome, and the resulting dysbiosis induces chronic inflammation and tissue destruction.

Over 700 microorganisms have been identified in the mouth, many of which have yet to be named or characterized. This immense diversity, coupled with the lack of consistent association between any one organism and gum disease, suggests that we have yet to identify the underlying cause of the disease. It is not sufficient to note that *Porphyromonas*, *Treponema*, and *Tannerella* are present, as there are so many additional contributing microbial factors involved. A team led by Peter Jorth at University of Texas, Austin, describes this process of PSD that results in gum disease. They compared the metatranscriptome of gums from patients with aggressive periodontitis, sampling both diseased and nondiseased sites from each individual (**Box 10.2**) (Jorth et al., 2014). Despite the fact that oral microbiomes varied significantly between individuals, the metabolic profiles of diseased sites were highly conserved. This result argues for a high level of functional redundancy, in which different microbes can be substituted for each other to cause disease. As long as there are species present that promote iron acquisition and synthesize lipopolysaccharides and flagella, disease results. These functions contribute to the inflammation observed with periodontitis, as

BOX 10.1. A MICROBIOME-BASED MODEL OF PERIODONTAL DISEASE

In chapter 4 we discussed a relatively new technology, spatial metagenomics, and its application to understanding both the members and the locations of the plaque-associated microbiome. This detailed understanding of how a community is structured has provided an elegant perspective on how members of a microbiome communicate with and respond to each other's presence. In **Figure 10.6A** we see a simplified model of how microbiota interactions can be compromised by the entry of a single keystone species into the plaque community. In this example, the keystone species is almost certain to be *Porphyromonas gingivalis*, whose presence results in inflammatory periodontitis through its disruption of the formerly eubiotic microbiome. Cascading interactions result as *P. gingivalis* communicates with other accessory pathogens, such as *Streptococcus gordonii*, whose presence results in a dysregulation of immune surveillance. The dysbiotic community increases in number, pathogens (green in the figure) overgrow and become more active, and tissue destruction ensues. **Figure 10.6B** provides a snapshot of some of the mechanisms through which the keystone species interacts with accessory pathogens, potentially leading to periodontal disease. These interactions include direct contact, such as metabolites binding to cell surface receptors, and indirect contact, such as the production of H_2O_2, hydrogen peroxide, which can trigger increased invasion of accessory pathogens into the host tissue. In some cases, the signaling results in increased community cohesion, such as the production of extracellular polymeric substances (EPS).

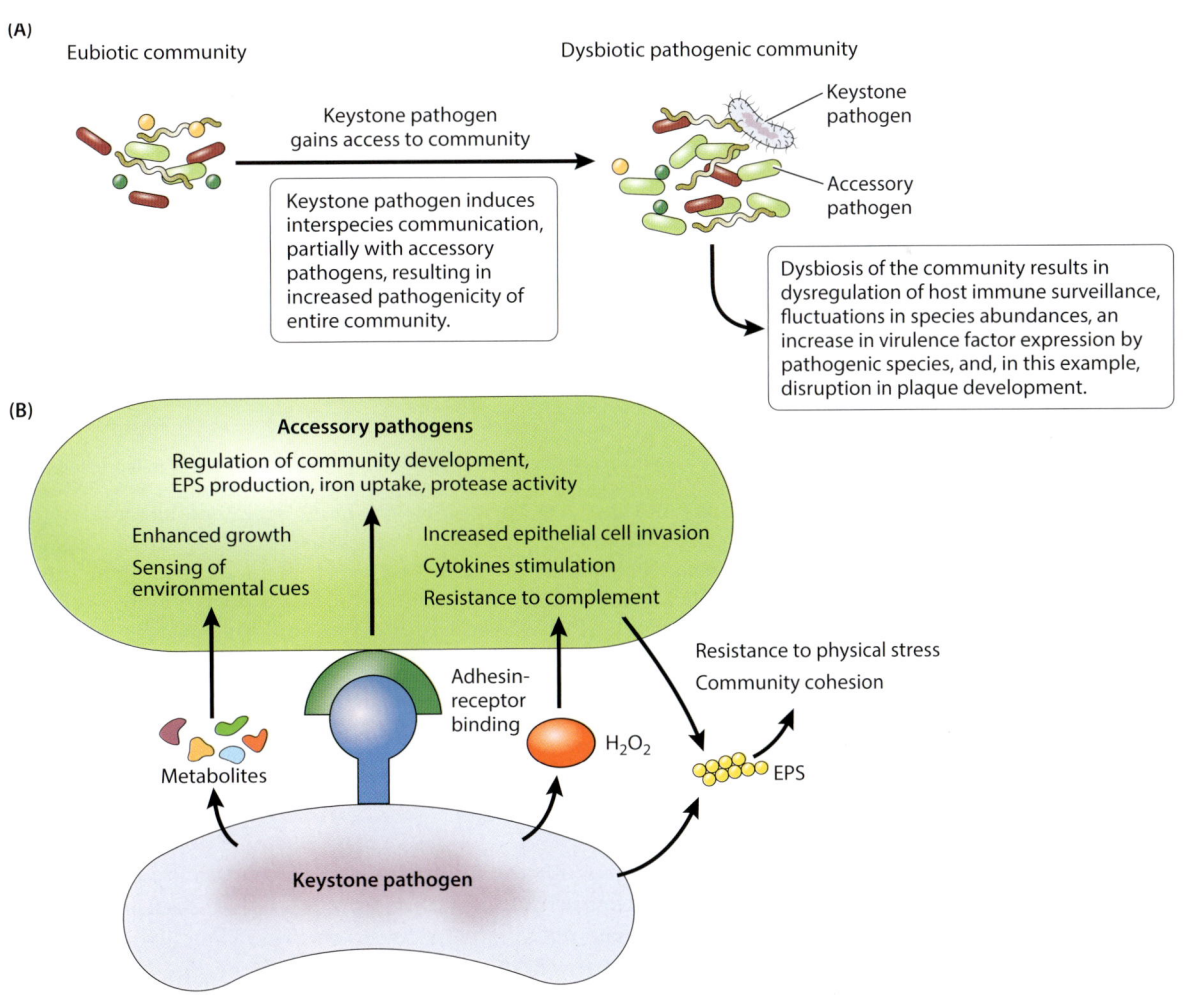

Figure 10.6 Periodontal Disease Progression (After Lamont and Hajishengallis, 2015.)

BOX 10.2. RESEARCH IN ACTION
Breaking the Single-Species Myth in Periodontal Disease

Jorth et al. (2014) were interested in identifying gene expression changes in the oral microbiome that accompanies periodontal disease. To that end, they designed and carried out a simple, yet elegant experiment.

❖ **Hypothesis.** The microbiome associated with periodontal disease has significantly altered patterns and levels of gene expression.

❖ **Methods.** They obtained oral microbiome samples from diseased versus healthy gum regions in the mouths of patients with periodontal disease and examined levels of diversity and patterns of gene expression.

❖ **Results.** Microbiomes associated with diseased gums show lower levels of alpha diversity and altered patterns of gene expression, relative to microbiomes from healthy gum regions (**Figure 10.7**). The researchers found that about 18% of the genes expressed were upregulated in the microbiomes from diseased gum regions.

❖ **Conclusions.** These results indicate that periodontal disease is associated with several significant changes in the oral microbiome, and it argues against a single-species view of periodontitis.

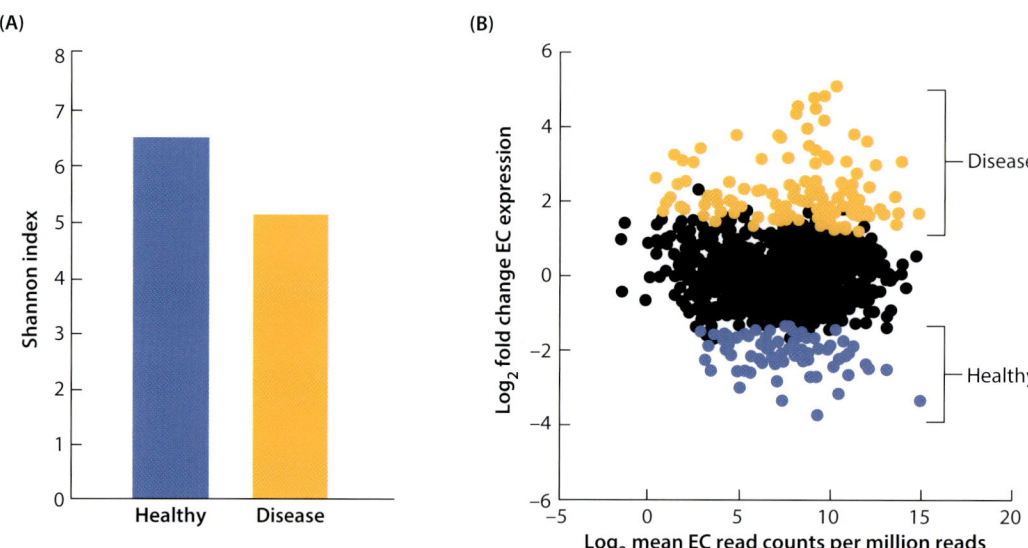

Figure 10.7 Periodontal Disease and the Oral Microbiome Dysbiotic oral microbiomes, found in individuals with periodontal disease, exhibit different levels of species diversity and patterns of gene expression. (A) provides a comparison of Shannon alpha diversity metrics for healthy (blue) and disease (orange) individuals. (B) provides a high-resolution, quantitative view of metabolism. Each dot corresponds to a unique enzyme gene family (EC) whose level of expression is increased in healthy individuals (blue dots) or disease individuals (orange dots). The black dots represent enzyme gene families whose expression levels are not associated with a diseased or healthy state. Levels of expression are indicated by the $\log_2$ fold change during disease plotted against the $\log_2$ mean read counts per million total reads for each gene family. (After Jorth et al., 2014.)

their metabolites provoke the immune system. Some of the pathogens present encode virulence factors that were found to be upregulated in diseased microbiomes. Surprisingly, these virulence features were found in organisms not previously considered to be involved in periodontal disease. As describe above, the PSD model suggests that it is not the bacterial species that matter, but rather what they bring to that particular microbial disturbance. This study provided a surprising view of periodontal disease progression; however, it was unable to definitively distinguish between cause and effect. It remained unclear whether the dysbiotic oral microbiome was causing gum disease or resulted from gum disease.

Another group (Yost et al., 2015) used a longitudinal study to further investigate cause and effect in gum disease. They compared metagenomic and metatranscriptomic data from healthy and diseased periodontal sites over time. They, too, observed that the microbiomes differed significantly between diseased and healthy sites. Certain genes were overexpressed in disease sites, including those involved in lipid A biosynthesis, iron transport, and cell motility. Several species showed upregulation of a variety of virulence genes. In comparison, those sites that did not develop gum disease maintained a nearly constant microbiota, with few changes in gene expression patterns. In healthy sites that went on to develop disease, those species that had previously been associated with periodontitis were present, including *Porphyromonas gingivalis*, *Prevotella intermedia*, and *Eubacterium nodatum*. However, additional species, normally associated with oral health, were detected, including *Streptococcus oralis*, *S. mitis*, and *Pseudomonas fluorescens*. Further, the patterns of gene expression for certain gene families, such as those involved in sulfur metabolism and proteolysis, were different in the microbiomes from healthy and diseased gum regions. These types of studies provide support for the PSD model, which will result in a dramatic shift in our search for appropriate therapeutics.

Shared Dysbiosis

Although we often focus on one particular microbiome, all of our microbiomes are interconnected. Dysbiosis of one part of the microbiome does not occur in isolation—it can trigger changes to other parts of the microbiome or other microbiomes. The gut and oral microbiomes have a particularly strong connection. Members of the oral microbiota can migrate into the digestive tract through saliva and food sources, although these oral microbes are generally poor colonizers of a healthy intestine, due to the drastically different conditions found in the gut. However, in several intestinal diseases, such as inflammatory bowel disease, liver disease, and colon cancer, the oral microbes survive in the intestine. As these orally derived microbes colonize, they activate the intestinal immune system and lead to chronic inflammation. Let's dive deeper into the development of one particular intestinal disease state, liver disease, which is linked to imbalances in the oral and gut microbiota.

Liver cancer is the sixth most common cancer worldwide. There is a strong link between oral health and liver cancer, with a 75% higher risk of developing liver cancer in individuals with poor oral health. The composition of the oral microbiota in patients with liver cancer differs from that of healthy people. Several taxa are found at elevated levels, such as *Clostridium*, *Oribacterium*, *Actinomyces*, and *Campylobacter*, while *Haemophilus*, *Streptococcus*, and *Pseudomonas* are reduced in number. In fact, *Clostridium* and *Oribacterium* are so commonly linked to liver cancer that they are used as biomarkers in liver cancer diagnosis. One mechanism that appears to be acting is the invasion of the intestinal tract by *Porphyromonas gingivalis* with a resulting increased intestinal mucosal permeability and insulin resistance. These changes permit subsequent invasion of gut bacteria into the liver and the triggering of inflammation, an early stage in cancer development (**Figure 10.8**).

The oral microbiota also appears to be associated with colorectal cancer (CRC), which is another quite common form of cancer, diagnosed in 1 in 25 individuals. In patients with CRC there is an increased abundance of the oral bacterium *Fusobacterium nucleatum*. The association is so strong that its presence is used as a biomarker for this disease. Further, in CRC disease progression, the biofilm that forms on the intestinal mucosa is similar to its oral counterpart. One mechanism proposed in CRC development involves *F. nucleatum* acting directly on cells in the colon, adhering to them, and activating a transcriptional pathway that leads to cancer. *F. nucleatum* also promotes the entry of normally noninvasive bacteria into epithelial cells, which further promotes tumor development.

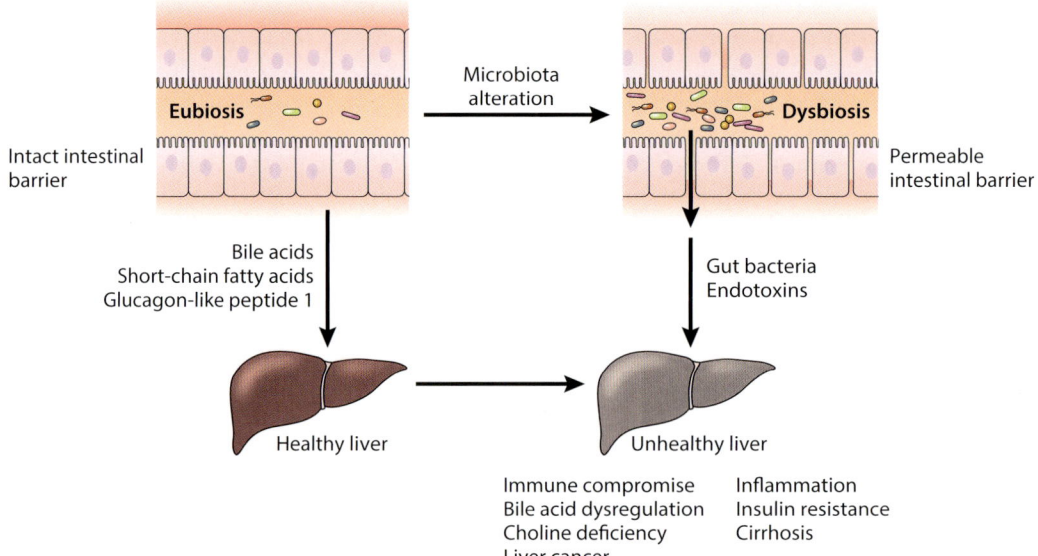

Figure 10.8 Gut Microbiome Dysbiosis and Liver Disease Perturbations of the gut microbiome not only impact the gut. In this illustration we see that alterations in the gut microbiome result in significant changes in intestinal cell wall permeability, which allows bacteria to translocate to distant sites, in this case the liver. The immune system responds to this invasion with inflammation in the liver tissue, and a cascade of health impacts result, such as dysregulation of bile acid production, insulin resistance, and ultimately, cirrhosis and/or liver cancer.

Is the Term *Dysbiosis* Still Useful?

Microbiome dysbiosis is easy to find. With such a broad definition—disruption of homeostasis—it is no wonder that we see it linked to most diseases. Some have attempted to limit the term to include only "major" changes in the microbiome. For example, Antharam and colleagues (2013) define dysbiosis as a profound alteration of a microbiome that results in lower species richness. Unfortunately, the finding that healthy and ill individuals have different levels of diversity is so common as to be effectively useless. Time series data may help refine the definition but do not get around the basic problem, which is that finding microbiota changes associated with a diseased state tells us simply that there is dysbiosis and provides no information about cause or effect.

Does this render the term *dysbiosis* useless? There are two scenarios to consider. The first, the **artifact situation**, occurs when dysbiosis is consistently associated with the cause of a disease (or by the disease itself) but is a consequence rather than cause of disease. The second, the causal situation, is when dysbiosis itself is a sole or major factor in disease production. If the first scenario holds, dysbiosis is simply a diagnostic tool. If the second, dysbiosis is a cause of a disease and can be the object of remediation efforts. This latter scenario is critical, and we certainly don't want to toss out the dysbiotic microbiome with the bathwater. However, for the term dysbiosis to be useful, the mechanisms by which dysbiosis causes disease must be pursued. So far, we have discovered very few examples where dysbiosis, itself, causes the disease and we understand the mechanisms by which it occurs. In fact, we often use dysbiosis as a placeholder, noting that the microbiome may be relevant to the disease condition. Our goal, of course, is to determine precisely how dysbiosis results in disease.

Another question to address is whether a focus on microbiome diversity is helpful if the goal is to identify disease causality. Should the focus be on function instead? Microbiota are assembled through both **stochastic** (random) and **deterministic** (selective) processes. The host selects for the presence of some microbes by providing an appropriate food source, pH, or temperature. However, the source of microbes that then experience that selection is largely due to chance interactions, such as whether there are pets in the household, or an individual's diet. Stochastic mechanisms introduce many random microbes to the host, which are then selected for or against through deterministic processes. Selection is likely to target specific functions rather than specific taxa. For example, your body needs butyrate producers, but it doesn't care whether it's *Faecalibacterium prausnitzii* or *Roseburia* producing butyrate. The large microbiota variability between microbiomes may actually be hiding a set of core functions encoded by whoever is present. In the end, uncovering the mechanisms of microbiome-caused disease will likely require both taxonomic and mechanistic data, which are, fortunately, becoming more and more abundant.

How can one distinguish between a eubiotic and dysbiotic state? Early attempts focused on quantifying the differences between healthy and dysbiotic microbiota to reveal the cause of such differences. The goal was to statistically determine the normal microflora. Even now, there have been attempts to develop indices of dysbiosis in order to quantify specific microbial contributions to a disease state. Indices require extremely large sample sizes, particularly given the levels of variation between even healthy microbiomes. For example, a sample of over 4,000 Europeans was examined, and the data revealed that associations between numerous disease states and variation in microbiome composition were not as clear as had been expected, and in some cases, microbiota that could be defined as dysbiotic were found to have positive effects on host health (Falony et al., 2019).

Koch's Postulates Applied to Microbiome-Health Associations

In 1876, Robert Koch proposed his famous postulates, which we first learned about in chapter 2. This approach to the study of infectious disease has dominated microbiological research and resulted in a fairly narrow perspective, the so-called one pathogen–one disease concept. However, we now understand that microbiomes are complex ecosystems and that infections are far more often due to an interplay between the host and numerous microbial constituents, which is completely outside the one pathogen–one disease paradigm. However, several modifications of Koch's postulates, in light of our new understanding of the microbiome, have been proposed (Neville et al., 2018). We will focus here on one modification that embraces the ecology of the microbiome as it seeks to identify disease causes, the so-called **ecological Koch's postulates** (Vonaesch et al., 2018).

This ecologically minded revision of Koch's original postulates is based on two major observations. First, not every individual infected with a pathogen will exhibit disease. Therefore, our prior focus on the target pathogen must now be expanded to include information on the host susceptibility, nutritional state, prior infections, and genetics as well as the resident microbiota and insults to the microbiome. The pathogens not only need to be present, but also must be able to establish themselves in a highly competitive niche in order for them to invade, persist, and cause disease. Second, there has been the discovery of several gastrointestinal illnesses in which there is no overt pathogen identified but where the entire microbiome appears to mediate disease. This raises the prospect that normal members of a microbiome may become invasive and cause disease under certain conditions. For example, if we turn to any of a number of intestinal diseases, such as IBD, CRC, or obesity, we find gut microbiome dysbiosis, increased levels of oxygen in the large intestine, and an increase in aerotolerant taxa, such as members of the Enterobacteriaceae, which don't need but can tolerate oxygen. These ecological disturbances result in decreased resilience of the microbiome and increased susceptibility of the host to pathogens and other, normally commensal, bacteria with potentially harmful properties (often called **pathobionts**). In short, these diseases are characterized by the fact that "the wrong bacteria are in the wrong proportions, in the wrong company or in the wrong place" (Vonaesch et al., 2018).

An overview of this ecological perspective of Koch's postulates is detailed in **Box 10.3**. According to the revised postulates, a dysbiosis may lead to a disease. For instance, a dysbiotic microbiome is observed in a diseased individual. This hypothesized cause of disease, a dysbiotic microbiome, is tested by transferring the microbiota

BOX 10.3. A COMPARISON OF THE ORIGINAL AND ECOLOGICAL KOCH'S POSTULATES

Original Koch's Postulates (Figure 10.9A) (Koch, 1876)

1. The microorganism must be present in all diseased individuals.
2. The microorganism must be isolated from the diseased host and be grown in a pure culture.
3. The reinoculation of a naive host with this pure culture must lead to the same disease as in the original host.
4. The microorganism must be recovered from the newly diseased host.

Ecological Koch's Postulates (Figure 10.9B) (Vonaesch et al., 2018)

1. The dysbiotic microbiota is found in similar composition / with similar characteristics in all affected individuals.
2. The dysbiotic microbiota can be retrieved from the affected host.
3. Introducing the dysbiotic microbiome into germ-free hosts leads, in combination with a similar environment (e.g., genetic makeup of the host, nutrition, age), to symptoms similar to those in the affected individual.
4. The dysbiotic microbiota composition remains stable in the newly affected host.

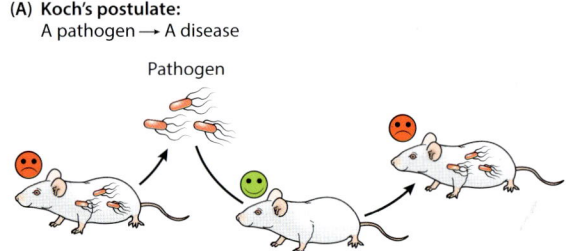

Figure 10.9 Application of Koch's Postulates (Original and Ecological) to Prove Causation in Microbiome-Based Disease (After Vonaesch et al., 2018.)

from the diseased individual into a healthy individual and monitoring changes in health. Let's take obesity as an example. We have discussed several times the highly influential study on obesity carried out by Jeff Gordon and his colleagues. They discovered that if the gut microbiome from an obese mouse is moved into the gut of a lean mouse, the perturbed mouse will gain weight (Turnbaugh et al., 2006). In this example, the mice are genetically identical and are housed under identical conditions. Thus, it is reasonable to deduce that the transferred gut microbiota is the cause of weight gain. A second example is provided by the condition of acute undernutrition. Children suffering from malnutrition have an altered gut microbiota and increased abundance of certain pathogens, including *E. coli*, *Campylobacter* spp., and *Giardia* spp. A study of malnutrition in Bangladeshi children revealed that a strain of *Bacteroides fragilis* was pathogenic when in a dysbiotic gut microbiome but was a healthful commensal in normally nourished children. In the dysbiotic state, the strain produces enterotoxin, which damages gut mucosal epithelial cells by altering their membrane permeability. Transplantation of dysbiotic microbiota into germ-free mice and then infecting the mice with the enterotoxin-producing *B. fragilis* strain led to malnutrition, but malnutrition did not occur when the pathogen was administered to mice harboring healthy microbiota (Wagner et al., 2016). This example illustrates that the whole ecosystem, rather than an isolated element, contributes to morbidity.

We are only just beginning to understand the complex relationships and interactions within the microbial ecosystems in the human body and, in particular, those interactions that result in disease. Dysbiosis has previously provided a unifying framework for microbiome studies, as it seeks to identify members of the microbiome that are associated with healthy and/or diseased states. However, it is increasingly clear that it is time to move beyond the concept of dysbiosis and into a more nuanced view based upon an ecological perspective of the microbiota and host interactions if we hope to translate our growing body of microbiome data into therapeutic interventions in disease.

CHECK YOUR UNDERSTANDING

1. A healthy microbiome composition
   a. is the same across all individuals.
   b. varies from person to person.
   c. is the same for your entire life.
   d. was discovered by NIH researchers as part of the Human Microbiome Project.

2. Microbiome diversity
   a. is a clear indicator of overall health.
   b. varies according to diet and lifestyle changes.
   c. decreases when people are unhealthy.
   d. is a good metric of homeostasis.

3. Probiotics may help to treat diseases associated with some microbiome composition changes. These include
   a. presence of bacteria associated with a disease.
   b. absence of health-associated bacteria.
   c. a major shift in community composition.
   d. presence of immune-modulating microbes.

4. What is the hypothesized role of *Lactobacillus* in the vaginal microbiome?
   a. Aid in lactose digestion
   b. Production and secretion of mucus
   c. Prevention of infection via colonization resistance
   d. Colonization of newborns' skin microbiomes

5. Paneth cells are located in the
   a. small intestine.
   b. large intestine.
   c. gallbladder.
   d. appendix.

6. Which vitamin, the one human body can't produce itself, is synthesized in the small intestine?
   a. B_{12}
   b. C
   c. B_{14}
   d. D

7. The low oxygen level in the large intestine is important because
   a. it promotes the growth of aerobic bacteria.
   b. it selects for bacteria that produce their own oxygen.
   c. it selects for anaerobic bacteria that ferment carbohydrates.
   d. it selects for anaerobic pathogens.

8. Colonization resistance
   a. helps prevent new microbes from establishing themselves in an environment.
   b. ensures that your microbiome composition is constantly changing.
   c. is the host's hostility towards resident microbes.
   d. is the same as antibiotic production.

9. The polymicrobial synergy and dysbiosis model refers to
   a. dysbiosis caused by a specific set of microbes.
   b. a "core" microbiome being responsible for overall health.
   c. disease states being triggered by microbiome dysbiosis, not specific bacteria.
   d. decreasing microbiome diversity linked to better overall health.

10. The conservation of disease-state metabolomic profiles in periodontitis patients suggests
    a. there is a high level of functional redundancy between microbe species.
    b. periodontitis is likely caused by a single type of bacteria.
    c. 16S sequencing would be a good approach to understanding periodontitis risk.
    d. which microbes are present is more important than the metabolites they produce.

11. Approximately how many individuals were included in the first phase of the Human Microbiome Project?
    a. 50
    b. 300
    c. 1,000
    d. 500

12. Why might a more diverse microbiome be more resilient to environmental changes?
    a. Antibiotics are more likely to be effective against diverse communities.
    b. Diverse microbes produce more metabolites.

c. Having more bacteria in a system allows them to reproduce more quickly.
d. Diverse communities likely have multiple species that can fill the same niche if one is eliminated.

13. One example of increased microbiome diversity leading to a negative health outcome is
    a. *C. difficile* infections.
    b. bacterial vaginosis.
    c. GI tract imbalance.
    d. mycorrhizal interactions.

14. Which bacteria create a barrier to vaginal infection?
    a. *C. difficile*
    b. *E. coli*
    c. *Lactobacillus*
    d. *Bifidobacterium*

15. Which type of cells secrete mucin to exclude bacteria from the large intestine, airways, and reproductive tract?
    a. Paneth cells
    b. Goblet cells
    c. T cells
    d. B cells

16. Which condition is a type of inflammatory bowel disease that is associated with disruption of the Paneth cells?
    a. IBS
    b. Colon cancer
    c. Crohn's disease
    d. Periodontitis

17. In the small intestine, overgrowth of microbes is prevented by all of the following *except*
    a. antibacterial secretion from the pancreas.
    b. gastric acid production.
    c. antibody immunoglobulin secretion.
    d. the mucus escalator.

18. Oxygen levels in the lumen of the large intestine are kept low by bacteria through a pathway known as
    a. β-oxidation.
    b. the electron transport chain.
    c. phosphorylation.
    d. photosynthesis.

19. Organisms' alteration of their environment is known as
    a. niche preemption.
    b. nourishing immunity.
    c. niche modification.
    d. sterilizing immunity.

20. All of the following are associated with colorectal cancer *except*
    a. *Fusobacterium*.
    b. *Porphyromonas*.
    c. *Escherichia*.
    d. *Enterobacter*.

21. Which model describes a disease triggered by a dysbiotic microbiome as opposed to specific pathogens?
    a. Polymicrobial synergy and dysbiosis
    b. Loss of health-associated bacteria
    c. Limited pathogens category
    d. Colonization resistance

22. All of the following are associated with periodontitis *except*
    a. *Escherichia coli*.
    b. *Porphyromonas gingivalis*.
    c. *Treponema denticola*.
    d. *Tannerella forsythia*.

23. A disease biomarker
    a. can be used to treat a disease state.
    b. is used to diagnose disease.
    c. indicates that a disease is being treated.
    d. always is associated with a specific disease state.

24. In genetically identical mice, transferring microbiota from an obese individual to a lean individual will cause the lean individual to
    a. stop eating.
    b. lose more weight.
    c. attack the obese individual.
    d. gain weight.

Answers: 1B, 2B, 3B, 4C, 5A, 6A, 7C, 8A, 9C, 10A, 11B, 12D, 13B, 14C, 15B, 16C, 17D, 18A, 19C, 20C, 21A, 22A, 23B, 24D

DIVING DEEPER

1. What are some of the reasons why developing a specific, useful definition of dysbiosis and eubiosis is challenging?
2. What is one example that shows why high microbiome diversity is not always a positive?
3. What types of microbiome-disease associations were proposed by Gilbert et al. and by Duvallet et al.?
4. How is breastfeeding a form of host control of the microbiome?

5. What are Paneth cells, and how do they alter the microbiome?
6. How are microbiota in the colon affected by oxygen levels?
7. How is host health affected when oxygen levels in the colon increase?
8. What is the polymicrobial synergy and dysbiosis model of periodontitis?
9. Why are oral microbiota not normally good colonizers of the gut microbiome?
10. How are oral microbiota involved in liver cancer and colorectal cancer?
11. What are the two processes by which the microbiome is assembled?
12. What are Koch's postulates applied to microbiome research?

DISCUSSING AND REFLECTING

1. In this chapter, the Duvallet group at MIT suggested that narrow-spectrum antimicrobials may be effective at preventing diseases associated with specific microbes. Why might we prefer narrow- to broad-spectrum antimicrobials? What are the benefits and drawbacks of each?

2. As we learn more about the microbiome, many researchers are approaching host-microbe interactions through the lens of ecology. Understanding which selection pressures are acting upon both host and microbe can give us important insight into what role these microbes are playing in human health. Aside from those we've discussed in this chapter, what are some selection pressures bacteria must overcome to thrive in the human environment? How may these have shaped your microbiome community?

3. Reflection. The following poem is "Host" by Jarod Anderson. Keeping in mind the ecological perspective on the microbiome we've used this chapter, reflect on how many bacteria have made their homes in your ecosystem. Microbiome community composition is clearly important to our overall health; has this information changed your perspective on a "healthy" lifestyle?

> Host
> To invading germs, you are a jungle full of hungry tigers.
> To your gut bacteria, you are a warm orchard of perpetual bounty.
> To your eyelash mites, you are a walking fortress and a mountaintop pasture.
> How many generations have you hosted?
> What do they name the wilderness of you?
> (Anderson, 2020)

RECOMMENDED READINGS

Popular Science Reviews

Brüssow, H. (2019). Problems with the Concept of Gut Microbiota Dysbiosis. *Microbial Biotechnology*, 13(2), 423–434. https://doi.org/10.1111/1751-7915.13479

Maroney, D. (2021, December 3). Is Your Modern Lifestyle Destroying Your Gut Microbiome? This Probiotic Strain Could Fix It! *Discover*. https://www.discovermagazine.com/lifestyle/is-your-modern-lifestyle-destroying-your-gut-microbiome

Popular Science Book

Cowell, R. (2020). *Dysbiosis: A Study of Underlying Causes* (first edition). Nova Science.

Scientific Review

Wei, S., Bahl, M. I., Baunwall, S. M. D., Hvas, C. L., & Licht, T. R. (2021). Determining Gut Microbial Dysbiosis: A Review of Applied Indexes for Assessment of Intestinal Microbiota Imbalances. *Applied and Environmental Microbiology*, 87(11), e00395-21. https://doi.org/10.1128/AEM.00395-21

The Microbiome and Obesity

11

CHAPTER CONTENTS

11.1 The Obesity Epidemic
11.2 The Microbiome of Obesity
11.3 Metabolic Markers of Weight Loss
11.4 The Gut Microbiome and Weight Loss
11.5 Diet-Based Approaches

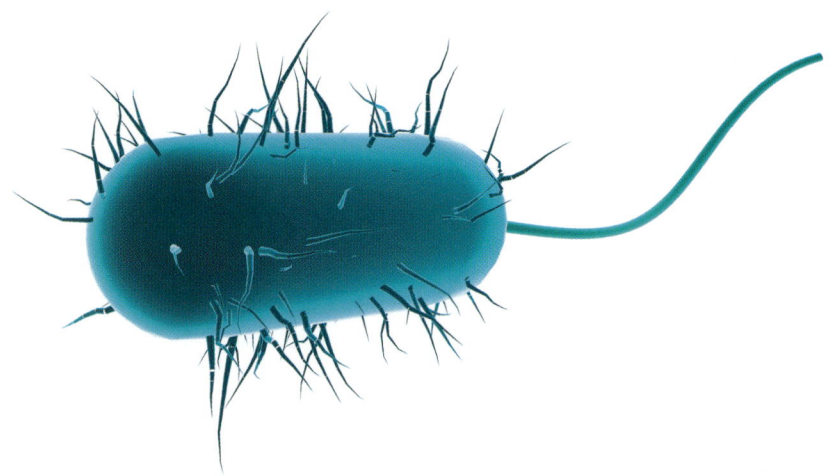

If you want to greet me, get in line. Everyone wants to interview me, bestow awards on me, and I'm very popular on social media. Who am I? Oh, I assumed you already knew. I am *Akkermansia muciniphila*. For you word nerds, you might be able to guess that my moniker means that I love mucin—yes, that slimy gunk that lines your large intestine. I dine on mucin day and night. Now, you might think that would make me one of the bad guys—mucin keeps my bad cousins from slipping out of your gut in their inopportune quests to wreak havoc in your body. However, my mucin digestion is special; it actually stimulates more mucin production, so there is always just the right amount. I also release key metabolites, such as the short-chain fatty acids propionate and acetate that help you stay healthy and lean. If that weren't enough, I have been granted the prestigious moniker of *keystone species*, as even with my low abundance in your gut, I have an oversized impact on your weight! (Photo from iStock.com/grechina)

"When you eat, you're not just nourishing your body, you're feeding the trillions of microbes that live inside your gut."
—Tim Spector (Leitch, 2021)

For the past 50 years, physicians have considered most obesity to result from eating too much and exercising too little. "Eat less, move more" or "calories in, calories out" were, and in many cases remain, the mantras to weight loss. And yet, even for those who have valiantly stuck to their diets and walked their 10,000 steps each day, the weight simply won't come off. Why is that? Why are some individuals prone to weight gain, while others are not? Why do some people lose weight regardless of what they eat, while others can't drop an ounce regardless of how few calories they consume? Here we will start by gaining a perspective on the enormity of the problem. We will then turn to the changes in the microbiome in obese individuals, which involve disruptions not only in which microbes are present, but also in the metabolites being produced. Finally, we will explore how we hope to employ the microbiome itself in our weight loss efforts in the future.

11.1 THE OBESITY EPIDEMIC

It should not be news to anyone that the United States, indeed the world, faces an obesity epidemic. How do we know? Let's employ a simple, albeit incomplete, measure of the amount of fat in our body—the body mass index, or BMI. The following equation shows you how to calculate your BMI, and it is followed by a sample calculation.

$$\text{BMI calculation: BMI} = 703 \times (\text{weight in pounds}) \div (\text{height in inches})^2$$

BMI sample calculation: If I weigh 150 pounds and stand 5 feet 6 inches tall (i.e., 66 inches), then my BMI $= 703 \times 150 \div 662 = 703 \times 150 \div 4{,}356 = 24.21$.

Now find your BMI on **Figure 11.1**, which provides the calculated BMIs over a range of weight and height measures, with BMIs color coded by the following five categories: underweight, healthy, overweight, obese, and extremely obese. The sample BMI we calculated above just barely falls within the healthy range on this chart.

Obesity is generally defined as having a BMI greater than 30, and using this metric, it is estimated that more than one-third of the world's population is either overweight (BMI of 25–30) or obese. Worse yet, these numbers are rapidly rising, and by 2030 there will be over 1 billion obese individuals. However, BMI is a single measurement and can't provide you with a full picture of what is a healthy versus unhealthy weight. Using BMI alone, you would have no insight into your metabolic health or any risk factors associated with weight gain. Additionally, BMI was devel-

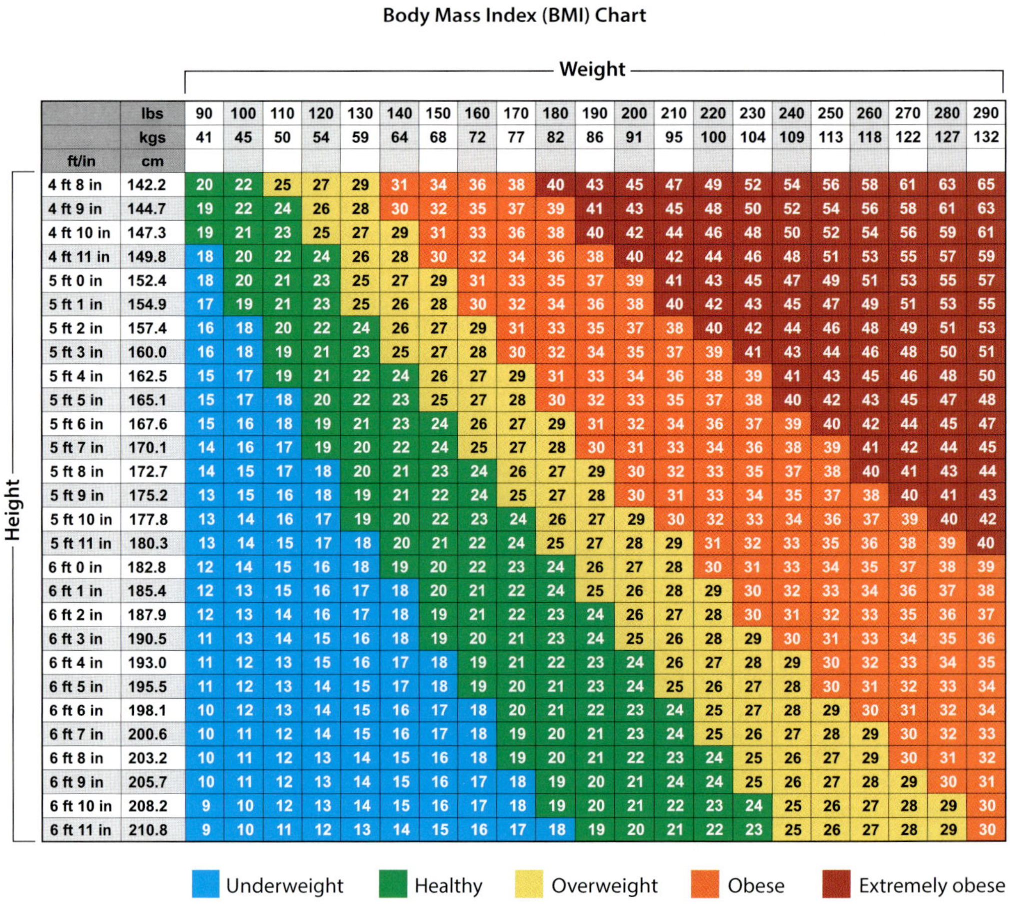

	lbs	90	100	110	120	130	140	150	160	170	180	190	200	210	220	230	240	250	260	270	280	290
	kgs	41	45	50	54	59	64	68	72	77	82	86	91	95	100	104	109	113	118	122	127	132
ft/in	cm																					
4 ft 8 in	142.2	20	22	25	27	29	31	34	36	38	40	43	45	47	49	52	54	56	58	61	63	65
4 ft 9 in	144.7	19	22	24	26	28	30	32	35	37	39	41	43	45	48	50	52	54	56	58	61	63
4 ft 10 in	147.3	19	21	23	25	27	29	31	33	36	38	40	42	44	46	48	50	52	54	56	59	61
4 ft 11 in	149.8	18	20	22	24	26	28	30	32	34	36	38	40	42	44	46	48	51	53	55	57	59
5 ft 0 in	152.4	18	20	21	23	25	27	29	31	33	35	37	39	41	43	45	47	49	51	53	55	57
5 ft 1 in	154.9	17	19	21	23	25	26	28	30	32	34	36	38	40	42	43	45	47	49	51	53	55
5 ft 2 in	157.4	16	18	20	22	24	26	27	29	31	33	35	37	38	40	42	44	46	48	49	51	53
5 ft 3 in	160.0	16	18	19	21	23	25	27	28	30	32	34	35	37	39	41	43	44	46	48	50	51
5 ft 4 in	162.5	15	17	19	21	22	24	26	27	29	31	33	34	36	38	39	41	43	45	46	48	50
5 ft 5 in	165.1	15	17	18	20	22	23	25	27	28	30	32	33	35	37	38	40	42	43	45	47	48
5 ft 6 in	167.6	15	16	18	19	21	23	24	26	27	29	31	32	34	36	37	39	40	42	44	45	47
5 ft 7 in	170.1	14	16	17	19	20	22	24	25	27	28	30	31	33	34	36	38	39	41	42	44	45
5 ft 8 in	172.7	14	15	17	18	20	21	23	24	26	27	29	30	32	33	35	37	38	40	41	43	44
5 ft 9 in	175.2	13	15	16	18	19	21	22	24	25	27	28	30	31	33	34	35	37	38	40	41	43
5 ft 10 in	177.8	13	14	16	17	19	20	22	23	24	26	27	29	30	32	33	34	36	37	39	40	42
5 ft 11 in	180.3	13	14	15	17	18	20	21	22	24	25	27	28	29	31	32	33	35	36	38	39	40
6 ft 0 in	182.8	12	14	15	16	18	19	20	22	23	24	26	27	28	30	31	33	34	35	37	38	39
6 ft 1 in	185.4	12	13	15	16	17	18	20	21	22	24	25	26	28	29	30	32	33	34	36	37	38
6 ft 2 in	187.9	12	13	14	16	17	18	19	21	22	23	24	26	27	28	30	31	32	33	35	36	37
6 ft 3 in	190.5	11	13	14	15	16	18	19	20	21	23	24	25	26	28	29	30	31	33	34	35	36
6 ft 4 in	193.0	11	12	13	15	16	17	18	19	21	22	23	24	26	27	28	29	30	32	33	34	35
6 ft 5 in	195.5	11	12	13	14	15	17	18	19	20	21	23	24	25	26	27	28	30	31	32	33	34
6 ft 6 in	198.1	10	12	13	14	15	16	17	18	20	21	22	23	24	25	27	28	29	30	31	32	34
6 ft 7 in	200.6	10	11	12	14	15	16	17	18	19	20	21	23	24	25	26	27	28	29	30	32	33
6 ft 8 in	203.2	10	11	12	13	14	15	16	18	19	20	21	22	23	24	25	26	27	29	30	31	32
6 ft 9 in	205.7	10	11	12	13	14	15	16	17	18	19	20	21	24	24	25	26	27	28	29	30	31
6 ft 10 in	208.2	9	10	12	13	14	15	16	17	18	19	20	21	22	23	24	25	26	27	28	29	30
6 ft 11 in	210.8	9	10	11	12	13	14	15	16	17	18	19	20	21	22	23	25	26	27	28	29	30

Figure 11.1 BMI and Obesity The BMI is one of the most common metrics of weight health. You can readily find your BMI on a chart, such as the one provided here. BMI helps to inform you whether you are at a healthy weight or fall into the overweight or underweight category. (Photo from Abhijeet Bhosale/Shutterstock)

oped using data from white populations, making it even less accurate for individuals with different body compositions, such as percentage body fat, which can vary by race or ethnic group (Fontaine et al., 2003).

We used to consider obesity the result of certain lifestyle choices, and it is true that a sedentary lifestyle and a higher intake of energy-rich, nutrient-poor foods, such as processed snacks, cakes, ice cream, and pastries, are strongly linked to obesity. However, environmental, genetic, and hormonal factors all play a role. There are hundreds of variants in the human genome that predispose us to obesity. These genetic differences can range from a single gene playing a strong role in increasing a person's likelihood of becoming obese to a combination of many different genes, each of which has a modest effect. Studies involving twins have shown that obesity has a high **heritability rate**, which is an indication of how much variation in a trait, such as body weight, is caused by your genes rather than the environment. The heritability of body weight is between 40% and 75%, which means that, yes, your genes do matter. However, there is considerable room for external factors to play a role. And, you guessed it, the microbiome is gaining more and more credit for its contributions to modulating body weight.

Losing Weight Is Hard to Do

A mere one in six adults who have lost weight are able to keep it off long-term. The reason weight loss is so challenging is quite simple: there is a mismatch between our biology and the environments in which we find ourselves. Many people in affluent, industrialized nations have access to highly processed food and lead a sedentary lifestyle. Those two factors clearly play a role in influencing weight gain. Additionally, obesity is strongly correlated with poverty in these industrialized nations, due to unhealthy food being cheaper and easier to access than healthier options. The term *food desert* describes geographic areas where access to healthy food options is limited due to the scarcity of grocery stores within convenient traveling distance. But equally important is the fact that our bodies have been programmed since our earliest primate ancestors' time to put on fat in order to survive famine. We are designed to preserve this fat by slowing down our metabolism while simultaneously signaling hunger. Even our microbiome has evolved to extract the maximum amount of energy possible from the food we consume. Those ancient humans who stored enough fat to make it through the lean winter months survived. The thinner individuals were selected against. Well, even though our environment and lifestyle have changed, our bodies (including our microbiomes) still think we should be fattening up for that proverbial tough winter to come.

Obesity is not simply a matter of weight gain. It is associated with a cluster of metabolic conditions referred to as **metabolic syndrome** (**MetS**), illustrated in **Figure 11.2**, which increase your risk of cardiovascular disease, stroke, and diabetes. One such metabolic condition is **glucose intolerance**, a set of physiological changes that can result in increased levels of blood sugar. Another is **insulin resistance**, in which cells in certain tissues, such as the liver, aren't able to respond to insulin levels and thus don't take up glucose from your blood properly. Your pancreas attempts to make more insulin to help glucose enter these cells. Over time, the pancreas produces increasing amounts of insulin to regulate blood sugar, until all cells are insulin-resistant and the pancreas cannot produce enough insulin for blood sugar levels to remain in a healthy range. A third component of MetS is **lipid imbalance**, which refers to levels and proportions of **low-density lipoprotein cholesterol** (**LDL-C**), **high-density lipoprotein** (**HDL**), and **triglycerides**. LDL-C and HDL are types of cholesterol that are important for maintaining cell membrane structure and synthesizing some hormones, but high levels of LDL-C can block arteries and cause heart attack or stroke. Triglycerides are used for energy storage, but like LDL-C, when too much is produced, they can increase your chance of heart attack or stroke. Finally, blood pressure rises, resulting in **hypertension**, or a higher-than-normal blood pressure. All of these health consequences are combined with a state of low-grade, chronic, systemic inflammation that feeds directly back into insulin levels, glucose levels, and the like.

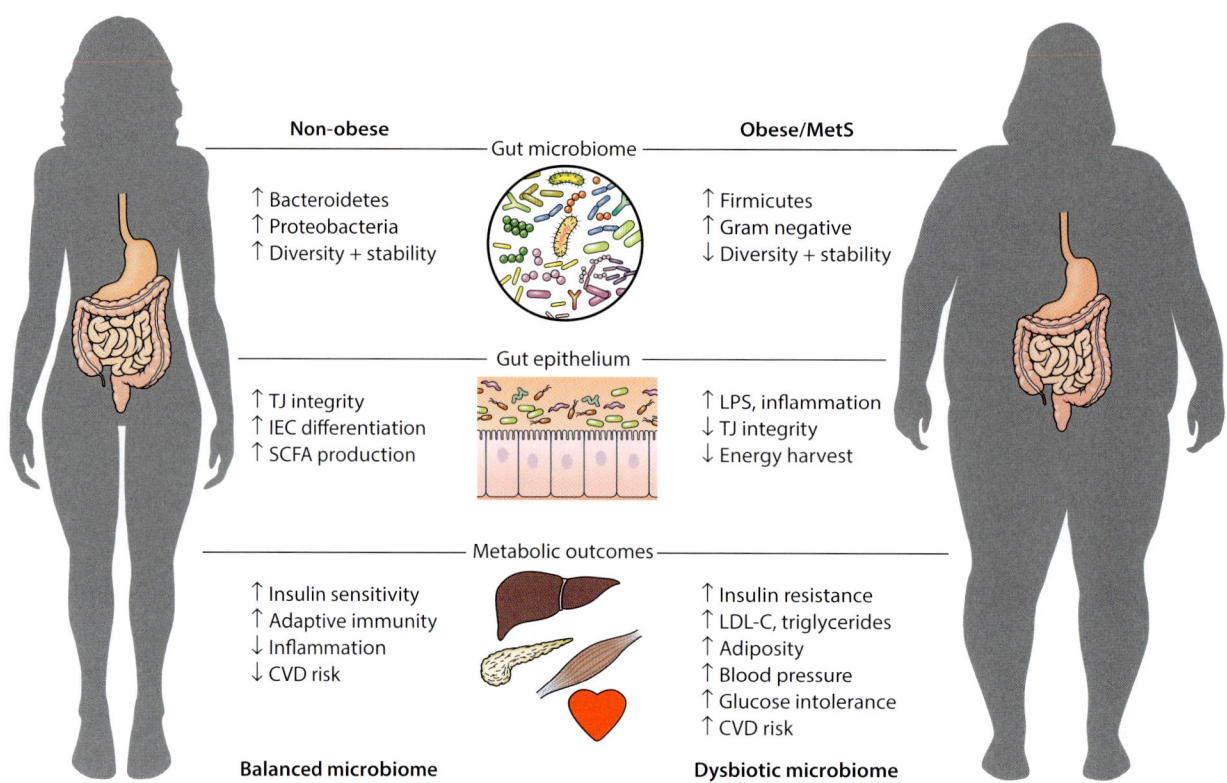

Figure 11.2 The Role of the Microbiome in Metabolic Syndrome and Obesity A comparison of how the gut microbiome impacts health with respect to obesity and metabolic syndrome (MetS). Changes in the microbiota result in impacts on the integrity of the tight junctions (TJs) in the lining of the large intestine, as well as a decrease in the levels of short-chain fatty acids (SCFAs) specifically and energy harvest in general. Two outcomes on the host's metabolism are increased levels of fat deposition, or adiposity, and increases in insulin resistance. (IEC = intestinal epithelial cells, CVD = cardiovascular disease, LPS = lipopolysaccharides, LDL-C = low-density lipoprotein cholesterol)

We now understand that a Westernized diet, one rich in highly processed foods and lacking fiber, impacts the gut microbiota, resulting in a significant risk factor for MetS (see Figure 11.2). The dysbiotic gut microbiota lacks some of the key fiber-fermenting microbes, resulting in a drop in short-chain fatty acid (SCFA) production, the promotion of adiposity (body fat), an increase in fat deposition, and insulin resistance, perhaps by altering energy extraction during digestion. There is an increase in gram-negative bacteria, which have lipopolysaccharides on their outer membranes that increase the permeability of the tight junctions, which normally create an intercellular barrier between gut epithelial cells. Tight junction integrity is crucial in preventing pathogens from penetrating host tissues and eliciting an immune response.

11.2 THE MICROBIOME OF OBESITY

Much of the early work on the relationship between obesity and the gut microbiome employed animal studies. Let's recall the research of Dr. Jeffrey Gordon, who we first met in chapter 2. He was interested in studying what happens when mice are raised without any microbes, that is, they are germ-free. These mice weigh less than conventional mice, even when they are fed more food. You see, without a microbiome, nondigestible food, which normally feeds our microbiome, passes through the mouse GI system undigested. The result is that the host doesn't gain any benefit from the undigestible fraction of its feed. In contrast, conventional mice extract what they can from their feed and then rely on their gut microbiota to extract even more energy from the undigestible bits, providing the host with some of this extra energy.

Dr. Gordon and Ruth Ley, who was his postdoctoral research fellow at the time, used germ-free mice to research whether the microbiome impacts weight. Their first study focused on mice with a mutation resulting in obesity, which they used to study the difference in the microbiomes of lean and obese mice (Ley et al., 2005). Leptin-deficient mice cannot produce the hormone **leptin**, which conveys satiety. Thus, leptin-deficient mice eat excessively and eventually become obese. Gordon and Ley showed that these mice acquire a different microbiota than lean mice, even when sharing the same mothers and consuming the same diet. Analyzing the microbial communities present in the feces of wild-type and leptin-deficient mice revealed that obese mice had changes in their microbiota compositions. Namely, obese mice had high levels of Firmicutes and low levels of Bacteroidetes, suggesting that weight can impact the composition of the microbiome. This obesity-associated microbiota provided an increase in energy-harvesting functions.

This result led to a second study in which gut microbiota from the leptin-deficient mice were transplanted into guts of germ-free mice without the obesity mutation, to determine whether the microbiome could affect the weight of the mice. Researchers also transplanted microbiota from lean mice into some of the germ-free mice. This control ensured that researchers were isolating the effect of the obese or lean microbiota, and not the effect of other aspects of the protocol. The result was that mice given the gut microbiota from obese mice gained more weight than those who received microbiota from lean mice.

In subsequent studies, Gordon's team turned to humans. They transferred the gut microbiota from human twins, one of whom was lean and the other obese, into germ-free mice (see Figure 2.18) (Ridaura et al., 2013). Transplanting gut microbiota from humans into mice allows researchers to study the microbial composition found in humans while controlling factors, such as diet and lifestyle, that cannot be controlled for in a human study. This approach is the best way to study the effects of the human microbiome on aspects of health in a controlled manner, although it is obviously limited, relying on mice instead of humans. The mice given the obese twin's microbiota gained more weight, while those receiving the lean twin's microbiota remained lean. Although all the mice ate the same food, the gut microbes from the obese twin were able to harvest more energy from the food, leading to weight gain. These studies created quite a stir, as they suggested that our microbiomes create the condition of obesity and that if we can change our microbiota, perhaps we can lose weight.

Obese versus Lean Gut Microbiomes

We have known for some time that the microbiomes of obese and lean people differ. In general, the gut microbiomes of obese individuals show a decreased microbial richness and diversity and alterations in the abundance of particular taxa, including Firmicutes, Bacteroidetes, Proteobacteria, and Actinobacteria. Similar to the findings in mice, a hallmark of the gut microbiome in obese individuals is that there is an increase in the number of Firmicutes and a pronounced decrease in the abundance of Bacteroidetes, referred to as the **F/B ratio**. Firmicutes encode genes that enable them to release more energy from the foods they ferment, which results in an increase in the storage of excess energy in **adipose** (fat) tissue in the host. Changes to the F/B ratio result in suppressed production of a key molecule called fasting-induced adipose factor (see Figure 11.8), which results in increased fat storage and decreased secretion of hormones that signal satiety, such as leptin. In other words, fasting-induced adipose factor not only results in more fat stored, it causes you to feel hungry so you eat more and more fat is stored. What a vicious cycle.

Although the F/B ratio was the original framework for understanding weight-associated differences in the microbiome, more-recent work has allowed scientists to narrow in on specific genera associated with weight. A gut microbiota dominated by *Prevotella* tends to be associated with individuals who eat a fiber-rich diet and are lean, while an abundance of *Bacteroides* is associated with a meat-based diet and obesity. With respect to weight loss, those individuals with more *Prevotella* are better able to lose weight.

Figure 11.3 shows the metabolic impact of gut dysbiosis in obesity. The changes in metabolite production are numerous and quite damaging, resulting in an increase in immune cell activation, a disruption to the mucin-coated lining of the intestine, and a reduction in the intestinal wall barrier. We will identify several taxa involved in each of these metabolic disruption mechanisms and then turn to their metabolic impact. Let's start by examining the bacteria that are most often associated with the microbiomes of obese versus lean individuals. One promising microbe is *Akkermansia muciniphila*, which is abundant in the gut microbiota of healthy individuals, comprising up to 5% of the gut microbiome. *A. muciniphila* feeds on mucin, which might make you think it is one of the bad gut microbes. We know that the mucin layer helps protect the integrity of the gut wall barrier, preventing microbes from moving from the gut to other parts of the body. However, digestion of mucin by this species actually stimulates more mucin production, so there is always the right amount present. The metabolites produced by *A. muciniphila* include the SCFAs propionate and acetate, which are implicated in a well-balanced microbiome and a healthy host. Finally, this species has been identified as a **keystone species** in the human gut. That means that it plays an oversized role, relative to its abundance, in ensuring your weight is optimal.

Another research group found that members of the *Lactobacillus* genus were correlated with obesity (Crovesy et al., 2017). For example, the abundance of *L. paracasei* is inversely related to obesity, while a positive correlation is found for the abundance of *L. reuteri* and *L. gasseri*. These findings reveal the species-specific (or even strain-specific) nature of obesity-related bacteria. Even closely related bacteria, such as those members of the genus *Lactobacillus*, may have quite different effects on metabolic syndrome and obesity. Naming **strains** is a way of identifying types of bacteria even more precisely than by species level. Strains are subtypes of a species that have slight genetic differences but that are still similar enough to be members of the same species. They're labeled with a combination of numbers and letters after the species name, such as *Bifidobacterium longum* APC1472.

Another significant change in the gut microbiota of obese individuals is an increase in the number of Proteobacteria present. In fact, Proteobacteria was one of the most consistently reported abundant phyla in the gut microbiota of obese individuals. *Proteus mirabilis* and *E. coli*, both species of Proteobacteria, have been shown to be involved in increased levels of inflammation in the gut. They reduce the rate at which mucus is produced by the cells lining the gut. Having less mucus can result in damage

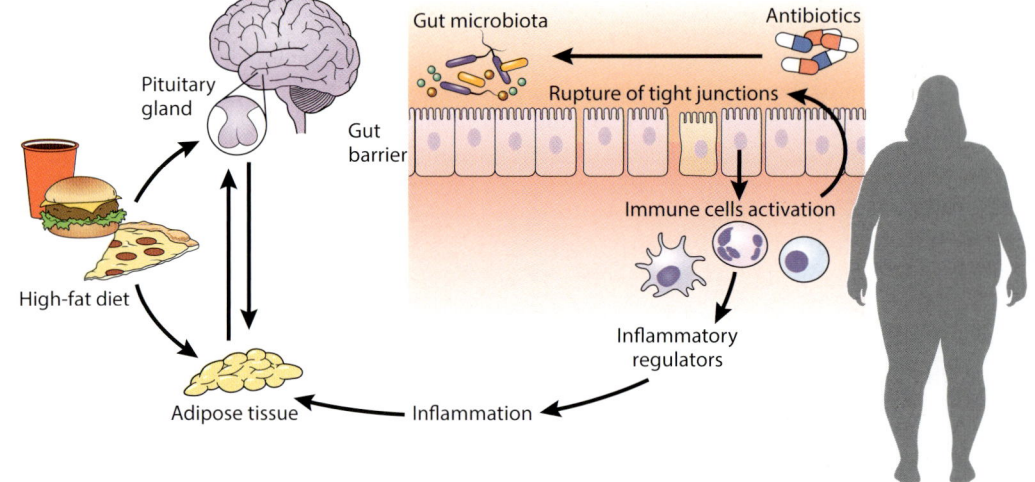

Figure 11.3 Gut Dysbiosis in Obesity You have likely heard that what you eat plays a role in obesity. But what about other environmental changes, like using antibiotics? Disrupting your gut microbiome can have devastating consequences to the metabolic pathways that regulate the storage of energy from your food as adipose tissue.

to the gut barrier and low-grade inflammation. Damage to the gut barrier also opens the door to the invasion of opportunistic pathogens and even higher levels of inflammation, which we discussed in chapter 7.

On the flip side, we now recognize numerous bacteria associated with leanness. *Bifidobacterium* is particularly well known for its effect in lowering cholesterol levels in the blood. You may recall *Bifidobacterium* from chapter 7 as it is a dominant genus in the healthy microbiome of babies and helps break down nutrients in breast milk. Several species of *Bifidobacterium* are commonly employed as probiotics and have a track record of safety and result in lower levels of cholesterol in blood.

WHR is a calculation of the ratio of your waist to your hip circumference and provides an indication of how much fat you store and where. Not all excess weight is equally unhealthy. Individuals with an apple-shaped body, that is, those carrying more weight around the waist, have a greater risk of heart disease and type 2 diabetes than do pear-shaped individuals, that is, those who carry more weight in their hips and thighs (**Figure 11.4**).

Roseburia, *Prevotella*, and *Ruminococcus* are consistently found to be associated with leanness. *Roseburia* is a major producer of butyrate, a short-chain fatty acid that helps to maintain gut barrier function. *Prevotella copri* (Bacteroidetes phylum) was found to improve glucose tolerance. Individuals with a higher *Prevotella*-to-*Bacteroides* ratio are better able to lose weight when consuming a high-fiber diet. *Ruminococcus bromii* is considered to be a keystone species of the gut microbiome due to its ability to degrade resistant starches such as amylose, which is found in lentils, corn, quinoa, potatoes, and more. *Dysosmobacter welbionis* is a newly isolated human commensal bacterium that has been shown to prevent diet-induced obesity and metabolic disorders in mice (Le Roy et al., 2022). Patrice Cani and colleagues showed that 70% of healthy humans host a particular strain of *D. welbionis* (J115^T) (Depommier et al., 2019). Absence of this strain is associated with higher BMI and excess blood glucose. When the strain is added to the mouse gut microbiome, those mice lose weight and store less fat while also showing improved glucose tolerance and lower levels of inflammation.

Microbial Guilds at Work in Obesity

The term **microbial guild** has been used to describe the interactions between taxa in an ecological unit such as the gut environment, where competition, cooperation, and selection mold the taxa present into a higher-level structure, namely the guild. Each guild has a shared metabolic outcome, and each member contributes to that outcome

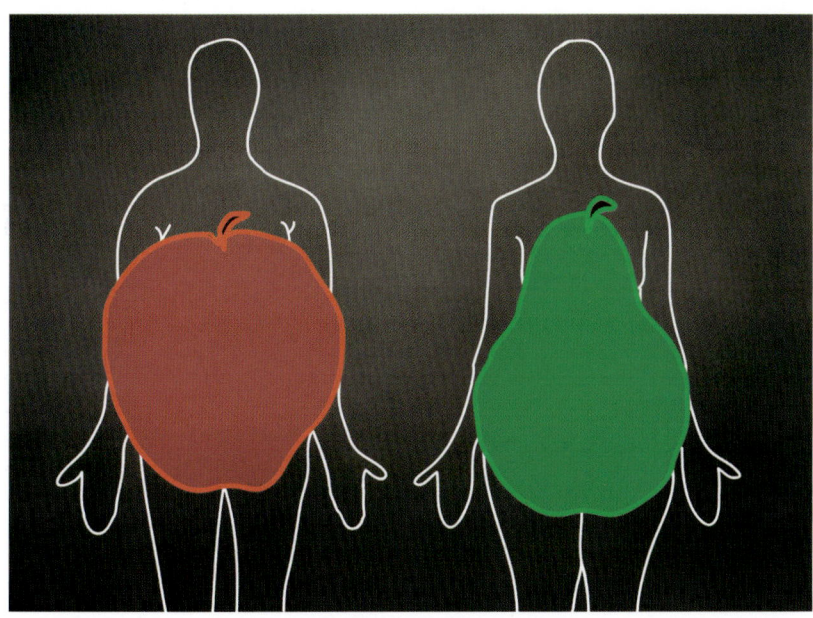

Figure 11.4 A Pear versus an Apple in Human Health Considerations Assessing your position on a scale of lean to obese may simply require a picture. There are clear associations between obesity and body shape. Those individuals with larger hips and thighs (the pear shape) tend to have reduced risk of death from all causes. In contrast, those who tend to gain weight around their midsection (the apple shape) are associated with higher risks of death. The shape is not causing the elevated risks; rather, the shape is correlated with factors that play a role in overall health. (Photo from iStock.com/sssimone)

in a specific manner. Members of a guild may have quite varied taxonomic profiles, but they thrive because of their combined presence. You can think of a guild like a company, where each employee has a specific role to play in achieving the company's mission, and without the contributions of all employees, the company will fall apart.

Some researchers have questioned the value of species-level identification in determining the taxon associations with obesity we described above and argue that a guild-based description is more valuable (van der Vossen et al., 2022). These scientists argue that the metabolic activities of microbes depend upon *where* they find themselves and *what* other microbes are present. Factors such as pH, bile acids, and substrate availability all play critical roles in determining which microbes will thrive, and which will not. These factors are, in turn, dependent on what other microbes are present. This means that the functions of our gut microbiota depend on the other microbes present. The absence of one member may result in other members being forced to utilize alternative metabolic pathways.

Van der Vossen and colleagues would argue that obese individuals tend to have lower diversity in their microbiomes, specifically alpha diversity. Recall from chapter 5 that alpha diversity is a measure of diversity within a sample, reflecting species richness and/or evenness. Alpha diversity is shaped by complex factors, including natural selection, variation in the local environment, ingestion of microbes, and other factors related to the ecology of the gut environment. Diversity by itself is not the heart of the matter; rather, it reflects a well-developed ecological network of species that have evolved to break down food, produce nutrients, and survive in the gut. In adult humans, *Bacteroides*-rich gut microbiomes tend to have lower diversity and far less complex metabolic networks. These microbiomes tend to be enriched in species that create a risk factor for obesity, including Enterobacteriaceae, *Fusobacterium*, *Streptococcus*, *Ruminococcus gnavus*, and/or various *Bacteroides* species.

A fascinating exploration of this idea was provided by Wu et al. (2022), who studied the gut microbiome metagenomes in individuals with type 2 diabetes consuming a high-fiber diet. Wu and colleagues identified associations between pairs of microbial genomes that revealed the presence of several competing guilds. **Figure 11.5** shows a visualization of the relationships within and between these guilds, based upon the correlations observed during the trial. What is not clear from this complex visualization of relationships is that these guilds seesawed in their abundance during the intervention, but the associations between the members of each guild remained. All the species in one guild increased in abundance while the species in another guild became less abundant, or vice versa, depending on the specific host.

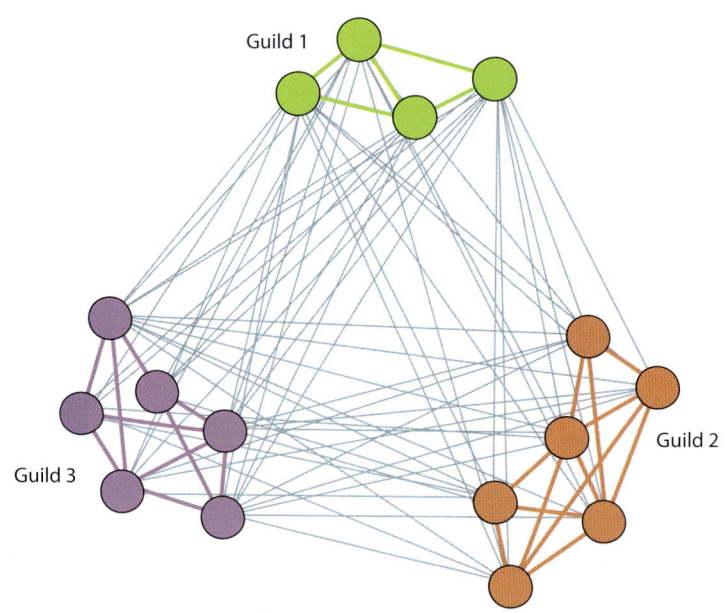

Figure 11.5 Microbial Guilds in the Gut Microbiome A guild refers to a group of species that utilizes resources in a similar way. In this figure, there are three guilds identified, which emerged in individuals subjected to a high-fiber diet. The blue connecting lines indicate two species with similar functions found in different guilds. Once a guild is established, it is hard for new species to invade, as they will compete for resources already being consumed efficiently by the corresponding member of the resident guild.

The importance of the guild-based assessment of microbiomes is that it requires us to consider the entire composition of the microbiome, rather than a subset of what we might think are key species. Although studies have identified key species, such as *A. muciniphila*, that have an outsized effect on human health, they are dependent on other members of the guild to function properly. Examining the effect of the microbiome through the guild framework provides important insights because it considers the interactions of numerous species in the gut. This perspective with respect to the association between the microbiome and human disease remains a tantalizing harbinger of how microbiomes may serve as **disease markers** in the future. Disease markers indicate that a disease may develop. For example, cholesterol is a disease marker for heart disease. While the markers for many diseases are currently known, understanding how microbiome composition can indicate the potential for sickness is a diagnostic science still in its infancy.

Not Just Your Gut Microbes

A surprising observation has implicated the oral microbiome in obesity: the oral microbiome composition of obese individuals differs from that in lean individuals. A large cohort study was undertaken in which the oral microbiomes of overweight and lean individuals were compared (Tam et al., 2018). An increased BMI is associated with a drop in the diversity in saliva bacterial communities. Lean individuals generally have a higher proportion of Bacteroidetes in their oral microbiome compared to obese individuals, while obese individuals tend to have a higher abundance of Firmicutes, often with a higher ratio of Firmicutes to Bacteroidetes, which is considered a key marker of potential dysbiosis associated with obesity. Obese individuals generally exhibit a higher ratio of Firmicutes to Bacteroidetes in their gut microbiomes compared to lean individuals, meaning they have a larger proportion of Firmicutes bacteria relative to Bacteroidetes bacteria, which is often considered a potential marker for obesity. There are other differences between the oral microbiomes of obese and lean individuals, such as an increase in *Porphyromonas gingivalis*, which may contribute to an altered amino acid metabolism that then plays a role in the host's immune response and increases the risk of metabolic syndrome. It appears that it is not simply a shift in the Firmicute/Bacteroidetes ratio that implicates the oral microbiome in obesity!

Meisel et al. (2021) showed that obesity is associated with **chronic periodontitis** (**CP**) in which the gums become inflamed. CP is also associated with an altered oral microbiome, including an increased abundance of *Eubacterium nodatum*, *Aggregatibacter actinomycetemcomitans*, and *Fusobacterium nucleatum*. (Refer back to chapter 3 for a review of the geography of the oral microbiome.) Olsen and Yamazaki (2019) revealed that the oral and gut microbiome compositions are connected and found that alterations in one impact the composition of the other. When the researchers administered the oral pathogen *Porphyromonas gingivalis*, Bacteroidetes levels plummeted in the mouse gut microbiota. The impact of that oral pathogen on the gut Bacteroidetes was followed by an increase in insulin resistance and inflammation. Another study showed that oral administration of *P. gingivalis* resulted in increased intestinal permeability, which allowed more lipopolysaccharides (LPS) and more bacteria to move through the intestinal barrier (Feng et al., 2020). The oral microbiome appears to play a significant role in modulating the gut microbiome, which can impact the overall metabolism of the host.

11.3 METABOLIC MARKERS OF WEIGHT LOSS

Members of the Enterobacteriaceae family, known as enteric bacteria, are key taxa in the production of SCFAs, which result from the fermentation of hard-to-digest carbohydrates. In particular, these bacteria produce high levels of the SCFAs acetate, propionate, and butyrate, which are crucial for the regulation of host energy homeostasis. These fatty acids can turn on or off the **inflammatory cascade**, which is intimately connected to obesity. The inflammatory cascade is the body's response to tissue injury or infection and involves coordinated communication between immune cells and blood vessels. This response has three components: (1) the trigger (infection or tissue damage); (2) the response (production of inflammatory sensors, such as mast cells and macrophages, and inflammatory mediators, such as cytokines and chemokines); and (3) the tissues impacted (gut epithelial cells, etc.). Chronic inflammatory conditions are of particular interest because they coincide with other diseases such as obesity, type 2 diabetes, atherosclerosis, neurodegenerative diseases, and cancer.

However, the role of SCFAs is complex and multifaceted. For example, acetate plays a role in the regulation of both the gut microbiota and obesity. On the one hand, it can induce the host to secrete hormones that help to prevent obesity. On the other hand, it is involved in cholesterol synthesis, resulting in higher serum cholesterol levels, which can increase the risk of obesity. Propionate plays a role in weight loss by promoting intestinal **lipolysis**, which is the breakdown of fat. Butyrate is a key energy source for the intestine and an inhibitor of proinflammatory cytokines. Like propionate, butyrate stimulates lipolysis, which results in greater energy consumption. Butyrate has also been shown to reduce the levels of lipopolysaccharides in the bloodstream, thereby reducing the negative LPS-related effects. **Figure 11.6** provides a summary of the production of SCFAs in the gut microbiome and their impact on host metabolism. Supplementation of these acids in mouse studies has shown that they protect against weight gain, inhibit inflammation, and increase insulin sensitivity.

Some consider SCFAs to be the silver bullet of weight loss. These metabolites help to inhibit fat production while also increasing levels of satiety hormones and increasing energy expenditure. They accomplish these feats by directly binding to receptors found in gut epithelium, stimulating the production of certain gut hormones that promote lipolysis. However, the potential benefits of SCFAs may be limited. When present in excess, SCFAs are removed from the body as waste. These higher levels of fecal SCFAs are associated with obesity. The gut microbiota of individuals with obesity produce twice the SCFAs that lean individuals do. Overall, SCFAs seem to be a double-edged sword. They protect the host from obesity, but excess SCFAs can promote obesity through the production of excess energy.

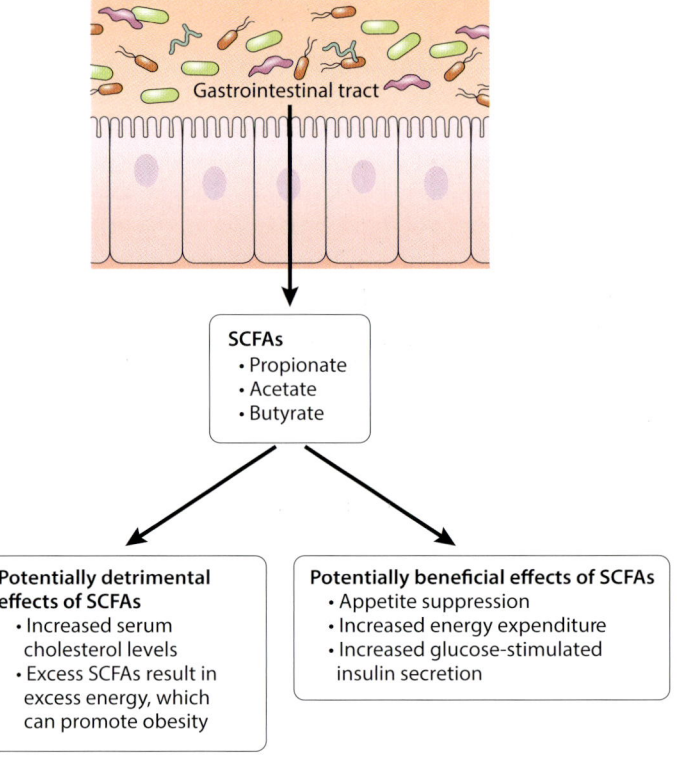

Figure 11.6 The Double-Edged Sword of Short-Chain Fatty Acids Some of the microbiota in our gut produce short-chain fatty acids as a by-product of fiber fermentation. These acids are food for intestinal epithelial cells and serve as signals that impact a variety of factors related to obesity, such as appetite suppression and levels of insulin secretion.

Rendering Bile Acids Impotent

Bile acids (BAs) are steroids synthesized from cholesterol. The amino acids taurine or glycine can be bonded to them, resulting in conjugated BAs, also known as bile salts. Bile salts play a key role in the regulation of fat absorption while also serving as potent antimicrobials within the intestine. The microbiota have evolved strategies to modify bile salts to limit their antimicrobial actions. Certain members of the gut microbiota produce bile salt hydrolases (BSHs), which can cleave off amino acids, resulting in deconjugated BAs. Other bacteria can then further transform the bile acids, resulting in reduced toxicity, which in turn affects the survival of various bacteria in the gut. New research has revealed that the deconjugation process and its resulting impact on microbial survival is far more complex than this simple description implies (Foley et al., 2021). According to Casey Theriot from North Carolina State College and one of the authors of the study, "The assumed relationship was that probiotic bacteria like *Lactobacillus* have BSHs that just deconjugate the bile acid, rendering it less toxic and allowing the bacteria to survive. But the reality is a lot more complex—these enzymes are more specific than we thought. Depending on which BSH is there and which bile acid it acts on, you can shape the gut in different ways, making it more or less hospitable to bacteria or pathogens" (North Carolina State University News, 2021).

Theriot and her colleagues studied the mechanics of how BSHs reduced bile toxicity. They grew two strains of *Lactobacillus* in the presence of several BAs. Depending on which amino acid had been added to the BA, different toxicity against the *Lactobacillus* was observed. Through a series of experiments, Theriot and her team revealed that the toxicity of BAs is dependent not only on whether it is conjugated or not, but also on which strain of bacteria is present, the type of BA, and which bile salt hydrolases are being produced.

Theriot's team genetically modified the *Lactobacillus* strains to produce different types of BSHs and found that the combination of BSHs produced by a strain affected its ability to survive different types of BAs, which would in turn affect its ability to survive in the gut microbiome. "These BSH enzymes have diverse properties. Bacteria pick up and drop off enzymes regularly—sometimes they pick up enzymes that will help them survive . . . or they could pick up an enzyme that will hurt competition. If we're going to try and design the gut microbiota in the future, we really have to understand all the players—bacteria, enzymes and bile acids—and their situational relationships," Theriot says. "This work is a big first step in that direction" (North Carolina State University News, 2021).

Triggers of Fat Storage

Fat storage is another function impacted by our gut microbiota. Early work showed that when germ-free mice were colonized with a fecal microbiota, they gained body fat and their insulin sensitivity decreased. Increased energy resulted from the presence of a more diverse set of microbes able to digest a larger number of polysaccharides and also able to signal the host to store the resulting energy as fat. In other words, the microbiome affects both how much energy we get from the foods we eat and how that energy is stored.

The relationship between the gut microbiome and fat storage is far more complex than simply the microbiome's effect on energy production. **Figure 11.7** shows how the gut microbiota can trigger inflammation resulting in adipose tissue dysfunction leading to increased fat storage and obesity. Certain microbes produce LPS, which can activate Toll-like receptors (TLRs) on macrophages that trigger the production of cytokines, such as IL-8, that result in the signaling of inflammatory pathways that impair insulin signaling and promote adipose tissue inflammation. As part of the inflammatory response, adipose tissue secretes hormones and cytokines, such as tumor necrosis factor alpha (TNF-α) and interleukin-6 (IL-6), that can modulate the gut microbiota's composition and function. These bidirectional interactions result in a vicious cycle that can lead to obesity.

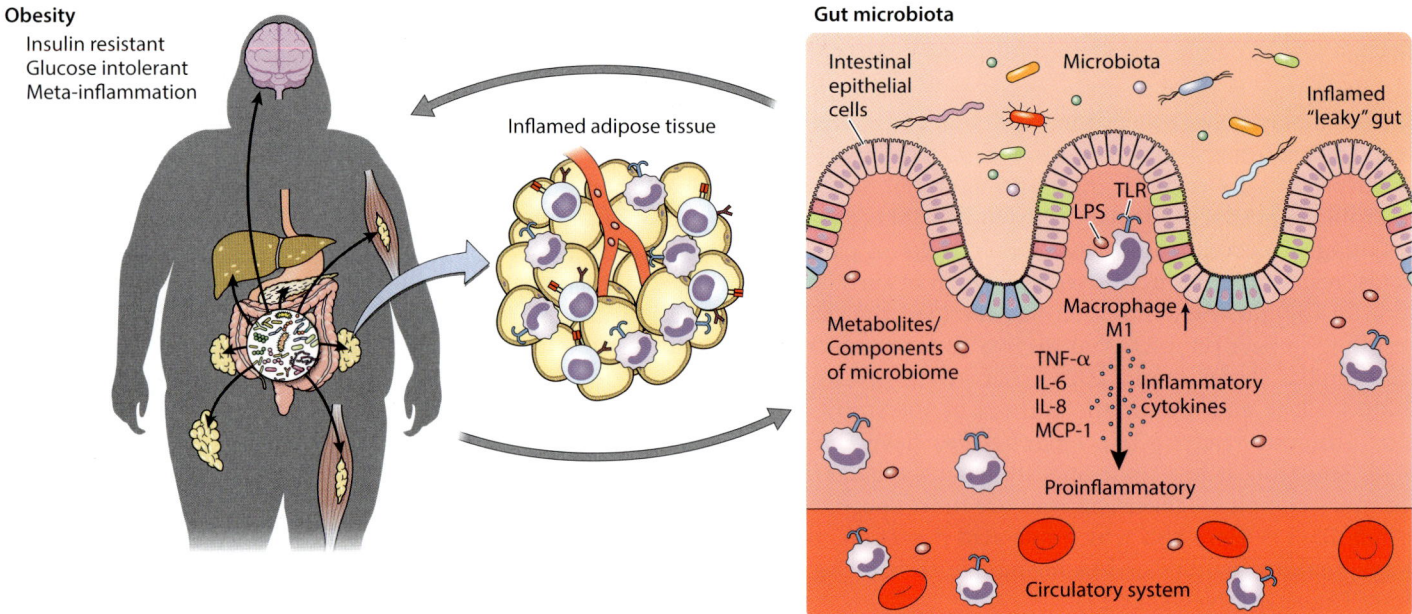

Figure 11.7 The Bidirectional Interactions between the Gut Microbiome and Adipose Tissue
The gut microbiota and adipose tissue communicate via metabolites, hormones, and the immune system. Certain metabolites produced by the gut microbiota, such as lipopolysaccharides (LPS), activate receptors on macrophages, such as Toll-like receptors (TLRs), which trigger the production of inflammatory cytokines, such as IL-8 inflammatory pathways that impair insulin signaling and promote inflammation of adipose tissue. In turn, adipose tissue secretes hormones and cytokines, such as tumor necrosis factor (TNF) and IL-6 that can alter the microbiota of the gut and its production of metabolites. This bidirectional interaction creates a cycle that leads to dysfunction of the adipose tissue and obesity. (After Wang et al., 2023.)

There is yet another mechanism involved in fat storage, which involves a family of **microRNAs** (**miRs**). These noncoding RNAs are involved in regulating gene expression and are highly expressed in fat tissue. Noncoding RNAs are molecules of nucleic acid that cells use to communicate but that are never translated into protein. A team of researchers led by Anthony Virtue and Sam McCright at the University of Pennsylvania were interested in exploring how changes in levels of certain microbial metabolites affect obesity (Virtue et al., 2019). They discovered that a family of miRs produced by the host promoted the progression of obesity. When the researchers looked at the gut microbiome, they found that certain metabolites influence the expression of miRs in fat cells. In lean mice, these circulating metabolites influence how energy is expended and what portion is directed to fat production and storage. They accomplish this by inhibiting one particular miR (miR-181). In mice fed a high-fat diet, the levels of one specific metabolite, indole, decreased, resulting in an abnormally high activity of miR-181, which in turn drove obesity, insulin resistance, and fatty tissue inflammation. To validate their findings in humans, the researchers analyzed the changes in indole levels in 38 children, including 19 with obesity. All children with obesity had lower levels of indole in their blood than children without obesity, suggesting that this microbial metabolite also plays a role in human obesity.

Mediating Low-Grade Inflammation

Part and parcel of obesity is chronic low-grade inflammation, and our gut microbiota is intimately involved in mediating inflammation. As you may recall from earlier in this chapter, obese individuals typically have elevated levels of Proteobacteria, in particular enteric bacteria, which can lead to a thinning of the mucin layer that coats gut epithelial cells and leads to a decrease in protective bacteria, such as *Roseburia*. When the intestinal barrier is compromised, the gut microbiota and its metabolites can

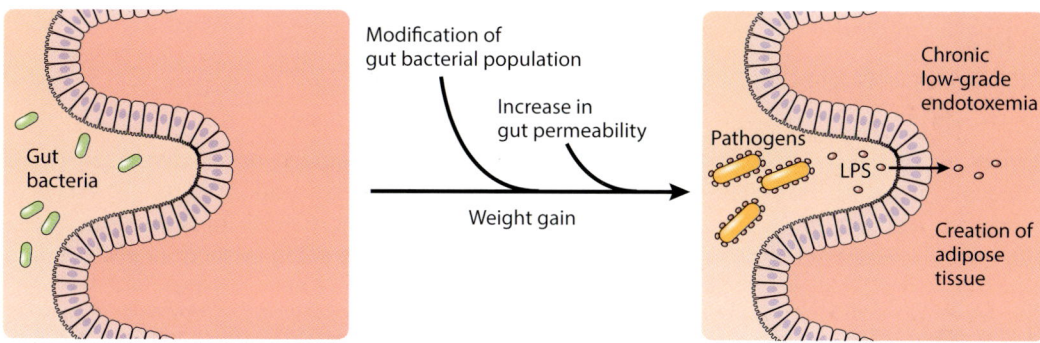

Figure 11.8 Microbe-Mediated Inflammation Associated with Obesity Obesity is accompanied by a change in the gut microbiota, including an increase in pathogenic bacteria. These pathogens can trigger a weakening of the intestinal epithelial barrier. One outcome is that as the pathogens die, the lipopolysaccharides in their cell membranes accumulate and are able to pass into the bloodstream, resulting in chronic endotoxemia. High levels of circulating LPS are associated with a broad range of diseases, including obesity, autoimmunity, and metabolic syndrome.

travel outside of the intestine, where they can signal the immune system to produce inflammation in numerous organs. **Figure 11.8** shows how an increase in pathogens in a dysbiotic gut microbiome can result in increased levels of the endotoxin LPS, which triggers the immune cells to produce inflammation and stimulates creation of adipose tissue. Infusing LPS into mice induces both obesity and inflammation (Scheithauer et al., 2020). When obese mice lose weight, the levels of LPS-producing bacteria decrease in the gut, and there is a corresponding decrease in levels of circulating LPS and lower levels of inflammation.

11.4 THE GUT MICROBIOME AND WEIGHT LOSS

We do not yet know the composition of gut microbiota that can help you gain or lose weight. In a more general sense, a more diverse microbiome is a healthier microbiome. However, which species (and even strains within a species) are present appears to matter as well. One highly revealing review of clinical studies examined how well gut microbiome composition predicted weight loss in a variety of weight loss dietary interventions (**Box 11.1**) (Hernández-Calderón et al., 2022). Although there was variation in the results from one study to another, certain taxa were revealed to be highly correlated with weight loss success under dietary interventions.

When one goes on a diet to lose weight, the gut microbiome changes, and there are increases in microbiota diversity and decreases in levels of systemic inflammation. As might be predicted, an individual on a carbohydrate- or fat-restricted diet shows a decrease in Firmicutes and an increase in Bacteroidetes abundance. Numerous microbial metabolites, including bile acids, aromatic amino acids, SCFAs, and branched-chain amino acids, play roles in the altered gut microbiome metabolic pathways associated with obesity.

For one overriding question—Which bacteria promote weight loss—two of the most promising answers are *Akkermansia muciniphila* and *Christensenella minuta*. They are found in abundance in lean individuals. You may recall from earlier in the chapter that *A. muciniphila* can feed on the mucin lining the gut and helps to strengthen the intestinal barrier. It also produces acetate, which helps regulate body fat and appetite. *C. minuta* is abundant in the microbiomes of lean individuals. It is also a bacterium that is acquired from your family; in other words, you are more likely to host this bacterium in your gut if your relatives have it too. *C. minuta* plays a role in limiting fat accumulation, and one study showed that the levels of "good" cholesterol (HDL) were higher when these bacteria were present (Tavella et al., 2021). Further, the presence of *Christensenella* is correlated with the presence of

BOX 11.1. RESEARCH IN ACTION
Microbiome Predictors—Biomarkers for Weight Loss Success

- ❖ **Hypothesis.** Differences in gut microbial profile play a role in the success of health interventions and may serve to predict whether an individual will lose weight under a calorie-restricted diet.

- ❖ **Methods.** This study examined the data generated in a large sample of clinical trials focused on assessing the correlation between success in dietary interventions and the gut microbiome composition.

- ❖ **Results.** Figure 11.9 provides a summary of the taxa identified in the clinical trials examined. Taxa are identified as positively or negatively predictive of weight loss success under a variety of weight loss interventions, including high-fiber diets, calorie restriction, intermittent fasting, and low-carbohydrate diets. (A) lists taxa positively or negatively associated with waist circumference and fat mass. (B) lists taxa associated with weight. (C) lists taxa associated with weight-associated parameters such as glucose and triglyceride levels.

- ❖ **Conclusions.** The presence of particular members of the microbiota may serve as biomarkers to indicate the likely success of weight loss interventions. Although there remains significant variation in the results obtained from the clinical trials, patterns are beginning to emerge, suggesting that we may soon be in a position to use a baseline microbiome composition to predict which dietary intervention is most likely to succeed in an individual.

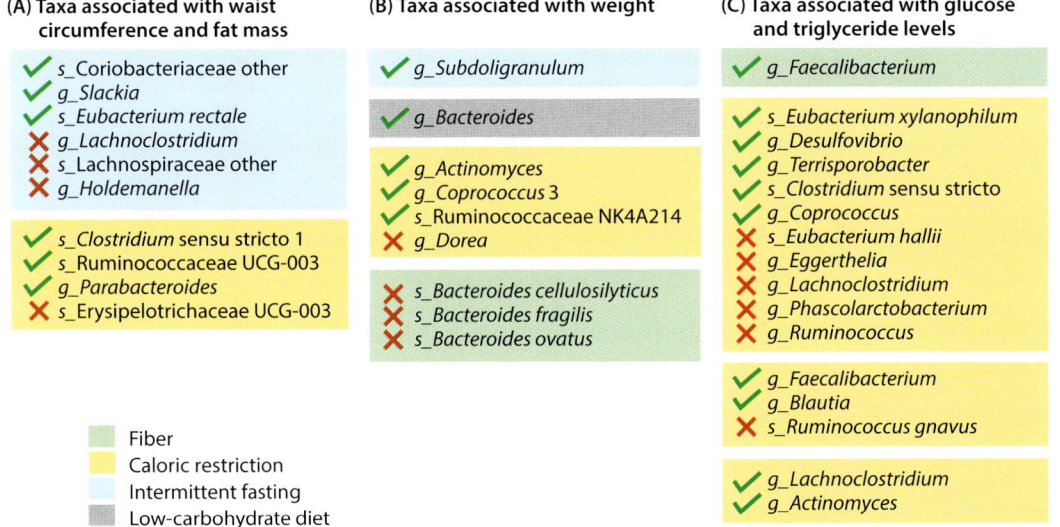

Figure 11.9 Overview of Microbiota-Based Weight Loss Biomarkers Several microbial taxa have been tested for their predictive value in relation to the success of weight loss programs. Success is divided into (A) reduced waist circumference and fat mass, (B) reduced weight, and (C) improved metabolism, shown by glucose and triglyceride levels. The results are also color coded by program type (e.g., calorie restriction, fasting). A green check mark indicates each taxa whose presence predicted success; a red X indicates each taxa whose absence predicted success. (After Hernández-Calderón, Wiedemann and Benítez-Páez, 2022.)

another beneficial bacterium, *Oscillospira*, which was identified as a biomarker of health. Thus, *Christensenella* may be a keystone member of a healthy human gut microbiome.

However, simply ingesting these microbes is not a weight loss panacea, and having certain gut bacteria won't necessarily cause you to lose weight. Rather, it is the impact of their activities, such as production of SCFAs, which determines how much energy you gain from your diet and how much weight you gain, as well as how hungry you feel. For example, obese individuals tend to have lower levels of gut microbes that produce butyrate. These microbes are involved in fermenting resistant starches, which produces SCFAs, such as butyrate. Numerous studies have revealed that butyrate protects us from inflammation and helps regulate our food intake by stimulating the release of satiety signals, which tell us we aren't hungry.

11.5 DIET-BASED APPROACHES

We have previously discussed how the composition of the gut microbiome is impacted by a variety of factors such as age, genetics, diet, lifestyle, and medications, particularly antibiotics. We are just beginning to explore how we can potentially use this knowledge to purposefully impact our own microbiomes. In other words, can we use the gut microbiome as a clinical tool in gastrointestinal disease? Let's focus on how we might manipulate our microbiome composition as a means to decrease obesity.

Let's start with the simplest approach, changing our diet with the goal of manipulating our gut microbiome composition. We have mentioned numerous times that a Western diet is associated with a distinctly different composition of the gut microbiota, compared with a plant-based diet, and that its consumption increases the risk of obesity dramatically. But what is it in a plant-based diet that is so critical to weight gain or loss? One of the simplest approaches to studying the effect of specific compounds on the gut microbiome is to extract them from raw food and test their impact on the microbiome. Let's discuss several of the more compelling compounds found to date, including phytochemicals, polysaccharides, fermented foods, and pro- and prebiotics.

Phytochemicals

Plant-derived chemicals, known as phytochemicals, can impact the gut microbiota composition and, through those impacts, help in weight loss and mediate metabolic diseases (**Figure 11.10**). There are numerous types of phytochemicals, such as polyphenols

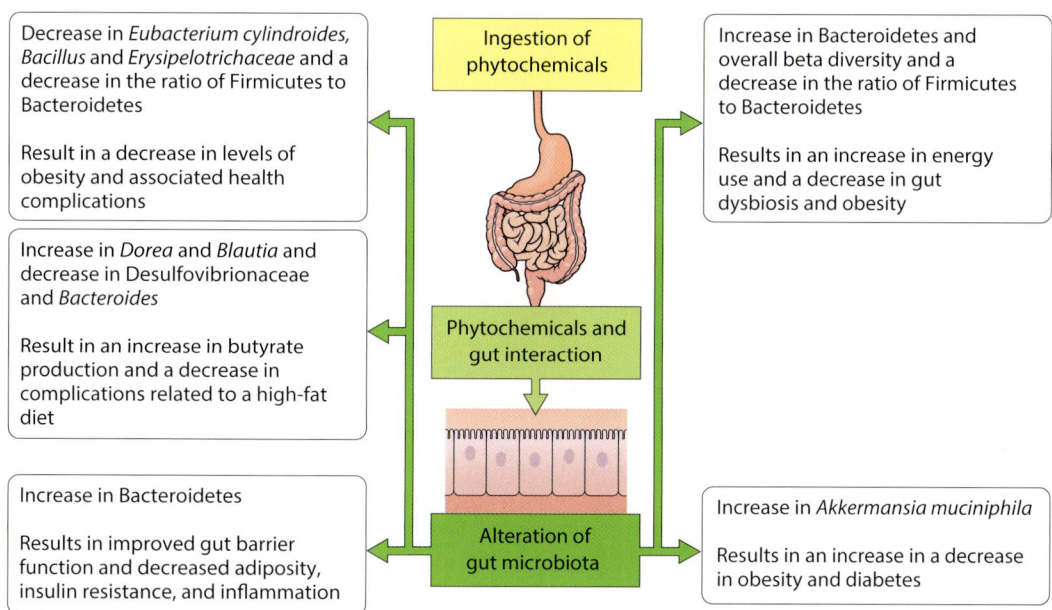

Figure 11.10 Impact of Diet on Microbiome Composition Ingesting plant-based phytochemicals has an enormous impact on the members of the gut microbiota. Some relative changes in various members of the microbiota are listed, followed by the associated health impacts of those changes. In some cases, specific species are impacted, such as when levels of *Eubacterium cylindroides* decrease, while in other cases changes occur at the genus (e.g., *Bacillus*) or phylum (e.g., Bacteroidetes) level. (After Santhiravel et al., 2022.)

and carotenoids, that play a role in protecting the plant from infectious agents. **Polyphenols** are a large family (over 800 types) of phytochemicals that have far-reaching activities, such as improving digestion, protecting against heart disease, and influencing the health of the central nervous system. They are bound to fermentable fibers and are abundant in fruits, vegetables, and cereals and are released during bacterial fermentation. They have strong anti-inflammatory effects, which play a key role in colon health. Although the concentration of polyphenols, even in a polyphenol-based diet, is low, their impact is enormous. Polyphenols are released in the host gut and are converted by bacteria to key metabolites. The health effect of polyphenols is largely due to their role in supporting our "good" or "healthful" bacteria. More specifically, as the result of a diet rich in polyphenols, levels of *Lactobacillus*, *Bifidobacterium*, *Akkermansia*, and *Bacteroides* are higher. In addition, pathogen levels are lower, which means that endotoxin levels remain low and more short-chain fatty acids are produced, resulting in lower levels of inflammation (Mithul Aravind et al., 2021).

Plant dietary fiber and polysaccharides (such as cellulose and pectin) are also powerful gut microbiota modulators. These plant-derived carbohydrates are a food source for our "healthful" gut microbes to digest and have promising potential for preventing and alleviating gut inflammation through their ability to modulate the microbiota. Plant-based polysaccharides exhibit diverse biological functions, including antioxidative, anti-inflammatory, immunomodulatory, antitumor, hypoglycemic, and microbiota modulation capabilities. Their presence selects for key beneficial bacteria in the large intestine, helps to increase the movement of food through the large intestine, and simultaneously produces key metabolites such as SCFAs. Consumption of plant fiber has been shown to have a direct impact on the diversity of the gut microbiome, with a predictable increase in the numbers of SCFA-producing bacteria in the human gut. As we have noted several times previously, SCFAs are the main players in the interplay of diet, microbiota, and host health.

Fermented Foods

Fermented foods are those produced by controlled microbial growth, meaning that components of the foods are converted by microbial enzymatic activity. Examples include cheese, yogurt, kombucha, kefir, and kimchi. The process of fermentation occurs when bacteria and yeast break down carbohydrates into alcohol or organic acids. A recent Stanford School of Medicine study showed that consumption of fermented foods increases the diversity of the microbiome and decreases inflammation (Wastyk et al., 2021). This study reported that being on a fermented-foods diet or a high-fiber diet produced very different results. Adults who consumed a fermented-foods diet for 10 weeks had an increase in overall microbial diversity, whereas adults who consumed a high-fiber diet generally had no change in microbial diversity.

A stunning finding of the Stanford study was that all the participants in the fermented-foods group, and not those in the higher-fiber group, showed reduced levels of inflammation. Specifically, it was found that four types of immune cells showed less activation in those who ate fermented foods for a mere 10 weeks. Additionally, the levels of 19 inflammatory proteins were also reduced. Among these 19 inflammatory proteins was interleukin-6, which is associated with inflammatory conditions like type 2 diabetes and rheumatoid arthritis as well as chronic stress. Since every participant who consumed the fermented-foods diet showed a reduction in inflammatory markers, this suggests that eating more fermented foods may have applications to people with autoimmune disease and chronic diseases of aging that are in part driven by inflammation.

Probiotics

Probiotics are defined by the World Health Organization as "live microorganisms that confer a beneficial effect to the host when administered in appropriate amounts." They offer yet another therapeutic intervention aimed at altering the gut microbiome composition. Probiotics act to modify the gut microbiota and enhance the gut epithelial barrier while also producing antimicrobials, such as lactic acid, and modulating the immune system. Probiotic bacteria inhibit the growth of harmful bacterial species in the gut, in part by competing for adhesion to the epithelial cells, and stimulate the growth of other beneficial strains.

Numerous studies have shown that probiotic supplementation may help reduce body weight, lower levels of fat deposition, and decrease an individual's BMI. The mechanisms involved include improving the gut barrier integrity, reducing the production of endotoxins, and decreasing systemic inflammation. Some argue that the healthful impact of probiotics is due primarily to the resulting increased production of SCFAs, which promotes epithelial barrier repair and the overall health of the intestine.

Most single probiotics are unable to permanently colonize the gut. However, studies have shown that ingesting a complex of bacterial species can result in far better results. One multistrain, multispecies probiotic, VSL#3 (three strains of *Bifidobacterium* and four of *Lactobacillus*), when administered to young adults on a high-fat diet prevented weight gain (Cheng et al., 2020). Studies employing a mouse model of obesity showed that the probiotic strain *Lactobacillus casei* Shirota resulted in lower toxin levels. Other such studies revealed that *Bifidobacterium animalis* ssp. *lactis* 420 helped to limit invasion of pathogens through the mucin boundary layer in the gut. There are many promising probiotic strains of bacteria that may help reduce the risk of, or even treat, metabolic disease, especially when used in combination. However, probiotics are not risk-free, as shown by the use of a potentially detrimental species, *Bifidobacterium bifidum*, in probiotics for infants. As we consider the benefits of probiotics, we must fully test and study these strains to understand any other impacts they may have on health.

Recent studies have found that *Akkermansia muciniphila* is associated with leanness and, when administered as a probiotic, can produce several health benefits. Patrice Cani led a clinical study that involved participants taking a daily dose of *A. muciniphila* for 3 months, while maintaining their normal diets and levels of activity (Depommier et al., 2019). The study revealed that the probiotic supplement helped reduce various factors associated with heart disease, such as fat storage and insulin resistance. The supplement also changed the participants' blood markers that signal inflammation. This was the first clinical study to demonstrate that a probiotic administration of what is a member of the normal human gut microbiome may help reduce metabolic syndrome. According to Cani, "Although diet and physical activity are the major cornerstones for managing cardiovascular disease, our findings pave the way for using next-generation beneficial microbes such as *Akkermansia* and/or specific bacterial components to play a role in improving metabolic health in obese and overweight human subjects" (Depommier, et al., 2019).

Bifidobacterium longum APC1472 was the focus of another study comparing the effect of adding this strain to the diets of obese mice and humans (Schellekens et al., 2020). Probiotic supplementation of *B. longum* APC1472 resulted in a drop in body weight, a reduction in fat levels, and an increase in glucose tolerance in mice fed a high-fat diet. In obese humans, although probiotic supplementation did not change primary BMI or **waist-to-hip ratio** (**WHR**), a positive effect on fasting blood glucose levels was found.

Prebiotics

Prebiotics are defined as "non-digestible food ingredients that beneficially affect the host by selectively stimulating the growth and/or activity of one or a limited number of bacterial species already established in the colon, and thus improve the host health" (Linares et al., 2016). In short, prebiotics are fibers that are indigestible by us and that promote the growth of a beneficial gut microbiome. Common prebiotic substances include fructooligosaccharides, galactooligosaccharides, and human milk oligosaccharides, which you may recall from chapter 7 prebiotics can promote the absorption of ions and trace elements, such as calcium, iron, and magnesium, and

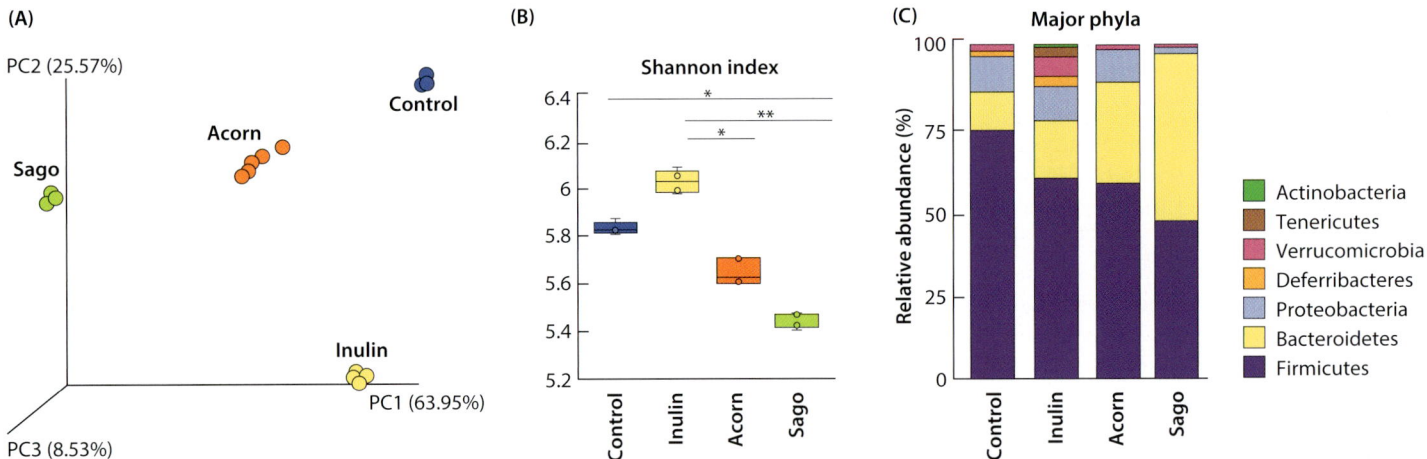

Figure 11.11 Prebiotics Modulate the Diversity and Composition of Gut Microbiome in Mice Fed a High-Fiber Diet For 5 weeks mice were fed diets enriched in one of three prebiotics (inulin, sago, or acorn) and then compared with a control group fed a standard diet. (A) Principal coordinate analysis of the beta diversity of the mouse gut microbiomes. (B) Shannon index representing the alpha diversity of the mouse gut microbiomes. (C) Relative abundances of the major phyla in the mouse gut microbiomes. (After Ahmadi et al., 2019.)

help regulate the immune system by modulating cytokine production through microbial metabolic products. Prebiotics selectively stimulate the growth of *Bifidobacteria* and *Lactobacillus* species, and these taxa, in turn, support the growth of a diversity of other members of the gut microbiome.

Ahmadi et al. (2019) fed mice diets enriched in prebiotics (sago, inulin, and acorn) for a period of 5 weeks. They saw significant changes in a variety of microbiome diversity measures (**Figure 11.11**), including changes in the major bacterial phyla present, the levels of alpha diversity in each treatment group, and well-separated treatment clusters in a principal coordinate analysis of beta diversity. They also noted an improvement in glucose tolerance, a reduction in insulin resistance, and changes in energy signaling networks in the brain. Several clinical trials have also shown a strong impact of prebiotics and an alteration of the gut microbiota and decreased obesity. Prebiotics result in increased production of SCFAs, which regulates the feeling of satiety and improves the secretion of hormones in the gut involved in insulin secretion, glycemic regulation, and satiation. The altered gut microbiota also impacts lipid metabolism.

Fecal Microbiota Transplantation

As we learned in chapter 2, **fecal microbiota transplant** (**FMT**) involves the use of a fecal suspension from a healthy donor, which is transferred into the colon of a patient. It has been a treatment option as far back as the 4th century in China, where it was used for several conditions, including diarrhea and food poisoning. FMT has made a resurgence in modern medicine and is now considered a first-line therapeutic treatment for *Clostridioides difficile* overgrowth. The Food and Drug Administration (FDA) has approved this as the single use of FMT to date. The FDA may also approve the use of FMT for life-threatening, emergency conditions through the use of what it calls an investigational new drug application. Despite FMT's potential, its uses are currently limited because there is a risk of acquiring an infection through the process, and there have been patient deaths as a result of these infections. However, two preliminary human studies have suggested the utility of FMT in treating obesity (Vrieze et al., 2012; Kootte et al., 2017). In both studies, FMT resulted in an improvement in insulin sensitivity, albeit transiently. However, we simply don't have enough information about this application, and there appear to be risks involved in its use. For example, one woman developed new-onset obesity after an FMT to treat a *C. difficile*

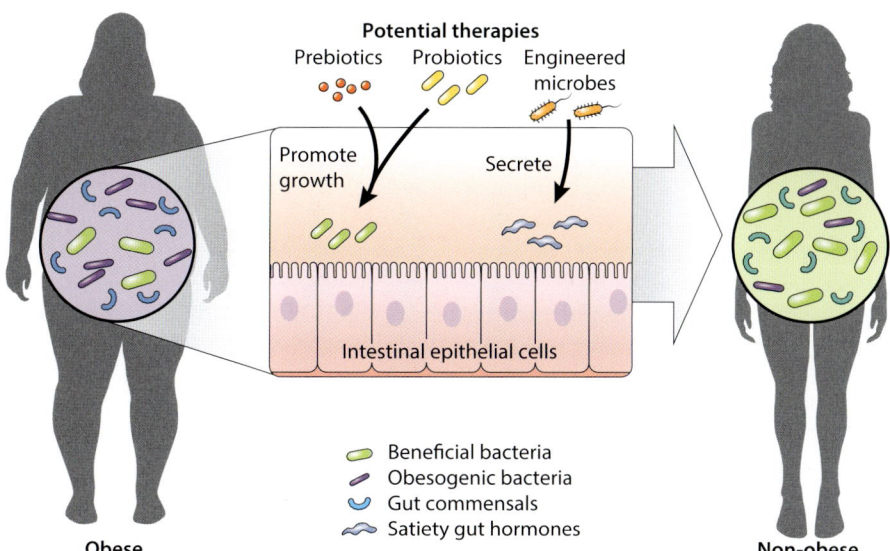

Figure 11.12 Potential Microbiome-Based Therapies to Treat Obesity One future approach to modulating obesity is through the promotion of healthful bacteria using combinations of prebiotics, probiotics, and synthetic microbes. Beneficial bacteria reduce inflammation by lowering the production of lipopolysaccharides and producing metabolites that aid in the satiety response.

infection. It turned out that the donor was overweight. There is simply not enough known about the long-term risks of FMT at this point.

This chapter has provided an overview of several of the microbiome-based strategies being explored for the prevention and/or treatment of obesity. What we find is that each approach appears to succeed by reducing obesity through similar functions, primarily by promoting the growth of beneficial bacteria (**Figure 11.12**). The presence of these bacterial "friends" results in a reduction in inflammation, often by decreasing the production of LPS, and an increase in production of bacterial metabolites that tell us we feel full, such as short-chain fatty acids. Future efforts will include producing engineered strains of bacteria capable of achieving all of these goals and that may be more readily retained in our gut through the consumption of particular foods or use of certain drugs. The future is, indeed, bright for taming the obesity epidemic!

CHECK YOUR UNDERSTANDING

1. BMI is an accurate tool for measuring weight-based health.
   a. True
   b. False

2. All of the following are linked to obesity *except*
   a. sedentary lifestyle.
   b. genetic markers.
   c. microbiome composition.
   d. laziness.

3. Glucose intolerance is
   a. a metabolic condition causing high levels of blood sugar.
   b. the inability to produce insulin at normal levels.
   c. having a strong distaste for sugar.
   d. unrelated to obesity.

4. Insulin is produced in the
   a. liver.
   b. kidneys.
   c. pancreas.
   d. thyroid.

5. Leptin
   a. is a signaling hormone conveying danger.
   b. helps the gut communicate with the immune system.
   c. is a hormone conveying satiety.
   d. is not present in germ-free mice.

6. Compared with lean individuals, the microbiomes of obese individuals
   a. show decreased species richness and diversity.
   b. produce fewer signaling hormones.
   c. contain more bacteria.
   d. promote weight loss.

7. Metabolic disruptions caused by gut microbiome dysbiosis can cause all of the following *except*
   a. increased immune cell activation.
   b. disruption of the mucin coating of the intestine.
   c. weakening of the intestinal wall barrier.
   d. loss of sight.

8. The most consistently reported phylum in the gut microbiota of obese individuals is
   a. Proteobacteria.
   b. Firmicutes.
   c. Acidobacteriota.
   d. Pseudomonadota.

9. A microbial guild is
   a. a microbiota structure composed of numerous species.
   b. composed of a single bacterial species.
   c. generally found only in the GI tract.
   d. a marketplace for probiotics.

10. Supplementing short-chain fatty acids in mouse diets has produced a decrease in all of the following *except*
    a. weight gain.
    b. microbiome diversity.
    c. inflammation.
    d. insulin resistance.

11. Bile acids are produced from
    a. cholesterol.
    b. proteins.
    c. microbes.
    d. blood.

12. The heritability rate of a given trait
    a. indicates how likely any individual is to display the trait.
    b. is based on how much genetics influence trait variation.
    c. explains environmental influence over a trait.
    d. determines how much microbiome composition will impact trait expression.

13. Which hormone do mice and humans use to signal satiety and regulate appetite?
    a. Testosterone
    b. Somatotropin
    c. Cortisol
    d. Leptin

14. Obesity-associated microbes often provide an increase in
    a. Mental acuity.
    b. metabolism speed.
    c. energy harvesting.
    d. sexual selection.

15. The ratio of _____ may be associated with increased weight gain, according to Jeffrey Gordon's research.
    a. *Firmicutes*:*Bacteroidetes*
    b. *Klebsiella*:*Bacteroides*
    c. *Fusobacterium*:*Firmicutes*
    d. *Acinetobacter*:*Klebsiella*

16. Which bacterial genus tends to encode genes that enable them to extract more energy from foods they ferment?
    a. *Bacteroides*

b. *Klebsiella*
c. *Acinetobacter*
d. *Lactobacillus*

17. A keystone species
    a. is an ecosystem engineer.
    b. plays a large role in its environment relative to its population.
    c. suppresses the growth of other ecosystem members.
    d. always includes a secondary consumer.

18. Which bacterial species was used in the first clinical study to demonstrate the impact of probiotics on metabolic syndrome?
    a. *Escherichia coli*
    b. *Staphylococcus aureus*
    c. *Akkermansia muciniphila*
    d. *Bifidobacterium longum*

19. Which classification is more precise than the species level?
    a. Strain
    b. Genus
    c. Kingdom
    d. Class

20. What bacterial genus, which helps newborns break down nutrients in breast milk, has also been shown to be associated with leanness?
    a. *Lactobacillus*
    b. *Streptococcus*
    c. *Acetobacter*
    d. *Bifidobacterium*

21. The following bacterial genera have been associated with leanness *except*
    a. *Roseburia*.
    b. *Prevotella*.
    c. *Acetobacter*.
    d. *Ruminococcus*.

22. When on a high-fiber diet, individuals with a high ratio of _____ demonstrated an increased ability to lose weight.
    a. *Prevotella*:*Bacteroides*
    b. *Bacteroidetes*:Firmicutes
    c. *Klebsiella*:Firmicutes
    d. *Bacteroides*:*Acetobacter*

23. The breakdown of fat occurs through a process known as
    a. glycolysis.
    b. lipolysis.
    c. fermentation.
    d. oxidation.

24. This group of compounds have been described by some as the "silver bullet to weight loss," but much more research is needed to fully understand their role in the microbiome and body.
    a. Lipoic acids
    b. Short-chain fatty acids
    c. Interleukin-21 and 22
    d. Spice melanges

Answers: 1B, 2D, 3A, 4C, 5C, 6A, 7D, 8A, 9A, 10B, 11A, 12B, 13D, 14C, 15A, 16D, 17B, 18C, 19A, 20D, 21C, 22A, 23B, 24D

DIVING DEEPER

1. What is the heritability rate of obesity? What other factors can influence an individual's weight?
2. From an evolutionary perspective, why was weight gain beneficial?
3. What changes occur in metabolic syndrome?
4. What microbial phylum is more plentiful in lean individuals than in those with obesity?
5. What function does *Akkermansia muciniphila* perform in the gut?
6. What is a microbial guild? What are the advantages of studying microbial guilds instead of focusing on individual species or other taxa?
7. What is LPS? What happens when LPS circulates through the body?
8. What are the effects of acetate, butyrate, and propionate?
9. Why is having higher levels of short-chain fatty acids not always better?
10. How do bile acids alter the composition of the gut microbiome?
11. How do some bacteria survive bile acids?
12. How do high or low levels of the metabolite indole affect miR-181, and as a result, how does miR-181 affect metabolic health?
13. When the intestinal barrier is weakened, what happens to LPS levels?
14. How is the oral microbiome different in obese than in lean individuals?
15. Some differences in a child's microbiome are associated with an increased likelihood of obesity later in childhood. At what age can they be detected?

16. What are some bacteria that are associated with an increased likelihood of losing weight?
17. What have been some effects of a diet high in fermented foods?
18. What are probiotics, and how are they different from prebiotics?
19. Are single-strain or multi-strain probiotics more effective?
20. Name three effects of prebiotic treatments that studies have observed.
21. What is the only FDA-approved use of fecal microbiota transplant (FMT) currently?
22. What are the risks of FMT?
23. How does the gut microbiome change over time?
24. Look up the term postbiotics and explain what they might do with respect to the gut microbiome?

DISCUSSING AND REFLECTING

1. Much of this chapter focuses on how your microbes can change your body's physiology. What are some dietary and therapeutic interventions you can use to flip the script and impact your microbiome's composition?
2. A future chapter will discuss the impact of the built environment on our microbiome and how its composition has changed from that of our prehistoric ancestors. Using only the information we've learned so far, such as the impact of diet and birth mode on community composition, generate some hypotheses about what our ancestors' microbiomes would have looked like.
3. Reflection. Many people, whether they are obese, lean, or somewhere in between, are self-conscious about their weight and are often critical of others. Does understanding the role your microbiome plays in regulating your body's weight and composition change your perspective on this? What other traits are we critical of in ourselves and others that may be due to our microbes?

RECOMMENDED READINGS

Popular Science Review
Berger, M. W. (2022, June 6). *How the Microbiome Influences Weight and Obesity*. Everyday Health. https://www.everydayhealth.com/digestive-health/what-does-the-microbiome-have-to-do-with-weight-and-obesity/

Popular Science Book
Sonnenburg, J. L., & Sonnenburg, E. D. (2016). *The Good Gut: Taking Control of Your Weight, Your Mood, and Your Long-term Health*. Penguin Books.

Scientific Reviews
Aoun, A., Darwish, F., & Hamod, N. (2020). The Influence of the Gut Microbiome on Obesity in Adults and the Role of Probiotics, Prebiotics, and Synbiotics for Weight Loss. *Preventive Nutrition and Food Science, 25*(2), 113–123. https://doi.org/10.3746/pnf.2020.25.2.113

Geng, J., Ni, Q., Sun, W., Li, L., & Feng, X. (2022). The Links between Gut Microbiota and Obesity and Obesity Related Diseases. *Biomedicine & Pharmacotherapy, 147*, 112678. https://doi.org/10.1016/j.biopha.2022.112678

Allergic Diseases and the Microbiome

12

CHAPTER CONTENTS

- 12.1 The Allergic Response
- 12.2 A Critical Window of Immune Training to Prevent Allergic Disease
- 12.3 Epigenetic Changes and Allergic Disease
- 12.4 Impact of the Maternal Microbiome on Allergic Reactions
- 12.5 The Impact of the Environment in Allergic Disease
- 12.6 The Old Friends and Biodiversity Hypotheses
- 12.7 The Role of Antibiotics in Allergic Disease
- 12.8 The Impact of the Microbiome on Allergic Disease
- 12.9 Microbiome-Based Therapeutics for Allergic Diseases
- 12.10 A Circle of Causality

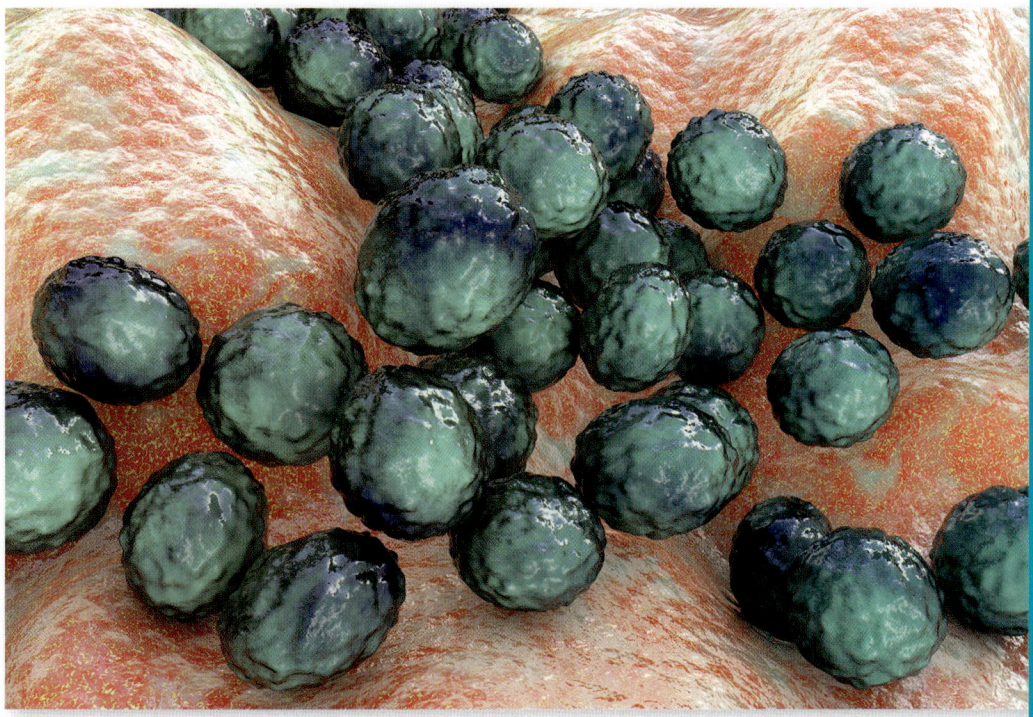

You may think that with the moniker *Staphylococcus*, I am one of those horrifying pathogens like my close relative, *S. aureus*, also known as MRSA. However, please understand that I am nothing like my kin. I am one of the nice guys, an abundant and beneficial member of your skin microbiome. I play a key role in keeping the skin microbiome in homeostasis, and I am often able to outcompete *S. aureus* when it attempts to invade the skin microbiome. I have several magical powers that permit me to regulate immunological reactions, such as the triggering of inflammation and the production of inflammatory cytokines. These are all critical defense mechanisms that I employ to protect you against eczema. (Photo from iStock.com/Dr_Microbe)

There has been an explosion in the number of children in the United States with food allergies, hay fever, and asthma. These inflammatory disorders involve immune reactions to what are normally harmless environmental triggers, or allergens, such as peanuts or plant pollen. As you will learn, our microbiome plays a key role in training our immune system to tolerate many of these allergens. However, when a newborn lacks exposure to a diverse array of microbes and their metabolites, inflammatory diseases run rampant. We will begin this chapter with an introduction to the molecular and physiological mechanisms underlying an allergic response. We will then turn our attention to the roles of early-life exposure to microbes and dysbiosis of the gut and lung microbiomes in the development of allergic diseases, with a particular focus on asthma.

"The lesson here is that it is possible that raising children in an overly hygienic environment could have a long-lasting detrimental impact on the development of their immune systems."

—Justin Sonnenburg
(Sonnenburg & Sonnenburg, 2016)

12.1 THE ALLERGIC RESPONSE

Your immune system is in constant motion—identifying and then attacking invading pathogens. However, there are times when it wages war on things that are not pathogens, such as allergens, and that is what we call an **allergic response**. Numerous external agents can trigger an allergic response, such as peanuts, animal fur, and pollen. These triggers are called **allergens**. When your immune system is triggered by an allergen, it responds by releasing **antibodies**, such as **immunoglobulin E** (**IgE**). As we learned in chapter 9, these highly specific proteins deliver a message to other immune cells, such as **mast cells**, which are white blood cells in connective tissue, that stop allergens in their tracks, resulting in the mast cell releasing **histamine**, which causes dilation of capillaries and contraction of smooth muscle. During an actual infection, histamine allows immune cells to leave the bloodstream and fight pathogens in soft tissue, but in the case of an allergy, where there is no invader, this release of histamine leads to harmful side effects. Histamine, and several other chemicals, are at the heart of creating what we experience as an allergic response, including wheezing, sneezing, overproduction of mucus, and swelling of mucosal surfaces. Although this type of immune response is common to all allergies, the antibodies that trigger the response are specific to each allergen type, which explains why one person might be allergic to pollen but not to cat fur. You can come into contact with allergens through the eyes, nose, mouth, or skin, which can result in skin inflammation, clogged sinuses, or stomach problems.

The Allergic Cascade

The more common allergic diseases are mediated by the antibody IgE and include asthma, allergic rhinitis (hay fever), allergic conjunctivitis (eye allergy), atopic dermatitis (eczema), and food allergies. These allergies trigger an **allergic cascade**, which is a series of predictable immune reactions to the presence of specific allergens (**Figure 12.1**). Allergens trigger B cells to produce IgE antibodies, which then bind to mast cells and trigger them to release histamines (a process called degranulation). Diverse groups of cell types contribute to IgE-dependent disease, including epithelial cells, **antigen-presenting cells** (**APCs**), T and B cells, and mast cells, as well as a whole group of white blood cells: basophils, eosinophils, and neutrophils. The role of epithelial cells in allergies extends far beyond providing a simple mechanical barrier that protects against invading pathogens. They are responsible for integrating innate and adaptive immune responses, which are central to allergic reactions.

Let's dive into the allergic cascade in more detail (**Figure 12.2**). Allergic reactions are initially mediated by antigen-presenting cells (APCs) such as dendritic cells (DCs), which capture and present allergens to T cells. DCs also help determine how a T cell responds to an allergen: DCs can trigger naive T cells (Th0) to differentiate into **type 1 T helper cells** (**Th1**), which are key to fighting bacterial infections; **type 2 T helper cells** (**Th2**), which promote an allergic response; or type 17 T helper cells (Th17), as well

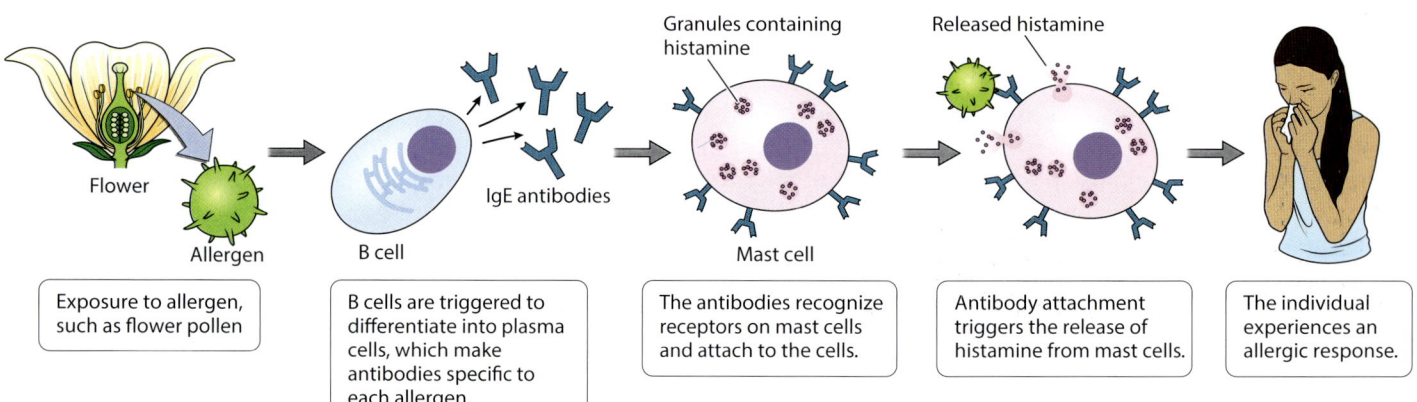

Figure 12.1 The Allergic Cascade Allergy is the result of an aberrant immune response towards harmless allergens, also known as antigens. Allergens trigger B cells to produce IgE, which then binds to mast cells and triggers their degranulation, which results in the release of histamine.

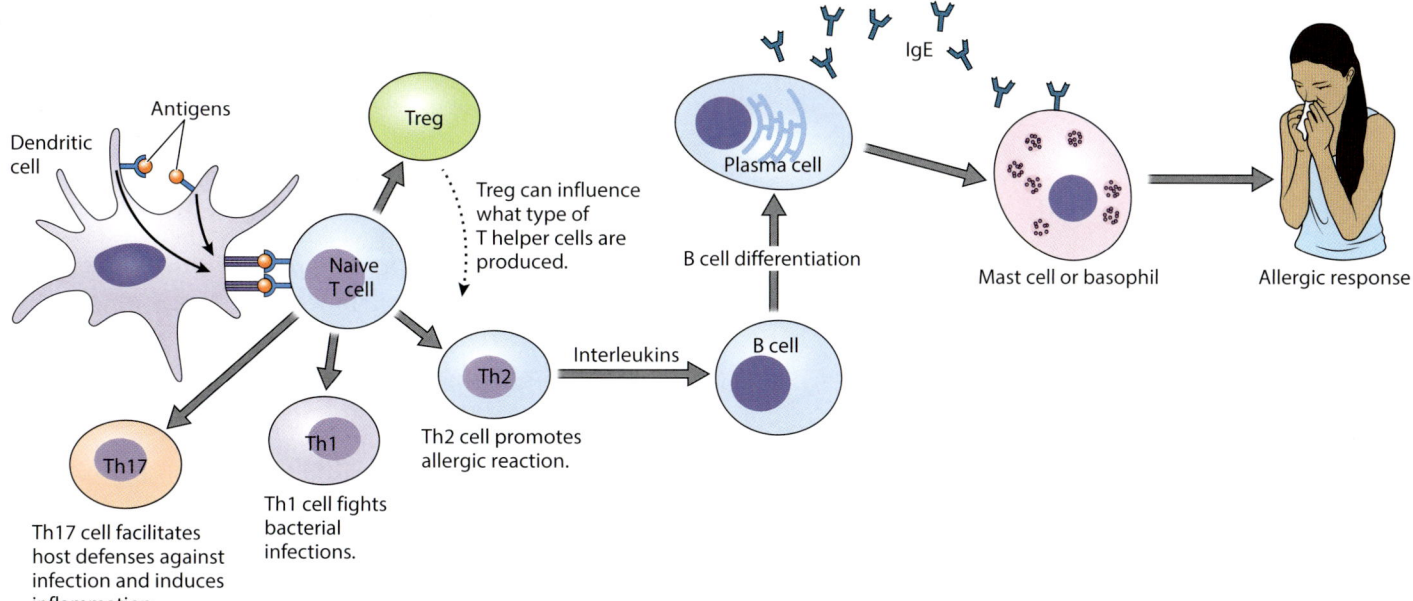

Figure 12.2 T and B Cell Involvement in an Allergic Response Naive T cells can be induced to differentiate into three main types of T cells: regulatory cells (Tregs) and type 1 (Th1) and type 2 (Th2) T helper cells. In allergic individuals, antigen presentation results in the differentiation of naive T cells predominately into Th2 cells. They secrete interleukins that trigger the differentiation of B cells into plasma cells. Plasma cells proliferate and produce an excessive amount of allergen-specific IgE that binds to mast cells and basophils.

as several other T cell types. Th2 responses in humans evolved to be protective against parasitic worm infections, but in the absence of these infections (such as in developed countries), the Th2 cells respond unnecessarily to harmless antigens, resulting in allergic responses. Th2 cells promote allergic responses by secreting cytokines, such as interleukins, which trigger the production of IgE antibodies by B cells. Th17 cells facilitate host defense against pathogen infection and help maintain the mucosal barrier, but they also induce tissue inflammation and contribute to autoimmune diseases. B cells differentiate into plasma cells, proliferate, and produce IgE that binds to mast cells and **basophils**, another type of white blood cell intimately involved in an allergic response. The production of IgE results in a sensitized immune system. That means that future exposure to that allergen results in a full-blown allergic response, such as release of histamine and inflammation. In skin tissue the response results in hives, swelling, and eczema. When allergens are inhaled, hay fever symptoms result, such as itchy eyes, runny nose, and wheezing. Ingestion of food allergens can result in nausea, vomiting, and diarrhea. The most severe reaction is anaphylaxis, which involves both the respiratory and cardiovascular systems and can result in death.

12.2 A CRITICAL WINDOW OF IMMUNE TRAINING TO PREVENT ALLERGIC DISEASE

There has been a substantial increase in allergic diseases over the past few decades that cannot be explained solely by genetics. It is becoming clear that some of this increase is due to a lack of exposure to a diversity of microbes during two critical phases of immune system development: the periods of fetal development and the first 2 years of life. Newborns and infants raised in large families and with household pets or farm animals are exposed to an enormous diversity of potential allergens in the so-called **farm effect**. Microbial exposure during this critical period results in an immune system educated to consider these potential allergens as friend rather than foe.

Upon repeated exposures later in life, an individual's immune system does not perceive those substances as threats and thus does not trigger an immune response. We will explore the farm effect in more detail later in this chapter.

More and more children do not have those early-life exposures to allergens. Families are smaller, there is an increased emphasis on hygiene, and kids spend less time outdoors and are exposed to ever-increasing amounts of antibiotics and antibacterial soaps. When these children are later exposed to allergens outside of the critical window of immune training, an immune response is mounted. Unfortunately, this immune hypersensitivity goes well beyond responses to the presence of pathogens and includes responses to harmless substances like peanuts, pollen, or dust.

12.3 EPIGENETIC CHANGES AND ALLERGIC DISEASE

One way our environment and our microbiota can impact our susceptibility to allergies is through changes in gene expression. Although we tend to think of the way our genes shape us as set in stone, gene expression is the result of many factors, including our environment and lifestyle. Our genetic "recipe book" may stay the same, but our environment plays a major role in which recipes get cooked and when. Our DNA is surrounded by RNA and proteins (mainly histones) that package and condense it into a structure that is called **chromatin**. **Epigenetic changes** are DNA modifications that impact the expression of genes. **Figure 12.3** provides an overview of the primary categories of epigenetic modification, including histone modification, DNA methylation,

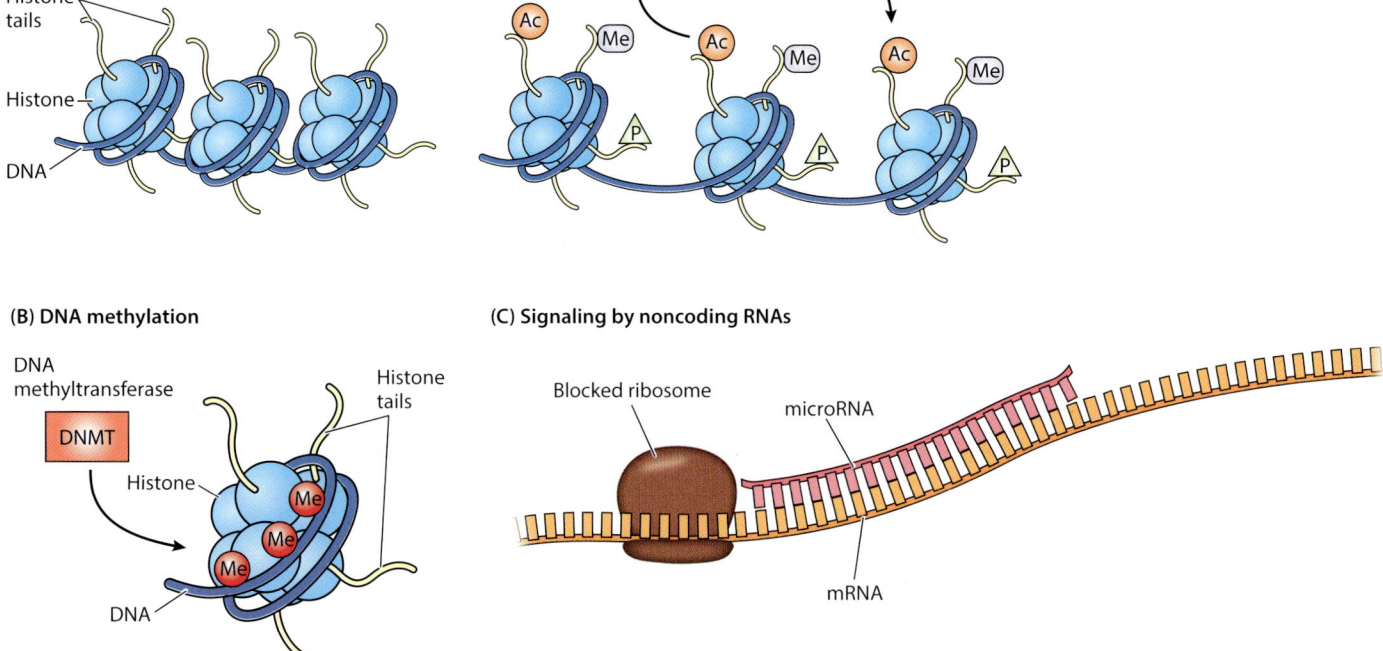

Figure 12.3 Overview of Epigenetic Modification (A) Histone modification includes the process of DNA acetylation (addition of an acetyl group, Ac) by histone acetyltransferase (HAT), which allows chromatin to relax, thus making it accessible for transcription. In contrast, deacetylation by histone deacetylase (HDAC) condenses the chromatin, preventing DNA transcription. (B) DNA methylation involves the activity of the enzyme DNA methyltransferase (DNMT), which adds methyl groups (Me) to DNA, promoting DNA condensation and inhibiting gene expression. (C) Noncoding RNAs, such as microRNA, are responsible for most epigenetic regulation of gene expression, which occurs by preventing translation of transcripts or promoting their degradation (silencing).

and signaling by **noncoding RNAs** (RNAs that don't encode proteins). These changes affect the transcriptional accessibility of genes and thus impact the regulation of gene expression.

In general, exposure to different organisms, such as bacteria or parasitic worms, can induce epigenetic changes that then directly impact an immune response. Additionally, certain environmental exposures during pregnancy, such as diet, smoking, and antibiotic use, can result in the modification of fetal T cell function through epigenetic mechanisms. These changes impact the expression of genes associated with immune responses. The window of opportunity for immune system development extends into the first 2 years of life, when these same environmental exposures are almost always accompanied by epigenetic changes.

One study of the impact of epigenetic changes that trigger an allergic response involved the sampling of cells from the nasal cavity of allergic individuals. These cells were screened for an association of epigenetic modifications with allergic disease. The study revealed that nasal cells harvested from allergic individuals shared many epigenetic signatures (van Breugel et al., 2022). For example, methylation levels in the nasal epithelial cells were lower in allergic individuals. In fact, the correlation was so strong that methylation status is now considered a biomarker of allergic disease. Another example is provided by the administration of *Lactobacillus reuteri* to pregnant mothers, which results in epigenetic changes that affect gene expression patterns in their offspring. Finally, production of short-chain fatty acids (SCFAs) by gut microbes can alter histone acetylation levels, thus causing epigenetic changes in regulatory T cells (Tregs) and mast cell regulation. We learned above that Th0 cells can differentiate into Th1 and Th2 cells. Th0 cells can also mature into Treg cells, which, as their name implies, regulate other immune cells. Tregs can dampen immune responses and prevent autoimmunity, the condition in which immune cells attack the host's cells.

12.4 IMPACT OF THE MATERNAL MICROBIOME ON ALLERGIC REACTIONS

Allergic responses are often seen as accidents—your immune system misinterpreting a harmless protein, such as that in cow's milk, as a threat. The immune system responds as if the threat were real, and you end up breaking out in hives, sneezing, or in the worst case, in anaphylactic shock. Why do some individuals' immune systems make frequent mistakes, while others never do? Some of those mistakes are the result of your genome, with genetics explaining some 30% to 90% of allergic disease. However, it should come as no surprise to you by now that studies of the microbiome are beginning to highlight the importance of the environment, starting as early as in the womb, as accessory factors in an allergic immune response.

Maternal diet is a significant factor in asthma development in offspring. As we learned in chapter 7, metabolites from the maternal gut microbiome, such as SCFAs, are passed from mother to fetus through the placenta and impact fetal immune development. A low-fiber diet during pregnancy results in decreased bacterial diversity and lower levels of SCFA production by the maternal gut microbiome, which leads to inhibited IgA and IgG production in the fetus. In contrast, a high-fiber diet during pregnancy results in increased levels of SCFA, and this lowers the risk of asthmatic symptoms in the newborn's first year of life (Thorburn et al., 2015).

This dietary effect is so highly predictive of asthma that an index was developed (**Figure 12.4**) (Venter et al., 2022) that measures food items that are able to prevent allergic responses, such as yogurt and vegetables, versus those that result in an increased allergic response, including red meat, French fries, and fruit juice. A higher score on the maternal

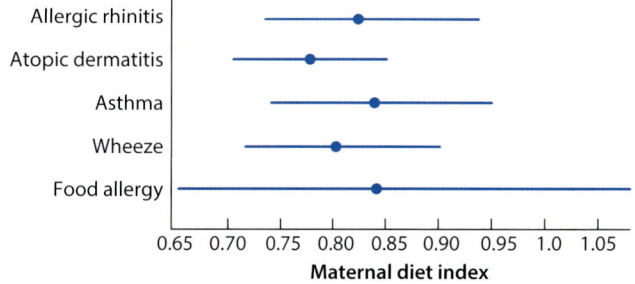

Figure 12.4 Maternal Diet Index Predicts the Risk of Allergic Disease A maternal diet index was produced by tracking the diets of large numbers of mothers while also tracking the appearance of allergic diseases in their offspring until age 4. The association between the diet index and prevalence of disease was highly significant. (After Venter et al., 2022.)

diet index indicates that the mother ate a diet that is better at preventing allergies. The correlation between the diet index value measured and subsequent risk for disease in offspring is stunningly high. For every unit increase in the diet metric, the odds of developing allergic diseases are decreased 16% for asthma, 23% for eczema, and 18% for hay fever (Venter et al., 2022).

One highly significant factor in determining the risk for allergic disease is maternal exposure to particular environments. Children born into families who live on farms experience significantly lower levels of allergic disease. The environment provides an increase in oral, respiratory, and skin-based exposure to several groups of bacteria, in particular *Lactobacillus* and *Bifidobacterium* species. The presence of these taxa results in increased maternal recruitment and functionality of Tregs, which regulate the immune system and help prevent reactions to harmless allergens. Women on farms are also more likely to drink unpasteurized milk, rich in beneficial microbes that confer protection to their offspring from allergic disease. The child then shows higher levels of cytokine production and reduced levels of IgE, suggesting a shift from Th2 to Th1 immunity. The mechanisms underlying these changes remain elusive. However, it appears likely that the maternal gut microbiome can affect systemic immune activity in the fetus, resulting in less inflammatory response by macrophages and DCs. The maternal microbiota itself can also decrease the risk of allergic disease. For example, nasal administration of the farm-associated bacterium *Acinetobacter lwoffii* in pregnant mice resulted in progeny that were protected from asthma (Alashkar Alhamwe et al., 2022). The effect of *A. lwoffii* was mediated by stabilization of histone H4 acetylation, which relaxes the conformation of DNA, making it more accessible for processing activities, such as transcription.

12.5 THE IMPACT OF THE ENVIRONMENT IN ALLERGIC DISEASE

During and immediately following birth, a newborn is exposed to an enormous diversity of microbes, which go on to potentially seed the gut with a dynamic community of microorganisms. The communication between this nascent microbiome and the newborn's immature immune system is shaped by numerous factors within the gut and through environmental exposures until the microbiome matures to its more stable, adult version by about 3 years of age. Healthy and allergic newborns and young infants have significant differences in the composition and diversity of their gut microbiome. **Table 12.1** lists examples of activities that impact the newborn's gut microbiota and contribute to an increased or decreased risk of allergic disease. Several of these activities have a particularly high impact on the risk of allergic disease, including delivery by C-section, hospital versus home delivery, antibiotic exposure, lack of pets such as dogs, lack of farm exposure, and smaller family sizes (Sbihi et al., 2019).

Some of the more compelling data implicating environmental microbes in allergic disease come from mouse studies, where a highly controlled environment can reveal the role external microbes play in populating the host microbiome and impacting immune response. In one such study, environmental factors, including the laboratory from which the mice were obtained, the types of cages in which the mice were housed, and diet, correlated with the resulting mouse gut microbiome composition (Dickson et al., 2018). This study shows how just about every aspect of the environment and lifestyle can impact the gut microbiota. Another study involved transferring germ-free embryos into wild mice. The offspring adopted the microbiomes of their surrogate mothers as well as their immune responses (Rosshart et al., 2019). These highly controlled experiments demonstrate that, at least in these subjects, exposure to environmental microbes has a greater influence on the microbiota than does the host genome.

Table 12.1 Early-life Environmental Factors Affecting Gut Microbiota Composition and Allergic Diseases Risk

RISK FACTORS	PROTECTIVE FACTORS
C-section delivery	Vaginal delivery
Premature hospital delivery	Home delivery
Antibiotics	Farm exposure
	Household pets
	Daycare exposure/older siblings
	Maternal high-fiber diet

Source: Sbihi et al., 2019

The Response of Gut Microbes to Environmental Triggers of Allergic Disease

One way gut microbes modulate the host immune response to environmental triggers is via interactions with receptors located on the host intestinal epithelial cells. These **pattern recognition receptors** (**PRRs**) recognize conserved structures known as **microbe-associated molecular patterns** (**MAMPs**) that are presented on the bacterial cell surface or released into the lumen (**Figure 12.5**). Basically, PRRs allow intestinal epithelial cells to detect common indicators of bacterial cell presence and coordinate an immune response. One very common PRR is the **Toll-like receptor** (**TLR**) part of the innate immune system of vertebrates. TLRs recognize and bind MAMPs and then initiate and transfer signals to other immune cells. The signals, in essence, amplify the TLRs' call for help from immune cells, which then produce inflammatory factors. **Leucine-rich repeats** (**LRRs**) are found in the extracellular region of TLRs, and they are involved in the recognition of MAMPS. The intracellular region, known as the TLR domain, binds to proteins in the cytoplasmic region (adapter proteins). Finally, the death domain is involved in initiating cell death or in signaling an immune response.

We have discussed several MAMPs previously, such as the lipopolysaccharides (LPS) found on the surface of certain pathogenic bacteria. In fact, LPS exposure during the few weeks before and after birth is protective against Th2-mediated inflammation in a murine model of asthma. Another example is polysaccharide A, which is present in the cell capsules of *Bacteroides fragilis*, which has beneficial effects on the immune system. Specific MAMPs are recognized by specific TLRs. MAMPs include lipids, proteins, and nucleic acids with unique molecular structures that are not found in host cells but are generally essential for a microbe's survival. Thus, this component of the innate immune system recognizes MAMPs and acts to distinguish self from non-self and respond to pathogen presence.

Members of the gut microbiome, in addition to their MAMP-mediated immune effects, also act indirectly with the production of metabolites, such as SCFAs, that affect host physiology and immune function. In addition to providing energy for epithelial cells, SCFAs protect the host against numerous immune and metabolic diseases. For example, the production of acetate by members of the gut microbiome has been shown to protect against asthma. Numerous studies have shown that a high-fiber diet, which results in numerous beneficial bacterial fermentation products, has been shown to result in improved asthma symptoms by impacting Treg production.

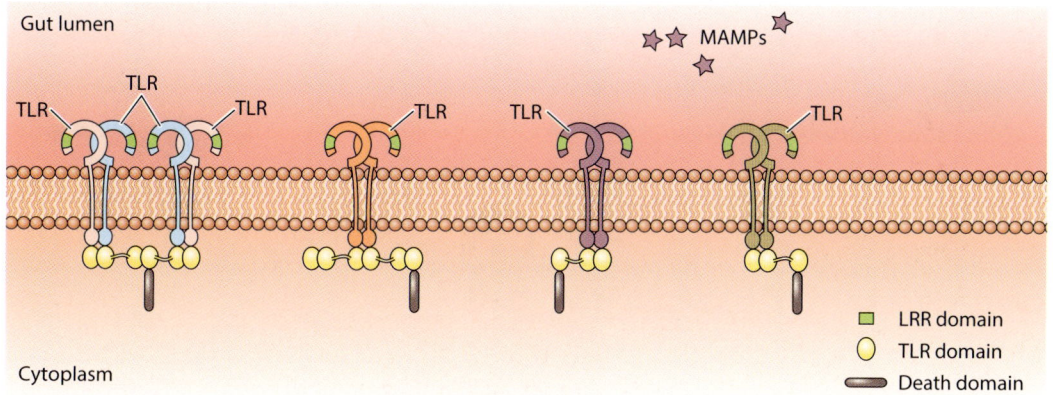

Figure 12.5 The Recognition of MAMPs by Cell Surface Receptors Pattern recognition receptors can recognize one or more microbe-associated molecular patterns (MAMPs) through the external leucine-rich repeat (LRR) domain. The internal Toll-like receptor (TLR) domain is involved in signal transduction.

The Farm Effect

In the 1990s, a Swiss physician noticed that children who lived on farms experienced far fewer respiratory allergies than nonfarm kids. The farm effect dovetailed with the **hygiene hypothesis**, a concept we have discussed repeatedly, and this provided a plausible explanation for the rapid rise of allergic diseases in developed countries. Today, this hypothesis centers on the microbiome and suggests that a variety of environmental and lifestyle factors converge to disrupt the body's natural community of microbes, and this prevents critical cross talk between the microbiota and innate immune cells that would help establish immune tolerance.

From an early age, farm children inhale a range of substances that children in urban settings rarely encounter, such as the microbes and their metabolites found on farm animals (**Figure 12.6**). For example, levels of certain MAMPs, such as LPS, are 7 times higher in dust samples from farm homes. Children from these homes suffer far lower levels of allergic diseases than do urban children. Researchers have administered farm dust to young mice and shown that it prevents airway hyperreactivity. Conversely, children in urban settings with asthma have decreased bacterial richness in their gut microbiomes, specifically lower levels of Firmicutes and Bacteroidetes. One study examined the bacteria from cowsheds and determined that two of the bacterial species detected (*Acinetobacter lwoffii* and *Lactococcus lactis*) impacted the levels of DCs in mice. Exposing mice to these strains intranasally resulted in a weakened responsiveness to a diversity of environmental allergens. Thus, exposure to farm-associated microbes can protect against allergies, and these same microbes can make their way into the gut microbiomes of children and appear to prevent the development of allergic diseases.

One of the more compelling studies into the farm effect focused on a comparison of microbiota among Amish and Hutterite farmers (Stein et al., 2016). These two populations have remained largely reproductively isolated since their immigration from Europe during the Protestant Reformation of the 1500s. Certain features of their lifestyles are extremely similar, such as having large families, their diets, prolonged breastfeeding, and abundant indoor pets. However, two key differences exist between these communities, which involve the manner in which they farm and the role of children in the farming activities. The Amish employ traditional farming methods, which include the use of farm animals for transportation and field work and the

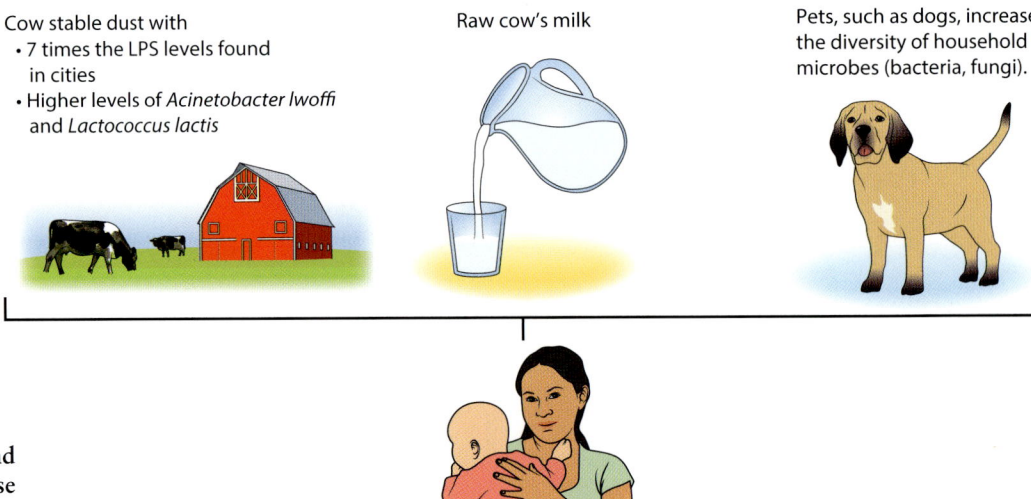

Figure 12.6 The Farm Effect and Protection from Allergic Disease The development of allergy and asthma is influenced by numerous farm factors, many of which are discussed as protective, especially if they act during pregnancy and childhood. (After Mayerhofer & Pali-Schöll, 2021.)

Cow stable dust with
- 7 times the LPS levels found in cities
- Higher levels of *Acinetobacter lwoffi* and *Lactococcus lactis*

Raw cow's milk

Pets, such as dogs, increase the diversity of household microbes (bacteria, fungi).

Pregnant women or newborns who are exposed to a diverse set of metabolites not found in urban settings are provided with protection from asthma and allergies, known as the farm effect.

distribution of farm chores among the entire family. The Hutterites, in contrast, employ more modern farming practices that do not involve animals. Their children do not come in contact with farm animals as Amish children do. The prevalence of asthma is 4 times lower among Amish than Hutterite children. To explore the cause of this difference, one study examined house dust in the respective homes and revealed that dust from Amish homes had far higher levels of the bacterial endotoxin LPS (Stein et al., 2016).

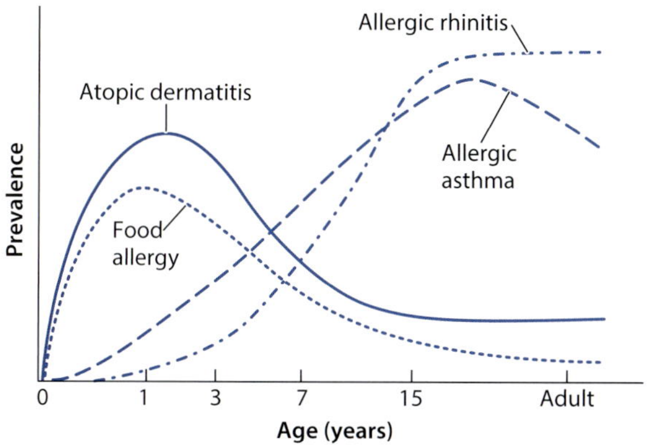

The Atopic March

Numerous additional studies show that what children consume during their first few years can influence their risk of later developing allergic conditions, in what is known as the **atopic march**, which speaks to the fact that children who develop an allergic disease, such as eczema, allergic rhinitis, asthma, or food allergy, are likely to develop other additional allergic diseases (**Figure 12.7**). One long-term study of the impact of diet on food allergies revealed that for each additional food type introduced by 6 months of age, the child's risk of developing food allergies dropped about 10% (Venter et al., 2020). To delay or possibly prevent allergies and asthma in children, the American Academy of Allergy, Asthma & Immunology recommends introducing a range of healthy solid foods starting at age 4 months, including common allergens, such as egg, dairy, peanut, tree nuts, fish, and shellfish. One allergy researcher recommends, "Every new food helps. Feed a new food every time the baby is happy" (Venter et al., 2020). Additionally, the consumption of processed foods should be minimized. Processing removes healthy fiber and reduces microbial content, which can help preserve food but can also deplete beneficial bacteria. Another suggestion is to spend time outdoors. Growing evidence suggests that in addition to diet diversity, much of the farm-related protection against asthma and allergies comes from early exposure to nature and animals. Getting a pet may also help, though that evidence is less clear. The take-home message is that exposing a newborn to a diversity of environments and environmental microbes appears to be one avenue for reducing the incidence of allergic disease.

Figure 12.7 Atopic March The atopic march is the predictable appearance of specific allergic diseases over time. Atopic dermatitis is generally one of the first atopic responses to appear, closely followed by food allergies, allergic asthma, and allergic rhinitis. Depending on the severity of atopic dermatitis in infancy, one can predict the subsequent severity of the other three conditions. (After Tsuge et al., 2021.)

12.6 THE OLD FRIENDS AND BIODIVERSITY HYPOTHESES

The **hygiene hypothesis** posits that early-life exposure to microbial pathogens and experiencing certain infections is required to educate or train the immune system. These childhood microbial exposures teach the immune system to distinguish harmless substances from those that trigger allergic diseases. In 2003 Graham Rook, now an emeritus professor at University College London, proposed a hygiene hypothesis revision called "**old friends**," which states that rather than focusing on the reduced exposure to pathogens, we should be considering those commensals that we have evolved with and that need to be tolerated by us (Rook, 2003). As shown in **Figure 12.8**, Rook proposed that the immune system requires input from our old friends, that is, the commensal microbiota we inherit from our mothers at birth or are exposed to through family members, nature, and pets that help to diversify the commensal microbiota. Early in our evolutionary history, it is likely, many infections were able to persist as relatively harmless. These "tolerated" pathogens played a key role in the education of the newborn's immune system. The Western lifestyle has resulted in the elimination of these previously common childhood infections, as well as a more limited exposure to microbes in the natural environment. One unfortunate outcome of our modernization is an increase in autoimmune disease.

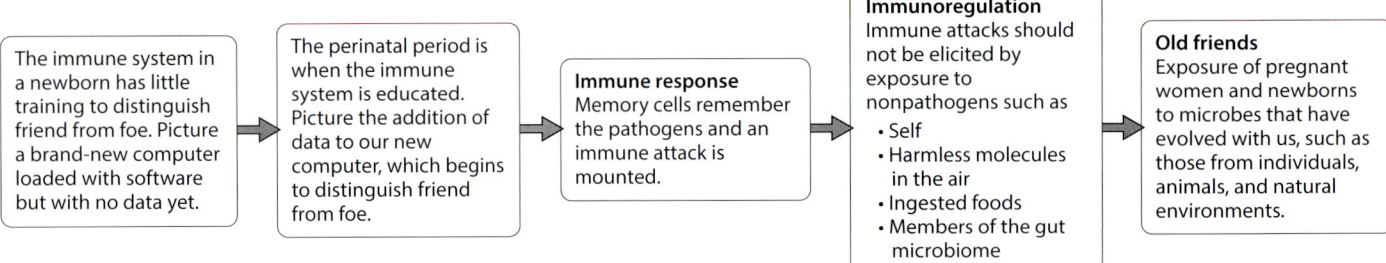

Figure 12.8 The Mechanisms Underlying the Old Friends Hypothesis The microbiota with which we have coevolved serve the role of educating our immune system, resulting in immune regulation that recognizes self, harmless allergens, and gut contents and ensures that background inflammation is inhibited when not required. (After Rook et al., 2014.)

A second revision to the hygiene hypothesis focuses, yet again, on the decline of microbial diversity. The **biodiversity hypothesis** proposes that early-life contact with natural environments results in a more diverse microbiome, which promotes immune balance (Haahtela, 2019). The human and environmental microbes work together to ensure immune tolerance. In attempting to understand the causes of allergic disease, exploring the determinants of immune tolerance is the key. This hypothesis lies at the center of a national campaign to combat allergic disease that ran in Finland from 2008 through 2018 (Haahtela, 2019). Standardized diagnostics were implemented to determine severe versus mild allergies. Severe asthma became the healthcare focus, while children with mild allergies were encouraged to learn how to tolerate these allergens and employ strategies of diverse food exposure early in life to limit these allergies in later life. The key messages to the public were to "endorse health, not allergy; strengthen tolerance, adopt a new attitude to allergy, and avoid allergens only if necessary; recognize and treat severe allergies early to prevent exacerbations; and improve air quality and decrease smoking" (Haahtela, 2019). There are encouraging signs this paradigm shift is working. Asthma-related visits to the hospitals for children have dropped by 62%. The levels of severe asthma, which were 20% before the program, now stand at 2.5%. This novel approach to thinking about and acting upon allergic disease appears to be having a significant impact on how people in Finland experience allergies.

12.7 THE ROLE OF ANTIBIOTICS IN ALLERGIC DISEASE

The administration of antibiotics in newborns and infants is highly correlated with an elevated risk of developing allergic diseases. Newborns who receive antibiotics in the first week of life have reduced levels of *Bifidobacterium* and increased levels of *Enterococcus* in their stool. By one month, these same antibiotic-treated babies have elevated levels of inflammatory Enterobacteriaceae (Sbihi et al., 2019). Children born by C-section to mothers who received antibiotics prior to or during childbirth experience the same reduction in *Bifidobacterium* species and possess even lower levels of gut microbiome diversity. For children prescribed antibiotics during their first year of life, there is a whopping 200% increased risk of developing asthma.

The more antibiotics an infant is exposed to, the more likely they are to develop asthma. Indeed, there is a 20% increase in asthma risk for each additional antibiotic prescription. This same increased risk applies to food allergies. **Figure 12.9** provides a model for how both an overly hygienic environment and heavy antibiotic use results in decreased microbial stimulation and reduced Th1 response. The presence of antibiotics disrupts the newborn's gut microbiome colonization, skewing the newborn's immune response by repressing the Treg response, which results in an elevation in Th2 cells.

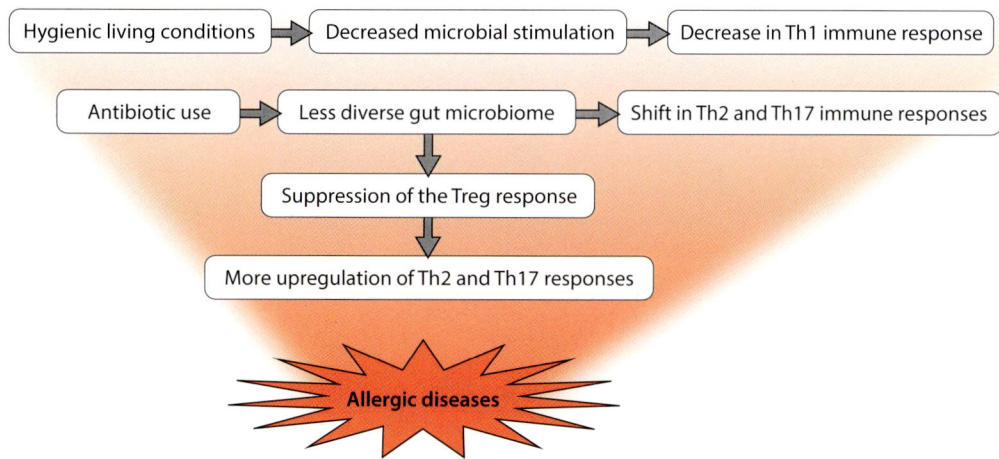

Figure 12.9 A Model for the Antibiotic-Mediated Induction of Allergic Diseases (After Kuo et al., 2013.)

Numerous studies have examined the impact of antibiotic use on microbiome dysbiosis and the correlations that exist between dysbiosis and rates of allergic disease. One such study employed mouse pups exposed to therapeutic doses of the antibiotics azithromycin or amoxicillin (Borbet et al., 2022). The mice were then challenged with an allergen, in this case dust mites. Mice with early-life azithromycin exposure had increased levels of IgE. To test that the dysbiotic microbiota was the cause of this increase, germ-free mice were given fecal microbiota transplantation (FMT) from the antibiotic-treated donor mice. These FMT mice did not show an altered response to mite exposure; however, their offspring showed elevated IgE levels and an altered reactivity in their airways following mite exposure. What might appear to be a genetic predisposition to allergies, in this case, is actually due to an altered gut microbiota.

As we become better antibiotic stewards, meaning that we limit antibiotic use to when it is medically necessary, to reduce the spread of antimicrobial resistance, one beneficial outcome may very well be a decrease in the risk of childhood allergies. Patrick et al. (2020) assessed the association between antibiotic prescriptions, gut microbiome composition, and asthma incidence levels in a cohort of over 2,000 children. Their findings suggest that there has been a reduction in the incidence of childhood asthma recently that might be the result of prudent use of antibiotics during infancy, which preserves the diversity of the infant's gut microbiome.

12.8 THE IMPACT OF THE MICROBIOME ON ALLERGIC DISEASE

We have clearly established the human microbiome as serving a critical role in determining allergic disease, due, in part, to its extensive communication and interaction with the immune system. One hypothesis regarding the risk of allergic disease focuses squarely on the impact of gut microbiome dysbiosis (**Figure 12.10**). In this proposal, allergic disease results from the failure of the gut microbiome to produce sufficient butyrate. Without butyrate, naive T cells cannot differentiate into Tregs, resulting in an inability of the immune system to dampen an immune response. Across the life of an individual, numerous environmental factors, including maternal microbiome, mode of birth, diet, and medications, all contribute to the composition and metabolic productivity of the human microbiome with significant consequences for host immune function. In this section we will focus on how the microbiome can directly

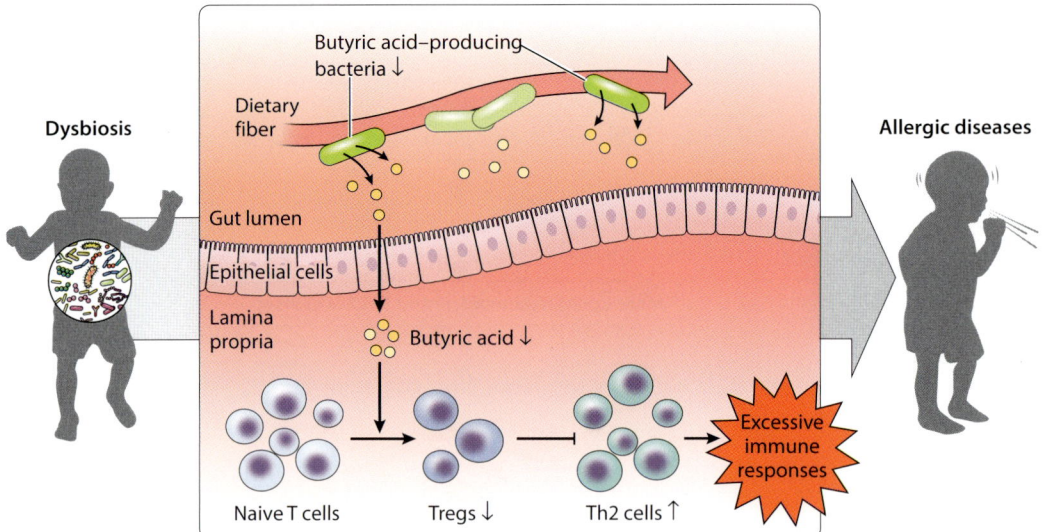

Figure 12.10 Dysbiosis and the Onset of Allergic Diseases Gut microbiome dysbiosis results from a variety of factors. There is a disruption in levels of the short-chain fatty acid butyric acid, which then results in fewer naive T cells differentiating into Tregs, which impairs the ability of the immune system to suppress excessive inflammatory responses.

impact the immune system and promote three highly common classes of allergic disease: asthma, atopic dermatitis, and food allergy.

Asthma

Asthma is one of the most common chronic inflammatory airway diseases, affecting roughly 25 million Americans, and is characterized by coughing, wheezing, and difficulty breathing in response to pollen, exercise, cold air, and many other common environmental triggers. An asthma attack results in inflammation and swelling of the air passages. The airways narrow, constricting airflow and resulting in shortness of breath, chest pain, and wheezing. Although asthma can affect individuals at any age, it often appears before age 6. The timing of its appearance and the levels of immune hypersensitivity depend on an individual's genetic susceptibility, viral and bacterial infections, and exposure to certain allergens, tobacco smoke, and air pollution.

Asthma is a heterogeneous disease, which means it has a variety of root causes and responses by the body. We now understand that genetic and immunological factors shape asthma but that the microbiome also plays a role and may underpin the heterogeneous nature of its expression in different individuals. The likelihood of developing asthma and the severity of the disease are heavily influenced by early-life exposure to microbes, with significant benefit arising from early exposure to a diversity of microbes. They are also impacted by the patterns of respiratory tract colonization and the occurrence of acute viral infections.

We are continuously exposed to inhaled agents, such as pollen and smoke, and if an individual is genetically predisposed to asthma, contact between these agents and airway mucosal surfaces may trigger an immune response. The airway epithelium plays a key role in the development of asthma. An asthmatic reaction is characterized by activation of **eosinophils**, which are white blood cells involved in the inflammatory response, as well as hypersecretion of mucus, proliferation of mucin-producing goblet cells, airway hypersensitivity to allergens, and breathlessness (**Figure 12.11**). During an asthmatic attack, the Th2-mediated immune response is dominant, resulting in the release of numerous cytokines. These cytokines trigger histamine release from mast cells, an accumulation of eosinophils in the airway, and an increase in mucus production and goblet cell proliferation.

A healthy respiratory tract has a dynamic yet relatively low-density microbial community dominated by Bacteroidetes, Actinobacteria, and Firmicutes. There is a

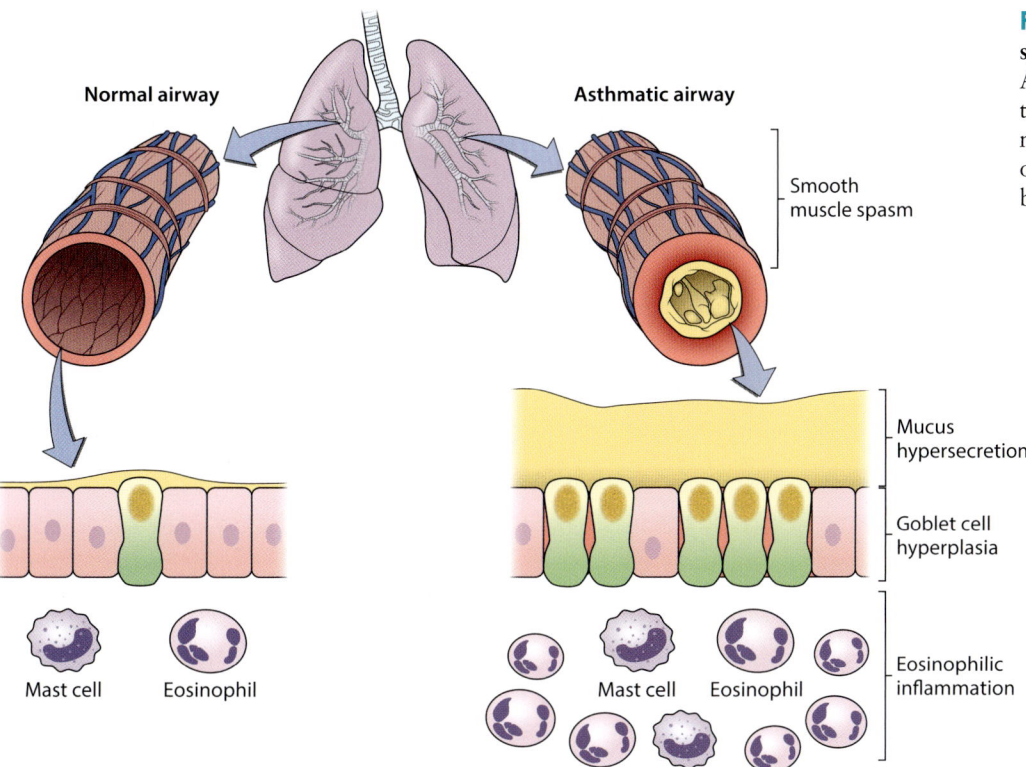

Figure 12.11 Physiological Responses in an Asthmatic Attack An asthmatic reaction is characterized by inflammation, excess mucus production, sensitivity of the airways, and shortness of breath.

short window of time following birth during which the newborn's respiratory and gut microbiomes are colonized. Disruption of this process results in lung microbiome dysbiosis and the risk for development of several respiratory diseases. As discussed above, a variety of factors affect the microbial colonization of our microbiota, including that of the respiratory tract. These include mother's health status, her diet, transfer of maternal antibodies and microbial metabolites through the placenta, mode of delivery, breastfeeding, exposure to smoking and air pollution, and the use of antibiotics and other drugs before birth and as a newborn.

Asthmatics have an altered microbiome composition in the respiratory tract, with increased numbers of Proteobacteria, especially *Haemophilus, Moraxella, Streptococcus*, and *Neisseria*, and reduced numbers of Bacteroidetes and Fusobacteriota. This microbial community is associated with hyper-responsiveness of airway epithelial cells to allergens, which readily trigger an inflammatory response and other asthma symptoms. When these asthma-related bacteria are detected in the lungs within the first few months of life, especially following viral infections, a child is more likely to develop asthma by age 6.

THE GUT-LUNG AXIS The **gut-lung axis** was so named because asthmatic children, in addition to alterations in their lung microbiome, often also show dysbiosis of the gut microbiome. During the first few months of life, there is a shift in the levels of several gut bacteria, including an increase in *Lachnospira* and *Clostridium neonatale* and a decrease of *Veillonella, Faecalibacterium*, and *Rothia*, suggesting that these bacteria play a role in protecting against or promoting the development of childhood asthma. Levels of key microbial metabolites also change; for instance, there is a drop in acetate production. Further, it turns out that the gut-lung axis works in both directions. For example, stimulation of mouse lungs with an allergen such as lipopolysaccharides results in an increase in bacterial density in the gut, suggesting that respiratory exposure to microbial metabolites stimulates the gut microbiome (Zhang et al., 2020). The mechanisms that mediate this bidirectional communication are not yet understood.

Microbial gut colonization and proliferation of several bacterial taxa, including members of the genera *Bifidobacterium, Lactobacillus*, and *Akkermansia*, among

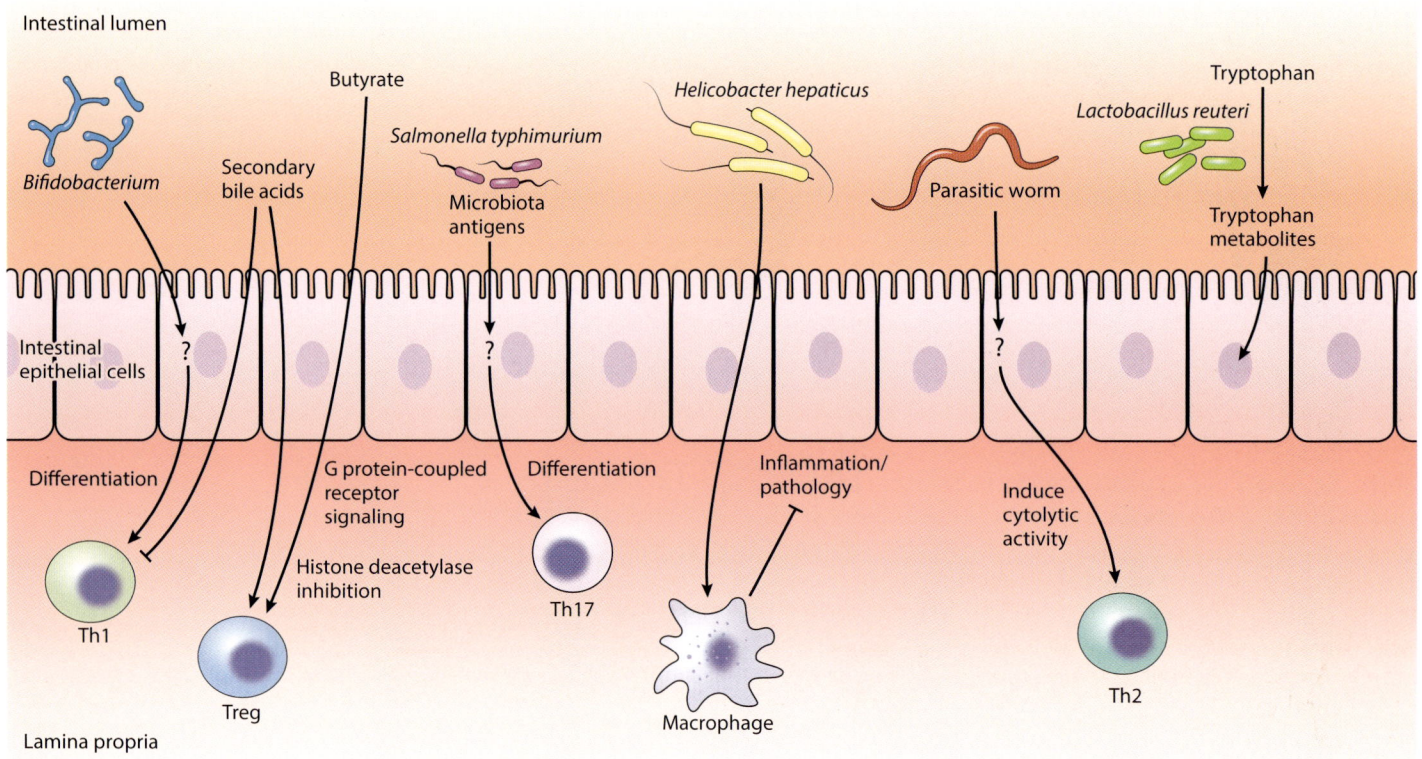

Figure 12.12 Microbial Impact on Intestinal Immune Cell Differentiation Our gut microbiome provides signals that trigger naive T cells to differentiate into a diverse family of immune cells, including Tregs and T helper cells. Some bacteria, such as *Bifidobacterium*, signal production of type 1 T helper cells (Th1). The presence of short-chain fatty acids, such as butyrate, signals production of Tregs. The presence of parasitic worms signals production of Th2 cells, while certain pathogenic antigens, such as LPS, signal production of Th17 cells. (After Geuking and Burkhard, 2020.)

others, is associated with protection against asthma (Zhang et al., 2020). Specific bacteria are able to modulate the balance between a variety of immune cells, as depicted in **Figure 12.12**. Furthermore, the gut microbiome produces key metabolites that directly impact immune function. For example, members of the genus *Clostridium* produce propionate, which appears to play a role in the activation of Tregs and a reduction in Th2-based inflammation.

The lung microbiomes of asthmatics show an abundance of Proteobacteria, including *Haemophilus* and *Neisseria*. These taxa are known to cause respiratory diseases, such as pneumonia. Members of this phylum produce LPS, which, as discussed above, is recognized by epithelial cells and results in the triggering of inflammation and other allergic responses. This increase in LPS levels can also be seen in newborns and is associated with the later development of wheezing and asthma. Not only does lung microbiome dysbiosis increase the risk of asthma, it also impacts the severity of viral infections. Colonization of the air passages with Proteobacteria in newborns increases the risk and severity of childhood pneumonia and bronchitis. Infants with these conditions also exhibit heightened inflammatory immune responses.

EARLY EXPOSURE OF THE LUNG TO ENVIRONMENTAL MICROBES The impact of the microbiota on asthma is most clearly seen in studies with germ-free mice. These mice are born with a heightened susceptibility to lung allergens. They clearly illustrate the window of opportunity during which the microbiome trains the immune system and influences allergic disease susceptibility. Once their airway and gut microbiomes are colonized, within several weeks following birth, that susceptibility is reduced. Exposing mice to "farm dust" results in a reduction of their allergic airway responses. Both

Acinetobacter lwoffii F78 and *Lactococcus lactis* G121 have been isolated in farm dust and, when introduced into neonatal mice, can protect against asthmatic symptoms.

Exposure to environmental viruses also influences the risk of developing asthma and its consequent severity. Infection of a newborn with rhinovirus or respiratory syncytial virus (RSV) is associated with a 10-fold increase in asthma risk in young children. The timing of these common respiratory infections and the specific viral strains involved impact the odds of developing asthma. Infections in younger babies, under 3 months old, result in a more significant risk for asthma development later in childhood. Further, the severity of the infection is also key, as babies with severe infections have a far higher risk of developing asthma. These infections appear to push the immune system towards a Th2 response.

Atopic Dermatitis

Atopic dermatitis, or **eczema**, is an inflammatory skin condition that causes your skin to become itchy, dry, and bumpy. Over time, rashes develop, along with scaly patches and sometimes skin infection. The condition weakens the skin's barrier, which is responsible for both protecting the body from outside elements and helping the skin retain moisture. There is often a cycle of flare-ups when the disease is worse and periods of remission when the skin lesions clear up.

The cause of eczema remains unknown, but genes, the immune system, and environmental exposures all play a role in disease development and severity. There are two major components to the development of eczema: dysfunction of the skin epithelial cells and alterations in immune responses. As is the case with asthma, the rapid rise in the prevalence of eczema is believed to be due, at least in part, to excessive cleanliness and lack of exposure to a diversity of environmental microbes early in life, resulting in a failure in the proper education of the host's immune system. On healthy skin, diversity of the microbiota is high, while in the case of eczema the microbial balance is lost.

We have focused our attention frequently in this textbook on the effect of early-life microbial exposure, our gut microbiome, and its impact on immune system functions. However, it is becoming clear that this same early-life microbial exposure is just as important, or even more so, in immune function of the skin. For example, colonizing newborn mice with the beneficial, commensal skin bacteria *Staphylococcus epidermidis* results in production of a large number of Tregs specific to this species, and subsequently, the immune system mounts far less inflammation when exposed to it later in life. In contrast, a delay in *S. epidermidis* exposure results in an increase in skin inflammation in response to this otherwise "healthy" bacterium (Paller et al., 2019).

Although the skin provides a physical barrier to block harmful agents from entering the body, skin epithelial cells and skin microbiota are in direct contact, unlike in the gut and lung, where mucus separates the microbiota from the epithelial cell surfaces. Furthermore, the skin provides an enormous surface area over which there are intimate interactions between the epithelial cells and members of the skin microbiome. To ensure control over these microbes, the skin produces a diverse ensemble of antimicrobials, including antimicrobial peptides and proteins, lipids, and a pH barrier. An army of immune cells are constantly patrolling the skin's surface to ensure the physical barrier remains intact. Failure to reinforce this barrier can result in pathogens penetrating the epidermis.

Dysbiosis of the skin microbiome precedes the development of eczema. In particular, an abundance of *Staphylococcus aureus* is highly predictive of eczema. The presence of *S. aureus* in bedroom dust is associated with an increased risk and severity of eczema, while exposure to farm dust, which are often enriched with *S. sciuri* and *Bacillus licheniformis* reduces eczema risk (Vercelli, 2023). There is a complex relationship between *S. aureus* and the skin that involves both host and pathogen factors. Healthy individuals' skin barrier and antimicrobial protein production create a hostile environment for *S. aureus*, limiting its ability to survive on the skin. However, individuals with eczema have impaired skin defense mechanisms, so *S. aureus* is able to proliferate. *S. aureus* has evolved mechanisms that allow it to adhere to the skin, invade the skin barrier, and modulate the host immune system, all of which triggers proinflammatory mechanisms (**Figure 12.13**). This potent pathogen produces a rich tool kit of bacterial virulence factors, including **superantigens**, which are a type of immune stimulatory molecule that enhances Th2 production and subverts the activity of Treg cells. *S. aureus* also produces molecules that permit it to adhere to skin, including **clumping factors A and B**, which are proteins anchored in the bacterial cell wall that recognize host proteins, allowing the pathogen to firmly attach to the epidermis and form biofilms. *S aureus* also produces and releases **α-toxin**, also known as α-hemolysin, which creates pores in host cell membranes, and a diversity of proteases, which facilitate dissolution of the outer skin layer. These virulence factors further compromise the skin's barrier function.

There is no cure for eczema. However, there are several treatment options that work to varying degrees in affected individuals. These range from home remedies, such as oatmeal baths and exposure to sunlight, to over-the-counter options such as hydrocortisone and lotions. When eczema leads to skin infections, antibiotics are typically used. However, these drugs target not only the pathogen, often *S. aureus*, but also the numerous beneficial members of the skin microbiome. Further, their use selects for ever-more-resistant strains of pathogens, creating a vicious cycle in which the increasing use of drugs results in their increasing failure rates.

Given the relatively new knowledge that members of the skin microbiome play a role in maintaining the integrity of the skin barrier and augmenting host defenses against skin pathogens, a new form of treatment is being developed. Known as **biotherapy**, this procedure involves the supplementation of the skin microbiomes of individuals with

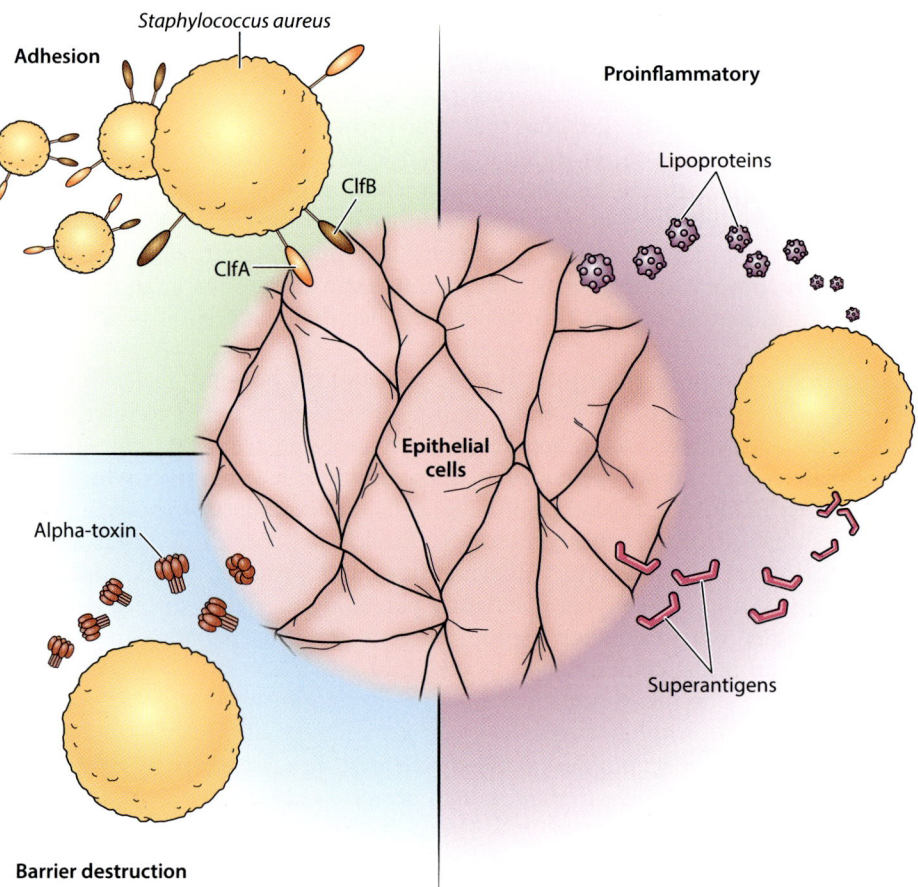

Figure 12.13 Pathogen-Specific Factors Involved in Atopic Dermatitis *Staphylococcus aureus* adheres to human skin through the production of external cell surface proteins. Their presence results in disturbance to the skin barrier due to physical, chemical, and inflammatory mechanisms. ClfA and ClfB = clumping factors A and B. (After Paller et al., 2019.)

eczema with strains known to have anti-pathogenic properties. Biotherapy takes advantage of the natural abilities of skin microbes to prevent pathogen invasion through a variety of mechanisms, including colonization resistance. In this procedure, the skin microbiome of a patient with eczema is first screened for anti-pathogenic properties. For example, *S. epidermidis* and *S. hominis*, members of the healthy skin microbiome, secrete potent antimicrobials that inhibit the growth and biofilm formation of *S. aureus*. Strains that show high activity against *S. aureus* are then selected for further testing. This generates a lead strain, which is then compared with a placebo strain for its impact on eczema symptoms. Larger cohort studies are then carried out until a candidate is chosen for application to high-risk children to prevent eczema.

Food Allergy

A new theory about the mechanistic basis of food allergy is emerging. If key beneficial microorganisms are missing from the gut microbiome, the intestinal barrier is disrupted and weakened, which permits food allergens to penetrate the epithelial boundary, reach the bloodstream, and trigger allergic responses. In support of this hypothesis is the observation that although many food allergens show few biochemical similarities, they do share the ability to defy digestion and remain intact in the digestive tract. "That seems to be what makes peanuts the champion—its ability to resist degradation in the gut," says Cathryn Nagler, a scientist who studies food allergies (Landhuis, 2020).

The gut microbiome can regulate food allergies in several ways. There is a continuous stream of communication between microbes and the immune system via a biochemical language that involves components of microbial cells, such as LPS, as well as the metabolites the microbes produce, such as SCFAs. The response by the immune system upon recognizing these signals includes an increase in mucus production, a decrease in gut permeability, differentiation of Treg cells, and cytokine production (Berni Canani et al., 2019). These actions result in a decrease in intestinal permeability.

STRUCTURE OF THE GUT MICROBIOME–IMMUNE SYSTEM AXIS Let's look at these activities in a bit more detail. Dietary fiber undergoes metabolism into SCFAs. These metabolites bind to intestinal epithelial cell surface receptors, which produce signals that help establish eubiosis in the gut microbiome and reduce host inflammatory responses (**Figure 12.14**). SCFAs impact host gene expression through epigenetic mechanisms,

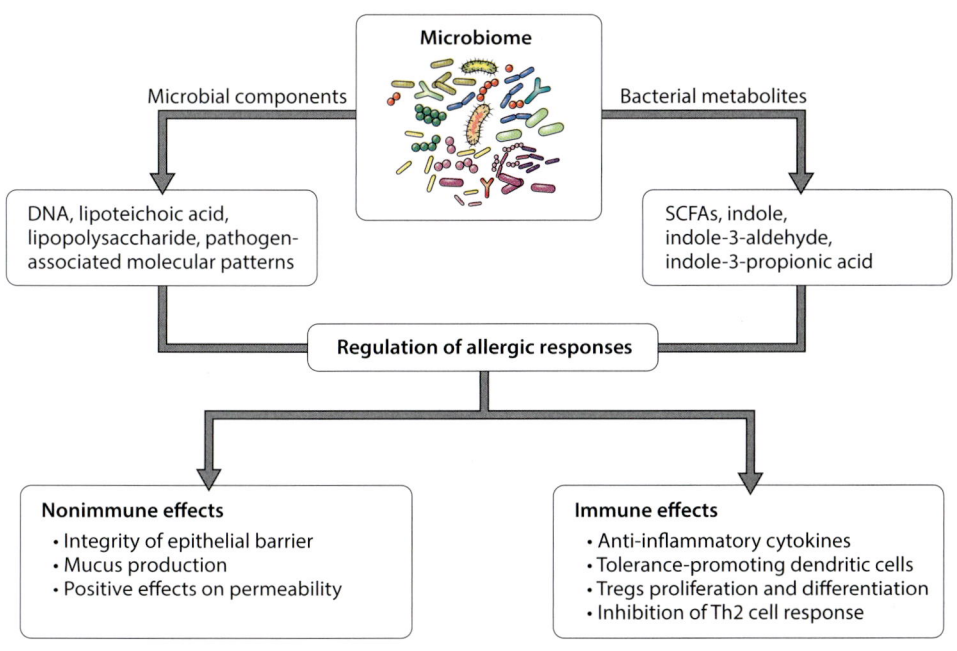

Figure 12.14 The Structure of the Gut Microbiome–Immune System Axis Communication is bidirectional within the gut microbiome–immune system axis. Microbes present immune cells with cell surface proteins and metabolites that trigger immune activity. By modulating the communication pathway, it may be possible to counteract allergic responses to foods. The gut microbiota can protect the epithelial boundary and thereby decrease gut permeability. A positive modulation of this axis can counteract food allergic response, both by nonimmune outcomes, such as increased epithelial integrity that results in decreased gut permeability and increased mucus production, and by immune-related outcomes, such as the production of Treg cells and anti-inflammatory cytokines and suppression of Th2 and Th17 cellular responses. (After Berni Canani et al., 2019.)

including the inhibition of histone deacetylases, resulting in changes in the immune response. For example, butyrate promotes immune tolerance mechanisms, such as the production of cytokines that modulate Th2 inflammation. In fact, the microbiota of children who later develop food allergies have lower levels of butyrate production. One particular allergy—to cow's milk—has been linked to variations in the production of SCFAs. Enrichment of the microbiota with butyrate-producing microbiota can alleviate that allergic response.

Much of what we understand about the mechanisms underpinning food allergies comes from studies done in germ-free mice. In one study, Clostridia and *Bacteroides* were introduced separately into germ-free mice, and Clostridia, but not *Bacteroides*, prevented an allergic response (Suther et al., 2020). Mice colonized with Clostridia produced more Tregs, which resulted in a dampening of an immune response. The microbiomes of these same mice produced more IL-22, which resulted in an improvement of the gut barrier. The presence of Clostridia increases production of SCFAs, whose presence promotes epithelial integrity and homeostasis.

The path to either immune tolerance or food allergy appears to be tied to that same window of opportunity for microbial exposures discussed earlier in this chapter. Factors that determine that the food tolerance path will be followed include a full-term birth and vaginal delivery, breastfeeding, a maternal diet rich in fiber, environmental exposure, and an infant diet rich in fiber and fermented foods (**Figure 12.15**). The result is higher levels of Treg production and immune tolerance. All of these factors contribute to gut eubiosis, thus laying the foundation for long-lasting protection against food allergies later in life. Conversely, factors that lead to the path to food allergy include C-section delivery, a mother eating a low-fiber diet, lack of environmental exposure, a low-fiber diet in infants, and exposure to antibiotics. The result is an increase in populations of Th1 cells and cytokine production. All of these factors lead to gut dysbiosis with an increase in the presence of bacterial pathogens in the gut, lower levels of immunomodulatory factors, and increased gut permeability.

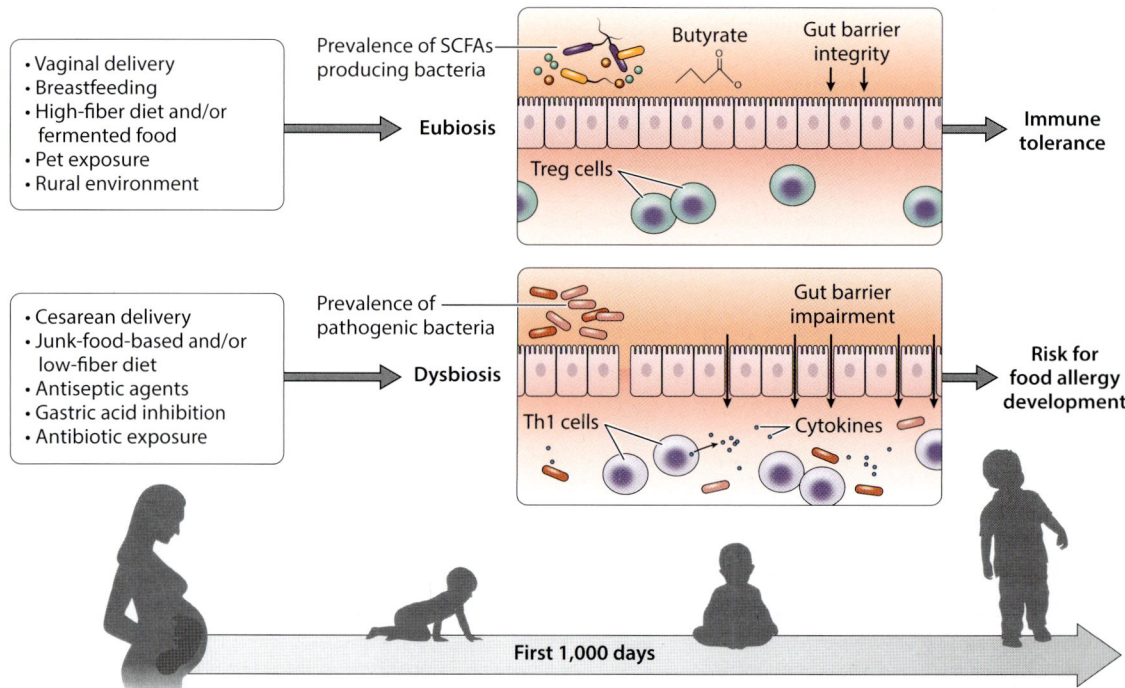

Figure 12.15 The Path to Either Immune Tolerance or Food Allergy The top developmental path leads to food tolerance, while the bottom path leads to food allergy. (After Di Costanzo et al., 2020.)

Low levels of gut microbial richness in newborns are associated with an increased likelihood of food allergies. However, this association vanishes in slightly older infants. Analyzing feces of healthy and allergic babies reveals differences in their gut microbiomes. In young children who outgrow their allergies, their gut microbiomes shortly after birth are enriched in bacterial diversity, including the presence of Clostridia. This protective effect is not seen in older babies, suggesting that the microbial protection may be limited to the first few months of life. In germ-free mice, colonization early, but not later, in life results in the suppression of IgE production and prevents food allergy. Clearly, at least in mice, the microbiome affects early immune development, which impacts food allergies later in life.

12.9 MICROBIOME-BASED THERAPEUTICS FOR ALLERGIC DISEASES

Our rapidly improving understanding of the microbiome provides an opportunity to develop an entirely new approach to identifying and preventing and/or treating allergic disease. Current therapeutic approaches include the use of antibiotics, probiotics, and prebiotics, and synbiotics and postbiotics are under development to treat food allergies and asthma (**Table 12.2**). Let's explore some microbiome-based therapeutics.

Oral Immunotherapy

One rational, and common, response to a food allergy is to simply avoid consuming that food. Over the past decade or so, physicians have developed a novel therapeutic approach, known as oral immunotherapy. **Food allergy oral immunotherapy** (**OIT**) has the goal of desensitizing an individual to a food allergen, resulting in immune tolerance to that allergen. OIT involves exposing the individual to increasing amounts of the target food allergen to determine the highest dose tolerated. An escalating treatment dose is then administered for a period ranging from months to years. When treatment is complete, if successful, the individual will tolerate the food allergen at a level that is protective upon accidental exposure. The mechanisms underlying successful OIT appear to be based upon Treg cellular responses, with effects including suppression of an inflammatory response, reduced production of eosinophils and mast cells, and an increased ratio of Th1 to Th2.

There is only one FDA-approved OIT treatment for food allergies to date, Palforzia, which is an oral immunotherapy against peanut allergy (Vickery et al., 2021). This approach can be stressful, as it requires eating food that can trigger an allergic response. Further, it increases tolerance but doesn't fix the underlying condition, and it doesn't work for everyone. When it does succeed, it gives the individual the ability to safely consume only small doses of the food in question. However, for some individuals, such as those with extreme peanut allergies, this modest gain can be life-altering.

Table 12.2 Potential Anti-Allergy Effects of Probiotics, Prebiotics, Synbiotics, and Postbiotics for Gut Microbiome Manipulation

PROBIOTICS	PREBIOTICS	POSTBIOTICS	SYNBIOTICS
Regulates intestinal flora and its metabolites	Stimulates growth of bacterial strains	Enhances survival and adhesion of probiotics	Stimulates growth of beneficial microbiota
Enhances epithelial integrity	Regulates intestinal flora metabolites	Improves epithelial barrier function	Enhances epithelial barrier function
Modulates Th1/Th2 balance	Influences immune response	Modulates immune response	Modulates immune response
Promotes Treg differentiation		Regulates inflammatory response	Promotes colonization of probiotics

Source: Shi et al., 2022

Synbiotics

Synbiotics are compounds made from probiotics and substrates that select for some of the more healthful bacteria in the gut microbiome (Shi et al., 2022). Synbiotic development is in its infancy, and as yet, there is no direct evidence that synbiotics are effective in alleviating food allergies in humans, save for one particular combination of plant sugars and a strain of *Bifidobacterium breve*, which resulted in a gut microbiota composition in children with cow's milk allergy such that it resembled those of nonallergic infants (Fox et al., 2019). The synbiotic appears to help regulate the balance of an immune response, resulting in reduced levels of cytokine production.

Fecal Microbiota Transplantation

Although most often associated with treatments for *Clostridioides difficile* recurrent infections, **fecal microbiota transplantation** (**FMT**) is being explored to treat allergic disease as well. FMT involves the transfer of microbial communities from a donor into a recipient individual. Studies in germ-free mice have shown that FMT from healthy infants protects mice against allergic responses to a cow's milk allergen, while transfer from children with milk allergies does not (Feehley et al., 2019). The healthy versus allergic transplanted mice showed differences in their gut microbiome compositions and had unique transcriptome signatures in gut epithelial cells. One bacterial species in particular, *Anaerostipes caccae*, protected against cow's milk allergy. Another recent study involved FMT from healthy donors into individuals with severe peanut allergy (Abdel-Gadir et al., 2019). The recipients were able to safely eat small amounts of peanuts for up to 4 months posttreatment. The participants' blood samples showed increases in Treg cells, which are associated with immune tolerance, along with reductions in the cytokine IL-13 and Th2 cells, which are associated with allergy. However, as we have discussed, there are risks to FMT, such as the introduction of life-threatening pathogens, so more research needs to be done to determine whether FMT can be used safely.

Faecalibacterium, *Lachnospira*, *Veillonella*, and *Rothia*

Understanding the key role served by the microbiome in asthma has focused attention on therapeutic options that directly target microbiome dysbiosis. Scientists have identified four genera of bacteria—*Faecalibacterium, Lachnospira, Veillonella,* and *Rothia*, commonly referred to as **FLVR**—that play a key role in reducing asthma risk levels (Arrieta et al., 2015). If a neonate has the FLVR bacteria in their gut microbiome by 3 months, they have a very low risk of developing asthma later in life, while babies with low levels of the FLVR four are at high risk of developing asthma. Studies in germ-free mice showed that administration of the FLVR species results in significantly lower levels of lung inflammation and asthma symptoms. The driving therapeutic vision is that safe formulations of FLVR bacteria could help establish a normal microbiota before the appearance of food allergies. In an ideal situation, FLVR treatments would be tailored to meet the needs of each child. According to Stuart Turvey, a coauthor of the study, "It opens the door for a way to maybe prevent asthma. That's the Holy Grail" (Van Evra, 2017).

In the same vein, probiotics could help regulate the gut and lung microbiome axis. Asthmatic patients who received *Bifidobacterium lactis* in concert with conventional treatments had reduced asthma symptoms and showed an increased resilience of their gut microbiome (Liu et al., 2021). Levels of potentially beneficial species, such as *B. animalis* and *B. longum*, were increased as well. Finally, the gut metabolome of those receiving the probiotic showed increased levels of key bioactive metabolites. It is likely that administration of the probiotic caused a synergistic effect with conventional therapy to alleviate asthmatic symptoms.

Susan Lynch, director of the University of California San Francisco Benioff Center for Microbiome Medicine, and her team have identified microbial metabolites that signal the Treg dysfunction often associated with the development of asthma at a later date (Levan et al., 2019). "We've considered that engineering the gut microbiome during this key window of immune training could have a long-term beneficial impact on the health status of high-risk children." They have created a therapy in which FLVR is administered just after birth for newborns at high risk for asthma. The microbial mix is composed of bacteria that are able to modulate the immune response, and their bacterial genomes encode functions that are usually absent from the gut microbiomes of babies at high risk for asthma. The hope is that shaping the newly developing gut microbiome and the metabolites it produces will allow the microbiome to appropriately train the immune system early in life and ultimately influence the microbiome and immune functions over the long term.

Inulin

Another microbiome-based approach in asthma therapies involves the dietary fiber **inulin**, a prebiotic soluble form of fiber. Its application can impact the composition of the gut microbiota and alter the biomarkers that are used to demonstrate inulin degradation and immune modification. One particularly revealing study showed that one of the hallmarks of asthma, the presence of eosinophils in sputum, was significantly diminished following inulin supplementation (Brehm & Pfefferle, 2019). Thus, a simple dietary intervention has the power to influence both the gut microbiome composition and function and the immune system. Cellulose has also been tested for its ability to alleviate asthma symptoms (Wen et al., 2022). In a mouse model of asthma, a diet rich in cellulose resulted in reduced levels of lung inflammation and asthmatic symptoms. The gut microbiome composition was impacted by the presence of a new fungal family and a unique genus of bacteria, *Romboutsia*, which is involved in lipid metabolism. This study was surprising, as levels and types of short-chain fatty acids are generally considered one of the key metabolic indicators of allergic risk, while this genus appears to regulate lipid metabolism.

Microbiome-based therapies are rapidly being developed, ranging from prebiotics and probiotics to live biotherapeutics, synbiotics, and postbiotics. Although progress has been made, further work is required to optimize methods to identify candidate microbes, develop appropriate preclinical validation models, and eventually progress to personalized targeting of microbiome-based medicines. Allergic diseases provide an exciting and amenable avenue for these explorations, and it is likely that we will soon see a panoply of options emerge.

12.10 A CIRCLE OF CAUSALITY

In this chapter we have seen how the microbiome of the gut, skin, and lungs play a role in the rapid increase in allergic disease we have experienced over the past 30 years. One way to consider this relationship between cause and effect is known as the **circle of causality**. Cyclic causality refers to the interconnection of events that result in a cycle of outcomes, where the first event triggers the second, which triggers the third, which triggers the first event, and so on. In **Figure 12.16** we see this concept applied to human urbanization, microbiome changes, immune system changes, and increases in allergic disease. Developed by Haahtela (2019), this perspective provides a simplified view of what is now understood to be a highly complex series of interactions. In this illustration a person is exposed to an urban lifestyle which results in reduced connections with natural microflora and a dysbiotic microbiome. That dysbiosis results in triggers to the immune system that support inflammation, such as increased levels of Th2 and Th17 cells (inflammation promoting) and reduced levels of Th1 and Treg cells (inflammation depressing). Inflammation is triggered resulting in asthmatic or allergic symptoms.

Let's walk through a series of questions and answers that Haahtela suggests a physician might use to help understand how this complex system of causes and effects result in imbalances in immune responses and the symptoms of allergies.

Figure 12.16 The Circle of Causality Urbanization has led to a dramatic increase in inflammatory diseases. The circle of causality illustrated here emphasizes the complexity of the causes as well as their interrelatedness. (After Haahtela, 2019.)

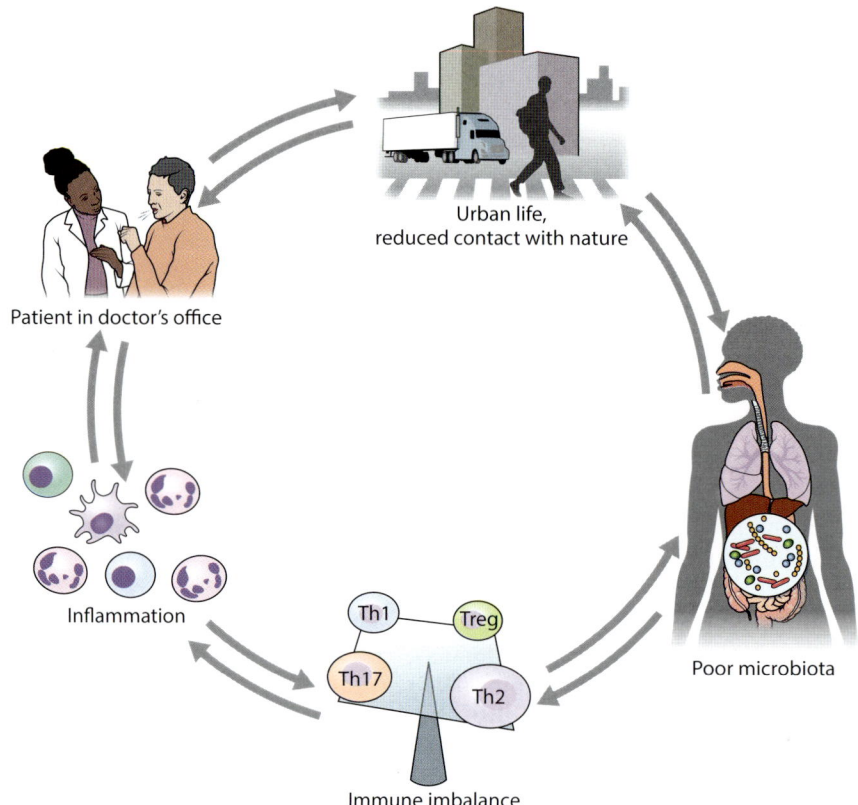

1. *Why does one sneeze, cough, and itch when exposed to an allergen?* These symptoms are due to an inappropriate and excessive inflammatory defense response. In short, the individual cannot cope with their environment.

2. *What causes the inflammatory response?* The immune system is not able to distinguish between dangerous and nondangerous signals. The immune system immediately pushes inflammatory cells to the epithelium to reject whatever is creating the signal, which might be simply a grain of pollen. The immune system attempts to neutralize this harmless pollen. In short, there is an imbalance in the immune responses.

3. *What is behind the immune imbalance?* Alterations in exposure to commensal microbes during a critical window of opportunity in infants are clearly involved. We used to think of an allergic response as a hyperactive immune response, meaning that, for example, pollen results in hyper-responsiveness to the pollen. However, we now know that immune responses in nonallergic individuals are modified upon exposure to the antigen, suggesting that health is a dynamic state, requiring a complex balance of immune responses.

4. *What impoverishes the human microbiome in the gut, skin, and respiratory tract?* A lack of exposure to the environment, particularly microbial exposure, is involved, as is the diet.

5. *Finally, what is causing loss of biodiversity, driving modern life, and pushing people to the cities?* Population overgrowth, technological innovations, and massive exploitation of natural resources have resulted in a dramatic cultural shift. Urban lifestyles dominate, and the human microbiome and its impact on immune responses are fundamentally altered.

This example is clearly an oversimplification, and there is a network of interactions, rather than a simple circle. However, the circle of causality illustrates our increasing awareness of the role played by the microbiome, influenced by environment and lifestyle, in contributing to what was previously perceived to be genetic and epigenetic disease states. This illustration might serve as a paradigm for how our microbiome, our lifestyles, our genetics, and environmental factors can act in concert to create allergic disease.

CHECK YOUR UNDERSTANDING

1. Allergies are caused by a reaction from your _____ system.
   a. immune
   b. neurological
   c. digestive
   d. thyroid

2. Bacteria that colonize your microbiome in your first 2 years of life can indicate likelihood of developing allergies.
   a. True
   b. False

3. External agents that can trigger an allergic response are known as
   a. antibodies.
   b. antigens.
   c. allergies.
   d. pollen.

4. IgE is produced by plasma cells, which themselves are a specialized type of
   a. T cell.
   b. B cell.
   c. natural killer cell.
   d. antigen.

5. The farm effect describes increased exposure to _____ during early development.
   a. diverse potential allergens
   b. isolated communities
   c. pollen
   d. other children

6. Modifications that affect gene expression without modifying the nucleotide sequences are known as
   a. chromatin.
   b. epigenetic changes.
   c. histones.
   d. transcriptomic regulation.

7. The following are all examples of epigenetic modifications *except*
   a. DNA methylation.
   b. noncoding RNA signaling.
   c. base excision repair.
   d. histone modification.

8. The introduction of _____ to pregnant mothers has been shown to alter DNA methylation patterns in their offspring.
   a. *Escherichia coli*
   b. *Lactobacillus reuteri*
   c. *Klebsiella pneumoniae*
   d. *Staphylococcus aureus*

9. A diet high in _____ during pregnancy can positively increase SCFA production, lowering the risk of asthmatic symptoms in a newborn's first year of life
   a. fiber
   b. fats
   c. sugars
   d. protein

10. At what age does a child's microbiome generally mature to a more stable, adult-like system?
    a. 5 years
    b. 1 year
    c. 3 years
    d. 10 years

11. Children raised on farms have lower incidence rates of asthma than children not raised on farms.
    a. True
    b. False

12. The term *old friends* has been used to describe
    a. tolerated pathogenic bacteria that have coevolved with humans.
    b. potential allergens.
    c. childhood pets.
    d. pathogenic bacteria introduced from the environment.

13. Newborns who receive antibiotic treatments are less likely to develop allergic diseases.
    a. True
    b. False

14. Antibiotics prescribed by pediatricians
    a. should never be taken.
    b. should target as few bacterial species as possible.
    c. often select for individual bacterial species.
    d. show no correlation with asthma development.

15. Asthma is a heterogeneous disease, which means that it
    a. has a single cause.
    b. has no known cause.
    c. can be caused by several different factors
    d. often has severe symptoms.

16. _____ are specialized white blood cells that activate to cause an asthmatic reaction.
    a. Macrophages
    b. T cells
    c. B cells
    d. Eosinophils

17. Asthmatic children with lung microbiome dysbiosis often show dysbiosis of the _____ microbiome.
    a. gut
    b. skin
    c. urogenital
    d. oral

18. Exposure to _____ is highly predictive of eczema development.
    a. *Staphylococcus aureus*
    b. *Escherichia coli*
    c. *Lactobacillus reuteri*
    d. *Klebsiella pneumoniae*

19. The following conditions may increase risk of developing food allergies *except*
    a. C-section delivery.
    b. the pregnant individual being on a low-fiber diet.
    c. lack of allergen exposure.
    d. high alpha diversity in the gut microbiome.

20. Synbiotics combine live bacteria with
    a. vitamins and minerals.
    b. prebiotics.
    c. allergens.
    d. probiotics.

21. Which procedure is best known for its role in treating recurrent *C. difficile* infections?
    a. Fecal microbiota transplantation
    b. Appendectomy
    c. Murine microbiota transplantation
    d. Microbiome cytokine supplementation

22. Inulin is a type of
    a. probiotic.
    b. cytokine.
    c. prebiotic.
    d. hormone.

23. Which bacterial genus is involved in lipid metabolism and has shown increased abundance in the gut microbiome of individuals rich in cellulose?
    a. *Escherichia*
    b. *Klebsiella*
    c. *Romboutsia*
    d. *Bifidobacterium*

Answers: 1A, 2A, 3C, 4B, 5A, 6B, 7C, 8B, 9A, 10C, 11A, 12A, 13B, 14B, 15C, 16D, 17A, 18A, 19D, 20B, 21A, 22C, 23C

DIVING DEEPER

1. List five examples of common allergens.
2. Briefly summarize the cascade of events that precipitate an allergic response.
3. Describe the farm effect.
4. What are epigenetic changes? Provide two examples.
5. How do microbiome changes affect fetal IgA and IgG production?
6. List five factors that when present at birth or in early childhood can impact an individual's likelihood of developing allergies.
7. Describe an experiment demonstrating the impact of environment versus genetics on microbiome composition.
8. How does the hygiene hypothesis support the observation of the farm effect? List some examples of lifestyle factors that support the hygiene hypothesis.
9. Why does the use of antibiotics in early childhood affect the risk of developing allergies?
10. List the steps involved in an asthmatic reaction.
11. How does the dust in your bedroom affect your risk for developing eczema?
12. What are synbiotics and how do they work?
13. What is the goal of FMT, and what are the risks of using it?
14. Why does the addition of inulin to a diet affect gut microbiome composition?
15. Describe the circle of causality.

DISCUSSING AND REFLECTING

1. Throughout the chapter we discussed risk factors for a variety of different immune system dysfunctions, from food allergy to eczema. Despite the symptoms, and often mechanisms, of these conditions being quite different, the microbiome-related risk factors are often quite similar. Describe the overlapping risk factors we discussed in the chapter. How can they be the root cause of such diverse symptoms?

2. Choose an allergic response described in this chapter, and create a circle of causality for it following the example provided in section 12.10.

3. Reflection. A recurring theme throughout this chapter is how the farm effect, hygiene hypothesis, and biodiversity hypothesis affect an individual's risk of developing autoimmune disorders. Reflect on your own childhood—what risk factors were present, and how might they have affected your developing microbiome?

RECOMMENDED READINGS

Popular Science Review

Landhuis, E. (2020, May 23). Gut Microbes May Be Key to Solving Food Allergies. *Scientific American*. https://www.scientificamerican.com/article/gut-microbes-may-be-key-to-solving-food-allergies/

Popular Science Book

Ho, V. (2022). *The Healthy Baby Gut Guide: Prevent Allergies, Build Immunity and Strengthen Microbiome Health from Day One*. Greystone Books. https://www.librarything.com/ner/detail/44687/The-Healthy-Baby-Gut-Guide-Prevent-Allergies-Build-Immunity-and-Strengthen-Microbiome-Health-from-Day-One

Scientific Review

Kloepfer, K. M., McCauley, K. E., & Kirjavainen, P. V. (2022). The Microbiome as a Gateway to Prevention of Allergic Disease Development. *Journal of Allergy and Clinical Immunology: In Practice*, *10*(9), 2195–2204. https://doi.org/10.1016/j.jaip.2022.05.033

Our Evolving Microbiome

13

CHAPTER CONTENTS

- **13.1** Where Did Our Microbiome Come From?
- **13.2** The Industrialization of Our Microbiome
- **13.3** The Microbiome and the Missing-Heritability Problem
- **13.4** Reacquiring a More Healthful Microbiota

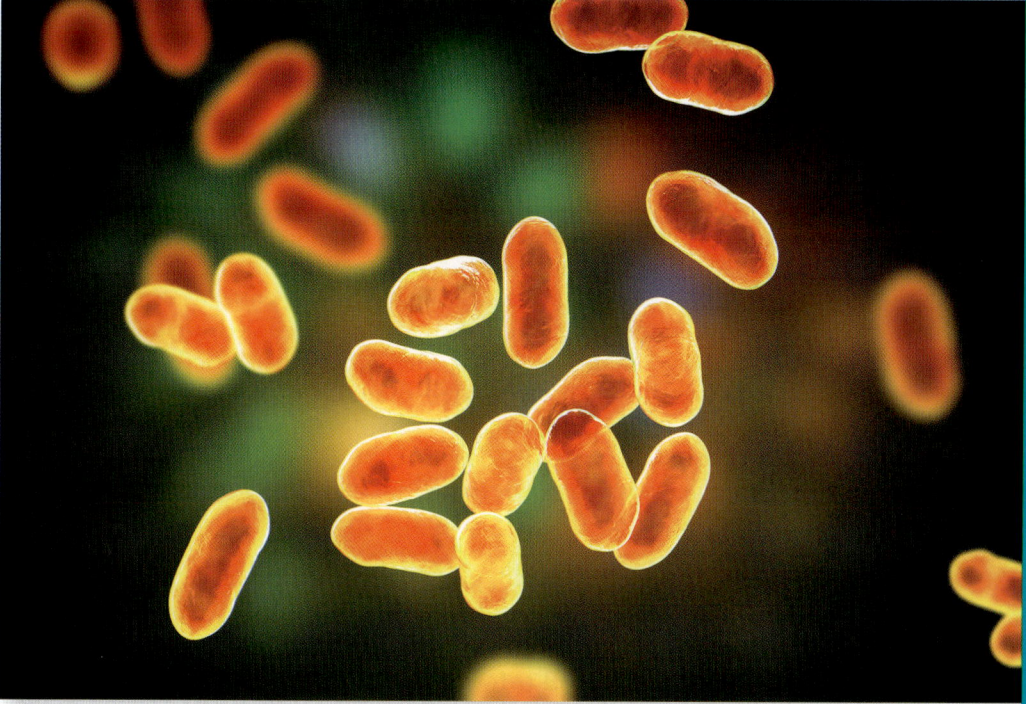

My name is *Prevotella copri*, and my claim to fame is that for those who eat a more traditional human diet, like hunter-gatherers, I am the most abundant member of the microbiome. Seriously, I can make up 40% of the cells in the gut microbiome. If you have succumbed to fast food and few veggies, then I have likely left the stage, and who knows who took my place? I love high-fiber foods, and when I am around, good things happen—think leaner and with lower blood sugar. (Photo from iStock.com/Dr_Microbe)

Trillions of microbes have accompanied us as we evolved from our distant primate ancestors, adapting with us every step of the way. Comparing our microbiome to that of our closest relatives can help us theorize about what the microbiome of our ancestors looked like, and how it evolved over time. We have lost some ancient strains that still inhabit our primate cousins, the great apes, which may help explain several human-specific diseases. In this chapter we will explore how our gut microbiome has changed over distant, and more recent, time. Comparisons of our gut microbiome with those of our closest living, and extinct, relatives reveal some of the anticipated differences and, even more surprising, the extraordinary similarities in microbiota we still share. One of the more perplexing observations is that our gut microbes are more likely to be inherited from our ancestors than acquired from the environment, notwithstanding the enormous impact of diet in shaping the bacteria in our gut on a daily basis! Let's dive right into that incongruence.

> "I often think about the long and winding road from organic compounds floating in the so-called primordial soup to humans. Lately I've been wondering if microbes helped drive the bus."
>
> —Susan Erdman (Erdman, 2017)

13.1 WHERE DID OUR MICROBIOME COME FROM?

Howard Ochman led a team of evolutionary biologists at the University of Texas at Austin in their quest to assess the relative roles of inheritance versus environment in determining the members of the human gut microbiome. The team proposed an intriguing hypothesis: our gut bacteria are descended from the ancestral microbiota that evolved in the common ancestor of humans and our close relatives, the great apes, which include chimpanzees, gorillas, and orangutans (**Figure 13.1**). These lineages diverged about 15 million years ago, and Ochman's team hypothesized that they share not only a common ancestral primate, but an ancestral microbiome as well. If this idea is correct and these primate hosts and their microbial symbionts have coevolved over millions of years, then their evolutionary histories will be congruent. That means that if you examine the branching patterns of the phylogenetic trees of the primate hosts, they should match the branching patterns of the phylogenetic trees of the members of their microbiomes. We discussed phylogeny first in chapter 1 and then again in chapter 5 if you need a refresher. Briefly, phylogenetic trees are visualizations of how closely related different organisms are. When the branching pattern of a host's phylogenetic tree matches that of its microbiome, this suggests that the members of the microbiome and the host have coevolved. When the branching patterns do not match, it suggests that the microbes have moved between different hosts.

To test this idea, Ochman's team sampled feces from hundreds of chimpanzees and bonobos (our closest living relatives), gorillas, and humans and sequenced a gene found in all Bacteria and Archaea, **gyrB** (Moeller et al., 2016). This particular gene is often used in generating bacterial phylogenies because, like the 16S rRNA gene (think back to chapter 5), it is highly conserved in evolution; however, it evolves more rapidly than the 16S rRNA gene and thus provides more information when close relatives are compared. The team used the resulting DNA sequences to generate a phylogeny of the *gyrB* gene, which served as a proxy for the phylogeny of the microbes present in these primate guts (**Figure 13.2**). By comparing the microbe and host trees, the team found that most of the branching patterns matched between phylogenies. For example, if we examine the microbial tree, all strains from one host are most closely related to other strains of that same host; a microbe from a gorilla clusters with microbes from other gorilla microbiomes. This pattern was true for all hosts studied. Furthermore, the branching patterns of microbes from each host matched the branching patterns of their host phylogeny. For example, bonobos and chimpanzees are very closely related, and their microbes were also found to be closely related. This indicates that many of the prominent members of the primate gut microbiomes appear to be long-term residents of their hosts, over an exceedingly long (i.e., evolutionary) time frame! In other words, to explain the fact that the microbe branching patterns match those of their hosts, one must posit that the host and symbionts have coevolved, in this case over a period of 16 million years or so. This observation flies squarely in the face of all that we have learned in prior chapters about how adaptable our microbiome is to its current environment. Recall that the levels of diversity and the precise taxa in our gut microbiome rapidly adjust to novel diets and different modes of birth and are decimated with antibiotic usage. How can both sets of observations be true?

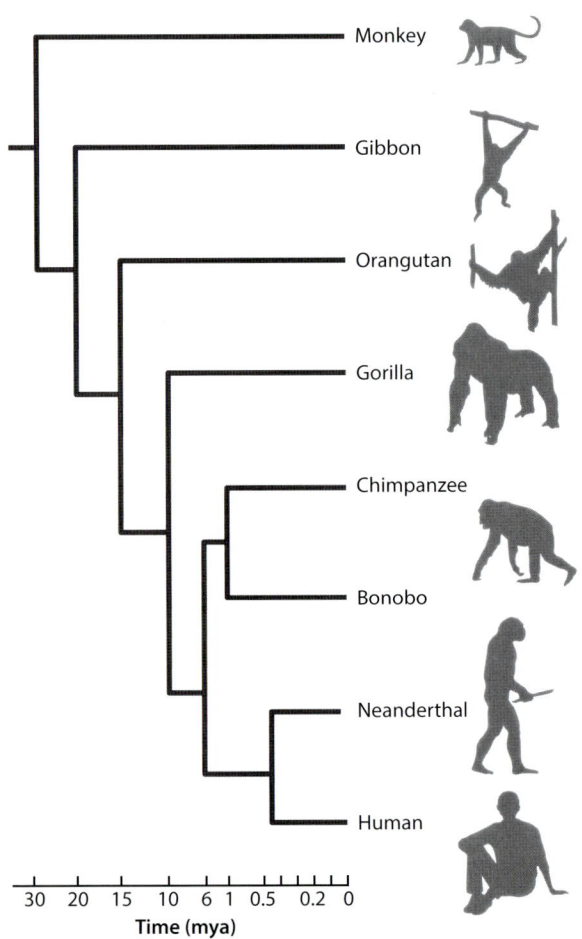

Figure 13.1 Evolutionary Relationships of Primates Humans' closest living relatives are chimpanzees and bonobos (genus Pan), which diverged from a common ancestor about 6 million years ago (mya), and gorillas, with whom we share a common ancestor from about 8-10 mya. Our closest relative is Neanderthals, who lived alongside humans for hundreds of thousands of years before going extinct. (After Sousa et al., 2017.)

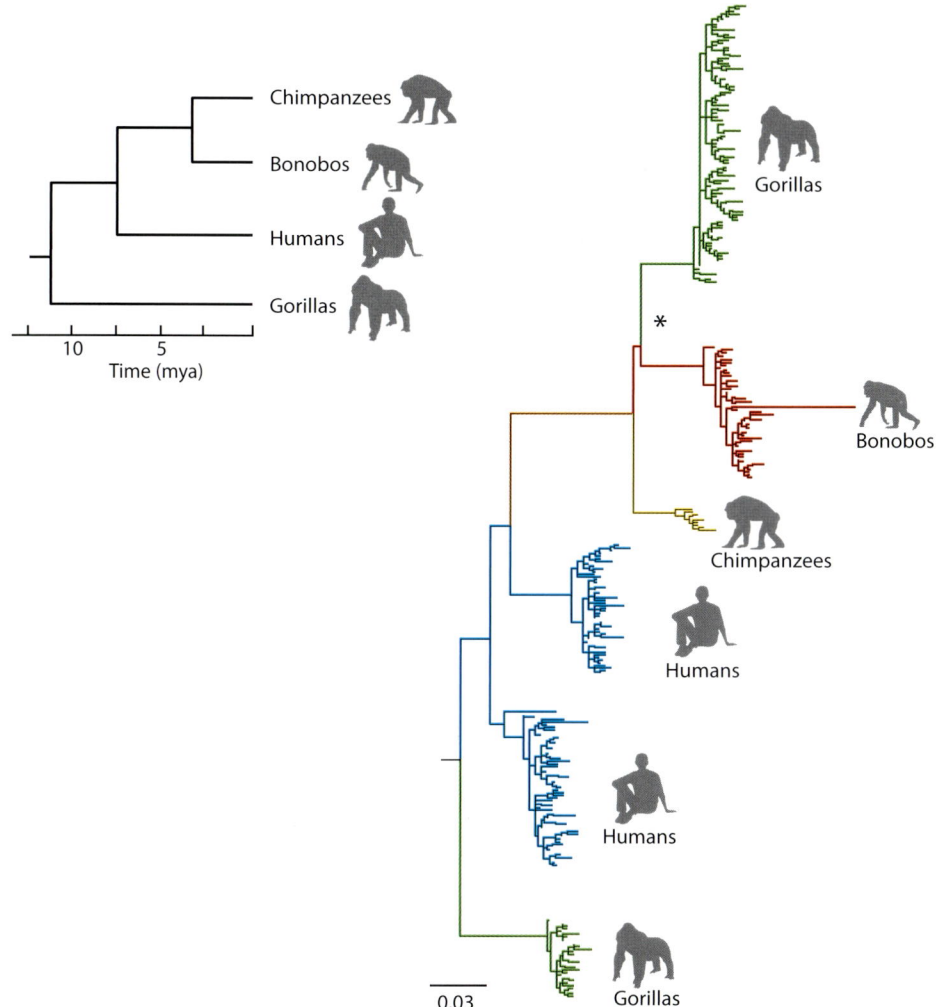

Figure 13.2 Congruence between Primate Gut Microbiota and Host Phylogenies The phylogeny on the left shows the evolutionary relationships of humans and their closest extant relatives. The phylogeny on the right is based upon the *gyrB* gene sequences obtained from *Bifidobacterium adolescentis* strains recovered from fecal samples for humans and African apes. The asterisk indicates the proposed horizontal transfer event of a *B. adolescentis* strain from bonobos into gorillas. Many of the *Bifidobacterium* strains found in the gorilla microbiome are native to that species, and distinct from the species of bacteria found in humans, bonobos, and chimpanzees, until the indicated transfer event. (After Moeller et al., 2016.)

Ochman's team found that most of the major bacterial lineages match the corresponding phylogenies of the host species, save for some rare examples of horizontal gene transfer between primate species (see Figure 13.2). The authors concluded that as the great apes diverged from a common primate ancestor, their microbiomes evolved in parallel to adapt to the varied diets and lifestyles these primates have today. As Ochman notes, "It's surprising that our gut microbes, which we could get from many sources in the environment, have actually been coevolving inside us for such a long time" (University of California—Berkeley, 2016).

An earlier study by the same group identified the specific microbiota changes that occurred during the diversification of African apes (Moeller et al., 2014). **Figure 13.3** shows the major changes in microbiota frequency within the gut microbiomes of various African apes. These data reflect 35 changes in abundance that have occurred since the divergences of the great apes. Nearly half of those changes occurred during the split between the human lineage and the chimpanzee and bonobo lineages. Several of these microbiome compositional changes have known functional implications. For example, *Bacteroides* is fivefold more abundant in humans, which is unsurprising as it is associated with diets rich in fats and proteins. Chimpanzees eat mostly fruit, leaves, nuts, and minimal meat, unlike humans, who need more *Bacteroides* to break down a diet heavy in animal products. In contrast, two taxa that promote the breakdown of plant polysaccharides, the archaean *Methanobrevibacter smithii* and the bacterium *Fibrobacter*, are both less abundant in humans than chimps.

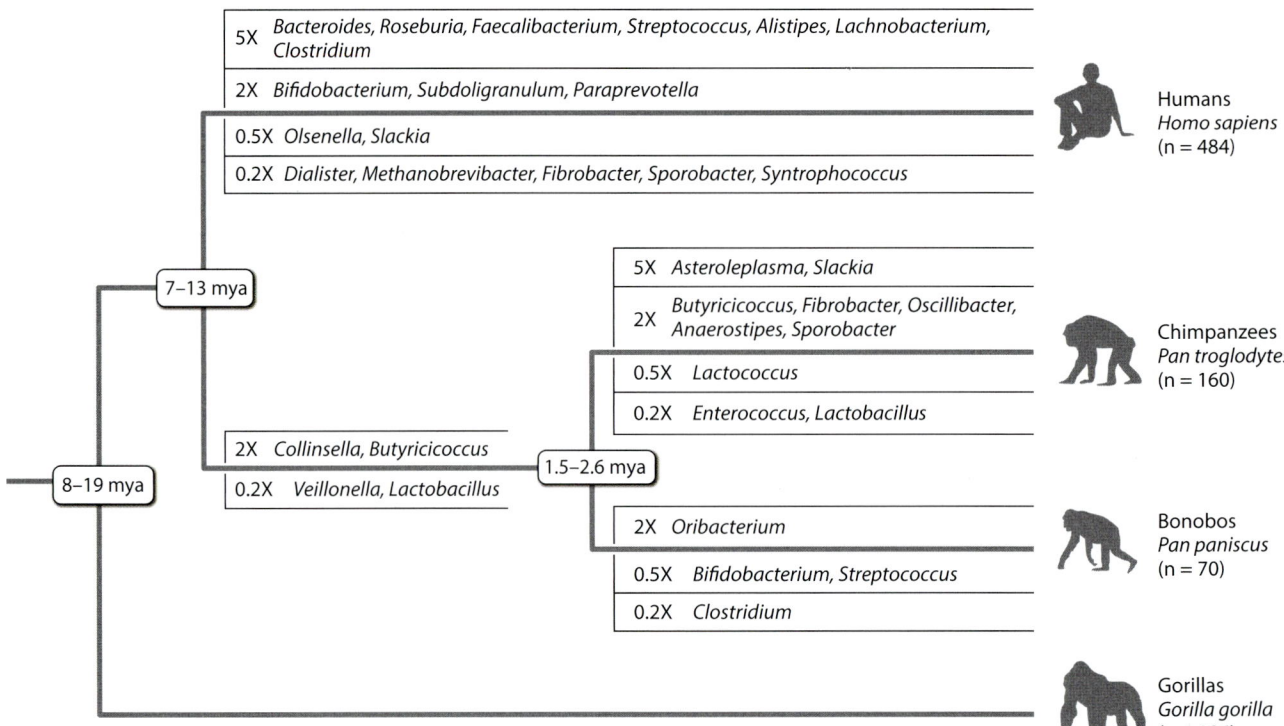

Figure 13.3 Changes in the Gut Microbiome during African Ape Diversification Humans, chimpanzees, bonobos, and gorillas share a common ancestor with an ancestral microbiome. By studying the differences in the composition of the microbiomes of these primates, scientists can make inferences about how the microbiome has evolved in each species over time. For example, the "5X" in the top row indicates that the abundance of *Bacteroides* increased fivefold as humans evolved from their most recent common ancestor with chimpanzees and bonobos. (After Moeller et al., 2014.)

Ochman and his colleagues similarly compared DNA sequences of the *gyrB* gene obtained from fecal samples for several distinct populations of humans, specifically those living in the industrialized northeastern United States and those from rural communities in Malawi, Africa. Based upon the resulting gene phylogeny, the authors estimated that the bacterial strains in these human populations appear to have diverged about 1.7 mya. Interestingly, that period corresponds with when the first human migrations out of Africa occurred! How cool is that? In fact, it is possible that the two clusters of *gyrB* sequences observed for humans in Figure 13.2 are the result of that early migration event out of Africa, though that remains conjecture at this point. However, what is clear is that our gut bacteria may help to inform us about early human migrations. Even more revealing, the Malawian individuals showed much higher levels of gut microbiome diversity, levels more like those found in chimpanzees, bonobos, and gorillas. This observation corresponds with the reduction in gut microbiome diversity observed in humans living in modernized societies. We will return to this subject shortly.

Our Closest Living Relatives

Let's narrow our comparative focus a bit. Our closest living relatives are the chimpanzees and bonobos with whom we share a common ancestor from a mere 6 mya. Our genomes have remained quite similar, with 99% of the gene content retained. Ochman's team next asked whether the genetic similarity we share with chimpanzees extends to our gut microbiota (Nishida & Ochman, 2019). By comparing

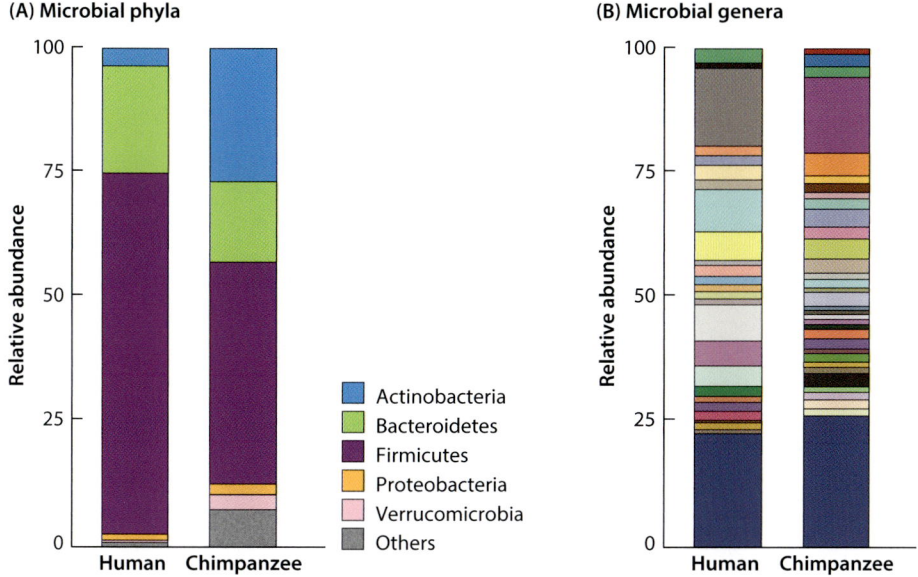

Figure 13.4 Gut Microbiota in Human and Chimpanzee A comparison of the phylum-level (A) and genus-level (B) taxa of human and chimpanzee gut microbiomes. (After Nishida and Ochman, 2019.)

the microbiota of human and chimpanzee, the team identified which taxa were likely present when these species diverged.

As you might already suspect, the human and chimpanzee gut microbiota are highly similar with respect to bacterial phyla (Figure 13.4A). In both hosts, their gut microbiota are dominated by Firmicutes and Bacteroidetes. Other phyla, such as Actinobacteria, Proteobacteria, and Verrucomicrobia, occur in both, but their frequencies are more variable. For example, 25% of the chimpanzee gut microbiome is composed of Actinobacteria, compared with <1% in the human gut microbiome. If we compare genera (Figure 13.4B), we begin to see significant numbers of taxa unique to either humans or chimps. Of the approximately 250 genera found in the two host species, over 20% are unique to one host, and 80% show a strong bias towards one or the other.

Ochman et al. used microbiota abundance data in a principal component analysis (PCA—a technique described in chapter 5 and revealed that chimpanzee and human gut microbial communities are host specific (Figure 13.5A). They then created separate PCA plots for the human and chimpanzee microbiomes and showed how distinct the three clusters (enterotypes) are within each species (Figure 13.5B). **Enterotypes** are defined as microbial communities adapted to particular environments, and both humans and chimps share the same three enterotypes, shown in red, yellow, and blue ovals in Figure 13.5. Thus, humans and chimps share most of the same genera of bacteria in their gut microbiota. However, the precise species within those genera differ, likely due to the varied selective pressures encountered in their respective host guts. Finally, and most exciting of all, the enterotype organization preceded the divergence of chimpanzees and humans. Despite the differences between microbiota species, the fact that chimpanzees and humans have the same enterotypes may help us understand the incongruence we were concerned about earlier when we learned that, even though our microbiomes can rapidly respond to diet changes and the use of antibiotics, we retain a microbiome that we inherited from our distant primate ancestors. What has been retained are the general ecological conditions of our guts, resulting in stable, shared ecotypes. However, that does not mean that our microbiomes have been static. On the contrary, they have evolved to be human and chimpanzee specific. The precise details of which species are present and the functions they carry out differ as the microbiomes adapt to our changing lifestyles and diets. So, we see at the same time an evolutionary stability paired with individual dynamism in our gut microbiomes.

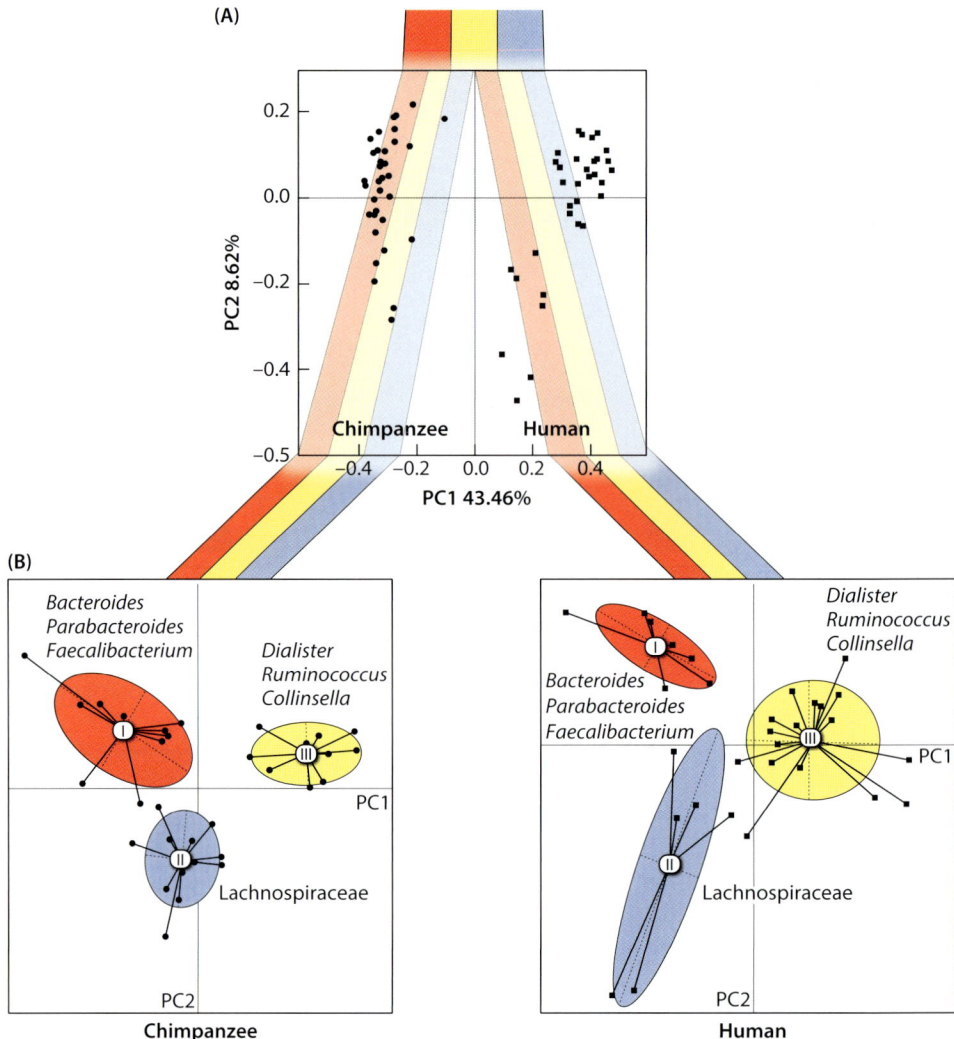

Figure 13.5 Stratification of Human and Chimpanzee Gut Microbiomes into Compositionally Similar Enterotypes (A) The human and chimpanzee gut microbiomes are well separated in a PCA analysis of *gyrB* sequence variation. (B) The presence of three enterotypes, or clusters—I (red), II (blue), and III (yellow)—arises before the human and chimpanzee divergence and becomes quite distinct in present-day populations. (After Nishida and Ochman, 2019.)

Looking to Our Evolutionary Siblings

The time span since our split with chimpanzees is about 6 million years. Another group of scientists turned to an even closer human relation, *Homo neanderthalensis*, to get a more fine-tuned picture of how our microbiota has evolved. **Neanderthals** are our closest evolutionary siblings. We shared a common ancestor about 500,000 years ago, although the two species intermixed well after that split. The Neanderthals (*Homo neanderthalensis*) belonged to the same genus as us (*Homo*) and lived contemporaneously with our species (*Homo sapiens*) until the former went extinct some 35,000 years ago. Studies of their gut and oral microbiomes have revealed some fascinating insights into the evolutionary histories of our own microbiome.

Coprolites

One way to study the microbiomes of extinct animals involves the examination of **coprolites**—fossilized feces. Feces are chock-full of microbes, which means they normally decompose quickly and, thus, are rarely found except in cold, dry places, such as caves. However, human samples have been found that date back to around 22,000 years ago in latrine deposits. **Figure 13.6** provides a particularly striking example, the Lloyds Bank coprolite, which earned its fame for both its size—measuring 20 cm (7.87 in.) × 5 cm (1.97 in.)—and having been deposited by a Viking some 1,200 years ago.

A large, multinational team of scientists examined coprolite samples from a latrine site in Spain where Neanderthals lived about 50,000 years ago (Buguliskis, 2021). The group identified over 10 million ancient bacterial DNA sequences, which represent many of the same microbes thriving in our own gut today. They identified genera known to be important to our health, including *Blautia, Dorea, Roseburia, Ruminococcus*, and *Faecalibacterium*. As you may recall from chapter 3, these taxa produce short-chain fatty acids (SCFAs) by degrading dietary fiber. *Bifidobacterium* was also present, which we know is critical in regulating our immune system, particularly in neonates, information which is hopefully familiar from chapter 7.

Dental Plaque

Another group, also interested in a comparison between Neanderthal and human microbiomes and led by Christina Warinner at Harvard University, focused on the one other

Figure 13.6 Viking Coprolite The Lloyds Bank coprolite was found at a Viking site at Coppergate, York, England. (Photo from Linda Spashett Storye book, CC BY 2.5, via Wikimedia Commons.)

material that can capture a microbiome when fossilized—dental plaque. Plaque forms when bacteria and food mix with saliva. Saliva is rich with calcium, which will solidify on teeth, layer after layer, particularly where the teeth meet the gums, as shown for a fossilized tooth in **Figure 13.7A**. As the plaque is laid down, microbes inhabit this environment in a particular sequence and create a complex, highly structured biofilm (**Figure 13.7B**). Much as there's a particular ecological progression of plant species needed to reestablish a forest after a wildfire, starting with grasses and shrubs and ending with tall trees, distinct bacterial species establish themselves in order and perform similar types of roles in creating the intricate ecosystem that is a plaque biofilm. These structural and functional requirements ensure the stability of the oral microbiome over time. Aware of this stability, the researchers sought to define the core oral microbiome for the common ancestor of humans and Neanderthals. To this end, they sampled fossilized dental biofilms from ancient human and Neanderthal samples spanning the past 100,000 years and compared the microbiota detected with those found in the oral microbiome of present-day humans, chimpanzees, gorillas, and howler monkeys (a more distant primate relative) (Yates et al., 2021).

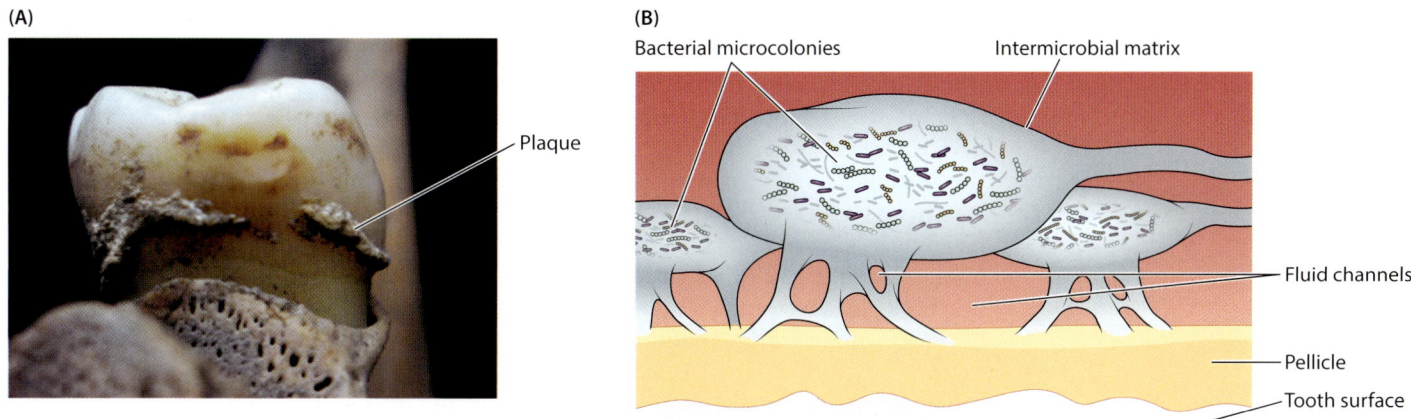

Figure 13.7 Ancient Dental Plaque (A) Neanderthal jaw showing tartar, which is a hardened form of dental plaque, on the molar teeth. (B) An illustration of dental plaque structure on the tooth surface. Over time this structure hardens and fossilizes into the tartar you see on the Neanderthal jaw. (A photo from Sawafuji et al., 2020; CC-BY.)

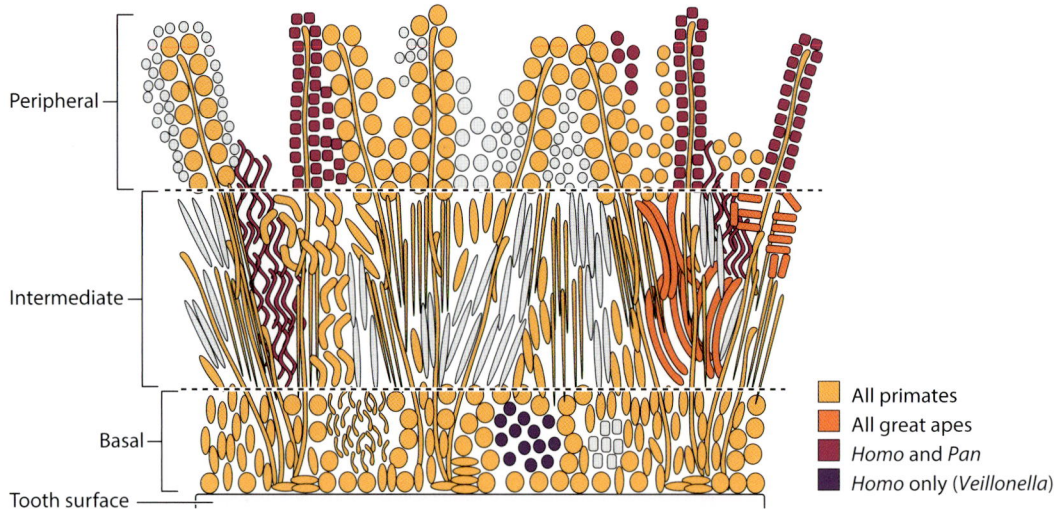

Figure 13.8 Structure of the Oral Microbiome of African Hominids An illustration of the hominid oral microbiome shows a complex structural and functional organization. As this microbiome is established, it results in a complex biofilm covering the teeth (solid line at bottom), with compositionally distinct basal, intermediate, and peripheral regions (separated by dashed lines). (After Yates et al., 2021.)

Figure 13.8 illustrates the complex structural organization of the core hominid plaque microbiome. The taxa shared among primates are indicated in light orange, and you can see that these make up the majority of the plaque biofilm community. Only one genus, *Veillonella*, is specific to humans, and an additional three genera distinguish *Pan* and *Homo* from gorilla and howler monkey. As predicted by the researchers of this study, it seems likely that one reason for the existence of this long-lived core is due to the functional and structural requirements of dental plaque formation, a function that was critical enough to have been conserved since howler monkeys diverged from great apes over 25 mya.

One key finding of this research was that humans possess oral bacteria that are adapted to break down starch. When starch is a regular part of the host's diet, species in the *Streptococcus* genus express genes that allow them to use the enzymes found in saliva to ferment the starches for energy. Modern and ancient humans, as well as our closest extinct kin, the Neanderthals, possess such starch-adapted strains in their plaque. The other primates have virtually no streptococci that can break down starch. According to Christina Warinner, "It seems to be a very human specific evolutionary trait that our *Streptococcus* acquired the ability to do this." The researchers concluded that humans adapted to eating starch-rich foods much earlier than had previously been thought, some 100,000 years ago. Starch-rich foods were clearly important after the introduction of farming, some 12,000 years ago. However, this study suggests starch was a key component of our diet well before the agrarian revolution. One interesting observation is that this addition of starch to the diet occurred just as the human brain began rapidly growing in size. You may not realize what a caloric beast our brain is, consuming some 25% of our glucose supply every day. For the human brain to significantly expand, an additional energy supply was required. It appears that our transition to starch-rich food provided just what was needed to permit rapid brain expansion in the genus *Homo*. Check out **Box 13.1** to learn more about this uniquely human evolutionary innovation.

Oral Microbiome

Another surprising outcome of the Yates et al. (2021) study is the insight provided into how little we know about many members of this core oral microbiome. Clearly

BOX 13.1 EXPANSION OF THE HUMAN BRAIN MAY HAVE BEEN TRIGGERED BY A STARCH-RICH DIET

One of the key characteristics of *Homo sapiens* is our brain size, doubled over the past few million years (**Figure 13.9**). This increase was previously attributed to the invention of stone tools and cooperative hunting. As the old story goes, as our human ancestors became better hunters, they had access to the energy-rich foods required to fuel the growth of larger brains. However, this explanation has troubled some, as our expanding brain required energy-dense foods, such as glucose, and meat is not a good source of glucose.

Warinner and colleagues (2015) studied the oral microbiomes of Neanderthals, ancient humans, and nonhuman primates and learned that preagricultural humans and Neanderthals had highly similar microbiota and, particularly intriguing, both harbored *Streptococcus* that are able to bind human salivary α-amylase, whose activity releases sugars from starchy food sources. The presence of these bacteria on the teeth of both Neanderthals and ancient modern humans, but not chimpanzees or bonobos, suggests that members of the genus *Homo* were eating starchy foods earlier than had been thought.

These findings argue for the importance of starch in the diet when human brains were still expanding. Further, because the salivary α-amylase enzyme is more efficient at digesting cooked starch, it has been suggested that the use of fires for cooking was common before 0.5 mya. Who would have thought that the presence of amylase-binding microbes could inform us about how one of the defining traits of humans, large brain size, might have evolved?

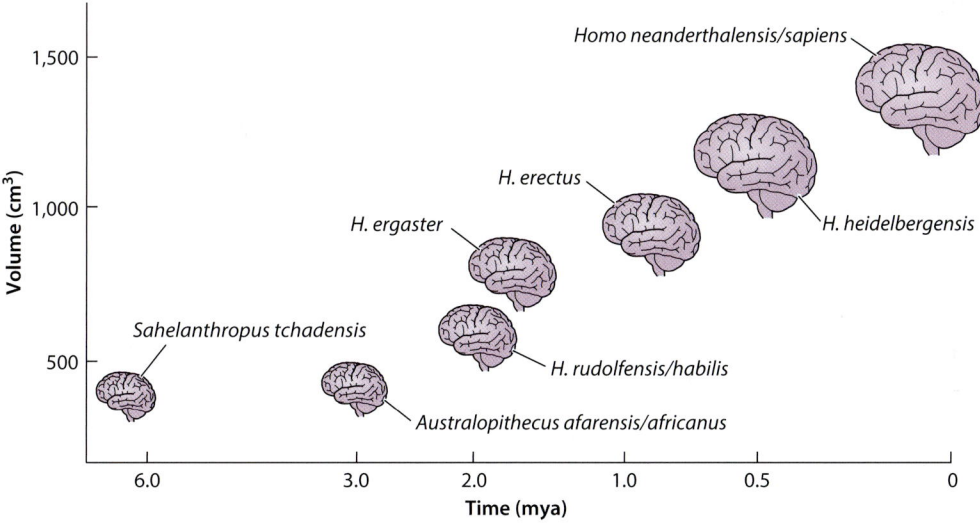

Figure 13.9 Increase in Human Brain Size Over Time (After Tattersall, 2008.)

a core that is maintained for over 25 million years must serve important and beneficial roles to the host, but as yet, many species in the oral microbiome are nameless. **Figure 13.10** provides a PCA plot of metabolic functions inferred from the primate oral metagenome. We see a clear distinction between the great apes and the humans in the functions encoded, suggesting that although members of the core are shared, their specific functional contributions have changed over time. However, when we compare the Neanderthals and modern humans, we find they are almost indistinguishable in metabolic functions. As we've previously discussed in chapter 3, many microbiome researchers are more focused on the functional role of the microbes present than on their species designations. These functional roles are often more highly conserved than specific community members and can provide important insight into selective pressures experienced by the microbiome and, even more exciting, potential clinical intervention routes.

The microbial-encoded genes that appear to drive the separation of the oral microbiomes of *Homo* and nonhuman primates consistently relate to carbohydrate

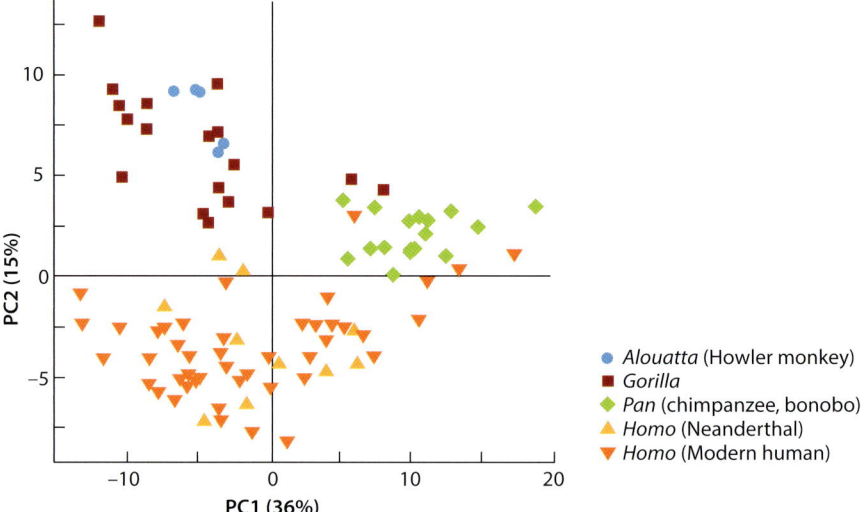

Figure 13.10 Metabolic Functions of Oral Microbiomes Differ between Homo and Other African Primates A principal component analysis is provided for oral microbiome gene functions, plotted based upon the host genus. *Homo* clusters separately from the nonhuman African hominids and the howler monkey, suggesting a functionally distinct oral microbiome. (After Yates et al., 2021.)

fermentation. The genes involved are usually possessed by species of *Streptococcus* (*S. mitis*, *S. sanguinis*, and *S. salivarius*) that are dominant in *Homo* and absent in chimpanzees. In contrast, the oral microbiomes of nonhuman primates are dominated by three different *Streptococcus* species (*S. anginosus*, *S. mutans*, and *S. pyogenes*. The human *Streptococcus* species cluster is notable for its members' ability to bind to salivary **α-amylase**. Humans express high levels of amylase, and it has been argued that increases in this salivary enzyme resulted from a dietary shift that included more starch-rich foods. This combination of the increased production of α-amylase and an increase in microbes that can use α-amylase to break down starch may have provided early humans with the necessary energy to survive and support a larger brain. It appears that members of the oral microbiome played a key role in driving human evolution, in particular, the emergence of that *Homo* specific trait, an enlarged brain.

Microbes from the Middle Ages

A team of archeologists were exploring a site in the Belgian town of Namur when they encountered wooden barrels that had been sitting for some 700 years. It turned out that the barrels didn't hold wine or ale, but rather feces. They were part of a latrine system that included both human and animal waste. Samples of the fecal matter revealed numerous microbes. Even more surprising, the archeologists were able to culture several microbial species directly from these ancient feces. The most abundant bacterial phyla identified include Proteobacteria, Firmicutes, Actinobacteria, and Chlamydiae. Several genera of pathogenic bacteria were found, including *Bartonella* and *Bordetella*, as well as several eukaryotic parasites, including the roundworms *Trichuris* and *Ascaris* (Appelt et al., 2014). **Table 13.1** identifies the numerous human pathogens detected in these medieval samples, all of which still plague us today.

Bacteriophages

Bacteriophages were also detected in the latrine barrels. The viral sequences were dominated by phages similar to those found today in both stool and soil. Even more intriguing, the gene content of these viruses was highly conserved, with functions detected in the 700-year-old samples similar to those encoded by phage in modern human guts. However, one significant difference was noted—the level of viral taxonomic diversity was higher in the medieval samples. This trend of decreasing diversity in the modern human gut microbiota is found for all types of microbes examined to date, something we'll discuss in further detail later in this chapter.

One of the more notable functions encoded in the bacteriophages was totally unexpected: resistance to antibiotics. Genes homologous to those currently circulating in bacteria today were found in the viral genomes, even though these viruses were

Table 13.1 Human Pathogens Detected in Medieval Samples

PATHOGENS	DISEASES
Species	
Bartonella henselae	Cat scratch disease, bacillary angiomatosis
Bartonella quintana	Trench fever, bacillary angiomatosis
Bordetella parapertussis	Respiratory tract infections
Bordetella bronchiseptica	Respiratory tract infections
Bordetella pertussis	Pertussis (whooping cough)
Brucella abortus	Brucellosis
Coxiella burnetii	Q-fever
Granulibacter bethesdensis	Associated with chronic granulomatous disease
Leptospira borgpetersenii	Leptospirosis
Mycobacterium abscessus	Chronic lung disease, post-traumatic wound infections
Mycobaterium tuberculosis	Tuberculosis
OPPORTUNISTIC PATHOGENS	
Species	
Burkholderia gladioli	Pulmonary infections
Clostridium botulinum	Botulism
Legionella drancourtii	Pneumonia, flu-like illnesses
Legionella pneumophila	Legionnaires' disease
Parachlamydia acanthamoebae	Lower respiratory tract infections

Source: Appelt et al., 2014

present long before antibiotics were used in medicine. However, we should note that antibiotics are commonly produced by fungi and bacteria in nature, meaning that these genes may have been a response to naturally produced antibiotics. Alternatively, it is also possible that antibiotic resistance genes evolved first to serve roles other than resisting antibiotics and only later, when humans began mass production of antibiotics, provided this additional benefit to their bacterial hosts. The leader of this study, Christelle Desnues, notes, "Our evidence demonstrates that bacteriophages represent an ancient reservoir of resistance genes, and that this dates at least as far back as the Middle Ages." Further, she adds, "Persistence of metabolic functionalities across centuries may reinforce the crucial role of the viral community in the human gastrointestinal tract" (American Society for Microbiology, 2014). This should serve as an appropriate reminder that, although we have made outstanding progress in our efforts to explore the human microbiome, we have really just begun, and at least for the viruses, microbial eukaryotes, and archaeans, we have hardly scratched the surface in terms of revealing their taxonomic diversity and ecological roles in human health and disease.

13.2 THE INDUSTRIALIZATION OF OUR MICROBIOME

Let's move forward to the present day, and consider what we can learn by examining the microbiomes of societies that retain a more traditional, hunter-gatherer lifestyle, such as the Hadza that live in the central Rift Valley in Tanzania. Over the past 100 years, this community has garnered increasing attention, with tourist visits and food donations that clearly impact their formerly remote situation. However, their overall lifestyle has remained relatively constant over thousands of years. Examining their gut microbiome may help us understand how ancient humans once ate, and perhaps even suggest changes to how we eat today.

A research team led by Justin Sonnenburg, a microbiologist at Stanford University, collected several hundred fecal samples from individuals in the Hadza community (Smits et al., 2017). As hunter-gatherers, their diet changes from one season to the next. During the rainy season, they consume honey and foraged berries, while during the dry season, they hunt for game (**Figure 13.11**). Sonnenburg's team resampled the same individuals during both seasons and found that their gut microbiome compositions were seasonally distinct. In fact, 70% of the gut bacteria present during the dry season disappeared during the wet season. Shifts in abundance occurred primarily in three bacterial families: Prevotellaceae, Succinivibrionaceae, and Spirochaetaceae. At the phylum level, members of the Firmicutes were stable across seasons, while half of the *Bacteroides* taxa were lost during the wet season only to reappear during the dry season. This study reveals, once again, how quickly the microbes in our gut can change in response to shifts in diet.

In a subsequent study from Sonnenburg's group (Olm et al., 2022), researchers compared 18 populations of hunter-gatherers, including the Hadza. Striking similarities were observed when these additional populations were examined. The hunter-gatherers had far more bacteria that ferment complex plant carbohydrates, such as fiber, than is found in the gut microbiomes of those eating a Western diet, and there was an overall higher level of taxonomic diversity. As we have mentioned several times previously, "industrialization" has resulted in the loss of what were once dominant members of our ancient microbiome. Although change is not necessarily bad, in this case it appears that this loss may have long-term impacts on our health. Or, as Dr. Sonnenburg notes, "In the industrialized world, we now may have lost many of those bacteria. It really looks like an ecosystem in disrepair."

We are slowly learning about the changes in our microbiota that have accompanied the transition from the ancient hunter-gatherer lifestyle to the more industrialized version found in Western societies. However, we should not, at least at this moment, use this limited knowledge to dictate how best to change our diets to improve our health. Alyssa Crittenden, an expert on the Hadza diet, suggests, "We need to be very careful about adopting any diet that portends to be mimicking our evolutionary past. . . . The microbiome is influenced by our whole world, not just what we eat." In other words, it isn't just what you eat that matters, but also how you prepare your food and where it came from that contributes to your microbiota. Attempting to re-create a Hadza diet in an urban setting is simply not possible. What is clear,

(A)

(B)

Figure 13.11 Hadza People The Hadza people in Tanzania survive by gathering during the rainy season (A) and hunting during the dry season (B), which leads to seasonal shifts in their microbiomes. (Photos from iStock.com/Katiekk2)

however, is that a high-fiber diet is probably ideal. But, according to Sonnenburg, co-author of "The Good Gut", who is refreshingly direct, "That's about the extent of what we can tell people.... All the rest of what you might hear is a bunch of hooey" (Brown, 2017).

Another research team carried out a similar sort of study, only this time focused on people from three different lifestyles, including traditional hunter-gatherers (Matsés, a remote population from the Peruvian Amazon), traditional agriculturalists (in Tunapuco, a community from the Andean highlands), and urban-industrialized people in the US (Obregon-Tito et al., 2015). The two rural populations have completely different lifestyles. The Matsés are hunter-gatherers who rely on gathered tubers and plantains. They eat some fish and occasional game meat. In contrast, Tunapuco is a rural agriculturalist community where the farmers supplement their diet with small farm animals. Their main sources of nutrition include plant tubers, fruits, and small amounts of dairy, rice, and bread. The college town of Norman, Oklahoma, has a conventional urban community, whose members eat processed foods, bread, and prepared meals. They consume significant amounts of dairy products, particularly milk and cheese.

Fecal samples were obtained from the three communities and the gut microbiota compared. A PCA plot reveals that the urban and traditional communities have distinct microbiota (**Figure 13.12**). The authors included data generated from several other studies in their analysis and revealed that while hunter-gatherers (Matsés and Hadza) and traditional agriculturalists (in Tunapuco, Venezuela, and Malawi) have very different diets and lifestyles, their microbiomes are much more similar to each other than they are to urban-industrialized people (in Oklahoma). Furthermore, 70% of the individuals examined have gut microbiome profiles consistent with their lifestyles and diets. For example, members of hunter-gatherer populations in different areas have very similar gut microbiomes, as do rural populations in Africa and South America. In other words, lifestyles and diets truly do matter!

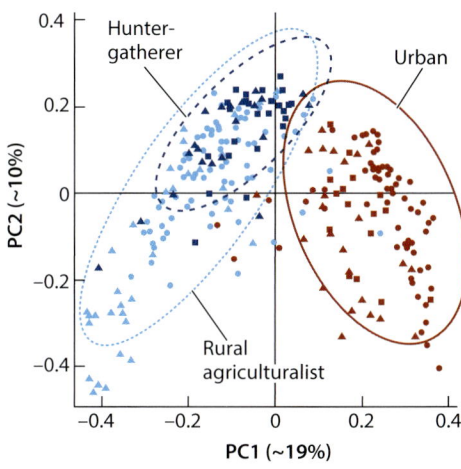

Figure 13.12 Diversity of Gut Microbiomes in Three Types of Human Societies A principal component analysis of beta diversity (Bray-Curtis distances) at the genus level is provided for the gut microbiome samples obtained from individuals in hunter-gatherer (dark blue marks), rural agriculturalist (light blue marks), and urban/industrialized (red marks) societies. The ellipses correspond to 95% confidence boundaries for each of the three categories. (After Obregon-Tito et al., 2015.)

The Hygiene Hypothesis

The studies of our evolutionary ancestors and modern-day hunter gatherers have revealed what is now accepted as a significant trend, that those living in industrialized areas host lower levels of gut microbiome diversity compared with those in more rural settings. Let's think about what might account for this pattern. Ever since Robert Koch developed the germ theory of disease, modern societies have focused on actions that limit exposure to pathogens. From the development of chlorinated water to the use of antibiotics, we have become more and more intent on, and indeed have succeeded in, eliminating many germs from our lives. **Figure 13.13A** shows a dramatic illustration of the correlation between the introduction of filtered and then chlorinated water and the plummeting rates of typhoid fever, a disease spread through water, food, and sewage, in Pittsburgh, Pennsylvania, in the early 1900s. And yet, despite the obvious benefits of filtration and chlorination, such as increased lifespan and reduction in infant mortality, the industrialized world has seen equally dramatic increases in allergies and autoimmune diseases. Just as the frequencies of measles, mumps, tuberculosis, hepatitis, and rheumatic fever plummeted, the incidence of asthma, multiple sclerosis, Crohn's disease, and type 1 diabetes skyrocketed (**Figure 13.13B,C**).

Several ideas have been put forth to explain why autoimmune diseases have increased so precipitously, including our changing diets and exposure to environmental contaminants resulting from industrialization. One of the most interesting suggestions is the **hygiene hypothesis**, which we introduced in chapter 12, which argues that our focus on cleanliness is the problem. This fascinating idea was first formulated in 1976, when a physician, J. W. Gerrard, working in Saskatchewan, Canada, noticed that white communities had higher levels of allergic diseases than did indigenous

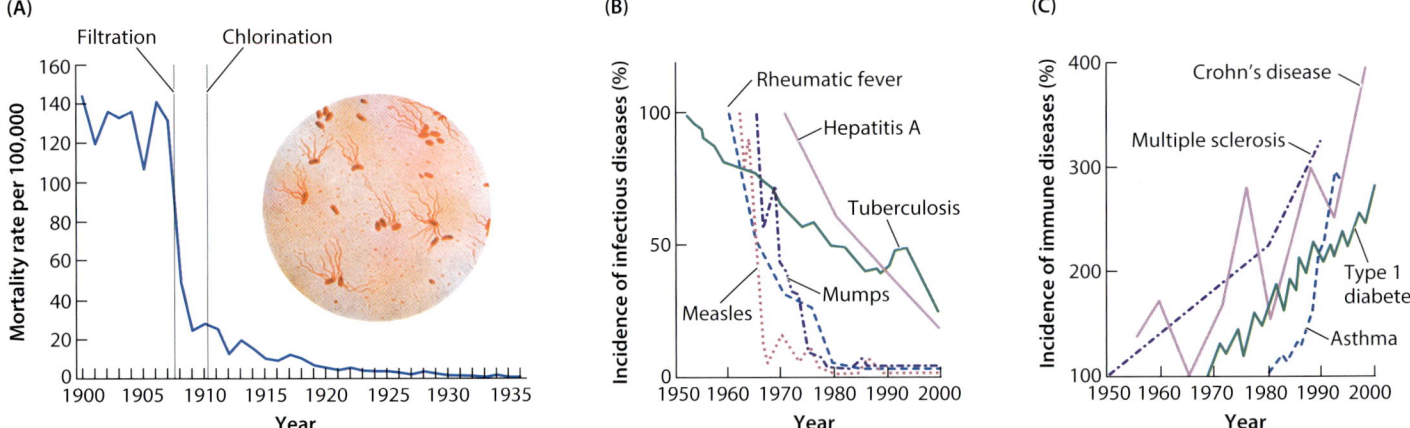

Figure 13.13 The impact of improved hygiene on the frequency of infectious and immune diseases (A) Between 1908 and 1910 there was the introduction of both filtration and chlorination of water sources. These two treatments essentially eradicated one of the most prevalent diseases of the nineteenth and early twentieth centuries, typhoid fever, which is caused by contamination by the bacterial pathogen *Salmonella typhi* (pictured in the inset). (B) As improvements in hygiene and health practices resulted in dramatic reductions in rates of infectious disease, (C) there was a corresponding increase in immune disorders. (A graphs after Cutler and Miller, 2005, photo insert courtesy CDC; B,C graphs after Bach, 2002.)

communities. He suggested "atopic disease is the price paid by some members of the white community for their relative freedom from diseases due to viruses, bacteria and helminths" (Gerrard et al., 1976).

This trend was named the hygiene hypothesis by another physician, D. P. Strachan, who observed the risk of hay fever was higher in children from small families. He suggested that "allergic diseases were prevented by infection in early childhood, transmitted by unhygienic contact with older siblings, or acquired prenatally from a mother infected by contact with her older children" (Strachan, 1989). Numerous subsequent studies have revealed that early childhood exposure to bacteria is essential for immune system training so that we tolerate the members of our microbiomes. If this doesn't happen, our immune system will overreact when exposed to many environmental triggers, in a phenomenon known as **allergic reactions**.

The Disappearing-Microbiota Hypothesis

The hygiene hypothesis lacks a mechanistic explanation for how exposure to pathogens specifically impacts our immune system and thus leads to a decreased chance of developing an allergic disease. However, as scientists learned about our intimate exposure to the microbes found in our microbiomes, a new theory was proposed to explain the increased incidence of autoimmune diseases. The **disappearing-microbiota hypothesis** argues that introduction of clean water, Cesarean births, increased toxic pollution, and the dramatic increase in antibiotic use have resulted in a significant shift in the balance between the microbial species and their relative abundances in our gut.

The disappearing-microbiota hypothesis was first elaborated by Martin Blaser from the New York University School of Medicine, who has spent decades studying the link between *Helicobacter pylori*, which can cause ulcers, and human diseases. "I came to this hypothesis through my work on *Helicobacter*, which is clearly disappearing. But the disappearance [of human microbiome diversity] seems to have begun even before *Helicobacter* was discovered, and not because people are treating ulcers" (Hunter, 2012). It is commonplace to treat stomach ulcers with antibiotics to eliminate *H. pylori* from the stomach.

Blaser suggests that numerous factors are involved in the disappearance of our microbiota, including sanitation practices and antibiotics. "We know that chlorination of water impedes the spread of pathogens, but another thought is that it impedes the spread of commensals," he explained. "Antibiotics are wonder drugs, but everyone assumed they would be free, with no biological cost. When you start learning about our microbiome, it's not too hard to imagine courses of antibiotics leading to extinctions, and when [the commensals] are gone, they're gone. It was assumed that everything bounces back when the course is over, but there is more and more evidence that this is not the case" (Hunter, 2012). Blaser also suggested that this change has been gradual and, since there are so many species of microbes in our gut and our microbiota is so variable, that it has been challenging to identify until recently. "If you have thousands of species, you may not see it at first, but our hypothesis is that it is cumulative" (Hunter, 2012).

Antibiotics

Antibiotics have been essential to modern medical practices since the discovery of penicillin by Alexander Fleming in 1928. Often described as "wonder drugs," they have prevented countless deaths from bacterial infections. However, these drugs exhibit "broad-spectrum" killing, meaning that they are capable of inhibiting the growth of, or killing, most bacteria. In fact, that was one of the selection criteria used in their initial discovery; the more bacterial pathogens an antibiotic could kill, the better. We didn't know then that while antibiotics are quite effective in eliminating pathogenic bacteria, they can also devastate the healthy members of our microbiomes. This is the formerly hidden cost of the use of antibiotics.

Much of the microbiome diversity we have lost can be explained by the increased use of antibiotics, but not just, or even primarily, in human health. In fact, more than 70% of the antibiotics produced are used in agriculture. Soon after penicillin was commercially produced, farmers found that low doses of antibiotics in animal feed resulted in larger animals, and the younger these animals were when exposed to the drugs, the larger they became. This same phenomenon was observed in swine, chicken, and cows and resulted in the widespread use of antibiotic-supplemented feed.

If you have read the prior chapters in this textbook, you are already well aware of why this expanded use of antibiotics turned out to be a very bad idea. Imagine an animal's microbiota under the long-term action of antibiotics. First, having more antibiotics introduced into the environment means greater selection for antibiotic resistance. This resistance can spread from environmental bacteria to pathogens quite readily. Second, antibiotic residues remain in many food sources—produce as well as meats—and can impact your microbiota when the food is eaten.

According to Blaser, the wonders of modern medical practices, such as C-sections and antibiotics, which have had such positive impacts on childhood survival rates, have simultaneously impacted that critical period when immunological training and metabolic patterning are established. "There's a developmental trajectory that's set in the first couple years of life," says Blaser. "This is just when the intergenerational transfer [of microbiota] is going on. It's been going on for millions of years. We're thwarting that" (Hunter, 2012). We now understand that the use of antibiotics can be devastating to our microbiomes, but particularly so during the first 2 years of life. Children under 2 years of age who are treated with antibiotics are far more likely to have allergies, asthma, and inflammatory bowel disease. In fact, we now understand that following a course of antibiotics, the microbiota of these children may never recover completely.

Diet

We learned in prior chapters that our diet also influences our microbiota. Western diets rich in saturated fats, sugar, and refined carbohydrates have reduced microbial community complexity. Imagine a gut microbiome exposed to little of the indigestible fiber that many of the members of our ancient gut microbiomes thrived upon. These members either adapt and utilize another energy source, such as the mucin lining our gut, or go extinct. The loss of key species can also result in extinction of other microbes that depend upon them. In other words, the loss of one important species could potentially lead to the disappearance of numerous other dependent species. Over time, each loss results in slight modifications to our physiology and our remaining microbes. This loss is not just unfortunate for the microbes; it can harm our health and quality of life. For example, if individuals from countries with lower risks of certain chronic diseases move to Western countries, their risks of developing those diseases will rapidly increase to the levels of their new country of residence. The conveniences of a modern society certainly come with their own risks.

The Consequences of the Missing Microbiota

Blaser proposed not just that this loss of microbiome diversity is occurring in individuals, but that our highly hygiene-focused lifestyles have led to a systemic loss of microbes that has snowballed over time. Western civilization has seen a gradual loss of conserved microbiota as birth parents have fewer microbes to pass on to their offspring, who in turn have fewer microbes to pass on to their offspring. **Figure 13.14** highlights the outcome of this progressive loss. This vertical loss (from parents to children) coupled with a lack of horizontal transmission of microbes (from the environment into the host), due to our increasingly hygienic lifestyles, has resulted in an ever more diminished microbiota. Meanwhile, there is an increase in the fraction of opportunistic pathogens in our environmentally acquired microbes.

Let's examine the potential impact of just such a vertical loss, only this time focused on our stomach, rather than our gut, microbiome. Until recently our stomachs were considered essentially sterile. The presence of stomach acids, bile reflux from the small intestine, and peristaltic action were thought to prevent the long-term survival of gastric microbes. In 1982, two scientists, Barry Marshall and J. Robin Warren, disrupted that theory with the discovery that *Helicobacter pylori* not only survives in our stomach, but is, rather than stress, the principal cause of gastritis and peptic ulcers. This discovery was monumental and completely unexpected, and Marshall and Warren were awarded a Nobel Prize for providing this fundamental shift in our understanding of the cause of peptic ulcers.

We now know that *H. pylori* has been a member of our stomach microbiome for at least several hundred thousand years. In fact, when our species migrated out of Africa, *H. pylori* came along for the ride. Now, our relationship with *H. pylori* has been a rocky one to be sure. Although it is found in half of the human stomachs on the planet, its precise role has never been clear. It survives in our harsh acidic stomach environment by producing and secreting urease, which converts urea to ammonia. The ammonia it creates enables *H. pylori* to neutralize stomach acidity and thus make the stomach a more hospitable environment for itself. Once established in the stomach, *H. pylori* can attach to epithelial cells and influence their behavior, including the release of nutrients and cytokines. These changes result in an inflammatory response that ultimately results in chronic gastritis, or inflammation of the stomach lining.

Our gastric microbiota is not limited to *Helicobacter*. In fact, we now know that gastric microbial density varies based upon site, local pH, and environmental factors such as diet and medications. Further, and curiously, the resident microbes vary depending upon whether *H. pylori* is present. When *H. pylori* is not present, the most prominent phyla in the stomach are Proteobacteria, Firmicutes, Bacteroidetes, Actinobacteria, and Fusobacteriota, with the most common genera being *Streptococcus*, *Prevotella*, *Veillonella*, and *Rothia*. When *H. pylori* is present, it is the dominant species, comprising about 75% or more of the microbiome, and the remaining gastric microbiota abundances shift and *Bacteroides* and *Fusobacteria* are absent.

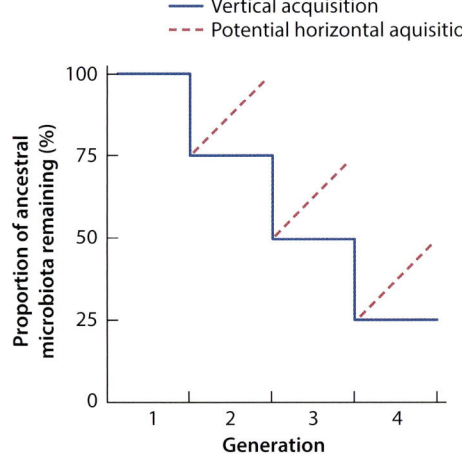

Figure 13.14 The Effect of Maternal Status on the Resident Microbiota of the Next Generation There is a progressive loss of microbiota in the human gut microbiome over time (solid line). During this same period, there is an increase in human population sizes, more social crowding, and less hygienic food and water sources, resulting in the introduction of more opportunistic pathogens into the gut microbiome via horizontal transfer (dashed lines). The progressive loss of vertically transmitted microorganisms from mother to child, without horizontal replacement of the "healthy" bacteria, represents a cumulative birth cohort phenomenon as envisioned in the disappearing-microbiota hypothesis. (After Blaser and Falkow, 2009.)

Blaser and his team made a highly intriguing observation, which is that the abundance of *H. pylori* in the gastric microbiome has been decreasing over the past 100 years in industrialized populations (**Figure 13.15**). This disappearance is associated with a decline in gastric cancer. However, another set of conditions is concurrently increasing, including esophageal reflux, Barrett's esophagus (when the flat pink lining of the esophagus becomes damaged by acid reflux), and adenocarcinoma (a form of cancer of glandular tissue). Are these reciprocal phenomena? *H. pylori* has long been considered a human pathogen, but perhaps it is naturally a commensal that only occasionally causes disease, and in its absence we are seeing an increase in diseases that it formerly helped to hold at bay. Let's dive a bit deeper into this story.

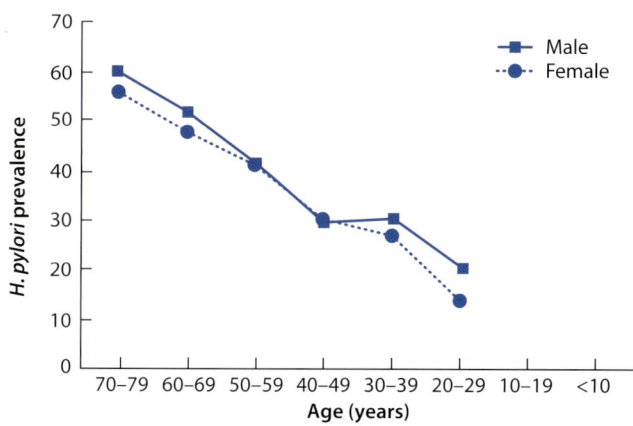

Figure 13.15 *Helicobacter pylori* Prevalence in the United States by Age A study of *H. pylori* prevalence in individuals of different ages found that younger individuals were less likely to have the species, showing that *H. pylori* has been disappearing from the microbiome over time. (After Blaser and Falkow, 2009.)

The stomach produces the hormones ghrelin and leptin. Ghrelin is known as the "hunger hormone" and stimulates food intake, while leptin informs us that we are full. Antibiotic treatment of young children eliminates *H. pylori* and results in increased ghrelin levels, telling the brain that they're hungry and should eat, even if they aren't hungry. It is possible that in the absence of *H. pylori*, regulation of fat production and storage is altered and this has contributed to, or even caused, some of the current epidemic of early-life obesity.

The presence of *H. pylori* also impacts the populations of the T cells and B cells in our stomach. Individuals lacking *H. pylori* have a diminished immune system and reduction in stomach cytokines. Those who retain the species have a rich immune system in the stomach and have lower risks of childhood allergies and asthma. There is a very strong correlation between the loss of *H. pylori* and the increases in childhood autoimmune diseases.

Let's look at another example. *Streptococcus pneumoniae* causes numerous medical conditions, including upper respiratory tract infections, pneumonia, and meningitis. Scientists have worked hard to develop vaccines active against this key human pathogen. The result is a number of highly effective vaccines, whose use has resulted in a reduction of disease in certain high-risk groups. However, *S. pneumoniae* is a perfectly pleasant member of your nasopharynx microbiome, and although the vaccines are designed to target the most pathogenic strains, the vaccinated individual is also inadvertently protected against colonization by some nonpathogenic strains. Hmmm, we are now well aware of what happens when there is an empty niche—microbes will compete to take up residence. To our detriment, one of the more successful competitors is *Staphylococcus aureus*. The combination of the increased incidence of *S. aureus* in our nasopharynx and the alarming rise in **methicillin-resistant *S. aureus*** (**MRSA**) explains the observed increase in staph infection levels observed over the past 30 years. Yet again, our most well-intentioned activities, such as the development of a vaccine to prevent respiratory diseases, has resulted in an increase in our community of a potentially highly virulent replacement.

Recolonizing a Vacated Niche

There was a popular ecological saying in the twentieth century that "everything is everywhere, but the environment selects" (O'Malley, 2007). As our microbiota disappears, vacant niches are created, and the door is open to any microbe able to fill that ecological role. The community of microorganisms interact directly and indirectly with their host, for better or worse. If those interactions disappear, the host and the microbial community will be impacted. Let's imagine two scenarios. First, there are transient microbes that don't interact directly with the host cells; second, there are transient microbes that do interact directly with the host cells. Consider the impact on the host immune response of these two scenarios. Clearly, the absence of microbes that never interacted with the host, and thus never elicited an immune response, will have a far less significant impact on the host than the loss of the microbes that do. In most cases, microbes and the host have coevolved to create a particular set of actions and reactions. For example, gram-negative bacteria are covered with lipopolysaccharides. The interactions between the host epithelial cells and the microbial lipopolysaccharides can trigger an immune response. This is great in the neonate, as the immune system is being trained. However, it is not quite so good if there is a constant trigger to inflammation, such as is seen in obesity and other inflammatory diseases.

Figure 13.16 provides an illustration of four possible types of host-microbiota interactions, which include (A) no microbes present, resulting in host-based signaling only (our germ-free mouse example); (B) little to no specific interaction with the host (our transient microbes example); (C) specific signaling between the microbes and the host (as with many of the members of the gut microbiome); and (D) specific interactions between select microbes (*Helicobacter pylori* is one example) and select host cells. In germ-free animals (A), the host regulatory mechanisms respond differently than they do when a microbiome is present (B). A second type of interaction might involve a more complex consortium of microbes that recognize multiple host cell types and tissues and activate complex neural, hormonal, and immunological pathways of the host (D). Clearly the disappearance of even one member of such a consortium can have far-reaching impacts on the host physiology. An example of this situation might be the microbial cross-feeding observed in the infant gut, which is dependent upon the food consumed. In this case lactic-acid-producing bacteria consume the sugars in breast milk and produce lactate and acetate SCFAs. These metabolites are then used by butyrate-producing bacteria as their food source, and their metabolites are critical for proper functioning of the microbiota-gut-brain axis. Lose one of these consortium members, and the functional pathways may fall apart. If there is no microbe that can fill the void, the host's immune, neural, and hormonal equilibrium will be fundamentally changed.

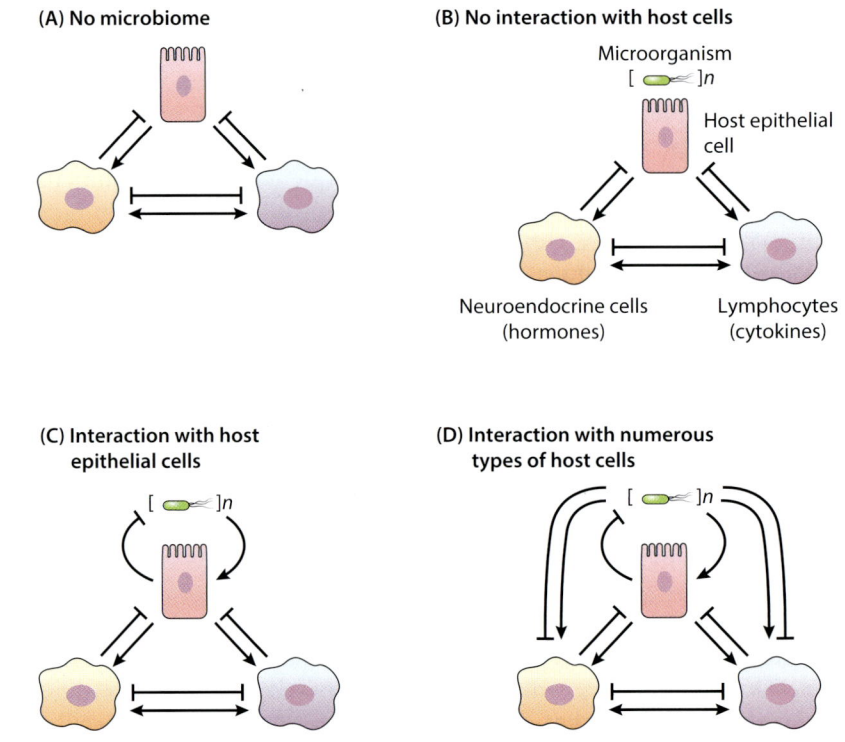

Figure 13.16 Four Potential Types of Host-Microbiota Interactions (A) In the absence of a gut microbiome, the host cells maintain their own signaling patterns. (B) Numerous bacterial taxa reside in the intestinal lumen and do not specifically interact with host cells, resulting in no impact on the hormones and cytokines produced by the host. (C) Those taxa that directly interact with epithelial cells are involved in signaling between the host and the bacteria. (D) Some bacteria have evolved to interact with numerous types of host cells, such as epithelial, immunological, and neuroendocrine cells. *Helicobacter pylori* is one species that has this type of bacterial-host interaction. (After Blaser and Falkow, 2009.)

13.3 THE MICROBIOME AND THE MISSING-HERITABILITY PROBLEM

Human genome-wide association studies (**GWASs**) provide a powerful tool for investigating the correlations that exist between specific genes and diseases (or other traits). This method involves searching the genomes of a large sample of people for **single nucleotide polymorphisms**, or **SNPs** (pronounced like *snips*). Scientists are searching for SNPs that are associated with a certain disease (or trait). One example of the importance of SNPs in human medicine is seen in **sickle cell disease** (**SCD**), which includes several inherited red blood cell disorders. Red blood cells are normally round and move easily through the blood vessels, including the smallest capillaries. In an individual with SCD, the hemoglobin is abnormal, which results in red blood cells that are C-shaped, resembling a farm tool known as a sickle (**Figure 13.17**). Sickle cells have a shortened lifetime, resulting in a shortage of red blood cells. Their sickle shapes cause them to get stuck in smaller blood vessels, causing pain and other complications, such as stroke. SCD is caused by a SNP, or mutation, in the β-globin gene. Using SNPs can help investigators identify genes that may be involved in specific diseases.

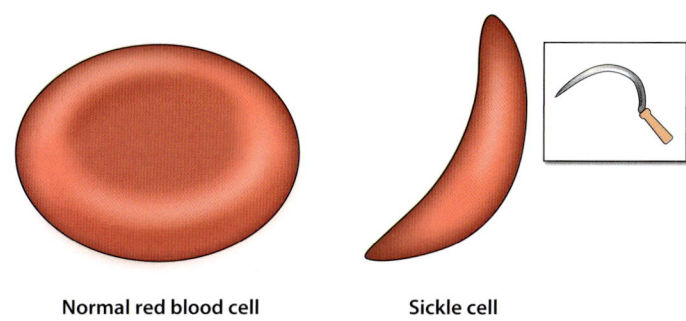

Figure 13.17 Sickle-Shaped Red Blood Cells Sickle cell disease got its name from the resemblance between the misshapen red blood cell associated with this disease and the agricultural tool known as the sickle.

This same principle is now being applied to search for correlations between SNPs and the composition of our microbiomes. In other words, does our genome dictate, in part, what taxa we host in our microbiomes? **Box 13.2** provides a description of one of the more powerful applications of the GWAS method to study genetic associations with the microbiome. This area of study is so new that at this point we simply can't say much about any of the associations detected, yet. However, there is another way in which our microbiome may play a critical role in helping us to interpret prior GWASs. Most such studies have estimated the heritability of a trait to be far lower than what we find if we just examine how common the trait is in families, meaning that genetics alone doesn't fully explain the presence or absence of a disease or trait. One fairly simple example is found in human height. Studies of families show that most of the variation (80%) in human height is associated with genetic effects, rather than environmental factors. In contrast, GWASs reveal that about 50 SNPs are associated with height but explain only 5% of the variation in height observed. This discrepancy between heritability estimates from family (80%) and from GWASs (5%) is common for other traits and diseases and is referred to as the **missing-heritability problem**.

The human microbiome may potentially help explain this discrepancy between heritability estimates from family and GWASs (Sandoval-Mota et al., 2019). As we learn more about the important influences of our microbiome on our immune responses, hormone production, neurological state, and gene expression patterns, the variation in one's microbiome should be included in determining heritability. The authors argue that the missing-heritability problem may be at least partially resolved by considering the impact of the microbiome. Regardless of how well the microbiome may resolve our hidden heritability, the use of GWASs is quickly becoming a powerful tool for investigating the relationships between specific microbial taxa in our microbiomes and human diseases.

13.4 REACQUIRING A MORE HEALTHFUL MICROBIOTA

We can all agree that our microbiota has been altered in industrialized countries. So, what do we do? The first step is to focus on antibiotics from a new perspective. We

> **BOX 13.2. RESEARCH IN ACTION**
> Lactose Digestion and the Microbiome—An Evolutionary Link
>
> ❖ **Hypothesis.** The human genome impacts which microbes are present in the microbiome.
>
> ❖ **Methods.** Single nucleotide polymorphisms (SNPs) and microbiota data were obtained for 6,000 individuals. SNPs that showed a correlation with the presence or absence of specific microbial taxa were identified.
>
> ❖ **Results.** Researchers identified 567 SNPs associated with specific taxa. Those in the *LCT* locus, which encodes lactase, are associated with *Bifidobacterium*, while those in the *ABO* locus, which encodes a type of transferase, are associated with *Faecalicatena lactaris*. SNPs in the *Med13L* locus, which is involved in a suite of development decisions, are associated with *Enterococcus faecalis*.
>
> ❖ **Conclusions.** The *LCT* locus encodes an enzyme called lactase, which helps to digest lactose, which is one of the most strongly selected traits in recent human evolutionary history (for about the past 10,000 years) and is believed to have been associated with domestication of animals and the use of animal milk as food. The association of *LCT* SNPs with the presence of *Bifidobacterium* in the gut microbiome is the most highly significant result of this study, and it suggests a long-lived and sustained interaction between *Bifidobacterium* and humans. Individuals who are lactose intolerant show higher levels of *Bifidobacterium*, as this gene metabolizes both human and bovine milk sugars. The *ABO* gene encodes a glycosyltransferase, which contributes to the determination of the ABO blood groups. This gene was correlated with the presence of *Bacteroides* and *Faecalibacterium*. Although this link is not fully worked out, *F. lactaris* is a mucin-degrading commensal that can digest blood antigens. The association between *E. faecalis* and a SNP at *Med13L* may be explained by the suspected link between *E. faecalis* and development of colorectal cancer, which is mediated in part by the *Med13L* gene. *E. faecalis* produces free radicals that can result in point mutations and chromosomal instability in colorectal cells.

must ensure that society understands how grave the consequences of antibiotic overuse and misuse are. Their use must be restricted, not only to protect their efficacy for the truly ill, but also to ensure that the microbes in our local environments are not selected to be more virulent and ever more highly resistant to antibiotics. We must also recognize that, in particular, the use of antibiotics early in life has far-reaching negative impacts on the individual, and the risk-to-reward ratio must be more carefully and deliberately weighed. For instance, is it worth waiting a week for strep throat to resolve itself, or should we rush to treat it with antibiotics? Finally, there should be a high priority given to the development of narrow-spectrum antibiotics that are able to specifically target the pathogen of interest while leaving the microbiome relatively undisturbed.

There are other medical changes that should be considered. The World Health Organization has stated that C-section should be employed only when medically necessary. Research is underway to identify methods to potentially seed the newborn with birthing parent's vaginal flora when C-sections are required. We must also find ways to encourage and support breastfeeding, which plays a critical role in establishing the infant microbiota and developing the newborn immune system.

Although systemic changes in the medical system are needed to support microbiome diversity, an action each of us can take is to simply change our diets. The critical role of fiber in supporting a healthy and diverse microbiome is well documented, and yet it continues to be nearly absent from many of our dinner plates. Plant fibers that are indigestible by humans are a powerful prebiotic, meaning that their presence nurtures beneficial microbes, which, as they metabolize the fibers, concurrently produce short-chain fatty acids that have numerous health benefits.

(A) 　(B)

Figure 13.18 Global Microbiome Conservancy The Global Microbiome Conservancy project has created a long term storage facility using ultracold freezers (A) to preserve microbial samples from across the planet, such as this remote location in Limi Valley, upper Humla (Nepal) (B). (A © Global Microbiome Conservancy/Photo by C. Corzett; B photo from Ganga Raj Sunuwar/Shutterstock)

One group of scientists have created a microbial "seed bank" that is essentially a collection of large ultracold freezers (**Figure 13.18A**), the **Global Microbiome Conservancy**, into which are being deposited cultures of microbial taxa that are missing from the Western gut microbiota (Karmacharya, 2022). The team gathers microbiome samples from diverse and underrepresented populations around the world, such as from the desolate Limi Valley in Nepal (**Figure 13.18B**) (Karmacharya, 2022). This bank may serve as a crucial resource for microbial diversity as we seek to both preserve species diversity and understand the health benefits these species may provide.

The future of microbiome-mediated approaches to human health is promising. We are rapidly learning how to harness the power of our microbiome to treat debilitating diseases. The earliest successes were seen in reducing the rates of recurrent *Clostridioides difficile* infection. However, the potential of microbiota-based biotherapeutics could extend well beyond infectious diseases, reaching neurological disorders and metabolic and autoimmune diseases. Novel therapeutic approaches, such as fecal microbiota transplantation, phage therapy, and probiotics, are receiving the research and clinical study required to ensure their safety and efficacy. We will further discuss these promising ways to alter the microbiome in chapter 15 The future for microbiome-based therapies is only getting brighter! According to Bernat Olle, the CEO of Vedanta Biosciences and a leader in the field of microbiome-based therapeutics, "I don't think there's any other field of medicine today that holds as much promise for the future of medicine as the microbiome" (Adams & Yakowicz, 2020).

CHECK YOUR UNDERSTANDING

1. The common ancestor of great apes and humans likely had a microbiome.
   a. True
   b. False

2. We can determine whether the microbiomes of humans and great apes share an evolutionary history if
   a. the microbiomes of humans and chimpanzees have at least 95% species composition similarity.
   b. ape microbiomes also play a role in immune system education and regulation.
   c. the phylogenetic trees of microbiomes and hosts match.
   d. Human microbiomes permit humans to eat the same foods as the great apes

3. In Ochman's study, what character was used to determine the relatedness of different species in the microbiome?
   a. The 16S rRNA gene
   b. Their nutritional requirements
   c. Their pathogen status
   d. Their whole genome sequences

4. A difference between human and ape microbiomes is that humans have a higher abundance of
   a. *Methanobrevibacter smithii.*
   b. *Bacteroides.*
   c. *Fibrobacter.*
   d. *Acinetobacter.*

5. Humans and chimpanzees diverged
   a. 200,000 years ago.
   b. 2 million years ago.
   c. 6 million years ago.
   d. 12 million years ago.

6. About _____ of the genera that occur at appreciable frequencies in the gut microbiome are unique to either the human or chimpanzee.
   a. 3%
   b. 20%
   c. 45%
   d. 72%

7. Microbial communities adapted to particular environments are known as
   a. subtypes.
   b. genera types.
   c. microbiota states.
   d. enterotypes.

8. A relatively stable, consistent microbiota is a feature unique to humans and has not been found in any other species.
   a. True
   b. False

9. *Homo sapiens* and *Homo neanderthalensis* diverged roughly
   a. 100,000 years ago.
   b. 500,000 years ago.
   c. 2 million years ago.
   d. 6 million years ago.

10. Which two of the following are substances that let us study the microbiota of extinct species?
    a. Plaque
    b. Food remnants
    c. Remains of the host
    d. Coprolites

11. A unique ability of the human oral microbiome is
    a. short-chain fatty acid production.
    b. immune system education.
    c. starch digestion.
    d. breakdown of fiber.

12. For humans, a large portion of energy (25%) is used by the
    a. muscles.
    b. brain.
    c. heart.
    d. lungs.

13. The core of the oral microbiome has been conserved over time.
    a. True
    b. False

14. Humans digest carbohydrates using
    a. butyrate.
    b. α-amylase.
    c. acetate.
    d. glucose phosphorylase.

15. The genus that is responsible for helping with carbohydrate digestion in humans is
    a. *Staphylococcus.*
    b. *Escherichia.*
    c. *Streptococcus.*
    d. *Pseudomonas.*

16. Hadza microbiota composition undergoes regular changes
    a. due to medication use.
    b. across seasons.
    c. as the result of industrialization.
    d. due to sickness.

17. Hadza microbiota are different from those of humans in industrialized areas because they
    a. have more bacteria that digest complex carbohydrates.
    b. have lower overall diversity.
    c. are consistent over time.
    d. do not vary significantly between individuals.

18. In a study of hunter-gatherer, agriculturalist, and urban groups, what percentage of individuals had a microbiome consistent with their group's lifestyle?
    a. 10%
    b. 25%
    c. 47%
    d. 70%

19. Since the 1950s, which disease has increased in prevalence in the US?
    a. Hepatitis A
    b. Measles
    c. Rheumatic fever
    d. Multiple sclerosis

20. The theory that our increases in sanitation and antibiotic use led to a decrease in the gut commensals, negatively impacting the immune system's training early in life and increasing the risk of developing autoimmune disorders, is known as the
    a. hygiene hypothesis.
    b. absent commensals hypothesis.
    c. disappearing-microbiota hypothesis.
    d. gut-immune axis.

21. When was penicillin, the first antibiotic, discovered?
    a. 1907
    b. 1928
    c. 1941
    d. 1959

22. Broad-spectrum antibiotics can inhibit or kill
    a. only the species causing the infection.
    b. many types of bacteria, including gut commensals.
    c. all bacteria and archaeans.
    d. many types of microbes, including bacteria, archaeans, and viruses.

23. Although many individuals in Western societies have lost important microbiome diversity, it is easy to regain this diversity from people or other sources in their society.
    a. True
    b. False

24. *Helicobacter pylori*
    a. decreases stomach acidity.
    b. is the cause of all peptic ulcers.
    c. is not a gut commensal.
    d. is a small portion of the stomach microbiome.

25. When a microbe disappears from the microbiome, which of the following is not a potential outcome?
    a. A host pathway or system is significantly altered.
    b. A pathogen fills the empty niche.
    c. Important microbial metabolites are produced.
    d. There is no effect on host health.

26. All of the following are public health changes that can support gut microbiota diversity *except*
    a. performing C-sections only when necessary.
    b. limiting antibiotic use.
    c. considering novel therapeutic approaches, such as fecal microbiota transplantation.
    d. discovering more broad-spectrum antibiotics.

27. Studies that search for correlations between human genetics and microbiota composition are called
    a. SNP searches.
    b. microbiota-genetics association studies.
    c. genome-wide association studies.
    d. heritability factor studies.

28. Some traits have been shown to be more heritable through generations than SNP associations can explain. This difference may be explained by
    a. the fact that all families have different levels of heritability.
    b. the microbiome, which may be influenced by family and heritability.
    c. medications used by multiple family members.
    d. the fact that SNPs in the human genome are difficult to detect, so genetic data are often incomplete.

29. Only a subset of diseases can be explained by single nucleotide polymorphisms.
    a. True
    b. False

Answers: 1A, 2C, 3C, 4B, 5C, 6B, 7D, 8B, 9B, 10A & D, 11C, 12B, 13A, 14B, 15C, 16B, 17A, 18D, 19D, 20C, 21B, 22B, 23B, 24A, 25C, 26D, 27C, 28B, 29A

DIVING DEEPER

1. Why might the fact that the *garbs* gene is evolving more quickly than the gene for 16s rRNA be relevant to the Ochman study to determine if host and microbe co-evolve?
2. What conclusions can be drawn from the fact that phylogenetic relationships of microbiome species match the phylogenetic relationships of their primate hosts?
3. What are enterotypes, and when did they emerge?
4. How can we study the microbiome of Neanderthals?
5. How are the oral microbiomes of humans and Neanderthals different from those of other primates, and how was this alteration of the oral microbiome important for human evolution?
6. Why was it surprising to find antibiotic resistance genes in samples from the Middle Ages?
7. Refer back to chapter 3; how does diet impact your microbiome?
8. What is the most significant difference between the diets of hunter-gatherer societies and Western Eurocentric societies?
9. Summarize the hygiene hypothesis.
10. How does the use of antibiotics affect our microbiome?
11. When was the last time you used antibiotics? Did they resolve your symptoms?
12. List some of the selection pressures microbes in your body must endure in order to successfully colonize.
13. What are the consequences of a vacant niche in your microbiome?
14. Think back to chapter 6; why might the World Health Organization connect C-section births and breastfeeding to diminishing microbiome communities?
15. What is the significance of the GWAS method described in this section?
16. What are sources of heritability besides the genome?

DISCUSSING AND REFLECTING

1. In chapters 3 and 6 you learned how your microbiome plays a role in training your immune system to recognize harmful pathogens. The end of this chapter introduces the idea of the hygiene hypothesis and provides a correlative dataset linking increasing antibiotic usage and focus on hygiene with an increase in autoimmune diseases. Combining this with your knowledge from previous reading, what might be the mechanistic explanation for this connection?
2. Describe the differences between prebiotic and probiotic therapies, using information from previous chapters. What are some of their advantages and disadvantages? Would you consider taking a prebiotic or probiotic to reintroduce ancestral microbes to your microbiome? Why or why not?
3. Reflection. Previous chapters have focused on ways that your microbiome changes based on factors such as diet, birthing mode, and lifestyle. Here, we've focused more on your ancestral microbiome, which persists across generations. Many students learn early on that their fundamental traits come from their parents' genes—but this is an oversimplification. How do you feel knowing that the microbes you've inherited from your ancestors may have played as important a role in your development as your genes?

RECOMMENDED READINGS

Popular Science Review

Yong, E. (2016). How Miraculous Microbes Help Us Evolve Better, Faster, Stronger: Invisible Yet Crucial, Our Microbial Partners Add a Gene-Swapping Plot Twist to Evolutionary Theory. *Smithsonian Magazine*, July 26.

Scientific Reviews

Gibbons, A. (2016). Microbes in Our Gut Have Been with Us for Millions of Years. *Science*. https://doi.org/10.1126/science.aag0679

Davenport, E. R., Sanders, J. G., Song, S .J., et al. (2017). The Human Microbiome in Evolution. *BMC Biology*, *15*, 127. https://doi.org/10.1186/s12915-017-0454-7

The Microbiome of the Built Environment

14

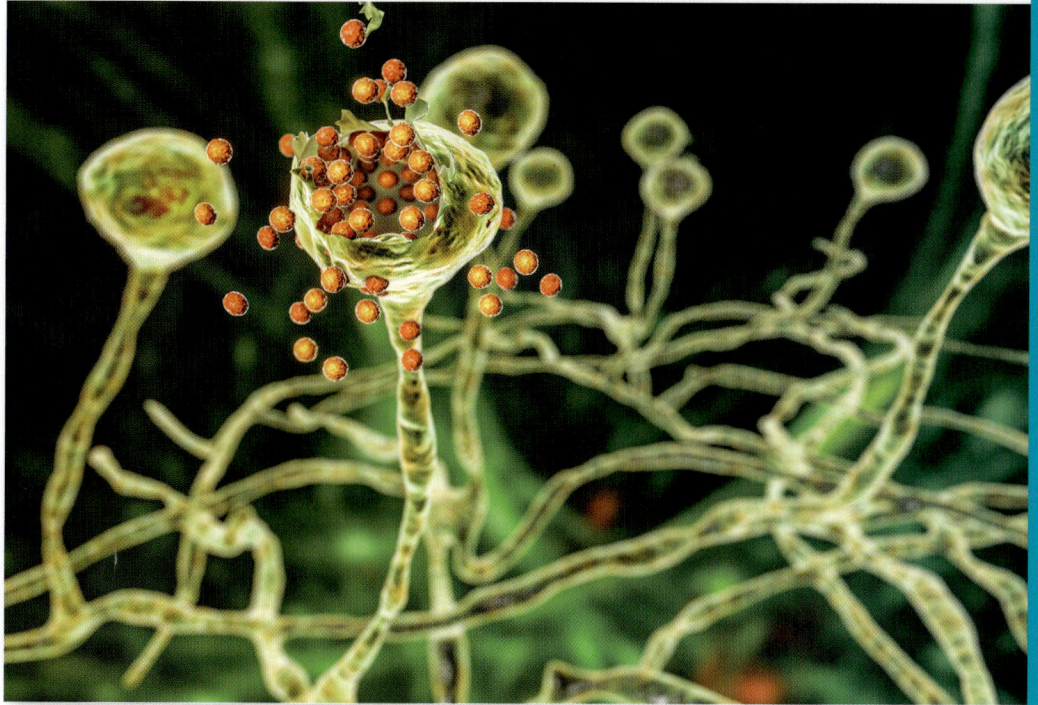

CHAPTER CONTENTS

- 14.1 What Is the Built Environment?
- 14.2 What Is the Microbiome of the Built Environment?
- 14.3 Microbiology of the Built Environment
- 14.4 BE Factors That Influence the MoBE
- 14.5 The Impact of the MoBE on Health
- 14.6 Tracking Microbes in the Built Environment
- 14.7 Microbial Metabolomics and the BE
- 14.8 The Future MoBE

My common name is toxic mold, which pretty much says it all. From a human perspective I am a particularly nasty creature. I love to grow on cellulose—think hay, cereals, and fiberboard—and when the humidity is just right, I blossom! When this happens in your home, my presence is blamed for respiratory problems. To be fair, I don't cause asthma and allergy; I just aggravate the symptoms. Some blame me for sick building syndrome, but I take exception to that claim. I am just trying to make a living, and before the built environment, I was perfectly happy growing outdoors! (Photo from iStock.com/Dr_Microbe)

The built environment refers to everything humans create, including our homes, offices, public buildings, cars, roads, and public transport, but also drinking water, sewage treatment plants, swimming pools, golf courses, and the like. In essence, it is anywhere on Earth where humans have even a moderate presence. In these spaces you will find unique microbial assemblages unlike those found in pristine natural environments, such as the Arctic tundra or deep ocean floor. We call these human-associated microbial communities the **microbiomes of the built environment** (**MoBE**). Few of us venture outside our built environments, which means that our microbial exposure is dominated by whatever microbes exist in these spaces. Although we have begun to identify the microbes present, we still know little about their community compositions, their ecological roles, or their impacts on human health. Are these microbes even alive, or just ghosts we find through their molecular signatures, like fingerprints? Can they promote disease, or do they protect us from illness? Can we harness beneficial microbes while simultaneously eradicating harmful ones? We will begin this chapter by considering how different the MoBE is from microbiomes found in pristine natural spaces and will consider the impact of these

"My home is my microbial castle."
—Roberto Kolter (Kolter, 2017)

Figure 14.1 A Chimpanzee Resting in a Nest of Leaves Chimpanzees and bonobos create their own "built environment" using twigs and leaves. (Photo from alterfalter/Shutterstock)

differences on our health. We will conclude by exploring how this knowledge might inform the future design of homes, workplaces, and even cities to ensure we experience a full and rich exposure to the microbes that are so essential for our well-being and health.

14.1 WHAT IS THE BUILT ENVIRONMENT?

The **built environment** (**BE**) dates to the very origin of our species, some 200,000 years ago. Before emerging onto the savannah, our ancestors very likely created temporary resting spots, much like the leaf nests chimpanzees and bonobos create to nap in (**Figure 14.1**). By swabbing abandoned nests, scientists have learned that a mere 3.5% of the bacterial species present in the leaf litter came from the chimps' own skin, saliva, or feces, and macroscopic parasites, such as ticks and fleas, are scarce (Thoemmes et al., 2018). The majority of species identified are members of the local environmental, predominantly soil, microbiomes. This suggests that these primitive BEs created little to no microbial separation between the natural environment and its occupants. Compare that with samples from human beds today, where over 35% of the microbes present are from our own bodies. To be fair, chimps and bonobos create a new bed each night, so, unless we are washing our sheets daily, it makes sense that our microbial impact is considerably greater in our beds than found for chimps and their nests.

The Earliest Human-Built Environments

For the several million years before our ancestors emerged from the forest habitat and transitioned to life on the savannah, they would have been predominantly exposed to leaf- and soil-associated microbes (Thoemmes et al., 2018). However, as human populations increased and our ancestors began to gather in larger, more dense colonies, those leaf- and soil-associated species began to be accompanied by increasing numbers of pathogens. Human settlements provided an opportunity for pathogens to spread amongst the inhabitants and their environment and evolve new pathogenic traits. Fast-forward to present time, and the dozens of pathogens known to afflict chimpanzees have been transformed into more than 1,400 microbial species that routinely impact human health (Dunn & Thoemmes, 2022). The never-ending process of the MoBE adapting to human presence and activities is well underway.

The Earliest Homes Appear

Archaeologists suggest that the first structures to resemble our modern idea of a home appeared some 12,000 years ago, in the region around the Mediterranean Sea. The Natufians constructed small circular huts, which marked a revolution in human culture and the advent of a true built environment (**Figure 14.2**). People began living in semi-permanent settlements that were more than simple shelters; they were designed to serve as the focal points of many human activities. The advent of homes marks a significant transition in the human relationship with the outdoor environment. To put it simply, we began limiting our exposure to environmental microbes when we first built homes, and the apparent culmination of that transition is that today many of us spend no time in pristine natural environments. Our microbiomes reflect that transition and are composed predominately of human-associated microbes.

Figure 14.2 Natufians: At the Origins of Agriculture and Sedentary Life Image of a Natufian structure in Beidha. Some of these structures date back to 8500 BC. Unearthing these ancient structures has revealed a variety of flint and bone tools as well as personal items, such as limestone beads and pots. (Photo from Anton_Ivanov/Shutterstock)

14.2 WHAT IS THE MICROBIOME OF THE BUILT ENVIRONMENT?

Once homes were invented, their structures and the human activities within them began to impact the resident microbes, ultimately creating the first MoBE. One example is seen in the adobe constructions of ancestral Puebloans of the American Southwest. These native Americans were agriculturalists who lived across the northern regions of the Southwest from the very beginnings of North American civilization until the Spaniard invasions in the 1540s. Their homes were complex structures embedded in clusters that created small cities of pueblos (**Figure 14.3**). The interior of their homes had a surprisingly high level of *Mycobacterium tuberculosis*, the causative agent of tuberculosis (TB), likely due to the reduced exposure of indoor surfaces to direct sunlight and low ventilation rates (Frobisher & Fuerst, 1983). With the Puebloans we see the first signs of how the BE directly impacts the MoBE, which then has a direct link to human health, in this case putatively elevated levels of TB. However, despite living in these settlements, most individuals continued to interact directly with the outdoors regularly, unlike people today.

Figure 14.3 Ancient Adobe Pueblo Structures A typical pueblo structure is composed of adobe blocks forming the walls of each room. These homes were often up to five stories tall with wooden ladders providing vertical access. Several families might have inhabited each structure. (Photo from iStock.com/bboserup)

Health Impacts of the MoBE

As population density in cities increased and humans spent more of their time in BEs, these spaces increasingly impacted our health. As early as the nineteenth century, we realized that population growth, overcrowding, and poor ventilation can lead to disease (**Figure 14.4**). One prime example is the bubonic plague, which resulted in several devastating pandemics over the course of human history. Ever since the Justinianic Plague of 541, which started in central Africa and spread to Egypt and the Mediterranean, plague has reemerged in areas where humans are densely packed together and there is a lack of basic hygiene, resulting in open sewage, rotting food, and stagnant water. The primary vector of this disease, the flea, and its natural host, the rat, are free to flourish under these conditions, and more and more humans get infected with the infectious agent, *Yersinia pestis*.

Controlling the MoBE

In response to rampant infectious disease, we slowly began to develop ways to mitigate its spread. We started to design buildings to be less hospitable to microbes and developed sophisticated systems to handle the air we breathe, the water we drink, and the surfaces we touch, such that microbial pathogens are excluded to the greatest degree possible. Through these actions we have shaped those microbes of our BEs. However, recall from chapter 1 that microbes live in some of the most unimaginable environments possible, such as deep ocean hydrothermal vents, and we can perhaps agree that it is a bit presumptuous on our part to imagine that we could actually prevent a microbe from adapting to the meager range of environments we inhabit. Even as we try to make our environments as sterile as possible in the name of hygiene, some microbiota will find ways around our defenses. Nevertheless, our impact on our MoBE has been substantial. Over the past 75 years or so, researchers have investigated the sources of, survival of, and, eventually, means of controlling the MoBE. We have learned about the associations between fungal spores in air and allergy symptoms. We tracked the microbial emissions from human respiration, air-conditioning systems, and open windows in our homes to reveal the dominant modes of transmission of respiratory infectious disease. With the advent of omic technologies, we can even examine the functional contributions of the MoBE.

Figure 14.4 Urbanization and Industrialization Lead to Crowding and Increases in Infectious Disease Urban centers become hotbeds for infectious disease due to overcrowding, lack of adequate ventilation, and primitive sewage systems. (Wellcome Library, London, CC BY 4.0, via Wikimedia Commons)

Physical Factors Influence the MoBE

Commercial buildings, as compared with private residences, host a far greater density of occupants, all of which are in contact with shared surfaces, such as handrails, door handles, elevator buttons, light switches, and the like. These are the primary means of microbial movement aside from ventilation systems. Commercial buildings tend to have sophisticated air-handling systems with particle filtration, which help limit microbial disbursement. However, the use of air-conditioning coils and humidification systems, which modulate the air's temperature and humidity, results in aeration of water and creation of damp spaces that support increased growth. These air-handling systems also increase the entry of outdoor air, and its associated microbes, as well as increasing movement of microbes between spaces in the building. Bringing outside air into a building boosts the diversity of the building's microbiota. Figure 14.5 illustrates the physical factors that influence microbial presence and movement in the BE, including airflow, filtration efficiency, and disinfection processes, as well as the biological factors involved, such as microbial growth (Stamper et al., 2016).

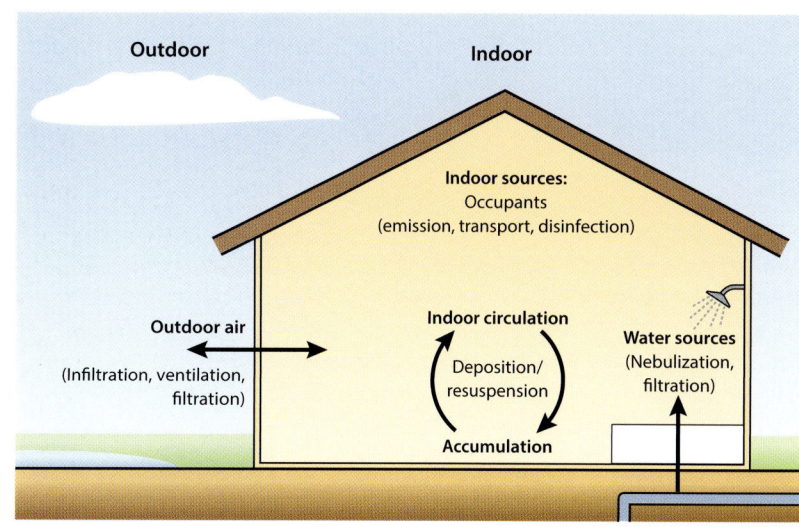

Figure 14.5 Physical Factors Involved in Microbial Transport in the MoBE

Early Studies of the MoBE

Let's consider one of the more influential early studies of the MoBE. The goal was to characterize the MoBE in homes spanning a range of modernization within the Amazon River basin (Ruiz-Calderon et al., 2016) (**Figure 14.6**). The targeted homes were located in a traditional hunter-gatherer village in Checherta, a rural village in Puerto

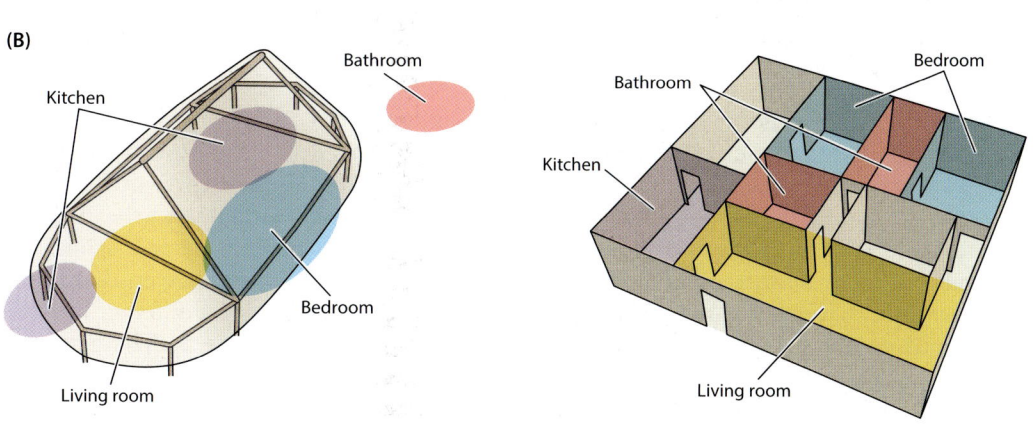

Figure 14.6 Representative Structures across the Range of Modernization within the Amazon River Basin (A) Photos of representative structures found in four communities along an urbanization gradient: Checherta (jungle), Puerto Almendras (rural), Iquitos (town), and Manaus (city). (B) Typical floor plans of houses in Checherta (left) and Manaus (right). (A photos courtesy of Humberto E. Cavallín Calanche; B after Ruiz-Calderon et al., 2016.)

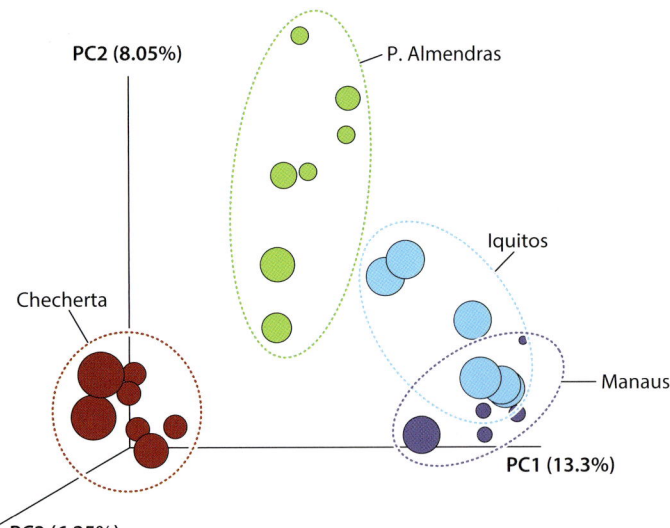

Figure 14.7 Microbial Community Structure across the Urbanization Gradient Buildings that represent an urbanization gradient were sampled and the measure of beta diversity was presented in a principal coordinates plot. The colors represent different communities, and each point of the same color reflects a different location in the home, such as kitchen, bathroom, or sleeping quarters. The size of the point indicates the relative level of phylogenetic diversity obtained. (After Ruiz-Calderon et al., 2016.)

Almendras, the town of Iquitos, and, finally, the modern Brazilian city of Manaus. The designs of the homes spanning this urbanization gradient varied considerably, most significantly in terms of their increasing compartmentalization and decreasing openness to the outdoors. While a home in Checherta was typically one large, open room that was used as a bedroom, kitchen, and living space, homes in Manaus had separate rooms specifically for these purposes. Not surprisingly, the microbial communities differed significantly across these four types of homes (**Figure 14.7**). Further, the study revealed those microbes residing on the walls were most predictive of the type of structure. The MoBE represented the inhabitants, rather than the outdoor environments, in homes in the more urban areas. The authors reached the conclusion that "urbanized spaces uniquely increase the content of human-associated microbes—which could increase transmission of potential pathogens—and decrease exposure to the environmental microbes with which humans have coevolved."

Human Microbial Clouds

One of the most eye-opening findings in these early studies was that we, the people, are the sources for most of the microbes that reside in our structures. An average human sheds 10 million bacteria per hour, and we travel within this "**microbial cloud**," leaving trails of microbes wherever we roam (Meadow et al., 2015). Several studies examined the impact of human presence in sterile spaces (Arnold, 2014, Becerik-Gerber, et al., 2022). The unique combination of bacteria identified often made it possible to distinguish unique individuals solely from their microbial traces, much like a fingerprint does. Moreover, when ventilation is limited, the microbial seeding effect is even more pronounced. This all may sound obvious after having read the prior chapters in this book, but most scientists at that time simply didn't pay much attention to this matter.

As we create BEs that are cleaner and more insulated, we have increased the abundance of members of our own microbial clouds in the MoBE and diminished the natural diversity of environmental microbes formerly present. This fact appears to have important health consequences, bringing us back, yet again, to the hygiene hypothesis, which we have brought up repeatedly in this book and which suggests that living in environments that are too clean results in weakened immune system development and thus an increase is autoimmune disorders. We will return to this hypothesis later, but first we will explore the ways human design and architectural choices impact the MoBE.

14.3 MICROBIOLOGY OF THE BUILT ENVIRONMENT

Most humans spend up to 90% of their time indoors and, as we noted above, while inside, shed and respire millions of microbes per hour. Thus, it isn't surprising that this microbial matter has a significant impact on the MoBE. The presence, density, and activity of humans, as well as their pets and plants, all significantly impact the diversity and distribution of the MoBE (**Figure 14.8**). When new occupants enter a space, the microbes they release can colonize a space in a matter of hours (Lax et al., 2015). Locations that experience more human traffic, such as hallways, have different constellations of microbes than do those with lower traffic, such as closets. The activities humans perform in these areas affect what species are shed. For example, bathrooms are more tightly connected with the skin microbiota of their human visitors.

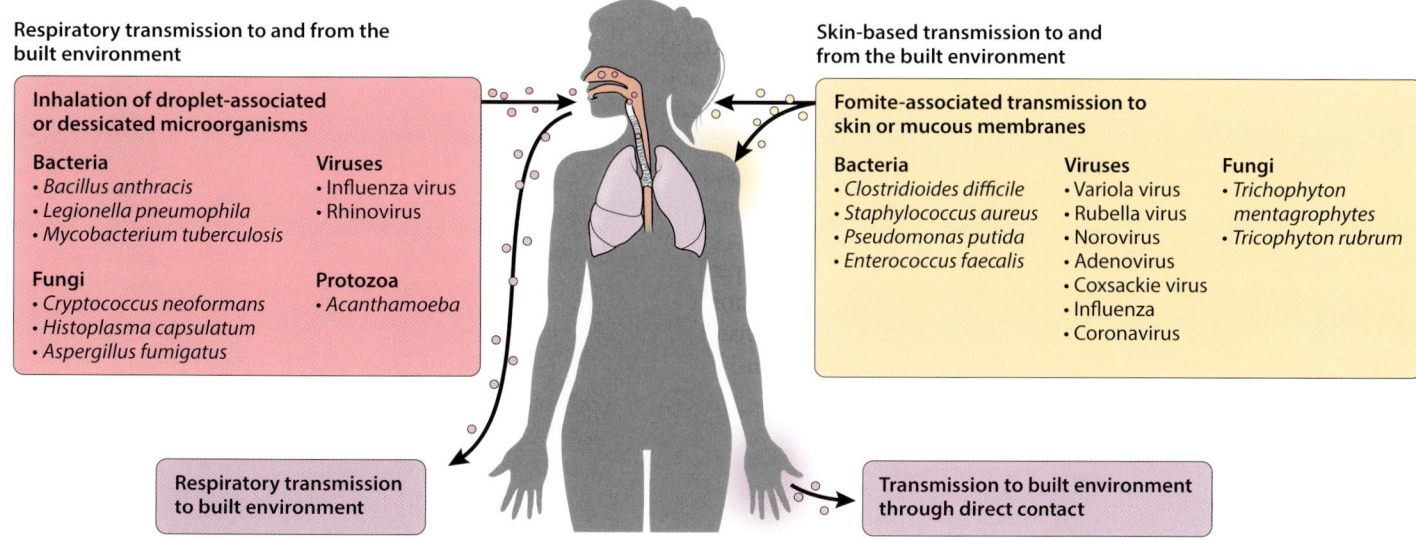

Figure 14.8 Biological Aspects of Microbial Transport in the MoBE (After Gilbert and Stephens, 2018.)

Constituents of the MoBE

One of the very first metagenomic-based microbial surveys of the BE involved bacterial biofilms associated with shower curtains (Kelley et al., 2004). These biofilms included many alpha proteobacteria, such as *Sphingomonas* and *Methylobacterium*, which have close relatives that are known opportunistic pathogens. Norm Pace, one of the scientists who conducted the study, presented his findings at a conference, and even among an audience of microbiologists, there was an audible gasp when Norm described the abundance of microbes they had identified. "I scraped a little bit of soap scum on the shower curtain, put it under the microscope and went: 'Wooah!'" he says. "The sample teemed with bacterial life." Other studies have revealed that lung pathogens can be found in the air hovering over a hot tub! "I would not get into a public hot tub. I would not get into a private hot tub, frankly," says Pace (Aldhous, 2004).

Numerous studies over the past few decades have continued to identify constituents of the MoBE, which allows scientists to compare the communities present in different buildings, such as urban versus rural homes or public schools versus hospitals. One such study cataloged the microbial diversity detected in hospitals, subways, and homes and revealed the abundance of human-associated bacteria, such as the well-known pathogenic *Staphylococcus*, as well as benign commensals, such as *Propionibacterium* (Gilbert & Stephens, 2018). Another study examined the microbial communities from inside and outside 20 shopping malls in China (An et al., 2023). They discovered that human-associated microbiota are enriched indoors, including higher levels of human pathogens and higher antibiotic resistance than outdoors.

The Skin and Oral Microbiomes Contribute Most to the MoBE

It bears repeating that most microbes found in the BE seem to originate from human respiratory and skin microbiomes, such as those listed in Figure 14.8. The MoBE commonly harbors human skin colonizers, such as *Staphylococcus* species, but also fungi and viruses. The **Home Microbiome Project** followed seven families and their pets over the course of 6 weeks. Participants' skin was sampled, as well as surfaces located throughout their homes (Lax et al., 2014). These data revealed that occupants have a major impact on the MoBE. Three of the families moved during the study, and within hours, their new residences harbored the same MoBE as their prior homes. Microbes were shared most often in samples from hands, while noses had more unique microbial profiles, supporting the idea that touching surfaces is a central way

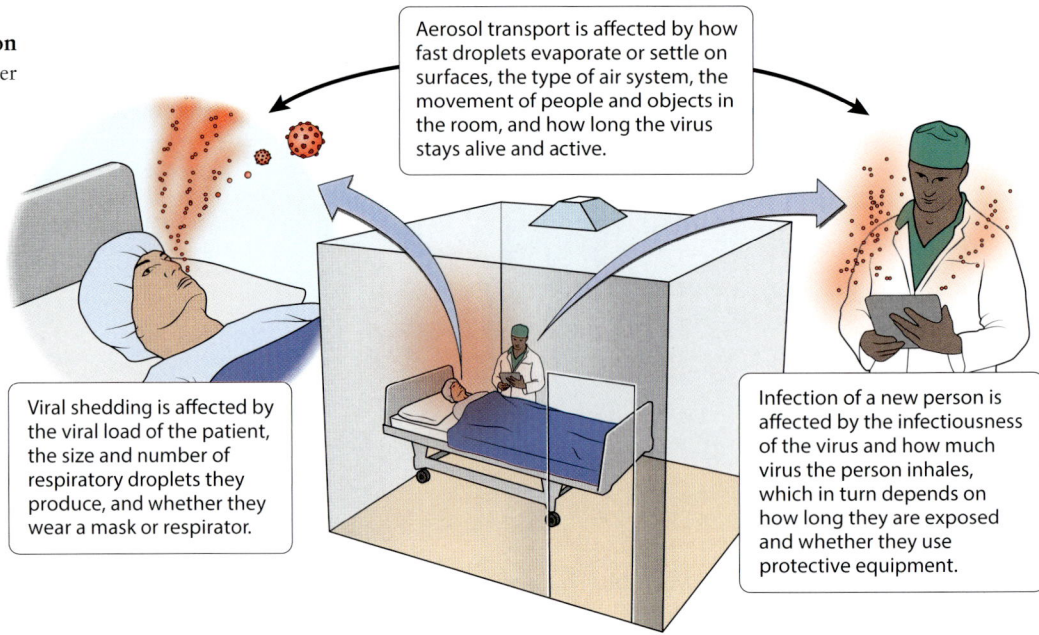

Figure 14.9 Indoor Transmission of the SARS-CoV-2 Virus (After Li et al., 2022.)

humans impact the MoBE. One of the leaders of this study, Jack Gilbert, suggests that the MoBE might serve as a forensic tool in the future. He noted that if he were to take an unidentified sample, "we could easily predict which family it came from."

The Virome

Compared with the knowledge about bacteria and fungi, far less information is available about the viral components of the BE, also known as the **virome**. The main sources of viruses are the same as for other microbes, including humans, pets, plants, dust, and ventilation systems. These viromes harbor beneficial viruses, such as bacteriophages, and also known pathogens, such as the influenza virus. Respiratory viruses, including SARS-CoV-2 and influenza, are spread primarily through droplets expelled when breathing or talking, and, thus, indoor spaces are a primary source of transmission, including through air and person-to-person contact (**Figure 14.9**). As many of us are already well aware, there are other transmission routes for SARS-CoV-2. Due to their ability to survive for days on some surfaces, the SARS-CoV-2 virus and other enveloped viruses such as herpes and influenza viruses are transmitted on surfaces readily as well (Li, Lester et al., 2022).

Plant Microbiomes

Yet another avenue for microbial contributions to the BE is through the **phyllosphere** (**Figure 14.10**), which refers to the aboveground microbiome associated with plants. As urban centers have grown, access to plant-associated microbes has been restricted primarily to those we consume. In fact, plant-based foods are one of the main ways the human microbiome is exposed to plant microbiomes. The phyllosphere is slowly emerging as a key player in human and plant health, as the presence of plants in our BEs helps to increase overall microbial diversity and provides avenues for novel exposures to potential allergens that contribute to immune system training.

It is not solely the fact that plants enrich the MoBE that makes them so important to humans. These microbes can act as protection against opportunistic human pathogens. The microbes of the phyllosphere help to keep down potential pathogen densities in the BE through several mechanisms such as contributing to colonization resistance and the production of antimicrobials. Understanding the roles that plant microbiomes serve can provide an important avenue to establishing a healthy microbiome in built environments.

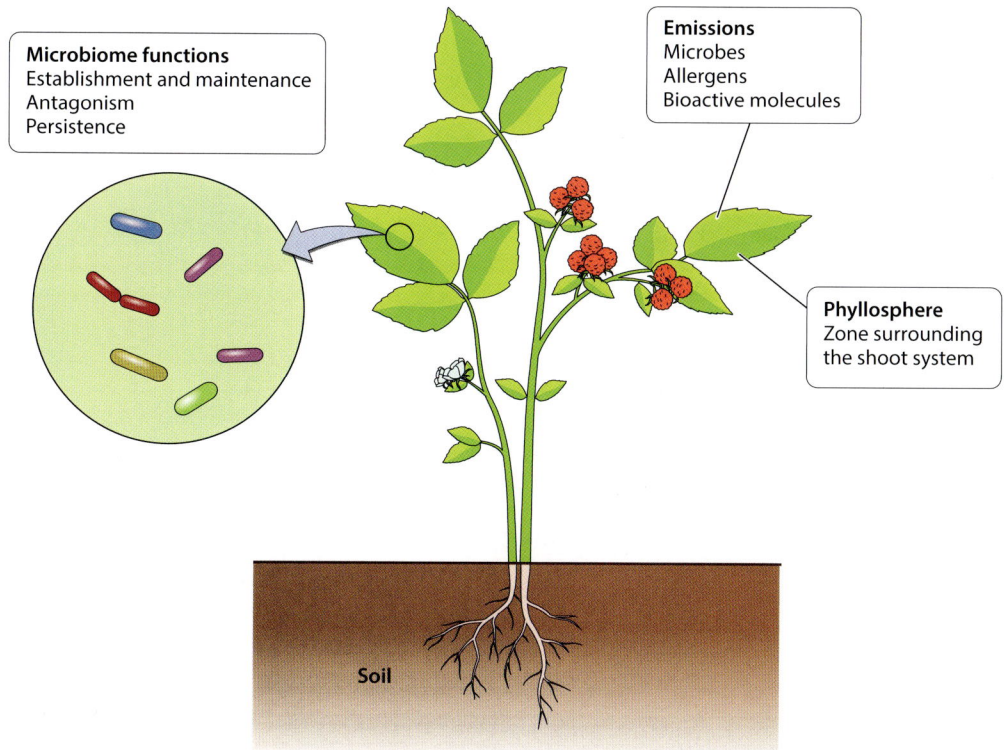

Figure 14.10 **Plant Microbiomes Include the Phyllosphere and the Rhizosphere** The above ground portion of the plant, the phyllosphere, provides a rich source of microbial diversity in the BE, which helps to minimize pathogen invasion.

14.4 BE FACTORS THAT INFLUENCE THE MoBE

Numerous properties of the BE influence the members of the MoBE, including building materials and heating and cooling systems, as well as the occupants and their activities (**Figure 14.11**). Building layout directly affects how individuals use and move through the space and, thus, how microbes are dispersed. Ventilation appears to be the primary determinant of microbial movement, which is significantly influenced by building design. Room temperature dictates which organisms can survive and thrive and alters the level of human and animal microbial shedding. The relative humidity of a space dictates the levels of aerosolization of microbial cells and spores. Carpets and rugs create microenvironments of high relative humidity that can increase the growth, prolonged survival, and transfer of microorganisms from surfaces to individuals. Numerous functional items in a building, such as sofas, rugs, and chairs, provide nutrients that support microbial growth. Windows permit the inhibitory effects of sunlight and permit air exchange that aids in the reduction of potentially contaminated air, while simultaneously serving as access points for outdoor microbes. Humans, pets, pests, and plants are constantly exchanging microbial partners, contributing to the diversity of the MoBE.

Rooms with different functions tend to have different microbiota. **Figure 14.12** shows the results of a survey of microbial taxa identified in samples from various locations in the BE. Offices harbor higher levels of soil-dwelling bacteria, presumably due to the increased traffic in and out of them. Ventilated offices show an increase in species better suited to hot dry air, such as *Deinococcus*. Restrooms are dominated by members of the human skin microbiota, including *Lactobacillus* and *Staphylococcus*. One study focused on comparing the microbes in dust sampled from different rooms in several buildings and revealed that bacterial communities in restrooms were particularly distinct from those in all other types of spaces (Kembel et al., 2014). A typical bathroom has a single door and harbors a higher diversity of human occupants, at least in office buildings. In a similar fashion, buildings with different functions, such as apartment buildings, food-processing plants, hospitals, and museums, all host their own unique microbial constituents.

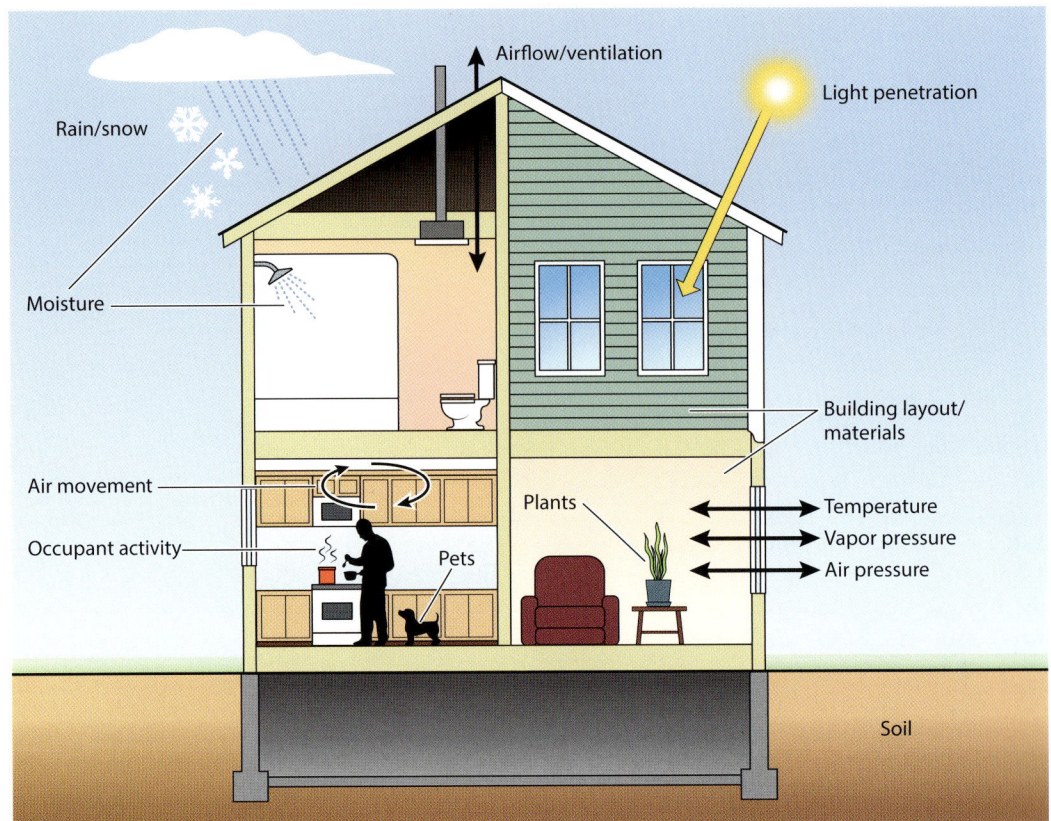

Figure 14.11 Attributes of the Built Environment That Impact the MoBE A building's attributes, such as layout, ventilation, and airflow, coupled with the building envelope, which creates a distinct internal environment, impact the resulting MoBE. Various features of the BE, such as the levels of sunlight, moisture, and other ambient household conditions, select for different microbiota. Building materials provide nutrients, and pockets of high relative humidity can support bacterial and mold growth, while windows and the human, animal, and plant inhabitants introduce novel microbes.

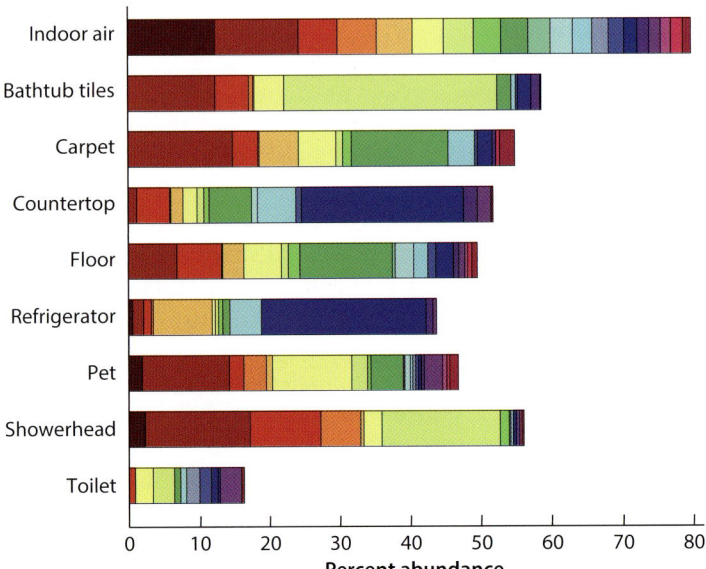

Figure 14.12 Relative Abundance of Bacterial Communities on BE Surfaces and in the Air Microbial communities vary based upon location in the BE. In this graph, the different colors refer to different taxa identified in samples from a variety of sources within the structures. (After Miletto and Lindow, 2015.)

Humidity and Mold

All types of building materials are subject to microbial colonization. One particularly nasty microbe, at least with respect to human health, **Stachybotrys**, also known as toxic black mold, can thrive on certain building materials, particularly when damp (Figure 14.13). Building materials offer substrates and nutrients to support a variety of microbial growth. For example, cellulose-based materials like wood, readily metabolized by numerous microbes, are more likely to be colonized than inorganic materials, such as mortar and concrete. Most building materials are porous and have rough surfaces, which help dust and its microbial partners to adhere. In addition, porous material will retain more moisture, a key ingredient in microbial growth. In fact, to counter this growth-supportive feature, many indoor surface materials incorporate metal nanoparticles to discourage microbial growth.

Ventilation and Microbial Spread

We have previously mentioned the importance of ventilation on the MoBE. This feature has a direct impact on temperature, levels of humidity and carbon dioxide, and airflow rates, all of which influence which species can grow on a surface and how fast. The most direct impact of airflow on the MoBE is to either encourage or exclude the entry of outdoor microbes. Conventional air filters can remove fungi and bacteria, but not viruses, such as influenza and SARS-CoV-2. Natural ventilation, such as through open windows or entrances to balconies and decks, provides more entry to external microbes. However, such natural ventilation also introduces potentially harmful allergens, which can create their own set of challenges.

Light Influences Which Microbes Survive

Light, visible or UV, can impact which microbes will survive in a building, and it tends to result in fewer human-associated bacteria in the MoBE. Sunlight and UV can inactivate viruses but are far less effective against bacteria and fungi. Architects can employ the disinfecting capacity of sunlight by creating structures that enhance exposure to direct sunlight in rooms that are the usual harbingers of microbial contamination, such as lobbies and bathrooms.

Indoor Plumbing

The incorporation of indoor plumbing, a key innovation in the improved functional utility of buildings, has also created the ideal ecological opportunity for waterborne pathogens, such as *Legionella*, to colonize. Microbial growth is enhanced in numerous locations in indoor plumbing, such as cold and hot water reservoirs, faucets, showers, and water filters. Most plumbing systems have a period in which water sits, which provides optimal growth conditions for some microbes. *Legionella* and *Mycobacterium* readily spread through drains, faucets, and showerheads, resulting in their aerosolization and leading to further spread throughout the building. Certain high-end filtration systems are effective in reducing bacterial load in filtered water; however, such systems are costly and require continued maintenance. Electronic, or motion-sensor, faucets seemingly had a huge impact in lowering microbial spread, as they prevent spread of pathogens

Figure 14.13 Toxic Black Mold Growing on Building Materials (Photo from Infrogmation of New Orleans, CC BY-SA 2.0, via Wikimedia Commons)

between hands and faucets. However, we now know that these same faucets enable the increased spread of *Legionella*. It appears that the more parts there are to a faucet (such as the many valves of an automatic faucet), the more surfaces there are for bacteria to grow on.

Cleaning Practices Impact the MoBE

Cleaning practices can have an impact on the MoBE and vary from residential to commercial buildings. In many homes, cleaning involves the use of disinfectants such as bleach or detergents to clean kitchens and bathrooms, while sweeping, vacuuming, and dusting are the principal activities in most other rooms. However, particularly with the advent of COVID-19, our cleaning efforts have gone so far that we might literally be cleaning the MoBE out of our homes! As we have learned repeatedly in this book, the presence of a diverse, rich microbiota is a key factor in determining health. With such clean homes, are our babies and infants getting the level of microbial exposure they need to achieve the best possible health outcomes?

Commercial buildings generally go much further in their attempts to eliminate the presence of microbes. These actions include filtration, to remove airborne microbes, and application of UV irradiation and chemicals, to disinfect surfaces. In particular, the use of UV irradiation has become a standard cleaning method in hospitals, where its use has significantly reduced the numbers of **healthcare-associated infections** (**HAI**) (Anderson et al., 2014). These "kill them all" approaches to cleaning may make sense in a hospital environment, but due to their lack of selectivity, they result in the elimination of the overwhelming majority of beneficial microorganisms as well. New approaches to more-targeted cleaning methods are needed but not yet in sight.

14.5 THE IMPACT OF THE MoBE ON HEALTH

We now recognize that most humans spend an enormous amount of their lives exposed to the MoBE, with relatively scant input of microbes from more natural settings. But does this lack of environmental microbiome exposure really matter to our health? According to numerous studies, the answer is a resounding yes.

Sick Building Syndrome

Since the late 1970s, public health agencies have been inundated with building-associated complaints collectively called the **sick building syndrome** (**SBS**) (**Figure 14.14**), in which occupants of a built structure experience negative health impacts linked to how much time they spend in the building, and for which no cause of disease can be

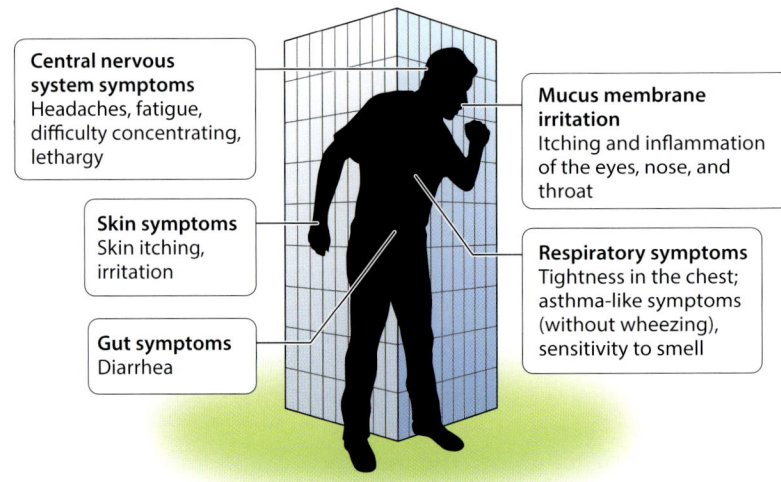

Figure 14.14 Sick Building Syndrome Symptoms (After Yaguang, 2011.)

identified. Symptoms of SBS include irritation of the mucous membranes of the eyes, nose, and throat, and symptoms, such as fatigue, headaches, and inability to concentrate, and dermal symptoms, such as itching and rashes on the skin, face, or scalp, all of which are associated with the length of building occupancy.

Traditionally, the MoBE was not considered in SBS investigations. However, a recent study found a link between SBS in schoolchildren and the MoBE in their classrooms. Fu et al. (2021) identified eight SBS-associated bacterial and fungal genera that were correlated with SBS symptoms experienced by some of these children. Further, the study revealed that in these schools, the relative humidity, the presence of visible dampness and mold, and the levels of dust potentially affected the abundance of microorganisms and occupant health. SBS is a complex condition that appears to be triggered by ventilation issues. As our understanding of the MoBE increases, it is likely that some of the SBS may also be due to the MoBE itself.

The Farm Effect

Let's examine one of the earliest and most influential studies of this matter, which focused on two distinct groups of American farmers, the Amish and the Hutterites. Both groups emigrated to the US from Europe during the Protestant Reformation of the sixteenth century and have remained socially and reproductively isolated ever since. They share somewhat similar lifestyles, including large families; diets rich in raw milk, fat, and salt; and low rates of obesity. However, they differ in their housing situations and in their approaches to farming. The Amish are traditional farmers, often dairy farmers, who use horses for field work and transportation and live in single-family homes. Children are active in caring for farm animals and participating in farming activities from a very early age. The Hutterites, in sharp contrast, are industrialized farmers who live communally. They employ modern technology in their farming and focus more on crop production. Their children are removed from many farming practices and proximity to farm animals.

Levels of several chronic allergic diseases, such as asthma, differ substantially in these two groups, with low levels in the Amish (5.2%) and fourfold higher levels in the Hutterites (21.3%), which we mentioned briefly in chapter 12 (Stein et al., 2016) (**Box 14.1**). This difference was surprising given their shared genetic ancestry and lifestyles. However, blood samples revealed that Amish children had more neutrophils, crucial to fighting infections, and fewer eosinophils, which promote allergic inflammation. There were also differences in gene expression profiles in these immune cells, with enhanced activation of innate immunity genes in Amish, but not Hutterite, children.

Samples of house dust from both communities provide one potential explanation for the differences in immune cells observed in their blood samples. Levels of bacterial endotoxins in dust from Amish homes were nearly sevenfold higher. Mice exposed to dust from Amish homes were protected from asthma-like responses to allergens, while those exposed to dust from Hutterite homes were not. When we consider the observed differences in asthma levels among children in these two communities, it appears that the Amish children, who have barns situated close to their homes and interact frequently with farm animals, gain an environmental exposure that confers protection from asthma, and it does so by triggering the innate immune system of these young children at a key period of immune development.

Compared with farm children, those in urban and suburban BEs are exposed primarily to members of the human microbiome. BEs are designed to create a strict separation from the natural world, resulting in conditions that have imposed strong selective pressures on the microbes that are able to colonize and persist inside. In addition, it has been shown that some of the adaptations needed for survival in the BE result in an increase in the accumulation of antimicrobial resistance and virulence genes by the members of the microbiome (Lax et al., 2017). Thus, children in the BE now come into contact with a much smaller subset of the microbial diversity that exists in nature, which has consequences for immune development.

BOX 14.1. RESEARCH IN ACTION
The Amish Advantage—How Dust Exposure Reduces Asthma Risk

❖ **Hypothesis.** Comparing two communities with different farming practices will reveal why they harbor significantly different levels of allergic disease.

❖ **Methods.** Blood samples were obtained from Hutterite and Amish children along with dust samples from their homes. The blood was examined for the presence of immune cells and their patterns of gene expression. Dust samples were examined for microbiome composition and used in inoculation of mice to determine the impact of dust on the production of asthma symptoms.

❖ **Results.** Figure 14.15A shows sixfold higher levels of bacterial endotoxins in dust samples from the Amish (blue) versus Hutterite (red) homes. Figure 14.15B compares the gene expression levels of immune cells of the two groups, showing higher expression in the Hutterites (1,360 genes) in red, and higher expression in the Amish (1,449 genes) in blue.

❖ **Conclusions.** The environment encountered by Amish children is responsible for the lower levels of asthma detected, in part due to the exposure to immune-stimulating dust particles.

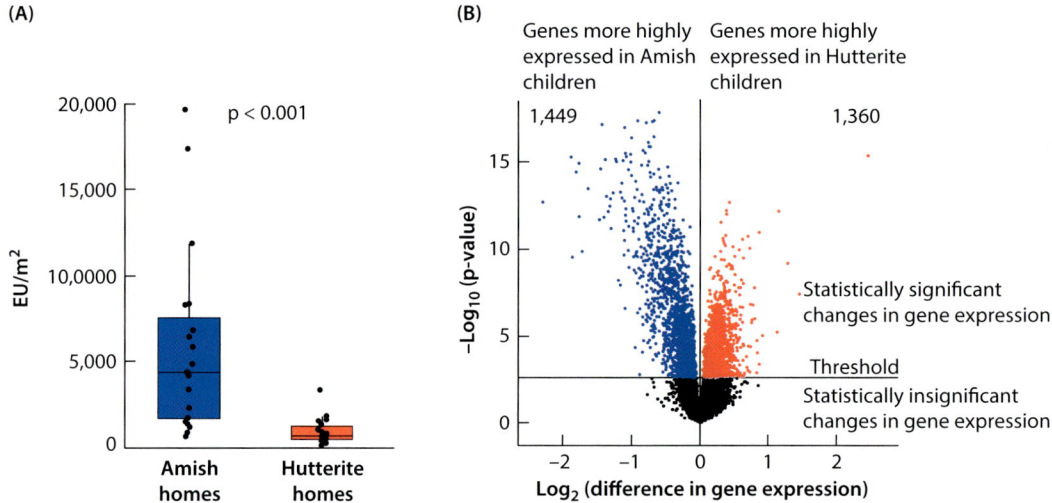

Figure 14.15 Comparing Amish and Hutterite Children (A) Dust samples from Amish and Hutterite homes show dramatically different levels of bacterial endotoxin (measured in endotoxin units per square meter, EU/m^2). (B) This may underlie the significant differences in leukocyte gene expression measured in the children. The authors chose to illustrate these gene expression differences using a volcano plot, which is useful for identifying events that differ significantly between groups. In this case the blue and red dots indicate increased expression of genes in leukocyte cells in Amish (1,449 genes) versus Hutterite (1,360 genes) children. The most significant points are found at the top of the plot, where it is clear that blue dots predominate over red dots. (After Stein et al., 2016.)

14.6 TRACKING MICROBES IN THE BUILT ENVIRONMENT

Interest in tracking microbes in the BE goes back as far as 1847, when Ignaz Semmelweis, an obstetrician, uncovered a microbial transmission route that led from the hospital autopsy chamber directly into the maternity wards. The inadvertent movement of pathogens by physicians from deceased patients to pregnant mothers was transmitting a deadly disease, postpartum sepsis, and resulting in high death rates. Semmelweis mandated handwashing for all physicians entering the maternity ward, which represents the first recorded intervention for a nosocomial (or hospital-acquired) infection.

Since then, most work on tracking microbes has focused on either hospitals or food manufacturing, with the goal of preventing outbreaks of specific pathogens and determining their routes of infection. One prominent example is the discovery of the bacteria that causes **Legionnaires' disease**, *Legionella pneumophila*, whose name is based upon an outbreak that occurred at an American Legion convention in 1976. The symptoms of Legionnaires' disease resemble other forms of pneumonia, including coughing, shortness of breath, fever, muscle aches, and headaches. Of those who contract Legionnaires' disease, 10% die from it, with the elderly and immunocompromised at the greatest risk. Although *L. pneumophila* can be found in natural water sources, it thrives in the BE, particularly in the cooling towers of hotels and commercial buildings, where it gets circulated through the air-conditioning systems (**Figure 14.16**). Understanding how this pathogen reaches the hotel guests or hospital inhabitants has resulted in a suite of changes with respect to the placement of cooling systems and their routine maintenance. And yet, over 50,000 individuals in the US contract this disease each year. Numerous sanitation practices have emerged in the built environment as pathogens have been tracked from their sources. For example, black mold (*Stachybotrys*), which can cause respiratory illness, is controlled with ultraviolet sanitation systems, while *Clostridioides difficile*, which causes gastrointestinal illness, is eliminated from hospital rooms by introducing hydrogen peroxide vapor.

Tracking Hospital Pathogens

Understanding the spread of infectious disease in hospitals is of particular importance because they harbor a high concentration of human pathogens and host a large population of susceptible patients. Significant work has been done to understand and limit the prevalence of hospital-acquired infections. As early as 1859, Florence Nightingale argued the virtue of open windows in her hospital wards to increase the rate of patient healing and to decrease the death rate (**Figure 14.17**). She was spot-on: opening windows can shape the MoBE in ways that benefit patient health by both increasing microbial diversity and reducing the levels of human pathogens. Today, scientists and doctors continue to consider how best to design and clean hospitals. One study focused on the path of airway contamination in hospitals (Kembel et al., 2012).

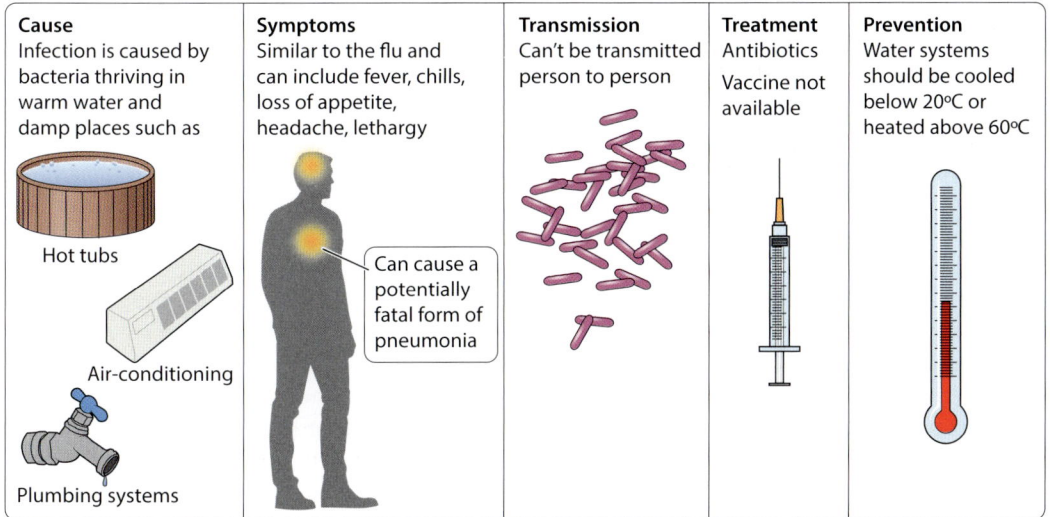

Figure 14.16 Legionnaires' Disease: The Source, Symptoms, and Treatments (After Piecková, 2017, and Bunbury, 2015.)

Figure 14.17 Florence Nightingale Inspecting a Hospital Ward (Wellcome Library, London, CC BY 4.0, via Wikimedia Commons)

Samples were obtained from hospital rooms that differed in their ventilation processes. In rooms in which windows were opened, there was greater microbial diversity than in rooms with mechanical ventilation. In fact, the air in mechanically ventilated rooms hosted more human pathogens. While this one study doesn't prove that opening a window will improve a patient's health, it does reveal how air sources and circulation patterns can impact a room's microbial content, which can in turn affect the number of pathogens vulnerable patients are exposed to.

The Neonatal Intensive Care Unit

Another hospital-based study showed that there was a relationship between environmental microbes and babies in the **neonatal intensive care unit** (**NICU**) (Cason et al., 2021) (**Figure 14.18A**). As we learned in chapter 7, babies born vaginally are typically colonized with the bacteria of their birth parents' vaginal and gut microbiomes. However, premature infants are often born via C-section and are immediately whisked away to the NICU. These babies are often born underdeveloped and are frequently treated with antibiotics, which means the colonization of their microbiomes is further disrupted. Carolina Cason and her team carried out a longitudinal study of preterm newborns in the NICU by following their nasal microbiomes from birth (Cason et al., 2021). Their study revealed that bacterial pathogens, including *Staphylococcus* and *Streptococcus*, were detected at higher levels in the nasal passages of babies after 13 days of hospitalization in the NICU (**Figure 14.18B**).

In addition to the pathogens present, the newborns' resistomes were sampled, which refers to the presence of antibiotic resistance genes. No resistance genes were present in the microbes detected on the babies at birth. However, following admission to the NICU, the resistome emerged and became more abundant and diverse. These data suggest that microbiota of the hospital surfaces might be transported to respiratory mucous membranes. In this example, in the absence of the mother's close physical presence, the baby was exposed to primarily hospital microbes, which successfully colonized the nasal passages. It is not yet clear how this information might be incorporated into future designs of NICUs, but clearly there is a need to limit the spread of pathogens and increase the presence of the birth parent's microbes. Vaginal birth and constant skin-to-skin contact is how we evolved to gain our microbiomes

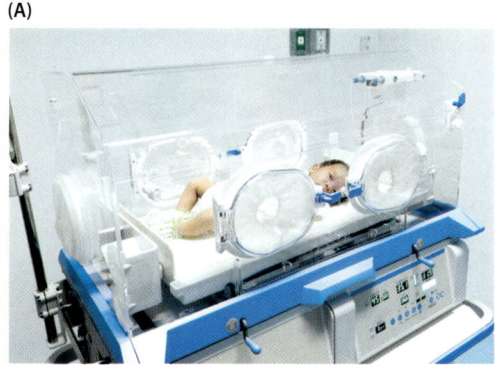

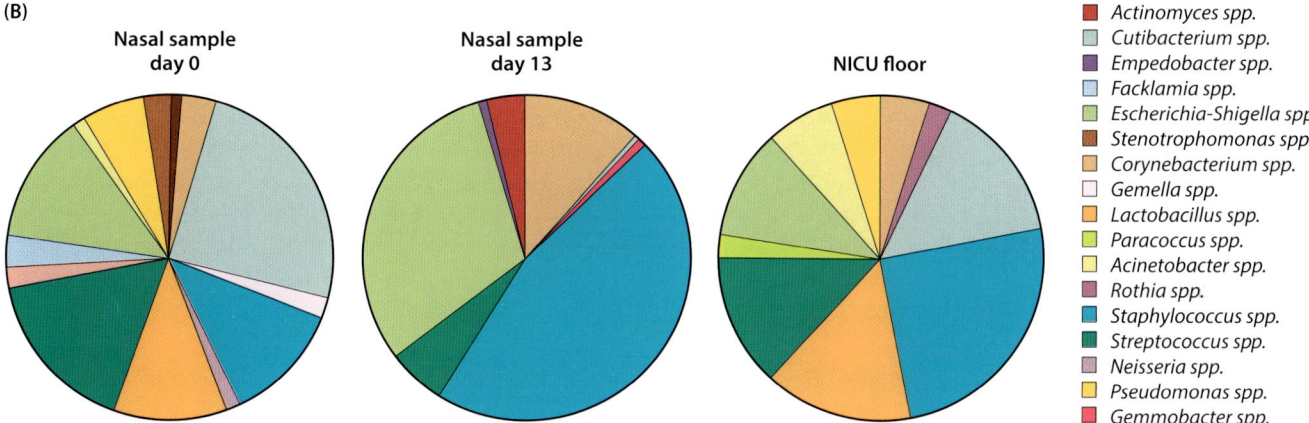

Figure 14.18 The Presumed Sterility of the Neonatal Unit in Hospitals (A) Although this image from a NICU suggests a highly sterile environment, the levels of pathogens in NICUs are often higher than in other hospital locations. (B) Newborn nasal swabs were collected at birth and again at 13 days in the NICU. Comparisons of the microbiota obtained from these samples and from the NICU floor reveal the source of numerous pathogenic microbes that ultimately make their way into the newborn's nasal microbiota. (A photo from Zapp2Photo/Shutterstock; B after Cason et al., 2021.)

as newborns, and we must be cognizant of the fact that changing that dynamic, particularly in the first few weeks of life, may extract a significant cost later in life, in terms of the newborn's long-term health.

The Hospital MoBE

Hospitals provide a special case for the MoBE. Far from being sterile, they harbor a variety of microorganisms, which have had to adapt to the extreme selective forces they experience in constantly sanitized spaces. Not only are there strict cleaning protocols, but hospitals also employ high levels of antibiotic administration and have a constant stream of new pathogens arriving with the patients. The result is a hospital microbiome dominated by human-associated bacteria with a far higher abundance of potential pathogens and far fewer beneficial bacteria. Those microbes that can survive are often better able to utilize the drugs administered and the cleansing solutions as food sources, and they often possess a rich resistome.

The Hospital Microbiome Project was designed to help us to understand how hospital microbiomes are populated. To this end, when the University of Chicago opened a new hospital, daily samples were collected from surfaces, air, staff, and patients. Over 15,000 samples were taken over a period of a year, allowing researchers to understand how its MoBE changed upon opening, admitting patients, and more. It was truly an impressive sampling feat and was one of the most in-depth microbially characterized environments on Earth at that time. As soon as the hospital opened its doors, the microbiota shifted from a predominance of soil bacteria, such as *Acinetobacter* and *Pseudomonas*, to human skin-associated microbes, such as *Staphylococcus* and *Streptococcus*, brought in by patients. Skin-associated microbiota quickly spread everywhere in the hospital: in the obvious places, such as common areas and surfaces in patients' rooms, but also in operating rooms and areas where patients rarely entered. Researchers were able to observe the impact of each individual patient admitted on the hospitals' microbial community, and vice versa. On a patient's first day in the hospital, microbes moved from surfaces in the room to the patient. By the next day, the movement was going in the other direction, with microbes from the patient adding to the microbial diversity of the room's surfaces. Within 24 hours, the patient's microbiome takes over the local hospital space.

Several findings from the Hospital Microbiome Project were quite surprising. The first involved the increase in microbial transmission due to heat and humidity. Over the summer months, staff members shared more bacteria with each other, perhaps because they were not as densely clothed. The second finding occurred when measures of the impact of treatments were examined. When antibiotics were given intravenously or orally, there was no impact on the skin microbiome. However, when a patient received a topical antibiotic, it wiped out the patient's skin microbes. The third finding related to the length of a patient's hospital stay. When patients remained in the hospital for periods of months, some of the more pathogenic microbes in their microbiomes, such as *Staphylococcus aureus* and *S. epidermidis*, acquired genes that could increase their ability to resist antibiotics and promote infection.

The Hospital Resistome

Although humans are similarly strong determinants of the MoBE in the hospital and nonhospital spaces, the difference that makes hospitals unique in their MoBE is that the abundance of potential pathogens is higher, the resistomes are richer, and the transfer of resistance genes is far more likely when antibiotics are being continuously administered to patients. The grave risk is that as levels of antibiotic resistance increase, hospitals have become reservoirs of **multidrug-resistant pathogens** (**MDRP**), representing a threat to patients, newborns, physicians, and visitors but also providing the opportunity for dispersal into the local community. The solution to this major human health challenge is not to continue business as usual. The indiscriminate use of sterilization procedures has resulted in strong selection for microbes that can withstand antibiotics, antiseptics, and cleaning products and, in fact, learn to consume them in some cases. Although some areas in a hospital truly do need to be relatively free of potential pathogens, particularly for immunocompromised patients, there are many other areas in which this need is far less or absent. In these cases, opening windows may provide an avenue to introduce environmental microbes and lower the density of pathogens. Additionally, introducing more plants would boost the diversity of the hospital microbiome. Both are simple solutions that have been proven to positively impact the MoBE. There are also efforts to engage in biocontrol of the hospital indoor microbiome.

14.7 MICROBIAL METABOLOMICS AND THE BE

While we now have a good understanding of how the BE shapes the types of microbes present, this is only part of the picture. More data are required to understand the lives of the microbes present—for example, how do we know whether a community is growing quickly, producing toxins, or struggling to survive? Microbes exist in one of four metabolic states: *growing*, *active*, *dormant*, and *deceased* (**Figure 14.19**). A microbe is growing if it is increasing in numbers and is considered active if it is taking up nutrients, whether dividing or not. In contrast, a microbe is in a dormant state when life processes stop but can be resurrected with appropriate stimuli, such as nutrients or temperature, and it is deceased when it cannot resume growth. These

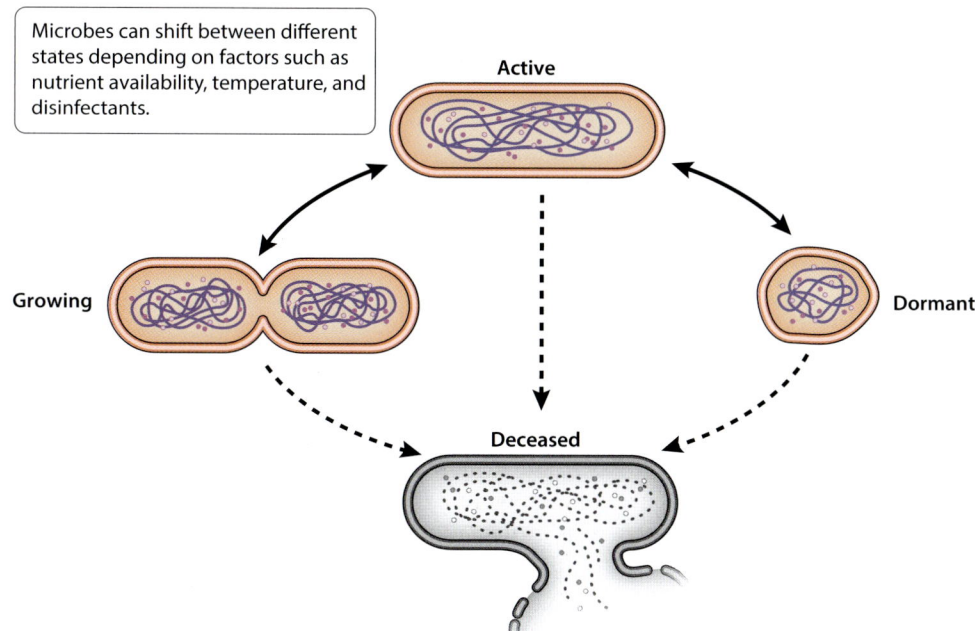

Figure 14.19 Microorganism Metabolic States Microorganisms exist in one of four metabolic states. Viable cells are (left) growing, which involves active dividing; (top) active, which involves metabolizing, but not necessarily dividing; or (right) dormant, in which state cells are not dividing or metabolizing but are not dead. Inviable cells (bottom) are deceased.

states correspond to different degrees of influence that microbes can have on the BE. Employing DNA-based screening methods will inform us about who is present, but not about what they are doing or whether those present are even alive. Assessing their metabolic activity requires the use of metabolomic methods that are not as commonly applied in studies of the MoBE. You may recall from chapters 4 and 5 that many scientists believe a functional profile of the microbiome, as revealed through metabolomics screens, has a greater impact on human health than the species composition.

Metabolites in the BE

Most of our knowledge of the MoBE is based upon census data, which record who is present and where they come from. However, a more limited set of research projects have attempted to determine the impact of the BE on the metabolic activity in the MoBE. In other words, how does the environment of the BE impact not just who is present, but what they do and what metabolites they produce? Under certain conditions, members of the MoBE can be highly active. For example, dust provides a rich substrate to support microbial growth. When moisture is present, fungal and bacterial spores present may germinate, leading to an increase in the production of metabolic products, such as chlorinated hydrocarbons, alcohols, and aldehydes, among several others. One group examined the metagenomic and metabolomic signatures of microbes growing on damp building materials over a 30-day period (**Figure 14.20**) (Lax et al., 2019). Their data revealed that although the bacterial and fungal diversity changed over the course of the experiment, the metabolic output did not, suggesting that regardless of who was growing, certain levels of metabolites were produced. This is a feature of the metabolome that we have encountered in the past. Very few studies have examined both the microbial and chemical signatures of the BE. Not surprisingly, surfaces that are frequently exposed to water, such as showers and sinks, are more metabolically active and chemically distinct than dry surfaces. One study examined the role microbes play in generating the chemical profiles obtained from kitchen sinks and shower stalls (Adams et al., 2017)

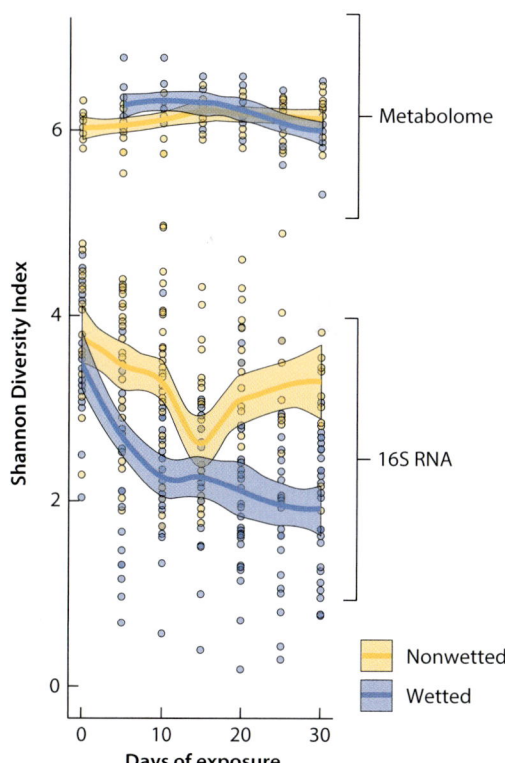

Figure 14.20 Impact of Wetness on Microbial and Metabolic Diversity Estimates Shannon diversity measures are plotted separately for wetted (blue) and nonwetted (yellow) building materials. Points represent individual samples, and the trend lines are given as moving averages of the means, while shaded regions around the trend lines indicate the standard error. Metrics are provided for both the microbiota (16S rRNA gene amplicons) and metabolomes. (After Lax et al., 2019.)

and found a rich microbial community in both locations, with 25% more bacteria than fungi.

Volatile Organic Compounds and the BE

Perhaps most critical with respect to human health are the **microbially produced volatile organic compounds** (**MVOCs**) (Figure 14.21). These are low-molecular-mass chemicals that are easily inhaled and which can negatively impact our metabolic, immune, and endocrine functions. MVOC levels increase in damp areas. As we noted above, some building materials promote microbial growth, and thus they promote production of MVOCs. Additionally, microbes can interact with and alter many chemicals that are not naturally produced and are commonly used in household goods, such as bisphenol A (BPA), used in the production of polycarbonate plastics, and perfluorooctane sulfonic acid (PFOS), commonly used in a wide range of industrial processes and found in many consumer products. These chemicals have been shown to negatively impact endocrine function and growth in humans (Heindel et al., 2017). Environmental microbes can metabolize these and other chemicals, sometimes creating more bioactive or bioavailable chemicals and sometimes reducing their toxicity (Vejdovszky et al., 2017).

The **House Observations of Microbial and Environmental Chemistry Project** focused on sampling a test house to determine the metabolites produced by the MoBE (Farmer et al., 2019). The target home was cleaned with bleach and then sampled to obtain a microbial and chemical census. Volunteers occupied the space and carried out normal household tasks, such as cooking and cleaning. After a month, a second census was taken to determine the impact of human activities. At the start of the study, the indoor microbiome was like that found outdoors. Within a month, the microbiome was dominated by members of the skin and gut microbiota of the volunteers.

Microbes produce a diversity of metabolites that remain poorly studied, particularly with respect to the BE. Not only do they produce their own molecules, but they can also modify those provided by the host or the BE. Over the course of one month, the chemical diversity in the test house dramatically increased, particularly on frequently touched surfaces, in the kitchen and in the bathroom. One example of how molecules can readily be distributed is provided by capsaicin, the molecule that makes peppers spicy. Even though the house was thoroughly cleaned at the start of the project, capsaicin was, nonetheless, already present. It was localized to surfaces in the kitchen. However, an oxidized version of the molecule began to show up on all the surfaces that were touched by the volunteers. Our liver metabolizes capsaicin, and it is likely that after the volunteers ingested capsaicin, it was modified and secreted through their skin. Researchers also found that volunteers' various medications, such

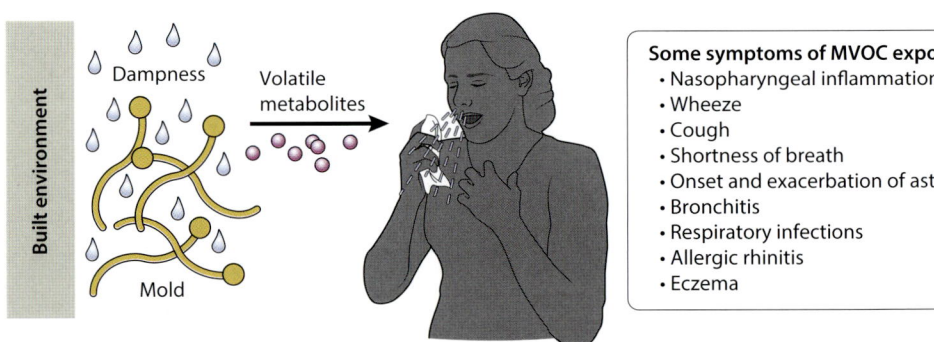

Figure 14.21 Effects of Volatile Microbial Metabolic Products on Human Health Buildings harbor not only the microbiota, but also their metabolites. Temperature, humidity, and light all impact which metabolites are present and in what concentrations. Some of the more troublesome metabolites include alcohols, aldehydes, ketones, and chlorinated hydrocarbons, and there is growing evidence that their presence impacts human health. MVOC = microbially-produced volatile organic compound (After Gilbert and Stephens, 2018.)

as antidepressants and back pain medication, left a chemical presence on the surfaces, where they could impact microbial communities and be chemically altered by microbes. One of the team members said, "One thing that completely blew my mind was the most pronounced trace humans left behind was coffee. Even though coffee was not part of scheduled indoor activities, we found multiple versions of a coffee-derived molecule all around the house, including some that originated from coffee and were then modified by microbe. . . .. What are the health effects of those molecules? We have no idea."

14.8 THE FUTURE MoBE

The world is becoming more urban and more highly populated every day. In part, this transformation from rural to urban living is because urban environments can provide better access to sanitation services, such as bathrooms and waste removal, as well as to more readily available clean drinking water, electricity, and nutritious food. However, city dwellers pay a steep price for those conveniences. The increasingly sterile environments we live in, the increased presence of antibiotics, and more highly processed foods have all been shown to influence what microbes an individual is exposed to, with numerous impacts on our health.

It is predicted that by 2050, 68% of the world's population will live in urban settings. Africa will see the greatest impact, with a whopping 300% increase in urbanization over the next 40 years. One real-life impact of our increasing urbanization was seen with the rapid spread of the SARS-CoV-2 virus. More than ever before, we are realizing that if we wish for a different outcome in future pandemics, and if we hope to decrease the incidence of allergic disease, we must make changes to our BE now. These changes will require input from a truly multidisciplinary team that engages architects, biologists, chemists, and physicians, not to mention social workers, business leaders, and politicians. Many of the issues raised in this chapter required data-driven solutions, which include space configuration, occupant densities, materials selection, window placements, lighting spectra, and ventilation strategies. Let's briefly touch upon what the COVID-19 pandemic taught us about the future of our BEs (Megahed & Ghoneim, 2020).

COVID-19 and the MoBE

Our lives after the COVID pandemic will never be the same. Our values, lifestyles, and habits have all been impacted, and in the future, our buildings will mirror many of these changes. Architects should design our BEs to limit viral spread. This design approach will require decisions based upon biological and chemical data, applying ecology and evolutionary principles, and incorporating the best information about BE structure, materials, and ventilation equipment. In this example, the desired result is buildings and cities that are greener, smarter, and more sustainable while also antiviral in design.

Build Back Better

"Build back better" has taken on a new meaning as it is applied to efforts to reenvision post-pandemic buildings, neighborhoods, and cities (Frumkin, 2021). Architects are calling for innovative HVAC systems to limit viral spread. Ecologists are proposing to green cities by adding plants to roofs and public areas, which will populate the MoBE with a more diverse array of microbes (**Figure 14.22**). Transportation experts are encouraging the use of public transportation instead of cars, which not only is great for limiting toxic emissions, but also permits greater sharing of our microbiomes as we sit in buses and trains. Smarter homes, that is, with internet-based systems to control temperature and the use of appliances and such, may help us improve our health by including mechanisms to monitor health indices, such as blood pressure. Social justice advocates are calling for a path to correct historical injustices in the BE that contribute to health inequities. Maimunah Mohd Sharif, Executive Director of the United Nations

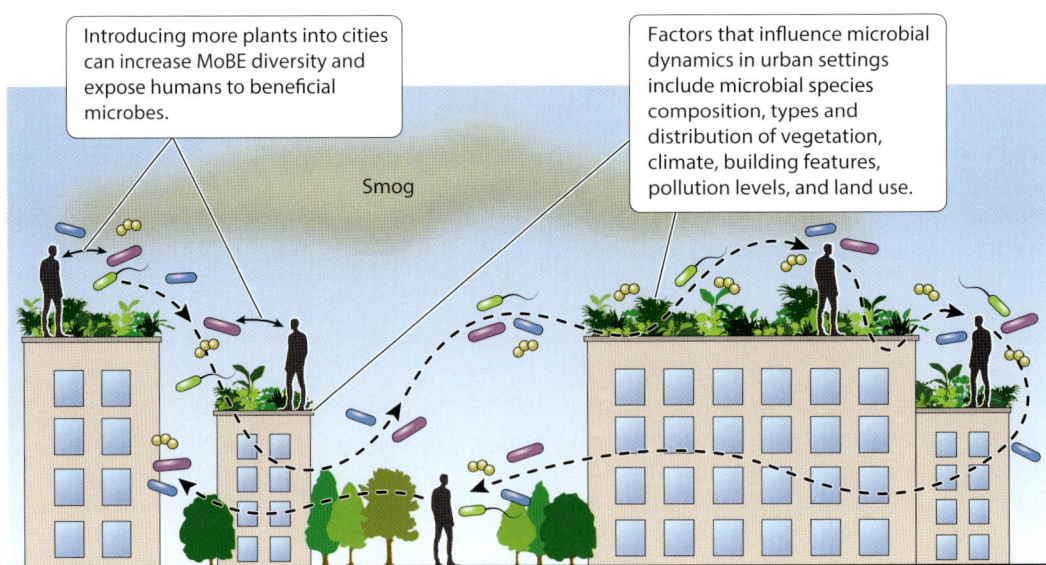

Figure 14.22 Manipulating the MoBE through Green Spaces Attention is being paid to the use of green spaces, such as the rooftop gardens shown here, to increase the abundance of health-promoting microbes in the BE. It is a complicated equation with numerous factors to consider, including which microbial species are selected for and which plants will host them. The dashed arrows represent the movements of microbes. (After Robinson et al., 2018.)

Human Settlements Programme, reflected on the "long-term failure to address fundamental inequalities and guarantee basic human rights. The post-COVID-19 response," he wrote, "will require these failures to be addressed and all urban residents provided with basic services—especially health care and housing—to ensure everyone can live with dignity and be prepared for the next global crisis" (Florida et al., 2020). Neal Gorenflo, an advocate of "shareable cities," argues that post-COVID cities should become "more self-governing, and thus less dependent upon state and national governments, more self-sufficient both financially and materially, more democratic, with more residents engaged in urban governance, and better at cooperating and sharing resources with each other" (Frumkin, 2021). One of the more positive outcomes of this pandemic is that we appear to finally have the will and the opportunity to bring a new way of thinking about the BE, envisioning one that more closely addresses human needs, becomes more self-sustainable, and harnesses technologies that make our buildings and cities healthier and more resilient than they have ever been before. In short, we anticipate nothing less than a revolution in the future of our BE (Tompkins, 2020).

The Rewilding Hypothesis

We learned in chapter 12 about the old friends and biodiversity hypotheses, and how the revisions they offer to the original hygiene hypothesis aid our understanding of the cause of the rapidly increasing levels of allergic disease observed over the past 50 years. As we consider the impact of the MoBE on health, it should come as no surprise that one approach has been to consider "rewilding" our microbiomes. The **microbiome rewilding hypothesis** (Figure 14.23) starts with an acceptance of the old friends and biodiversity hypotheses as explanations of the existing challenge of allergic disease (Mills et al., 2017). The old friends component argues that urbanization has resulted in a loss of our exposure to those microbes we evolved with, both in terms of our inner microbiomes, such as those found in our gut, on our skin, or in our mouths, and the outer microbiomes we were exposed to in the environment, such as those found in the soil, in animals, and on plants. The resulting loss of biodiversity due to urbanization (the biodiversity hypothesis) then resulted in an increase in immune dysregulation. The solution, the rewilding hypothesis, proposes a mechanism to rediscover our old friends by reestablishing the biodiversity that has been lost as we have become an urban species.

Nature-Based Solutions

Nature-based solutions have been proposed to address some of the environmental and societal challenges in efforts to increase diversity in the MoBE in urban areas.

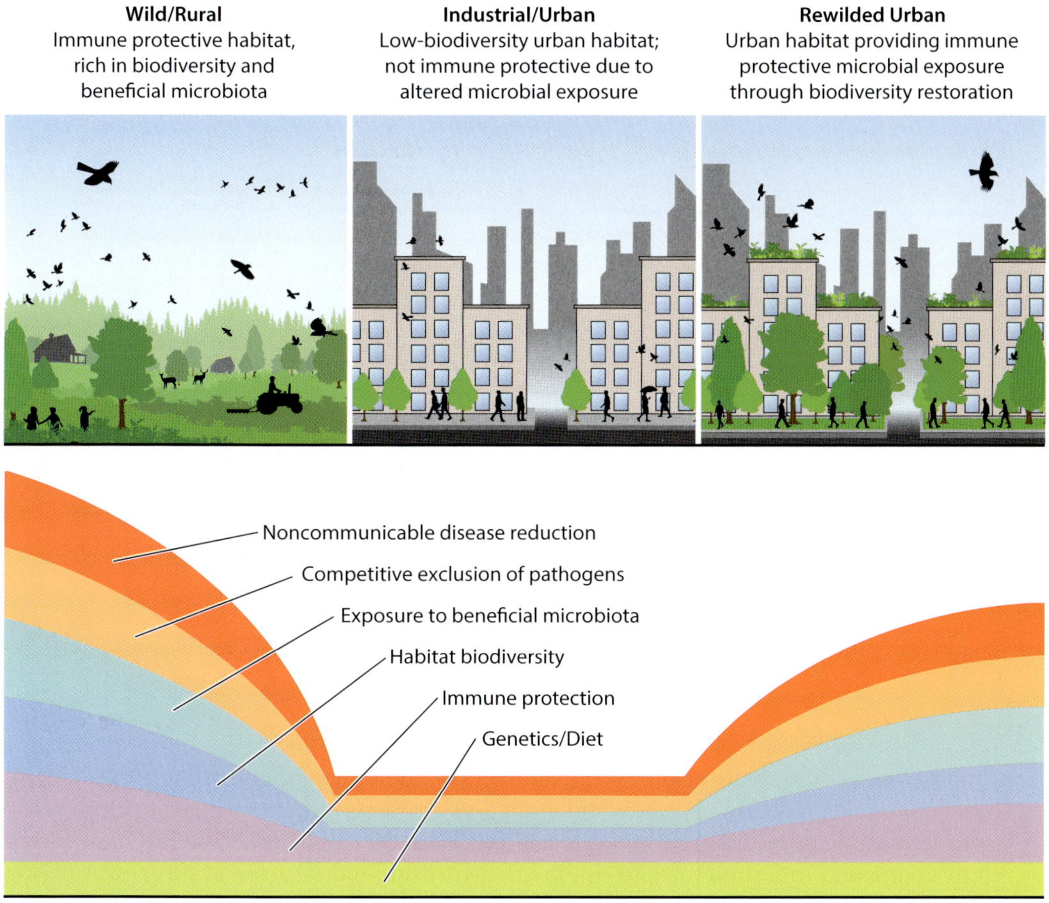

Figure 14.23 The Microbiome Rewilding Hypothesis The rewilding hypothesis argues for the use of green spaces to recover microbiota diversity in the built environment, which will directly impact human health. The thickness of each colored layer represents how much that factor contributes to human health. (After Mills et al., 2019.)

Exposure to plant and soil microbiomes in the indoor and outdoor BE is core to restoring the diversity of beneficial microbes we need to properly train our immune system and reduce immune-based diseases, a concept we've previously explored in chapters 12 and 13. A diverse population of plants brings an extraordinary increase in the microbial diversity to the local environment. Some of the benefits of having plants in the BE include competitive exclusion of pathogens, exposure to beneficial microorganisms, immune protection, and increased habitat biodiversity. **Table 14.1** provides a list of some of the potential impacts of urban regeneration based upon our current knowledge of the MoBE.

Attempts at rewilding urban settings are helping us to re-establish a green infrastructure within the built environment. Natural ecosystems are resilient and provide a range of ecosystem services, even in urban settings, including lining roads with trees and ensuring parks, wetlands, and forests are available. One study revealed that the incorporation of green roofs can result in significant reductions in air pollution (Tomson et al., 2021). Another demonstrated that plant-based restoration efforts can relatively rapidly (in this case, within 10 years) return a natural environmental microbiome to its native state (Gellie et al., 2017).

There remain significant gaps in our knowledge that must be filled before we can consider employing the microbiome rewilding hypothesis as a cornerstone for urban renewal and immune health interventions (**Figure 14.24**) (Mills et al., 2017). Although the positive relationships between human health and exposure to "green spaces" are clear, the mechanisms, both ecological and physiological, that underlie this relationship are not, and more research is needed to understand this complex and

Table 14.1 Main Outcomes and Potential Impacts of an Urban Regeneration Based on Knowledge of the Microbiome of Built Environments

OUTCOMES	POTENTIAL IMPACTS
Microbiome-inspired green infrastructure	Regenerate nature
	Preserve and improve biodiversity
Greener cities with an increased contact with nature, healthier urban environment	Tackle autoimmune and noncommunicable diseases
Smarter mass transit systems	Limit the spread of AMR[b], healthier transit BE
Wise use of antimicrobial products, new approaches in buildings design	Limit the spread of AMR, introduce beneficial microbial communities
Smart technologies to monitor the microbiome of BE and to generate FAIR[a] data	Microbiome perturbations prediction and prevention of disease outbreak
Stakeholders' active engagement	Scientific innovations accepted
Introduction of governance policies to ensure a healthier urban environment	Scientific innovations implemented

[a]findable, accessible, interoperable, and reusable

[b]antimicrobial resistance

Source: Bruno et al., 2022

multidimensional relationship. For example, do different types of urban green spaces, such as manicured lawns versus unkempt lots or restored native plant communities, confer significantly different health benefits? What mechanisms allow the human and environmental microbiomes to interact? Must individuals spend time wandering through parks, or is an open window sufficient? Equally important to understand, does restoration return the environmental microbiome to urban green spaces? Or does the precise microbiome not matter, and simply increased diversity is sufficient to positively impact our health? In addition, some of the benefits of urban greening come with their

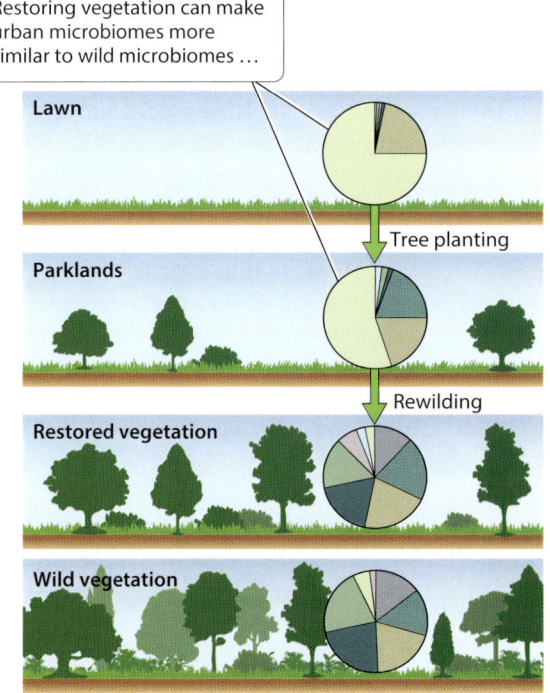
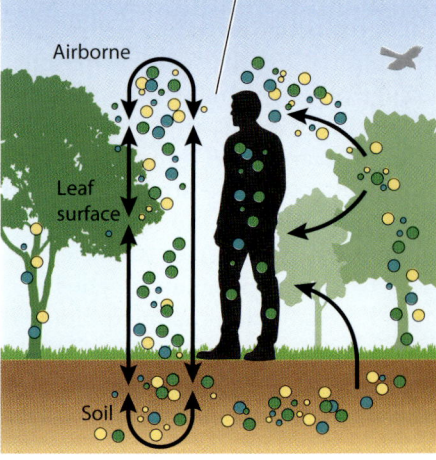

Figure 14.24 Some Key Knowledge Gaps of the Microbiome Rewilding Hypothesis The rewilding hypothesis sounds deceptively simple. Grow plants and the wild microbiota will return and human health will improve. However, there are many unknowns that remain, such as does a greener building actually result in positive impacts to human health? (After Mills et al., 2017.)

own set of challenges. For example, access to urban green spaces results in an increase in physical activity and lower levels of depression. However, spending more time outdoors exposes individuals to allergens that can trigger allergic responses.

Imagine a home, neighborhood, or city designed to maximize human health in the twenty-first century. Biological and social scientists, physicians, and architects must come together to consider truly novel ways to design, construct, and manage buildings and cities in a way that promotes increased microbial diversity and increased exposure of humans to this diversity. This is not really so farfetched. Biologists and architects both look at the world in terms of networks. An architect might ask how people move through and interact with a building, while a biologist considers how microbes move and interact in a particular space. Bringing these differing perspectives together to tackle this enormous challenge may result in truly incredible outcomes. **Figure 14.25** brings together the various factors we have addressed in this chapter that are involved in creating a more healthful built environment. These factors include biological components, which are actions that promote microbial diversity through the presence of pets, open windows, and diverse plantings, coupled with architectural components, designs that limit surface moisture to restrict undesired microbial growth, and ventilation patterns that reduce disease transmission. Design inspired by the natural world will almost certainly play a significant role in the future of the BE. The goal is to create greener and sustainable cities, as we learn from ecologists and environmental experts who provide input on how to enhance the sustainability of city design. Future cities may be designed to better meet challenges related to climate change, as well as societal and economic demands. And yes, much of this change is focused on re-creating a highly diverse microbiome composed of our old microbial friends, who will continue to serve their critical role of helping us to be as healthy as we can be.

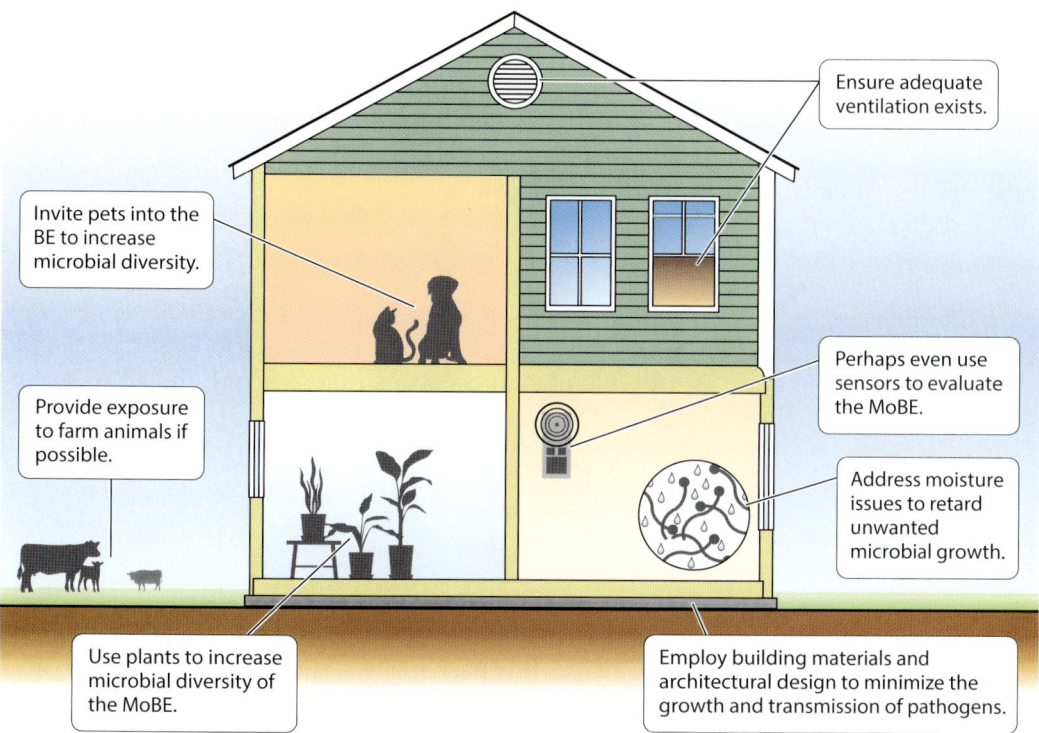

Figure 14.25 Shaping the Indoor Microbiome for Better Health Outcomes Our rapidly expanding knowledge linking exposure to microbial diversity with positive impacts on human health argues that we should find ways to use our indoor microbiomes in a therapeutic fashion. Simple acts, such as introducing pets, reducing surface moisture, and increasing the presence of plants in our homes and offices, will move us towards producing a more healthful environment. (After Gilbert and Stephens, 2018.)

CHECK YOUR UNDERSTANDING

1. In both the built environment of humans in Western societies and the built environment of chimpanzees, the majority of microbes found are from their human or chimpanzee hosts.
   a. True
   b. False

2. Spending more time in a built environment and less time in nature led to an increased risk of acquiring pathogens. An example of this is Puebloan homes, which have been shown to have had high levels of pathogens that cause
   a. MRSA.
   b. scarlet fever.
   c. tuberculosis.
   d. diarrhea

3. Which feature of commercial buildings limits microbial growth?
   a. Particle filtration systems
   b. Air-conditioning
   c. Ventilation systems that circulate air
   d. Common surfaces, such as doorknobs

4. How many bacteria per hour does an average human shed in their "microbial cloud"?
   a. 70,000
   b. 400,000
   c. 10 million
   d. 3 billion

5. Which type of home had the greatest abundance of microbes from humans who lived there?
   a. Town of Iquitos
   b. Brazilian city of Manaus
   c. Hunter-gatherer village in Checherta
   d. Rural village in Puerto Almendras

6. New occupants can introduce microbes that colonize a space within a matter of
   a. minutes.
   b. hours.
   c. days.
   d. weeks.

7. Studies of various parts of the built environment, including public transportation, hospitals, and malls, have found that
   a. most bacteria are from the outdoors.
   b. air vents are the primary way bacteria enter indoor spaces.
   c. humans are the primary source of microbes in these spaces.
   d. pathogen levels are lower in these spaces than in nature.

8. The Home Microbiome Project showed that homes have unique microbial compositions.
   a. True
   b. False

9. The least understood aspect of the BE is
   a. bacteria.
   b. fungi.
   c. resistance genes.
   d. viruses.

10. The above ground microbiome associated with plants is known as the
    a. phyllosphere.
    b. plantbiotica.
    c. soil sphere.
    d. plant-associated microbiota.

11. Which of the following correctly pairs a room and the main elements of its microbial profile?
    a. Bathrooms; gut microbiota
    b. Offices; soil microbes
    c. Living rooms; skin microbes
    d. Kitchens; plant microbes

12. Which of these thrives on damp building materials?
    a. Archaea
    b. *Mycobacterium tuberculosis*
    c. Black mold
    d. Methicillin-resistant *Staphylococcus aureus*

13. Sunlight can
    a. kill many types of bacteria.
    b. limit the growth of fungi.
    c. inactivate viruses.
    d. increase bacterial growth.

14. Plumbing systems can increase the spread of pathogens, such as
    a. *Mycobacterium*.
    b. *Staphylococcus*.
    c. *Streptococcus*.
    d. *Enterococcus*.

15. When multiple people experience health effects associated with time spent in a building without a specific cause, this phenomenon is known as
    a. built environment disorder.
    b. sick building syndrome.
    c. building-related illness.
    d. diseased commercial space.

16. What is a potential explanation for the increased prevalence of asthma seen in Hutterites compared with the Amish?
    a. Diet
    b. Large families
    c. Farming methods
    d. Breastfeeding

17. The first work involving the MoBE involved Dr. Ignaz Semmelweis understanding pathogen transmission from the
    a. infectious disease ward to the rest of the hospital.
    b. autopsy chamber to the maternity ward.
    c. community to the hospital.
    d. emergency room to the pediatric ward.

18. The BE increased the spread of Legionnaires' disease, as it is spread through
    a. air droplets.
    b. ventilation systems.
    c. human contact.
    d. the room furnishings.

19. A hospital room with an open window has fewer pathogens than one ventilated mechanically.
    a. True
    b. False

20. The antibiotic genes present in an environment are known as a
    a. resistance collective.
    b. resistome.
    c. AMR group.
    d. resistibiota.

21. The Hospital Microbiome Project found that the longer a patient stays in the hospital, the more resistance genes the pathogens in their microbiome acquire.
    a. True
    b. False

22. The metabolic states of microbes are
    a. alive and deceased.
    b. active, sporulated, and deceased.
    c. growing, active, dormant, and deceased.
    d. dividing, active, homeostatic, and deceased.

23. Microbially produced _____ are a class of small organic compounds that can be inhaled by humans and have negative health effects not yet fully understood.
    a. polyfluorinated carbons
    b. volatile organic compounds
    c. biopolycarbonates
    d. short-chain fatty acids

24. A change that has been proposed to improve the BE of cities after the COVID pandemic is
    a. less public transportation.
    b. decreased use of HVAC systems.
    c. systems to increase humidity in homes.
    d. increasing the abundance of plants.

Answers: 1B, 2C, 3A, 4C, 5B, 6B, 7C, 8A, 9D, 10A, 11B, 12C, 13C, 14A, 15B, 16C, 17B, 18B, 19A, 20B, 21A, 22C, 23B, 24D

DIVING DEEPER

1. How have the built environments most humans interact with changed over time?
2. What factors allowed infectious diseases like the plague to spread quickly?
3. What are five aspects of building design that can influence the type and abundance of microbes present?
4. In terms of the MoBE, what are the benefits and drawbacks of air ventilation systems found in commercial buildings compared with natural ventilation, such as an open window?
5. Why do hospitals commonly use UV light to sanitize? What are some potential downsides of cleaning in this way?
6. What are four aspects of the room you're working in that affect its MoBE?
7. How do the traditional farming practices of the Amish appear to decrease Amish children's likelihood of developing asthma?
8. How is Legionnaires' disease an example of the built environment affecting human health?
9. What are three ways hospitals could change the built environment to reduce healthcare-associated infections?
10. Summarize the key findings of the House Observations of Microbial and Environmental Chemistry Project.
11. What are three examples of how the BE shapes transmission and risk of certain diseases?
12. What are some of the changes proposed by scientists to improve the MoBE as we reinvest in cities after the COVID pandemic?

DISCUSSING AND REFLECTING

1. Consider the environment of your university. What aspects of your university's design positively impact the MoBE and improve the health of students, and which aspects have a negative impact? What changes would you make to your campus to improve the MoBE?

2. Both in the past and in the modern era, humans have lived in different types of built environments. Compare the built environment you interact with daily to a built environment that is very different from your own, such as Puebloan homes. What are the benefits provided by each BE, and what are the downsides?

3. Reflection. After learning about how the built environment shapes our microbiomes, and vice versa, how do you feel about the impact of your BE on your health and well-being? Does learning about the MoBE change how you view the spaces you interact with, such as your home, your workplace, stores, gyms, hospitals, and more?

RECOMMENDED READINGS

Popular Science Review

Finbow, A. (2019, September 19). How Microbiomes Could Save the Planet. In the former blog network of *Scientific American*. https://blogs.scientificamerican.com/observations/how-microbiomes-could-save-the-planet/

Popular Science Book

Goldhagen, S. W. (2017). *Welcome to Your World: How the Built Environment Shapes Our Lives* (1st ed.). Harper, an imprint of Harper Collins.

Scientific Review

Ciric, L. (2022). Microbes in the Built Environment. *Scientific Reports*, *12*(1), Article 1. https://doi.org/10.1038/s41598-022-12254-w

Taking Charge of Your Microbiome

15

CHAPTER CONTENTS

- **15.1** Is Your Gut Microbiome Healthy?
- **15.2** Seeking Professional Advice
- **15.3** The DIY Approach
- **15.4** Defining a Healthy Microbiome
- **15.5** The Healthy Plate
- **15.6** Microbiome-Based Therapeutics
- **15.7** Microbiome Recovery Following Antibiotic Use
- **15.8** Let Food Be Thy Medicine

Well, hello there! I am *Candida albicans*, a mild-mannered fungus who makes a quiet home in your gut microbiome with my billion, nay, trillion bacterial neighbors. My numbers are generally small, but I play an oversized role in training your immune system. Now, if you disturb my community, say with antibiotics, I instantly transform into a sneaky, shape—shifting troublemaker, ready to cause mayhem. Think of me as an overly dramatic roommate: harmless most of the time, but capable of turning your world upside down if I don't get my way. When pushed, I overgrow and crowd out some of my neighbors. The result is oral thrush, yeast infections, or gut inflammation. So, don't rock my boat if you know what's good for ya! (Photo from iStock.com/iLexx)

> "The single greatest predictor of a healthy gut microbiome is the diversity of plants in one's diet."
> —Will Bulsiewicz (Bulsiewicz, 2020)

We have learned so much about our microbiome in our journey through this textbook. Congratulations for sticking with it! So, now what do we do with all this knowledge? My hope is that you use what you have learned to positively impact your own health, and that of others around you. To this end, we need to translate some of what we have learned into actionable items. And yet, there is so much information to sort through, and much of it leads to statements like "we need more data" from the scientists involved. How can we possibly know how to act on what we have learned? This chapter is devoted to addressing precisely that question, as well as other related questions about how we can best understand and improve our own microbiomes. For example, is your gut microbiome reasonably healthy? Should you have it analyzed? Should you be on a microbiome recovery diet? If you must take an antibiotic, what is the best strategy to limit the drug's impact on your microbiome? There are endless questions that arise, and we will target several of the more pressing ones and find answers supported by science. We will frequently draw from what we have learned in prior chapters and use this knowledge to inform our decisions as we take charge of our own microbiomes.

15.1 IS YOUR GUT MICROBIOME HEALTHY?

Imagine a lovely meadow, rich with grasses, flowers, butterflies, and birds. In fact, you don't have to imagine it; **Figure 15.1A** provides just such a scene. Take a moment to gaze at this image. How do you feel? Perhaps your mind feels a bit calmer, or your heart rate slows. Maybe you wish you could stroll through the meadow and take in the sights, scents, and sounds and feel the earth beneath your feet. Clearly there is something healthful and rich in this meadow image, but what do we actually mean by that? Now look at **Figure 15.1B** and you find the same community, only this time illustrated by a complex series of arrows and connections, also known as an ecological network. As you look more closely at this network, you begin to realize that the number and types of interactions between the plants, insects, birds, amphibians, and mammals, just to name a few of the macroscopic members, are almost unfathomable. A healthy **ecosystem** is one that hosts a high level of biological diversity, along with its **abiotic** components, such that it is resilient in the face of change and can return to a healthy state when perturbed. This is precisely what we mean when we say a gut microbiome is healthy. We mean that it has a rich network of microbial interactions that creates resilience in the face of perturbations, such as a dose of antibiotics or a change in your diet.

We learned in chapter 3 that the gut microbiome plays an inordinately large role in our health, but in particular in our gut health. Our GI system functions by digesting food, absorbing the nutrients released, and expelling the waste products. But how do you know whether it's working well? From a physician's perspective, a healthy gut meets certain criteria (**Table 15.1**), including effectively digesting food and readily

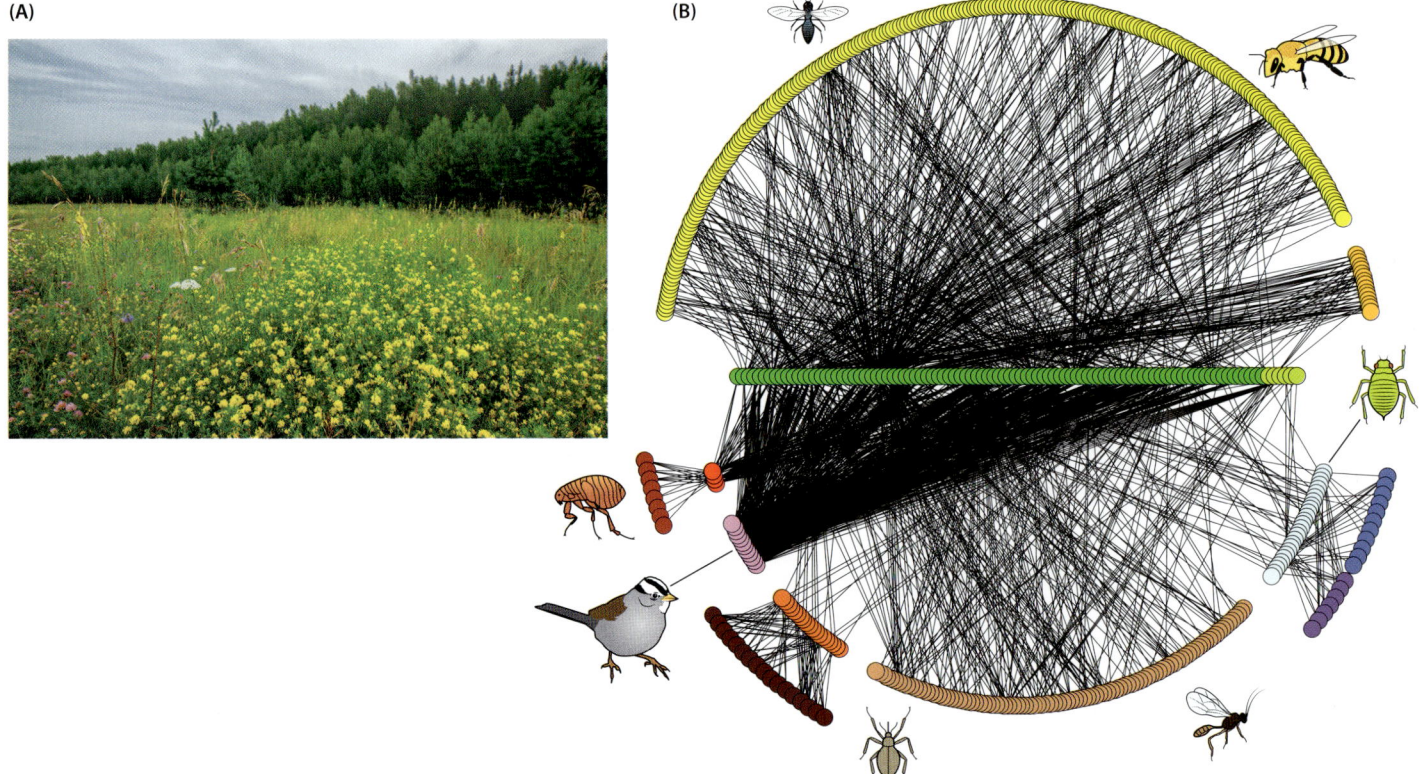

Figure 15.1 Ecological Networks Driving Meadow Diversity (A) A meadow appears deceptively tranquil and static. (B) Layer onto this image the ecological interactions that drive just a fraction of the animal diversity in this meadow, and we realize that the interactions are literally innumerable. (A photo from Alex prokopenko/Shutterstock; B modified from Pocock, et al., 2012. *Science* 335: 973–977. Reprinted with permission from AAAS; after Bohan, et al., 2013. *Adv Ecol Res* 49: 1–67.)

Table 15.1 Five Criteria for a Healthy Gut

CRITERIA	SIGNS OF GASTROINTESTINAL HEALTH
Effective digestion and absorption of food	Normal nutritional status and effective absorption of food, water and minerals
	Regular bowel movement, normal transit time and no abdominal pain
	Normal stool consistency and rare nausea, vomiting, diarrhea, constipating and bloating
Absence of GI illness	No detectable GI illness
Normal and stable intestinal microbiota	Normal composition and vitality of the gut microbiome
Effective immune status	Normal levels of immunoglobulin A, normal numbers and normal activity of immune cells
	No mucosal hypersensitivity
Status of well-being	Normal quality of life

Source: Dr. Jayaram Menon, 2017

absorbing nutrients, hosting a normal and stable intestinal microbiota, and producing effective immune responses, along with the absence of GI tract disease. From an individual's perspective, you might assess your gut health by asking the following questions: Do you feel that what you eat is being digested properly? Do you feel satiated after a meal, or bloated and gaseous? What is the quality and quantity of your poop? The last question is really at the heart of our DIY gut health check. You know your gut is healthy if you have a regular pattern of bowel movements, which may range from three times a day to three times a week. Your gut should be able to move food through its system within about 48 hours, and when fecal matter emerges, it should be smooth in texture, form soft to firm sausage shapes that are passed in a single or few pieces, and then sink in the toilet. Not to get too personal, but I imagine that most of us have room for improvement in this department, and we will return to this topic shortly.

On the flip side, you may have an unhealthy gut microbiota if you experience stomach discomfort, fatigue, food cravings, skin irritation, allergies, or mood issues. I wager that most individuals in the US suffer from one or more of these common conditions at some point in their lives, perhaps even most of the time, especially since, according to the CDC, 6 in 10 US adults have at least one chronic illness. As you learned in prior chapters, there is compelling evidence that a dysbiotic gut microbiome is correlated with these conditions and is very likely partially, if not fully, responsible for all the symptoms mentioned above. As you have seen throughout this book, a healthy gut microbiome is critical for our physical and emotional health. But what do we do if we believe our gut microbiome is out of kilter?

15.2 SEEKING PROFESSIONAL ADVICE

Consulting a Medical Professional

For some, the first thing you might consider is visiting your general practitioner (GP), who has a rich tool kit of tests on hand to explore your gut microbiome health. In addition to discussing your symptoms, they might choose to determine the flow rate of food through your GI tract, assess the quality and quantity of your stool, or discuss the details of your diet and lifestyle. They may also want to identify the composition of your gut microbiota with one or more molecular screens (16S rRNA gene amplicons or metagenomics) or assess the microbial functions present (with metabolomics or proteomics). We explored these technologies in detail in chapters 4 through 6. Once the physician has the results of all these assays in hand, what happens next?

Herein lies the challenge. We have repeatedly stated that there is no such thing as a single "healthy" gut microbiome. In fact, we employ the term *normal*, rather than *healthy*, to make it clear that healthy individuals have a diversity of gut microbiome constituents and functions, and one person's healthy gut microbiome may look very different from another's. Your GP must determine where you fall in this distribution of good (or bad) gut microbiome health. In other words, does your gut microbiome appear normal? And if they note dysbiosis, your doctor must then determine how it is related to the original symptoms that inspired your visit in the first place.

Gut Microbiome Index

Knowledge of who is present in your gut and what they are doing is providing a new power in addressing your health complaints and determining whether your gut microbiome may be involved. To aid in this determination, there are gut health indexes, such as the **Gut Microbiome Health Index** (**GMHI**) (Gupta et al., 2020), which provides a summary statistic that may be useful to a physician in detecting dysbiosis. Briefly, the GMHI is based upon 50 microbial species associated with normal gut ecosystems, which were identified through an analysis of an immense sample of human stool metagenomes from both healthy and nonhealthy individuals (i.e., those with disease or abnormal body weight). A higher GMHI score indicates a higher richness and relative abundance of species associated with health. To determine how effective the GMHI was at assessing a person's microbiome health, it was used in a blind test and proved able to correctly identify whether stool samples had been taken from healthy or nonhealthy individuals 74% of the time (**Box 15.1**). Considering how young this field of science is, and how much variation exists between microbiome samples, that is decent predictive power. It remains challenging to assess what is a healthy gut microbiome, and what is not, but the GMHI is a foundation upon which more accurate models can be built as we collect ever more data.

Let's pause to address a critical concern with the use of a gut health index. The microbiome samples used to generate the index are highly biased, with the majority of data sourced from highly industrialized countries. In industrialized societies, C-section births, hyper-cleanliness, and increased antibiotic use are commonplace and are factors known to disrupt microbiome diversity. While an individual from an industrialized society may be disease-free, it doesn't necessarily mean that their microbiome composition is normal. For example, it has been suggested that beneficial species are becoming extinct in developing countries, owing to increased industrialization (Sonnenburg & Sonnenburg, 2019). One example of our disappearing microbes is provided by *Eubacterium coprostanoligenes*, which is most commonly found in populations living in less-industrialized communities. Sistiaga et al. (2019) determined that *E. coprostanoligenes* can degrade cholesterol in a unique way and prevent it from cycling through the bloodstream (Sistiaga et al., 2019). The resulting metabolite, coprostanol, is then secreted in the feces. Individuals from industrialized societies who do not have the same levels of *E. coprostanoligenes*, and thus do not convert cholesterol as readily, accumulate cholesterol in the bloodstream, which contributes to heart disease. Thus, this bias towards employing microbiome samples from more highly developed countries in calculating a microbiome health index may be hiding even more beneficial microbes rendered extinct by industrialization. At present, there is a dearth of data on the microbiome composition of numerous regions, including South America, Southeast Asia, and Africa. Because our current microbiome datasets do not reflect the world at large, our understanding of the microbiome is limited, and our working knowledge of a normal microbiome will surely evolve over time as researchers work to expand these datasets.

Let's return to our physician, who has a patient who has a GMHI of -3.5 and complains about frequent diarrhea and discomfort in his gut area. The patient has been treated repeatedly with antibiotics to relieve those symptoms, to no avail. Glancing at Figure 15.2, you see that an individual with such a low GMHI score is more likely to harbor disease. The physician prescribes a gut microbiome metagenomic

BOX 15.1. RESEARCH IN ACTION
Gut Microbiome Health Index—A Predictor of Disease Probability

❖ **Hypothesis.** One can use a gut microbiota signature to predict the health status of an individual.

❖ **Methods.** The Gut Microbiome Health Index (GMHI) employs the detection of 50 microbial species, which were identified through a metagenomic analysis of 4,347 human stool samples obtained from healthy and nonhealthy individuals.

❖ **Results. Figure 15.2** provides a graph of the GMHI distribution. The values on the y-axis are GMHI index numbers, which range from just under +6 to just above –6. The distribution of GMHI values is provided separately for healthy and nonhealthy individuals. Although the range of GMHI values is quite broad for both health classes, there are clear, and significant, differences between them. If you were told your GMHI value was 4, you might feel pretty positive about your gut health, as the value of 4 falls well within the healthy gut values. However, if your GMHI value was –2, you might be concerned, as that value is more often found in nonhealthy guts. The resulting metric, when applied to a large metagenomic dataset, was able to predict disease presence in the volunteers who provided their stool samples and completed detailed health surveys. In fact, it was correct in the identification of disease in 73.7% of the individuals employed to calibrate the metric.

❖ **Conclusions.** The GMHI provides a simple metric to quantify the probability of disease in an individual, using a metagenomic analysis obtained from a single gut microbiome sample.

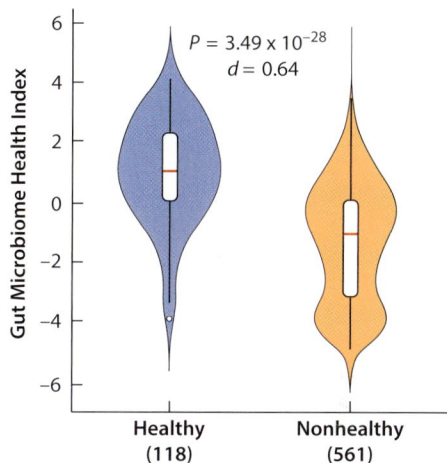

Figure 15.2 The Gut Microbiome Health Index Distribution These violin plots show the distributions of the GMHI in healthy and nonhealthy individuals. The plots summarize the data using six metrics. The box plots (internal white oblongs) provide the minimum, first quartile, median, third quartile, and maximum values of the GMHI. The probability density functions (blue and orange distributions) show the density of each of the GMHI values, with a wider distribution indicating that the values occur more frequently. The two plots show a clear distinction between the healthy and nonhealthy values; however, there is considerable overlap in the values. (After Gupta et al., 2020.)

---

analysis and examines the list of species returned. She notes an overabundance of *Clostridioides difficile*. Knowing that recurrent *C. difficile* infections (CDIs) are a marker of disease and that once dominant in the gut, they are exceedingly hard to eliminate, she might raise the option of fecal microbiota transplantation, which you will recall from chapter 2 refers to the transfer of fecal matter from a healthy donor into the patient. This is a superb option for certain individuals, as it is over 90% successful in knocking back CDI and relieving many of the attendant symptoms. However, it is, currently, the only microbiome-based FDA-approved therapy. Even more challenging, there are, as yet, very few GPs with experience addressing gut microbiome dysbiosis and even fewer trained in interpreting the results from a gut metagenomic analysis. The good news is that this situation is rapidly improving, so there is hope for a plethora of novel microbiome-based treatments in the near future and well-trained GPs to support patients in their journey towards health.

Consulting a Naturopath

Perhaps you would rather not visit a GP and, instead, prefer to turn to a nutritionist or naturopath. One of their first steps might be to suggest those same gut microbiome assays described above. However, they might also explore your food sensitivities or provide instructions for a specialized diet. According to Rosia Parrish, a **naturopathic** doctor based in Boulder, Colorado, "The very first step in healing the gut is to identify and remove the offending foods." Further, once the naturopath has identified the constituents of your microbiota, they will very likely recommend a diet tailored to your needs, which might include probiotics, prebiotics, fish oil, and more. Addressing your lifestyle habits can also help. "Balancing other aspects of health can restore your gut to optimal functioning," says Parrish. "It's amazing how much stress plays a role in digestion, as well as sleep." This option is quite attractive to many, as it is clear that both diet and lifestyle have an enormous impact on your gut microbiome health. But how does the naturopath know which food components provide the best diet for you? Do they simply provide a list of gut-healthy foods and activities and encourage you to follow this generic regime? Unfortunately, this area of medicine is so new that we simply don't have science-supported answers that permit us to create personalized diets, yet. We can certainly offer generic diet and lifestyle advice known to improve the diversity of the gut microbiota in some individuals in industrialized societies. However, we anticipate reaching a point in the near future where we will be able to identify specific prebiotic foods to support your own particular microbiome.

15.3 THE DIY APPROACH

Assessing Stool Quality

Perhaps you don't want to enlist the help of a professional or you have limited access to microbiome-related healthcare. There are several free, at-home options for you to consider. Let's start with one that is incredibly simple and that we introduced at the start of this chapter, examining your own poop. To quickly assess your gut health, you can compare your stool over the course of a few days or weeks with the images in the Bristol stool chart (**Figure 15.3**). This will provide an entry into understanding whether you have good gut health. Although a medical professional is required to interpret your stool, you can at least learn whether you have what is typically interpreted to be stool types that represent good gut health, such as types 3 and 4 in the Bristol chart. For example, if you produce stool of types 6 and 7, this suggests an imbalance in your gut microbiome. Perhaps this is due to a bacterial pathogen that

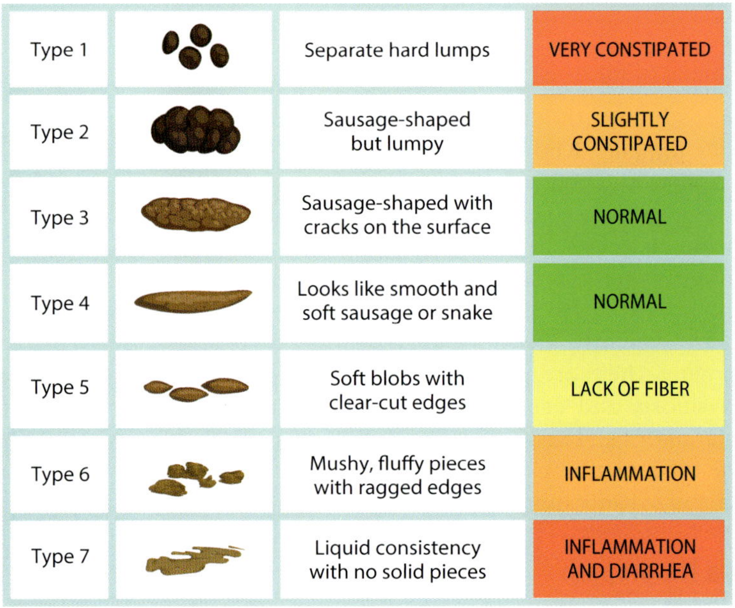

Type 1		Separate hard lumps	VERY CONSTIPATED
Type 2		Sausage-shaped but lumpy	SLIGHTLY CONSTIPATED
Type 3		Sausage-shaped with cracks on the surface	NORMAL
Type 4		Looks like smooth and soft sausage or snake	NORMAL
Type 5		Soft blobs with clear-cut edges	LACK OF FIBER
Type 6		Mushy, fluffy pieces with ragged edges	INFLAMMATION
Type 7		Liquid consistency with no solid pieces	INFLAMMATION AND DIARRHEA

Figure 15.3 Gut Health Check with the Bristol Stool Chart (Image from Juliam.no/Shutterstock)

has invaded your gut and taken up residence, resulting in diarrhea. A type 5 stool might suggest a lack of sufficient fiber to feed your healthful microbes. These are only very rough health indicators, but examining your stool on a regular basis could help you decide whether you might want to pursue a visit with a professional. One benefit of this method is that it puts you in charge. You have chosen to investigate your gut health, you can do so in the privacy of your home, or bathroom, and you can experiment with different food sources and lifestyle choices to assess their impact on your stool quality and your state of health.

Assessing Gut Transit Time

There is a second and equally simple DIY test, in which you measure the gut **transit time** of food you eat. Think of it as a measure of how long it takes food to go from the table to the toilet, which is relevant to health because it is linked to host and microbiota metabolism (**Figure 15.4**). What you eat directly affects the motility of food through your gut. Fiber-rich foods can increase the fecal bulk and increase gut transit times. Your diet also dictates what substrates are available to the gut microbiota. Based upon what you eat, the microbes in your gut produce metabolites, including short-chain fatty acids (SCFAs), secondary bile acids, tryptamine, histamine, H_2, or CH_4, which can stimulate gut motility. Further, the gut transit time impacts the composition and metabolism of the gut microbiota, which impacts the gut environment, particularly pH. In addition, host factors, such as age, gender, and physical activity, impact gut transit time.

Ideally, transit times should be between 12 and 48 hours. If the transit time is too slow, that is, more than 72 hours, you may have an imbalance of gut microbiota, a buildup of toxins, constipation, gas, or an infection. Alternatively, if transit time is too fast, that is, less than 10 hours, food is passing through your GI tract too quickly, suggesting that you may not be absorbing nutrients from your food properly. Fast transit times can be a sign of nutrient deficiency and several other serious medical conditions, such as inflammatory bowel disease (IBD), ulcerative colitis, or celiac or Crohn's disease.

To carry out this test, you choose a food marker, which you won't eat for a week prior to this test. This might be sweet corn, sesame seeds, or red beet root. When you are ready, eat only the marker food, an hour before other foods. Record the time and date. Over the next few days, look for the food marker in your feces, and record the time and date you first see it. Certain foods will naturally move more slowly or quickly through your digestive system. This is because transit time will depend on their fiber and water content, as well what else is consumed that day. Many variables affect transit time, so you will want to take repeated measures and determine an average. Recall our discussions of experimental design in chapter 4; your goal is to

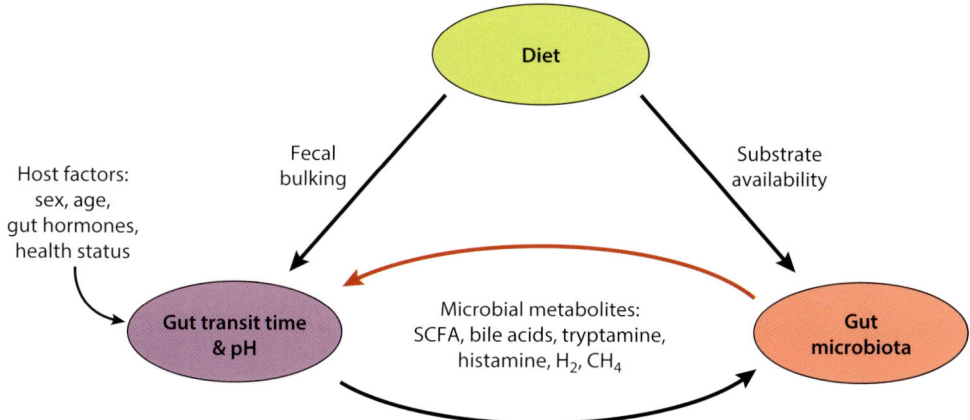

Figure 15.4 Complex Relationship between Gut Microbiome, Diet, and Gut Transit Time Diet and gut microbiota directly impact the gut transit time. (After Procházková et al., 2023)

measure enough food transit times to ensure that you have a robust dataset from which to determine your next step.

Studies of the mechanistic links between transit time and gut microbiome composition are underway. In one such study, the abundances of several key bacterial species were compared between individuals with fast versus slow gut transit times (**Figure 15.5**) (Asnicar et al., 2021). Individuals with faster food transit times tended to have an abundance of Firmicutes, such as *Eubacterium rectale*. Those with longer transit times had a larger abundance of all species except for *E. rectale*, which is a saccharolytic bacterium that breaks down sugars to produce energy. In guts with longer transit times, its abundance may decrease as protein digestion becomes more dominant. Levels of *Akkermansia muciniphila* are higher in guts with longer gut transit times. When there is a long transit time, some bacteria may run out of their preferred carbohydrates and feed on leftovers, which impacts their metabolite production. Even worse, without enough dietary fiber, bacteria feed on the mucus layer of our intestinal cells. This lowers the ability of the mucus layer to provide protection against invading pathogens and increases the risk of DNA mutation and, ultimately, developing colorectal cancer. Knowing your gut transit time and adjusting it through dietary interventions can be a powerful approach to improving gut health.

Assessing Gut Microbiome Composition

Yet another DIY approach involves the use of an at-home gut microbiome health check. Check your latest news feed, and you may very well find this method advertised. There is a buzz in the air, growing ever louder, about these DIY gut microbiome health kits,

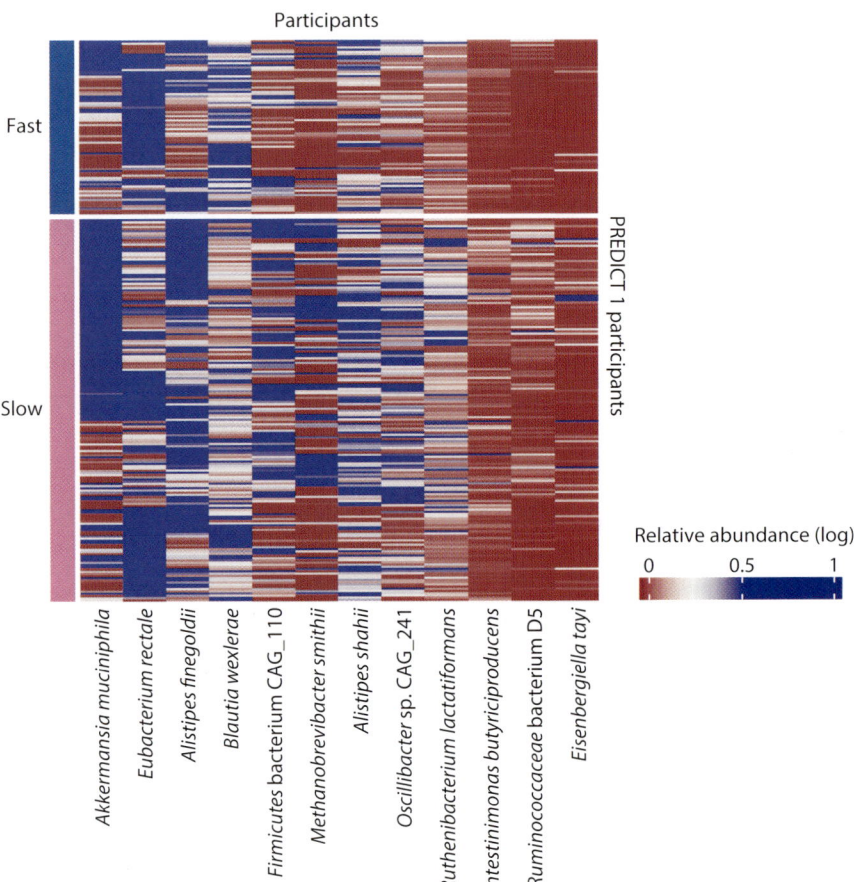

Figure 15.5 Gut Transit Time and Microbiota Associations Abundance of select microbial species in participants with either a fast (top) or slow (bottom) gut transit time. (From F. Asnicar et al., 2021. *Gut*, Sep;70(9):1665–1674. Image Copyright Zoe Limited.)

and the movement is rapidly expanding to include vaginal and oral microbiome tests, and even kits for our pets! The advertisements testify to a rich new frontier where knowledge about who is in your gut, or mouth, or pet, gives you power over your own, or even your pet's, health. I certainly agree that knowledge is power, and knowing who resides in your gut brings you one step closer to understanding whether that ecosystem is healthy. However, my reservation is in what you do with this information once it is provided. But I'm getting ahead of myself. Let's learn about the tests themselves first, and then discuss how we translate this knowledge into actions.

We will focus solely on gut microbiome health tests here. They are simple, are relatively inexpensive, and have a longer commercial track record, having been available since 2012. The procedure involves purchasing a kit; obtaining your sample, usually stool; and sending it to the testing company, who apply their molecular biology magic to produce a gut health report (**Figure 15.6**). Recall that we went over the molecular procedures involved in identifying microbes from a fecal sample in chapter 4. Most commercial kits employ universal 16S rRNA primers to create and sequence prokaryote amplicons in your sample. By including additional primers that anneal to the 18S rRNA, the screen will also identify any fungal members of the microbiome, such as yeast, or intestinal parasites, such as roundworms or tapeworms that might be present.

The advantage of the amplicon approach is that you will obtain a comprehensive list of the taxa and their relative abundance found in your stool at one moment in time. One disadvantage is that, while you learn who is present, you don't know

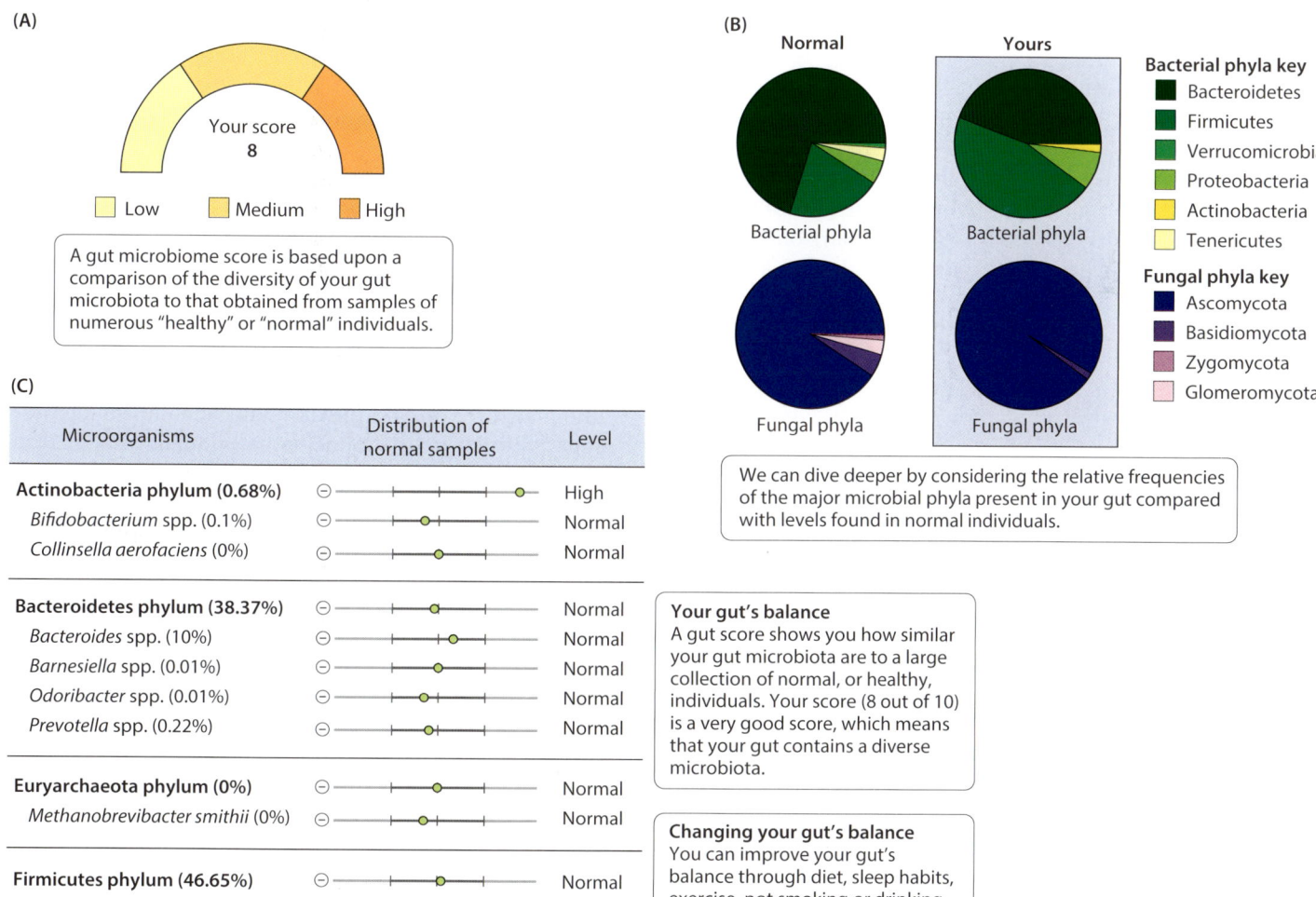

Figure 15.6 A Gut Health Report Summary Score

whether they are alive and, if alive, what they do. To overcome this drawback, and for a significantly higher price, you can opt for a metatranscriptomics assay, in which the RNA present in your stool sample is sequenced. The advantage here is that it identifies which genes are active in your sample, rather than simply who is there. With this type of dataset, we begin to learn something about the functional capacities present. Finally, there are methods that are rarely offered commercially, metabolomics and proteomics, that inform us about the metabolites or proteins being produced by our gut microbes.

These tests should describe the members of your gut microbiota and should provide guidance on what that list of species means in terms of your health. They should inform you about your microbiota's role in a healthy gut microbiome and whether they are species that are harbingers of an unhealthy gut ecosystem. The latter bit is essential, because without context, you're just left with a long list of unpronounceable bacteria names. Useful test results should detail the following:

- the overall microbial species diversity, such as an alpha diversity estimate, which we learned about in chapter 5
- the microbial phyla present and their relative abundance
- the individual microbial species and strains present and their relative abundance

Figure 15.6 provides generic examples of what gut microbiome reports might include. They usually start with an overall gut health score, which is 8 of 10 in the example provided (Figure 15.6A). That score is determined by calculating the species diversity in the submitted stool sample and comparing that with values available from a large collection of individuals. It is a relative measure of the sample's species richness. The person in this example has a reasonable level of species diversity. That's great, but what does it actually mean for this person's health? We generally assume that higher diversity is correlated with better gut health, but is that true for this individual? We will return to these questions shortly.

Let's turn to the next data summary often provided, which is shown in pie charts in Figure 15.6B. Here you find the phylum-level bacterial diversity in this sample, and a comparison to what is determined to be normal levels. In this sample, the individual's phylum-level diversity would be rated as "okay," and by examining the pie charts of their phylum-level diversity, you can see why they didn't ace their gut health exam. In this example, the stool sample had higher numbers of Firmicutes than is usual, while the diversity of fungi present was low. The report then provides a summary of the presence or absence of some of the more critical bacterial taxa found in a normal gut (Figure 15.6C). In this report you find the individual's numbers (colored dots) placed relative to the range for each taxon found in individuals with healthy guts (black bars). In this report, four bacterial phyla are identified, and additional detail is provided about eight genera or species. One phylum (Actinobacteria) is present at levels higher than what is found in stool samples from a healthy person.

The report then provides a set of recommendations to improve your gut microbiome score, such as through changes in diet. These recommendations are quite generic. For example, it is a fact that the typical American should increase their fiber intake and decrease their red meat consumption; one doesn't need a microbiome health check to understand those facts.

The most important part of providing a test is making it actionable. If your test results don't provide new insight into how you can improve your health, then the test isn't useful. Although some features of your microbiome are fixed, others can be impacted by long-term adjustments to your diet and lifestyle. Your gut health report should explain how to do this. What is often suggested in a report is a set of supplement recommendations that offer products made by the manufacturer of the microbiome gut health kit. Why might that bother me? Well, the issue is that you have been sold on purchasing a microbiome gut health kit because it brings the latest science to your fingertips. You are purchasing the molecular magic that occurs behind the curtain. That is fantastic and not something you can achieve on your own, unless you are a microbiome scientist. However, these kits are not advertised as a fun educational opportunity that exposes you to microbiome science; they promise to help you understand and improve your health. Will the recommendations provided in the reports actually help improve your gut health? The proposed outcome is for you to purchase probiotic and prebiotic supplements that are, in most cases, not clinically proven to have an impact on your gut microbiome health. To be sure, several of the companies in this industry are working on generating those clinical data. However, relevant data do not yet exist, meaning that these supplement recommendations are not based upon hard science.

It is important to recognize that even though these kits are advertised as relatively inexpensive, they remain prohibitive for most individuals in the world. The price varies greatly, generally ranging from about fifty dollars to several hundred dollars, depending on the company and service provided, and they are currently not covered by insurance companies. There are efforts to make them more affordable. One such effort involved the creation of a not-for-profit organization, the **American Gut Project**, which provided free gut health checks for individuals willing to have their anonymized data used in scientific studies. That project is currently closed to new submissions, but they continue to fundraise to support the original mission. There should be many more such programs made available so that anyone who wishes to explore this facet of their health is free to do so.

So, if you're fortunate enough to be able to afford this kind of testing, are the results worth the cost? Let's return to the data generated. The microbiome health metric assigned is based upon a comparison to a very large sample of supposedly healthy gut microbiomes. However, as we have repeatedly stated, we do not, as yet, know what a healthy gut microbiome looks like. Think about the meadow we viewed in Figure 15.1. Now step back and think about other meadows you may have seen. I imagine that each of us might have a different vision of what a healthy meadow looks like, a vision that might bear little resemblance to the one shown in Figure 15.1. For example, all of the meadows shown in **Figure 15.7** are considered healthy, and yet

(A)

(B)

(C)

Figure 15.7 Snapshots of Meadow Diversity (A photo from Robin Gyorgyfalvy, Public domain, via Wikimedia Commons; B photo from Auritulus Cinereus, CC BY 2.0, via Wikimedia Commons, C photo from Schwäbin, via Wikimedia Commons)

they look extremely different in terms of their species distributions. The bottom line is that if the right functions are present, regardless of the individual members, then a meadow can be healthy. The same holds true for your gut microbiome. Even if you have a species distribution that is highly skewed relative to the normal gut microbiome, whatever that means, it does not necessarily follow that your gut microbiome is unhealthy, or dysbiotic.

The fact that every microbiome is unique makes it hard to know whether your particular composition and abundances of key species are good or not. For example, too much *C. difficile* is bad, and yet it is often present in the gut. What amount of *C. difficile* is so bad that illness may result? It is also the case that Proteobacteria species are usually considered undesirable. But, again, there is no clear consensus on how much of these species is too much or why one microbiome composition might be better than another. The gut health metric we discussed earlier, the GMHI, is only able to accurately assess a person's health status based on their microbiome sample 74% of the time, meaning that one in four people would be given an incorrect assessment. This metric also uses some of the best testing methods, datasets, and analyses available, and there is no guarantee that commercial tests are equally rigorous and thorough.

Even assuming you received an accurate assessment of your gut health, could you then use that information to positively improve your well-being? Many microbiome scientists might argue that the data provided by these gut health test kits are of limited value in terms of providing the specific knowledge required to fundamentally alter your health. In part, this is due to the fact that microbiome science is still a new field and there is so much more we don't know than what we do know. Let's take your results at face value and use that information to attempt to improve your gut microbiome health score. Using probiotics or changing your diet can alter the diversity of your gut microbiome and how it functions, but the responses to a particular lifestyle change will be specific to you. Unique gut microbial communities, coupled with person-specific genetic, health, diet, lifestyle, and other factors, make it virtually impossible to prescribe a universal solution.

Knowing all the caveats associated with gut health testing kits, you might fairly ask, Should anyone bother getting a gut microbiome test at all? For some, the answer may be that they are excited to be contributing their own body's data as we fill in the details of our understanding of the microbiome. In short, they crave being a part of science in a direct and tangible way. Others may want to track their microbiomes over time to assess changes they intend to make in their diet or lifestyle. There is power in following how your microbiome responds to changes, regardless of where it starts out. You may choose to think of a gut microbiome test as a personalized experiment. We do not yet understand what microbiome changes mean or how to "manage" our microbial partners, but surely learning who is present is a great first step. According to one microbiome expert, Dr. Rashmi Sinha at the National Cancer Institute, "you won't learn much about your health from the test because the field is still in its infancy.... It's not ready for prime time." Another researcher is even more pessimistic: "You'll get an enormous amount of data that is basically uninterpretable, though there are people who will be very happy to take your money and tell you they can interpret it." In short, a microbiome testing kit may not provide you with specific, evidence-based ways to improve your gut health, but if you can afford it, you may find the results interesting or informative. The decision, ultimately, is yours.

15.4 DEFINING A HEALTHY MICROBIOME

Most microbiome researchers would agree that high species diversity and balance among major bacterial phyla are the strongest indicators of microbiome health. This is because both of those conditions increase the stability of the gut ecosystem, which allows it to bounce back from disruptions, such as stress or malnutrition. A diverse microbiome also possesses a more varied range of functions, as more microbes means more specialization. Recall the analogy we raised in chapter 3, in which we envisioned a microbiome as a baseball team. In every team there is a pitcher, catcher, shortstop, and the like. But the individuals assuming each of these roles on different teams will be unique. The same holds true for your microbiome; you need fiber-digesting microbes, but there are numerous species that can get that job done. So, what are the key positions that must be filled for a healthy microbiome? A baseball team requires a minimum of nine players to cover the field. How many players are needed for a healthy gut? **Figure 15.8** provides a list of the genera of a normal human GI tract microbiome, based upon 20 years of research (Ruan et al., 2020). **Figure 15.9** shows the correlation between select species and a suite of health indicators in the gut. These combined images provide a way to begin to tease out which genera are involved in certain factors of gut health. If you were to take a microbiome gut health test, you could compare your results directly with these lists of taxa to determine how many your gut harbors, which is an indication of the health of your gut microbiome.

Fiber-Fermenting Bacteria

Another feature of the microbiome that we can all agree is critical is the presence of bacteria that can ferment fiber that our bodies cannot digest on their own. This is not just for the very obvious reason that we can extract more energy from our food with bacterial fermentation, although that is true, but rather because the metabolites generated during this process are like the baseball commissioner's trophy of the microbial world series. These bacteria belch out short-chain fatty acids as by-products,

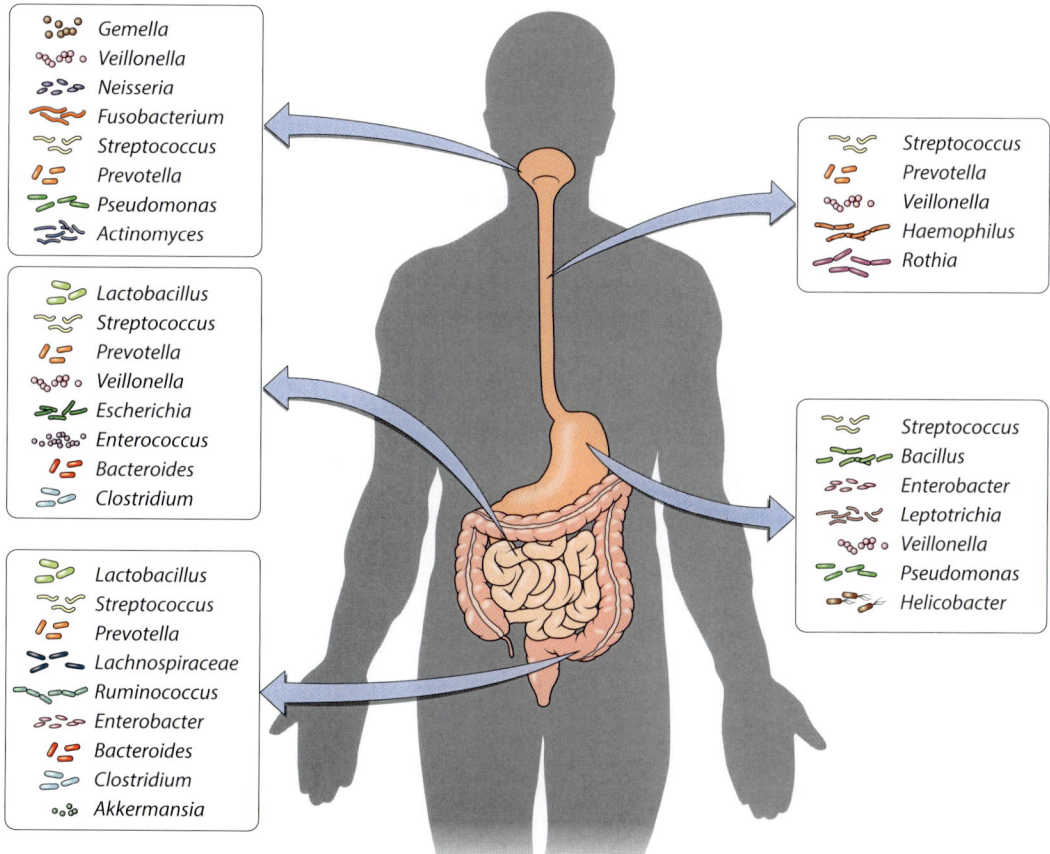

Figure 15.8 Human GI Tract Microbiome Composition in a Healthy Individual The microbiota composition varies along the GI tract. The more-abundant taxa are indicated for the mouth, esophagus, stomach, and small and large intestines. (After Ruan et al., 2020.)

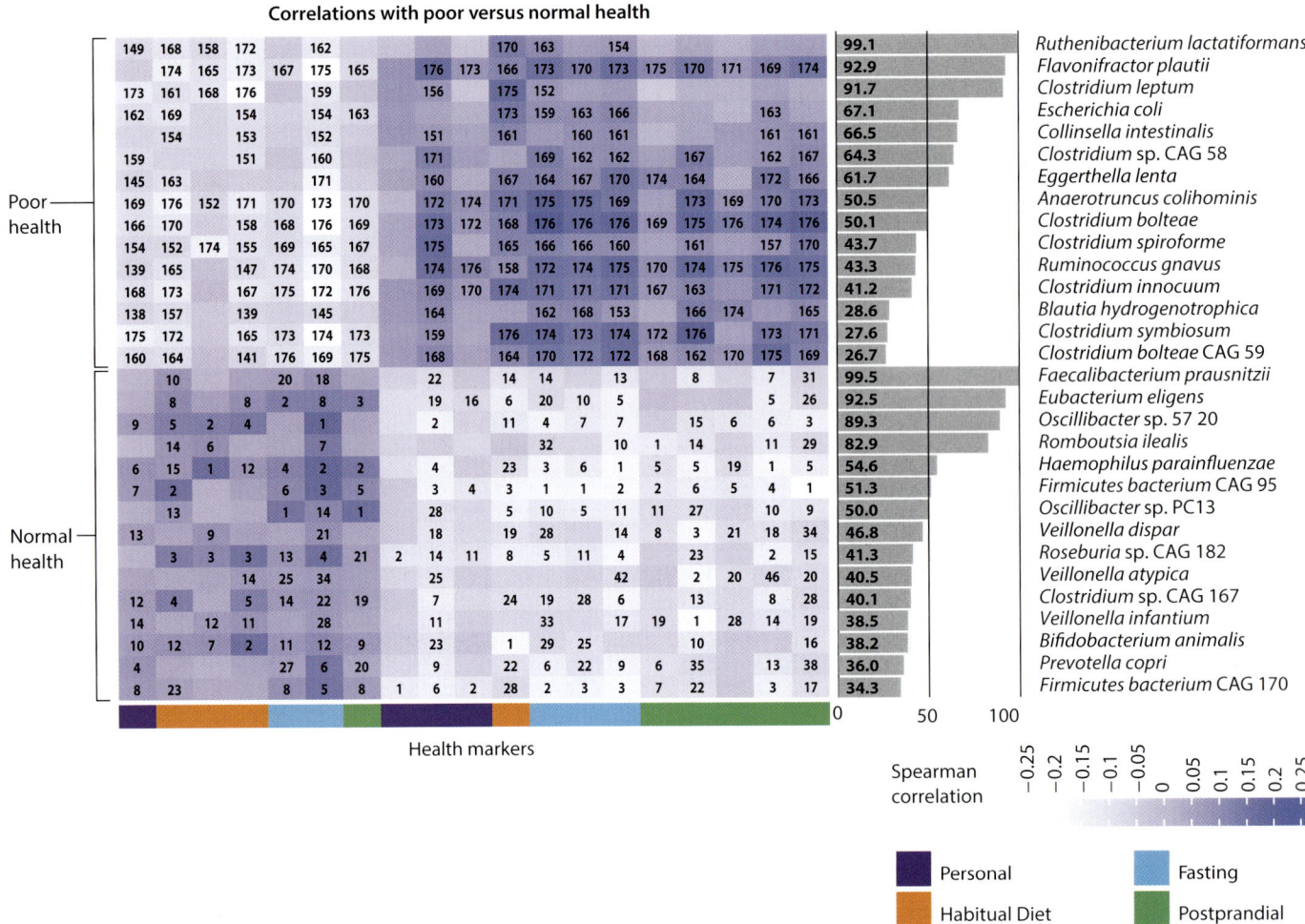

Figure 15.9 Correlations between Gut Microbiota and Markers of Nutritional and Cardiometabolic Health A set of 30 microbial species were strongly linked with 19 markers of nutritional and cardiometabolic health. The species are listed along the right margin and include 15 associated with poor health (top) and 15 associated with normal health (bottom). For each species, Spearman correlation metrics are provided. These metrics indicate the strength of the correlation between the species and each of the dietary/health markers indicated along the bottom axis and representing the following: personal—food indexes reflecting different levels of healthy diets; habitual diet—indicators of cardiometabolic health; fasting—circulating metabolites connected with cardiometabolic risk; and postprandial—metabolic markers occurring after a meal. Positive correlations are indicated in light shading, while negative correlations are noted in darker shading. (From Asnicar, F., Berry, S. E., Valdes, A. M., et al., 2021. Microbiome connections with host metabolism and habitual diet from 1,098 deeply phenotyped individuals. *Nat Med* 27, 321–332, with permission from Springer Nature. Image Copyright Zoe Limited.)

including the now-famous three we have repeatedly mentioned in prior chapters: butyrate, propionate, and acetate. These compounds fuel the epithelial cells lining the large intestine, reinforce the intestinal wall, and prevent inflammation. Butyrate, in particular, is a key metabolite of the gut microbiome that is considered a strong indicator of microbiome health, as it offers protection against inflammatory disease. The main butyrate-producing bacteria in the human gut are *Faecalibacterium prausnitzii*, *Clostridium leptum*, *Eubacterium rectale*, and *Roseburia* spp. A microbiome health test will be able to tell you whether these key butyrate-producing bacteria are present in your gut—a valuable indication of gut health.

It is critical to note that the gut microbiome health checks we described previously represent just one snapshot of this ecosystem. To truly begin to understand the ecology of our gut microbiome requires repeated sampling of our feces, our gut transit times, and our gut microbiota. Even better, we might wish to sample before and after some treatment, such as a change in our diet, lifestyle, or medication. Of course, determining the precise microbiota is preferred, but the commercial cost of such an experiment remains prohibitive. More important, you likely would still not have enough data to make sense of what you find. This is not due to a failure of the technology or the scientists behind it, but rather due to the complexity of ecological studies and the number of players (in this case microbes) involved.

15.5 THE HEALTHY PLATE

Our DIY investigations have informed us either that our microbiome is in great shape or that it needs improvement. Diet is one area over which we have a reasonable level of control, and it makes sense to do whatever possible to ensure that we are feeding our gut microbiomes the best fuel available. There are untold numbers of microbiome-friendly diets; however, the key is that we make decisions that work for us and our microbiome and that we can maintain over the long haul. One source of information on creating a microbiome-friendly diet is shown in **Figure 15.10**, Harvard's Healthy Eating Plate tool. The information provided in this tool and at the associated website (https://www.hsph.harvard.edu/nutritionsource/healthy-eating-plate/) is straightforward, is based upon the very best scientific knowledge, and can be readily adapted to meet our specific dietary needs.

If you choose to pursue a microbiome-healthy diet, you should consider addressing the following goals as you construct it.

1. *Include a diverse range of foods.* Each of the hundreds of species in your gut microbiome has a specific role in that ecosystem and requires different nutrients for growth. A diet consisting of a diverse plate of food types can lead to a more

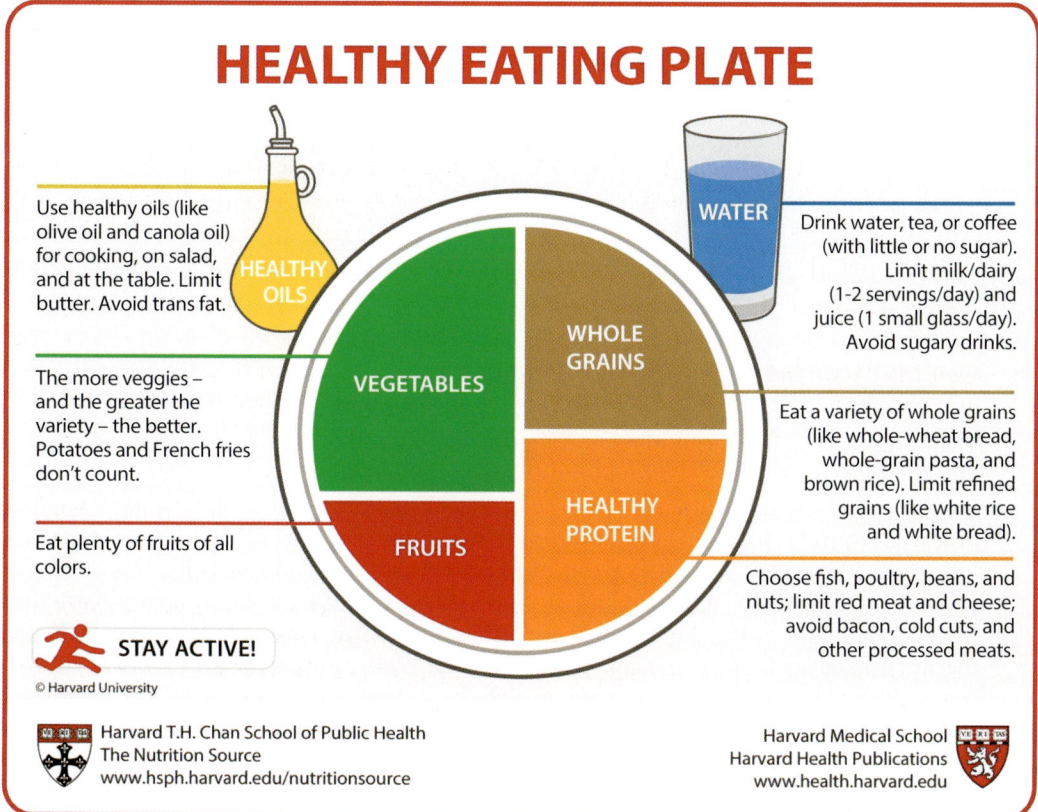

Figure 15.10 Harvard's Healthy Eating Plate (Copyright © 2011, Harvard University)

species-diverse microbiome. However, if you follow a Western diet, which often obtains nutrients from only 15 plants and 5 animal species, you are limiting that diversity from the start. The good news is that you can easily expand your dietary horizon and have fun with novel food choices while supporting your microbiome's health, and your own. Check out Harvard's Healthy Eating Plate list of the healthiest food sources and start experimenting!

2. *Include a diversity and large portions of vegetables, legumes, and fruits.* These ingredients are superb sources of nutrients for a healthy microbiome. They are high in resistant starches and your body can't digest them, so they are intact when they reach your large intestine, where they feed beneficial bacteria. Fiber-rich foods that support a healthy microbiome include artichokes, broccoli, peas, lentils, whole grains, unripened bananas, and apples. These are the energy source of choice for *Bifidobacterium*, which, you will recall, is one of the more highly beneficial bacterial taxa.

3. *Include fermented foods.* Fermentation means that the sugars in these foods have been broken down by microbes. Some examples are yogurt, kimchi, sauerkraut, kombucha, and tempeh. Those who consume yogurt on a regular basis have more lactobacilli and less Enterobacteriaceae in their gut microbiome, while kimchi promotes the growth of beneficial commensal bacteria, such as Bifidobacterium and Lactobacillus, while decreasing levels of harmful species.

4. *Include prebiotic foods.* Prebiotics are foods that are made up of fiber and complex carbohydrates that the host cannot digest. They serve as food for the beneficial bacteria in the gut. The highest levels of fibers are found in raw garlic, onions, asparagus, dandelion greens, bananas, whole grains, beans, and seaweed. Studies have shown that prebiotics can promote the growth of beneficial bacteria, such as *Bifidobacterium*. You get prebiotics from your food or using supplements; however, it is not yet clear whether supplements provide the same beneficial impact on your microbiome.

5. *Include probiotic foods.* Probiotic foods contain live bacteria that can impact the gut microbiome. They include kefir, yogurt, tempeh, kombucha, kimchi, and miso. Again, there are supplements you might purchase, but you can also include probiotics in your diet by simply choosing to include fermented foods.

6. *Include polyphenols.* Polyphenols are compounds in plants that include flavonoids, phenolic acid, and polyphenolic amides. They appear to improve digestion and protect against heart disease and certain cancers. These compounds are not absorbed by the small intestine, so they end up in the large intestine, where they are used as energy sources by some of our beneficial bacteria. Examples of foods rich in polyphenols include dark chocolate, red wine, grape skins, blueberries, almonds, and onions.

15.6 MICROBIOME-BASED THERAPEUTICS

Coupling our increasing understanding of the gut microbiome with the application of ecological principles is providing exciting opportunities for the development of microbiome-based therapeutics. There is a well-justified hope that we will rapidly progress from the sole current microbiome-based therapy, fecal microbiota transplantation (FMT), to the administration of precisely defined and clinically validated microbes and consortia, which refers to two or more species living in symbiosis, that protect against disease. **Figure 15.11** identifies the major types of microbiome-based therapeutics under investigation, which include diet, prebiotics, probiotics, synbiotics, antibiotics, microbiota-derived metabolites and proteins, and FMT (Gulliver et al., 2022). Most of these types of therapies are so new that there are, as yet, no products or therapies available to the public. However, there are three therapeutic approaches—probiotics, prebiotics, and synbiotics—that are commercially available, and the public can make informed decisions about if and when to incorporate them into their microbiome management efforts.

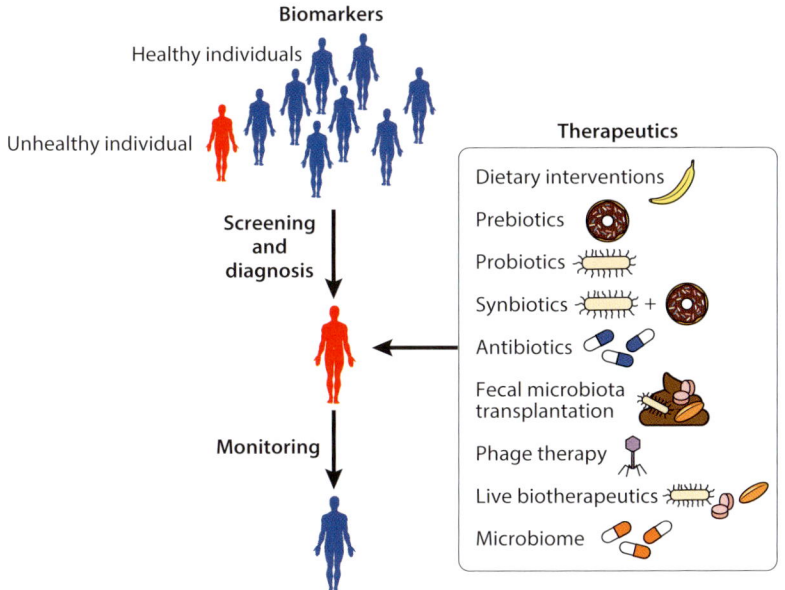

Figure 15.11 The Future of Microbiome-Based Therapeutics Overview of the different uses of the microbiome for medicine, including use as biomarkers (on the left), where patients are screened and monitored, and nine current forms of therapeutics (on the right). *Live biotherapeutics* refers to a recently introduced category of probiotics that have a defined therapeutic action. These probiotics come with health claims that must be approved by the FDA for use in the US. (After Gulliver et al., 2022.)

Probiotics

We have mentioned probiotics repeatedly in this textbook, so you are already familiar with them, and perhaps you are one of the 4 million individuals in the US who take them as a dietary supplement. According to the World Health Organization (WHO), **probiotics** are live microorganisms (mainly bacteria and yeast) that can confer health benefits when consumed in adequate amounts and that are available in probiotic-containing foods as well as supplements. What sets probiotic supplements apart from foods are the claims that manufacturers make about their impact on your gut microbiome in both healthy and impaired states. Whether you believe the health claims or not, they are the drivers in a multibillion-dollar industry, with sales projected to exceed $65 billion soon. However, it remains challenging for the consumer to make sound judgments on whether to consume probiotics and, if so, which ones to choose. Let's focus on probiotic supplements, since probiotic foods have been addressed in the previous section, The Healthy Plate.

Dr. Allan Walker, professor at Harvard Medical School, notes that studies about the efficacy of probiotics are conflicting. However, he feels there are situations where probiotics may be useful. "Probiotics can be most effective at both ends of the age spectrum, because that's when your microbes aren't as robust as they normally are," Walker explains. "You can influence this huge bacterial colonization process more effectively with probiotics during these periods." However, he argues, "If you're dealing with a healthy adult or older child who isn't on antibiotics, I don't think giving a probiotic is going to be that effective in generally helping their health."

The study of probiotics is actually an ancient field of science, one that starts with the acknowledgment by ancient Romans and Greeks that certain fermented foods have beneficial effects on health. However, it was not until the early 1900s that scientific research into probiotics truly began. A Polish physician, Dr. Józef Brudziński, carried out a study in which he created a suspension of *Bacillus lactis aërogenes* and successfully treated infants with acute diarrhea (Krawczyk & Banaszkiewicz, 2021). A more well-known early contributor to this field is Élie Metchnikoff, a Russian scientist, who noted that the health of certain Russian peasants was related to their consumption of yogurt containing strains of *Lactobacillus acidophilus*. From these

Table 15.2 Probiotic Selection Criteria

CRITERION	REQUIRED PROPERTIES
Safety	Human or animal origin
	Isolated from GI tract of healthy individuals
	History of safe use
	Precise diagnostic identification (phenotype and genotype traits)
	Absence of data regarding an association with infective disease
Functionality	Competitiveness with respect to the microbiota inhabiting the intestinal ecosystem
	Competitiveness with respect to microbial species inhabiting the intestinal ecosystem (including closely related species)
	Antagonistic activity towards pathogens (e.g., *Helicobacter pylori, Salmonella* spp., *Listeria monocytogenes, Clostridioides difficile*)
Technological usability	Easy production of high biomass amounts and high productivity of cultures
	Viability and stability of the desired properties of probiotic bacteria during the fixing process (freezing, freeze-drying), preparation, and distribution of probiotic products

Sources: Markowiak and Śliżewska 2017, FAO 2002, and EFSA 2005

simple beginnings has emerged a multibillion-dollar industry that permeates every aspect of our conversations about gut microbiome health.

According to the WHO, each probiotic strain must be evaluated for safety, function, and usability (**Table 15.2**). A probiotic must be identified as a strain, not a species, and one with properties that, when it is present in the gut in sufficient numbers and for long enough, has a beneficial effect on the body. Each strain is evaluated for its safety, based upon its origin, a history of safe use, and its antibiotic resistance profile. The strain must possess features that permit it to survive in the competitive environment of the gut microbiome, which might include the ability to survive in the low pH of the stomach, resist the action of enzymes and bile salts present in the GI tract, or outcompete pathogens. Finally, probiotic strains must be able to survive the storage and distribution processes. The most common probiotic microorganisms used by humans are listed in **Table 15.3**.

Table 15.3 Probiotic Strains

TYPE *LACTOBACILLUS*	TYPE *BIFIDOBACTERIUM*	OTHER LACTIC ACID BACTERIA	OTHER MICROORGANISMS
L. acidophilus [a,c]	B. adolescentis [a]	Enterococcus faecium [a]	Bacillus clausii [a,c]
L. amylovorus [b,c]	B. animalis [a,c]	Lactococcus lactis [b,c]	E. coli Nissle 1917 [a]
L. casei [a,b,c]	B. bifidum [a]	Streptococcus thermophilus [a,c]	Saccharomyces cerevisiae (boulardi) [a,c]
L. gasseri [a,c]	B. breve [b]		
L. helveticus [a,c]	B. infantis [a,c]		
L. johnsonii [b,c]	B. longum [a,c]		
L. pentosus [b,c]			
L. plantarum [b,c]			
L. reuteri [a,c]			
L. rhamnosus [a,b,c]			

[a] mostly as pharmaceutical products

[b] mostly as food additives

[c] QPS (qualified presumption of safety) microorganisms

Sources: Markowiak and Śliżewska, 2017; EFSA 2011, 2013, and 2017

Probiotics are regulated as foods and must have the **GRAS (Generally Regarded As Safe)** status, which is determined by the **FDA (Food and Drug Administration)**. However, because they are supplements and not drugs, there is no requirement for them to be tested using the gold standard of clinical trials. It also means that the FDA does not regulate the production of these products. This means that a probiotic product may not contain the strain indicated, be in the amounts described on the label, or even be alive. Further, the FDA does not investigate any health benefit claims made on these products.

People choose to use probiotic supplements because they believe they will help them to improve one or more aspects of their health. Several thousand clinical studies have been conducted on probiotics and have yielded a wide range of outcomes with respect to their use as therapeutics in treating disease. A systematic meta-review of many of these studies (Markowiak & Śliżewska, 2017) reports the following conclusions:

1. Certain probiotic strains have been shown to be effective in inducing remission of ulcerative colitis.
2. Probiotic strains have been used successfully in the treatment of lactose intolerance and irritable bowel syndrome and the prevention of colon cancer.
3. There are strains that have an impact on the urogenital system, preventing and treating urinary tract infections and bacterial vaginosis.
4. Probiotic strains have been shown to have a positive impact on blood cholesterol levels and are linked to the prevention of obesity, diabetes, and cardiovascular diseases.
5. Numerous studies have shown a positive impact of several probiotic strains in the treatment of a variety of diarrheal diseases.
6. Probiotic strains have been shown to arrest some viral infections, presumably due to their impact on increasing levels of IgA antibodies.
7. The regular use of a probiotic strain can result in the reduction of several types of respiratory tract infections.
8. Probiotics have been shown to reduce the levels of Enterobacteriaceae in the guts of yogurt consumers.

You will notice that all the clinical study results listed above focused on the use of probiotics in already dysbiotic gut microbiomes. However, these microbiomes are usually already depleted of key microbes, which may provide the probiotic strain with a selective advantage such that it can invade and remain in the gut long enough for its beneficial activity to emerge. Just as Professor Walker noted above, studies of the use of probiotics in individuals who have healthy gut microbiomes suggest that probiotics have little or no effect. They are unable to invade the competitive microbiome environment. So, depending on your starting point, in terms of gut microbiome health, you should consider whether probiotics may provide you with a therapeutic option. If you choose to pursue this option, it is highly challenging to identify which probiotic strains and what dosages are likely to have the impact on your gut microbiome that you wish. Because of these challenges, search engines are being developed to provide guidance to physicians as they seek appropriate probiotics for their patients. If you take a probiotic or are considering taking one, type the strain name into a search engine to see whether it has been studied in clinical trials. As mentioned above, that is the gold standard used by the FDA to approve drugs for use in specific therapeutic applications. The search may help you decide whether a particular probiotic strain is right for you.

There is little evidence that probiotics cause harm, and certainly they may be beneficial for individuals with specific diseases, but they do not appear to be beneficial for use in maintaining health. Nassim Nicholas Taleb, an essayist and mathematician, provides an appropriate summary, "We ingest probiotics because we don't eat enough 'dirt' anymore." Since there are so many ways for an individual to take in healthy bacteria, including spending more time outdoors and eating a rich variety of foods, you might want to steer away from this industry for now unless you have a medical condition, at least until there is more data about probiotic use in maintaining health.

Prebiotics

We have previously described prebiotics as food ingredients that humans are unable to digest on their own and that promote the growth of beneficial microorganisms in the gut microbiome. These food sources have enormous potential for altering the gut microbiome composition. However, the changes that might occur in a microbiome following prebiotic consumption are not easily predicted prior to their use. Most **prebiotics** are carbohydrates, and their physiological properties determine their potential benefit to the host's microbiome and health. Prebiotics reach the colon intact, where they are fermented by **saccharolytic** bacteria (e.g., *Bifidobacterium*), which are bacteria that are able to ferment, or break down, complex sugars into simple sugars and energy. The production of this energy and the simple sugar metabolites impacts both the composition of the gut microbiota and its metabolic activity.

Fruit, vegetables, and cereals have the potential to provide prebiotics, as they generally possess resistant starches, which are not digested by humans and require fermentation by microbes in our large intestine. Some of the more common prebiotic sources include tomatoes, bananas, berries, garlic, onions, green vegetables, and oats. There are also artificially produced prebiotics, such as lactulose, galactooligosaccharides, fructooligosaccharides, and maltooligosaccharides. It is common knowledge that most people in Westernized societies don't eat enough fiber: Americans consume about 16 grams of fiber per day, while the WHO recommendation is 25 to 38 grams per day. Clearly there is room for improvement—and one that requires a very simple dietary adjustment.

According to Wang (2009), there are several criteria used in the selection of prebiotics for commercial development:

1. They should not be digested by the host's enzymes.

2. They should be fermented by beneficial bacteria.

3. Their fermentation should lead to increases in SCFAs.

4. Their bacterial-growth-stimulating action should be associated with improved health.

5. They must be able to survive until they reach the intestine.

For some, prebiotics represent the most powerful tool at our disposal if we want to support our good gut bacteria. They represent a healthy addition to a regular diet, they do no harm, and they have beneficial properties that can impede the development of a variety of infectious diseases and cancers (Akutko & Stawarski, 2021): they can reduce the prevalence of several types of diarrhea, reduce symptoms associated with IBD, prevent colon cancer, reduce some risk factors for heart disease, promote weight loss, and prevent obesity.

These claims are made based upon several hundred clinical trials that focused on specific health end points, such as allergies, constipation, and IBD (Gibson et al., 2017). It is challenging to relate these outcomes to a specific individual's health needs, because the state of the science is so young and the factors involved—microbiome composition, prebiotic use, duration of use, host health, genetics, and physiology—are complex and interdependent. However, there are so many tasty ways to incorporate prebiotic foods into your diet that, regardless of the specific health benefit you seek, prebiotics will potentially provide key resources to the beneficial bacteria in your gut, help create a more varied and healthful diet, and do no harm!

Synbiotics

Synbiotics are defined as mixtures of probiotics strains and prebiotic substances that provide a benefit to the host (**Figure 15.12**). The prebiotic components not only improve the survival of the probiotics, but also stimulate the proliferation of the resident good microbiota. **Table 15.4** provides several examples of some of the more common synbiotic combinations. Although few of us are likely familiar with synbiotics, commercial claims related to their use include that they can improve digestion and help in the treatment of common GI disease. Further, there are claims that synbiotics may also help to keep your heart healthy, relieve inflammation, and promote weight loss.

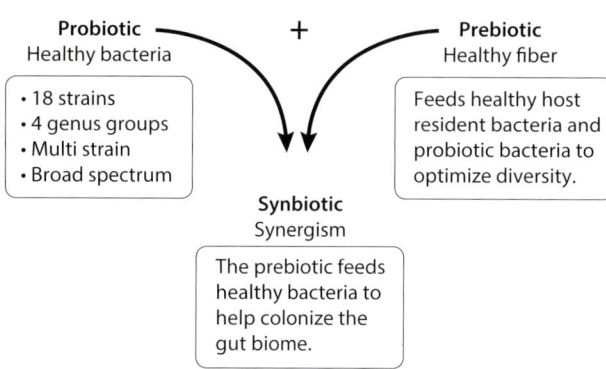

Figure 15.12 Synbiotics Created from Probiotics and Prebiotics

After decades of research on the efficacy of probiotics and prebiotics, the results of clinical trials are still highly conflicting. In part, this is because there are numerous probiotic strains, which cannot be compared across trials and which require their own specific dosing regimens. However, it is also due to the varying ability of probiotic microbes or prebiotic substances to actually effect changes in the GI tract. Many probiotic strains were not originally isolated from humans and may lack the characteristics required to successfully compete in the human gut. One of the primary advantages of the synbiotic concept is that by providing a strain with the substrate that supports its growth, one may be able to circumvent some of these ecological challenges.

One of the most successful synbiotic trials was for a combination of prebiotics and probiotics that could reduce infant sepsis (Panigrahi et al., 2017). The study employed a randomized controlled design and employed over 5,000 participants, some of whom were given an oral synbiotic of *Lactiplantibacillus plantarum* strain

Table 15.4 Examples of Prebiotics and Synbiotics

PREBIOTICS	SYNBIOTICS
FOS[a]	*Lactobacillus* genus bacteria + inulin
GOS[b]	*Lactobacillus*, *Streptococcus* and *Bifidobacterium* genus bacteria + FOS
Inulin	*Lactobacillus*, *Bifidobacterium*, *Enterococcus* genus bacteria + FOS
XOS[c]	*Lactobacillus* and *Bifidobacterium* genus bacteria + oligofructose
Lactitol	*Lactobacillus* and *Bifidobacterium* genus bacteria + inulin
Lactosucrose	
Lactulose	
Soy oligosaccharides	
TOS[d]	

[a] FOS = fructooligosaccharides
[b] GOS = galactooligosaccharides
[c] XOS = xylooligosaccharides
[d] TOS = transgalactooligosaccharides

Sources: Markowiak and Śliżewska, 2017; Crittenden and Playne 2009; Olveira and González-Molero, 2016; Sáez-Lara, et al. 2016

ATCC 202195 and fructooligosaccharides. Levels of sepsis in the treatment group were significantly lower. Unfortunately, the protocol did not include determining the fecal microbiota, so it isn't possible to determine what caused the lower rates of infection. Another example of a successful clinical trial involved an oral synbiotic that include *Lacticaseibacillus rhamnosus* strain GG coupled with arginine. The goal was to combat *Streptococcus mutans*, which causes dental cavities. Arginine is used by commensal oral microbes as a source of energy, resulting in the production of ammonia, which increases the pH, which inhibits the growth of *S. mutans*.

Synbiotics are formulated as either **complementary supplements** or **synergistic supplements**. *Complementary* refers to the combination of probiotics and prebiotics that provides an added health benefit. Most commercial synbiotics are complementary. *Synergistic* refers to the design in which the probiotic strain is stimulated, or its persistence is enhanced, by the prebiotic it is paired with. There are few examples of synergistic synbiotics so far.

It is quite easy to create your own synbiotics. For example, sauerkraut and kimchi prepared with onions and garlic (sources of prebiotics) undergo fermentation that results in an increase in the number of beneficial organisms (sources of probiotics). If you prefer to employ a commercial product, look for a reputable brand that has active cultures with high bacterial counts and a reasonable level of probiotic strain diversity. The product should also contain plant-based fiber to provide food for the probiotics. According to Daniel Ramón Vidal, an expert in synbiotic development, "Synbiotics are an example of the interaction in the human microbiome ecosystem and, by definition, they are more versatile than separate probiotics or prebiotics."

15.7 MICROBIOME RECOVERY FOLLOWING ANTIBIOTIC USE

One of the most highly cited impacts on our microbiome is the use of antibiotics—and for good reason, as over 35% of the US population have used antibiotics directly at some point in their lives, and even more have experienced antibiotics indirectly, such as in fetal exposure during pregnancy or at birth or by ingesting antibiotic residues in foods. We learned in chapter 2 that the use of antibiotics has a significant, negative impact on gut microbiome biodiversity. Further, microbiomes vary in their ability to recover their former diversity. Can we use microbiome-based therapies to help in the recovery of the gut microbiome following antibiotic-mediated disruption?

Chng et al. (2020) report on an analysis of clinical trials focused on the recovery of gut microbiomes following antibiotic treatment. In this meta-analysis, five cohorts of individuals were examined, and in each cohort, individuals were identified as either "recoverers," those whose guts rebounded following antibiotic use, or "nonrecoverers" whose microbiomes did not recover readily. The team identified 21 bacterial species positively correlated with recoverer microbiomes and labeled these as recovery-associated bacteria (RABs). RABs tend to specialize in degrading fiber and complex carbohydrates, which suggests that they are **keystone species** in the gut microbiome and that they promote rapid recovery of both ecological diversity and microbial biomass following antibiotics. Chng and colleagues noted that the recovery of a microbiome from antibiotic disruption appears to be contingent on the presence of a set of beneficial microbes and does not seem to be inhibited by the presence of nonbeneficial bacteria.

To examine the ecological network involved in recovery, the team looked for co-occurring species. In this analysis they identified taxa as primary, secondary, and ter-

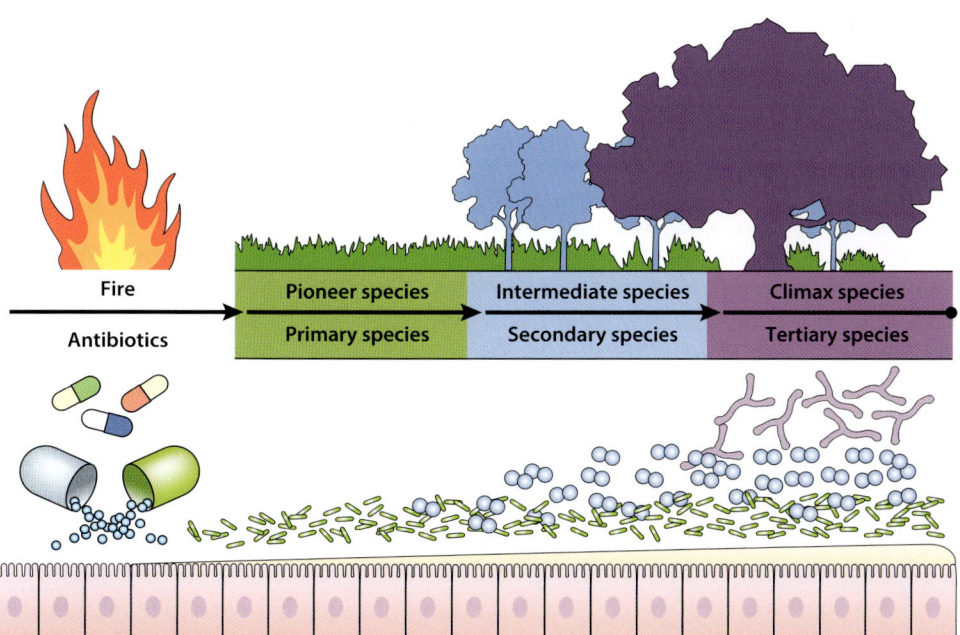

Figure 15.13 Secondary Succession Following an Ecological Disturbance Ecological succession is the process by which the mix of species and habitat in an area changes over time. Here we compare the highly predictable succession that follows a forest fire to the corresponding stages of recovery for a gut microbiota following the use of antibiotics. Just like the plant community after a fire, the gut microbiota shows a transition from primary species (e.g., complex polysaccharide degraders) to secondary species (SCFA producers) and then tertiary species (bacteria that require a recovered ecosystem to thrive). The mucin layer recovery is indicated by the light yellow shading over the epithelial cells. (After Gibbons et al., 2020.)

tiary. Primary, or pioneer, species are essential for the presence of other taxa and are crucial for ecosystem recovery—consider them as similar to keystone species. Secondary species are involved primarily in the production of SCFAs, while tertiary species are dependent upon many other taxa and may be useful in diagnosing the recovery process. RABs were exclusively either primary or tertiary species. Specific secondary species appeared not to be critical in the recovery process. This meta-analysis provides an ecological perspective on antibiotic recovery that appears to follow a well-known ecological process known as **secondary succession**. Secondary succession is the predictable chain of species recovery in an ecological habitat following a dramatic disturbance, for example a fire destroying an old-growth forest (**Figure 15.13**). The fire destroys the community. The first phase of recovery is the appearance of **pioneer** species, which are fast-growing grasses and weeds in our forest example. These species are followed by the invasion of a series of intermediate species, such as shrubs and perennials. The ecosystem recovery culminates in the re-establishment of the climax species, which is the presence of canopy trees in our example.

Chng et al. (2020) argue that the gut microbiome follows a similar ecological recovery process following a disturbance such as the use of antibiotics. Fast-growing, facultative anaerobes are the pioneer species. They are followed by aerotolerant fiber- and mucin-degrading species. The ecosystem has been restored once the dominance of strict anaerobic fiber degraders is regained, along with a species that relies upon them. RABs serve as crucial keystone species and are key for gut microbiome ecosystem recovery (Gibbons, 2020).

15.8 LET FOOD BE THY MEDICINE

This fascinating study seems like an appropriate place to end our investigations in the human microbiome. What we are reminded of is that our microbiomes are complex ecosystems, with species interdependent upon each other's presence and functions in the creation of a healthy ecosystem. The therapeutic applications of this knowledge are immense. In this particular case we learn that there are keystone species whose presence is vital for the natural process of microbiome self-remediation to occur. In our DIY push, we can imagine identifying probiotics and prebiotics to use in creation of our own individualized synbiotics that will hasten the process of gut microbiome dysbiosis recovery. The field is simply too young to provide the scientific details on how to create these synbiotics at this moment, but it is poised to translate the emerging knowledge into personalized, highly effective natural therapies that will help us recover from drug treatments, bolster weakened immune functions, impact our moods, and improve our energy levels.

What I love about the field of microbiome science is that it places the primary power to heal in the hands of the people—unlike our usual system in which we rely on nurses and physicians to heal us. We can make informed choices that directly impact our microbiome and thus indirectly impact our own physical and mental health. We understand that our brain and our gut are in constant communication and that this intimate connection is directly influenced by the foods that we eat—through which we nurture or discourage the beneficial microbes inhabiting our bodies. As the medical community learns more about the biology of these interactions, novel therapeutic options will naturally emerge. In the meantime, you are in the driver's seat, and you can steer your body towards the very best, science-based interventions based upon the diet and lifestyle choices you make. Or, in the words credited to Hippocrates of Kos, a Greek physician born in 460 BC and the founder of modern medicine, "All disease begins in the gut" (**Figure 15.14**). Hippocrates was convinced that food was our most potent medicine, saying, "Let food be thy medicine and medicine be thy food." He went even further and stated what may be so very obvious and yet remains so challenging and key to this conversation: "Before you heal someone, ask him if he's willing to give up the things that make him sick." Are you ready to address the needs of your microbiome? Are you ready to make food and lifestyle choices that we know can make a difference to your health? The rewards are unfathomable, and it is all up to you!

Figure 15.14 Hippocrates and the Microbiome Hippocrates is quoted as saying "all disease begins in the gut," and over 2,000 years later we are beginning to appreciate his sentiment.

CHECK YOUR UNDERSTANDING

1. A health microbiome, much like a health ecosystem,
   a. has as much diversity as possible.
   b. has few interactions between species.
   c. is able to return to a health state when perturbed.
   d. includes many abiotic components.

2. The Gut Microbiome Health Index
   a. is a metric based on 10 key species found in healthy individuals.
   b. can correctly identify a person's health status based on their stool 74% of the time.
   c. ranges from −5 to +5.
   d. uses a lower score to indicate a healthier gut microbiome.

3. Which important species can convert cholesterol into a metabolite that can be removed from the body in feces?
   a. *Bacteroides fragilis*
   b. *Methanobrevibacter smithii*
   c. *Akkermansia muciniphila*
   d. *Eubacterium coprostanoligenes*

4. There is a strong body of evidence that can help individuals find a diet that is personalized for their unique health needs.
   a. True
   b. False

5. On the Bristol stool chart, what type(s) is/are healthy?
   a. Types 3 and 4
   b. Type 6
   c. Types 4 and 5
   d. Type 2

6. A healthy gut transit time is
   a. 36–82 hours.
   b. 48–72 hours.
   c. 12–48 hours.
   d. 3–12 hours.

7. Commercial gut microbiome health tests have been available since 2012.
   a. True
   b. False

8. Most commercial gut microbiome tests use
   a. whole genome sequencing.
   b. proteomics.
   c. LC-MS.
   d. 16S and/or 18S rRNA primers.

9. Some commercial tests may provide metatranscriptomics assays, which can provide you with information about the
   a. whole genome of bacteria present.
   b. functions of your active bacteria.
   c. variation between individual members of a species.
   d. spatial location of bacteria in your GI tract.

10. Ideally, a quality commercial gut microbiome test would provide you with
    a. information about different subspecies.
    b. a holistic look at different parts of your microbiome, such as the oral, skin, and gut microbiomes.
    c. interpretation of the results and context about what they mean for your health.
    d. raw sequencing files so you can perform your own bioinformatics analysis.

11. A limitation of recommendations provided by gut microbiome test kits is that they are often
    a. not evidence based.
    b. generic.
    c. too difficult to implement.
    d. ignored by consumers.

12. What nonprofit offers to help people who want gut microbiome testing but can't afford it, by providing testing in exchange for being able to use the data in scientific studies?
    a. Gut Microbiome for All Initiative
    b. American Gut Project
    c. Microbiota International
    d. Universal Gut Assessment Project

13. Which of the following does not produce butyrate?
    a. *Roseburia* spp.
    b. *Faecalibacterium prausnitzii*
    c. *Bacteroides fragilis*
    d. *Eubacterium rectale*

14. All of the following help the gut microbiome, except for _____, which should be limited.
    a. fermented foods
    b. beans
    c. prebiotic foods
    d. refined grains

15. High levels of polyphenols are found in
    a. red wine.
    b. chicken.
    c. brownies.
    d. cheese.

16. The only FDA-approved microbiome therapy is
    a. probiotics.
    b. prebiotics.
    c. FMT.
    d. synbiotics.

17. According to Dr. Allan Walker, probiotics are the least likely to help
    a. healthy adults.
    b. children.
    c. people on antibiotics.
    d. the elderly.

18. Which of the following could be a probiotic?
    a. Bacteroidetes
    b. *Bacillus lactis aërogenes*
    c. *Methanobrevibacter smithii*
    d. Firmicutes

19. Probiotics are
    a. FDA approved on an individual basis.
    b. regulated like drugs by the FDA, with a schedule rating.
    c. regulated like food by the FDA, with a GRAS rating.
    d. evaluated through clinical trials.

20. Most prebiotics are
    a. proteins.
    b. lipids.
    c. nucleic acids.
    d. carbohydrates.

21. Synbiotics include both probiotic and prebiotic components.
    a. True
    b. False

22. What percentage of Americans have taken antibiotics at least once?
    a. 7%
    b. 15%
    c. 35%
    d. 82%

23. Keystone species that help the microbiome recover after exposure to antibiotics were identified by Chng et al. by
    a. comparing recoverers and nonrecoverers.
    b. doing a controlled study in mice.
    c. completing metatranscriptomics analysis to determine the functions present.
    d. performing clinical testing with probiotics of these keystone species.

24. Secondary species are
    a. necessary for any other taxa to recover.
    b. dependent on the return of many other taxa.
    c. not essential for microbiome recovery.
    d. SCFA producers.

Answers: 1C, 2B, 3D, 4B, 5A, 6C, 7A, 8D, 9B, 10C, 11B, 12B, 13C, 14D, 15A, 16C, 17A, 18B, 19C, 20D, 21A, 22C, 23A, 24D

DIVING DEEPER

1. What tests can a general practitioner run to assess your gut microbiome?

2. How can you test your gut transit time?

3. What information can a commercial gut microbiome kit using rRNA primers provide? What are the limitations of this type of testing?

4. Why is it difficult for a gut microbiome test to determine whether an individual's gut microbiome is healthy?

5. What are reasons why someone may choose to get a gut microbiome test?

6. What are the two main functions of a healthy gut microbiome?

7. What are some of the recommendations about diet that will improve your gut health?

8. Why are polyphenols beneficial to the gut microbiome and your health?

9. How must probiotic strains be tested before they get put into foods?

10. Probiotics are considered Generally Regarded As Safe (GRAS) by the FDA. What does this status mean, and what kinds of testing and regulation are mandatory or not mandatory?

11. In what applications have studies shown probiotics to be helpful? Why are they helpful in these cases?

12. What criteria are used to select prebiotics for commercial development?

13. Why might synbiotics be more beneficial than probiotics alone?

14. What are the phases of microbiome recovery after a perturbance?

DISCUSSING AND REFLECTING

1. After learning about the potential benefits and drawbacks of different methods of microbiome testing, what types of assessment would you consider doing and why?

2. Have you taken a probiotic, prebiotic, or other therapeutic? What would make you consider trying one of these microbiome therapies, and how would you evaluate the different commercial options available?

3. Reflection. After learning about the functions of the gut microbiome and the different ways to assess it, how do you feel about your gut microbiome? Do you feel like it's healthy? Do you feel like you understand your own gut microbiome, or do you realize how little you know about your own gut? Are there any changes you plan on making to alter your gut microbiome?

RECOMMENDED READINGS

Popular Science Review

Jabr, F. (2017, July 1). Do Probiotics Really Work? *Scientific American.* https://doi.org/10.1038/scientificamerican0717-26

Popular Science Book

Watson, R. R., & Preedy, V. R. (2015). Probiotics, Prebiotics, and Synbiotics: Bioactive Foods in Health Promotion. Academic Press.

Scientific Review

Spacova, I., Dodiya, H. B., Happel, A.-U., Strain, C., Vandenheuvel, D., Wang, X., & Reid, G. (2020). Future of Probiotics and Prebiotics and the Implications for Early Career Researchers. *Frontiers in Microbiology, 11.* https://www.frontiersin.org/articles/10.3389/fmicb.2020.01400

Glossary

A

abiogenesis The study of the creation of life from nonlife.

abiotic Not associated with, or derived from, living organisms.

abstract The portion of a research article that provides a summary of its main contents.

acetate A salt formed by the combination of acetic acid with a base.

acquired immune system See *adaptive immune system*.

adaptive (acquired) immune system A part of the immune system that develops a specific response to a foreign substance, such as a pathogen, which it remembers and can respond to rapidly in future exposures.

adaptive trait A genetic trait that gives an organism an advantage in a specific environment that is passed on to its offspring and becomes more common in a population.

adenine One of two purine nucleotides (the other being guanine) that make up DNA.

adenosine triphosphate (ATP) A molecule that uses the energy stored in its phosphate bonds to power chemical reactions; the "currency" of the cell.

adherens A protein complex that creates junctions that hold cells together.

adipose Connective tissue composed of fat.

aliquot A portion of a larger whole.

allergen A substance that can cause an allergic reaction, such as pollen or dust.

allergic cascade A predictable series of physiological responses to the detection of an allergen by the immune system.

allergic reaction An immune response to a substance that most people tolerate.

allergic response A physiological response triggered by the detection of an allergen by the immune system, such as wheezing, sneezing, and overproduction of mucus.

alpha diversity The species diversity within a community.

α-amylase An enzyme that breaks down starch into smaller sugar subunits.

α-toxin A protein produced by *Staphylococcus aureus* that causes disease by damaging host tissues.

American Gut Project A citizen science project focused on the human gut microbiome.

amplification The process of exponentially increasing the amount of a specific segment of DNA.

animalcule A term coined by Antonie van Leeuwenhoek for microscopic organisms.

anoxic Containing no oxygen; anaerobic, such as conditions on early Earth.

antibiotic An antimicrobial substance that inhibits or kills the growth of bacteria.

antibody A protein produced by the immune system that identifies and neutralizes foreign substances, such as pathogens.

antigen A toxin or other foreign substance that induces an immune response in the body.

antigen-presenting cell (APC) An immune cell that initiates the adaptive immune response by presenting antigens on its surface, which triggers recognition by T cells.

APC See *antigen-presenting cell*.

Archaea A domain of single-celled organisms, often found in extreme environments.

artifact In metagenomics, a collection of sequence files and their metadata that is used in analysis.

artifact situation An experimental result that appears to be genuine but is caused by a flaw in the experimental design.

asthma A chronic lung condition that causes inflammation of the lung airways.

astrocytes A type of glial cell that holds neurons in place and helps them work properly.

atopic dermatitis A chronic skin condition that causes inflammation, redness, and irritation of the skin.

atopic march The predictable progression of allergic diseases that begins in infancy.

atopy A predisposition to developing allergic diseases.

ATP See *adenosine triphosphate*.

autoantigen A protein that naturally occurs in the body and is identified as foreign by the immune system.

autoclave A steam-based sterilizer that kills bacteria, viruses, fungi, and spores.

autoimmune response The mounting of an attack when the immune system mistakes an antigen produced in the body as foreign.

autoimmunity A condition in which the immune system mistakenly attacks the body's own tissues.

autonomic nervous system A part of the peripheral nervous system that controls involuntary processes, such as heart rate and respiration.

autophagy The body's cellular-level recycling system, which breaks down damaged or abnormal proteins in a cell's cytoplasm.

autotroph An organism that makes its own energy by using inorganic compounds as a food source.

axon A portion of a nerve cell, or neuron, that carries nerve impulses away from the cell.

B

B cell A type of white blood cell that produces antibodies.

B lymphocyte A type of white blood cell that produces antibodies; also called B cells.

BE See *built environment*.

bacterial vaginosis (BV) An infection that results from changes to the vaginal microbiome.

bacteriophage A virus that infects and replicates only in bacteria.

barcode In biology, a short sequence of DNA that is used to identify a species.

base In biology, one of four chemical building blocks that make up the genetic code.

Basic Local Alignment Search Tool (BLAST) A program that finds similarity between sequences of nucleotides or proteins.

basophil A type of white blood cell that produces granules with enzymes that are released during an allergic reaction.

BBB See *blood-brain barrier*.

BDNF See *brain-derived neurotrophic factor*.

beta diversity The species diversity between communities.

β-oxidation A metabolic process in which fatty acids are broken down to generate energy.

bimodal A distribution with two peaks.

binomial A system of naming organisms that includes the genus name followed by the species name.

biodiversity hypothesis A proposal suggesting that reduced contact with environmental microbes in early life can negatively impact human health.

biodiversity The variety of life on Earth.

biofilm A layer of microbes that grows on the surface of a structure.

biomolecule A molecule produced by living organisms, such as carbohydrates, lipids, nucleic acids, and proteins.

biosphere The regions of the planet, including the surface and the atmosphere, occupied by living organisms.

biotherapy The use of substances produced by living organisms to treat disease.

BLAST See *Basic Local Alignment Search Tool*.

blood-brain barrier (BBB) A diffusion barrier that protects the brain from harmful substances

BMI See *body mass index*.

body mass index (BMI) A measure of body fat based upon weight and height.

bone marrow A substance inside bones that that produces blood cells.

box and whiskers plot A graph summarizing data; the shape of the plot shows how the data is distributed.

brain-derived neurotrophic factor (BDNF) A secreted protein that promotes the survival of nerve cells by playing a role in their growth and maturation.

Bray-Curtis (BC) dissimilarity A statistical measure used to quantify the dissimilarity in species composition between two samples.

built environment (BE) Human-made structures and conditions that provide people with living, working, and recreational spaces.

butyrate Metabolite produced when "good" bacteria in the large intestine help the body break down dietary fiber.

BV See *bacterial vaginosis*.

by-product An outcome of cellular respiration, often considered a waste product.

C

Cambrian Explosion The unexpected appearance of diverse animal and plant life forms in the fossil record approximately 540 million years ago.

cancer A disease caused by uncontrolled division of abnormal cells.

carbohydrate fermentation A metabolic process by which microorganisms break down carbohydrates into simpler substances, like acids or gases.

cardiometabolic disease (CMD) A group of conditions that affect cardiovascular and metabolic health.

CD-HIT See *Cluster Database at High Identity with Tolerance*.

cell-mediated immunity An immune response in which macrophages and killer T cells destroy host cells infected with a pathogen.

cellulose The most abundant organic molecule on the planet and a common component of plant cell walls.

cellulose-degrading bacterium Bacterium that possess the enzymes that enable degradation of cellulose.

Central Dogma of Molecular Biology A theory that describes how a cell's genetic information, DNA, is transcribed into RNA, which is then used to produce proteins.

central nervous system The portion of the nervous system made up of the brain and spinal cord.

chemoautotroph An organism that uses chemicals for energy; commonly found in environments rich in inorganic compounds, such as near deep-sea hydrothermal vents.

chemolithoautotrophy An autotrophic organism that obtains energy by oxidizing inorganic compounds (like hydrogen or sulfur) and fixing carbon dioxide to produce organic molecules.

chemolithotroph An organism that obtains energy from minerals and chemicals, such as those that spew from hydrothermal vents, and releases compounds that other microorganisms then use for food.

chromatin The packaged form of DNA, known as a chromosome, consisting of DNA, RNA, and proteins that is found in the nucleus.

chromosome structure An organized form of the genetic material, DNA, and its associated proteins, such as histones.

chronic periodontitis (CP) A bacterial infection that results in inflammation of the gums; this condition is associated with an altered oral microbiome.

circle of causality A series of events that influence each other in a cyclical manner.

clumping factors A and B Virulence factors on the surface of *Staphylococcus aureus*.

Cluster Database at High Identity with Tolerance (CD-HIT) A computer program that compares DNA or protein sequences and identifies groups based upon sequence identity.

CMD See *cardiometabolic disease*.

codon bias The preferential or non-random use of synonymous codons.

codon A sequence of three nucleotides that corresponds to a specific amino acid or signals the start or stop of protein synthesis.

colitis A chronic digestive disease that results in inflammation of the inner lining of the colon.

colonization resistance The ability of a host's microflora to prevent invasion of pathogenic microbes.

colostrum The first form of milk released by the mammary glands after giving birth.

community state type (CST) One of the five most common types of vaginal microbiomes, some of which may be predictive of certain vaginal diseases.

competitive exclusion The ability of one species to prevent entry of another species into a community, such as a microbiome.

complementary supplement A nonprescription medicine, such as vitamins.

contig Contiguous sequences created by alignment of genomic or metagenomic data.

coprolites Fossilized fecal matter.

coprophagy The consumption of feces.

correlated A relationship in which one variable affects or depends on another.

CP See *chronic periodontitis*.

Crohn's disease A chronic inflammatory disease that causes inflammation of the inner wall of the intestine.

cross-sectional study A study designed to compare two groups.

crown species A species-level designation for a member of a clade that includes all living members (the crown) and their most recent common ancestors, such as extinct species of plants, animals or fungi.

crypt Tubelike gland in the lining of the large intestine that contains stem cells that develop into epithelial cells.

CST See *community state type*.

culture The growth of a life form, generally a microorganism, in a nutrient source.

culturing A process that results in the multiplication of microorganisms under defined laboratory conditions.

cyanobacteria A phylum of bacteria also known as blue-green algae.

cytokine A secreted protein that signals immune cells to activate.

cytosine One of two pyrimidine nucleotides (the other being thymine) that make up DNA.

D

data transformation The process of cleaning and structuring data, such as DNA sequences and their metadata, for subsequent analysis.

de novo A biological process or entity that starts anew.

de-identified Data that has been cleansed of personal information to prevent the identification of an individual.

delta toxin A protein produced by *Staphylococcus aureus* that degrades the cellular membrane, causing cell lysis and subsequent cell death.

demultiplex The process of extracting individual signals from within a stream of data.

dental caries (cavities) A disease that occurs when decay-causing bacteria in the mouth produce acid that damages tooth surface structure.

dental plaque A sticky film of bacteria that constantly forms on teeth.

deoxyribonucleic acid (DNA) A molecule that contains the genetic information required to build and maintain a cellular organism.

deoxyribose A five-carbon sugar molecule that is a component of the backbone of DNA.

deterministic A mathematical model that produces the same result each time it is programmed with the same parameters.

dietary carbohydrate Carbohydrate that serves as the body's main fuel source.

digital object identifier (DOI) A unique string of numbers and letters that serves as a permanent web address for a document.

disappearing-microbiota hypothesis A proposal that the composition of the human gut microbiome has changed over the past few hundred years, resulting in impacts to human health.

discussion The portion of a research article in which the authors provide an interpretation and explanation of the significance of its findings.

disease marker An indicator that a disease is present.

distribution A statistical function that shows the possible values for a variable and how often they occur.

diversity index A measure that reflects how many different species there are in a community.

DNA See *deoxyribonucleic acid*.

DOI See *digital object identifier*.

double helix The term used to describe the physical structure of DNA.

dysbiosis An imbalance between the types of microflora present in a microbiome; thought to contribute to a range of health conditions.

E

Easy Microbiome Analysis Platform (EzMAP) A computer program that provides user-friendly microbiome analysis tools.

ecological core The group of microbial taxa that play essential roles in maintaining the ecological stability and balance within a microbial community.

ecological Koch's postulates The application of Koch's postulates to microbiome community associations to identify the cause of a disease.

ecosystem A community of organisms interacting with each other and their environment.

eczema A chronic condition that results in inflammation and irritation of the skin.

effector A part of the body that can respond to a stimulus sent from the central nervous system.

18S rRNA gene Homologous to the 16S rRNA gene, a DNA sequence that encodes the small eukaryotic ribosomal subunit.

endocrine pathway A series of events that involves hormone release into the circulatory system, which then signals other parts of the body.

endosymbiont A form of symbiosis in which a symbiont lives inside its host.

ENS See *enteric nervous system*.

enteric nervous system (ENS) A network of neurons that controls the digestive system.

enterotype A classification of gut microbiota based on the microbial species present.

enzyme A substance produced by a living organism which acts as a catalyst to bring about a specific biochemical reaction.

eosinophil A type of white blood cell that serves in the innate defense system by recognizing molecules produced by pathogens.

epigenetic change A reversible chemical modification of DNA that results in altered gene expression.

ethical considerations A set of principles that guide research practices and ensures participants are given informed consent and protected from harm.

eubiosis A balanced microbiome.

eukaryote An organism consisting of a cell or cells in which the genetic material is DNA in the form of chromosomes contained within a distinct nucleus.

evenness The relative abundance of different species in a community.

evolutionary tree See *phylogenetic tree*.

evolved dependence A process through which a host trait evolves to require the presence of one or more microorganisms.

experiment A procedure carried out under controlled conditions to test a hypothesis or establish a fact.

experimental control An element of an experiment that remains unchanged by other variables, which serves as a benchmark against which other elements are compared.

experimental treatment Variable explicitly controlled by the experimenter.

extant A species, genus, or other taxonomic group that currently exists.

extinction The process in which a species, genus, or other taxonomic group ceases to exist.

extremophile A microorganism, especially an archaean, that lives in conditions of extreme temperature, acidity, alkalinity, or chemical concentration.

EzMAP See *Easy Microbiome Analysis Platform*.

F

F/B ratio The ratio of Firmicute to Bacteroides in the human gut microbiome.

facultative anaerobe An organism able to metabolize with or without oxygen present.

family A taxonomic rank falling between order and genus in the Linnaean hierarchical system.

farm effect The observation that infants raised on farms have a reduced risk of allergic disease.

FASTA A format for representing DNA sequences, in which bases are designated by a single-letter code.

FastTree A computer program that generates a phylogenetic tree of 16S rRNA sequences.

fatty acid Carboxylic acid consisting of a hydrocarbon chain and a terminal carboxyl group, especially any of those occurring as esters in fats and oils.

fecal microbiota transplant (FMT) A medical procedure in which fecal matter from a healthy donor is delivered into a patient for therapeutic purposes.

fetal programming hypothesis The proposal that there is a critical window, the period between conception and the first 1,000 days after birth, when factors such as birth mode, antibiotics, diet, and exposure to pets results in long-term impacts on the programming of tolerance in the immune system.

FDA See *Food and Drug Administration*.

fiber Substances, such as cellulose, that are resistant to the action of digestive enzymes.

Firmicutes A phylum of Gram-positive bacteria that have a particularly thick cell wall.

five prime (5′) Refers to the end where a phosphate group is attached to the fifth carbon atom of the sugar (deoxyribose in DNA or ribose in RNA) in the nucleotide.

5′ See *five prime*.

FMT See *fecal microbiota transplant*.

food allergy oral immunotherapy (OIT) A procedure that involves exposing an individual to increasing amounts of a target food allergen to in order to desensitize them to it.

Food and Drug Administration (FDA) An agency that protects public health by regulating the safety of numerous products and services, such as drugs and medical devices.

forward and reverse primers Primers that correspond to which strand of the DNA they anneal to.

functional core The set of metabolic functions consistently present in a microbiome regardless of the specific species present.

Fungi A kingdom of eukaryotic organisms distinct from plants, animals, and bacteria, which includes yeasts, molds, and mushrooms.

G

GABA See *gamma-aminobutyric acid*.

gamma-aminobutyric acid (GABA) A neurotransmitter that regulates brain activities involved in mood and anxiety.

GasPak A commercially available method used in the production of an anaerobic environment.

gastrointestinal (GI) tract The pathway through which food and liquids travel through as they are digested.

GBA See *gut-brain axis*.

GenBank The NIH genetic database that contains a collection of all publicly available DNA sequences.

Generally Regarded As Safe (GRAS) A designation by the Food and Drug Administration that indicates that a substance added to food is safe when used in the manner intended.

genomics A branch of biology focused on the study of genomes, which are the complete set of DNA, including all of an organism's genes.

genus A rank in the classification hierarchy that groups species that are closely related based on shared characteristics.

germ A microorganism causing disease; a pathogenic agent (such as a bacterium or virus).

germ theory of disease The theory that certain infectious diseases are caused by pathogenic microorganisms invading a host.

Global Microbiome Conservancy A not-for-profit organization that preserves microbial samples from the human gut microbiome.

glucose intolerance A metabolic condition that results from an inability to properly process glucose.

glutamate The most abundant excitatory neurotransmitter in the brain.

glycan See *polysaccharide*.

glycosidic linkage A characteristic chemical bond that links monosaccharides.

GMHI See *Gut Microbiome Health Index*.

gnotobiotic Refers to a microbe-free organism.

goblet cell A type of cell found in the gastrointestinal, respiratory, and reproductive tract epithelial layers that secretes mucin, which forms mucus.

GOE See *Great Oxidation Event*.

Google Scholar A web search engine that provides a simple way to search scholarly literature.

GRAS See *Generally Regarded As Safe*.

Great Oxidation Event (GOE) The appearance of oxygen (O_2) in Earth's atmosphere due to biological activities.

great plate count anomaly The observation that microbial culture methods underestimate the actual number of species in a sample.

Greengenes An online database that collects and provides access to 16S rRNA sequences.

guanine One of two purine nucleotides (the other being adenine) that make up DNA.

Gut Microbiome Health Index (GMHI) A formula that predicts the likelihood of disease based upon a gut microbiome sample.

gut-brain axis (GBA) A network of nerves that links the enteric and central nervous systems.

gut-liver axis The two-way communication system that links the gut and its microbiome to the liver.

gut-lung axis The two-way communication route between the lungs and the gut microbiome that involves the exchange of microbes, their metabolites, or immune signals.

GWAS See *human genome-wide association study*.

gyrB A gene that encodes a subunit of a bacterial protein, DNA gyrase, whose DNA sequence is often used in phylogenetic studies of bacteria.

H

HAI See *healthcare-associated infection*.

halotolerant An organism that is adapted to high salt concentrations.

HDL See *high-density lipoprotein*.

healthcare-associated infection (HAI) An infection that occurs while an individual is receiving medical care.

healthy gut phageome A community of bacteriophages that are common in healthy adults and that are thought to be essential for the proper functioning of a healthy gut microbiome.

helper T cell A type of immune cell that senses an infection and activates other immune cells to fight it.

hematopoietic stem cell An immature cell found in bone marrow that can develop into all types of blood cells.

heritability rate A statistic used to estimate the degree to which a trait's presence is due to inherited genetic factors rather than the environment.

heterotroph An organism that obtains it energy by ingesting organic matter.

high microbiome diversity Occurs when numerous microbial taxa are present.

high-density lipoprotein (HDL) A form of lipid, also called "good" cholesterol, that circulates in the blood.

histamine A signaling compound released by cells in response to injury and allergic reactions.

histogram A bar graph used to represent the frequency distribution of data points of one variable.

HMOs See *human milk oligosaccharides*.

HMP See *Human Microbiome Project*.

holobiont A community of a host and its microbial symbionts that form a discrete ecological entity.

Home Microbiome Project A study that explored how humans and their pets impact the microorganisms in their homes.

homeostasis The tendency toward a relatively stable equilibrium between interdependent elements, especially as maintained by physiological processes.

homogenization The process of making a mixture completely uniform in composition.

homolog A gene that shares a common ancestor with another gene.

host range The variety of species that a particular pathogen, parasite, or organism can infect or associate with.

host-adapted core The set of microbial taxa that have evolved and adapted specifically to the human body and its unique environments.

House Observations of Microbial and Environmental Chemistry Project A study that explored how activities in a home, such as cleaning and cooking, impact the chemical composition of indoor air.

housekeeping task A routine task required for life in a bacterium or other microbe.

HPA See *hypothalamic-pituitary-adrenal axis*.

human genome-wide association study (GWAS) An approach that examines whole genome sequences from a large sample of a population to identify genetic variants associated with diseases.

Human Microbiome Project (HMP) A research initiative sponsored by the National Institutes of Health to characterize the normal human microflora.

human milk oligosaccharides (HMOs) A group of complex sugars found in breast milk.

humoral immune response A response that involves the production of antibodies by B cells to target pathogens found outside the body's cells, like bacterial and fungi.

Hungate technique A specialized method used for cultivating anaerobic microorganisms.

hydrothermal vent An opening in the sea floor out of which heated mineral-rich water flows.

hygiene hypothesis The proposal that exposure to infectious agents is required to ensure the development of a healthy gut microbiome and immune system.

hypertension A condition in which the pressure of the blood against the artery walls is consistently high.

hypothalamic-pituitary-adrenal (HPA) axis A neuroendocrine system that controls the body's response to stress.

hypothesis An idea that is proposed and tested to determine if it is true.

hypoxia A state of insufficient oxygen reaching body tissues.

I

IBD See *Inflammatory bowel disease*.

if/then A logical structure that expresses a conditional relationship between two statements or events.

IgE See *immunoglobulin E*.

IgG See *immunoglobulin G*.

ILC See *innate lymphoid cell*.

immune pathway The processes by which the gut microbiota influences the development and function of the immune system.

immunoglobulin E (IgE) Antibody produced by the immune system in response to an allergen.

immunoglobulin G (IgG) The most common antibody that protects against infections; crosses from mother to placenta during fetal development.

independent variable In a study, one variable that is not influenced by any other variables.

Inflammatory bowel disease (IBD) A chronic condition that causes inflammation in the digestive tract.

inflammatory cascade The body's response to tissue injury or infection that involves the production of histamines and cytokines.

innate immune system The body's defense system that passively prevents pathogen invasion; tears and mucus are part of this system.

innate lymphoid cell (ILC) An innate immune cell that does not possess antigen-specific receptors and plays a role in numerous responses, such as regulation of adaptive immunity, protection against pathogens, and cellular repair.

insulin resistance A condition resulting from an inability of tissues to take up glucose from the blood.

integration The process by which the central nervous system takes in sensory stimuli that leads to a response, such as muscle movement.

interquartile range A measure of the spread of the middle half of the data.

introduction The portion of a research article in which the question under study is described and the background information of the subject area is provided.

inulin A prebiotic form of plant fiber that is fermented by key gut bacteria and is used to treat asthma.

ion Atom or molecule that has gained or lost one or more electrons, resulting in a net electrical charge.

irreproducibility Unable to be reproduced.

K

keystone species A species that, for its abundance, has a disproportionately large impact on the ecosystem in which it lives.

killer T cell (KTC) A type of white blood cell that identifies and destroys certain cells, such as cancer cells or cells infected with a virus.

kingdom A taxonomic rank falling between domain and phylum in the Linnaean hierarchical system.

Koch's postulates A set of criteria that establish whether a particular organism is the cause of a particular disease.

KTC See *killer T cell*.

L

Last Universal Common Ancestor (LUCA) The extinct ancestral protocell from which all species descended.

LDL-C See *low-density lipoprotein cholesterol*.

Legionnaires' disease A form of pneumonia caused by *Legionella* bacteria.

leptin A hormone produced by fat cells that regulates appetite and fat storage.

leucine-rich repeat (LRR) A stretch of leucine in proteins that provides a framework for protein interactions.

lipid imbalance A condition in which there are abnormal levels of lipids in the blood.

lipolysis The breakdown of fat.

lipophilic Substances that are attracted to, or have an affinity for, lipids

lipopolysaccharide (LPS) The outer membrane components of Gram-negative bacteria.

longitudinal study A study designed to track changes in an individual or group over time.

low microbiome diversity Occurs when relatively few microbial taxa are present.

low-density lipoprotein cholesterol (LDL-C) A form of lipid, also called "bad" cholesterol, that circulates in the blood.

LPS See *lipopolysachharide*.

LRR See *leucine-rich repeat*.

LUCA See *Last Universal Common Ancestor*.

lumen The channel within a tubular organ, such as the intestine.

lymphatic A part of the immune system that involves the lymph nodes and that defends against infections and helps keep body fluid levels in balance.

lysis The process of breaking down a cell wall or membrane.

M

macrophage A type of white blood cell that surrounds and kills microorganisms.

major depressive disorder (MDD) A serious mood disorder that involves a persistent feeling of sadness or emptiness.

MAMP See *microbe-associated molecular pattern*.

mass spectrometry A tool used to determine the mass-to-charge ratio of a molecule.

mast cell A type of white blood cell, primarily found in connective tissue, that releases immune mediators, such as cytokines and histamine.

maximum value The largest value in a data set.

MDD See *major depressive disorder*.

median value The middle value in a data set.

membrane argument An argument concerning the origin of life that emphasizes the role of cellular structure.

metabolic module Conserved sequence of chemical transformations.

metabolic syndrome (MetS) A cluster of physiological conditions that is associated with an increased risk for heart disease, type 2 diabetes, and stroke.

metabolism The chemical processes that occur within a cell that transform food into energy.

metabolite An intermediate or end-product of metabolism, such as the production of butyrate by bacteria as they digest certain starches.

metabolome The metabolites produced by members of a microbiome.

metabolomics The large-scale study of small molecules, commonly known as metabolites, within cells, biofluids, tissues, or organisms.

Metabolomics Standards Initiative A group of experts in metabolomics who created a series of reporting standards for the field.

metadata A set of data that describes and gives information about other data.

metagenome The complete genetic content of the members of a microbiome.

metagenomic sequence library A DNA library that contains fragments from every organism present in the sampled microbiome.

metagenomics The study of all of the genetic material isolated from an environmental sample.

metaproteomics The study of the collective protein composition of the microbiome or environmental sample.

metatranscriptome The complete set of RNA transcripts produced by the collective genomes (or metagenome) of microorganisms within a particular environment or community.

metatranscriptomics The study of the RNA transcripts expressed within the community of cells or organisms.

methanogenesis The use of hydrogen gas as a food source and release of methane as a waste product.

methicillin-resistant *S. aureus* (MRSA) Strains of *Staphylococcus aureus* that are resistant to multiple antibiotics, including methicillin.

methods The portion of a research article that explains how data was generated or collected and how it was analyzed.

MetS See *metabolic syndrome*.

MGBA See *microbiota-gut-brain axis*.

Mian An online metagenomics platform designed to provide a diverse set of bioinformatics tools for metagenomic data analysis.

miasma theory The theory that certain diseases originate from "bad air," or particles, rising from decomposing matter.

micelle A spontaneously created aggregate of molecules that possess both hydrophilic and hydrophobic components.

microbe An organism that is too small to be seen by the naked eye.

microbe-associated molecular pattern (MAMP) A protein found on the cell surface of pathogens that is recognized by immune cells.

microbial cloud A collection of microscopic biological particles, including bacteria and viruses, that emanate from the human body.

microbial ecologist A scientist who studies how microorganisms interact with each other and with their environment.

microbial guild A group of microbial species that use the same resources in a similar way.

microbially produced volatile organic compound (MVOC) A gaseous metabolite produced by a microorganism during metabolism.

microbiology The scientific study of the biology of microorganisms.

microbiome Microorganisms and their collective genetic material, metabolites, and interactions with the host and environment.

MicrobiomeAnalyst A microbiome-focused metagenomics analysis program specifically designed for those with no experience using computer code.

microbiome rewilding hypothesis A proposal that posits that by restoring urban green spaces to a more natural state, one high in microbial diversity, we can improve human health.

microbiomes of the built environment (MoBE) The communities of microorganisms that reside in human-made structures and spaces.

microbiota A list of the species in a microbiome.

microbiota-gut-brain axis (MGBA) The communication system of the brain and the gut microbiota.

microglial cell A type of immune cell found in the central nervous system.

microRNA (miR) Non-coding RNA involved in regulating gene expression.

milk-oriented microbiome (MOM) The gut microbiome of a breast-fed baby.

minimum value The smallest value in a data set.

miR See *microRNA*.

missing-heritability problem The discrepancy between heritability estimates obtained from heritability studies, such as GWAS, and those from family and identical twin studies.

mitochondrion An organelle found in large numbers in most eukaryotic cells, in which the biochemical processes of respiration and energy production occur.

MoBE See *microbiomes of the built environment*.

model organism A nonhuman species that is studied extensively to understand biological phenomena.

molar The unit of concentration equal to the number of moles per liter of a solution.

molecular networking The grouping and connecting of mass spectrometry data based upon similarity.

MOM See *milk-oriented microbiome*.

Monera A historical kingdom in hierarchical classification that is made up of prokaryotes.

monocyte A type of white blood cell that travels to foreign cells and tissues and transforms into macrophages or dendritic cells.

monosaccharide A simple sugar that is the building block for more complex sugars.

mothur An open-source software package that is frequently used in microbiome research to analyze microbial DNA.

motor function of the CNS The control of voluntary movement.

MRSA See methicillin-resistant *S. aureus*.

mucus escalator The self-clearing mechanism of the respiratory system that moves mucus up and out of the lungs.

multidrug-resistant pathogen (MDRP) A bacterium that is resistant to multiple antibiotics.

multi-omics Biological analysis of multiple "omics" datasets.

MVOC See *microbially produced volatile organic compound*.

myeloid cell A family of immune cells, including monocytes and mast cells, that are produced in the bone marrow.

N

NAST A multiple sequence alignment server.

National Institutes of Health (NIH) A US federal agency responsible for supporting and conducting biomedical research and training research scientists.

natural killer (NK) cell A type of immune cell that serves in the innate immune response by controlling tumor cells and cells infected with pathogens.

natural log The base-e log, where e is the natural exponential.

natural selection The evolutionary process by which organisms that are better adapted to their environment tend to survive and reproduce, thus leaving more offspring.

naturopathic An approach to medicine that relies on natural substances, such as herbs, and physical means, such as acupuncture.

Neanderthal An extinct species of human that lived some 400,000 to 40,000 years ago.

neonatal intensive care unit (NICU) A specialized ward in a hospital that cares for premature and critically ill newborns.

nerve fiber Projections of nerve cells (or neurons) that carry electrical impulses away from the cell.

nerve impulse An electrical signal that moves along a neuron to the brain, where it creates sensations, produces thoughts, or to adds to memory.

network workspace In the metagenomics platform QIITA, the space where data analysis commands are entered.

neural pathway A series of connected neurons through which electrical signals are sent to various components of the nervous system.

neurodivergent A term that refers to a broad range of conditions that impact how the brain processes information.

neuroendocrine system A network of nerves and glands that produce and release hormones into the circulatory system.

neurogenesis The generation of new neural tissue in the brain.

neurotransmitter A chemical signal released by a neuron that causes nerve impulses to be transferred to a target, such as a muscle or gland.

neurotrophic factor A molecule that helps neurons grow and maintain themselves.

neurotrophic signal Protein that binds to receptors on target neural cells and regulates neuronal differentiation and growth.

neutrophil The most abundant white blood cell type and the first responders to pathogen infection.

niche modification The creation of new niches by the organisms in a community.

niche preemption A process by which organisms occupy specific ecological niches, which prevents other organisms from occupying them.

NICU See *neonatal intensive care unit*.

NIH See *National Institutes of Health*.

nitrogenous base See *base*.

NK See *natural killer cell*.

noncoding region The DNA located between coding regions, which does not directly contribute to protein synthesis.

normal Conforming to an average type, or pattern.

normal distribution A probability distribution that is symmetrical around its mean.

normal microbiome The consensus microbiome in individuals without known disease.

nucleobase See *base*.

nucleoside A component of DNA or RNA that consists of a purine or pyrimidine base linked to a sugar.

nucleotide An organic compound consisting of a nucleoside linked to a phosphate group, which forms the basic structural unit of nucleic acids, such as DNA.

nucleus A membrane-enclosed organelle in a eukaryotic cell that contains the genetic material.

null hypothesis The hypothesis that there is no difference between variables.

O

obligate anaerobe An organism that lives only in environments that lack oxygen.

OIT See *food allergy oral immunotherapy*.

old friends hypothesis Pathogenic microbes that humans coevolved with and that are considered necessary for proper immune system development.

open reading frame A portion of a DNA molecule that, when translated into amino acids, contains no stop codons.

open-access The practice of making digital information freely available to the public.

open-source software Computer code designed to be freely accessible to the public.

operational taxonomic unit (OTU) A set of DNA sequences that share a sequence identity above a given threshold.

organic molecule Material relating to, or derived from, living matter.

OTU See *operational taxonomic unit*.

outcome The results of an experiment.

ozone layer A thin layer of Earth's atmosphere that absorbs most of the sun's harmful ultraviolet light.

P

p-value A number representing the probability of obtaining the observed data under the null hypothesis.

Paneth cell Secretory epithelial cell located in crypts in the intestines that produces antimicrobial peptides.

passive immunity A temporary protection against infectious agents that occurs when antibodies are introduced to the body rather than produced by it.

pathobiont Organism of the microbiome that can cause disease under certain conditions.

pathogen trapping A defense mechanism that employs various methods to trap and eliminate pathogens.

pattern recognition receptor (PRR) A protein that detects molecules found on the cell surface of pathogens, which leads to an innate immune response.

PCA See *principal component analysis*.

PCoA See *principal coordinate analysis*.

PCR See *polymerase chain reaction*.

PCR amplification Sometimes called "molecular photocopying," a rapid, inexpensive method used to "amplify," or copy, small segments of DNA.

PD See *phylogenetic diversity*.

peer review The process by which a research manuscript is examined by experts in the field to ensure scientific quality.

periodontal (gum) disease A bacterial infection that results in inflammation of the gums and tissues that support the teeth.

peripheral nervous system (PNS) Includes the nerves that branch out from the brain and spinal cord.

phage See *bacteriophage*.

phagocytosis The ingestion of bacteria or dead cells by a type of white blood cell.

phenotype The term applied to an organism's observable characteristics.

phosphate group A functional group made up of a phosphorus atom bonded to four oxygen atoms.

photosynthesis The process by which green plants and some other organisms use sunlight to synthesize food from carbon dioxide and water.

phyllosphere The aboveground portion of a plant that is in inhabited by microbes.

phylogenetic diversity (PD) A measure of biodiversity, based on phylogeny.

phylogenetic tree A branching diagram that shows the evolutionary history of a group of organisms or molecules.

phylogeny See *phylogenetic tree*.

pioneer The first organism to colonize an area, such as microbes that enter a newborn's gut during birth.

PNS See *peripheral nervous system*.

polymerase chain reaction (PCR) An amplification method that mimics how DNA is copied in nature.

polyphenol A plant antioxidant that helps to improve digestion and brain health by neutralizing free radicals.

polysaccharide A carbohydrate that consists of several monosaccharides linked together.

power The probability that a test of significance will pick up on an effect that is present.

power analysis A calculation of the probability that a test of significance will pick up on an effect that is present given a particular sample size.

prebiotic Non-digestible foods that promote the growth of beneficial microorganisms in the large intestine.

predatory journal A publication that uses deceptive practices to trick researchers into publishing in exchange for money.

primary literature Scholarly publication that presents original research.

primer A short, single-stranded nucleic acid sequence used as a starting point for DNA synthesis.

primer pad region A non-template sequence incorporated into a primer used in amplification processes.

principal component analysis (PCA) A method used to reduce the dimensionality of large data sets.

principal coordinate analysis (PCoA) An analysis approach reduces the dimensionality of a large dataset without losing key information.

priority Refers to an advantage gained when a microbe arrives first in an ecosystem, such as the human gut, and can thus prevent subsequent microbes from colonizing it.

probiotic Live microorganisms that provide health benefits when consumed in sufficient numbers.

process specific to human-associated adaptations Adaptations that permit a microorganism to live within a human host, such as adhesion or metabolism.

prokaryote An organism that lacks a nucleus and other organelles, such as bacteria or archaea.

propionate A short-chain fatty acid produced during the fermentation of fiber.

protein A large organic molecule composed of one or more long chains of amino acids.

Proteobacteria A phylum of Gram-negative bacteria.

protist A eukaryotic organism that is not an animal, plant, bacteria or fungus.

protocell A primitive cell composed of a lipid membrane.

PRR See *pattern recognition receptor*.

psychobiotic A microorganism that confers mental health benefits when ingested.

PubMed An online search engine for biomedical or life sciences literature.

Q

QC See *quality control*.

QIIME 2 See *Quantitative Insights into Microbial Ecology*.

QIIMP See *Quick and Intuitive Interactive Metadata Portal*.

QS See *quality score*.

quality control (QC) The inspection of DNA sequence reads for contamination or low quality.

quality score (QS) A measure of the probability that a base is identified incorrectly.

Quantitative Insights into Microbial Ecology (QIIME 2) A microbiome bioinformatics tool.

Quick and Intuitive Interactive Metadata Portal (QIIMP) An online application that provides user-friendly access to metagenomic data analysis tools.

R

rarefaction A plot of the number of species as a function the number of samples.

rDNA See *ribosomal DNA*.

references The portion of a research article that provides the sources used during the preparation of the manuscript.

regulatory pathway Process in which a stimulus leads to a change in the expression or activity of a particular gene product, which in turn alters the expression or activity of another/other gene product(s), which in turn could regulate another level of activity.

regulatory T cell (Treg) A type of white blood cell that controls how the immune system responds to both foreign and internal substances.

replication argument An argument concerning the origin of life that emphasizes the role of genetics and inheritance.

resilience The ability to return to an original state of equilibrium after being subjected to a perturbation.

resistance The ability not to be affected by something, especially adversely.

resistant starch A type of carbohydrate that is not digested in the small intestine but is fermented in the large intestine.

respiration The process by which cells break down chemical energy from nutrients into a more useable form, such as adenosine triphosphate.

results The portion of a research article where the findings of a study are provided.

retention rate The number of individuals or organisms that will remain for the duration of a study.

ribosomal DNA (rDNA) Genes that encode ribosomal RNA, which is used to make proteins.

roll-tube A modification of the Hungate method designed to maintain strict anaerobic conditions during the cultivation process.

S

saccharolytic The ability to break down sugars for energy.

satiety signal Signals that arise in the gastrointestinal tract during eating that influence eating behavior.

SBS See *sick building syndrome*.

SCD See *sickle cell disease*.

SCFA See *short-chain fatty acid*.

sebaceous gland A small gland in the skin that secretes sebum, which lubricates the skin and hair.

second brain Another name for the enteric nervous system.

second quartile The value below which the lower 25% of the data are contained.

secondary literature Scientific publications that include review articles and specialty books, written with scientists in mind.

secondary succession The predictable chain of species recovery in an ecological habitat following a dramatic disturbance, such as a forest fire.

selection A process by which genetic factors determine which organisms survive in certain environments.

self-tolerance The ability of the immune system to recognize and ignore antigens produced by the body.

semiconservative replication The process by which DNA replicates in cells, ensuring that each new DNA molecule consists of one original (parent) strand and one newly-synthesized (daughter) strand.

sensory receptor A specialized cell that converts physical stimuli into electrical signals that the brain can process.

sequence alignment A way of arranging sequences to identify regions of similarity.

sequence annotation The process of adding descriptive information about a DNA, RNA, or protein sequence to identify specific features and their structure or function.

sequencing The process of determining the order of nucleotides in a stretch of DNA.

serial endosymbiosis A series of symbiotic events.

serotonin A chemical that carries messages between nerve cells in the brain and throughout the body.

sexually transmitted infection (STI) An infection passed from one person to another through blood, semen, or vaginal fluids during sex with an infected partner.

short-chain fatty acid (SCFA) Fatty acid with fewer than six carbon atoms.

shotgun sequencing A procedure that involves randomly breaking up a genome into pieces that are sequenced individually.

sick building syndrome (SBS) The situation in which occupants of a built structure experience negative health impacts linked to how much time they spend in the building, with no cause of disease identified.

sickle cell disease (SCD) A form of anemia characterized by a mutated form of hemoglobin that results in misshapen red blood cells.

SILVA An online resource for aligned ribosomal RNA sequence data.

single nucleotide polymorphism (SNP) A variation in a single nucleotide between two individuals, often used to inform GWAS studies.

single-line identifier A description (greater-than symbol, >) at the start of a DNA sequence in the FASTA format.

16S rRNA gene A DNA sequence that encodes the RNA component of the 30S subunit of a prokaryotic ribosome.

SNP See *single nucleotide polymorphism*.

somatic nervous system A part of the peripheral nervous system that controls voluntary movement.

spatial omics The study of biological molecules (such as RNA, DNA, proteins, or metabolites) in their natural spatial context within tissues or cells.

spatial transcriptomics The study of the level and location of gene activity in a tissue sample.

species The basic taxonomic rank in the Linnaean hierarchical system, denoted with the genus and species (e.g., *Homo sapiens*).

species accumulation curve The expected number of species observed as a function of sampling effort.

species richness The number of different species represented in a community.

spontaneous generation See *miasma theory*.

spore A minute, typically one-celled, reproductive unit capable of giving rise to a new individual without sexual fusion; a characteristic of lower plants, fungi, and protozoans.

stability A property of a microbiome that allows it to maintain its functions under external conditions.

Stachybotrys Also called toxic black mold; a genus of mold commonly found in wet, indoor environments.

stack plot A graph that displays multiple sets of data vertically.

starch A mixture of carbohydrates that is the primary form of stored energy in plants.

statistical power The probability that a statistical test will correctly reject a false null hypothesis.

statistical significance The probability of finding a given deviation from the null hypothesis in a sample.

sterilized Made free from bacteria or other living microorganisms.

STIs See *sexually transmitted infection*.

stochastic A mathematical model that produces different results each time it is programmed with the same parameters.

strain A genetic variant, or subtype, within a species.

stromatolite Mound of layers of lime-secreting bacteria and trapped sediment.

superantigen A type of immune stimulatory molecule that can cause an excessive activation of the immune system.

symbiosis A mutually beneficial interaction between two organisms that live in close physical association.

synbiotic A mixture of probiotics and the prebiotics that support their growth.

synergistic supplement A supplement or dietary compound that has a greater benefit when taken in combination with another supplement or dietary compound than it would on its own.

syntrophic The phenomenon of one species living off the metabolites of another.

systemic Not localized to one part of the body.

T

T cell A type of white blood cell that originates from stem cells in bone marrow and that protects the body from infection.

T lymphocyte A type of white blood cell that protects the body from infection; also called T cells.

Taq polymerase A specially engineered polymerase, originally derived from bacteria, that lives in superheated environments, such as hot springs.

target gene amplicon sequencing A process in which the target gene is amplified so that it can be sequenced.

temporal core The set of microbial taxa in a microbiome that remain consistently present over time, despite fluctuations in the environment or other members of the microbiome.

tertiary literature Science-focused publications including textbooks (such as this one) and encyclopedias, written with the general public in mind.

testable hypothesis A hypothesis that can be proved or disproved as a result of experimentation.

Th1 See *type 1 T helper cell*.

Th2 See *type 2 T helper cell*.

thermocycler A tool used to amplify DNA using polymerase chain reaction.

third quartile The value above which the upper 25% of the data are contained.

three prime (3′) The end of a DNA strand that ends with a phosphate group.

3′ See *three prime*.

thymine One of two pyrimidine nucleotides (the other being cytosine) that make up DNA.

tipping point A critical threshold at which small changes result in significant shifts in the composition and/or function of a microbiome.

title The portion of a research article that provides a simple and informative description of the research presented.

TLR See *Toll-like receptor*.

ToL See *tree of life*

Toll-like receptor (TLR) A receptor found on a cell's surface that recognizes molecules from pathogens and damaged cells and triggers an innate immune response.

transit time The time it takes for food to move through the gastrointestinal system.

tree of life (ToL) The concept, introduced by Darwin, that all living and extinct life forms are connected to a common ancestor.

Treg See *regulatory T cell*.

triglyceride A type of fat in the blood that is formed from a glycerol molecule with three fatty acids attached.

trophic interaction The transfer of energy between organisms in an ecosystem.

type 1 diabetes An autoimmune disease in which the immune system attacks cells in the pancreas that produce insulin.

type 1 T helper cell (Th1) An immune cell that releases a molecule that activates macrophages, which help eliminate foreign substances from the body.

type 2 T helper cell (Th2) An immune cell that mediates antibody-based responses against allergens and toxins.

U

UniFrac See *unique fraction*.

UniProt A database of all known protein sequence data linked to their functional information.

unique fraction (UniFrac) A metric that measures the phylogenetic distance between sets of taxa in a phylogenetic tree.

urinary tract infection (UTI) An infection in any part of the urinary tract.

UTI See *urinary tract infection*.

V

vaginal microbiome transplant (VMT) The transfer of vaginal fluid from a healthy person to a person with an imbalanced vaginal microbiome.

vagus nerve The major nerve of the parasympathetic nervous system.

VAMPS See *Visualization and Analysis of Microbial Population Structures*.

variable Any characteristic that can take on different values.

virome A collection of viruses that live within a particular environment, such as a host.

Visualization and Analysis of Microbial Population Structures (VAMPS) A software platform that permits researchers using marker gene sequence data to analyze the diversity of microbial communities and the relationships between them.

VMT See *vaginal microbiome transplant*.

vortex A device used to quickly mix small volumes of liquids in test tubes, microcentrifuge tubes, or small containers.

W

waist-to-hip ratio (WHR) A measure of the ratio of the circumference of the waist to that of the hips.

white blood cell A part of the immune response that includes lymphocytes and granulocytes.

WHR See *waist-to-hip ratio*.

References

A

Abdel-Gadir, A., Stephen-Victor, E., Gerber, G. K., Rivas, M. N., Wang, S., Harb, H., Wang, L., Li, N., Crestani, E., Spielman, S., Secor, W., Biehl, H., Dibendetto, N., Dong, X., Umetsu, D. T., Bry, L., Rachid, R., & Chatila, T. A. (2019). Microbiota Therapy Acts via a Regulatory T Cell MyD88/RORγt Pathway to Suppress Food Allergy. *Nature Medicine*, 25(7), 1164–1174. https://doi.org/10.1038/s41591-019-0461-z

Abed, J., Emgård, J. E. M., Zamir, G., Faroja, M., Almogy, G., Grenov, A., Sol, A., . . . & Bachrach, G. (2016). Fap2 Mediates *Fusobacterium nucleatum* Colorectal Adenocarcinoma Enrichment by Binding to Tumor-Expressed Gal-GalNAc. *Cell Host & Microbe*, 20(2), 215–225.

Adams, R. I., Lymperopoulou, D. S., Misztal, P. K., De Cassia Pessotti, R., Behie, S. W., Tian, Y., Goldstein, A. H., Lindow, S. E., Nazaroff, W. W., Taylor, J. W., Traxler, M. F., & Bruns, T. D. (2017). Microbes and Associated Soluble and Volatile Chemicals on Periodically Wet Household Surfaces. *Microbiome*, 5(1), 128. https://doi.org/10.1186/s40168-017-0347-6

Adams, S., & Yakowicz, W. (2020). Drugs from Bugs: Why Gates, Zuck, and Benioff think the Next Blockbusters Will Come from Inside Your Gut. *Forbes*. (Updated Feb 7, 2020, 03:15 pm EST.)

Ahmadi, S., Nagpal, R., Wang, S., Gagliano, J., Kitzman, D. W., Soleimanian-Zad, S., Sheikh-Zeinoddin, M., Read, R., & Yadav, H. (2019). Prebiotics from Acorn and Sago Prevent High-Fat-Diet-Induced Insulin Resistance via Microbiome-Gut-Brain Axis Modulation. *Journal of Nutritional Biochemistry*, 67, 1–13. https://doi.org/10.1016/j.jnutbio.2019.01.003

Ahmadi, S., Nagpal, R. K., Wang, S., & Yadav, H. (2018). New Prebiotics to Ameliorate High-Fat Diet-Induced Obesity and Diabetes via Modulation of Microbiome-Gut-Brain Axis. *Diabetes*, 67(Suppl. 1), 264-LB. https://doi.org/10.2337/db18-264-LB

Airhart, M. (2016, August 3). Some Bacteria Have Lived in the Human Gut Since Before We Were Human. The University of Texas at Austin, College of Natural Sciences, podcast. https://cns.utexas.edu/news/podcast/some-bacteria-have-lived-human-gut-we-were-human

Akutko, K., & Stawarski, A. (2021). Probiotics, Prebiotics and Synbiotics in Inflammatory Bowel Diseases. *Journal of Clinical Medicine*, 10(11), Article 11. https://doi.org/10.3390/jcm10112466

Alashkar Alhamwe, B., Gao, Z., Alhamdan, F., Harb, H., Pichene, M., Garnier, A., El Andari, J., Kaufmann, A., Graumann, P. L., Kesper, D., Daviaud, C., Garn, H., Tost, J., Potaczek, D. P., Blaser, M. J., & Renz, H. (2022). Intranasal Administration of *Acinetobacter lwoffii* in a Murine Model of Asthma Induces IL-6-Mediated Protection Associated with Cecal Microbiota Changes. *Allergy*, 78(5). https://doi.org/10.1111/all.15606

Aldhous, P. (2004). Biohazard Lurks in Bathrooms. *Nature*. https://doi.org/10.1038/news040216-2

Allaband, C., McDonald, D., Vázquez-Baeza, Y., Minich, J. J., Tripathi, A., Brenner, D. A., Loomba, R., Smarr, L., Sandborn, W. J., Schnabl, B., Dorrestein, P., Zarrinpar, A., & Knight, R. (2019). Microbiome 101: Studying, Analyzing, and Interpreting Gut Microbiome Data for Clinicians. *Clinical Gastroenterology and Hepatology*, 17(2), 218–230. https://doi.org/10.1016/j.cgh.2018.09.017

Almeida, A., Mitchell, A. L., Boland, M., Forster, S. C., Gloor, G. B., Tarkowska, A., Lawley, T. D., & Finn, R. D. (2019). A new genomic blueprint of the human gut microbiota. *Nature*, 568(7753), 499–504. https://doi.org/10.1038/s41586-019-0965-1

Almeida, A., Nayfach, S., Boland, M., Strozzi, F., Beracochea, M., Shi, Z. J., Pollard, K. S., Sakharova, E., Parks, D. H., Hugenholtz, P., Segata, N., Kyrpides, N. C., & Finn, R. D. (2021). A Unified Catalog of 204,938 Reference Genomes from the Human Gut Microbiome. *Nature Biotechnology*, 39(1), Article 1. https://doi.org/10.1038/s41587-020-0603-3

Alsharairi, N. A. (2020). The Infant Gut Microbiota and Risk of Asthma: The Effect of Maternal Nutrition during Pregnancy and Lactation. *Microorganisms*, 8(8), 1119. https://doi.org/10.3390/microorganisms8081119

Alsukhon, J., Chatfield, A., Zoratti, E., Kim, H., Sitarik, A., Havstad, S., Johnson, C., Joseph, C., & Wegienka, G. (2020). Prenatal Pet Exposure and Total Serum IgE Trajectory at 10 Years of Age. *Annals of Allergy, Asthma & Immunology*, 125(5), S19. https://doi.org/10.1016/j.anai.2020.08.076

Altschul, S. F., Gish, W., Miller, W., Myers, E. W., & Lipman, D. J. (1990). Basic Local Alignment Search Tool. *Journal of Molecular Biology*, 215(3), 403–410. https://doi.org/10.1016/S0022-2836(05)80360-2

American Society for Microbiology. (2014, February 27). Fossilized Human Feces from 14th Century Contain Antibiotic Resistance Genes. ScienceDaily. www.sciencedaily.com/releases/2014/02/140227164534.htm

An, X.-L., Xu, J.-X., Xu, M.-R., Zhao, C.-X., Li, H., Zhu, Y.-G., & Su, J.-Q. (2023). Dynamics of Microbial Community and Potential Microbial Pollutants in Shopping Malls. *mSystems*, e00576-22. https://doi.org/10.1128/msystems.00576-22

Anderson, D. J., Podgorny, K., Berríos-Torres, S. I., Bratzler, D. W., Dellinger, E. P., Greene, L., Nyquist, A.-C., Saiman, L., Yokoe, D. S., Maragakis, L. L., & Kaye, K. S. (2014). Strategies to Prevent Surgical Site Infections in Acute Care Hospitals: 2014 Update. *Infection Control & Hospital Epidemiology*, 35(6), 605–627. https://doi.org/10.1086/676022

Anderson, J. K. (2020). "Host" *Field Guide to the Haunted Forest*. Independently published.

Anderson, M. W., & Schrijver, I. (2010, May 25). Next Generation DNA Sequencing and the Future of Genomic Medicine. *Genes*, 1, 38–69. https://doi.org/10.3390/genes1010038

Antharam, V. C., Li, E. C., Ishmael, A., Sharma, A., Mai, V., Rand, K. H., & Wang, G. P. (2013). Intestinal Dysbiosis and Depletion of Butyrogenic Bacteria in Clostridium difficile Infection and Nosocomial Diarrhea. *Journal of Clinical Microbiology*, 51(9), 2884–2892. https://doi.org/10.1128/JCM.00845-13

Appanna, V. D. (2018). The Human Microbiome: The Origin. In V. D. Appanna, *Human Microbes-The Power Within*. Springer.

Appelt, S., Armougom, F., Le Bailly, M., Robert, C., & Drancourt, M. (2014). Polyphasic analysis of a middle ages coprolite microbiota, Belgium. *PLOS ONE*, 9(2): e88376. https://journals.plos.org/plosone/article?id=10.1371/journal.pone.0088376

Arnold, C. (2014, July 1). Rethinking sterile: the hospital microbiome. *Environmental Health Perspectives* 122(7), A182–187. https://ehp.niehs.nih.gov/doi/10.1289/ehp.122-A182.

Arrieta, M.-C., Stiemsma, L. T., Dimitriu, P. A., Thorson, L., Russell, S., Yurist-Doutsch, S., Kuzeljevic, B., Gold, M. J., Britton, H. M., Lefebvre, D. L., Subbarao, P., Mandhane, P., Becker, A., McNagny, K. M., Sears, M. R., Kollmann, T., the CHILD Study Investigators, Mohn, W. W., Turvey, S. E., & Brett Finlay, B. (2015). Early Infancy Microbial and Metabolic Alterations Affect Risk of Childhood Asthma. *Science Translational Medicine*, 7(307). https://doi.org/10.1126/scitranslmed.aab2271

Asnicar, F., Berry, S. E., Valdes, A. M., . . . & Segata, N. (2021). Microbiome Connections with Host Metabolism and Habitual Diet from 1,098 Deeply Phenotyped Individuals. *Nature Medicine*, 27, 321–332.

Asnicar, F., Leeming, E. R., Dimidi, E., Mazidi, M., Franks, P. W., Al Khatib, H., Valdes, A. M., Davies, R., Bakker, E., Francis, L., Chan, A., Gibson, R., Hadjigeorgiou, G., Wolf, J., Spector, T. D., Segata, N., & Berry, S. E. (2021, September). Blue Poo: Impact of Gut Transit Time on the Gut Microbiome Using a Novel Marker. *Gut*, 70(9), 1665–1674.

Aun, M., Bonamichi-Santos, R., Arantes-Costa, F. M., Kalil, J., & Giavina-Bianchi, P. (2017). Animal Models of Asthma: Utility and Limitations. *Journal of Asthma and Allergy*, 10, 293–301. https://doi.org/10.2147/JAA.S121092

Australian Academy of Science. (2019). What We Can Learn from Prehistoric Poo. Curious. https://www.science.org.au/curious/people-medicine/what-we-can-learn-prehistoric-poo

Azad, M. B., Konya, T., Maughan, H., Guttman, D. S., Field, C. J., Sears, M. R., Becker, A. B., Scott, J. A., & Kozyrskyj, A. L. (2013). Infant Gut Microbiota and the Hygiene Hypothesis of Allergic Disease: Impact of Household Pets and Siblings on Microbiota Composition and Diversity. *Allergy, Asthma, and Clinical Immunology*, 9(1), 15.

B

Bach, J.-F. (2002). The Effect of Infections on Susceptibility to Autoimmune and Allergic Diseases. *New England Journal of Medicine*, 347(12), 911–920.

Baker, H. (1743). *The Microscope Made Easy*. London, England: Robert Dodsley. https://digital.sciencehistory.org/works/gkw2jwq.

Barr, J. J., Auro, R., Furlan, M., Whiteson, K. L., Erb, M. L., Pogliano, J., Stotland, A., Wolkowicz, R., Cutting, A. S., Doran, K. S., Salamon, P., Youle, M., & Rohwer, F. (2013). Bacteriophage Adhering to Mucus Provide a Non–Host-Derived Immunity. *Proceedings of the National Academy of Sciences*, 110(26), 10771–10776. https://doi.org/10.1073/pnas.1305923110

Baumann-Dudenhoeffer, A. M., D'Souza, A. W., Tarr, P. I., Warner, B. B., & Dantas, G. (2018). Infant Diet and Maternal Gestational Weight Gain Predict Early Metabolic Maturation of Gut Microbiomes. *Nature Medicine*, 24(12), 1822–1829. https://doi.org/10.1038/s41591-018-0216-2

Becerik-Gerber, B., Lucas, G., Aryal, A., . . . & Zhu, R. 2022. The field of human building interaction for convergent research and innovation for intelligent built environments. *Scientific Reports*, 12, 22092. https://doi.org/10.1038/s41598-022-25047-y

Belkaid, Y., & Hand, T. W. (2014). Role of the Microbiota in Immunity and Inflammation. *Cell*, 157(1), 121–141. https://doi.org/10.3389/fnut.2022.1006747

Benson, D. A., Cavanaugh, M., Clark, K., Karsch-Mizrachi, I., Lipman, D. J., Ostell, J., & Sayers, E. W. (2013). GenBank. *Nucleic Acids Research*, 41(Database issue), D36–42. https://doi.org/10.1093/nar/gks1195

Berni Canani, R., Paparo, L., Nocerino, R., Di Scala, C., Della Gatta, G., Maddalena, Y., Buono, A., Bruno, C., Voto, L., & Ercolini, D. (2019). Gut Microbiome as Target for Innovative Strategies against Food Allergy. *Frontiers in Immunology*, 10, 191. https://doi.org/10.3389/fimmu.2019.00191

Bescos, R., Ashworth, A., Cutler, C., Brookes, Z. L., Belfield, L., Rodiles, A., Casas-Agustench, P., Farnham, G., Liddle, L., Burleigh, M., White, D., Easton, C., & Hickson, M. (2020). Effects of Chlorhexidine Mouthwash on the Oral Microbiome. *Scientific Reports*, 10(1), 5254. https://doi.org/10.1038/s41598-020-61912-4

Blacher, E., Levy, M., Tatirovsky, E., & Elinav, E. (2017). Microbiome-Modulated Metabolites at the Interface of Host Immunity. *Journal of Immunology*, 198(2), 572–580. https://doi.org/10.4049/jimmunol.1601247

Blakeslee, S. (1996, Oct 15) *Scientist at Work: Carl R. Woese, Microbial Life's Steadfast Champion*. The New York Times. https://www.nytimes.com/1996/10/15/science/microbial-life-s-steadfast-champion.html

Blaser, M. J. (2014). *Missing Microbes: How the Overuse of Antibiotics Is Fueling Our Modern Plagues* (1st ed.). Henry Holt and Company.

Blaser, M., & Falkow, S. (2009). What Are the Consequences of the Disappearing Human Microbiota? *Nature Reviews Microbiology*, 7, 887–894.

Bojanova, D. P., & Bordenstein, S. R. (2016). Fecal Transplants: What Is Being Transferred? *PLOS Biology*, 14(7), e1002503. https://doi.org/10.1371/journal.pbio.1002503

Bolyen, E., Rideout, J. R., Dillon, M. R., Bokulich, N. A., Abnet, C. C., Al-Ghalith, G. A., Alexander, H., Alm, E. J., Arumugam, M., Asnicar, F., Bai, Y., Bisanz, J. E., Bittinger, K., Brejnrod, A., Brislawn, C. J., Brown, C. T., Callahan, B. J., Caraballo-Rodríguez, A. M., Chase, J., . . . & Caporaso, J. G. (2019). Reproducible, Interactive, Scalable and Extensible Microbiome Data Science Using QIIME 2. *Nature Biotechnology*, 37(8), 852–857. https://doi.org/10.1038/s41587-019-0209-9

Borbet, T. C., Pawline, M. B., Zhang, X., Wipperman, M. F., Reuter, S., Maher, T., Li, J., Iizumi, T., Gao, Z., Daniele, M., Taube, C., Koralov, S. B., Müller, A., & Blaser, M. J. (2022). Influence of the Early-Life Gut Microbiota on the Immune Responses to an Inhaled Allergen. *Mucosal Immunology*, 15(5), 1000–1011. https://doi.org/10.1038/s41385-022-00544-5

Brand, S. (2014, July 15). Quote of the Day. Mayo Clinic: In the Loop. https://intheloop.mayoclinic.org/2014/07/15/quote-of-the-day-61/

Bravo, J. A., Forsythe, P., Chew, M. V., Escaravage, E., Savignac, H. M., Dinan, T. G., Bienenstock, J., & Cryan, J. F. (2011).

Ingestion of *Lactobacillus* Strain Regulates Emotional Behavior and Central GABA Receptor Expression in a Mouse via the Vagus Nerve. *Proceedings of the National Academy of Sciences*, 108(38), 16050–16055. https://doi.org/10.1073/pnas.1102999108

Bray, J. R., & Curtis, J. T. (1957). An Ordination of the Upland Forest Communities of Southern Wisconsin. *Ecological Monographs*, 27(4), 325–349. https://doi.org/10.2307/1942268

Brown, K. V. (2017, August 24). What Present-Day Hunter-Gatherers Can Tell Us about the Bacteria in Our Gut. Gizmodo. https://gizmodo.com/what-present-day-hunter-gatherers-can-tell-us-about-the-1798372037

Bruno, A., Fumagalli, S., Ghisleni, G., & Labra, M. (2022). The Microbiome of the Built Environment: The Nexus for Urban Regeneration for the Cities of Tomorrow. *Microorganisms*, 10(12), 2311. https://doi.org/10.3390/microorganisms10122311

Buguliskis, J. S. (2021, February 8). Did Neanderthals Contribute to Our Gut Microbiome? *GEN - Genetic Engineering and Biotechnology News*. https://www.genengnews.com/news/did-neanderthals-contribute-to-our-gut-microbiome/

Bulsiewicz, W. (2020). *Fiber Fueled: The Plant-Based Gut Health Program for Losing Weight, Restoring Your Health, and Optimizing Your Microbiome*. Avery.

Bunbury, K. (2015). Legionnaires Outbreak: Are Your Clients Properly Insured? Environmental Risk Managers, Moline (USA).

Byndloss, M. X., Pernitzsch, S. R., & Bäumler, A. J. (2018). Healthy Hosts Rule within: Ecological Forces Shaping the Gut Microbiota. *Mucosal Immunology*, 11(5), Article 5. https://doi.org/10.1038/s41385-018-0010-y

C

Callaway, E. (2019, September 18). C-Section Babies Are Missing Key Microbes. *Nature*. https://www.scientificamerican.com/article/c-section-babies-are-missing-key-microbes/

Camarillo-Guerrero, L. F., Almeida, A., Rangel-Pineros, G., Finn, R. D., & Lawley, T. D. (2021). Massive Expansion of Human Gut Bacteriophage Diversity. *Cell*, 184(4), 1098–1109.e9. https://doi.org/10.1016/j.cell.2021.01.029

Canchy, L., Kerob, D., Demessant, A., & Amici, J.-M. (2023). Wound Healing and Microbiome, an Unexpected Relationship. *Journal of the European Academy of Dermatology and Venereology*, 37(Suppl. 3), 7–15.

Canfield, D. E. (2005) The early history of atmospheric oxygen: homage to Rober A. Garrels. *Annual Review of Earth and Planetary Sciences* 33, 1–36.

Caporaso, J. G., Kuczynski, J., Stombaugh, J., Bittinger, K., Bushman, F. D., Costello, E. K., Fierer, N., Peña, A. G., Goodrich, J. K., Gordon, J. I., Huttley, G. A., Kelley, S. T., Knights, D., Koenig, J. E., Ley, R. E., Lozupone, C. A., McDonald, D., Muegge, B. D., Pirrung, M., . . . & Knight, R. (2010). QIIME Allows Analysis of High-Throughput Community Sequencing Data. *Nature Methods*, 7(5), 335–336. https://doi.org/10.1038/nmeth.f.303

Careaga, M., Murai, T., & Bauman, M. D. (2017). Maternal Immune Activation and Autism Spectrum Disorder: From Rodents to Nonhuman and Human Primates. *Biological Psychiatry*, 81(5), 391–401. https://doi.org/10.1016/j.biopsych.2016.10.020

Cason, C., D'Accolti, M., Campisciano, G., Soffritti, I., Ponis, G., Mazzacane, S., Maggiore, A., Risso, F. M., Comar, M., & Caselli, E. (2021). Microbial Contamination in Hospital Environment Has the Potential to Colonize Preterm Newborns' Nasal Cavities. *Pathogens*, 10(5), 615. https://doi.org/10.3390/pathogens10050615

Cassano, O. (2022, June 14). What Are the Vaginal Microbiome Community State Types? *Evvy*. https://www.evvy.com/blog/community-state-types

Catlett, J. L., Carr, S., Cashman, M., Smith, M. D., Walter, M., Sakkaff, Z., Kelley, C., Pierobon, M., Cohen, M. B., & Buan, N. R. (2022). Metabolic Synergy between Human Symbionts *Bacteroides* and *Methanobrevibacter*. *Microbiology Spectrum*, 10(3), e0106722. https://doi.org/10.1128/spectrum

Cavalazzi, B., Lemelle, L., Simionovici, A., Cady, S. L., Russell, M. J., Bailo, E., Canteri, R., Enrico, E., Manceau, A., Maris, A., Salomé, M., Thomassot, E., Bouden, N., Tucoulou, R., & Hofmann, A. (2021). Cellular Remains in a ~3.42-Billion-Year-Old Subseafloor Hydrothermal Environment. *Science Advances*, 7(29), eabf3963. https://doi.org/10.1126/sciadv.abf3963

Chakrabarti, A., Geurts, L., Hoyles, L., Iozzo, P., Kraneveld, A. D., La Fata, G., Miani, M., Patterson, E., Pot, B., Shortt, C., & Vauzour, D. (2022). The Microbiota–Gut–Brain Axis: Pathways to Better Brain Health. Perspectives on What We Know, What We Need to Investigate and How to Put Knowledge into Practice. *Cellular and Molecular Life Sciences*, 79(2), 80. https://doi.org/10.1007/s00018-021-04060-w

Chen, C., Song, X., Wei, W., Zhong, H., Dai, J., Lan, Z., Li, F., Yu, X., Feng, Q., Wang, Z., Xie, H., Chen, X., Zeng, C., Wen, B., Zeng, L., Du, H., Tang, H., Xu, C., Xia, Y., . . . & Jia, H. (2017). The Microbiota Continuum along the Female Reproductive Tract and Its Relation to Uterine-Related Diseases. *Nature Communications*, 8(1), Article 1. https://doi.org/10.1038/s41467-017-00901-0

Chen, J., Wright, K., Davis, J. M., Jeraldo, P., Marietta, E. V., Murray, J., Nelson, H., Matteson, E. L., & Taneja, V. (2016). An Expansion of Rare Lineage Intestinal Microbes Characterizes Rheumatoid Arthritis. *Genome Medicine*, 8(1), 43. https://doi.org/10.1186/s13073-016-0299-7

Chen, S. G., Stribinskis, V., Rane, M. J., Demuth, D. R., Gozal, E., Roberts, A. M., Jagadapillai, R., Liu, R., Choe, K., Shivakumar, B., Son, F., Jin, S., Kerber, R., Adame, A., Masliah, E., & Friedland, R. P. (2016). Exposure to the Functional Bacterial Amyloid Protein Curli Enhances Alpha-Synuclein Aggregation in Aged Fischer 344 Rats and *Caenorhabditis elegans*. *Scientific Reports*, 6(1), 34477. https://doi.org/10.1038/srep34477

Chen, X., Lu, Y., Chen, T., & Li, R. (2021). The Female Vaginal Microbiome in Health and Bacterial Vaginosis. *Frontiers in Cellular and Infection Microbiology*, 11, 631972. https://doi.org/10.3389/fcimb.2021.631972

Cheng, F.-S., Pan, D., Chang, B., Jiang, M., & Sang, L.-X. (2020). Probiotic Mixture VSL#3: An Overview of Basic and Clinical Studies in Chronic Diseases. *World Journal of Clinical Cases*, 8(8), 1361–1384. https://doi.org/10.12998/wjcc.v8.i8.1361

Chichlowski, M., Shah, N., Wampler, J. L., Wu, S. S., & Vanderhoof, J. A. (2020). *Bifidobacterium longum* Subspecies *infantis* (B. infantis) in Pediatric Nutrition: Current State of Knowledge. *Nutrients*, 12(6), 1581. https://doi.org/10.3390/nu12061581

Chng, K. R., Ghosh, T. S., Tan, Y. H., . . . & Nagarajan, N. (2020). Metagenome-Wide Association Analysis Identifies Microbial Determinants of Post-Antibiotic Ecological Recovery in the Gut. *Nature Ecology & Evolution*, 4, 1256–1267. https://doi.org/10.1038/s41559-020-1236-0

Chong, J., Liu, P., Zhou, G., & Xia, J. (2020). Using MicrobiomeAnalyst for Comprehensive Statistical, Functional, and Meta-analysis of Microbiome Data. *Nature Protocols*, 15(3), Article 3. https://doi.org/10.1038/s41596-019-0264-1

Christoff, A. P., Sereia, A. F., Hernandes, C., & de Oliveira, L. F. (2019). Uncovering the Hidden Microbiota in Hospital and Built Environments: New Approaches and Solutions. *Experimental Biology and Medicine*, 244(6), 534–542. https://doi.org/10.1177/1535370218821857

Chudzik, A., Orzyłowska, A., Rola, R., & Stanisz, G. J. (2021). Probiotics, Prebiotics and Postbiotics on Mitigation of Depression Symptoms: Modulation of the Brain-Gut-Microbiome Axis. *Biomolecules*, 11(7), 1000. https://doi.org/10.3390/biom11071000.

Churko, J. M., Mantalas, G. L., Snyder, M. P., & Wu, J. C. (2013). Overview of High Throughput Sequencing Technologies to Elucidate Molecular Pathways in Cardiovascular Diseases. *Circulation Research*, 112(12), 1613–1623. https://doi.org/10.1161/CIRCRESAHA.113.300939

Crittenden, R., & Playne, M. J. (2009). Prebiotics. In Y. K. Lee & S. Salminen (Eds.), *Handbook of Probiotics and Prebiotics* (pp. 535–561). John Wiley & Sons.

Crovesy, L., Ostrowski, M., Ferreira, D. M. T. P., Rosado, E. L., & Soares-Mota, M. (2017). Effect of *Lactobacillus* on Body Weight and Body Fat in Overweight Subjects: A Systematic Review of Randomized Controlled Clinical Trials. *International Journal of Obesity*, 41(11), 1607–1614. https://doi.org/10.1038/ijo.2017.161

Crowe, S. A., Døssing, L. N., Beukes, N. J., Bau, M., Kruger, S. J., Frei, R., & Canfield, D. E. (2013). Atmospheric oxygenation three billion years ago. *Nature* 501, 535–538.

Curatola, G. P., & Reverand, D. (2017). *The Mouth-Body Connection: A 28-Day Program to Create a Healthy Mouth, Reduce Inflammation, and Prevent Disease Throughout the Body*. Center Street.

Cutler, D., & Miller, G. (2005). The Role of Public Health Improvements in Health Advances: The Twentieth-Century United States. *Demography*, 42(1), 1–22. https://doi.org/10.1353/dem.2005.0002

D

Darwin, C. (1859). *On the Origin of Species by Means of Natural Selection, or, The Preservation of Favoured Races in the Struggle for Life* (1st ed., pp. 1–564). John Murray. https://www.biodiversitylibrary.org/item/135954

Dekaborush, E., Suryavanshi, M. V., Chettri, D., & Verma, A. K. (2020). Human Microbiome: An Academic Update on Human Body Site Specific Surveillance and Its Possible Role. *Archives of Microbiology*, 202(8), 2147–2167. https://doi.org/10.1007/s00203-020-01931-x

de la Fuente-Nunez, C., Torres Meneguetti, B., Luiz Franco, O., & Lu, T. K. (2018). Neuromicrobiology: How Microbes Influence the Brain. *ACS Chemical Neuroscience*, 9(2), 141–150.

Delannoy-Bruno, O., Desai, C., Raman, A. S., Chen, R. Y., Hibberd, M. C., Cheng, J., Han, N., Castillo, J. J., Couture, G., Lebrilla, C. B., Barve, R. A., Lombard, V., Henrissat, B., Leyn, S. A., Rodionov, D. A., Osterman, A. L., Hayashi, D. K., Meynier, A., Vinoy, S., . . . & Gordon, J. I. (2021). Evaluating Microbiome-Directed Fibre Snacks in Gnotobiotic Mice and Humans. *Nature*, 595(7865), 91–95. https://doi.org/10.1038/s41586-021-03671-4

Depommier, C., Everard, A., Druart, C., Plovier, H., Van Hul, M., Vieira-Silva, S., Falony, G., Raes, J., Maiter, D., Delzenne, N. M., de Barsy, M., Loumaye, A., Hermans, M. P., Thissen, J.-P., de Vos, W. M., & Cani, P. D. (2019). Supplementation with *Akkermansia muciniphila* in Overweight and Obese Human Volunteers: A Proof-of-Concept Exploratory Study. *Nature Medicine*, 25(7), 1096–1103. https://doi.org/10.1038/s41591-019-0495-2

DeSantis, T. Z., Hugenholtz, P., Keller, K., Brodie, E. L., Larsen, N., Piceno, Y. M., Phan, R., & Andersen, G. L. (2006). NAST: A Multiple Sequence Alignment Server for Comparative Analysis of 16S rRNA Genes. *Nucleic Acids Research*, 34(web server issue), W394–399. https://doi.org/10.1093/nar/gkl244

Di Costanzo, M., Carucci, L., Berni Canani, R., & Biasucci, G. 2020. Gut Microbiome Modulation for Preventing and Treating Pediatric Food Allergies. *International Journal of Molecular Sciences*, 21(15), 5275. https://doi.org/10.3390/ijms21155275

Dickson, R. P., Erb-Downward, J. R., Falkowski, N. R., Hunter, E. M., Ashley, S. L., & Huffnagle, G. B. (2018). The Lung Microbiota of Healthy Mice Are Highly Variable, Cluster by Environment, and Reflect Variation in Baseline Lung Innate Immunity. *American Journal of Respiratory and Critical Care Medicine*, 198(4), 497–508. https://doi.org/10.1164/rccm.201711-2180OC

Dinan, T. G., Stanton, C., & Cryan, J. F. (2013). Psychobiotics: A Novel Class of Psychotropic. *Biological Psychiatry*, 74(10), 720–726. https://doi.org/10.1016/j.biopsych.2013.05.001

Dodd, M. S., Papineau, D., Grenne, T., Slack, J. F., Rittner, M., Pirajno, F., O'Neil, J., & Little, C. T. S. (2017). Evidence for Early Life in Earth's Oldest Hydrothermal Vent Precipitates. *Nature*, 543(7643). https://doi.org/10.1038/nature21377

Dogra, S. K., Doré, J., & Damak, S. (2020). Gut Microbiota Resilience: Definition, Link to Health and Strategies for Intervention. *Frontiers in Microbiology*, 11, 572921. https://doi.org/10.3389/fmicb.2020.572921

Dominguez-Bello, M. G., Costello, E. K., Contreras, M., Magris, M., Hidalgo, G., Fierer, N., & Knight, R. (2010). Delivery Mode Shapes the Acquisition and Structure of the Initial Microbiota across Multiple Body Habitats in Newborns. *Proceedings of the National Academy of Sciences*, 107(26), 11971–11975. https://doi.org/10.1073/pnas.1002601107

Dominguez-Bello, M. G., De Jesus-Laboy, K. M., Shen, N., Cox, L. M., Amir, A., Gonzalez, A., Bokulich, N. A., Song, S. J., Hoashi, M., Rivera-Vinas, J. I., Mendez, K., Knight, R., & Clemente, J. C. (2016). Partial Restoration of the Microbiota of Cesarean-Born Infants via Vaginal Microbial Transfer. *Nature Medicine*, 22(3), 250–253. https://doi.org/10.1038/nm.4039

Doolittle, W. F., & Booth, A. (2017). It's the Song, Not the Singer: An Exploration of Holobiosis and Evolutionary Theory. *Biology & Philosophy*, 32(1), 5–24. https://doi.org/10.1007/s10539-016-9542-2

Dunn, R. R., & Thoemmes, M. S. (2022, September 13). The Future of Microbiomes in the Built Environment. National Academy of Engineering. https://nae.edu/281130/The-Future-of-Microbiomes-in-the-Built-Environment

Dupraz, C. & Visscher, P. T. (2005). Microbial lithification in marine stromatolites and hypersaline mats. *Trends in Microbiology* 13, 29–438.

Duvallet, C., Gibbons, S. M., Gurry, T., Irizarry, R. A., & Alm, E. J. (2017). Meta-analysis of Gut Microbiome Studies Identifies Disease-Specific and Shared Responses. *Nature Communications*, 8(1), 1784. https://doi.org/10.1038/s41467-017-01973-8

E

Eiseman, B., Silen, W., Bascom, G. S., & Kauvar, A. J. (1958). Fecal Enema as an Adjunct in the Treatment of Pseudomembranous Enterocolitis. *Surgery*, 44(5), 854–859.

Erdman, S. (2017, June 9). Did the Microbiome Help Drive Human Evolution? BioSpace. https://www.biospace.com/article/did-the-microbiome-help-drive-human-evolution-/

Escherich, T. (1989). The Intestinal Bacteria of the Neonate and Breast-Fed Infant. *Reviews of Infectious Diseases*, 11(2), 352–356. https://doi.org/10.1093/clinids/11.2.352

European Food Safety Authority (EFSA). (2005). Opinion of the Scientific Committee on a Request from EFSA Related to a Generic Approach to the Safety Assessment by EFSA of Microorganisms Used in Food/Feed and the Production of Food/Feed Additives. *EFSA Journal*, 226, 1–12.

European Food Safety Authority (EFSA). (2013). Scientific Opinion on the Maintenance of the List of QPS Biological Agents Intentionally Added to Food and Feed (2013 update). *EFSA Journal*, 11, 1–108.

European Food Safety Authority (EFSA). (2017). Scientific Opinion on the Update of the List of QPS-Recommended Biological Agents Intentionally Added to Food or Feed as Notified to EFSA (2017 update). *EFSA Journal*, 15, 1–177.

European Food Safety Authority (EFSA). (2013). The European Union Summary Report on Trends and Sources of Zoonoses, Zoonotic Agents and Food-Borne Outbreaks in 2011. *EFSA Journal*, 11(4), (3129), 1–250.

F

Faith, D. P. (1992). Conservation Evaluation and Phylogenetic Diversity. *Biological Conservation*, 61(1), 1–10. https://doi.org/10.1016/0006-3207(92)91201-3

Fall, T., Lundholm, C., Örtqvist, A. K., Fall, K., Fang, F., Hedhammar, Å., Kämpe, O., Ingelsson, E., & Almqvist, C. (2015). Early Exposure to Dogs and Farm Animals and the Risk of Childhood Asthma. *JAMA Pediatrics*, 169(11), e153219. https://doi.org/10.1001/jamapediatrics.2015.3219

Falony, G., Joossens, M., Vieira-Silva, S., Wang, J., Darzi, Y., Faust, K., Kurilshikov, A., Bonder, M. J., Valles-Colomer, M., Vandeputte, D., Tito, R. Y., Chaffron, S., Rymenans, L., Verspecht, C., De Sutter, L., Lima-Mendez, G., D'hoe, K., Jonckheere, K., Homola, D., . . . & Raes, J. (2016). Population-Level Analysis of Gut Microbiome Variation. *Science*, 352(6285), 560–564. https://doi.org/10.1126/science.aad3503

Falony, G., Vandeputte, D., Caenepeel, C., Vieira-Silva, S., Daryoush, T., Vermeire, S., & Raes, J. (2019). The human microbiome in health and disease: hype or hope. *Acta Clinica Belgica*, 74(2), 53–64. https://doi.org/10.1080/17843286.2019.1583782

Farmer, D. K., Vance, M. E., Abbatt, J. P. D., Abeleira, A., Alves, M. R., Arata, C., Boedicker, E., Bourne, S., Cardoso-Saldaña, F., Corsi, R., DeCarlo, P. F., Goldstein, A. H., Grassian, V. H., Hildebrandt Ruiz, L., Jimenez, J. L., Kahan, T. F., Katz, E. F., Mattila, J. M., Nazaroff, W. W., ... & Zhou, Y. (2019). Overview of HOMEChem: House Observations of Microbial and Environmental Chemistry. *Environmental Science: Processes & Impacts*, 21(8), 1280–1300. https://doi.org/10.1039/C9EM00228F

Feehley, T., Plunkett, C. H., Bao, R., Choi Hong, S. M., Culleen, E., Belda-Ferre, P., Campbell, E., Aitoro, R., Nocerino, R., Paparo, L., Andrade, J., Antonopoulos, D. A., Berni Canani, R., & Nagler, C. R. (2019). Healthy Infants Harbor Intestinal Bacteria That Protect Against Food Allergy. *Nature Medicine*, 25(3), Article 3. https://doi.org/10.1038/s41591-018-0324-z

Feng, Y.-K., Wu, Q.-L., Peng, Y.-W., Liang, F.-Y., You, H.-J., Feng, Y.-W., Li, G., Li, X.-J., Liu, S.-H., Li, Y.-C., Zhang, Y., & Pei, Z. (2020). Oral *P. gingivalis* impairs gut permeability and mediates immune responses associated with neurodegeneration in LRRK2 R1441G mice. *Journal of Neuroinflammation*, 17(1), 347. https://doi.org/10.1186/s12974-020-02027-5

Fleming, A. (1929). On the antibacterial action of cultures of a penicillium, with special reference to their use in the isolation of *B. influenzae*. *Bulletin of the World Health Organization*, 79(8), 780–790.

Florida, R., Glaeser, E., Mohd Sharif, M., Bedi, K., Campanella, T. J., Chee, C. H., Doctoroff, D., Katz, B., Katz, R., Kotkin, J., Muggah, R., Sadik-Kahn, J., & Theil, S. (2020, May 1). How Life in Our Cities Will Look After the Coronavirus Pandemic. *Foreign Policy*. https://foreignpolicy.com/2020/05/01/future-of-cities-urban-life-after-coronavirus-pandemic/

Foley, M. H., O'Flaherty, S., Allen, G., Rivera, A. J., Stewart, A. K., Barrangou, R., & Theriot, C. M. (2021). *Lactobacillus* bile salt hydrolase substrate specificity governs bacterial fitness and host colonization. *Proceedings of the National Academy of Sciences USA*, 118(6), e2017709118. https://doi: 10.1073/pnas.2017709118.

Fontaine, K. R., Redden, D. T., Wang, C., Westfall, A. O., & Allison, D. B. (2003). Years of Life Lost Due to Obesity. *JAMA*, 289(2), 187. https://doi.org/10.1001/jama.289.2.187

Food and Agriculture Organization (FAO). (2002). Guidelines for the Evaluation of Probiotics in Food: Report of a Joint FAO/WHO Working Group on Drafting Guidelines for the Evaluation of Probiotics in Food. FAO, London, ON, Canada.

Forgie, A. J., Drall, K. M., Bourque, S. L., Field, C. J., Kozyrskyj, A. L., & Willing, B. P. (2020). The Impact of Maternal and Early Life Malnutrition on Health: A Diet-Microbe Perspective. *BMC Medicine*, 18(1), 135. https://doi.org/10.1186/s12916-020-01584-z

Foster, K. R., Schluter, J., Coyte, K. Z., & Rakoff-Nahoum, S. (2017). The Evolution of the Host Microbiome as an Ecosystem on a Leash. *Nature*, 548(7665), 43. https://doi.org/10.1038/nature23292

Fouhy, F., Ross, R. P., Fitzgerald, G. F., Stanton, C., & Cotter, P. D. (2012). Composition of the Early Intestinal Microbiota: Knowledge, Knowledge Gaps and the Use of High-Throughput Sequencing to Address These Gaps. *Gut Microbes*, 3(3), 203–220. https://doi.org/10.4161/gmic.20169

Fox, A. T., Wopereis, H., Van Ampting, M. T. J., Oude Nijhuis, M. M., Butt, A. M., Peroni, D. G., Vandenplas, Y., Candy, D. C. A., Shah, N., West, C. E., Garssen, J., Harthoorn, L. F., Knol, J., Michaelis, L. J., & ASSIGN Study Group. (2019). A Specific Synbiotic-Containing Amino Acid-Based Formula in Dietary Management of Cow's Milk Allergy: A Randomized Controlled Trial. *Clinical and Translational Allergy*, 9, 5. https://doi.org/10.1186/s13601-019-0241-3

Friedland, R. P. (2015). Mechanisms of Molecular Mimicry Involving the Microbiota in Neurodegeneration. *Journal of Alzheimer's Disease*, 45(2), 349–362. https://doi.org/10.3233/JAD-142841

Frobisher, M., & Fuerst, R. (1983). *Frobisher & Fuerst's Microbiology in Health & Disease* (15th ed). Saunders. http://books.google.com/books?id=PQprAAAAMAAJ

Fromentin, S., Forslund, S. K., Chechi, K., . . . & Pedersen, O. (2022). Microbiome and Metabolome Features of the Cardiometabolic Disease Spectrum. *Nature Medicine*, 28, 303–314.

Frumkin, H. (2021). COVID-19, the Built Environment, and Health. *Environmental Health Perspectives*, 129(7), 075001. https://doi.org/10.1289/EHP8888

Fu, X., Norbäck, D., Yuan, Q., Li, Y., Zhu, X., Hashim, J. H., Hashim, Z., Ali, F., Hu, Q., Deng, Y., & Sun, Y. (2021). Association between Indoor Microbiome Exposure and Sick Building Syndrome (SBS) in Junior High Schools of Johor Bahru, Malaysia. *Science of the Total Environment*, 753, 141904. https://doi.org/10.1016/j.scitotenv.2020.141904

G

Ganesan, K., Chung, S. K., Vanamala, J., & Xu, B. (2018). Causal Relationship between Diet-Induced Gut Microbiota Changes and Diabetes: A Novel Strategy to Transplant *Faecalibacterium prausnitzii* in Preventing Diabetes. *International Journal of Molecular Sciences*, 19(12), 3720. https://doi.org/10.3390/ijms19123720

Gardner, H. L., & Dukes, C. D. (1955). *Haemophilus vaginalis* Vaginitis: A Newly Defined Specific Infection Previously Classified Non-Specific Vaginitis. *American Journal of Obstetrics and Gynecology*, 69(5), 962–976.

Gassen, N. C., & Rein, T. (2019). Is There a Role of Autophagy in Depression and Antidepressant Action? *Frontiers in Psychiatry*, 10, 337. https://doi.org/10.3389/fpsyt.2019.00337

Gaynes, R. (2017). The Discovery of Penicillin—New Insights After More Than 75 Years of Clinical Use. *Emerging Infectious Diseases*, 23(5), 849–53. https:/doi: 10.3201/eid2305.161556

Gellie, N. J. C., Mills, J. G., Breed, M. F., & Lowe, A. J. (2017). Revegetation Rewilds the Soil Bacterial Microbiome of an Old Field. *Molecular Ecology*, 26(11), 2895–2904. https://doi.org/10.1111/mec.14081

Gerrard, J. W., Geddes, C. A., Reggin, P. L., Gerrard, C. D., & Horne, S. (1976). Serum IgE Levels in White and Metis Communities in Saskatchewan. *Annals of Allergy*, 37(2), 91–100.

Geuking, M. B., & Burkhard, R. (2020). Microbial Modulation of Intestinal T Helper Cell Responses and Implications for Disease and Therapy. *Mucosal Immunology*, 13(6), 855–866.

Gibbons, S. (2020). Keystone Taxa Indispensable for Microbiome Recovery. *Nature Microbiology*, 5, 1067–1068. https://doi.org/10.1038/s41564-020-0783-0

Gibson, G. R., Hutkins, R., Sanders, M. E., Prescott, S. L., Reimer, R. A., Salminen, S. J., Scott, K., Stanton, C., Swanson, K. S., Cani, P. D., Verbeke, K., & Reid, G. (2017). Expert Consensus Document: The International Scientific Association for Probiotics and Prebiotics (ISAPP) Consensus Statement on the Definition and Scope of Prebiotics. *Nature Reviews Gastroenterology*, 14, 491–502.

Gilbert J. A., & Dupont, C. L. (2011). Microbial metagenomics: beyond the genome. *Annual Review of Marine Science*, 3, 347–371.

Gilbert, J. A., & Lynch, S. V. (2019). Community Ecology as a Framework for Human Microbiome Research. *Nature Medicine*, 25(6), 884–889. https://doi.org/10.1038/s41591-019-0464-9

Gilbert, J. A., Quinn, R. A., Debelius, J., Xu, Z. Z., Morton, J., Garg, N., Jansson, J. K., Dorrestein, P. C., & Knight, R. (2016). Microbiome-Wide Association Studies Link Dynamic Microbial Consortia to Disease. *Nature*, 535(7610), 94–103. https://doi.org/10.1038/nature18850

Gilbert, J. A., & Stephens, B. (2018). Microbiology of the Built Environment. *Nature Reviews Microbiology*, 16(11), 661–670. https://doi.org/10.1038/s41579-018-0065-5

Gonzalez, A., Navas-Molina, J. A., Kosciolek, T., McDonald, D., Vázquez-Baeza, Y., Ackermann, G., DeReus, J., Janssen, S., Swafford, A. D., Orchanian, S. B., Sanders, J. G., Shorenstein, J., Holste, H., Petrus, S., Robbins-Pianka, A., Brislawn, C. J., Wang, M., Rideout, J. R., Bolyen, E., . . . & Knight, R. (2018). Qiita: Rapid, Web-Enabled Microbiome Meta-analysis. *Nature Methods*, 15(10), Article 10. https://doi.org/10.1038/s41592-018-0141-9

Guerin, E., & Hill, C. (2020). Shining Light on Human Gut Bacteriophages. *Frontiers in Cellular and Infection Microbiology*, 10, 481. https://doi.org/10.3389/fcimb.2020.00481 ALSO in chapter 7

Gulliver, E. L., Young, R. B., Chonwerawong, M., D'Adamo, G. L., Thomason, T., Widdop, J. T., Rutten, E. L., Rossetto Marcelino, V., Bryant, R. V., Costello, S. P., O'Brien, C. L., Hold, G. L., Giles, E. M., & Forster, S. C. (2022). Review Article: The Future of Microbiome-Based Therapeutics. *Alimentary Pharmacology & Therapeutics*, 56(2), 192–208. https://doi.org/10.1111/apt.17049

Guo, M., Miao, M., Wang, Y., Duan, M., Yang, F., Chen, Y., Yuan, W., & Zheng, H. (2020). Developmental Differences in the Intestinal Microbiota of Chinese 1-Year-Old Infants and 4-Year-Old Children. *Scientific Reports*, 10(1), 19470. https://doi.org/10.1038/s41598-020-76591-4

Gulliver, E. L., Young, R. B., Chonwerawong, M., D'Adamo, G. L., Thomason, T., Widdop, J. T., . . . & Forster, S. C. (2022). Review Article: The Future of Microbiome-Based Therapeutics. *Alimentary Pharmacology & Therapeutics*, 56, 192–208.

Gupta, V. K., Kim, M., Bakshi, U., Cunningham, K. Y., Davis, J. M., Lazaridis, K. N., Nelson, H., Chia, N., & Sung, J. (2020). A Predictive Index for Health Status Using Species-Level Gut Microbiome Profiling. *Nature Communications*, 11(1), Article 1. https://doi.org/10.1038/s41467-020-18476-8

György, P., Norris, R. F., & Rose, C. S. (1954). Bifidus Factor. I. A Variant of *Lactobacillus bifidus* Requiring a Special Growth Factor. *Archives of Biochemistry and Biophysics*, 48(1), 193–201. https://doi.org/10.1016/0003-9861(54)90323-9

H

Haahtela, T. (2019). A Biodiversity Hypothesis. *Allergy*, 74(8), 1445–1456. https://doi.org/10.1111/all.13763

Haahtela, T., Valovirta, E., Bousquet, J., & Mäkelä, M. (2017). The Finnish Allergy Programme 2008–2018 Works. *European Respiratory Journal*, 49(6), 1700470. https://doi.org/10.1183/13993003.00470-2017

Haitjema, C., Solomon, K., Henske, J., Theodorou, M., & O'Malley, M. (2014). Anaerobic Gut Fungi: Advances in Isolation, Culture, and Cellulolytic Enzyme Discovery for Biofuel Production. *Biotechnology and Bioengineering*. 111(8),1471–1482.

Hamady, M., Walker, J. J., Harris, J. K., Gold, N. J., & Knight, R. (2008). Error-Correcting Barcoded Primers for Pyrosequencing Hundreds of Samples in Multiplex. *Nature Methods*, 5(3), 235–237. https://doi.org/10.1038/nmeth.1184

Hämäläinen, H. (2021, August 27). Turn Data into Insight. *Cloudia*. https://cloudia.com/turn-data-into-insight/

Handelsman, J., Rondon, M. R., Brady, S. F., Clardy, J., & Goodman, R. M. (1998). Molecular Biological Access to the Chemistry of Unknown Soil Microbes: A New Frontier for Natural Products. *Chemistry & Biology*, 5(10), R245–R249. https://doi.org/10.1016/S1074-5521(98)90108-9

Harrison, G. P. (2021, October 29). *At Least Know This*. Quoted on Goodreads. https://www.goodreads.com/quotes/10688523-every-person-is-a-collective-a-vast-and-complex-gathering

Haynes, R. B. (2006). Forming Research Questions. *Journal of Clinical Epidemiology*, 59(9), 881–886. https://doi.org/10.1016/j.jclinepi.2006.06.006

Heindel, J. J., Skalla, L. A., Joubert, B. R., Dilworth, C. H., & Gray, K. A. (2017). Review of Developmental Origins of Health and Disease Publications in Environmental Epidemiology. *Reproductive Toxicology*, 68, 34–48. https://doi.org/10.1016/j.reprotox.2016.11.011

Helander, H. F., & Fändriks, L. (2014). Surface Area of the Digestive Tract—Revisited. *Scandinavian Journal of Gastroenterology*, 49(6), 681–689. https://doi.org/10.3109/00365521.2014.898326

Henker, J., Laass, M., Blokhin, B. M., Bolbot, Y. K., Maydannik, V. G., Elze, M., Wolff, C., & Schulze, J. (2007). The Probiotic *Escherichia coli* Strain Nissle 1917 (EcN) Stops Acute Diarrhoea in Infants and Toddlers. *European Journal of Pediatrics*, 166(4), 311–318. https://doi.org/10.1007/s00431-007-0419-x

Hernández-Calderón, P., Wiedemann, L., & Benítez-Páez, A. (2022). The Microbiota Composition Drives Personalized Nutrition: Gut Microbes as Predictive Biomarkers for the Success of Weight Loss Diets. *Frontiers in Nutrition*, 9, 1006747. https://doi.org/10.3389/fnut.2022.1006747

Ho, N. T., Li, F., Lee-Sarwar, K. A., Tun, H. M., Brown, B. P., Pannaraj, P. S., Bender, J. M., Azad, M. B., Thompson, A. L., Weiss, S. T., Azcarate-Peril, M. A., Litonjua, A. A., Kozyrskyj, A. L., Jaspan, H. B., Aldrovandi, G. M., & Kuhn, L. (2018). Meta-analysis of Effects of Exclusive Breastfeeding on Infant Gut Microbiota Across Populations. *Nature Communications*, 9(1), 4169. https://doi.org/10.1038/s41467-018-06473-x

Hooke, R. (1665). *Micrographia*. Printed by Jo. Martyn and J. Alleftry.

Huang, Y., Niu, B., Gao, Y., Fu, L., & Li, W. (2010). CD-HIT Suite: a web server for clustering and comparing biological sequences. *Bioinformatics*, Oxford, England. 26(5), 680–682. https:doi.org/10.1093/bioinformatics/btq003.

Hungate, R. E. (1950). The Anaerobic Mesophilic Cellulolytic Bacteria. *Bacteriological Reviews*, 14(1), 1–49. https://doi.org/10.1128/br.14.1.1-49.1950

Hunter, P. (2012). The Changing Hypothesis of the Gut: The Intestinal Microbiome Is Increasingly Seen as Vital to Human Health. *EMBO Reports*, 13(6), 498–500. https://doi.org/10.1038/embor.2012.68

Husain, S., Allotey, J., Drymoussi, Z., Wilks, M., Fernandez-Felix, B., Whiley, A., Dodds, J., Thangaratinam, S., McCourt, C., Prosdocimi, E., Wade, W., Tejada, B., Zamora, J., Khan, K., & Millar, M. (2019). Effects of Oral Probiotic Supplements on Vaginal Microbiota During Pregnancy: A Randomised, Double-Blind, Placebo-Controlled Trial with Microbiome Analysis. *BJOG: An International Journal of Obstetrics & Gynaecology*, 127(2), 275–284. https://doi.org/10.1111/1471-0528.15675

Huse, S. M., Mark Welch, D. B., Voorhis, A., Shipunova, A., Morrison, H. G., Eren, A. M., & Sogin, M. L. (2014). VAMPS: A Website for Visualization and Analysis of Microbial Population Structures. *BMC Bioinformatics*, 15(1), 41. https://doi.org/10.1186/1471-2105-15-41

I

Iddrisu, I., Monteagudo-Mera, A., Poveda, C., Pyle, S., Shahzad, M., Andrews, S., & Walton, G. E. (2021). Malnutrition and Gut Microbiota in Children. *Nutrients*, 13(8), 2727. https://doi.org/10.3390/nu13082727

J

Jarde, A., Lewis-Mikhael, A.-M., Moayyedi, P., Stearns, J. C., Collins, S. M., Beyene, J., & McDonald, S. D. (2018). Pregnancy Outcomes in Women Taking Probiotics or Prebiotics: A Systematic Review and Meta-analysis. *BMC Pregnancy and Childbirth*, 18(1), 14. https://doi.org/10.1186/s12884-017-1629-5

Jin, B. T. (2018). Mian: Interactive Web-Based 16S rRNA Operational Taxonomic Unit Table Data Visualization and Discovery Platform (p. 416073). *bioRxiv*. https://doi.org/10.1101/416073

Jo, J.-H., Kennedy, E. A., & Kong, H. H. (2016). Research Techniques Made Simple: Bacterial 16S Ribosomal RNA Gene Sequencing in Cutaneous Research. *Journal of Investigative Dermatology*, 136(3), e23–e27. https://doi.org/10.1016/j.jid.2015.11.023

Johnson, K. V.-A., & Foster, K. R. (2018). Why Does the Microbiome Affect Behaviour? *Nature Reviews Microbiology*, 16(10), 647–655. https://doi.org/10.1038/s41579-018-0014-3

Jorth, P., Turner, K. H., Gumus, P., Nizam, N., Buduneli, N., & Whiteley, M. (2014). Metatranscriptomics of the Human Oral Microbiome during Health and Disease. *mBio*, 5(2), e01012-14. https://doi.org/10.1128/mBio.01012-14

Judger, B., Batista, J., Gibson, J., Cunningham, P., Asara, J., & Watnick, P. (2022). *Vibrio cholerae* High Cell Density Quorum Sensing Activates the Host Intestinal Innate Immune Response. *Cell Reports*, 40(12), 111368. https://doi.org/10.1016/j.celrep.2022.111368

Junca, H., Pieper, D. H., & Medina, E. (2022). The Emerging Potential of Microbiome Transplantation on Human Health Interventions. *Computational and Structural Biotechnology Journal*, 20, 615–627. https://doi.org/10.1016/j.csbj.2022.01.009

K

Karmacharya, D. (2022, August 18). The Global Microbiome Conservancy—Uncovering the Mysteries of Important Bugs That Help Us Live a Healthy Life. Springer Nature Research Communities. https://healthcommunity.nature.com/posts/the-global-microbiome-conservancy-uncovering-the-mysteries-of-important-bugs-that-help-us-live-a-healthy-life

Kaur, K. K., Allahbadia, G., & Singh, M. (2021). The Role of Vaginal Microbiota Dysbiosis in Case of Gynaecological Diseases and the Potential Treatment with Antibiotics, Probiotics and Vaginal Microbiota Transplantation: How Practical Will It Be? *Frontiers in Obstetrics and Gynaecology*, 4, 2581–3226.

Kelley, S. T., Theisen, U., Angenent, L. T., St. Amand, A., & Pace, N. R. (2004). Molecular Analysis of Shower Curtain Biofilm Microbes. *Applied and Environmental Microbiology*, 70(7), 4187–4192. https://doi.org/10.1128/AEM.70.7.4187-4192.2004

Kembel, S. W., Jones, E., Kline, J., Northcutt, D., Stenson, J., Womack, A. M., Bohannan, B. J. M., Brown, G. Z., & Green, J. L. (2012). Architectural Design Influences the Diversity and Structure of the Built Environment Microbiome. *ISME Journal*, 6(8), 1469–1479. https://doi.org/10.1038/ismej.2011.211

Kembel, S. W., Meadow, J. F., O'Connor, T. K., Mhuireach, G., Northcutt, D., Kline, J., Moriyama, M., Brown, G. Z., Bohannan, B. J. M., & Green, J. L. (2014). Architectural Design Drives the Biogeography of Indoor Bacterial Communities. *PLOS ONE*, 9(1), e87093. https://doi.org/10.1371/journal.pone.0087093

Kim, S., Kwon, S.-H., Kam, T.-I., Panicker, N., Karuppagounder, S. S., Lee, S., Lee, J. H., Kim, W. R., Kook, M., Foss, C. A., Shen, C., Lee, H., Kulkarni, S., Pasricha, P. J., Lee, G., Pomper, M. G., Dawson, V. L., Dawson, T. M., & Ko, H. S. (2019). Transneuronal Propagation of Pathologic α-Synuclein from the Gut to the Brain Models Parkinson's Disease. *Neuron*, 103(4), 627–641.e7. https://doi.org/10.1016/j.neuron.2019.05.035

Koch, R. (1876). Die Ätiologie der Milzbrand-Krankheit, begründet auf die Entwicklungsgeschichte des *Bacillus Anthracis*. *Bacteriology Archive*, 277–310. Publication Server of Robert Koch Institute. https://doi.org/10.25646/5064

Kolter, R. (2017, October 30). My Home Is My Microbial Castle. *Small Things Considered*. https://schaechter.asmblog.org/schaechter/2017/10/my-home-is-my-microbial-castle.html

Kong, H. H. (2011). Skin Microbiome: Genomics-Based Insights into the Diversity and Role of Skin Microbes. *Trends in Molecular Medicine*, 17(6), 320–328.

Kootte, R. S., Levin, E., Salojärvi, J., Smits, L. P., Hartstra, A. V., Udayappan, S. D., Hermes, G., Bouter, K. E., Koopen, A. M., Holst, J. J., Knop, F. K., Blaak, E. E., Zhao, J., Smidt, H., Harms, A. C., Hankemeijer, T., Bergman, J. J. G. H. M., Romijn, H. A., Schaap, F. G., & Nieuwdorp, M. (2017). Improvement of Insulin Sensitivity After Lean Donor Feces in Metabolic Syndrome Is Driven by Baseline Intestinal Microbiota Composition. *Cell Metabolism*, 26(4), 611–619.e6. https://doi.org/10.1016/j.cmet.2017.09.008

Krawczyk, R. T., & Banaszkiewicz, A. (2021). Dr. Józef Brudziński—The True "Father of Probiotics." *Beneficial Microbes*, 12(3), 211–213. https://doi.org/10.3920/BM2020.0201

Kuo, C.-H., Kuo, H.-F., Huang, C.-H., Yang, S.-N., Lee, M.-S., & Hung, C.-H. (2013). Early Life Exposure to Antibiotics and the Risk of Childhood Allergic Diseases: An Update from the Perspective of the Hygiene Hypothesis. *Journal of Microbiology, Immunology and Infection*, 46(5), 320–329.

L

Lamont, R. J., & Hajishengallis, G. (2015). Polymicrobial Synergy and Dysbiosis in Inflammatory Disease. *Trends in Molecular Medicine*, 21(3), 172–183. https://doi.org/10.1016/j.molmed.2014.11.004

LaMonte, M. J., Gordon, J. H., Diaz-Moreno, P., Andrews, C. A., Shimbo, D., Hovey, K. M., Buck, M. J., & Wactawski-Wende, J. (2022). Oral Microbiome Is Associated with Incident Hypertension among Postmenopausal Women. *Journal of the American Heart Association*, 11(6), e021930. https://doi.org/10.1161/JAHA.121.021930

Landhuis, E. (2020, May 23). Gut Microbes May Be Key to Solving Food Allergies. *Scientific American*. https://www.scientificamerican.com/article/gut-microbes-may-be-key-to-solving-food-allergies/

Langille, J. (2019, October 23). Researchers Describe Gut Health's Influence on Brain Health. *Cornell Chronicle*. https://news.cornell.edu/stories/2019/10/researchers-describe-gut-healths-influence-brain-health

Laursen, M. F. (2021). Gut Microbiota Development: Influence of Diet from Infancy to Toddlerhood. *Annals of Nutrition and Metabolism*, 77(Suppl. 3), 21–34. https://doi.org/10.1159/000517912

Lax, S., Cardona, C., Zhao, D., Winton, V. J., Goodney, G., Gao, P., Gottel, N., Hartmann, E. M., Henry, C., Thomas, P. M., Kelley, S. T., Stephens, B., & Gilbert, J. A. (2019). Microbial and Metabolic Succession on Common Building Materials Under High Humidity Conditions. *Nature Communications*, 10(1), 1767. https://doi.org/10.1038/s41467-019-09764-z

Lax, S., Sangwan, N., Smith, D., Larsen, P., Handley, K. M., Richardson, M., Guyton, K., Krezalek, M., Shogan, B. D., Defazio, J., Flemming, I., Shakhsheer, B., Weber, S., Landon, E., Garcia-Houchins, S., Siegel, J., Alverdy, J., Knight, R., Stephens, B., & Gilbert, J. A. (2017). Bacterial colonization and succession in a newly opened hospital. *Science Translational Medicine*, 9, eaah6500. https://www.science.org/doi/10.1126/scitranslmed.aah6500

Lax, S., Nagler, C. R., & Gilbert, J. A. (2015). Our Interface with the Built Environment: Immunity and the Indoor Microbiota. *Trends in Immunology*, 36(3), 121–123. https://doi.org/10.1016/j.it.2015.01.001

Lax, S., Smith, D. P., Hampton-Marcell, J., Owens, S. M., Handley, K. M., Scott, N. M., Gibbons, S. M., Larsen, P., Shogan, B. D., Weiss, S., Metcalf, J. L., Ursell, L. K., Vázquez-Baeza, Y., Van Treuren, W., Hasan, N. A., Gibson, M. K., Colwell, R., Dantas, G., Knight, R., & Gilbert, J. A. (2014). Longitudinal Analysis of Microbial Interaction between Humans and the Indoor Environment. *Science*, 345(6200), 1048–1052. https://doi.org/10.1126/science.1254529

Lee, B. K., Magnusson, C., Gardner, R. M., Blomström, Å., Newschaffer, C. J., Burstyn, I., Karlsson, H., & Dalman, C. (2015). Maternal hospitalization with infection during pregnancy and risk of autism spectrum disorders. *Brain, Behavior, and Immunity*, 44, 100–105. https://doi.org/10.1016/j.bbi.2014.09.001

Leidy, J. (1853). *A Flora and Fauna within Living Animals* (Vol. 5). Smithsonian Institution. https://doi.org/10.5962/bhl.title.56319

Leidy, J. (1881). The Parasites of the Termites. *Journal of the Academy of Natural Sciences of Philadelphia*, 8(2), 425–436.

Leitch, C. (2021, January 12). Connecting Gut Microbes, Diet, and Health. *Labroots*. https://www.labroots.com/trending/microbiology/19594/connecting-gut-microbes-diet-health

LeMieux, J. (2020, October 2). Spatial: The Next Omics Frontier. *Genetic Engineering and Biotechnology News*, 40(10). https://www.genengnews.com/insights/spatial-the-next-omics-frontier/

Leonel, C., Sena, I. F. G., Silva, W. N., Prazeres, P. H. D. M., Fernandes, G. R., Mancha Agresti, P., Martins Drumond, M., Mintz, A., Azevedo, V. A. C., & Birbrair, A. (2019). *Staphylococcus epidermidis* role in the skin microenvironment. *Journal of Cellular and Molecular Medicine* 23(9), 5949–5955. https:// doi: 10.1111/jcmm.14415

LeRoy, T., Moens de Hase, E., Van Hul, M., Paquot, A., Pelicaen, R., Régnier, M., Depommier, C., Druart, C., Everard, A., Maiter, D., Delzenne, N.M., Bindels, L.B., de Barsy, M., Loumaye, A., Hermans, M.P., Thissen, J.P., Vieira-Silva, S., Falony, G., Raes, J., Muccioli, G.G., & Cani, P.D. (2022). *Dysosmobacter welbionis* is a newly isolated human commensal bacterium preventing diet-induced obesity and metabolic disorders in mice. *Gut*, 71(3), 534–543.

Levan, S. R., Stamnes, K. A., Lin, D. L., Panzer, A. R., Fukui, E., McCauley, K., Fujimura, K. E., McKean, M., Ownby, D. R., Zoratti, E. M., Boushey, H. A., Cabana, M. D., Johnson, C. C., & Lynch, S. V. (2019). Elevated Faecal 12,13-diHOME Concentration in Neonates at High Risk for Asthma Is Produced by Gut Bacteria and Impedes Immune Tolerance. *Nature Microbiology*, 4(11), 1851–1861. https://doi.org/10.1038/s41564-019-0498-2

Lev-Sagie, A., Goldman-Wohl, D., Cohen, Y., Dori-Bachash, M., Leshem, A., Mor, U., Strahilevitz, J., Moses, A. E., Shapiro, H., Yagel, S., & Elinav, E. (2019). Vaginal Microbiome Transplantation in Women with Intractable Bacterial Vaginosis. *Nature Medicine*, 25(10), 1500–1504. https://doi.org/10.1038/s41591-019-0600-6

Levy, M., Kolodziejczyk, A. A., Thaiss, C. A., & Elinav, E. (2017). Dysbiosis and the Immune System. *Nature Reviews Immunology*, 17(4), 219–232. https://doi.org/10.1038/nri.2017.7

Ley, R. E., Bäckhed, F., Turnbaugh, P., Lozupone, C. A., Knight, R. D., & Gordon, J. I. (2005). Obesity Alters Gut Microbial Ecology. *Proceedings of the National Academy of Sciences*, 102(31), 11070–11075. https://doi.org/10.1073/pnas.0504978102

Li, D., Liu, R., Wang, M., Peng, R., Fu, S., Fu, A., Le, J., Yao, Q., Yuan, T., Chi, H., Mu, X., Sun, T., Liu, H., Yan, P., Wang, S., Cheng, S., Deng, Z., Liu, Z., Wang, G., & Li, Y. (2022). 3β-Hydroxysteroid Dehydrogenase Expressed by Gut Microbes Degrades Testosterone and Is Linked to Depression in Males. *Cell Host & Microbe*, 30(3), 329–339. https://doi.org/10.1016/j.chom.2022.01.001

Li, W., & Godzik, A. (2006). Cd-hit: A Fast Program for Clustering and Comparing Large Sets of Protein or Nucleotide Sequences. *Bioinformatics*, 22(13), 1658–1659. https://doi.org/10.1093/bioinformatics/btl158

Li, X., Lester, D., Rosengarten, G., Aboltins, C., Patel, M., & Cole, I. (2022). A Spatiotemporally Resolved Infection Risk Model for Airborne Transmission of COVID-19 Variants in Indoor Spaces. *Science of the Total Environment*, 812, 152592. https://doi.org/10.1016/j.scitotenv.2021.152592

Linares, D. M., Ross, P., & Stanton, C. (2016). Beneficial Microbes: The Pharmacy in the Gut. *Bioengineered*, 7(1), 11–20. https://doi.org/10.1080/21655979.2015.1126015

Linehan, J. L., Harrison, O. J., Han, S.-J., Byrd, A. L., Vujkovic-Cvijin, I., Villarino, A. V., Sen, S. K., Shaik, J., Smelkinson, M., Tamoutounour, S., Collins, N., Bouladoux, N., Dzutsev, A., Rosshart, S. P., Arbuckle, J. H., Wang, C.-R., Kristie, T. M., Rehermann, B., Trinchieri, G., . . . & Belkaid, Y. (2018). Non-Classical Immunity Controls Microbiota Impact on Skin Immunity and Tissue Repair. *Cell*, 172(4), 784–796.e18. https://doi.org/10.1016/j.cell.2017.12.033

Litvak, Y., & Bäumler, A. J. (2019). Microbiota-Nourishing Immunity: A Guide to Understanding Our Microbial Self. *Immunity*, 51(2), 214–224. https://doi.org/10.1016/j.immuni.2019.07.002

Liu, A., Ma, T., Xu, N., Jin, H., Zhao, F., Kwok, L.-Y., Zhang, H., Zhang, S., & Sun, Z. (2021). Adjunctive Probiotics Alleviates Asthmatic Symptoms via Modulating the Gut Microbiome and Serum Metabolome. *Microbiology Spectrum*, 9(2), e00859-21. https://doi.org/10.1128/Spectrum.00859-21

Liu, G., Lu, J., Sun, W., Jia, G., Zhao, H., Chen, X., Kim, I. H., Zhang, R., & Wang, J. (2022). Tryptophan Supplementation Enhances Intestinal Health by Improving Gut Barrier Function, Alleviating Inflammation, and Modulating Intestinal Microbiome in Lipopolysaccharide-Challenged Piglets. *Frontiers in Microbiology*, 13, 919431. https://doi.org/10.3389/fmicb.2022.919431

Lozupone, C., & Knight, R. (2005). UniFrac: A New Phylogenetic Method for Comparing Microbial Communities. *Applied and Environmental Microbiology*, 71(12), 8228–8235. https://doi.org/10.1128/AEM.71.12.8228-8235.2005

Lucas, C., Barnich, N., & Nguyen, H. T. T. (2017). Microbiota, Inflammation and Colorectal Cancer. *International Journal of Molecular Sciences*, 18(6), Article 6. https://doi.org/10.3390/ijms18061310

Luna, P. C. (2020). Skin Microbiome as Years Go By. *American Journal of Clinical Dermatology*, 21(Suppl. 1), 12–17. https://doi.org/10.1007/s40257-020-00514-1

Lyons, T. W., Reinhard, C. T., & Planavsky, N. J. (2014). The rise of oxygen in Earth's early ocean and atmosphere. *Nature* 506, 307–315.

M

Macpherson, A. J., de Agüero, M. G., & Ganal-Vonarburg, S. C. (2017). How Nutrition and the Maternal Microbiota Shape the Neonatal Immune System. *Nature Reviews Immunology*, 17(8), Article 8. https://doi.org/10.1038/nri.2017.58

Madigan, M. T., & Martinko, M. (2006). *Brock Biology of Microorganisms* (11th ed.). Pearson Prentice Hall.

Maguire, M., & Maguire, G. (2019). Gut Dysbiosis, Leaky Gut, and Intestinal Epithelial Proliferation in Neurological Disorders: Towards the Development of a New Therapeutic Using Amino Acids, Prebiotics, Probiotics, and Postbiotics. *Reviews in the Neurosciences*, 30(2), 179–201. https://doi.org/10.1515/revneuro-2018-0024

Mahalanobis, P. (1933). Prasanta Mahalanobis quotes. *Today in Science History*. https://todayinsci.com/M/Mahalanobis_Prasanta/MahalanobisPrasanta-Quotations.htm

Maioli, T. U., Borras-Nogues, E., Torres, L., . . . & Chatel, J-M. (2021). Possible Benefits of *Faecalibacterium prausnitzii* For Obesity-Associated Gut Disorders. *Frontiers in Pharmacology*, 12, 740636. https://doi.org/10.3389/fphar.2021.740636

Mann, A. (2021). Making Headway with the Mysteries of Life's Origins. *Proceedings of the National Academy of Sciences*, 118(16), e2105383118. https://doi.org/10.1073/pnas.2105383118

Manus, M. B., Kuthyar, S., Perroni-Marañón, A. G., Núñez-de la Mora, A., & Amato, K. R. (2020). Infant Skin Bacterial Communities Vary by Skin Site and Infant Age across Populations in Mexico and the United States. *mSystems*, 5(6), e00834-20. https://doi.org/10.1128/mSystems.00834-20

Marietta, E. V., Murray, J. A., Luckey, D. H., Jeraldo, P. R., Lamba, A., Patel, R., Luthra, H. S., Mangalam, A., & Taneja, V. (2016). Suppression of Inflammatory Arthritis by Human Gut-Derived *Prevotella histicola* in Humanized Mice. *Arthritis & Rheumatology*, 68(12), 2878–2888. https://doi: 10.1002/art.39785.

Markowiak, P., & Śliżewska, K. (2017). Effects of Probiotics, Prebiotics, and Synbiotics on Human Health. *Nutrients*, 9(9), Article 1021. https://doi.org/10.3390/nu9091021

Marrs, T., Logan, K., Craven, J., Radulovic, S., McLean, W. H. I., Lack, G., Flohr, C., Perkin, M. R., & the EAT Study Team. (2019). Dog Ownership at Three Months of Age Is Associated with Protection Against Food Allergy. *Allergy*, 74(11), 2212–2219. https://doi.org/10.1111/all.13868

Matijašić, M., Meštrović, T., Čipčić Paljetak, H., Perić, M., Barešić, A., & Verbanac, D. (2020). Gut Microbiota Beyond Bacteria—Mycobiome, Virome, Archaeome, and Eukaryotic Parasites in IBD. *International Journal of Molecular Sciences*, 21(8), Article 2668. https://doi.org/10.3390/ijms21082668

Mayerhofer, H., & Pali-Schöll, I. (2021). The Farm Effect Revisited: From β-Lactoglobulin with Zinc in Cowshed Dust to Its Application. *Allergo Journal International*, 30, 135–140. https://doi.org/10.1007/s40629-021-00178-0

Mayneris-Perxachs, J., Castells-Nobau, A., Arnoriaga-Rodríguez, M., Martin, M., de la Vega-Correa, L., Zapata, C., Burokas, A., Blasco, G., Coll, C., Escrichs, A., Biarnés, C., Moreno-Navarrete, J. M., Puig, J., Garre-Olmo, J., Ramos, R., Pedraza, S., Brugada, R., Vilanova, J. C., Serena, J., . . . & Fernández-Real, J. M. (2022). Microbiota Alterations in Proline Metabolism Impact Depression. *Cell Metabolism*, 34(5), 681–701.e10. https://doi.org/10.1016/j.cmet.2022.04.001

Meadow, J. F., Altrichter, A. E., Bateman, A. C., Stenson, J., Brown, G., Green, J. L., & Bohannan, B. J. M. (2015). Humans Differ in Their Personal Microbial Cloud. *PeerJ*, 3, e1258. https://doi.org/10.7717/peerj.1258

Meckel, K. R., & Kiraly, D. D. (2020). Maternal Microbes Support Fetal Brain Wiring. *Nature*, 586(7828), 203–205. https://doi.org/10.1038/d41586-020-02650-6

Megahed, N. A., & Ghoneim, E. M. (2020). Antivirus-Built Environment: Lessons Learned from COVID-19 Pandemic. *Sustainable Cities and Society*, 61, 102350. https://doi.org/10.1016/j.scs.2020.102350

Meisel, P., Pink, C., Pitchika, V., Nauck, M., Völzke, H., & Kocher, T. (2021). Competing Interplay between Systemic and Periodontal Inflammation: Obesity Overrides the Impact of Oral Periphery. *Clinical Oral Investigations*, 25(4), 2045–2053. https://doi.org/10.1007/s00784-020-03514-y

Meliț, L. E., Mărginean, C. O., & Săsăran, M. O. (2022). The Yin-Yang Concept of Pediatric Obesity and Gut Microbiota. *Biomedicines*, 10(3), 645. https://doi.org/10.3390/biomedicines10030645

Menon, J. (2017, December 10). 1st ASEAN Conference on Healthy Ageing—Challenges, Successes and the Journey Ahead. Table 15.1.

Merriam-Webster. (2024, November). Statistics. Merriam-Webster Dictionary. https://www.merriam-webster.com/dictionary/statistics

Merriam-Webster. (2022, December). Microbiome. Merriam-Webster Dictionary. https://www.merriam-webster.com/dictionary/microbiome

Milani, C., Duranti, S., Bottacini, F., Casey, E., Turroni, F., Mahony, J., Belzer, C., Delgado Palacio, S., Arboleya Montes, S., Mancabelli, L., Lugli, G. A., Rodriguez, J. M., Bode, L., de Vos, W., Gueimonde, M., Margolles, A., van Sinderen, D., & Ventura, M. (2017). The First Microbial Colonizers of the Human Gut: Composition, Activities, and Health Implications of the Infant Gut Microbiota. *Microbiology and Molecular Biology Reviews*, 81(4), e00036-17. https://doi.org/10.1128/MMBR.00036-17

Miletto, M., & Lindow, S. E. (2015). Relative and Contextual Contribution of Different Sources to the Composition and Abundance of Indoor Air Bacteria in Residences. *Microbiome*, 3(1), Article 61. https://doi.org/10.1186/s40168-015-0128-z

Mills, H., Acquah, R., Tang, N., Cheung, L., Klenk, S., Glassen, R., Pirson, M., Albert, A., Hoang, D. T., & Van, T. N. (2022). The Use of Bacteria in Cancer Treatment: A Review from the Perspective of Cellular Microbiology. *Emergency Medicine International*, 2022, 8127137. https://doi.org/10.1155/2022/8127137

Mills, J. G., Brookes, J. D., Gellie, N. J. C., Liddicoat, C., Lowe, A. J., Sydnor, H. R., Thomas, T., Weinstein, P., Weyrich, L. S., & Breed, M. F. (2019). Relating Urban Biodiversity to Human Health with the "Holobiont" Concept. *Frontiers in Microbiology*, 26(10), 550. https://doi.org/10.3389/fmicb.2019.00550

Mills, J. G., Weinstein, P., Gellie, N.J.C., Weyrich, L.S., Lowe, A.J. & Breed, M.F. (2017). Urban habitat restoration provides a human health benefit through microbiome rewilding: the Microbiome Rewilding Hypothesis. *Restoration Ecology*, 25, 866–872. https://doi.org/10.1111/rec.12610

Miqdady, M., Al Mistarihi, J., Azaz, A., & Rawat, D. (2020). Prebiotics in the Infant Microbiome: The Past, Present, and Future. *Pediatric Gastroenterology, Hepatology & Nutrition*, 23(1), 1–14. https://doi.org/10.5223/pghn.2020.23.1.1

Mithul Aravind, S., Wichienchot S., Tsao R., Ramakrishnan S., & Chakkaravarthi S. (2021). Role of dietary polyphenols on gut microbiota, their metabolites and health benefits. *Food Research International*, 142, 110189. https://doi.org/10.1016/j.foodres.2021.110189

Moeller, A. H., Li, Y., Mpoudi Ngole, E., Ahuka-Mundeke, S., Lonsdorf, E. V., Pusey, A. E., Peeters, M., Hahn, B. H., & Ochman, H. (2014). Rapid changes in the gut microbiome during human evolution. *Proceedings of the National Academy of Sciences*, 111(46), 16431–16435. https://www.pnas.org/doi/10.1073/pnas.1419136111

Moeller, A. H., Caro-Quintero, A., Mjungu, D., Georgiev, A. V., Lonsdorf, E. V., Muller, M. N., Pusey, A. E., Peeters, M., Hahn, B. H., & Ochman, H. (2016). Cospeciation of gut microbiota with hominids. *Science*, 353(6297), 380–382. https://doi.org/10.1126/science.aaf3951

Morgan, R. L., Preidis, G. A., Kashyap, P. C., Weizman, A. V., Sadeghirad, B., & McMaster Probiotic, Prebiotic, and Synbiotic Work Group. (2020). Probiotics Reduce Mortality and Morbidity in Preterm, Low-Birth-Weight Infants: A Systematic Review and Network Meta-analysis of Randomized Trials. *Gastroenterology*, 159(2), 467–480. https://doi:10.1053/j.gastro.2020.05.096.

Moya-Pérez, A., Luczynski, P., Renes, I. B., Wang, S., Borre, Y., Ryan C. A., Knol, J., Stanton, C., Dinan, T. G., and Cryan, J. F. (2017). Intervention strategies for cesarean section-induced alterations in the microbiota-gut-brain axis. *Nutrition Reviews*, 75(4), 225–240.

Myhre, R., Brantsæter, A. L., Myking, S., Gjessing, H. K., Sengpiel, V., Meltzer, H. M., Haugen, M., & Jacobsson, B. (2011). Intake of probiotic food and risk of spontaneous preterm delivery. *The American Journal of Clinical Nutrition*, 93(1), 151–157. https://doi.org/10.3945/ajcn.110.004085.

N

Nakajima, A., Kaga, N., Nakanishi, Y., Ohno, H., Miyamoto, J., Kimura, I., Hori, S., Sasaki, T., Hiramatsu, K., Okumura, K., Miyake, S., Habu, S., & Watanabe, S. (2017). Maternal High Fiber Diet during Pregnancy and Lactation Influences Regulatory T Cell Differentiation in Offspring in Mice. *Journal of Immunology*, 199(10), 3516–3524. https://doi.org/10.4049/jimmunol.1700248

Nakamura, Y., Oscherwitz, J., Cease, K. B., Chan, S. M., Muñoz-Planillo, R., Hasegawa, M., Villaruz, A. E., Cheung, G. Y., McGavin, M. J., Travers, J. B., Otto, M., Inohara, N., & Núñez, G. (2013). Staphylococcus δ-Toxin Induces Allergic Skin Disease by Activating Mast Cells. *Nature*, 503(7476), 397–401. https://doi.org/10.1038/nature12655

Nakatsuji, T., Chiang, H.-I., Jiang, S. B., Nagarajan, H., Zengler, K., & Gallo, R. L. (2013). The Microbiome Extends to Subepidermal Compartments of Normal Skin. *Nature Communications*, 4, Article 1431. https://doi.org/10.1038/ncomms2441

Napolitano, M., & Covasa, M. (2020). Microbiota Transplant in the Treatment of Obesity and Diabetes: Current and Future Perspectives. *Frontiers in Microbiology*, 11, Article 590370. https://doi.org/10.3389/fmicb.2020.590370

Nath, S., Zilm, P., Jamieson, L., Kapellas, K., Goswami, N., Ketagoda, K., & Weyrich, L. S. (2021). Development and Characterization of an Oral Microbiome Transplant among Australians for the Treatment of Dental Caries and Periodontal Disease: A Study Protocol. *PLOS ONE*, 16(11), e0260433. https://doi.org/10.1371/journal.pone.0260433

Navarro-Tapia, E., Sebastiani, G., Sailer, S., Toledano, L. A., Serra-Delgado, M., García-Algar, Ó., & Andreu-Fernández, V. (2020). Probiotic Supplementation during the Perinatal and Infant Period: Effects on Gut Dysbiosis and Disease. *Nutrients*, 12(8), 2243. https://doi.org/10.3390/nu12082243

Neu, A. T., Allen, E. E., & Roy, K. (2021). Defining and quantifying the core microbiome: Challenges and prospects. *Proceedings of the National Academy of Sciences*, 118(51), e2104429118. doi: 10.1073/pnas.2104429118.

Neufeld, K.-A., & Foster, J. A. (2009). Effects of Gut Microbiota on the Brain: Implications for Psychiatry. *Journal of Psychiatry & Neuroscience*: JPN, 34(3), 230–231.

Neville, B. A., Forster, S. C., & Lawley, T. D. (2018). Commensal Koch's Postulates: Establishing Causation in Human Microbiota Research. *Current Opinion in Microbiology*, 42, 47–52. https://doi.org/10.1016/j.mib.2017.10.001

Ng, S., Chen, M., Kundu, S., Wang, X., Zhou, Z., Zheng, Z., Qing, W., Sheng, H., Wang, Y., He, Y., Bennett, P. R., MacIntyre, D. A., & Zhou, H. (2021). Large-Scale Characterisation of the Pregnancy Vaginal Microbiome and Sialidase Activity in a Low-Risk Chinese Population. *NPJ Biofilms and Microbiomes*, 7(1), Article 1. https://doi.org/10.1038/s41522-021-00261-0

Ni, J., Shen, T.-C. D., Chen, E. Z., Bittinger, K., Bailey, A., Roggiani, M., Sirota-Madi, A., Friedman, E. S., Chau, L., Lin, A., Nissim, I., Scott, J., Lauder, A., Hoffmann, C., Rivas, G., Albenberg, L., Baldassano, R. N., Braun, J., Xavier, R. J., . . . & Wu, G. D. (2017). A Role for Bacterial Urease in Gut Dysbiosis and Crohn's Disease. *Science Translational Medicine*, 9(416), eaah6888. https://doi.org/10.1126/scitranslmed.aah6888

Nishida, A. H., & Ochman, H. (2019). A Great-Ape View of the Gut Microbiome. *Nature Reviews Genetics*, 20(4), 195–206. https://doi.org/10.1038/s41576-018-0085-z

Nissle, A. (1925). Weiteres über Grundlagen und Praxis der Mutaflorbehandlung. *DMW - Deutsche Medizinische Wochenschrift*, 51(44), 1809–1813. https://doi.org/10.1055/s-0028-1137292

Nordqvist, M., Jacobsson, B., Brantsæter, A.-L., Myhre, R., Nilsson, S., & Sengpiel, V. (2018). Timing of Probiotic Milk Consumption during Pregnancy and Effects on the Incidence of Preeclampsia and Preterm Delivery: A Prospective Observational Cohort Study in Norway. *BMJ Open*, 8(1). https://doi.org/10.1136/bmjopen-2017-018021

North Carolina State University News. (2021). *Lactobacillus* Manipulates Bile Acids to Create Favorable Gut Environment. https://news.ncsu.edu/2021/02/lactobacillus-manipulates-bile-acids/

Nutt, D. (2020, December 2). Spatial Maps Give New View of Gut Microbiome. *Cornell Chronicle*. https://news.cornell.edu/stories/2020/12/spatial-maps-give-new-view-gut-microbiome

O

Obregon-Tito, A. J., Tito, R. Y., Metcalf, J., Sankaranarayanan, K., Clemente, J. C., Ursell, L. K., Zech Xu, Z., Van Treuren, W., Knight, R., Gaffney, P. M., Spicer, P., Lawson, P., Marin-Reyes, L., Trujillo-Villarroel, O., Foster, M., Guija-Poma, E., Troncoso-Corzo, L., Warinner, C., Ozga, A. T., & Lewis, C. M. (2015). Subsistence Strategies in Traditional Societies Distinguish Gut Microbiomes. *Nature Communications*, 6(1), Article 1. https://doi.org/10.1038/ncomms7505

Olesen, S. W., & Alm, E. J. (2016). Dysbiosis Is Not an Answer. *Nature Microbiology*, 1(12), Article 16228. https://doi.org/10.1038/nmicrobiol.2016.228

Olm, M. R., Bhattacharya, N., Crits-Christoph, A., Diamond, S., Lavy, A., Crusan, A., Thomas, B. C., & Banfield, J. F. (2022). Robust Variation in Infant Gut Microbiome Assembly across a Spectrum of Lifestyles. *Science*, 376(6598), 1220–1223. https://doi.org/10.1126/science.abg1671

Olsen, I., & Yamazaki, K. (2019). Can Oral Bacteria Affect the Microbiome of the Gut? *Journal of Oral Microbiology*, 11(1), 1586422. https://doi.org/10.1080/20002297.2019.1586422

Olveira, G., & González-Molero, I. (2016). An Update on Probiotics, Prebiotics and Symbiotics in Clinical Nutrition. *Endocrinología y Nutrición*, 63(6), 482–494. https://doi.org/10.1016/j.endonu.2016.09.010

O'Malley, M. A. (2007). The Nineteenth Century Roots of "Everything Is Everywhere." *Nature Reviews Microbiology*, 5(8), 647–651. https://doi.org/10.1038/nrmicro1711

P

Paller, A. S., Kong, H. H., Seed, P., Naik, S., Scharschmidt, T. C., Gallo, R. L., Luger, T., & Irvine, A. D. (2019). The Microbiome in Patients with Atopic Dermatitis. *Journal of Allergy and Clinical Immunology*, 143(1), 26–35. https://doi.org/10.1016/j.jaci.2018.11.015

Palmer, N. (2020). *The Regenerative Grower's Guide to Garden Amendments: Using Locally Sourced Materials to Make Mineral and Biological Extracts and Ferments*. Chelsea Green.

Pan, G., Wang, X., Wang, Y., Li, R., Li, G., He, Y., Liu, S., Luo, Y., Wang, L., & Lei, Z. (2021). *Helicobacter pylori* Promotes Gastric Cancer Progression by Upregulating Semaphorin 5A Expression via ERK/MMP9 Signaling. *Molecular Therapy: Oncolytics*, 22, 256–264. https://doi.org/10.1016/j.omto.2021.01.010

Panigrahi, P., Parida, S., Nanda, N. C., Satpathy, R., Pradhan, L., Chandel, D. S., Baccaglini, L., Mohapatra, A., Mohapatra, S. S., Misra, P. R., Chaudhry, R., Chen, H. H., Johnson, J. A., Morris, J. G., Paneth, N., & Gewolb, I. H. (2017). A Randomized Synbiotic Trial to Prevent Sepsis among Infants in Rural India. *Nature*, 548(7668), Article 7668. https://doi.org/10.1038/nature23480

Pastar, I., O'Neill, K., Padula, L., Head, C. R., Burgess, J. L., Chen, V., Garcia, D., Stojadinovic, O., Hower, S., Plano, G. V., Thaller, S. R., Tomic-Canic, M., & Strbo, N. (2020). *Staphylococcus epidermidis* Boosts Innate Immune Response by Activation of Gamma Delta T Cells and Induction of Perforin-2 in Human Skin. *Frontiers in Immunology*, 11, Article 550946. https://doi.org/10.3389/fimmu.2020.550946

Pasteur, L. (1864, April 23). On Spontaneous Generation. Lecture delivered at the Sorbonne Scientific Soirée, University of Paris. Retrieved from http://www.rc.usf.edu/~levineat/pasteur.pdf

Pasteur, L. (1862). *Mémoire sur les corpuscules organisés qui existent dans l'atmosphère: Examen de la doctrine des générations spontanée*. Wellcome Collection. https://wellcomecollection.org/works/njdg2696

Patra, V., Byrne, S. N., & Wolf, P. (2016). The Skin Microbiome: Is It Affected by UV-Induced Immune Suppression? *Frontiers in Microbiology*, 7, 1235. https://doi.org/10.3389/fmicb.2016.01235

Patrick, D. M., Sbihi, H., Dai, D. L. Y., Al Mamun, A., Rasali, D., Rose, C., Marra, F., Boutin, R. C. T., Petersen, C., Stiemsma, L. T., Winsor, G. L., Brinkman, F. S. L., Kozyrskyj, A. L., Azad, M. B., Becker, A. B., Mandhane, P. J., Moraes, T. J., Sears, M. R., Subbarao, P., . . . & Turvey, S. E. (2020). Decreasing Antibiotic Use, the Gut Microbiota, and Asthma Incidence in Children: Evidence from Population-Based and Prospective Cohort Studies. *Lancet Respiratory Medicine*, 8(11), 1094–1105. https://doi.org/10.1016/S2213-2600(20)30052-7

Pennisi, E. (2020). Meet the Psychobiome. *Science*, 368(6494), 570–573. https://doi.org/10.1126/science.368.6494.570

Pentz, J. T., Márquez-Zacarías, P., Bozdag, G. O., Burnetti, A., Yunker, P. J., Libby, E., & Ratcliff, W. C. (2020). Ecological Advantages and Evolutionary Limitations of Aggregative Multicellular Development. *Current Biology*, 30(21), 4155–4164.e6. https://doi.org/10.1016/j.cub.2020.08.006

Pérez, J. C. (2021). Fungi of the Human Gut Microbiota: Roles and Significance. *International Journal of Medical Microbiology*, 311(3), 151490. https://doi.org/10.1016/j.ijmm.2021.151490

Peterson, D., Bonham, K. S., Rowland, S., Pattanayak, C. W., RESONANCE Consortium, & Klepac-Ceraj, V. (2021). Comparative Analysis of 16S rRNA Gene and Metagenome Sequencing in Pediatric Gut Microbiomes. *Frontiers in Microbiology*, 12, 670336. https://doi.org/10.3389/fmicb.2021.670336

Phan, T. X., Nguyen, V. H., Duong, M. T., Hong, Y., Choy, H. E., & Min, J. J. (2015). Activation of Inflammasome by Attenuated *Salmonella typhimurium* in Bacteria-Mediated Cancer Therapy. *Microbiology and Immunology*, 59(11), 664–675. https://doi.org/10.1111/1348-0421.12333

Piecková, E. (2017). Indoor Microbial Aerosol and Its Health Effects: Microbial Exposure in Public Buildings—Viruses, Bacteria, and Fungi. In C. Viegas, et al. (Eds.), *Exposure to Microbiological Agents in Indoor and Occupational Environments* (pp. 237–252). Springer.

Pocock, M. J., Evans, D. M., & Memmott, J. (2012). The Robustness and Restoration of a Network of Ecological Networks. *Science*, 335(6071), 973–977. https://doi.org/10.1126/science.

Price, M. N., Dehal, P. S., & Arkin, A. P. (2009). FastTree: Computing Large Minimum Evolution Trees with Profiles Instead of a Distance Matrix. *Molecular Biology and Evolution*, 26(7), 1641–1650. https://doi.org/10.1093/molbev/msp077

Procházková, N., Falony, G., Dragsted, L. O., Licht, T. R., Raes, J., & Roager, H. M. (2023). Advancing Human Gut Microbiota Research by Considering Gut Transit Time. *Gut*, 72(1), 180–191. https://doi.org/10.1136/gutjnl-2021-325704

Proctor, D. M., Fukuyama, J. A., Loomer, P. M., Armitage, G. C., Lee, S. A., Davis, N. M., Ryder, M. I., Holmes, S. P., & Relman, D. A. (2018). A Spatial Gradient of Bacterial Diversity in the Human Oral Cavity Shaped by Salivary Flow. *Nature Communications*, 9(1), Article 1. https://doi.org/10.1038/s41467-018-02900-1

Proctor, L. M., Creasy, H. H., Fettweis, J. M., Lloyd-Price, J., Mahurkar, A., Zhou, W., Buck, G. A., Snyder, M. P., Strauss, J. F., Weinstock, G. M., White, O., Huttenhower, C., & Integrative HMP (iHMP) Research Network Consortium. (2019). The Integrative Human Microbiome Project. *Nature*, 569(7758), Article 7758. https://doi.org/10.1038/s41586-019-1238-8

Proctor, L. M., & iHMP Research Network Consortium. (2014). The Integrative Human Microbiome Project: Dynamic Analysis of Microbiome-Host Omics Profiles during Periods of Human Health and Disease. *Cell Host & Microbe*, 16(3), 276–289. https://doi.org/10.1016/j.chom.2014.08.014

Puniewska, M. (2017, July 9). How the Microbiome Could Transform Your Skin in Surprising Ways. Johnson & Johnson. https://www.jnj.com/innovation/how-the-microbiome-could-transform-your-skin-in-surprising-ways

Q

Quast, C., Pruesse, E., Yilmaz, P., Gerken, J., Schweer, T., Yarza, P., Peplies, J., & Glöckner, F. O. (2012). The SILVA Ribosomal RNA Gene Database Project: Improved Data Processing and Web-Based Tools. *Nucleic Acids Research*, 41(D1), D590–D596. https://doi.org/10.1093/nar/gks1219

R

Ranjan, R., Rani, A., Metwally, A., McGee, H. S., & Perkins, D. L. (2016). Analysis of the Microbiome: Advantages of Whole Genome Shotgun versus 16S Amplicon Sequencing. *Biochemical and Biophysical Research Communications*, 469(4), 967–977. https://doi.org/10.1016/j.bbrc.2015.12.083

Reardon, S. (2019). Do C-section Babies Need Mum's Microbes? Trials Tackle Controversial Idea. *Nature*, 572(7770), 423–424. https://doi.org/10.1038/d41586-019-02348-3

Redi, F. (1909). *Experiments on the Generation of Insects*. Open Court Publishing.

Rehel, J. (2015, retrieved 2024). Four Gut Bacteria Are Crucial to Asthma Prevention: Researchers. *Allergic Living*. https://www.allergicliving.com/2015/09/30/four-gut-bacteria-crucial-to-asthma-prevention-study-finds/

Relman, D. (2019, June 17). The Origins of Human Microbiota Research (A. Jagatia, Interviewer). *Nature*. https://www.nature.com/articles/d42859-019-00073-5

Reyman, M., van Houten, M. A., Watson, R. L., Chu, M. L. J. N., Arp, K., de Waal, W. J., Schiering, I., Plötz, F. B., Willems, R. J. L., van Schaik, W., Sanders, E. A. M., & Bogaert, D. (2022). Effects of Early-Life Antibiotics on the Developing Infant Gut Microbiome and Resistome: A Randomized Trial. *Nature Communications*, 13(1), Article 1. https://doi.org/10.1038/s41467-022-28525-z

Ridaura, V. K., Faith, J. J., Rey, F. E., Cheng, J., Duncan, A. E., Kau, A. L., Griffin, N. W., Lombard, V., Henrissat, B., Bain, J. R., Muehlbauer, M. J., Ilkayeva, O., Semenkovich, C. F., Funai, K., Hayashi, D. K., Lyle, B. J., Martini, M. C., Ursell, L. K., Clemente, J. C., . . . & Gordon, J. I. (2013). Gut Microbiota from Twins Discordant for Obesity Modulate Metabolism in Mice. *Science*, 341(6150), Article 1241214. https://doi.org/10.1126/science.1241214

Rivière, A., Selak, M., Lantin, D., Leroy, F., & De Vuyst, L. (2016). Bifidobacteria and Butyrate-Producing Colon Bacteria: Importance and Strategies for their Stimulation in the Human Gut. *Frontiers in Microbiology*, 7, Article 979. https://doi.org/10.3389/fmicb.2016.00979

Robinson, J., Mills, J., & Breed, M. (2018). Walking Ecosystems in Microbiome-Inspired Green Infrastructure: An Ecological Perspective on Enhancing Personal and Planetary Health. *Challenges*, 9(2), Article 40. https://doi.org/10.3390/challe9020040

Robitzski, D. (2022, March 17), Bacteria in the Lungs Can Regulate Autoimmunity in Rat Brains, *The Scientist*.

Rogers, S. A. (2002) Detoxify or Die. Prestige Publishing.

Rook, G. A.W., Martinelli, R., & Brunet, L. R. (2003). Innate immune responses to mycobacteria and the downregulation of atopic responses. *Current Opinion in Allergy and Clinical Immunology* 3(5), 37–342. https://pubmed.ncbi.nlm.nih.gov/14501431/

Rook, G. A. W., Raison, C. L., & Lowry, C. A. (2014). Microbial 'Old Friends', Immunoregulation and Socioeconomic Status. *Clinical and Experimental Immunology*, 177(1), 1–12. https://doi.org/10.1111/cei.12269

Rosshart, S. P., Herz, J., Vassallo, B. G., Hunter, A., Wall, M. K., Badger, J. H., McCulloch, J. A., Anastasakis, D. G., Sarshad, A. A., Leonardi, I., Collins, N., Blatter, J. A., Han, S.-J., Tamoutounour, S., Potapova, S., Foster St. Claire, M. B., Yuan, W., Sen, S. K., Dreier, M. S., . . . & Rehermann, B. (2019). Laboratory Mice Born to Wild Mice Have Natural Microbiota and Model Human Immune Responses. *Science*, 365(6452), Article eaaw4361. https://doi.org/10.1126/science.aaw4361

Ruan, W., Engevik, M. A., Spinler, J. K., & Versalovic, J. (2020). Healthy Human Gastrointestinal Microbiome: Composition and Function After a Decade of Exploration. *Digestive Diseases and Sciences*, 65(3), 695–705. https://doi.org/10.1007/s10620-020-06118-4

Ruiz-Calderon, J. F., Cavallin, H., Song, S. J., Novoselac, A., Pericchi, L. R., Hernandez, J. N., Rios, R., Branch, O. H., Pereira, H., Paulino, L. C., Blaser, M. J., Knight, R., & Dominguez-Bello, M. G. (2016). Walls Talk: Microbial Biogeography of Homes Spanning Urbanization. *Science Advances*, 2(2), Article e1501061. https://doi.org/10.1126/sciadv.1501061

S

Sáez-Lara, M. J., Robles-Sanchez, C., Ruiz-Ojeda, F. J., Plaza-Diaz, J., & Gil, A. (2016). Effects of Probiotics and Synbiotics on Obesity, Insulin Resistance Syndrome, Type 2 Diabetes, and Non-Alcoholic Fatty Liver Disease: A Review of Human Clinical Trials. *International Journal of Molecular Sciences*, 17(6), 928. https://doi.org/10.3390/ijms17060928

Sagan, L. (1967). On the Origin of Mitosing Cells. *Journal of Theoretical Biology*, 14(3), 225–274. https://doi.org/10.1016/0022-5193(67)90079-3

Salas Garcia, M. C., Schorr, A. R., Arnold, W., Fei, N., & Gilbert, J. A. (2020). Pets as a Novel Microbiome-Based Therapy. In M. R. Pastorinho & A. C. A. Sousa (Eds.), *Pets as Sentinels, Forecasters and Promoters of Human Health* (pp. 245–267). Springer. https://doi.org/10.1007/978-3-030-30734-9_11

Salliss, M. E., Maarsingh, J. D., Garza, C., Łaniewski, P., & Herbst-Kralovetz, M. M. (2021). Veillonellaceae Family Members Uniquely Alter the Cervical Metabolic Microenvironment in a Human Three-Dimensional Epithelial Model. *NPJ Biofilms and Microbiomes*, 7(1), Article 1. https://doi.org/10.1038/s41522-021-00229-0

Sandoval-Motta, S., Aldana, M., Martínez-Romero, E. & Frank, A. (2017). The Human Microbiome and the Missing Heritability Problem. *Frontiers in Genetics*, 8, 80. https://www.frontiersin.org/journals/genetics/articles/10.3389/fgene.2017.00080/full

Sant, J. (2019). Francesco Redi and Controlled Experiments. Scientus.org. http://www.scientus.org/Redi-Galileo.html

Santhiravel, S., Bekhit, A. E.-D. A., Mendis, E., Jacobs, J. L., Dunshea, F. R., Rajapakse, N., & Ponnampalam, E. N. (2022). The Impact of Plant Phytochemicals on the Gut Microbiota of Humans for a Balanced Life. *International Journal of Molecular Sciences*, 23(15), 8124. https://doi.org/10.3390/ijms23158124

Saturio, S., Nogacka, A. M., Alvarado-Jasso, G. M., Salazar, N., de los Reyes-Gavilán, C. G., Gueimonde, M., & Arboleya, S. (2021). Role of *Bifidobacteria* on Infant Health. *Microorganisms*, 9(12), 2415. https://doi.org/10.3390/microorganisms9122415

Sbihi, H., Boutin, R. C. T., Cutler, C., Suen, M., Finlay, B. B., & Turvey, S. E. (2019). Thinking Bigger: How Early-Life Environmental Exposures Shape the Gut Microbiome and Influence the Development of Asthma and Allergic Disease. *Allergy*, 74(11), 2103–2115. https://doi.org/10.1111/all.13812

Schaedler, R. W., Dubos, R., & Costello, R. (1965). Association of Germfree Mice with Bacteria Isolated from Normal Mice. *Journal of Experimental Medicine*, 122(1), 77–82. https://doi.org/10.1084/jem.122.1.77

Scheithauer, T. P. M., Rampanelli, E., Nieuwdorp, M., Vallance, B. A., Verchere, C. B., van Raalte, D. H., & Herrema, H. (2020). Gut Microbiota as a Trigger for Metabolic Inflammation in Obesity and Type 2 Diabetes. *Frontiers in Immunology*, 11, 571731. https://doi.org/10.3389/fimmu.2020.571731

Schellekens, H., Torres-Fuentes, C., van de Wouw, M., Long-Smith, C. M., Mitchell, A., Strain, C., Berding, K., Bastiaanssen, T. F. S., Rea, K., Golubeva, A. V., Arboleya, S., Verpaalen, M., Pusceddu, M. M., Murphy, A., Fouhy, F., Murphy, K., Ross, P., Roy, B. L., Stanton, C., & Cryan, J. F. (2020). *Bifidobacterium longum* Counters the Effects of Obesity: Partial Successful Translation from Rodent to Human. *EBioMedicine*, 63, 103176. https://doi.org/10.1016/j.ebiom.2020.103176

Schloss, P. D., Westcott, S. L., Ryabin, T., Hall, J. R., Hartmann, M., Hollister, E. B., Lesniewski, R. A., Oakley, B. B., Parks, D. H., Robinson, C. J., Sahl, J. W., Stres, B., Thallinger, G. G., Van Horn, D. J., & Weber, C. F. (2009). Introducing Mothur: Open-Source, Platform-Independent, Community-Supported Software for Describing and Comparing Microbial Communities. *Applied and Environmental Microbiology*, 75(23), 7537–7541. https://doi.org/10.1128/AEM.01541-09

Schonhoff, A. M., & Mazmanian, S. K. (2022). Lung Microbes Mediate Spinal-Cord Autoimmunity. *Nature*, 603(7899), 38–40. https://doi.org/10.1038/d41586-022-00468-x

Schröder, L., Kaiser, S., Flemer, B., Hamm, J., Hinrichsen, F., Bordoni, D., Rosenstiel, P., & Sommer, F. (2020). Nutritional Targeting of the Microbiome as Potential Therapy for Malnutrition and Chronic Inflammation. *Nutrients*, 12(10), Article 10. https://doi.org/10.3390/nu12102998

Scott, A. J., Alexander, J. L., Merrifield, C. A., Cunningham, D., Jobin, C., Brown, R., Alverdy, J., O'Keefe, S. J., Gaskins, H. R., Teare, J., Yu, J., Hughes, D. J., Verstraelen, H., Burton, J., O'Toole, P. W., Rosenberg, D. W., Marchesi, J. R., & Kinross, J. M. (2019). International Cancer Microbiome Consortium Consensus Statement on the Role of the Human Microbiome in Carcinogenesis. *Gut*, 68(9), 1624–1632. https://doi.org/10.1136/gutjnl-2019-318556

Sedghi, L., DiMassa, V., Harrington, A., Lynch, S. V., & Kapila, Y. L. (2021). The Oral Microbiome: Role of Key Organisms and Complex Networks in Oral Health and Disease. *Periodontology 2000*, 87(1), 107–131. https://doi.org/10.1111/prd.12393

Sedighi, M., Bialvaei, A. Z., Hamblin, M. R., Ohadi, E., Asadi, A., Halajzadeh, M., Lohrasbi, V., Mohammadzadeh, N., Amiriani, T., Krutova, M., Amini, A., & Kouhsari, E. (2019). Therapeutic Bacteria to Combat Cancer; Current Advances, Challenges, and Opportunities. *Cancer Medicine*, 8(6), 3167–3176. https://doi.org/10.1002/cam4.2148

Senghor, B., Sokhna, C., Ruimy, R., & Lagier, J.-C. (2018). Gut Microbiota Diversity According to Dietary Habits and Geographical Provenance. *Human Microbiome Journal*, 7–8, 1–9. https://doi.org/10.1016/j.humic.2018.01.001

Sgritta, M., Dooling, S. W., Buffington, S. A., Momin, E. N., Francis, M. B., Britton, R. A., & Costa-Mattioli, M. (2019). Mechanisms Underlying Microbial-Mediated Changes in Social Behavior in Mouse Models of Autism Spectrum Disorder. *Neuron*, 101(2), 246–259.e6. https://doi.org/10.1016/j.neuron.2018.11.018

Shanmugam, G., Lee, S.H. & Jeon, J. (2021). EzMAP: Easy Microbiome Analysis Platform. *BMC Bioinformatics* 22, 179. https://doi.org/10.1186/s12859-021-04106-7

Shao, Y., Forster, S. C., Tsaliki, E., Vervier, K., Strang, A., Simpson, N., Kumar, N., Stares, M. D., Rodger, A., Brocklehurst, P., Field, N., & Lawley, T. D. (2019). Stunted Microbiota and Opportunistic Pathogen Colonization in Caesarean-Section Birth. *Nature*, 574(7776), Article 7776. https://doi.org/10.1038/s41586-019-1560-1

Shapiro, H., Goldenberg, K., Ratiner, K., & Elinav, E. (2022). Smoking-Induced Microbial Dysbiosis in Health and Disease. *Clinical Science*, 136(18), 1371–1387. https://doi.org/10.1042/CS20210787

Sharma, H., Tal, R., Clark, N. A., & Segars, J. H. (2014). Microbiota and Pelvic Inflammatory Disease. *Seminars in Reproductive Medicine*, 32(1), 43–49. https://doi.org/10.1055/s-0033-1361822

Shi, B. (2017). Workshop 11. *Metagenomics Analysis*. CNSI 4338, UCLA.

Shi, H., Shi, Q., Grodner, B., Lenz, J. S., Zipfel, W. R., Brito, I. L., & De Vlaminck, I. (2020). Highly Multiplexed Spatial Mapping of Microbial Communities. *Nature*, 588(7839), Article 7839. https://doi.org/10.1038/s41586-020-2983-4

Shi, J., Wang, Y., Cheng, L., Wang, J., & Raghavan, V. (2022). Gut Microbiome Modulation by Probiotics, Prebiotics, Synbiotics and Postbiotics: A Novel Strategy in Food Allergy Prevention and Treatment. *Critical Reviews in Food Science and Nutrition*, 63(4), 1–17. https://doi.org/10.1080/10408398.2022.2160962

Shoubridge, A. P., Choo, J. M., Martin, A. M., Holmes, A. J., & Hannan, A. J. (2022). The Gut Microbiome and Mental Health: Advances in Research and Emerging Priorities. *Molecular Psychiatry*, 27(4), 1908–1919. https://doi.org/10.1038/s41380-022-01416-2

Silk, D. B., Davis, A., Vulevic, J., Tzortzis, G., & Gibson, G. R. (2009). Clinical Trial: The Effects of a Trans-Galactooligosaccharide Prebiotic on Faecal Microbiota and Symptoms in Irritable Bowel Syndrome. *Alimentary Pharmacology & Therapeutics*, 29(5), 508–518. https://doi.org/10.1111/j.1365-2036.2008.03911.x

Silva, M., Brunner, V., & Tschurtschenthaler, M. (2021). Microbiota and Colorectal Cancer: From Gut to Bedside. *Frontiers in Pharmacology*, 12, 760280. https://doi.org/10.3389/fphar.2021.760280

Silva, Y. P., Bernardi, A., & Frozza, R. L. (2020). The Role of Short-Chain Fatty Acids from Gut Microbiota in Gut-Brain Communication. *Frontiers in Endocrinology*, 11, 25. https://doi.org/10.3389/fendo.2020.00025

Sistiaga, A., Poyet, M., Groussin, M., Collins, M., & Summons, R. E. (2019). Analysis of Faecal Substrates Sheds Light into Coprostanol Origin, Preservation, and Diagenesis. *First Break*, 37(1), 49–53. https://doi.org/10.3997/2214-4609.201902866

Smilowitz, J. T., & Taft, D. H. (2020, June 1). Infographic: The Changing Infant Gut Microbiome. *Scientist*. https://www.the-scientist.com/infographics/infographic-the-changing-infant-gut-microbiome-67588

Smits, S. A., Leach, J., Sonnenburg, E. D., Gonzalez, C. G., Lichtman, J. S., Reid, G., Knight, R., Manjurano, A., Changalucha, J., Elias, J. E., Dominguez-Bello, M. G., & Sonnenburg, J. L. (2017). Seasonal Cycling in the Gut Microbiome of the Hadza Hunter-Gatherers of Tanzania. *Science*, 357(6353), 802–806. https://doi.org/10.1126/science.aan4834

Song, S. J., Wang, J., Martino, C., Jiang, L., Thompson, W. K., Shenhav, L., McDonald, D., Marotz, C., Harris, P. R., Hernandez, C. D., Henderson, N., Ackley, E., Nardella, D., Gillihan, C., Montacuti, V., Schweizer, W., Jay, M., Combellick, J., Sun, H., . . . & Dominguez-Bello, M. G. (2021). Naturalization of the Microbiota Developmental Trajectory of Cesarean-Born Neonates after Vaginal Seeding. *Med*, 2(8), 951–964.e5. https://doi.org/10.1016/j.medj.2021.05.003

Sonnenburg, J. L., & Sonnenburg, E. D. (2019). Vulnerability of the Industrialized Microbiota. *Science*, 366(6464), eaaw9255. https://doi.org/10.1126/science.aaw9255

Sousa, A. M. M., Meyer, K. A., Santpere, G., Gulden, F. O., & Sestan, N. (2017). Evolution of the Human Nervous System Function, Structure, and Development. *Cell*, 170(2), 226–247. https://doi.org/10.1016/j.cell.2017.06.036

Stamper, C. E., Hoisington, A. J., Gomez, O. M., Halweg-Edwards, A. L., Smith, D. G., Bates, K. L., Kinney, K. A., Postolache, T. T., Brenner, L. A., Rook, G. A. W., & Lowry, C. A. (2016). The Microbiome of the Built Environment and Human Behavior. In C. A. Lowry (Ed.), *International Review of Neurobiology* (Vol. 131, pp. 289–323). Elsevier. https://doi.org/10.1016/bs.irn.2016.07.006

Stein, M. M., Hrusch, C. L., Gozdz, J., Igartua, C., Pivniouk, V., Murray, S. E., Ledford, J. G., Marques dos Santos, M., Anderson, R. L., Metwali, N., Neilson, J. W., Maier, R. M., Gilbert, J. A., Holbreich, M., Thorne, P. S., Martinez, F. D., von Mutius, E., Vercelli, D., Ober, C., & Sperling, A. I. (2016). Innate Immunity and Asthma Risk in Amish and Hutterite Farm Children. *New England Journal of Medicine*, 375(5), 411–421. https://doi.org/10.1056/NEJMoa1508749

Strachan, D. P. (1989). Hay Fever, Hygiene, and Household Size. *British Medical Journal*, 299(6710), 1259–1260. https://doi.org/10.1136/bmj.299.6710.1259

Strait, J. E. (2017, March 3). The Father of the Microbiome. *The Source—Washington Magazine*. https://source.wustl.edu/2017/03/the-father-of-the-microbiome/

Superti, F., & De Seta, F. (2020). Warding Off Recurrent Yeast and Bacterial Vaginal Infections: Lactoferrin and Lactobacilli. *Microorganisms*, 8(1), 130. https://doi.org/10.3390/microorganisms8010130

Suther, C., Moore, M. D., Beigelman, A., & Zhou, Y. (2020). The Gut Microbiome and the Big Eight. *Nutrients*, 12(12), 3728. https://doi.org/10.3390/nu12123728

Szostak, J. W. (2017). The Narrow Road to the Deep Past: In Search of the Chemistry of the Origin of Life. *Angewandte Chemie International Edition*, 56(37), 11037–11043. https://doi.org/10.1002/anie.201704048

T

Tam, J., Hoffmann, T., Fischer, S., Bornstein, S., Gräßler, J., & Noack, B. (2018). Obesity Alters Composition and Diversity of the Oral Microbiota in Patients with Type 2 Diabetes Mellitus Independently of Glycemic Control. *PLOS ONE*, 13(10), e0204724. https://doi.org/10.1371/journal.pone.0204724

Tap, J., Mondot, S., Levenez, F., Pelletier, E., Caron, C., Furet, J.-P., Ugarte, E., Muñoz-Tamayo, R., Paslier, D. L. E., Nalin, R., Doré, J., & Leclerc, M. (2009). Towards the Human Intestinal Microbiota Phylogenetic Core. *Environmental Microbiology*, 11(10), 2574–2584. https://doi.org/10.1111/j.1462-2920.2009.01982.x

Tattersall, I. (2008). An Evolutionary Framework for the Acquisition of Symbolic Cognition by *Homo sapiens*. *Comparative Cognition & Behavior Reviews*, 3, 99–114. https://doi.org/10.3819/ccbr.2008.30007

Tavella, T., Rampelli, S., Guidarelli, G., Bazzocchi, A., Gasperini, C., Pujos-Guillot, E., Comte, B., Barone, M., Biagi, E., Candela, M., Nicoletti, C., Kadi, F., Battista, G., Salvioli, S., O'Toole, P. W., Franceschi, C., Brigidi, P., Turroni, S., & Santoro, A. (2021). Elevated Gut Microbiome Abundance of *Christensenellaceae*, *Porphyromonadaceae* and *Rikenellaceae* Is Associated with Reduced Visceral Adipose Tissue and Healthier Metabolic Profile in Italian Elderly. *Gut Microbes*, 13(1), 1880221. https://doi.org/10.1080/19490976.2021.1880221

Teresi, D. (2011, June 16). Discover Interview: Lynn Margulis Says She's Not Controversial, She's Right. *Discover Magazine*. https://www.discovermagazine.com/the-sciences/discover-interview-lynn-margulis-says-shes-not-controversial-shes-right

The UniProt Consortium, Bateman, A., Martin, M.-J., Orchard, S., Magrane, M., Agivetova, R., Ahmad, S., Alpi, E., Bowler-Barnett, E. H., Britto, R., Bursteinas, B., Bye-A-Jee, H., Coetzee, R., Cukura, A., Da Silva, A., Denny, P., Dogan, T., Ebenezer, T., Fan, J., . . . & Teodoro, D. (2021). UniProt: The Universal Protein Knowledgebase in 2021. *Nucleic Acids Research*, 49(D1), D480–D489. https://doi.org/10.1093/nar/gkaa1100

Thion, M. S., Low, D., Silvin, A., Chen, J., Grisel, P., Schulte-Schrepping, J., Blecher, R., Ulas, T., Squarzoni, P., Hoeffel, G., Coulpier, F., Siopi, E., David, F. S., Scholz, C., Shihui, F., Lum, J., Amoyo, A. A., Larbi, A., Poidinger, M., . . . & Garel, S. (2018). Microbiome Influences Prenatal and Adult Microglia in a Sex-Specific Manner. *Cell*, 172(3), 500–516.e16. https://doi.org/10.1016/j.cell.2017.11.042

Thoemmes, M. S., Stewart, F. A., Hernandez-Aguilar, R. A., Bertone, M. A., Baltzegar, D. A., Borski, R. J., Cohen, N., Coyle, K. P., Piel, A. K., & Dunn, R. R. (2018). Ecology of Sleeping: The Microbial and Arthropod Associates of Chimpanzee Beds. *Royal Society Open Science*, 5(5), 180382. https://doi.org/10.1098/rsos.180382

Thomas, L. (1978). *The Lives of a Cell: Notes of a Biology Watcher*. Penguin.

Thomson, C. A., & McCoy, K. D. (2021). The Role of Mom's Microbes during Pregnancy. *Scientist*. https://www.the-scientist.com/features/the-role-of-moms-microbes-during-pregnancy-68959

Thorburn, A. N., McKenzie, C. I., Shen, S., Stanley, D., Macia, L., Mason, L. J., Roberts, L. K., Wong, C. H. Y., Shim, R., Robert, R., Chevalier, N., Tan, J. K., Mariño, E., Moore, R. J., Wong, L., McConville, M. J., Tull, D. L., Wood, L. G., Murphy, V. E., . . .

& Mackay, C. R. (2015). Evidence That Asthma Is a Developmental Origin Disease Influenced by Maternal Diet and Bacterial Metabolites. *Nature Communications*, 6, 7320. https://doi.org/10.1038/ncomms8320

Thursby, E., & Juge, N. (2017). Introduction to the Human Gut Microbiota. *Biochemical Journal*, 474(11), 1823–1836. https://doi.org/10.1042/BCJ20160510

Tiffany, C. R., & Bäumler, A. J. (2019). Dysbiosis: From Fiction to Function. *American Journal of Physiology-Gastrointestinal and Liver Physiology*, 317(5), G602–G608. https://doi.org/10.1152/ajpgi.00230.2019

Tompkins, S. (2020, December 28). 4 Priorities for a Better Built Environment in the Post-COVID City. *World Economic Forum*. https://www.weforum.org/agenda/2020/12/4-priorities-better-built-environment-cities/

Tomson, M., Kumar, P., Barwise, Y., Perez, P., Forehead, H., French, K., Morawska, L., & Watts, J. F. (2021). Green Infrastructure for Air Quality Improvement in Street Canyons. *Environment International*, 146, 106288. https://doi.org/10.1016/j.envint.2020.106288

Tortora, G. J., Funke, B. R., & Case, C. L. (2003). *Microbiology: An Introduction* (8th ed.). Pearson Education.

Townsend, E. M., Kelly, L., Muscatt, G., Box, J. D., Hargraves, N., Lilley, D., & Jameson, E. (2021). The Human Gut Phageome: Origins and Roles in the Human Gut Microbiome. *Frontiers in Cellular and Infection Microbiology*, 11, 643214. https://doi.org/10.3389/fcimb.2021.643214

Tsuge, M., Ikeda, M., Matsumoto, N., Yorifuji, T., & Tsukahara, H. (2021). Current Insights into Atopic March. *Children*, 8(11), 1067. https://doi.org/10.3390/children8111067

Tuohy, K. M., & Scott, K. P. (2015). The Microbiota of the Human Gastrointestinal Tract: A Molecular View. In K. M. Tuohy & D. Del Rio (Eds.), *Diet-Microbe Interactions in the Gut* (pp. 1–15). Elsevier.

Turnbaugh, P., Ley, R., Mahowald, M., Magrini, V., Mardis, E. R., & Gordon, J. I. (2006). An obesity-associated gut microbiome with increased capacity for energy harvest. *Nature* 444, 1027–1031. https://doi.org/10.1038/nature05414

Turnbaugh, P. J. (2017). Microbes and Diet-Induced Obesity: Fast, Cheap, and Out of Control. *Cell Host Microbe* 21(3), 278–281. https://doi:10.1016/j.chom.2017.02.021.

U

University of California—Berkeley. (2016, July 21). Biologists Home In on Paleo Gut for Clues to Our Evolutionary History. ScienceDaily. https://www.sciencedaily.com/releases/2016/07/160721151457.htm

University of California—Davis Health. (2021, December 3). Probiotics improve nausea and vomiting in pregnancy, according to new study. https://health.ucdavis.edu/news/headlines/probiotics-improve-nausea-and-vomiting-in-pregnancy-according-to-new-study/2021/12.

V

Vacca, M., Celano, G., Calabrese, F. M., Portincasa, P., Gobbetti, M., & De Angelis, M. (2020). The Controversial Role of Human Gut Lachnospiraceae. *Microorganisms*, 8(4), Article 4. https://doi.org/10.3390/microorganisms8040573

van Breugel, M., Qi, C., Xu, Z., Pedersen, C.-E. T., Petoukhov, I., Vonk, J. M., Gehring, U., Berg, M., Bügel, M., Carpaij, O. A., Forno, E., Morin, A., Eliasen, A. U., Jiang, Y., van den Berge, M., Nawijn, M. C., Li, Y., Chen, W., Bont, L. J., . . . & Xu, C.-J. (2022). Nasal DNA Methylation at Three CpG Sites Predicts Childhood Allergic Disease. *Nature Communications*, 13(1), 7415. https://doi.org/10.1038/s41467-022-35088-6

van de Wijgert, J., & Verwijs, M. C. (2020). Lactobacilli-Containing Vaginal Probiotics to Cure or Prevent Bacterial or Fungal Vaginal Dysbiosis: A Systematic Review and Recommendations for Future Trial Designs. *BJOG: An International Journal of Obstetrics and Gynaecology*, 127(2), 287–299. https://doi.org/10.1111/1471-0528.15870

van der Hee, B., & Wells, J. M. (2021). Microbial Regulation of Host Physiology by Short-Chain Fatty Acids. *Trends in Microbiology*, 29(8), 700–712. https://doi.org/10.1016/j.tim.2021.02.001

van der Vossen, E. W. J., de Goffau, M. C., Levin, E., & Nieuwdorp, M. (2022). Recent Insights into the Role of Microbiome in the Pathogenesis of Obesity. *Therapeutic Advances in Gastroenterology*, 15, 17562848221115320. https://doi.org/10.1177/17562848221115320

Van Leeuwenhoek, A. (1677). Observations, Communicated to the Publisher by Mr. Antony van Leewenhoeck, in a Dutch Letter of the 9th Octob. 1676, Here English'd: Concerning Little Animals by Him Observed in Rain-Well-Sea- and Snow Water; as Also in Water Wherein Pepper Had Lain Infused. *Philosophical Transactions of the Royal Society of London*, 12(133), 821–831. https://doi.org/10.1098/rstl.1677.0003

Vejdovszky, K., Hahn, K., Braun, D., Warth, B., & Marko, D. (2017). Synergistic Estrogenic Effects of *Fusarium* and *Alternaria* Mycotoxins in Vitro. *Archives of Toxicology*, 91(3), 1447–1460. https://doi.org/10.1007/s00204-016-1795-7

Velegraki, A., Cafarchia, C., Gaitanis, G., Iatta, R., & Boekhout, T. (2015). *Malassezia* Infections in Humans and Animals: Pathophysiology, Detection, and Treatment. *PLOS Pathogens*, 11(1), e1004523. https://doi.org/10.1371/journal.ppat.1004523

Venter, C., Maslin, K., Holloway, J. W., Silveira, L. J., Fleischer, D. M., Dean, T., & Arshad, S. H. (2020). Different Measures of Diet Diversity during Infancy and the Association with Childhood Food Allergy in a UK Birth Cohort Study. *Journal of Allergy and Clinical Immunology: In Practice*, 8(6), 2017–2026. https://doi.org/10.1016/j.jaip.2020.01.029

Venter, C., Palumbo, M. P., Glueck, D. H., Sauder, K. A., O'Mahony, L., Fleischer, D. M., Ben-Abdallah, M., Ringham, B. M., & Dabelea, D. (2022). The Maternal Diet Index in Pregnancy Is Associated with Offspring Allergic Diseases: The Healthy Start study. *Allergy*, 77(1), 162–172. https://doi.org/10.1111/all.14949

Vickery, B. P., Vereda, A., Nilsson, C., du Toit, G., Shreffler, W. G., Burks, A. W., Jones, S. M., Fernández-Rivas, M., Blümchen, K., Hourihane, J. O'B., Beyer, K., Anagnostou, A., Assa'ad, A. H., Ben-Shoshan, M., Bird, J. A., Carr, T. F., Carr, W. W., Casale, T. B., Chong, H. J., . . . & Adelman, D. C. (2021). Continuous and Daily Oral Immunotherapy for Peanut Allergy: Results from a 2-Year Open-Label Follow-On Study. *Journal of Allergy and Clinical Immunology: In Practice*, 9(5), 1879–1889.e13. https://doi.org/10.1016/j.jaip.2020.12.029

Virtue, A. T., McCright, S. J., Wright, J. M., Jimenez, M. T., Mowel, W. K., Kotzin, J. J., Joannas, L., Basavappa, M. G., Spencer, S. P., Clark, M. L., Eisennagel, S. H., Williams, A., Levy, M., Manne, S., Henrickson, S. E., Wherry, E. J., Thaiss, C. A., Elinav, E., & Henao-Mejia, J. (2019). The Gut Microbiota Regulates White Adipose Tissue Inflammation and Obesity via a Family of microRNAs. *Science Translational Medicine*, 11(496), eaav1892. https://doi.org/10.1126/scitranslmed.aav1892

Vonaesch, P., Anderson, M., & Sansonetti, P. J. (2018). Pathogens, Microbiome and the Host: Emergence of the Ecological Koch's Postulates. *FEMS Microbiology Reviews*, 42(3), 273–292. https://doi.org/10.1093/femsre/fuy003

Vrieze, A., Van Nood, E., Holleman, F., Salojärvi, J., Kootte, R. S., Bartelsman, J. F. W. M., Dallinga-Thie, G. M., Ackermans, M. T., Serlie, M. J., Oozeer, R., Derrien, M., Druesne, A., Van Hylckama Vlieg, J. E. T., Bloks, V. W., Groen, A. K., Heilig, H. G. H. J., Zoetendal, E. G., Stroes, E. S., de Vos, W. M., . . . & Nieuwdorp, M. (2012). Transfer of Intestinal Microbiota from Lean Donors Increases Insulin Sensitivity in Individuals with Metabolic Syndrome. *Gastroenterology*, 143(4), 913–916.e7. https://doi.org/10.1053/j.gastro.2012.06.031

Vuong, H. E., Pronovost, G. N., Williams, D. W., Coley, E. J. L., Siegler, E. L., Qiu, A., Kazantsev, M., Wilson, C. J., Rendon, T., & Hsiao, E. Y. (2020). The Maternal Microbiome Modulates Fetal Neurodevelopment in Mice. *Nature*, 586(7828), 281–286. https://doi.org/10.1038/s41586-020-2745-3

W

Wadhwa, A., Kesavelu, D., Kumar, K., Chatterjee, P., Jog, P., Gopalan, S., Paul, R., Veligandla, K. C., Mehta, S., Mane, A., Pandit, S., Rathod, R., Jayan, S., & Mitra, M. (2022). Role of *Lactobacillus reuteri* DSM 17938 on Crying Time Reduction in Infantile Colic and Its Impact on Maternal Depression: A Real-Life Clinic-Based Study. *Clinics and Practice*, 12(1), 37–45. https://doi.org/10.3390/clinpract12010005

Wagner, V. E., Dey, N., Guruge, J., Hsiao, A., Ahern, P. P., Semenkovich, N. P., Blanton, L. V., Cheng, J., Griffin, N., Stappenbeck, T. S., Ilkayeva, O., Newgard, C. B., Petri, W., Haque, R., Ahmed, T., & Gordon, J. I. (2016). Effects of a Gut Pathobiont in a Gnotobiotic Mouse Model of Childhood Undernutrition. *Science Translational Medicine*, 8(366), 366ra164. https://doi.org/10.1126/scitranslmed.aah4669

Wallace, C. J. K., & Milev, R. V. (2021). The Efficacy, Safety, and Tolerability of Probiotics on Depression: Clinical Results from an Open-Label Pilot Study. *Frontiers in Psychiatry*, 12, 618279. https://doi.org/10.3389/fpsyt.2021.618279

Wang, C., Yi, Z., Jiao, Y., Shen, Z., Yang, F., & Zhu, S. (2023). Gut Microbiota and Adipose Tissue Microenvironment Interactions in Obesity. *Metabolites*, 13(7), 821. https://doi.org/10.3390/metabo13070821

Wang, G. P. (2015). Defining Functional Signatures of Dysbiosis in Periodontitis Progression. *Genome Medicine*, 7(1), 40. https://doi.org/10.1186/s13073-015-0165-z

Wang, J.-W., Kuo, C.-H., Kuo, F.-C., Wang, Y.-K., Hsu, W.-H., Yu, F.-J., Hu, H.-M., Hsu, P.-I., Wang, J.-Y., & Wu, D.-C. (2018). Fecal Microbiota Transplantation: Review and Update. *Journal of the Formosan Medical Association*, 118, 10.1016/j.jfma.2018.08.011

Wang, Y. (2009). Prebiotics: Present and Future in Food Science and Technology. *Food Research International*, 42(1), 8–12. https://doi.org/10.1016/j.foodres.2008.09.001

Ward, H. (1923). A Founder of American Parasitology, Joseph Leidy. *Journal of Parasitology*, 10(1), 1–21. Retrieved from H. W. Manter Laboratory Library Materials. https://digitalcommons.unl.edu/manterlibrary/19

Warinner, C., Speller, C., Collins, M. J., & Lewis, C. M. (2015). Ancient Human Microbiomes. *Journal of Human Evolution*, 79, 125–136. https://doi.org/10.1016/j.jhevol.2014.10.016

Wastyk, H. C., Fragiadakis, G. K., Perelman, D., Dahan, D., Merrill, B. D., Yu, F. B., Topf, M., Gonzalez, C. G., Van Treuren, W., Han, S., Robinson, J. L., Elias, J. E., Sonnenburg, E. D., Gardner, C. D., & Sonnenburg, J. L. (2021). Gut-Microbiota-Targeted Diets Modulate Human Immune Status. *Cell*, 184(16), 4137–4153.e14. https://doi.org/10.1016/j.cell.2021.06.019

Welch, J. L. M., Rossetti, B. J., Rieken, C. W., Dewhirst, F. E., & Borisy, G. G. (2016). Biogeography of a Human Oral Microbiome at the Micron Scale. *Proceedings of the National Academy of Sciences*, 113(6), E791–E800. https://doi.org/10.1073/pnas.1522149113

Wen, S., Yuan, G., Li, C., Xiong, Y., Zhong, X., & Li, X. (2022). High Cellulose Dietary Intake Relieves Asthma Inflammation through the Intestinal Microbiome in a Mouse Model. *PLOS ONE*, 17(3), e0263762. https://doi.org/10.1371/journal.pone.0263762

White, R.A. (2020). The Global Distribution of Modern Microbialites: Not So Uncommon After All. In V. Souza, A. Segura, & J. Foster (Eds.), *Astrobiology and Cuatro Ciénegas Basin as an Analog of Early Earth. Cuatro Ciénegas Basin: An Endangered Hyperdiverse Oasis*. Springer. https://doi.org/10.1007/978-3-030-46087-7_5

Willyard, C. (2021). How Gut Microbes Could Drive Brain Disorders. *Nature*. https://doi.org/10.1038/d41586-021-00260-3

Woese, C. R., & Fox, G. E. (1977). Phylogenetic Structure of the Prokaryotic Domain: The Primary Kingdoms. *Proceedings of the National Academy of Sciences*, 74(11), 5088–5090. https://doi.org/10.1073/pnas.74.11.5088

Wu, G. (2017). Engineering the Gut Microbiome with Good Bacteria May Help Treat Crohn's Disease. Penn Medicine News. https://www.pennmedicine.org/news/news-releases/2017/november/engineering-the-gut-microbiome-with-good-bacteria-may-help-treat-crohns-disease

Wu, G., Xu, T., Zhao, N., Lam, Y. Y., Ding, X., Wei, D., Fan, J., Shi, Y., Li, X., Li, M., Ji, S., Wang, X., Fu, H., Zhang, F., Shi, Y., Zhang, C., Peng, Y., & Zhao, L. (2022). Two Competing Guilds as a Core Microbiome Signature for Health Recovery. *Microbiology* [Preprint]. https://doi.org/10.1101/2022.05.02.490290

Wu, H.-J., & Wu, E. (2012). The Role of Gut Microbiota in Immune Homeostasis and Autoimmunity. *Gut Microbes*, 3(1), 4–14. https://doi.org/10.4161/gmic.19320

Y

Yaguang, S. (2011). Development and Characteristics of Central Business District under the Philosophy of Health. *Procedia Engineering*, 21, 258–266. https://doi.org/10.1016/j.proeng.2011.11.2021

Yang, J., Pu, J., Lu, S., Bai, X., Wu, Y., Jin, D., Cheng, Y., Zhang, G., Zhu, W., Luo, X., Rosselló-Móra, R., & Xu, J. (2020). Species-Level Analysis of Human Gut Microbiota with Metataxonomics. *Frontiers in Microbiology*, 11. https://doi.org/10.3389/fmicb.2020.02029

Yaseen, A., Mahafzah, A., Dababseh, D., Taim, D., Hamdan, A. A., Al-Fraihat, E., Hassona, Y., Şahin, G. Ö., Santi-Rocca, J., & Sallam, M. (2021). Oral Colonization by *Entamoeba gingivalis* and *Trichomonas tenax*: A PCR-Based Study in Health, Gingivitis, and Periodontitis. *Frontiers in Cellular and Infection Microbiology*, 11, 782805. https://doi.org/10.3389/fcimb.2021.782805

Yates, J. A. F., Velsko, I. M., Aron, F., Posth, C., Hofman, C. A., Austin, R. M., Parker, C. E., Mann, A. E., Nägele, K., Arthur, K. W., Arthur, J. W., Bauer, C. C., Crevecoeur, I., Cupillard, C., Curtis, M. C., Dalén, L., Díaz-Zorita Bonilla, M., Díez Fernández-Lomana, J. C., Drucker, D. G., . . . & Warinner, C. (2021). The Evolution and Changing Ecology of the African Hominid Oral Microbiome. *Proceedings of the National Academy of Sciences*, 118(20), e2021655118. https://doi.org/10.1073/pnas.2021655118

Yong, E. (2016). *I Contain Multitudes: The Microbes within Us and a Grander View of Life*. Ecco, an imprint of HarperCollins.

Yong, E. (2014). The Forest in Your Mouth. *National Geographic*. https://www.nationalgeographic.com/science/article/the-forest-in-your-mouth

Yost, S., Duran-Pinedo, A. E., Teles, R., Krishnan, K., & Frias-Lopez, J. (2015). Functional Signatures of Oral Dysbiosis during Periodontitis Progression Revealed by Microbial Metatranscriptome Analysis. *Genome Medicine*, 7(1), 27. https://doi.org/10.1186/s13073-015-0153-3

Z

Zaheer, R., Noyes, N., Ortega Polo, R., Cook, S. R., Marinier, E., Van Domselaar, G., Belk, K. E., Morley, P. S., & McAllister, T. A. (2018). Impact of Sequencing Depth on the Characterization of the Microbiome and Resistome. *Scientific Reports*, 8(1), Article 1. https://doi.org/10.1038/s41598-018-24280-8

Zeng, F., Wang, Z., Wang, Y., Zhou, J., & Chen, T. (2017). Large-scale 16S Gene Assembly Using Metagenomics Shotgun Sequences. *Bioinformatics*, 33(10), 1447–1456. https://doi.org/10.1093/bioinformatics/btx018

Zhang, D., Li, S., Wang, N., Tan, H.-Y., Zhang, Z., & Feng, Y. (2020). The Cross-Talk between Gut Microbiota and Lungs in Common Lung Diseases. *Frontiers in Microbiology*, 11, 301. https://doi.org/10.3389/fmicb.2020.00301

Zheng, D., Liwinski, T., & Elinav, E. (2020). Interaction between Microbiota and Immunity in Health and Disease. *Cell Research*, 30(6), 492–506. https://doi.org/10.1038/s41422-020-0332-7

Zheng, H., Liang, H., Wang, Y., Miao, M., Shi, T., Yang, F., Wang, X., Chen, Y., Feng, Y., Zhang, Y., & Fang, J. (2016). Altered Gut Microbiota Composition Associated with Eczema in Infants. *PLOS ONE*, 11(11), e0166026. https://doi.org/10.1371/journal.pone.0166026

Zhong, J., Wu, D., Zeng, Y., Wu, G., Zheng, N., Huang, W., Li, Y., Tao, X., Zhu, W., Sheng, L., Shen, X., Zhang, W., Zhu, R., & Li, H. (2022). The Microbial and Metabolic Signatures of Patients with Stable Coronary Artery Disease. *Microbiology Spectrum*, 10(6), e0246722. https://doi.org/10.1128/spectrum.02467-22

Zhu, G., Zhao, J., Zhang, H., Chen, W., & Wang, G. (2021). Administration of *Bifidobacterium breve* Improves the Brain Function of $A\beta_{1-42}$-Treated Mice via the Modulation of the Gut Microbiome. *Nutrients*, 13(5), 1602. https://doi.org/10.3390/nu13051602

Zhu, T. F., & Szostak, J. W. (2009). Coupled Growth and Division of Model Protocell Membranes. *Journal of the American Chemical Society*, 131(15), 5705–5713. https://doi.org/10.1021/ja900919c

Zoetendal, E. G., Raes, J., van den Bogert, B., Arumugam, M., Booijink, C. C., Troost, F. J., & de Vos, W. M. (2012). The Human Small Intestinal Microbiota Is Driven by Rapid Uptake and Conversion of Simple Carbohydrates. *ISME Journal*, 6, 1415–1426. https://doi.org/10.1038/ismej.2012.37

Index

A
abiogenesis, 4
abiotic components, 374
ABO gene, 340
abstract (article), 147, 149–51
abundance. *See* species abundance
acetate
    in gut microbiome, 60–62, 273, 278, 281–82, 285, 386
    health linked to, 386
    immune function, 184, 301
Acetobacter, 247
Acidianus virus, 16
acidic environment
    oral microbiome, 71
    skin microbiome, 74–75
acidophilic archaeans, 19
Acinetobacter, 247, 362
Acinetobacter lwoffii, 300, 302, 309
acne, 75–76
acquired (adaptive) immune system, 180–82, 237–38, 296–97
ACTH (adrenocorticotropin), 212
Actinobacteria
    colonization resistance, 62
    female reproductive tract microbiome, 176
    gut microbiome, 330–31
    infant microbiome, 188–90
    in leanness, 277
    oral microbiome, 69, 70
    primate microbiome, 325
    respiratory microbiome, 306–7
    skin microbiome, 73–74
    stomach microbiome, 336
Actinomyces, 70
active state, 362–63
adaptations
    niche, 45
    processes specific to, 55
adapter regions, 101–2, 105, 117
adaptive (acquired) immune system, 180–82, 237–38, 296–97
adaptive traits, 7
adenine, 99
adenocarcinoma, 337
adenosine triphosphate (ATP), 8, 60
adherens, 73
adiposity, 276, 277, 279, 283–84. *See also* obesity
adrenal glands, 212
adrenocorticotropin (ACTH), 212
aerobic bacteria, 37, 69, 71, 188, 190, 395
aerobic respiration, 7–8
Africa
    ape diversification, 323–24
    Malawi study, 324

agar, 37–38, 77
Aggregatibacter, 70
Aggregatibacter actinomycetemcomitans, 281
agriculture. *See* farming
Ahmadi, S., 290
air-borne microbes experiment, 30–31
air-handling systems, 349, 353–55, 359, 365
Akkermansia, 307–8
Akkermansia muciniphila
    gut microbiome, 273, 278, 281
    gut transit times linked to, 380
    infant microbiome, 194
    weight loss linked to, 285, 288, 289
Aktipis, Athena, 62
Alexander, Albert, 34
algorithms, similarity-searching, 120, 123
aliquot, 97
alkaliphilic archaeans, 19
allergens, 296
allergic cascade, 296–97
allergic diseases, 295–319. *See also specific disorder*
    antibiotics in, 304–5
    atopic march, 303
    circle of causality, 315–16
    environmental impact, 300–303
    epigenetic changes, 298–99
    farm effect, 297, 300, 302–3, 308–9, 357–58
    hygiene hypothesis, 245, 302–4, 333–34, 350, 366–69
    infant microbiome linked to, 198–99
    maternal microbiome linked to, 299–300
    microbiome in, 305–13
    prevention of, 297–98, 304
    treatment of, 313–15
allergic response, 295–97, 334
Alm, E. J., 254–55
alpha-aminoadipic acid, 246
alpha diversity, 126–29, 141, 255
    indices, 129–32, 166–69, 382
    infant microbiome, 196–97
    in obesity, 280
α-amylase, 330
α-synuclein, 227–28
α-toxin (α-hemolysin), 310
altered microbiome, 88. *See also* dysbiosis
Alzheimer's disease, 221, 228
Amazon River basin study, 349–50
Ambystoma mexicanum (axolotl), 87
American Academy of Allergy, Asthma & Immunology, 303
American Gut Project, 383

amino acids
    in bile salts, 283
    in Crohn's disease, 66–69
    in depression, 224–26
    origin of, 3–4
    triplets encoding, 119–20
Amish farmers (USA), 302–3, 357–58
amniotic sac, 179–80, 243
amplification, DNA, 101–6, 117–18
AMPs (antimicrobial proteins), 240, 241, 258
amylases, salivary, 60, 330
amylose, 279
anaerobic bacteria, 37–39, 65, 71, 190, 257–60, 395
anaerobic respiration, 7–8
Anaerostipes caccae, 314
anaphylaxis, 297
animalcules, 17, 28
animals. *See also specific animal*
    in built environment, 350, 353–54, 369
    infant microbiome affected by, 192–93, 297, 300, 302–3, 308–9, 357–58
    in research (*see* model organisms)
    in tree of life, 12–13, 19
annealing, 101–2, 117
anoxic environment, 7–8, 37–38
Antharam, V. C., 266
anthrax, 33
antibacterial mouthwash, 70
antibiotics
    agricultural use of, 335
    in allergic diseases, 304–5, 310, 312
    bactericidal, 35
    bacteriostatic, 35
    in children, 196–97, 300, 304–5, 335, 337, 360
    definition of, 34
    discovery of, 34–35
    dysbiosis caused by, 39, 57, 76, 196–97, 214, 241, 244, 260, 335, 337
    increased use of, 201, 245, 335, 365
    microbiome recovery from, 394–95
    naturally produced, 331
    new perspectives on, 339–40
    preventative use of, 261
    resistance to, 35, 85, 186–87, 310, 330–31, 335, 337, 360, 362
antibodies, 182, 184, 237–38, 242–45, 296
antidepressants, 220–21, 229
antigen-presenting cells (APCs), 241–42, 296–97
antigens, 237, 310
antimicrobial proteins (AMPs), 240, 241, 258

435

antimicrobials
    female reproductive tract, 177–79
    intestinal, 239
    skin, 309
antiseptics, 35
anxiety, 220
APCs (antigen-presenting cells), 241–42, 296–97
apple-shaped body, 279
Appleton, Joseph, 37
Archaea
    evolutionary divergence, 103
    extremophiles, 1, 3, 18–19
    gut microbiome, 58, 59
    oral microbiome, 69
    skin microbiome, 73–75
    in tree of life, 14–15, 17, 19
arginine, 73, 394
arthritis, rheumatoid, 246–47
artifact, 160–62
artifact situation, 266
aryl hydrocarbon receptor agonists, 214
ASD (autism spectrum disorder), 226–27
asparagine, 246
Aspergillus, 69, 78
asthma, 198, 306–9
    birth mode linked to, 86, 300, 304
    farm effect, 302–3, 308–9, 357–58
    gut dysbiosis, 245
    gut-lung axis, 307–8
    maternal microbiome linked to, 299–300
    physiological responses in, 306–7
    tolerance approach, 304
    treatment of, 314–15
astrocytes, 214, 221
atherosclerosis, 247
atmosphere
    anoxic, 7–8, 37–38
    early Earth, 2, 3
atopic dermatitis (eczema), 75, 198–99, 240, 300, 309–11
atopic march, 303
atopy, 198
ATP (adenosine triphosphate), 8, 60
authorship (article), 148–49
autism spectrum disorder (ASD), 226–27
autoantigens, 237
autoclave, 37
autoimmune diseases. *See also specific disorder*
    CNS, 241–42
    hygiene hypothesis, 245, 302–4, 333–34, 350, 366–69
    systemic, 246
autoimmunity, 182–84, 237
autonomic nervous system, 210, 217
autophagy, 222
autotrophs, 7
Axial Therapeutics, 226
axolotl (*Ambystoma mexicanum*), 87
axon (nerve fiber), 185, 217

B
babies. *See* infant microbiome; newborns
Bacillus, 78
Bacillus lactis aërogenes, 389
Bacillus licheniformis, 309
bacteria. *See also specific microbe*
    aerobic, 37, 69, 71, 188, 190, 395
    anaerobic, 37–39, 65, 71, 190, 257–60, 395
    antibiotics produced by, 331
    appendages, 69–70
    in built environment, 355, 363
    cellulose-degrading, 37
    in core microbiome, 54–55
    culturing methods, 18, 37–39, 42
    evolutionary divergence, 103
    filamentous, 43, 226–27
    first observations of, 17–18, 28
    immune responses to, 237 (*see also* immune system)
    recovery-associated, 394–95
    saccharolytic, 392
    tree of life, 15, 17, 19
    virulence factors, 264–65, 310
bacterial infections. *See specific infection*
    antibiotics for (*see* antibiotics)
bacterial vaginosis (BV), 179, 202, 255
bactericidal antibiotics, 35
bacteriocins, 177, 178
bacteriophages (phages)
    in built environment, 352
    characteristics of, 16
    gut microbiome, 63–65, 330–31
    infant microbiome, 189
    skin microbiome, 73
    therapeutic use of, 341
bacteriostatic antibiotics, 35
Bacteroides
    in birth mode study, 128–29, 187
    in colorectal cancer, 67
    in coronary heart disease, 247
    in depression, 225–26
    dietary impact on, 332
    in food allergy, 67
    infant microbiome, 189, 193, 194, 196, 198
    lung microbiome, 241
    in obesity, 278–80, 288
    primate microbiome, 323–24
    stomach microbiome, 336
    vaginal microbiome, 186
Bacteroides fragilis, 51, 189, 199, 268, 301
Bacteroides thetaiotaomicron, 60, 69
Bacteroidetes
    in colorectal cancer, 62–64
    core microbiome, 54
    in diabetes, 200
    female reproductive tract microbiome, 176, 183–84
    gut microbiome, 53, 59, 60, 67, 199–200, 302, 325
    infant microbiome, 189, 194, 195, 199
    in obesity, 199, 277, 281, 285
    oral microbiome, 69, 70, 281
    primate microbiome, 325
    respiratory microbiome, 306–7
    skin microbiome, 73–74
    stomach microbiome, 336
barcodes, 101–2, 117
bar plot, 164–66
Barrett's esophagus, 337
Bartonella, 330–31
BAs (bile acids), 283
baseball team metaphor, 54–55, 385
bases, 99
Basic Local Alignment Search Tool (BLAST), 120
basophils, 181, 238, 296–97
Bassi, Agostino, 32
bathrooms, 351, 353, 355–56, 363, 364
BBB (blood-brain barrier), 212–13, 216–17
BC (Bray-Curtis dissimilarity) metric, 135–36, 333
B cells, 182, 238
    in allergic cascade, 296–97
    development of, 236–37, 243–44
    in stomach, 337
BDNF (brain-derived neurotrophic factor), 221
BE. *See* built environment
beds, microbes in, 346
behavioral abnormalities, 185, 223
Belgium, 330–31
beta-alanine, 246
beta diversity, 135–39
    definition of, 126
    dysbiosis defined by, 255
    female reproductive tract, 176–77
    gut microbiome, 333
    visualisations, 140–41, 155–56
β-oxidation, 258–59
bias
    codon, 120
    research, 45, 376
Bifidobacterium
    in birth mode study, 129
    cholesterol-lowering effect, 279
    farm effect, 300
    in fossils, 327
    gut microbiome, 62, 64, 188, 198–99, 307–8, 388
    infant microbiome, 188, 190–201, 203–4, 256–57, 279, 304
    in lactose digestion, 340
    in microbiota-gut-brain axis, 219–21, 225
    in obesity, 288–90
    probiotic strains, 390, 393
Bifidobacterium adolescentis, 60, 323
Bifidobacterium animalis, 289, 314
Bifidobacterium bifidum, 192, 289
Bifidobacterium breve, 192, 229, 314
Bifidobacterium infantis, 175, 192, 203, 229
Bifidobacterium lactis, 314
Bifidobacterium longum, 203, 229, 256–57, 278, 289, 314
bifidus factor, 191
bile acids (BAs), 283
bile ducts, 241
bile salt hydrolases (BSHs), 283
bimodal distribution, 93–94
binominal (Latin) names, 12, 53–54
biodiversity, 126. *See also* diversity
biodiversity hypothesis, 304, 366–67
biofilms, 67, 70, 71, 265, 327, 351
bioinformatics, 38–41, 121
biomarkers
    disease, 281
    weight loss, 286
biomolecules, 5
biosphere, early Earth, 8–10
biotherapy, 310–11

birth, preterm, 46, 202–4, 243, 300, 360
birth mode
    colonization process during, 41, 186–87
    C-section (see Cesarean section)
    vaginal delivery, 186–87, 191
birth mode study
    data analysis, 116–17, 122–25, 128–32
    design of, 86–90, 95
    discussion, 157–58
    literature review, 148–58
    microbiome analysis program, 158–70
    research questions, 86, 121
    results, 154–57
    sample size, 125, 141, 150
    sampling methods, 97–98
bisphenol A (BPA), 364
black mold, 345, 355, 359
bladder cancer, 248
bladder microbiome, 21
Blaser, Martin, 334–37
BLAST (Basic Local Alignment Search Tool), 120
Blastocystis, 59
Blastocystis hominis, 115
Blautia, 225–26, 327
blood-brain barrier (BBB), 212–13, 216–17
blood pressure, 275
blood stem cells, 243
blood sugar, 212, 275
blue-green algae (Cyanobacteria), 8–9
B lymphocytes. See B cells
body mass index (BMI), 199, 274–75, 281
body odor, 73
body weight, 274–79. See also obesity
Bogaert, Debby, 196–97
bone marrow, 236
bonobos, 322–30, 346
Booth, A., 224
Bordetella, 330–31
bowel movements, 375. See also feces
box and whiskers plot, 134–35
box plot, 247–48
BPA (bisphenol A), 364
brain, 209–33
    development of, 185, 212–17
    expansion in humans, 328–29
    gut interactions, 211, 213–22 (see also microbiota-gut-brain axis)
brain-derived neurotrophic factor (BDNF), 221
Bray-Curtis dissimilarity (BC) metric, 135–36, 333
Brazil, 349–50
breastfeeding. See human milk
Bristol stool chart, 378
Brudziński, Józef, 389
Bry, Lynn, 43–44
BSHs (bile salt hydrolases), 283
bubonic plague, 348
build back better mantra, 365–66
built environment (BE), 345–72
    definition of, 345–46
    features of, 349, 353–56, 359–60, 369
    future improvements, 365–69
    history of, 346–48
    lifestyle based on (see Western lifestyle)
    metabolomics, 362–65
    microbe tracking in, 358–62
    microbiology of, 350–52 (see also microbiomes of the built environment)
butterfly gut amoeba, 29
butyrate
    gut microbiome, 60, 61, 193, 258, 267
    health linked to, 386
    immune function, 305, 308, 312, 338
    in obesity, 279, 281–82, 286–87
    in type 1 diabetes, 200
BV (bacterial vaginosis), 179, 202, 255
by-products, 22. See also metabolites; specific substance

C
Caenorhabditis elegans (nematode), 87
Callaway, Ewen, 186
Camarillo-Guerrero, L. F., 103
Cambrian Explosion, 11–12
Campylobacter, 268
cancer, 248–50
    adenocarcinoma, 337
    bladder, 248
    colorectal, 67, 249, 261, 265, 340
    liver, 265–66
    microbiome-based therapies, 250
    skin, 73–74
    stomach, 248–49, 337
Candida
    gut microbiome, 220
    oral microbiome, 69
    skin microbiome, 73, 78
Candida albicans, 373
    gut microbiome, 57, 59
    vaginal microbiome, 179
candidiasis, 179
Cani, Patrice, 289
capsaicin, 364
carbohydrates
    in cavity formation, 71
    dietary, 59–60, 285, 288, 332
    digestion of, 22, 188, 194, 281–82, 329–30, 392
    types of, 22, 59–60 (see also fiber; starches; sugars)
carbon dioxide, 3, 258, 355
cardiometabolic diseases (CMDs), 247–48. See also specific disease
carpets, 353–54
Cason, Carolina, 360–61
cavities (dental caries), 71, 394
CD-HIT, 154, 158
celiac disease, 245
cell density, 20–21
cell-mediated immunity, 237
cells
    culturing methods, 18, 37–39, 42
    eukaryotic (see eukaryotes)
    origin of, 4–6, 20
    plasma membranes, 18
    prokaryotic (see prokaryotes)
cellular respiration, 7–8
cellulose, 37, 59–60, 315, 345, 355
Center for Microbiome Innovation, 170
Central Dogma of Molecular Biology, 4–7
central nervous system (CNS), 185, 210, 241–42. See also brain
Cesarean section (C-section), 186–87

allergic disease linked to, 300, 304, 312
    avoidance of, 340
    dysbiosis linked to, 200–201
    NICU use after, 360
    studies of (see birth mode study)
    vaginal microbiome transplant after, 202, 340
c-FOS enzyme, 224–25
Chain, Ernst, 34
Chaos (formless) genus, 28
CHD (coronary heart disease), 247–48
Checherta (Peru), 349–50
chemical communication, gut-immune system, 63–64, 180, 239
chemical lysis, 101
chemoautotrophs, 8
chemolithoautotrophy, 7
chemolithotrophs, 2
chemoorganotrophs, 1
childbirth. See birth mode
chimpanzees, 322–30, 346
chlamydia, 33
Chlamydiae, 330–31
chloroplasts, 11
Chng, K. R., 394–95
cholesterol, 275, 279, 282, 283, 285, 376
Christensenella minuta, 285–86
chromatin, 298
chromosome structure, 4–5
chronic heart disease, 247
chronic periodontitis (CP), 281
Church, George, 109
circle of causality, 315–16
citations (article), 152
cities. See urban lifestyle
Citrobacter rodentium, 224–25
classification system, Linnaean, 12–17, 28, 53–54
cleaning practices, 333–35, 356, 359–62
Clostridiales, 78
Clostridioides difficile, 253, 369, 384
Clostridioides difficile infection
    disorders caused by, 57, 261
    FMT for, 39–40, 260, 290, 314, 341, 377
Clostridium
    in depression, 226
    immune function, 308, 312
    infant microbiome, 194, 198–99, 214–15
    in liver cancer, 265
Clostridium leptum, 386
Clostridium neonatale, 307
clumping factors A and B, 310
cluster, 154
CMDs (cardiometabolic diseases), 247–48. See also specific disease
CNS (central nervous system), 185, 210, 241–42. See also brain
codon bias, 120
codons, 119–20
coevolution, 219, 229–30, 322–23
coin toss example, 92
colic, 200, 203
colitis, 40, 65
    enterocolitis, 39–40, 200, 203
    ulcerative, 66–67, 248, 258–59

Collinsella, 246–47
colonization process
  during birth, 41, 186–87
  during infancy, 189–90, 196
  recolonization, 337–38
colonization resistance, 259
  gut microbiome, 36–37, 62–63, 196–97, 204, 259
  infant microbiome, 196–97
  lung microbiome, 240–41
  oral microbiome, 70
  skin microbiome, 74–75, 240–41, 309
colonocytes, 65
color coding, 123
colorectal cancer (CRC), 67, 249, 261, 265, 340
colostrum, 245
commensal interaction, 230, 259
community state types (CSTs), 177–78
competition, diversification driven by, 7
competitive exclusion, 179
complementary supplements, 394
confidence intervals, 140
confounding variables, 95
consortia, 388
constant regions, 103
contamination of samples, 118
contig (contiguous sequence), 119
control
  experimental, 77–79, 95–96, 125, 223, 247, 254, 277
  quality, 118
  sterility, 37–38, 77–78
cooling systems, 349, 353–55, 359, 365
coprolites, 326–27
coprophagy, 95, 226–27
coprostanol, 376
core microbiome, 54–57, 187–90, 254
coronary heart disease (CHD), 247–48
correlation, 86
corticotrophin-releasing hormone (CRH), 212
cortisol, 212
Corynebacterium, 70, 73–75
Corynebacterium accolens, 74–75
COVID-19 virus, 16, 356, 365–66
cow milk allergy, 312, 314
CP (chronic periodontitis), 281
CRC (colorectal cancer), 67, 249, 261, 265, 340
CRH (corticotrophin-releasing hormone), 212
Crick, 38
critical window, 196, 201, 235, 245, 297–98, 334–35, 357
Crittenden, Alyssa, 332
Crohn's disease, 66–69, 257–58
cross-feeding interactions, 188, 192, 338
cross-sectional study, 96
crown species, 15
Cryan, John, 227
crypts, 257
C-section. *See* Cesarean section
CSTs (community state types), 177–78
culturing
  fecal samples, 97
  methods, 18, 37–39, 76–78
  oral microbiome, 42

samples for (*see* sampling methods)
  skin microbiome, 78–80
Curatola, Gerry, 72
Cutibacterium, 73–75
Cutibacterium acnes, 73, 75–76
Cyanobacteria (blue-green algae), 8–9
cyclic causality, 315–16
cytokines, 217
  activation of, 237
  in allergic response, 297
  in autism spectrum disorder, 226–27
  in human milk, 244
  in MGBA pathway, 221
  in obesity, 283–84
  in rheumatoid arthritis, 247
cytosine, 99
cytotoxins, 182

D
Darwin, Charles, 12–13
data analysis, 88, 115–44
  microbiome, 121–25
  overview, 115–18
  quality control, 118
  reconstruction of genes and genomes, 118–20
  reporting on, 153–57 (*see also* literature)
  software, 121, 140, 145, 154, 158–70 (*see also specific program*)
  species diversity measures, 126–39
  species diversity visualisations, 139–41, 155–56, 163–64
data distribution, 93–94
data organization, 88
datasets. *See also specific study*
  genome reference, 119, 121–23, 154, 163, 376
  KOALA Birth Cohort Study, 195
  public (Qiita), 159–60
data transformation, 163–64
DCs (dendritic cells), 181, 239, 296
dead zones (ocean), 9
deceased state, 362–63
defensins, 256, 257
degranulation, 296
deidentification, 151
Deinococcus, 353
delta toxin, 75
dementia, 222
demineralization, 71
demultiplexing, 118
denaturing, 101–2
dendritic cells (DCs), 181, 239, 296
de novo reconstruction, 119
dental caries, 71, 394
dental plaque. *See also* oral microbiome
  creation of, 70
  early studies of, 17, 28, 37, 41–42
  fossilized, 327–28
deoxyribonucleic acid. *See* DNA
deoxyribose, 99
dependent variables, 89
depression, 220, 223–26, 229
dermatitis, atopic (eczema), 75, 198–99, 240, 300, 309–11
desensitization, 313
Desnues, Christelle, 331
deterministic (selection) process, 266–67

De Vlaminck, Iwijn, 110
diabetes
  type 1, 200, 245
  type 2, 46, 247, 280
diarrheal diseases, 33, 197–98, 253, 261. *See also specific disorder*
diet
  antibiotics in, 335
  body weight management, 287–91
  carbohydrates (*see* carbohydrates)
  evolved dependence, 229–30, 328–29
  fermented food, 288, 388–89
  fiber (*see* fiber)
  food additives, 258
  food allergies (*see* food allergies)
  food as medicine, 396
  gut transit time, 379–80
  healthy, 387–88
  human milk, 191–94
  in obesity, 199, 275, 278, 285–91
  oral microbiome impact, 70–71
  plants in, 323, 328, 332, 340, 352, 388, 394
  prebiotic foods (*see* prebiotics)
  during pregnancy, 182–84, 244, 299–300
  probiotic foods (*see* probiotics)
  professional advice on, 377–78
  solid food transition, 194, 257, 303
  Western, 287, 303, 332–33, 335–36, 340, 365, 388, 392
dietary emulsifiers, 258
dietary supplements
  prebiotics, 392–93
  probiotics, 389–92
  synbiotics, 393–94
digital object identifier (DOI), 148
Dinan, T. G., 228–29
Dinophysis algae, 9
disappearing-microbiota hypothesis, 334–37, 341
discovery metabolomics, 214
discussion section (article), 148, 149, 153, 157–58
disease. *See also specific disease*
  versus dysbiosis, 260, 266–69
  ecological perspective on, 258–59, 268, 374, 395
  environmental impact (*see* built environment)
  germ theory of, 31–32, 333
  humoral theory of, 254–55
  Koch's postulates about, 32–33
  microbiome associations, 260–69
  microbiome-based therapies, 39–41, 201–4, 228–29, 250 (*see also specific therapy*)
  most common and deadly, 33
  spontaneous generation theory of, 29–31
  susceptibility to, 196, 201, 235, 376–77 (*see also* immune system)
disease markers, 281
disinfectants, 35, 356
dissimilarity, measurement of, 135–39
distance matrix, 156
distribution of data, 93–94
diversity, 126
  biodiversity hypothesis, 304, 366–67

Cambrian Explosion, 11–12
cataloguing methods, 43
competition driving, 7
in core microbiome, 56–57, 189, 254
Cyanobacteria, 9
versus function, 266–67
as health indicator, 384–85
meadow metaphor, 374, 383–84
measurement of, 126–39, 141, 166–69, 255, 382
viruses, 16
visualisations, 139–41, 155–56, 163–64
diversity index, 129
DNA
amplification, 101–6, 117–18
chemistry of, 98–99
epigenetic changes, 298–99
extraction, 38–39, 99–101, 117
methylation, 298–99
noncoding, 119
origin of, 4
replication, 4–7, 99–100
ribosomal (*see* ribosomal DNA)
versus RNA, 107
shearing, 104
structure of, 38, 98–99
DNA isolation kits, 101
DNA methyltransferase (DNMT), 298
DNA sequence read, 117
DNA sequencing, 96, 98. *See also specific method or study*
contig (contiguous sequence), 119
cost of, 150
data analysis, 116–18
of fossils, 327
function prediction, 120
high-throughput, 104–6, 117
metagenomic library, 104
of primate microbiome, 322–24
purpose of, 38–39
ribosomal, 41–42
shotgun, 106
by synthesis (next-generation), 117–18
target gene amplicon, 101–2, 381 (*see also specific target*)
DNMT (DNA methyltransferase), 298
DOI (digital object identifier), 148
domains of life, 14–15
Dominguez-Bello, M. G., 148–70, 186, 202
Doolittle, W. F., 224
Dorea, 327
dormant state, 362–63
double helix, 99
Drosophila melanogaster (fruit fly), 87, 225
Dukes, C. D., 202
dust
farm, 308–9, 357–58 (*see also* farm effect)
house, 353–55, 357–58, 363
Duvallet, Claire, 261–62
dysbiosis, 253–71
circle of causality, 315–16
definition of, 58–59, 254–55, 266
versus disease, 260, 266–69
gut microbiome (*see* gut microbiome dysbiosis)
indices of, 267–69
infant microbiome, 175, 196–201 (*see also* birth mode)

lung microbiome, 241, 307, 308
maternal microbiome, 213, 226
oral microbiome, 71–72, 262–65
self-management of (*see* personal microbiome)
shared, 265–66
skin microbiome, 75–76, 240, 309–11
smoking-related, 46
vaginal microbiome, 179
dysentery, 36, 39
Dysosmobacter welbionis, 279

E
Earth
microbes as core of life on, 17
origin of life on, 2–12
tree of life on, 12–17, 19, 132
Earth Microbiome Project (EMP), 106
Easy Microbiome Analysis Platform (EzMAP), 170
Ebola virus, 16
ecological core, 55–56
ecological Koch's postulates, 267–69
ecological perspective, 255–60, 374, 395
ecosystem, 374, 386–87, 395
eczema (atopic dermatitis), 75, 198–99, 240, 300, 309–11
education, scientific, 86
EECs (enteroendocrine cells), 219
effectors, 211
Eggerthella, 246
Eiseman, Ben, 39–40
Elinav, Eran, 46
EMP (Earth Microbiome Project), 106
emulsifiers, 258
endocrine pathway (MGBA), 218–19
endocrine system, 211–12, 217
endosymbiont, 11
endosymbiosis, 10–12, 15, 20, 52, 223
endotoxins, 357–58
ENS (enteric nervous system), 211, 217, 219, 220
Entamoeba, 69
enteric bacteria. *See* Enterobacteriaceae
enteric nervous system (ENS), 211, 217, 219, 220
Enterobacter, 225, 261
Enterobacteriaceae
gut microbiome, 388
infant microbiome, 193, 194, 304
in obesity, 280–82, 284–85
in ulcerative colitis, 259
Enterococcus, 220, 304
Enterococcus faecalis, 145, 340
enteroendocrine cells (EECs), 219
enterotypes, 325–26
envelope, viral, 16–17
environment
in allergic disease, 300–303, 312
anoxic, 7–8, 37–38
built (*see* built environment)
early Earth, 2, 3
gene expression affected by, 298–300, 322, 357–58
health care, 186–87, 356, 359–62
infant microbiome impacted by, 194–95, 300–303
environmental metagenomics, 38–39, 41–43

enzymes. *See also specific enzyme*
carbohydrate-digesting, 22, 60
definition of, 22
eosinophils, 181, 237–38, 296, 306, 315
epigenetic changes, 298–99
epithelial cells, in allergic response, 296–97
EPS (extracellular polymeric substances), 263
error bars, 92–93
errors, process, 118, 121
Escherich, Theodor, 36
Escherichia coli, 209
classification of, 53–54
digestive function of, 60
discovery of, 36
disease caused by, 57
gut microbiome, 220, 258, 268
infant microbiome, 204
in microbiota-gut-brain axis, 220, 224–25
Nissle 1917 strain, 36–37, 39
in obesity, 278–79
O157:H7 strain, 53, 57
esophageal reflux, 337
esophagus, 21, 58. *See also* gut microbiome
ester bonds, 18
estrogen, 176–77, 256
ether bonds, 18
ethics, research, 150–51
Eubacterium, 67, 225–26
Eubacterium coprostanoligenes, 376
Eubacterium cylindroides, 287
Eubacterium nodatum, 265, 281
Eubacterium rectale, 60, 380, 386
eubiosis, 254, 260, 267
Eukarya, 14–15
eukaryotes
classification of, 15
definition of, 10
gut microbiome, 59
oral microbiome, 69–70
origin of, 10–12, 20
Euler's number, 130
evenness (species), 126. *See also* alpha diversity
evolutionary distance, 132
evolutionary divergence, 103
evolutionary tree (phylogeny), 12, 132, 138, 154, 322–23
evolved dependence, 229–30, 328–29
evolving microbiome, 321–44
ancestral origins, 322–31
healthful microbiota development, 339–41
industrialization, 331–38, 348, 376 (*see also* built environment; Western lifestyle)
missing-heritability problem, 339
experiment
design, 30–31, 88–89 (*see also* research)
methods, 97–106 (*see also* culturing; sampling methods)
experimental controls, 77–79, 95–96, 125, 223, 247, 254, 277
experimental data, *versus* metadata, 96–97
experimental treatment, 30
experimental variables, 89, 95
extant species, 7

extension phase of PCR, 102
extinction
　loss of microbiota, 334–38, 341
　mass events, 9
extracellular polymeric substances (EPS), 263
extremophiles, 1, 3, 18–19
EzMAP (Easy Microbiome Analysis Platform), 170

F
facultative anaerobic bacteria, 65, 258–60, 395
Faecalibacterium, 247, 307, 314–15, 327
Faecalibacterium prausnitzii, 194–95, 198–200, 267, 386
Faecalicatena lactaris, 340
FAIR principles, 90
Faith's phylogenetic diversity metric, 132–35
families, 12
farm effect, 297, 300, 302–3, 308–9, 357–58
farming
　antibiotics used in, 335
　introduction of, 328
FASTA file, 122–23
fasting-induced adipose factor, 277
FastTree, 154
fat
　absorption of, 283
　adiposity, 276, 277, 279, 283–84 (*see also* obesity)
　dietary, 285
　measurement of, 274–75
　production of, 60–61, 282
　storage of, 283–84
　subcutaneous layer, 73, 275
fat-loving (lipophilic) yeast, 73
fatty acids, 5
　free, 200
　short-chain (*see* short-chain fatty acids)
F/B ratio, 277, 281, 285
FDA (Food and Drug Administration), 290, 377, 391
fecal microbiota transplantation (FMT), 39–41, 253, 260, 261, 290–91, 305, 314, 341, 377, 388
feces
　amino acids in, 68–69
　consumption of, 95, 226–27
　fossilized, 326–27, 330–31
　obesity transmission through, 44
　quality assessment, 375, 378–79
　sampling methods, 97–98
female reproductive tract microbiome, 176–79. *See also* vaginal microbiome
　during pregnancy (*see* birth mode; pregnancy)
fermented food, 288, 388–89
fermenters, 60–61, 65, 385–87, 392
fetal microbiota transplant (FMT), 202
fetal programming hypothesis, 196
fetal stem cells, 217
fetus
　brain development, 185, 212–17
　delivery of (*see* birth mode; newborns)
　immune system development, 180–87, 195, 217, 242–44, 297–300
　mother's microbiome (*see* pregnancy)

FFAs (free fatty acids), 200
fiber (dietary), 59–60
　body weight link, 44, 277–80, 288, 321
　in gut health assessment, 379–80, 385–88
　immune function, 311–12
　maternal microbiome, 182–84, 244, 299–301
　prebiotic (*see* prebiotics)
　in Western diet, 332–33, 340, 392
Fibrobacter, 323–24
fight-or-flight response, 217
filamentous bacteria, 43, 226–27
filtration systems, 355–56
financial incentives, for research participants, 94
findings section (article), 148, 149, 151
Finland, 304
Firmicutes
　in colorectal cancer, 62–64
　in core microbiome, 54
　in diabetes, 200
　female reproductive tract microbiome, 176
　gut microbiome, 53, 59, 60, 62, 67, 188, 198–200, 302, 325, 330–31
　gut transit time, 380
　infant microbiome, 188–91, 194, 195, 198–99
　in obesity, 199, 277, 281, 285
　oral microbiome, 69, 70
　primate microbiome, 325
　respiratory microbiome, 306–7
　skin microbiome, 73–74
　stomach microbiome, 336
Fischbach, Michael, 70
five-kingdom tree of life, 12–13, 17, 19
5′ (five prime), 99
Fleming, Alexander, 34, 335
A Flora and Fauna within Living Animals (Leidy), 29
Florey, Howard, 34
fluorescent labels, 42–43, 104, 110, 117
FLVR, 314–15
fly larvae experiment, 29–30
FMT (fecal microbiota transplantation), 39–41, 253, 260, 261, 290–91, 305, 314, 341, 377, 388
FMT (fetal microbiota transplant), 202
folic acid, 216
food. *See* diet
food additives, 258
food allergies, 311–13
　antibiotics in, 304
　infant microbiome, 193, 198, 303
　physiological response to, 297
　treatment of, 313–15
food allergy oral immunotherapy (OIT), 313
Food and Drug Administration (FDA), 290, 377, 391
food desert, 275
forensics, microbiome used in, 352
formless *(Chaos)* genus, 28
forward primers, 101–2
fossils
　dental plaque, 327–28
　feces, 326–27, 330–31
　microbial, 2–3
Foster, Jane, 223

Foster, K. R., 230
Foster, Kevin, 256
free fatty acids (FFAs), 200
freezing of samples, 98
French Academy of Sciences, 30–31
Friedland, Robert, 228
fructooligosaccharides, 289, 392–94
fruit fly *(Drosophila melanogaster)*, 87, 225
functional core, 55–56, 329–30
functional redundancy, 262–65
function *versus* diversity, 266–67
funding, research, 94, 150
fungi. *See also specific microbe*
　antibiotics produced by, 331
　in built environment, 351, 355, 363
　gut microbiome, 59, 381
　immune responses to, 237 (*see also* immune system)
　oral microbiome, 69–70
　silkworm studies, 32
　skin microbiome, 73–74
　in tree of life, 12–13
furnishings, 353–54
Fusarium, 69
Fusobacteria
　infant microbiome, 189
　in obesity, 280
　respiratory microbiome, 207
　stomach microbiome, 336
Fusobacterium, 70, 261
Fusobacterium nucleatum, 249, 265, 281

G
GABA (gamma-aminobutyric acid), 220, 224–26
galactooligosaccharides, 289, 392–93
gamma-aminobutyric acid (GABA), 220, 224–26
ganglia, 210
garbage in, garbage out expression, 118
Gardner, H. L., 202
Gardnerella, 179
Gardnerella vaginalis, 202
GasPak™, 38
gastric acid, 257
gastric cancer, 248–49, 337
gastroenteritis, 259
gastrointestinal (GI) tract, 58. *See also specific organ*
　bowel movements, 375 (*see also* feces)
　digestion in, 60–61
　enteric nervous system, 211, 217, 219, 220
　immune function, 63–64, 180, 239, 245
　leash effect, 257–58
　microbes in (*see* gut microbiome)
　mucin (*see* mucin layer)
　transit time, 379–80
Gaussian (normal) distribution, 93
GBA (gut-brain axis), 211
GenBank, 120
Generally Regarded As Safe (GRAS) status, 391
general practitioner (GP), advice from, 375–78
genes
　expression of, 298–99, 322, 357–58
　reconstruction of, 118–20

genetic code, 120
genetic databases, 120, 154
genetic dictionary, 119–20
genetics
　allergic disease linked to, 299
　versus environment, 322
　infant microbiome impacted by, 188
　missing-heritability problem, 339
　in obesity, 43–44, 275
　primate evolution, 322–25
genetic variant (strain), 53, 278, 390–94
genome, reconstruction of, 118–20
genome reference sequences, 119, 121–23, 154, 163, 376
genome scaffold, 119
genome-wide association studies (GWASs), 339
genomics, 38–41
genus, 12, 53
　identification of, 121–24
germ-free mice, 40–41, 43–44, 87
　allergic disease studies, 300, 305, 308–9, 312–14
　host regulatory mechanisms, 338
　nervous system studies, 214–15, 223
　obesity studies, 43–44, 199, 268, 276–77, 283
germs, 34
germ theory of disease, 31–32, 333
Gerrard, J. W., 333–34
ghrelin, 337
Giardia intestinalis, 27, 268
Gilbert, Jack, 46–47, 260–61, 352
GI tract. *See* gastrointestinal tract
　microbes in (*see* gut microbiome)
Global Microbiome Conservancy, 341
glucose intolerance, 275
glucose production, 212
glutamate, 220
glycans. *See* polysaccharides
glycan degradation strategy, 60–61
glycine, 283
glycolysis, 258
glycoprotein, 70, 255–56
glycosidic linkages, 59
GMHI (Gut Microbiome Health Index), 376–77, 384
gnotobiotic (germ-free) animals, 40–41, 87. *See also* germ-free mice
goblet cells, 257
GOE (Great Oxidation Event), 9–10
gonorrhea, 85
Google Scholar, 90, 152
Gordon, Jeffrey, 43–44, 46, 268, 276–77
Gorenflo, Neal, 366
gorillas, 322–30
GP (general practitioner), advice from, 375–78
graphs, 141, 155
GRAS (Generally Regarded As Safe) status, 391
great apes, 322–27. *See also specific species*
Great Oxidation Event (GOE), 9–10
great plate count anomaly, 18
green cities, 365–69
Greengenes, 121, 154, 158, 163
gross change of microbiota diversity category, 261

growing state, 362–63
growth medium
　preparation of, 77
　species requirements, 18
　sterile, 37–38, 77–78
guanine, 99
gum disease (periodontitis), 69–72, 261–65, 281
gut-brain axis (GBA), 211
gut-liver axis, 241–42
gut-lung axis, 307–8
gut microbiome, 21, 58–69. *See also specific microbe*
　in allergic disease, 300–301, 305–6, 311–12
　birth mode impact (*see* birth mode)
　brain interactions, 211, 213–22 (*see also* microbiota-gut-brain axis)
　colonization resistance, 36–37, 62–63, 196–97, 204, 259
　common species in, 52–54
　definition of, 20
　dysbiosis (*see* gut microbiome dysbiosis)
　early observations of, 29
　ecological perspective on, 255–56, 259–60, 268, 374, 395
　example organisms, 27, 51, 85, 115, 145, 175, 209, 235, 273, 321, 373
　functions of, 43–44, 58
　genetic composition of, 58, 60
　health linked to, 37, 254, 374–75, 396
　immune system interactions, 63–64, 180, 239, 242, 244–50, 300–301, 305–6, 311–12
　infant, 187–94, 196–97, 300, 304–5
　lifestyle impacts, 331–38 (*see also* Western lifestyle)
　in neuropsychiatric disorders, 222–28
　in obesity (*see* obesity)
　oral microbiome connection, 266, 281
　phageome, 63–65
　during pregnancy, 179–80, 186, 299
　in primates, 322–26
　resilience of, 57
　self-management of (*see* personal microbiome)
　spatial omics, 110
　species count, 58–59
　therapies based on, 39–41, 228–29
　transplantation of, 268, 277
gut microbiome dysbiosis
　allergic diseases linked to, 198–99, 305–6
　antibiotic-associated (*see* antibiotics)
　intestinal disease linked to, 21, 64–69, 258–59, 267
　malnutrition linked to, 197–98
　obesity linked to, 21, 267, 278
　self-management of (*see* personal microbiome)
Gut Microbiome Health Index (GMHI), 376–77, 384
Gut Phage Database, 103
gut transit time, 379–80
GWASs (genome-wide association studies), 339
György, Paul, 191
gyrB, 322–24

H
Haahtela, T., 315
Hadza people (Tanzania), 331–32
Haemophilus, 265, 307–8
HAIs (healthcare-associated infections), 186–87, 356, 359–62
halophilic archaeans, 19
halotolerance, 73
Handelsman, Jo, 38–39
handwashing, 358–59
Harvard Healthy Eating Plate, 387–88
HAT (histone acetyltransferase), 298
hay fever, 297, 300, 334
HDL (high-density lipoprotein cholesterol), 275, 285
health, microbiome linked to, 21–22, 37, 56–57, 72, 254, 374–75, 396. *See also* normal microbiome
healthcare
　professional advice, 375–78
　self-management (*see* personal microbiome)
healthcare-associated infections (HAIs), 186–87, 356, 359–62
health inequities, 365–66
healthy gut criteria, 375, 385–86
healthy gut phageome, 64–65
heating systems, 349, 353–55, 359, 365
hedgehog structure, 42–43
Helicobacter pylori, 235, 248–49, 334, 336–38
helper T cells, 182, 226–27, 296–97, 308
hematopoietic stem cells, 243
heritability rate, 275. *See also* genetics
herpes, 352
Heterocephalus glaber (naked mole rat), 87
heterotrophs, 7
high-density lipoprotein cholesterol (HDL), 275, 285
high microbiome diversity, 56
high-pressure liquid chromatography (HPLC), 108
high-throughput DNA sequencing, 104–6, 117
Hippocrates, 396
histamine, 296
histogram, 93–94
histone acetyltransferase (HAT), 298
histone deacetylases, 312
history
　of human microbiome, 322–31
　of microbiology, 17–18, 27–38
　of microbiome research, 41–47
Hitchcock, Thomas, 76
HIV/AIDS, 33
HMOs (human milk oligosaccharides), 191–92
HMP (Human Microbiome Project), 45–47, 54, 254
holobiont, 51–83
　definition of, 20, 51–52, 254
　ecological perspective on, 255–60, 268, 374, 395
　genetic composition, 53
　host-microbiota interactions, 338
　species count, 53
Home Microbiome Project, 351–52. *See also* built environment

homeostasis, 58, 64–66, 259
  disruption of (see dysbiosis)
hominids. See also specific species
  microbiome in, 322–27
homogenization, 97, 99
homologs, 103
Homo neanderthalensis (Neanderthals), 71, 322, 326–30
Homo sapiens. See human
Hooke, Robert, 17
hormones, 211–12. See also specific hormone
  female, 176–77
hospital environment, 186–87, 356, 359–62
Hospital Microbiome Project, 361–62
host-adapted core, 55–56
host-microbiota interactions. See also holobiont
  types of, 338
host range, 16–17
hot springs, archaeans in, 19
housekeeping tasks, of core microbiome, 55
House Observations of Microbial and Environmental Chemistry Project, 364–65
howler monkeys, 327–30
HPA (hypothalamic-pituitary-adrenal) axis, 212, 217
HPLC (high-pressure liquid chromatography), 108
human (*Homo sapiens*)
  classification of, 12, 322–24
  evolution of, 322–30
  microbial clouds, 350
  microbiomes associated with (see microbiomes of the built environment)
  relationship with microbiome (see holobiont)
human genome-wide association studies (GWASs), 339
human microbiome. See also specific site
  altered, 88 (see also dysbiosis)
  core, 54–57, 187–90, 254
  definition of, 19
  evolution of (see evolving microbiome)
  functional role of, 45, 54–55, 329
  health linked to, 21–22, 37, 56–57, 72, 254, 374–75, 396 (see also normal microbiome)
  host relationship with (see holobiont)
  research on (see research; specific study)
  self-management of (see personal microbiome)
  size of, 20–21
  therapeutic use of, 39–41, 201–4, 228–29, 250, 291, 313–15, 341, 388–94 (see also specific therapy)
  uniqueness of, 22, 54–55, 156, 376, 384
Human Microbiome Project (HMP), 45–47, 54, 254
human milk, 191–94
  antibiotics in, 196–97
  composition of, 191–94, 244–45, 256, 289
  immune function, 244–45, 338, 340
  microbiome, 189–90, 256–57, 338
human milk oligosaccharides (HMOs), 191–92

human subjects, 91
  deidentification of, 151
  financial incentives for, 94
  metadata collected from, 97
humidity, 353–55, 362, 363
humoral immune responses, 237
humoral theory of disease, 254–55
Hungate, Robert, 37–38
Hungate technique, 38
hunter-gatherer lifestyle, 331–33
Hutterite farmers (USA), 302–3, 357–58
Hyalosphenia papilio, 29
hydrogen, in early Earth atmosphere, 3
hydrogen peroxide, 178, 359
hydrothermal vents, 1, 2, 348
hygiene hypothesis, 245, 302–4, 333–34, 350, 366–69
hygiene practices, 333–35, 348
hypertension, 275
hyperthermophiles, 1, 19
hypothalamic-pituitary-adrenal (HPA) axis, 212, 217
hypothalamus, 212
hypothesis, 30
  biodiversity, 304, 366–67
  disappearing-microbiota, 334–37, 341
  fetal programming, 196
  hygiene, 245, 302–4, 333–34, 350, 366–69
  microbiome rewilding, 366–69
  null, 88–89, 91, 135, 139
  old friends, 245, 303–4, 366–67
  testable, 76, 87–88
hypoxia, 258

I
IBD (inflammatory bowel disease), 46, 66–69, 246, 248, 261, 265. See also specific disorder
IBS (irritable bowel syndrome), 220–21, 229, 260
if/then format, 88
IgA (immunoglobulin A), 64, 241, 244–45, 256
IgE (immunoglobulin E), 192–93, 238, 296–97
IgG (immunoglobulin G), 243, 245
ILCs (innate lymphoid cells), 237
IL-6 (interleukin-6), 283–84, 288
IL-17 (interleukin-17), 226–27, 247
Illumina adapters, 101–2, 105
immune disorders, 187, 245–50. See also specific disease
immune pathway (MGBA), 219, 221
immune system, 235–52
  activation of, 237–38
  allergic response, 295–97 (see also allergic diseases; specific disorder)
  autophagy, 222–23
  components of, 180–82, 236–38
  development of, 21, 180–87, 194–96, 200, 201, 217, 242–45, 297–300, 312–13, 334–35, 357
  epigenetic changes, 298–99
  gut microbiome, 63–64, 180, 239, 242, 244–50, 300–301, 305–6, 311–12
  host-microbiota interactions, 338
  lung microbiome, 240–41, 306–9

  mucosal firewall, 239–40
  oral microbiome, 71–72
  during pregnancy, 180–82, 195, 217, 242–44, 299–300
  skin microbiome, 73–76, 240–41
  stomach microbiome, 337
immune tolerance, 304, 312, 313
immunoglobulin A (IgA), 64, 241, 244–45, 256
immunoglobulin E (IgE), 192–93, 238, 296–97
immunoglobulin G (IgG), 243, 245
immunotherapy, 313
independent variables, 89
indole, 284
industrialization, 331–38, 348, 376. See also built environment; Western lifestyle
infant formula, 192–93, 200, 201, 203–4
infant microbiome, 187–91
  antibiotics in, 196–97, 300, 304–5, 335, 337, 360
  dysbiosis, 175, 196–201 (see also birth mode)
  environmental impacts on, 194–95, 300–303
  immune function, 244–45, 297–303, 313, 334–35, 338, 357
  milk-oriented, 191–94, 256–57
  newborn (see birth mode; newborns)
  solid food transition, 194, 257, 303
inflammatory bowel disease (IBD), 46, 66–69, 246, 248, 261, 265. See also specific disorder
inflammatory cascade, in obesity, 282–85, 288, 338
inflammatory skin disorders, 240. See also specific disorder
influenza, 33, 352, 355
inheritance. See genetics
innate immune system, 180–82, 237–38, 296–97, 301
innate lymphoid cells (ILCs), 237
inorganic compounds, 3
insect gut microbiome, 29
insulin, 212
insulin resistance, 275–76
integrated HMP (iHMP), 45–46
integration, 211
interleukins, in allergic response, 297
interleukin-6 (IL-6), 283–84, 288
interleukin-17 (IL-17), 226–27, 247
internal capsule, 214
interquartile range, 135
intestines. See large intestine; small intestine
  microbes in (see gut microbiome)
intriguing association, 260
introduction (article), 147, 149, 152
inulin, 315
investigational new drug application, 290
ions, 108
Iquitos (Peru), 349–50
irreproducibility, 93
irritable bowel syndrome (IBS), 220–21, 229, 260

J
Johnson, K V.-A., 229–30
Jorth, Peter, 262–65

journals, 146–47, 152. *See also* literature
Justinianic Plague of 541, 348

K
keratinocytes, 74–75
keystone species. *See also specific microbe*
    female reproductive tract microbiome, 176
    gut microbiome, 273, 278, 281, 286, 384, 394–95
    loss of, 336
    oral microbiome, 71
    recovery-associated, 394–95
kidney failure, 247
killer T cells (KTCs), 182
kingdoms, 12
kitchens, 355–56, 363–64
Klebsiella, 225
Klebsiella pneumoniae, 221
Knight, Rob, 46, 137, 150
KOALA Birth Cohort Study, 195
Koch, Robert, 32–33, 224, 267, 333
Koch's postulates, 32–33, 255, 268
    depression study using, 224
    ecological, 267–69
KTCs (killer T cells), 182

L
Lachnospira, 307, 314–15
Lachnospiraceae, 194, 261
lactase, 340
lactate, 179
lactic acid, 176–78, 390
Lacticaseibacillus rhamnosus, 394
Lactobacillus
    in birth mode study, 128–29
    in built environment, 353
    in colorectal cancer, 67
    farm effect, 300
    gut microbiome, 62–64, 307–8, 388
    infant microbiome, 192, 194
    in microbiota-gut-brain axis, 219, 220, 225
    in obesity, 278, 283, 288–90
    oral microbiome, 70
    probiotic strains, 390, 393
    vaginal microbiome, 176–79, 186, 201, 256
Lactobacillus acidophilus, 229, 389
Lactobacillus brevis, 122–23, 129
Lactobacillus casei Shirota, 289
Lactobacillus crispatus, 177
Lactobacillus delbrueckii subsp. *bulgaricus*, 229
Lactobacillus gasseri, 177, 278
Lactobacillus helveticus, 229
Lactobacillus iners, 177–79
Lactobacillus jensenii, 177–78
Lactobacillus paracasei, 229, 278
Lactobacillus plantarum, 229
Lactobacillus reuteri, 60, 203, 226, 278, 299
Lactobacillus rhamnosus, 201
Lactococcus, 70
Lactococcus lactis, 302, 309
lactose digestion, 340
lactulose, 392
large intestine, 58

    bacteriophages in, 64–65
    digestion in, 60–61
    dysbiosis in, 64–69, 259, 276 (*see also* gut microbiome dysbiosis)
    immune function, 63–64, 180, 239, 245
    leash effect, 257–58
    microbes in, 59, 61–63, 273 (*see also* gut microbiome)
    mucin (*see* mucin layer)
Last Universal Common Ancestor (LUCA), 7
Latin (binomial) names, 12, 53–54
latrine deposits, 326–27, 330–31
Lawley, Trevor, 186–87
LDL-C (low-density lipoprotein cholesterol), 275
leaf-associated microbes, 346
leanness, 44, 277–79, 281, 289
learned immune response, 237–38
leash effect, 256–57
Leeuwenhoek, Antonie van, 17, 27–28, 37, 41
Legionella, 355–56
Legionella pneumophila, 359
Legionnaires' disease, 359
Leidy, Joseph, 29
leptin, 277, 337
leucine-rich repeats (LRRs), 301
Lewy bodies, 227–28
Ley, Ruth, 277
Li, D., 224
Li, Huiying, 76
life
    microbes as core of, 17
    origins of, 2–12
    tree of, 12–17, 19, 132
lifestyle
    hunter-gatherer, 331–33
    industrialization, 331–33, 348, 376 (*see also* built environment; Western lifestyle)
light, 355–56, 359
limited pathogens category, 261
Limi Valley (Nepal), 341
Linnaean classification system, 12–17, 28, 53–54
Linnaeus, Carolus, 12, 28
lipid imbalance, 275
lipolysis, 282
lipophilic yeast, 73
lipopolysaccharides (LPS), 221–22
    in allergic disease, 301–3, 308
    immune response, 338
    in lungs, 241
    in obesity, 276, 282–85
    in type 1 diabetes, 200
literature, 145–74
    article format, 147–49
    evaluation of, 89–90, 148–58
    peer review, 90, 146–47, 157
    primary, 89–90, 141, 146–48
    searching, 90, 152
    secondary, 146
    tertiary, 146
    types of, 146
Liu, Albert, 201
liver cancer, 265–66
liver microbiome, 241–42

Lloyds Bank coprolite, 326–27
longitudinal study, 96
loss of health-associated bacteria category, 261
loss of microbiota, 334–38, 341
low-density lipoprotein cholesterol (LDL-C), 275
low microbiome diversity, 56
Lozupone, C., 137
LPS. *See* lipopolysaccharides
LRRs (leucine-rich repeats), 301
LUCA (Last Universal Common Ancestor), 7
lumen, intestinal, 63–64
lung microbiome, 240–41, 306–9
lymphatic cells, 236, 243
lymph nodes, 236, 243
lymphocytes. *See* B cells; T cells
lymphoid progenitor cells, 238
Lynch, Susan, 314–15
lysis, 100–101
lysosomes, 222–23

M
macrophages, 181, 241, 243–45, 283–84
maggot experiment, 29–30
magnesium, 5
magnetic beads, 101
magnification, 28, 42
major depressive disorder (MDD), 220, 223–26, 229
major histocompatibility complex (MHC), 238
Malassezia, 73–75
Malawi, 324
malnutrition, 197–98, 244, 268
maltooligosaccharides, 392
MAMPs (microbe-associated molecular patterns), 301–2
Manaus (Brazil), 349–50
manuscripts, 146. *See also* literature
Margulis, Lynn, 11, 20, 52, 223
marker food, 379
Marshall, Barry, 336
mass extinction events, 9
mass spectrometry, 107–8
mast cells, 238, 296–97, 299
maternal microbiome. *See* birth mode; pregnancy
Matsés people (Peru), 333
maximum value, 134
Mayneris-Perxachs, J., 224–25
McCright, Sam, 284
MDD (major depressive disorder), 220, 223–26, 229
MDRP (multidrug-resistant pathogens), 362
meadow diversity metaphor, 374, 383–84
mean, 92
mechanical lysis, 100–101
median value, 134–35
medical professionals, advice from, 375–78
medieval microbiome, 330–31
Med13L gene, 340
Meisel, P., 281
melatonin, 212
membrane argument, 4–5
mental disorders. *See also specific disorder*
    microbiota-gut-brain axis in, 220–21
    psychobiotics for, 228–29

messenger RNA, 107
metabolic modules, 54
metabolic states, 362–63
metabolic syndrome (MetS), 275–76, 289
metabolism, 61
metabolites. *See also specific substance*
   in asthma, 314–15
   in built environment, 363–64
   in Crohn's disease, 66–69
   ecological perspective on, 256–58, 395
   gut transit time, 379
   health linked to, 22, 385–86
   immune function, 239, 241
   infant microbiome, 188–90, 192, 214–15
   maternal microbiome, 180, 182, 213, 242–43, 299
   in microbiota-gut-brain axis, 219, 221, 338
   in obesity, 278, 281–82
   of starch digestion (*see* short-chain fatty acids)
metabolome, 109
metabolomics, 106–7, 109, 214, 362–65, 375, 382
Metabolomics Standards Initiative, 109
Metabolomics Workbench, 109
metadata, 90, 96–97
metagenome, 104
metagenomics, 38–39, 41–43
   in built environment, 363
   DNA amplification, 101–6, 117–18
   DNA extraction, 99–101, 117
   personal screening, 375–77
   spatial transcriptomics, 109–10, 263
metagenomic sequence library, 104
metagenomic (shotgun) sequencing, 106
metaproteomics, 106–8, 375, 382
metatranscriptome, 107
metatranscriptomics, 106–7, 382
Metchnikoff, Élie, 389
Methanobrevibacter, 69
Methanobrevibacter smithii, 69, 323–24
methanogenesis, 3, 7
methanogens
   classification of, 14–15
   skin microbiome, 74–75
Methanopyrus kandleri, 3
methanotrophs, 9
methicillin-resistant *S. aureus* (MRSA), 295, 337
methods section (article), 147–48, 153, 154
Methylobacterium, 351
MetS (metabolic syndrome), 275–76, 289
MGBA. *See* microbiota-gut-brain axis
MHC (major histocompatibility complex), 238
Mian, 170
miasma theory (spontaneous generation), 29–31
mice. *See* mouse
micelles, 5
microbe(s)
   as core of life on Earth, 17
   definition of, 2–3
   metabolic states, 362–63

metabolites (*see* metabolites; *specific substance*)
   research on (*see* research)
microbe-associated molecular patterns (MAMPs), 301–2
microbial clouds, 350
microbial ecology, 37
microbial guilds, 279–81
microbial leash metaphor, 256–57
microbially produced volatile organic compounds (MVOCs), 364
microbial seed bank, 341
microbiology
   definition of, 32
   history of, 17–18, 27–38
   modern developments in, 38–41
microbiome
   in humans (*see* human microbiome)
   in primates, 322–27
microbiome analysis programs, 170. *See also* Qiita
MicrobiomeAnalyst, 170
microbiome rewilding hypothesis, 366–69
microbiomes of the built environment (MoBE), 345
   constituents of, 351–52
   control of, 348, 355–56
   early studies of, 349–50
   future, 365–69
   health impacts of, 348, 356–58
   history of, 347–48
   metabolomics, 362–65
   microbial transport in, 350–51
   physical factors influencing, 349, 353–56, 359–60, 369
   tracking methods, 358–62
microbiota
   definition of, 19–20, 52
   transfer of (*see* transplantation)
microbiota-gut-brain axis (MGBA), 209, 217–22, 338
   autophagy, 222
   endocrine pathway, 218–19
   immune pathway, 219, 221
   neural pathway, 219–21
Micrococcus, 78
microglial cells, 185, 213–14, 221, 241–42
microRNAs (miRs), 284, 298–99
Middles Ages, microbes from, 330–31
milk
   cow, allergy to, 312, 314
   human (*see* human milk)
   probiotic, 202
milk-oriented microbiome (MOM), 192
Miller, Stanley, 3–4
minimum value, 134
miRs (microRNAs), 284, 298–99
missing-heritability problem, 339
mitochondrion, 11
MoBE. *See* microbiomes of the built environment
model organisms, 87. *See also specific organism*
   Crohn's disease, 67
   germ-free, 40–41, 87 (*see also* germ-free mice)
   obesity studies, 276–77
   variables, 95

molar ratio, 61
mold
   Hooke's observations of, 17
   toxic, 345, 355, 359
molecular networking, 109
molecular photocopying, 101
molecular probing, 42–43
molecular therapies, 226–27
molecular tree of life, 13–14, 132
MOM (milk-oriented microbiome), 192
monera, 13
monocytes, 238
monosaccharides, 59
Moraxella, 307
Morganella, 225
mother's first gift, 186. *See also* birth mode; pregnancy
mother's milk. *See* human milk
mothur, 121, 140
motor function of the CNS, 211
mouse (*Mus musculus*), 87, 91
   allergic disease studies, 300–302, 305, 307–9, 312–15, 357
   autism spectrum disorder studies, 226
   brain studies, 214–16, 221, 223
   Crohn's disease studies, 67
   depression studies, 224–25
   germ-free (*see* germ-free mice)
   maternal microbiome, 180–71, 183–85, 244
   obesity studies, 43–44, 199, 268, 276–77, 279, 281, 283–85, 289, 290
   rheumatoid arthritis studies, 246
   spatial omics, 110
   variables, 95
mouth, 58. *See also* oral microbiome
mouth-body connection, 72
mouthwash, antibacterial, 70
MRSA (methicillin-resistant *S. aureus*), 295, 337
mucin layer
   host-microbiome interactions in, 256–57, 273, 380
   infant microbiome, 188, 194, 204
   in obesity, 278, 284–85
mucosal firewall, 239–40, 245
mucus escalator, 257
Mullis, Kary, 101
multicellularity, origin of, 10–12
multidrug-resistant pathogens (MDRP), 362
multi-omics, 109–10, 154, 224–25
multiple sclerosis, 221
multiplexing, 117–18
Mure, Nancy, 199
Mus musculus. *See* mouse
MVOCs (microbially produced volatile organic compounds), 364
Mycobacterium, 355
Mycobacterium neoaurum, 224
Mycobacterium tuberculosis, 347
myelination, 217
myeloid cells, 236, 243

N
Nagler, Cathryn, 311
naive T cells, 182, 296–97, 305
Nakajima, Akihito, 183
naked mole rat (*Heterocephalus glaber*), 87

Namur (Belgium), 330–31
nasopharynx microbiome, 21, 58, 337, 360–61
NAST, 154, 158
National Institutes of Health (NIH), 45, 254
Natufians, 346–47
natural environment, *versus* indoor, 346. *See also* built environment
natural killer T cells, 181, 237–38
natural log, 130
natural selection, 7
nature-based solutions, 366–69
naturopaths, 377–78
nausea, during pregnancy, 201–2
Neanderthals, 71, 322, 326–30
necrotizing enterocolitis, 200, 203
Neisser, Albert, 85
Neisseria, 307–8
Neisseria gonorrhoeae, 85
nematode (*Caenorhabditis elegans*), 87
neonatal intensive care unit (NICU), 360–61
neonates. *See* newborns
Nepal, 341
nerve fiber (axon), 185, 217
nerve impulses, 211
nervous system, 210–12. *See also* brain
Netherlands, 195
network workspace (Qiita), 161–64, 167
neural pathway (MGBA), 219–21
neural tube defects, 216
neurodegenerative disorders. *See also specific disorder*
    microbiota-gut-brain axis in, 221–22
neurodivergence, 226–27
neuroendocrine system, 211–12, 217–19
neurogenesis, 219
neuroinflammation, 214, 219, 222, 225
neuropeptides, 213
neuropsychiatric disorders, 222–28. *See also specific disorder*
neurotransmitters, 220
neurotrophic factors, 221
neurotrophic signals, 213
neutrophils, 181, 237–38, 296
newborns
    antibiotics in, 196–97, 300, 304–5, 335, 337, 360
    birthing process (*see* birth mode)
    dysbiotic microbiome, 196–200
    feeding (*see* human milk; infant formula)
    immune system, 21, 180–87, 194–95, 217, 242, 244–45, 297–304, 312–13, 334–35
    microbiome, 186–91, 300, 313
    preterm, 46, 202–4, 243, 300, 360
next-generation sequencing, 117–18
niche adaptation, 45
niche modification, 259–60
niche preemption, 259
NICU (neonatal intensive care unit), 360–61
Nightingale, Florence, 359–60
NIH (National Institutes of Health), 45, 254
nisin, 177
Nissle, Alfred, 36
nitric oxide, 70
nitrogen, metabolism of, 74–75
noncoding regions, 119
noncoding RNAs, 284, 298–99

normal (Gaussian) distribution, 93
normal microbiome
    core taxa, 54–56
    definition of, 52
    hallmarks of, 56–57
    research on, 45–47, 54
Norman (Oklahoma), 333
nosocomial infections, 186–87, 356, 359–60
nucleosides, 99
nucleotides, 98–99
    fluorescent labeling, 104, 117
    origin of, 4
    triplets of (codons), 119–20
null hypothesis, 88–89, 91, 135, 139
nutrient cycling, in oral microbiome, 70
nutrient medium. *See* growth medium
nutrition. *See also* diet
    microbiome in, 22, 197–98, 244, 268
nutritionists, 377–78

O
obesity, 273–94
    definition of, 274–75
    epidemic of, 274–76, 337
    genetic mutations in, 43–44
    gut microbiome in, 43–44, 60–61, 199–200, 268, 276–85, 338
    infants, 199–200
    versus leanness, 277–79, 281, 289
    microbial guilds in, 279–81
    oral microbiome in, 281
    treatment of, 40, 275–76, 285–91
    weight loss markers, 281–85
obligate anaerobic bacteria, 65, 257–59
observational studies, 86, 202
oceans
    dead zones, 9
    hydrothermal vents, 1, 2, 348
Ochman, Howard, 322–25
OIT (food allergy oral immunotherapy), 313
Oklahoma, 333
old friends hypothesis, 245, 303–4, 366–67
Olesen, S. W., 254–55
oligosaccharides, 191, 289, 392–94
Olle, Bernat, 341
Olsen, I., 281
omics, 106–10. *See also specific type*
one pathogen-one disease paradigm, 267
online data analysis programs, 121, 140, 145, 154, 158–70. *See also specific program*
open-access journals, 152
open reading frames (ORFs), 119–20
open-source software, 158
operational taxonomic units (OTUs), 121–24, 154–55, 162–63
oral immunotherapy, 313
oral microbiome, 69–72
    in built environment, 351–52
    cataloging of, 42–43
    colonization resistance, 70
    common species in, 69–70
    DIY assessment kits, 381
    dysbiosis, 71–72, 262–65
    early studies of, 17, 28, 37, 41–42
    evolution of, 70–71
    functions of, 70, 329–30
    gut microbiome connection, 266, 281

    in liver cancer, 265–66
    in obesity, 281
    in primates, 327–30
    site differences, 52
    species count, 69
    teeth (*see* dental plaque)
orangutans, 322–25
ORFs (open reading frames), 119–20
origin
    of human microbiome, 322–31
    of organic molecules, 3
    of species, 12–13
Oscillospira, 192, 286
OTUs (operational taxonomic units), 121–24, 154–55, 162–63
outcome, 86
outdoor environment, *versus* indoor, 346. *See also* built environment
outliers, 94
oxygen
    in early Earth atmosphere, 2, 7, 9–10
    environment without (anoxic), 7–8, 37–38
    in gut microbiome, 189–90, 257–60, 267, 395
    in large intestine, 65
    in oral microbiome, 69, 71
ozone layer, 2

P
Pace, Norm, 351
Palforzia, 313
pancreas, 275
Pandorina algae, 9
Paneth cells, 64–66, 257–58
parasitic worm infections, 297, 308, 330–31, 381
Parkinson, James, 227
Parkinson's disease (PD), 221, 227–28
Parrish, Rosia, 378
participants. *See* human subjects
Parvimonas, 261
passive immunity, 245
Pasteur, Louis, 30–32
pathobionts, 267
pathogens. *See also specific pathogen*
    in built environment, 346, 355–56 (*see also* built environment)
    defenses against (*see* colonization resistance; immune system)
pathogen trapping, 179
pattern recognition receptors (PRRs), 301
PCA (principal component analysis), 139–41, 169–71, 176, 325–26, 329–30, 333
PCoA (principal coordinate analysis), 156–57, 176–77
PCR (polymerase chain reaction), 19, 101, 117
PCR (polymerase chain reaction) amplification, 101–6, 117–18
PD (Parkinson's disease), 221, 227–28
PD (phylogenetic diversity), 132–35
peanut allergy, 313–14. *See also* food allergies
pear-shaped body, 279
peer review, 90, 146–47, 157
penicillin, 34, 335

Penicillius notatum, 34
Peptostreptococcus, 261
perfluorooctane sulfonic acid (PFOS), 364
perforins, 182, 240
periodontitis, 69–72, 261–65, 281
peripheral nervous system (PNS), 210
personal microbiome, 373–99
    antibiotic recovery, 394–95
    food as medicine, 396
    Gut Microbiome Health Index, 376–77, 384
    gut microbiome kits, 380–84
    gut transit time, 379–80
    health indicators, 384–87
    healthy diet, 387–88
    healthy gut criteria, 375, 385–86
    microbiome-based therapeutics, 388–94
    professional advice, 375–78
    stool quality assessment, 378–79
Peru, 333, 349–50
petri dishes, 77
pets
    in built environment, 350, 353–54, 369
    infant microbiome affected by, 192–93, 297, 300, 350, 353–54
    microbiome assessment kits, 381
Peyer's patches, 243
PFOS (perfluorooctane sulfonic acid), 364
phageome, gut, 63–65
phages. See bacteriophages
phagocytes, 237–38
phagocytosis, 11, 181
phenotypes, 13
phenylalanine, 226
phosphate groups, 99
phospholipid bilayer, 18
photosynthesis, 8–9
phyllosphere, 352–53
phylogenetic diversity (PD), 132–35
phylogenetic tree (phylogeny), 12, 132, 138, 154, 322–23
phylum-level diversity, 382
phytochemicals, 287–88
Picrophilus, 19
pili, 69–70
pilot study. See birth mode study
pioneer (primary) species, 394–95
pituitary gland, 212
placenta, 180, 184, 242, 299
plague, 29, 348
planetary habitability, 5
plants
    in built environment, 353–54, 362, 365–69
    chloroplasts, 11
    in diet, 323, 328, 332, 340, 352, 388, 394
    microbiomes, 352–53
    photosynthesis, 8–9
    phytochemicals, 287–88
    in tree of life, 12–13
plaque. See dental plaque
plasma membrane, 18
plumbing, 355–56
PNS (peripheral nervous system), 210
polymerase chain reaction (PCR), 19, 101, 117
polymerase chain reaction (PCR) amplification, 101–6, 117–18

polymicrobial synergy and dysbiosis (PSD), 261–65
polyphenols, 288, 388
polysaccharide A, 189, 301
polysaccharides (glycans), 59
    biological functions, 246, 288
    degradation strategy, 60–61
    glycan degradation strategy, 60–61
    infant microbiome, 188–89
    primate microbiome, 323–24
Porphyromonas, 70–72, 261
Porphyromonas gingivalis, 71–72, 262–65, 281
portal vein, 241
postbiotics, 313
postpartum sepsis, 358–59
power, statistical, 91, 141
power analysis, 94–95
prebiotics, 203, 228
    for allergies, 313, 315
    artificially produced, 392–93
    in breast milk, 191, 203
    definition of, 289, 392
    for depression, 229
    dietary sources, 340, 388, 392–93
    in infant formula, 203–4
    for obesity, 289–90
    synbiotics, 313, 393–94
predatory journals, 147
predictable association, 260
pregnancy
    delivery (see birth mode)
    dysbiosis during, 213, 226
    immune system during, 180–87, 195, 217, 242–44, 299–300
    inflammation and infection during, 243–44
    microbiome during, 46, 179–80, 212–17, 299–300, 336
    postpartum sepsis, 358–59
    probiotics during, 201–2
    vaginal sampling during, 97–98
preterm birth, 46, 202–4, 243, 300, 360
Prevotella, 179, 186, 277–79, 336
Prevotellaceae, 194, 332
Prevotella copri, 279, 321
Prevotella histicola, 247
Prevotella intermedia, 265
primary literature, 89–90, 141, 146–48
primary (pioneer) species, 394–95
primary starch degraders, 60–61
primates. See also specific species
    microbiome in, 322–27
primer pad region, 101–2
primers, 101–2, 117
primordial soup, 4
principal component analysis (PCA), 139–41, 169–71, 176, 325–26, 329–30, 333
principal coordinate analysis (PCoA), 156–57, 176–77
priority, 188
probability, 92
probiotics, 228, 341
    for allergies, 313–14
    brain effects, 220
    for cholesterol levels, 279
    definition of, 288, 389

    for depression, 229
    for diabetes, 200
    dietary supplements, 389–92
    foods containing, 388–89
    in infant formula, 193–94
    for inflammatory bowel disease, 261
    for obesity, 288–89
    during pregnancy, 201–2
    for preterm babies, 203
    for rheumatoid arthritis, 247
    synbiotics, 313, 393–94
    unregulated production of, 204
Proceedings of the National Academy of Sciences (PNAS), 148
process errors, 118, 121
proinflammatory cells, 221
prokaryotes
    classification of, 15
    definition of, 10
    origin of, 10
proline, 224–25
propionate, 60–62
    health linked to, 386
    immune function, 308
    infant microbiome, 193
    in obesity, 273, 278, 281–82
Propionibacterium, 73–74, 128–29, 351
proteases, 73, 310
proteins
    databases of, 120
    metaproteomics, 106–8, 375, 382
    misfolding, 227–28
    origin of, 4
    tree of life based on, 13–14
Proteobacteria
    in built environment, 351
    core microbiome, 54
    female reproductive tract microbiome, 176
    gut microbiome, 53, 188, 198, 200, 261–62, 325, 330–31, 384
    infant microbiome, 188–90, 198, 200
    in obesity, 277–79, 284–85
    oral microbiome, 69
    primate microbiome, 325
    respiratory microbiome, 307–8
    skin microbiome, 73–74
    stomach microbiome, 336
Proteus mirabilis, 278–79
protists
    in oral microbiome, 69–70
    in tree of life, 13
protocell, 5–7
PRRs (pattern recognition receptors), 301
PSD (polymicrobial synergy and dysbiosis), 261–65
pseudomembranous enterocolitis, 39–40
Pseudomonadota, 70
Pseudomonas, 265, 362
Pseudomonas aeruginosa, 221
Pseudomonas fluorescens, 265
psoriasis, 240
psychobiotics, 228–29
psychological disorders. See also specific disorder
    microbiota-gut-brain axis in, 220–21
    psychobiotics for, 228–29
Public Studies section (Qiita), 159–60

PubMed, 90
Puebloans (American Southwest), 347
Puerto Almendras (Peru), 349–50
pulmonary microbiome, 240–41, 306–9
p-value, 91–92
Pyrococcus furiosus, 1

Q
QIIME 2 (Quantitative Insights into Microbial Ecology), 121, 124, 140
QIIMP (Quick and Intuitive Interactive Metadata Portal), 97
Qiita, 145, 158–70
    alpha diversity analysis, 166–69
    basics, 158–63
    data transformation, 163–64
    PCA analysis, 169–71
    taxonomic distribution analysis, 164–66
quality control, in data analysis, 118
quality scores (QS), 118
quorum sensing, 230

R
RA (rheumatoid arthritis), 246–47
RABs (recovery-associated bacteria), 394–95
random (stochastic) process, 266–67
rarefaction (species accumulation) curves, 124–25, 169–70
rats
    autoimmune disease studies, 241–42
    depression studies, 224, 229
    naked mole rat, 87
    Parkinson's disease studies, 228
    plague spread by, 348
rDNA. *See* ribosomal DNA
receptor attachment specificity, 69–70
recolonization, 337–38
recovery-associated bacteria (RABs), 394–95
red blood cells, 339
Redi, Francesco, 29–30
references (article), 148, 152
reference sequences, 119, 121–23, 154, 163, 376
regulatory pathways, 54
regulatory T cells (Tregs), 182–84, 240, 244, 297, 299–301, 305, 308, 312, 314
Relman, David, 41–42
replication
    DNA, 4–7, 99–100
    viral, 16
replication argument, 4–7
reproductive tract, female, 176–79. *See also* vaginal microbiome
    during pregnancy (*see* birth mode; pregnancy)
research, 41–47, 85–113. *See also specific study*
    ethics in, 150–51
    funding for, 94, 150
    Human Microbiome Project, 45–47, 54, 254
    metagenomics, 41–43
    origins of, 17, 27–35, 43–44
    reports on (*see* literature)
    volume of, 45
research design, 86–88, 90–97
    analysis phase (*see* data analysis)
    bias in, 45, 376

examples of, 95–96, 151, 153 (*see also* birth mode study)
experimental phase, 97–106 (*see also* experiment)
hypothesis (*see* hypothesis)
literature review, 89–90, 146–58
power analysis, 94–95
sample size, 91–95, 125, 141, 150, 254, 267
subjects, 91 (*see also* human subjects; model organisms)
training in, 86
research questions, 85, 86, 116, 121
resilience, 54, 56–57
resistance, 56
    antibiotic, 35, 85, 186–87, 310, 330–31, 335, 337, 360, 362
    colonization (*see* colonization resistance)
resistant starch, 60–61, 229–30, 279, 286–87, 388
resistomes, 360, 362
respiration, cellular, 7–8
respiratory syncytial virus (RSV), 309
respiratory tract
    in built environment, 351, 356–57, 359–60
    infections, 33, 345
    microbiome, 240–41, 306–9, 337
restrooms, 351, 353, 355–56, 363, 364
results section (article), 148, 149, 153–54
retention rates, 94
reverse primers, 101–2
review articles, 146
rewilding hypothesis, 366–69
rheumatoid arthritis (RA), 246–47
rhinovirus, 309
ribosomal DNA (rDNA), 41
    sequencing of, 41–42, 183–84, 375, 381–82 (*see also specific method or study*)
    16S subunit, 18, 41, 133, 183–84
ribosomal RNA (rRNA)
    classification by, 14–15
    18S subunit, 14, 103
    reference data, 121–23, 154, 163, 376
    sequencing, 44, 96, 101–4, 121, 158, 163 (*see also specific study*)
    16S subunit, 14, 18, 41–42, 44, 96, 98, 101–4, 121, 132, 158, 163, 322, 375, 381
    tree of life based on, 14–15, 132
    variable regions, 103
ribosome, tree of life based on, 13–14, 132
richness (species), 126–27, 169–70, 266, 313, 382
RNA
    versus DNA, 107
    messenger, 107
    noncoding, 284, 298–99
    origin of, 4–7
    ribosomal (*see* ribosomal RNA)
    sequencing, 41–42, 107, 123
    tree of life based on, 14–15, 132
rock, ancient, 2–3, 9–10
Rogers, Sherry, 200
roll-tube, 37–38
Romboutsia, 315
Rook, Graham, 303

room temperature, 353–55, 362
Roseburia, 267, 279, 284–85, 327, 386
Roseomonas mucosa, 74–75
Rothia, 307, 314–15, 336
Royal Society of London, 28
rRNA. *See* ribosomal RNA (rRNA)
RSV (respiratory syncytial virus), 309
rugs, 353–54
Ruminococcaceae, 261
Ruminococcus, 192, 247, 279, 327
Ruminococcus bromii, 60, 279
Ruminococcus gnavus, 280
rust, 9–10

S
saccharolytic bacteria, 392
saliva, 281, 327
salivary amylases, 60, 330
Salmonella typhimurium, 259
sample size, 91–95, 125, 141, 150, 254, 267
sampling methods, 98. *See also* culturing
    DNA extraction, 99–101, 117
    fecal, 97–98
    quality control, 118
    skin, 78–80
    vaginal, 97–98
sanitation practices, 333–35, 356, 359–62
SARS-CoV-2 virus, 16, 352, 355, 365
satiety signal, 62, 277, 282
SBS (sick building syndrome), 345, 356–57
SCD (sickle cell disease), 339
SCFAs. *See* short-chain fatty acids
Schaedler, Russell, 40–41
Schistosoma haematobium, 248
scientific education, 86
scientific literature. *See* literature
seasonal patterns, 332
sebaceous glands, 73–76
sebum, 73–76
secondary literature, 146
secondary species, 394–95
secondary succession, 395
second brain (ENS), 211, 217, 219, 220
second quartile, 134
sedimentary rock, 9–10
seed bank (microbial), 341
selection, 187–88, 230, 266–67
selection (deterministic) process, 266–67
self-management. *See* personal microbiome
self-tolerance, 182
semaphorin 5, 248
semiconservative replication, 99–100
Semmelweis, Ignaz, 358–59
sensitized immune system, 297
sensorimotor deficits, 214
sensory receptors, 210–11
sequence alignment, 122–23
sequence annotation, 109
sequence functions, prediction of, 120
sequence ID (OTU), 162–63
sequence read, 117
serial endosymbiosis, 11
serotonin, 220
sexually transmitted infections (STIs), 178
Shannon diversity index, 130–32, 166–69, 363
shareable cities, 366
shared dysbiosis, 265–66

Sharif, Maimunah Mohd, 365–66
Shiga toxin, 53
Shigella, 36–37, 39
short-chain fatty acids (SCFAs), 22, 60. *See also specific acid*
 ecological perspective on, 256–58, 395
 function of, 61–62, 64
 health linked to, 385–86
 immune function, 239, 241, 246, 299, 301, 311–12
 infant microbiome, 190, 194, 200, 204, 256–57
 in microbiota-gut-brain axis, 219, 221–22, 229, 338
 in obesity, 199, 276, 278, 281–82, 286, 288
 oral microbiome, 70
 during pregnancy, 182–83, 187, 213–14, 244, 299
 production of, 65
shotgun sequencing, 106
siblings, 193, 300, 334
sick building syndrome (SBS), 345, 356–57
sickle cell disease (SCD), 339
significance, statistical, 91–94
silica columns, 101
silkworm studies, 31–32
SILVA, 121–23, 154
similarity-searching algorithms, 120, 123
single-line identifier, 122
single nucleotide polymorphisms (SNPs), 339–40
Sinha, Rashmi, 384
skin cancer, 73–74
skin infections, 74–75. *See also specific disorder*
skin microbiome, 21, 73–76
 in built environment, 351–53, 362
 colonization resistance, 74–75, 240–41, 309
 common species in, 73–74, 78, 295
 dysbiosis, 75–76, 240, 309–11
 functions of, 74–75, 240–41
 maternal, 186, 189–90
 newborn, 190–91
 as nutritional desert, 73
 sampling methods, 76–80
 species count, 73
small intestine, 58
 digestion in, 60
 immune function, 239
 leash effect, 257–58
 microbes in, 59 (*see also* gut microbiome)
smallpox, 16
smoking, 46
SNPs (single nucleotide polymorphisms), 339–40
software packages, data analysis, 121, 140, 145, 154, 158–70. *See also specific program*
soil-associated microbes, 346, 353, 362, 367
somatic (voluntary) nervous system, 210
Sonnenburg, Justin, 332–33
spatial transcriptomics, 109–10, 263
specialized functions, of core microbiome, 55, 329–30

species, 12, 53
 diversity of (*see* diversity)
 identification of, 121–24
 keystone (*see* keystone species)
 origin of, 12–13
species abundance
 definition of, 127
 measurement of, 124–29, 155, 164–65
 in personal microbiome, 382
 in primate microbiome, 325–26
species accumulation (rarefaction) curves, 124–25, 169–70
species evenness, 126. *See also* alpha diversity
species recovery, 395
species richness, 126–27, 169–70, 266, 313, 382
Sphingomonas, 351
spinal cord, 210
Spirochaetaceae, 332
Spirulin algae, 9
spontaneous generation (miasma theory), 29–31
spores, 40
stability, 56
Stachybotrys, 355, 359
stack plot, 127–29, 154–57, 164–65
standard deviation, 92
Staphylococcus, 295
 in birth mode study, 128–29
 in built environment, 351, 353, 362
 infant microbiome, 193
 nasal microbiome, 360
 in skin microbiome, 73–75, 78
Staphylococcus aureus
 enterocolitis caused by, 39
 infections, 33, 74–75, 240, 295, 309–11, 337
 penicillin mold in, 34
Staphylococcus epidermidis, 73–75, 240, 311
Staphylococcus hominis, 73, 311
Staphylococcus sciuri, 309
starches, 59–60
 digestion of, 22, 43–44, 328–29
 resistant, 60–61, 229–30, 279, 286–87, 388
start codon, 119–20
startle response, 214–16
statistical power, 91, 141
statistical significance, 91–94
statistical tests, 139–41
stem cells, 217, 243
sterility control, 37–38, 77–78
STIs (sexually transmitted infections), 178
stochastic (random) process, 266–67
stomach, 21, 58, 257
 immune system in, 337
 microbiome, 21, 59, 235, 336–37 (*see also* gut microbiome)
stomach cancer, 248–49, 337
stool. *See* feces
stop codon, 119–20
Strachan, D. P., 334
strains, 53, 278, 390–94
Streptococcus
 in cancer treatment, 250
 in hospital environment, 250

 in liver cancer, 265
 nasal microbiome, 360
 in obesity, 280
 oral microbiome, 70, 71, 328–30
 penicillin, 34
 respiratory microbiome, 307
 stomach microbiome, 336
Streptococcus anginosus, 330
Streptococcus gordonii, 263
Streptococcus mitis, 265, 330
Streptococcus mutans, 71, 330, 394
Streptococcus oralis, 265
Streptococcus pneumoniae, 337
Streptococcus pyogenes, 330
Streptococcus salivarius, 330
Streptococcus sanguinis, 330
Streptococcus thermophilus, 229
stress response, 212, 217, 220, 224–25, 257, 378
stromatolites, 2–3
subcutaneous fat layer, 73
subjects
 animal (*see* model organisms; *specific organism*)
 human (*see* human subjects)
subscription model, 152
Succinivibrionaceae, 332
sugars
 blood levels, 212, 275
 dietary, 22, 59–61, 392
sulfate-reducing microbes, 7–8
sunlight, 355
superantigens, 310
super bug, 85
surface area of intestines, 58
surprising associations, 260
Svedberg units, 14
swan-neck flask experiment, 30–31
sweat, 73
symbiosis, 11, 20, 22, 34, 322–23, 388
synbiotics, 313, 393–94
synergistic supplements, 394
synthesis (next-generation) sequencing, 117–18
syntrophy, 69
Szostak, Jack, 4–5

T
tagmentation, 104
Taleb, Nassim Nicholas, 391–92
Tannerella forsythia, 262
Tanzania, 331–32
Taq polymerase, 19, 102
targeted metabolomics, 109
target gene amplicon sequencing, 101–2, 381. *See also specific target*
target gene amplification, 105–6
taste receptors, 71
taurine, 283
taxonomic distribution analysis (Qiita), 164–66
taxonomic identities, assignment of, 121–24
taxonomy, 12–17, 28, 53–54
TB (tuberculosis), 347
T cells, 182, 236–38
 in allergic response, 296–97, 299, 305
 development of, 243–44
 helper, 182, 226–27, 296–97, 308

killer, 181–82
naive, 182, 305
regulatory (Tregs), 182–84, 240, 244, 297, 299–301, 305, 308, 312, 314
in stomach, 337
temperature
early Earth, 2
extremophiles, 1, 3, 18–19
PCR, 102
room, 353–55, 362
sampling protocols, 98
temporal core, 55–56
tertiary literature, 146
tertiary species, 394–95
testable hypothesis, 76, 87–88
testosterone, 224
therapies, microbiome-based, 39–41, 201–4, 228–29, 250. See also specific therapy
Theriot, Casey, 283
thermocycler, 102
thermophilic archaeans, 1, 19
thioalcohols, 73
Thion, Morgane, 185
third quartile, 134
Thomas, Lewis, 33
3′ (three prime), 99
throat, 21, 58, 337, 360–61. See also gut microbiome
thymine, 99, 107
thymus, 236
thymus-dependent cells. See T cells
tight junctions (TJs), 246, 276
tipping points, 56
title (article), 147–49
TJs (tight junctions), 246, 276
TLRs (toll-like receptors), 283–84, 301
T lymphocytes. See T cells
TNF-α (tumor necrosis factor alpha), 283–84
toll-like receptors (TLRs), 283–84, 301
tooth plaque. See dental plaque
toxic black mold, 345, 355, 359
training, scientific, 86
trans-galactooligosaccharide, 229
transit time (gut), 379–80
transplantation
fecal microbiota, 39–41, 253, 260, 261, 290–91, 305, 314, 341, 377, 388
fetal microbiota, 202
gut microbiome, 268, 277
vaginal microbiome, 202, 340
transportation, 365–66
tree of life (ToL), 12–17, 19, 132
Tregs (regulatory T cells), 182–84, 240, 244, 297, 299–301, 308, 312, 314
Treponema denticola, 262
Trichinella, 29
trichinosis, 29
Trichomonas, 69
triglycerides, 275
trophic interactions, 188
tryptophan, 214, 220, 226
t-test, 139
tuberculosis (TB), 347
tumor microbiomes, 248–49
tumor necrosis factor alpha (TNF-α), 283–84

Tunapuco people (Andes), 333
Turcimonas, 247
Turvey, Stuart, 314
twin studies, 199, 275, 277
type 1 diabetes, 200, 245
type 1 T helper cells (Th1), 296–97, 308
type 2 diabetes, 26, 247, 280
type 2 T helper cells (Th2), 296–97, 308
typhoid fever, 333

U
ulcerative colitis (UC), 66–67, 248, 258–59
ultrasound, 104
ultraviolet (UV) light, 355–56, 359
umbilical cord, 181, 242
UniFrac (unique fraction) metric, 137–39, 141, 176–77
UniProt, 120
United Nations Human Settlements Programme, 365–66
University of California, San Diego, 46–47
University of Chicago hospital, 361–62
untargeted metabolomics, 109
uracil, 107
urban lifestyle, 315–16, 348–50, 357, 365–69. See also built environment; Western lifestyle
urease, 67, 336
Urey, Harold, 3–4
urinary tract infections (UTIs), 177–78
UV (ultraviolet) light, 355–56, 359

V
vaccines, 337
vacuoles, 222–23
vaginal delivery, 186–87, 191. See also birth mode study
vaginal microbiome, 176–79, 201
community state types, 177–78
diversity in, 255
DIY assessment kits, 381
dysbiosis, 179
functions of, 178–79
leash effect, 256
vaginal microbiome transplant (VMT), 202, 340
vaginal sampling, 97–98
vaginosis, bacterial, 179, 202, 255
vagus nerve, 211, 219, 228
VAMPS (Visualization and Analysis of Microbial Population Structures), 170
van der vossen, E. W. J., 280
variability, data, 134–35, 140
variable regions, 103
variables, 89, 95
Variola virus, 16
Veillonella, 307, 314–15, 328, 336
Veillonellaceae, 179
ventilation, 349, 353–55, 359–60, 365
Vermes (worms) phylum, 28
Verrucomicrobia, 194, 325
vertical loss, 336–37
Vidal, Daniel Ramón, 394
Viking coprolite, 326–27
violin plot, 247–48, 377
virome, 352

Virtue, Anthony, 284
virulence factors, 264–65, 310
viruses. See also specific microbe
asthma linked to, 309
in built environment, 351–52, 355, 365–66
characteristics of, 16–17
genomes, 103
gut microbiome, 63–64, 330–31
immune responses to, 237, 245 (see also immune system)
infant microbiome, 189
skin microbiome, 73
visualisations
diversity estimates, 139–41, 155–56, 163–64
microbial guilds, 280
phylogenetic trees, 12, 154, 322–23
Visualization and Analysis of Microbial Population Structures (VAMPS), 170
vitamins, 22, 258
Vivomixx, 229
VMT (vaginal microbiome transplant), 202, 340
volatile organic compounds, microbially produced (MVOCs), 364
voluntary (somatic) nervous system, 210
vomiting, during pregnancy, 201–2
vortex, 99
VSL#3, 289
Vuong, H. E., 185, 214–16

W
Wadwah, Arun, 203
waist-to-hip ratio (WHR), 279, 289
Walker, Allan, 389, 391
Wang, G. P., 261–62
Wang, Y., 392
Warinner, Christina, 327–29
Warren, J. Robin, 336
Washington University (St. Louis), 46
waste molecules, 22. See also metabolites; specific substance
waterborne microbes, 355–56, 363
water supply, 333–35
Watson, 38
weight, body, 274–79. See also obesity
Weizmann Institute, 46
Welch, Jessica, 42
Western lifestyle, 287, 303
allergic diseases linked to, 315–16
diet, 287, 303, 332–33, 335–36, 340, 365, 388, 392
environment (see built environment)
healthful microbiota development, 340–41
microbiota loss caused by, 334–38, 341
white blood cells, 237–38, 296–97. See also specific cell type
White HMP, 45
whole-genome reconstruction, 119
WHO (World Health Organization), 340, 390
WHR (waist-to-hip ratio), 279, 289
windows, 355, 359–60, 362
Woese, Carl, 13–17, 41–42, 96, 132

workspace (Qiita), 161–64, 167
World Health Organization (WHO), 340, 390
worm infections, 297, 308, 330–31, 381
worms (Vermes) phylum, 28
wound healing, 74–75, 240
Wu, Gary, 67, 280

X
Xycrobe, 76

Y
Yamazaki, K., 281
Yang, J., 225
Yates, J.A.F., 328–30

yeast
   gut microbiome, 59, 381
   skin microbiome, 73
   vaginal microbiome, 179
Yellowstone National Park, 19
Yersinia pestis, 348
Yost, S., 265